A TEXTBOOK OF
ENGINEERING MECHANICS

by

DR. D.S. KUMAR

B.Sc. Engg. (Hons.), M.Sc. Engg, Ph.D, FIE

Formerly :

Professor Mech. Engg, Punjab Engg. College, Chandigarh,
Director Academic Affairs, Punjab Technical Univ., Jalandhar,
Principal S.U.S. College of Engg. & Technology, Tangori (Mohali) and
Advisor, Manav Rachna Educational Institutions, Faridabad

S.K. KATARIA & SONS®

Publisher of Engineering Books

4885/109, Prakash Mahal, Dr. Subhash Bhargav Lane,
Opposite Delhi Medical Association, Daryaganj, New Delhi–110002 (INDIA)
Phone: +91-11-43551243; Mobile: +91-9871775858, 7678425209
e-mail: katariabook@gmail.com, katariabooks@yahoo.com
Website: www.skkatariaandsons.com

(e-books, subscriptions and bulk buying also available on our website)

Published by :

S.K. KATARIA & SONS®

4885/109, Prakash Mahal, Dr. Subhash Bhargav Lane,
Opposite Delhi Medical Association, Daryaganj, New Delhi-110002 (INDIA)
Phone: +91-11-43551243; Mobile: +91-9871775858, 7678425209
e-mail: katariabook@gmail.com, katariabooks@yahoo.com
Website: www.skkatariaandsons.com

First Edition: August, 2008; Fourth Edition: July, 2012; Reprint: 2026

Laser Typesetting by : Plus Computers, Delhi

Preface

The Mahamaya Technical University, Noida offer a course titled **Engineering Mechanics**. This course forms a core subject which is taught to students of all disciplines of engineering in all the technical institutions affiliated to this university. The study of this subject is aimed at developing a thorough understanding of the basic concepts and principles of mechanics and their application to solve engineering problems.

The author had taught this subject for several years at Punjab Engg. College, Chandigarh and this monograph is essentially based on the lectures delivered by him. From the experience gained through useful class discussions and feed back, these class notes were revised to improve their clarity and necessary explanatory notes were added during each teaching semester.The subject matter has thus been thoroughly tested in the class room and found suitable.

The text presents an amalgam of numerous ideas and influences and is the outcome of the distillation of the available text books written by distinguished authors. Obviously then a question arises to 'Why another book on Engineering Mechanics'. The answer to this question is reflected by the following salient features incorporated in this book :

- Precise and logical presentation of the basic concepts and principles
- Text written in simple, lucid and easily understandable language
- Full use of line diagrams made to supplement the text and to explain a particular phenomenon as clearly as possible
- Step-wise solutions provided to a wide variety of problems of standard comparable to those set for engineering degree, AMIE, GATE and Engineering Service Examinations
- Learning objectives have been added to the introduction of each chapter to set the purpose and goal of the topics described therein.
- Each chapter ends with a review and summary of the material covered in that chapter; this is to help the students to comprehend the subject matter prior to sitting for any test or examination.
- There are also sections on objective questions keeping in view the recent trend of such questions being asked in the various university and competitive examinations.

The author strongly believes that this treatise on Engineering Mechanics will contribute effectively to the existing literature in the field, and will be a useful text for the students of engineering curriculum

Acknowledgments : On completion of the book **Engineering Mechanics**, I wish to acknowledge my thanks and gratitude to all those who have been associated with this work directly/indirectly. Acknowledgment is due and is hereby made to all the authors whose works have been freely consulted in the preparation of this text. I owe an enormous debt of gratitute to my colleagues and students for much valued assistance in the form of discussion and feed back. A word of thanks is due to the publishers too for their considerable patience and cooperation throughout. Above all, I with to place on record my sincerest thanks and apologies to my near and dear ones who willingly endured certain hardships as a result of my pre-occupation with this work. A special mention in this respect is reserved for my grand-children **Namrata, Ritvik, Arshia and Ankit.**

My efforts in writing this book will be amply rewarded if the readers send their constructive suggestions and objective criticism with a view to enhance the usefulness of the book. Further, I would gratefully acknowledge the contribution of the readers by way of bringing my attention to the errors and printing mistakes that might have inadvertently crept into the book.

D.S. Kumar

Contents

UNIT - I

Chapter-1

Basic Concepts

Chapter-2

Two-Dimensional Con-current Force Systems

CHAPTER

1

Basic Concepts

Learning objectives : Attention has been focussed in this chapter to introduce the reader to the following basic concepts of mechanics

- *matter, particle and body; space, time, motion and trajectory*
- *fundamental laws of motion*
- *scalar and vector quantities, force and its representation, system of forces*
- *principle of transmissibility*
- *system of units; dimensional homogenity*
- *vectors and their associated algebra*

Engineering mechanics is a basic subject which describes and predicts the effects of forces on rigid bodies. It essentially deals with the study of bodies at rest or in motion when subjected to external mechanical disturbances. A knowledge of its basic concepts and principles is a must for an engineer engaged in the design and construction of various types of machines and structures.

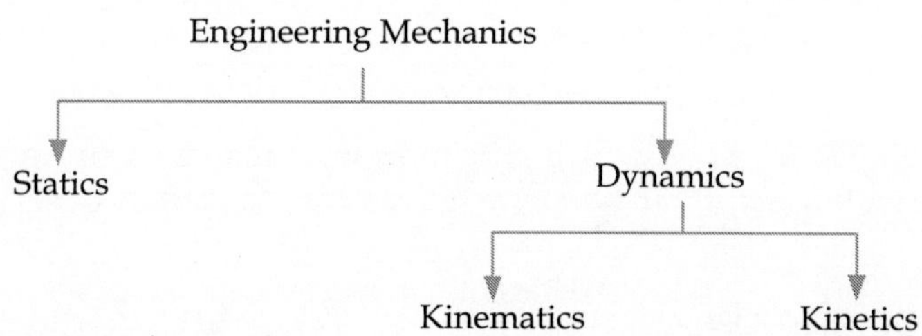

The subject of mechanics of a rigid body is divided into Statics and Dynamics. *Statics* deals with forces in terms of their distribution and effect on a body in equilibrium, *i.e.*, at absolute or relative rest. *Dynamics* deals with the study of bodies in motion. Dynamics is further divided into Kinematics and Kinetics. *Kinematics* is concerned with the description of motion of objects independent of causes of motion. Here study is made of motion interrelationship among position, velocity, acceleration and time without taking into account of the forces causing motion. In *Kinetics,* both the motion and its causes are considered. Kinetics relates the action of forces and the resulting motion.

1.1. MATTER, PARTICLE AND BODY

Matter is anything that occupies space, possesses mass and offers resistance to any external force. Iron, stone, wood salt, water and air etc represent matter.

A *particle* is an object that has infinitely small volume (occupies negligible space) but has a mass which can be considered to be concentrated at a point.

A *body* has a definite shape and consists of numbers of particles.

A body in which the distance between any two particles remains constant, *i.e.*, the size and shape of the body do not change, is called a *rigid* body. A rigid body exhibits no relative deformation between two parts when it is acted upon by a force system.

An *elastic* body undergoes deformation but regains its original shape after removal of the external forces.

1.2. SPACE, TIME, MOTION AND TRAJECTORY

Space is a region which extends in all directions and contains every thing in it, *e.g.*, sun, moon stars etc. In space, position of a body is located with respect to a reference system. The position of an aircraft in space is found with respect to the earth. On earth, position of a body is determined with respect to a permanent station on earth by taking measurements.

Time is a measure of succession of events. The unit of time is second which is a convenient fraction of the period by the earth's rotation, *i.e.*, $\frac{1}{86400}$ of an average solar day. According to SI units, second is taken to be a duration equal to 9 192 631 770 periods of vibration of cesium - 133 atom.

When a body changes its position with respect to other bodies, then body is said to be in motion. Thus, relative change in position is **motion**. It involves both space and time.

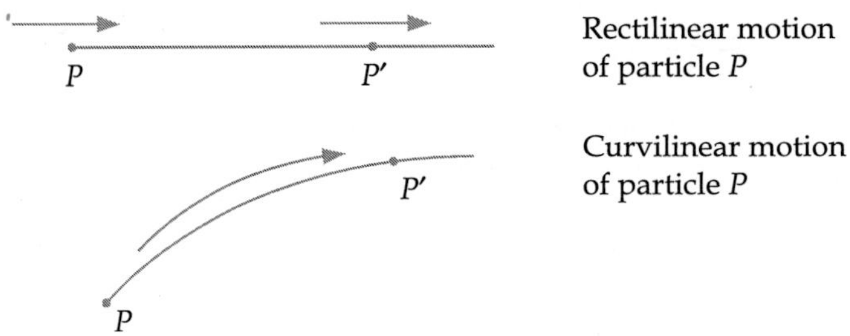

Fig. 1.1

Trajectory is path followed by a body during its motion. It may be a straight line or a curve. Straight line path is observed in *rectilinear motion*, whereas curved path is observed in *curvilinear motion*.

1.3. NEWTONS LAWS OF MOTION

First law: Every body continues in its state of rest or of uniform motion in a straight line if there is no unbalanced force acting upon it.

Newton's first law of motion gives the concept of inertia. Inertia is the property of matter by which a resistance is offered for change in the state of rest or uniform motion.

Second law: The rate of change of linear momentum is directly proportional to the impressed force and it takes place in the direction of the impressed force.

Newton's second law of motion connects rate of change of momentum and external force.

Third law: To every action, there is equal and opposite reaction.

Consider a ball resting on a floor (Fig. 1.2). At the point of contact, the ball exerts a force W (action) on the floor. The floor develops a reactive force R (reaction) which is equal in magnitude to action W but opposite in direction. Action and reaction act on different bodies and as such do not cancel each other.

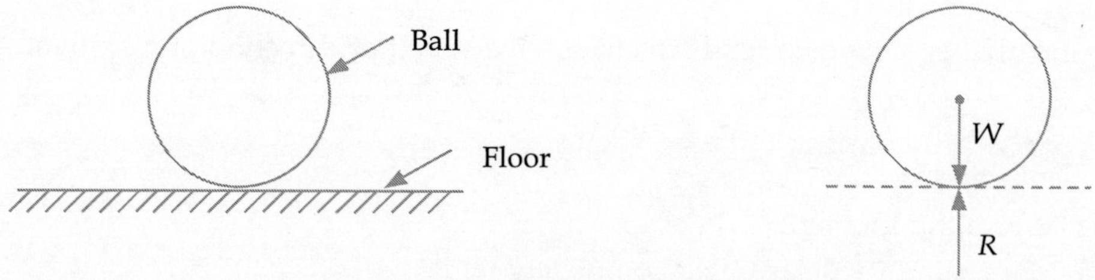

Fig. 1.2

Newton also formulated the **Law of gravitation**, which governs the mutual attraction between two isolated bodies. Newton's law of gravitation states that

"every body in the universe attracts every other body with a force directly proportional to the product of their mass and inversely proportional to the square of the distance separating them"

$$F = G\,\frac{m_1 m_2}{r^2}$$

where m_1 and m_2 are the masses of the two bodies, r is the distance separating them and G is the universal gravitation constant ($G = 6.67 \times 10^{-11}$ Nm^2/kg^2). The line joining the centres of bodies is then the line of action of the force of attraction.

1.4. SCALAR AND VECTOR QUANTITIES

A quantity is said to be *scalar* if it is completely defined by its magnitude alone. Mass, length, volume, temperature and time etc are scalar quantities.

Any quantity which possesses magnitude as well as direction is called a *vector* quantity. A vector quantity needs both magnitude and direction for its complete specification. Some examples of vector quantities are displacement, velocity, acceleration, momentum and force.

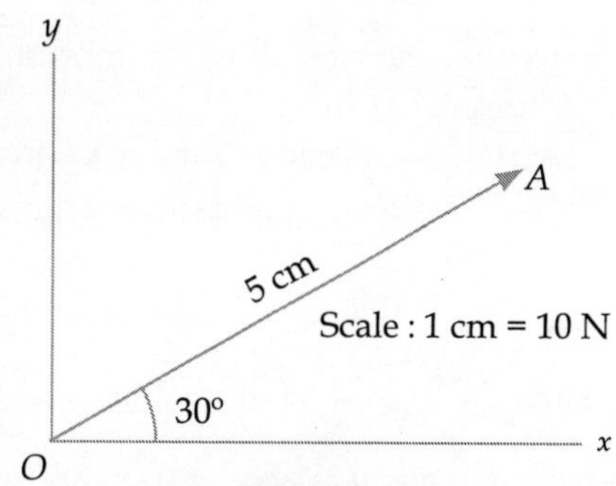

Fig. 1.3

A vector quantity is represented by a directed segment of a straight line. The length of the line represents the magnitude of the vector on some scale. The direction of the line indicates the direction of the vector, and arrowhead indicates the sense of direction. For example a force of 50 N acting in the x-y plane and making 30° with x-axis is represented by segment OA as shown in Fig. 1.3.

A vector is denoted by a letter of English alphabet with an arrow at its top such as $\vec{a}$, $\vec{A}$, $\vec{X}$ etc. The magnitude of any vector $\vec{A}$ is denoted by $|\vec{A}|$.

In Bow's method of force representation, a force is designated by writing two capital letters one on either side of it. With reference to Fig. 1.4, a force of $P_1 = 10$ kN is represented by AB and $P_2 = 15$ kN is represented by CD.

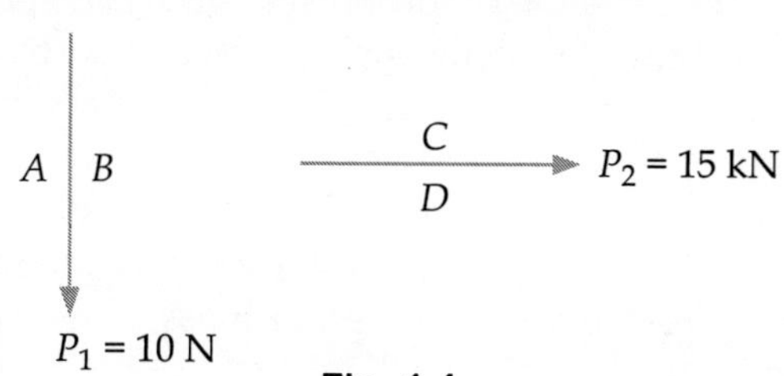

Fig. 1.4

1.5. MASS, FORCE AND WEIGHT

Mass is an indication of the quantity of matter present within a system. The more mass means more matter.

Force is an external agent which tends to change the speed or direction of a system. A force is applied whenever the system needs to be accelerated or decelerated. A force on a rigid body may produce one or both of the following effects:

(*i*) linear displacement

(*ii*) turning or rotating moment.

A force is a vector quantity determined completely by its

- magnitude
- point of application
- line of action, and
- direction.

With reference to Fig. 1.5, *AB* is the ladder that rests against a wall and a person weighing 700 N stands at point *C* on the ladder. For complete specification of the force exerted by the person on the ladder, the following characteristics need to be specified:

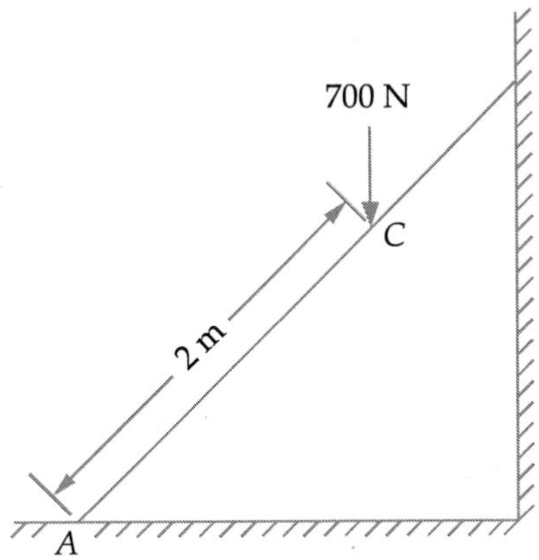

Fig. 1.5

- magnitude is 700 N
- the point of application is at a distance of 2 m from the floor along the ladder
- the line of action of force is vertical, and
- the direction is downward.

Newton's second law of motion states that force is proportional to rate of change of momentum. That is

$$\begin{aligned}\text{Force} &\propto \text{rate of change of momentum}\\ &\propto \text{rate of change of (mass x velocity)}\\ &\propto \text{mass x rate of change of velocity}\\ &\propto \text{mass x acceleration}\end{aligned}$$

Thus $$F \, \alpha \, m\,a\,; \qquad F = \frac{m\,a}{g_c}$$

where m is mass, a is acceleration and g_c is proportionality constant. The SI unit of force is newton (N) which represents the force required to accelerate 1 kg mass with an acceleration of 1 m/s^2

Thus $$1\text{ N} = \frac{1\text{ kg} \times 1\text{ m/s}^2}{g_c} \quad \text{and} \quad g_c = \frac{1\text{ kg m}}{\text{Ns}^2}$$

The constant of proportionality becomes unity for defining unit of force as newton and accordingly dropped. Such a practice corresponds to writing Newton's second law as

$$\text{Force} = \text{mass x acceleration}; \quad F = m\,a$$

Weight is a force which the system exerts due to gravitational acceleration. Weight is actually the force with which the system is attracted towards the earth. Where there is no gravity, there is no weight and this condition of the system is known as weightlessness. The weight of a system varies from place to place but mass remains constant. The mass of a given collection of matter cannot be altered by a change of location, a deformation of shape, a change of temperature, and through chemical reactions which the matter may undergo.

The weight (W) of system equals the product of mass (m) and local gravitational acceleration. The mass of a system remains constant whereas weight varies with change in value of gravitational acceleration from one place to another. The value of g at sea level is 9.8066 m/s^2 and is generally taken as 9.81 m/s^2.

The force system acting on a body may

– change its state of rest or motion
– accelerate or retard its motion
– change its shape or size
– turn or rotate it, and
– keep it in equilibrium.

1.6. TENSION AND COMPRESSION

A structural member is said to be in tension when it is subjected to two equal and opposite pulls and the member tends to elongate/increase in length. The stress induced in the bar is called tensile stress, and the tensile stress σ_t at any cross-section X-X is given by

$$\sigma_t = \frac{P}{A}$$

where A is the cross-sectional area of the bar.

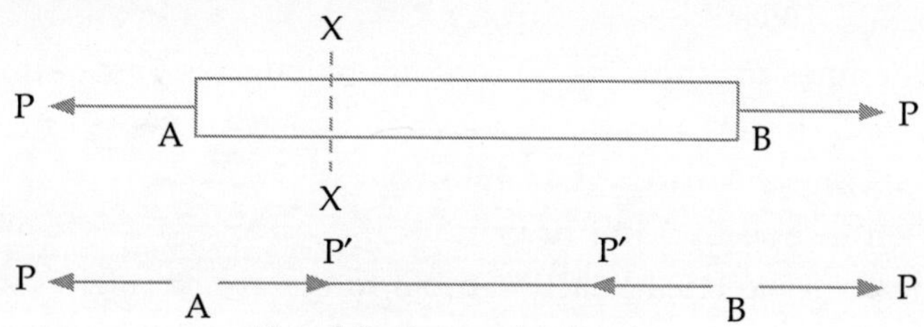

Fig. 1.6. Bar under tension

At every point of bar AB, equal and opposite forces P' (where $P' = P$) act. The force P' acting at point A balances the external force P; hence when considering the equilibrium of A, the arrow showing P' must be directed toward B. Likewise, when considering the equilibrium of B, the arrow points towards A.

The hoisting ropes used is cranes and passenger elevators are the elements subjected to tensile loading. An element under tension is called a *tie*.

If the structural member is subjected to two equal and opposite pushes, and the member tends to shorten/decrease in length, the member is said to be in compression. The stress so produced is called compressive stress, and the compressive stress σ_c at any cross-section X-X is given by

$$\sigma_c = \frac{P}{A}$$

Under equilibrium conditions of the bar, the resistance force P' (where $P' = P$) will be directed as shown in Fig. 1.7. A bar under compression is called a *strut*.

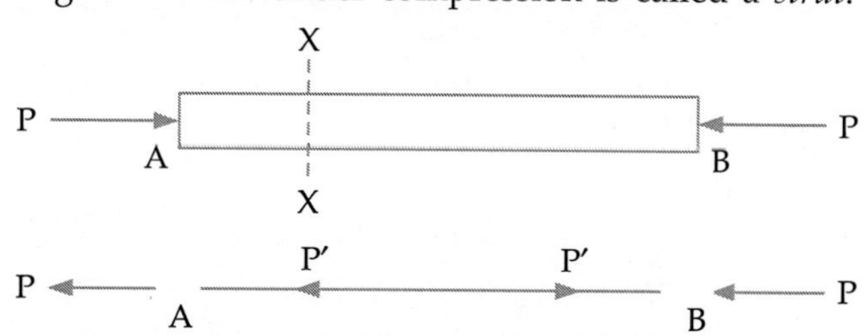

Fig. 1.7. Bar under compression

1.7. SYSTEM OF FORCES

When several forces of different magnitude and direction act upon a body, they constitute a force system.

Considering the plane in which forces are applied and depending upon the position of line of action, forces may be classified as shown follows:

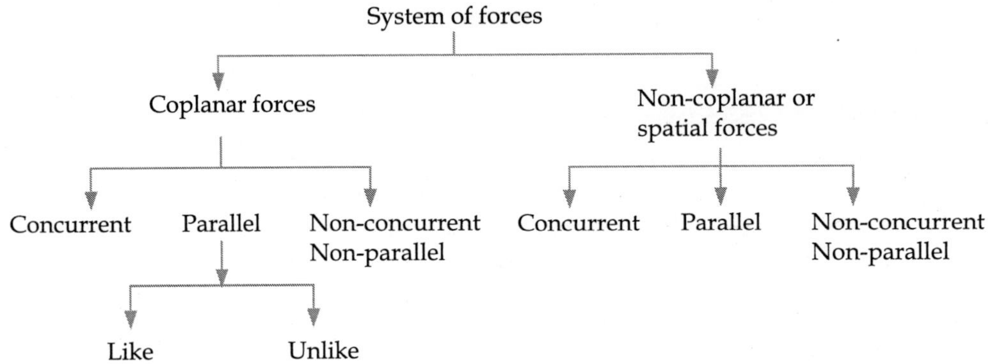

- *Collinear forces*: The lines of action of all forces lie along the same straight line.

Fig. 1.8

Example: Force on a rope in a tug of war.

- *Coplanar parallel forces* : The lines of action of all forces are parallel to each other and lie in a single plane.

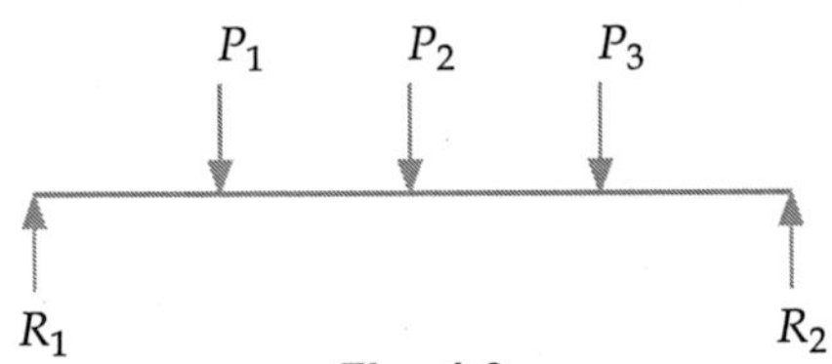

Fig. 1.9

Example: System of vertical loads (including reactions) acting on a beam.

- *Coplanar concurrent forces*: All forces lie in the same plane, have different directions but their lines of action act at one point (pass through a single point). The point where the lines of action of the forces meet is known as the point of concurrency of the force system.

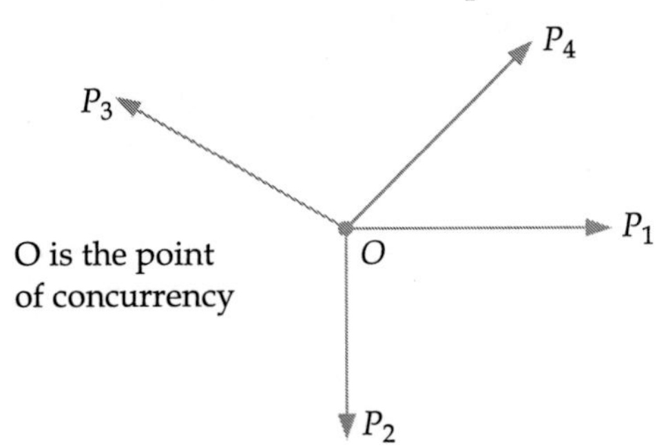

Fig. 1.10

Example: Forces on a rod resting against a wall.

- *Coplanar non-concurrent forces*: All forces lie in the same plane but their lines of action do not pass through a single point.

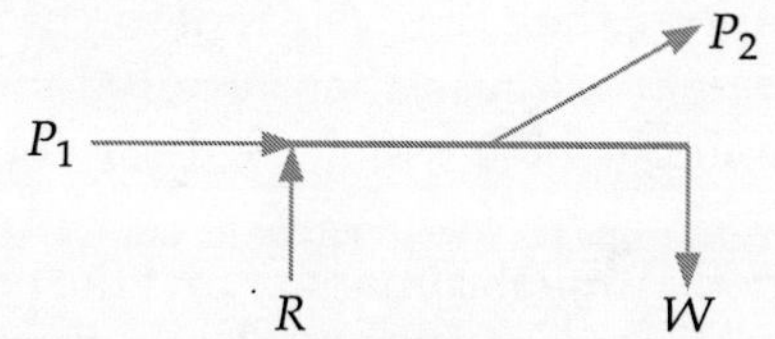

Fig. 1.11

Example: Forces on a ladder resting against wall and a person standing on a rung which is not at its centre of gravity.

- *Non-coplanar concurrent forces*: All forces do not lie in the same plane but their lines of action pass through a single point.

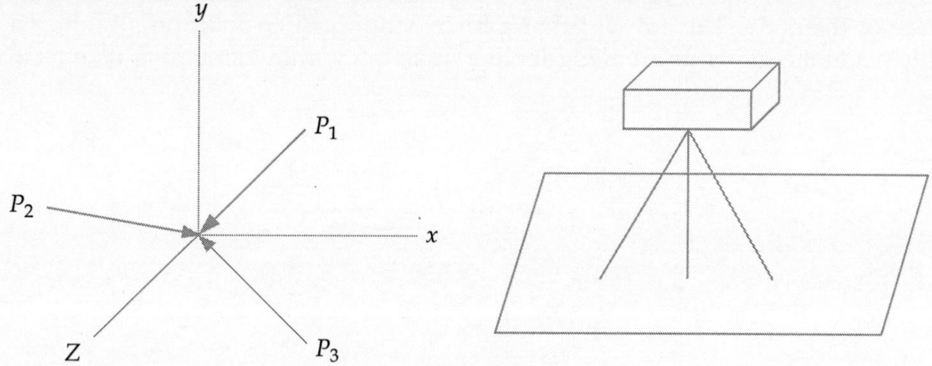

Fig. 1.12

Example: forces on a tripod carrying a camera.

- *Non-coplanar and non-concurrent forces:* All forces do not lie in the same plane, and their lines of action do not meet at a single point.

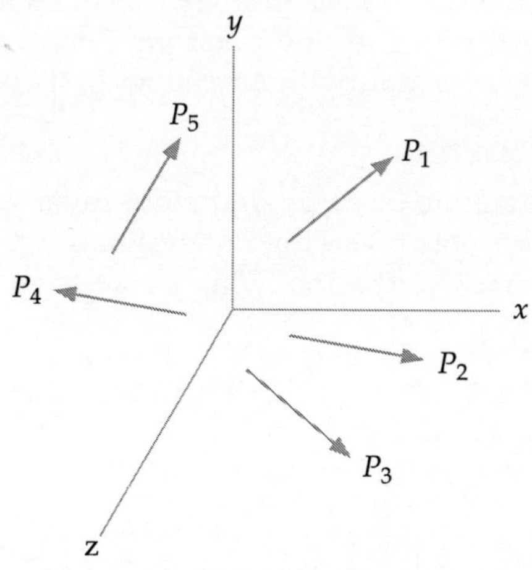

Fig. 1.13

Example: Forces acting on a moving bus.

1.8. EQUILIBRIUM, RESULTANT AND EQUILIBRANT

When two or more than two forces act on a body in such a way that the body remains in a state of rest or of uniform motion (no acceleration), then the system of forces is said to be in *equilibrium*.

When a body is acted upon by a system of forces, then vectorial sum of all the forces is known as *resultant*. Hence resultant refers to the single force which produces the same effect as is done by the combined effect of several forces.

A number of forces may act on a body in such a manner that the body is not in equilibrium. The resultant of several forces may cause a change of state of rest or of uniform motion. A single force may have to be applied to the body to bring it in equilibrium state. That single force is known as *equilibrant*. Equilibrant is equal and opposite to the resultant of several forces acting on the body.

1.9. PRINCIPLE OF TRANSMISSIBILITY

When the point of application of a force acting on a body is shifted to any other point on the line of action of the force without changing its direction, there occurs no change in the equilibrium state of the body. This implies that a force acting at any point on a body may also be considered to act at any other point along its line of action without changing its effect on the body.

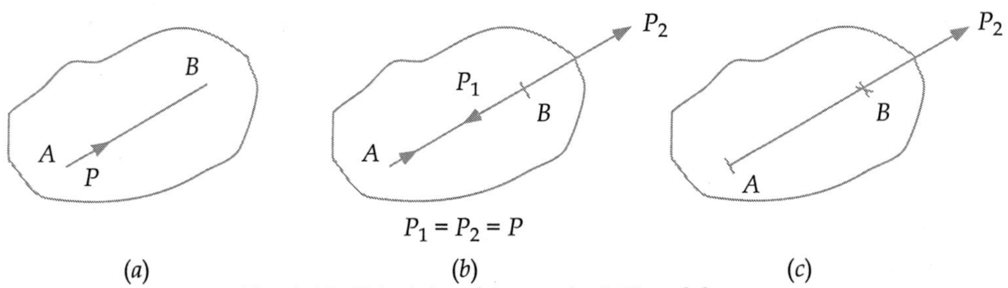

Fig 1.14. Principle of transmissibility of forces

Consider a force P acting at point A on rigid body (Fig 1.14). B is another point on the line of action of force P. At point B, apply two oppositely directed forces (P_1 and P_2) equal to and collinear with P. Such an application will in no way alter the action of given force P. At point A, forces P and P_1 are equal but opposite and accordingly cancel each other. That leaves a force $P_2 = P$ at B. This implies that a force acting at any point on a body may also be considered to act at any other point along its line of action without changing its effect on the body.

1.10. VECTORS AND VECTOR ALGEBRA

Vector approach is generally considered to be easier and more suitable while dealing with forces. For that, knowledge about basic concepts of vectors is essential. A brief review of vectors and their associated algebra has been made in the following paragraphs.

1.10.1. Vector and its representation

Vectors are mathematical expressions possessing magnitude and direction.

A vector quantity may be written either as thick bold letter (**A**) or a capital letter with arrow over it ($\vec{A}$). The plane capital letter (A) denotes the magnitude of that vector. Thus

$$\vec{A} = |\vec{A}| = A$$

where $\vec{A}$ is a vector quantity and A is its magnitude.

1.10.2. Components of a vector

The vector $\vec{A}$ represented by OP can be resolved into three components A_x, A_y and A_z along the rectangular axes.

$$\vec{A} = A_x i + A_y j + A_z k$$

where i, j and k are referred to as unit vectors along the positive x, y and z directions respectively. A unit vector is a vector having a unit magnitude or unit length.

Unit vector corresponding to vector $\vec{A}$

$$= \frac{\vec{A}}{|\vec{A}|} = \frac{\vec{A}}{\sqrt{A_x^2 + A_y^2 + A_z^2}}$$

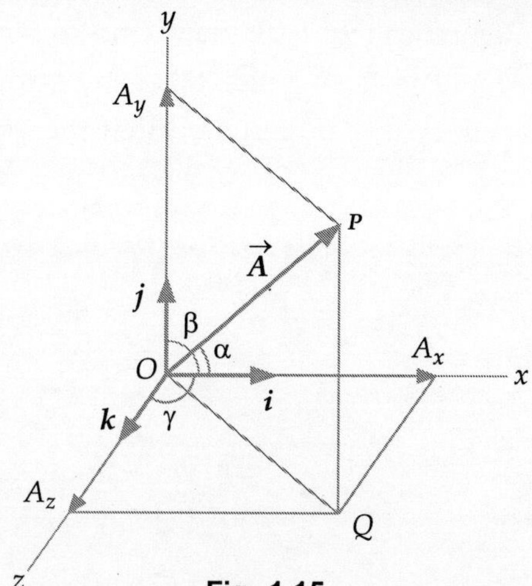

Fig. 1.15

The components A_x, A_y and A_z are called scalar components of the vector $\vec{A}$.

If the vector $\vec{A}$ makes angles α, β and γ with the x, y and z axes respectively, then

$$\vec{A} = (A \cos \alpha)\, i + (A \cos \beta)\, j + (A \cos \gamma)\, k$$

EXAMPLE 1.1.

A force of 25 N makes an angle of 30, 45 and 75 degree with the x, y and z-axis. What should be the corresponding force vector?

Solution :

$$\begin{aligned} \vec{F} &= F_x i + F_y j + F_z k \\ &= (F \cos \alpha)\, i + (F \cos \beta)\, j + (F \cos \gamma)\, k \\ &= (25 \cos 30)\, i + (25 \cos 45)\, j + (25 \cos 75)\, k \\ &= 21.65\, i + 17.68\, j + 6.47\, k \end{aligned}$$

1.10.3. Position vector of a given point

The position vector of a point P with respect to point O is the vector OP. It is denoted by $\vec{r}$ and is used to specify the position of point P relative to O. (Fig. 1.16)

$$\vec{r} = x\, i + y\, i + z\, k$$

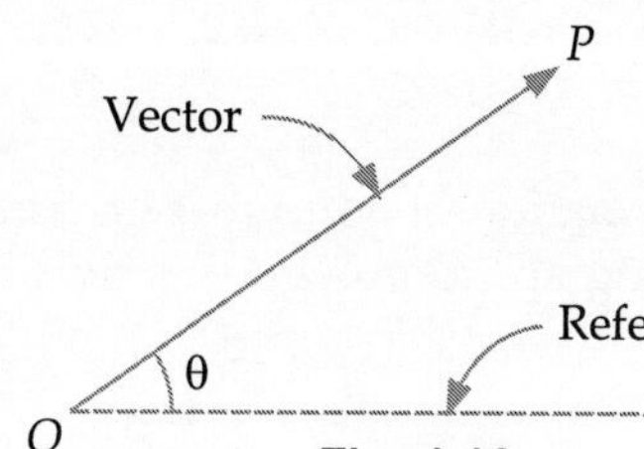

Fig. 1.16

A vector is generally represented by a line segment with an arrow at one end.

(*i*) the length of the line represents the magnitude of the vector quantity

(*ii*) the arrow mark specifies the sense of direction, *i.e.*, to or from a point

(*iii*) the orientation of the line shows the direction of the vector, *i.e.*, the angle which the vector makes with a reference line.

1.10.4. Addition and subtraction of vectors

The resultant of two vectors may be obtained according to parallelogram law or triangle law.

Reference Fig. 1.17, the resultant $\vec{R}$ of two vectors $\vec{A}$ and $\vec{B}$ is the diagonal of the parallelogram for which $\vec{A}$ and $\vec{B}$ are the adjacent sides.

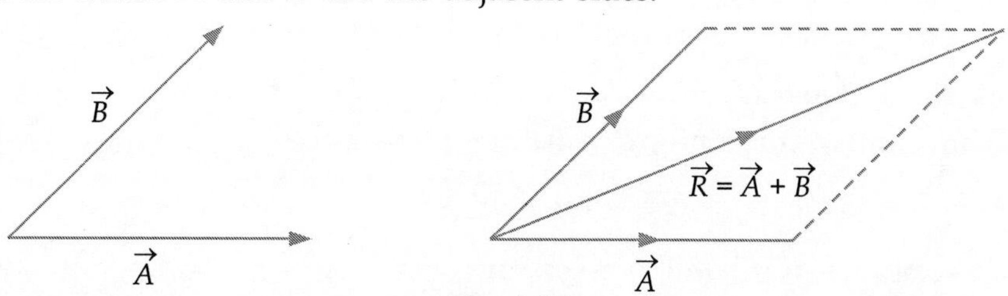

Fig. 1.17

Reference Fig. 1.18, the vectors $\vec{A}$ and $\vec{B}$ are represented by the two sides of a triangle taken in order. The closing side taken in opposite order represents the resultant $\vec{R}$ of the vector $\vec{A}$ and $\vec{B}$. Further, it may be noted that the vector addition is commutative. That is

$$\vec{R} = \vec{A} + \vec{B} = \vec{B} + \vec{A}$$

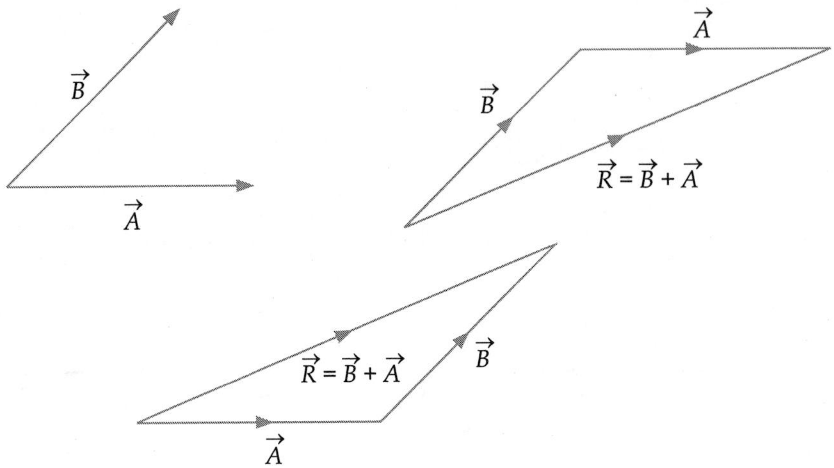

Fig. 1.18

The difference between vector $\vec{A}$ and $\vec{B}$ can be obtained by addition of vectors $\vec{A}$ and $-\vec{B}$ (Fig. 1.19).

$$\vec{A} - \vec{B} = \vec{A} + (-\vec{B})$$

The summation of a number of vectors can be obtained by the repeated use of parallelogram or by drawing space polygon.

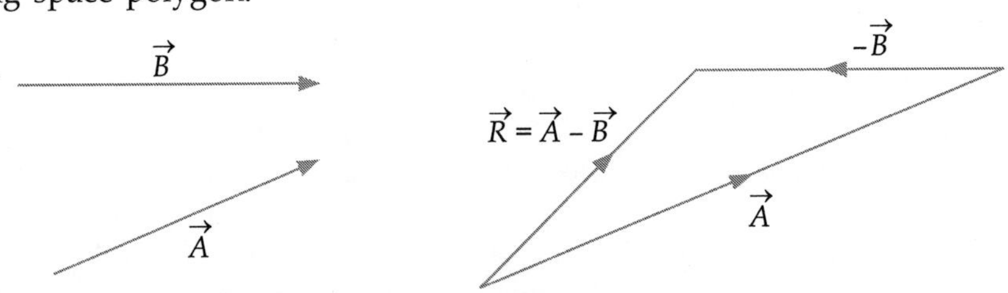

Fig. 1.19

EXAMPLE 1.2

Two vectors are prescribed as

$$\vec{A} = 3i + 2j \text{ and } \vec{B} = 2i - 3j$$

If $\vec{P} = \vec{A} + \vec{B}$ and $\vec{Q} = \vec{A} - \vec{B}$, then find

(*a*) magnitude of vectors $\vec{P}$ and $\vec{Q}$
(*b*) unit vectors along *P* and *Q*, and
(*c*) magnitude and unit vector along $\vec{S} = 3\vec{P} + 2\vec{Q}$

Solution : $\vec{P} = \vec{A} + \vec{B} = (3i + 2j) + (2i - 3j) = 5i - j$

Magnitude of vector $\vec{P} = \sqrt{5^2 + (-1)^2} = 5.1$ unit

Unit vector in the direction of $\vec{P} = \dfrac{5i - j}{5.1} = 0.98\, i - 0.196\, j$

$$\vec{Q} = \vec{A} - \vec{B} = (3\,i + 2\,j) - (2\,i - 3\,j) = i + 5\,j$$

Magnitude of vector $Q = \sqrt{1^2 + 5^2} = 5.1$ unit

Unit vector in the direction of $\vec{Q} = \dfrac{i + 5j}{5.1} = 0.196\, i + 0.98\, j$

(*b*) $\vec{S} = 3\vec{P} + 2\vec{Q} = 3\,(5\,i - j) + 2\,(i + 5\,j)$
$= (15\,i - 3\,j) + (2\,i + 10\,j) = 17\,i + 7\,j$

Magnitude of vector $\vec{S} = \sqrt{(17)^2 + (7)^2} = 18.38$ unit

Unit vector in the direction of $\vec{S} = \dfrac{17\,i + 7\,j}{18.38} = 0.925\, i + 0.381\, j$

1.10.5. Dot (scalar) product

The dot product of vectors $\vec{A}$ and $\vec{B}$ is a scalar quantity, and is defined as the product of the magnitude of the vectors and cosine of their included angle

$$\vec{A} \cdot \vec{B} = |\vec{A}|\,|\vec{B}| \cos\theta = AB \cos\theta$$

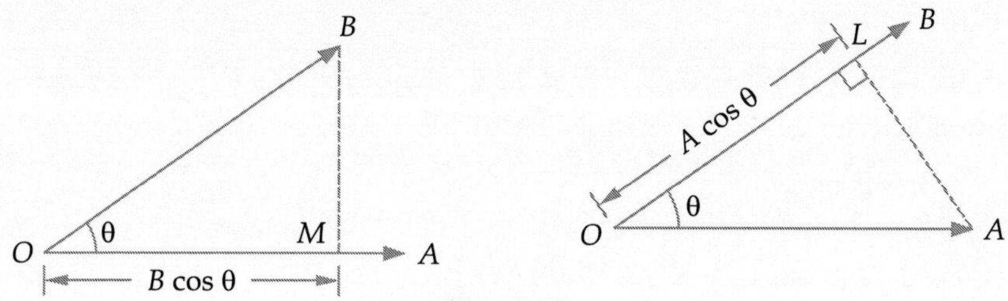

Fig. 1.20

It is to be noted that :

(*i*) When θ = 0°, i.e., when the vectors $\vec{A}$ and $\vec{B}$ are along the same direction

$$\vec{A} \cdot \vec{B} = AB \cos 0° = AB$$

(*ii*) When θ = 90°, i.e., when the vectors $\vec{A}$ and $\vec{B}$ are perpendicular to each other

$$\vec{A} \cdot \vec{B} = AB \cos 90° = 0$$

(*iii*) $\vec{A} \cdot \vec{B}$ = A times projection of $\vec{B}$ on $\vec{A}$

= B times projection of $\vec{A}$ on $\vec{B}$

(*iv*) In terms of components,

$$\vec{A} \cdot \vec{B} = (A_x i + A_y j + A_z k) \times (B_x i + B_y j + B_z k)$$
$$= A_x B_x + A_y B_y + A_z B_z$$

i, ***j*** and ***k*** are the unit vectors along *x*, *y* and *z* directions respectively, and

$i \cdot i = j \cdot j = k \cdot k = 1$ (∵ θ = 0° and cos 0° = 1)

$i \cdot j = j \cdot k = k \cdot i = 0$ (∵ θ = 90° and cos 90° = 0)

(*v*) The angle between the vectors $\vec{A}$ and $\vec{B}$ is

$$\cos\theta = \frac{\vec{A}\cdot\vec{B}}{|\vec{A}||\vec{B}|} = \frac{\vec{A}\cdot\vec{B}}{AB}$$

(*vi*) The dot product defines the term work in the field of dynamics.

1.10.6. Cross (vector) product

The cross product of vectors $\vec{A}$ and $\vec{B}$ is a vector quantity and is defined as the product of the magnitude of the vectors and the sine of the smaller angle between the vectors. Its direction is perpendicular to the plane containing the vectors.

If $\vec{n}$ is the unit vector which gives the direction of the resultant vector, then

$$\vec{R} = \vec{A}\times\vec{B} = |A|\,|B|\,\sin\theta\,\vec{n} = AB\sin\theta\,\vec{n}$$

If $\vec{A} = A_x i + A_y j + A_z k$ and $\vec{B} = B_x i + B_y j + B_z k$

then the cross product in terms of rectangular components is given by

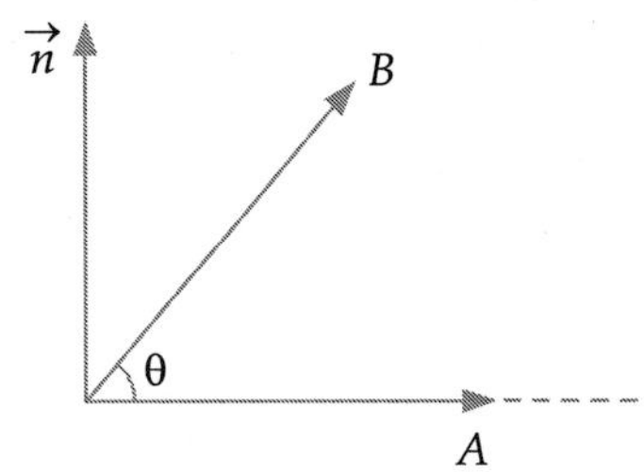

Fig. 1.21

$$\vec{A}\times\vec{B} = \begin{vmatrix} i & j & k \\ A_x & A_y & A_z \\ B_x & B_y & B_z \end{vmatrix}$$

$$= (A_yB_z - B_yA_z)\,i - (A_xB_z - B_xA_z)\,j + (A_xB_y - B_xA_y)\,k$$

Further it is to be noted that:

(*i*) $\vec{A}\times\vec{B} = 0$ if $\theta = \theta$ or π

(*ii*) $i \times i = j \times j = k \times k = 0$

$i \times j = k;\quad j \times k = i \quad$ and $\quad k \times i = j$

(*iii*) The cross product defines the term moment of force.

EXAMPLE 1.3

(*a*) Check whether the following set of vectors are perpendicular to each other or not

$$\vec{A} = 6i + 10j - 5k \quad \text{and} \quad \vec{B} = 5i + 2j + 10k$$

(*b*) Check whether the following set of vectors are parallel to each other or not

$$\vec{A} = 2i - 3j - k \quad \text{and} \quad \vec{B} = -6i + 9j + 3k$$

Solution: $\vec{A}\cdot\vec{B} = (6j + 10j - 5k)\cdot(5i + 2j + 10k)$

$$= 6\times 5 + 10\times 2 + (-5\times 10) = 0$$

Since the dot product of the given vectors is zero, the vectors are perpendicular to each other.

(*b*) $\vec{A}\times\vec{B} = (2i - 3j - k)\times(-6i + 9j + 3k)$

$$= \begin{vmatrix} i & j & k \\ 2 & -3 & -1 \\ -6 & 9 & 3 \end{vmatrix}$$

$$= i[-3\times 3 - 9\times(-1)] - j[2\times 3 - (-6)\times(-1)] + k[2\times 9 - (-6\times -3)]$$

$$= i(-9+9) - j(6-6) + k(18-18) = 0$$

Since the cross product of the given vectors is zero, the vectors are parallel to each other.

EXAMPLE 1.4

Two vectors $\vec{A}$ and $\vec{B}$ are given as

$$\vec{A} = 2i + 3j \quad \text{and} \quad \vec{B} = 3i - j$$

Determine : (*a*) the dot product and cross product of the vectors, (*b*) the angle between vectors $\vec{A}$ and $\vec{B}$ (*c*) the included angle between vector $\vec{A}$ and the vector resulting from the cross product.

Solution : The dot product is given as

$$\vec{A} \cdot \vec{B} = A_x B_x + A_y B_y$$
$$= 2 \times 3 + 3 \times (-1) = 6 - 3 = 3 \text{ units}$$

The cross product is expressed as

$$\vec{A} \times \vec{B} = \begin{vmatrix} i & j & k \\ A_x & A_y & A_z \\ B_x & B_y & B_z \end{vmatrix} = \begin{vmatrix} i & j & k \\ 2 & 3 & 0 \\ 3 & -1 & 0 \end{vmatrix}$$

$$= [3 \times 0 - (-1) \times 0]\, i - (2 \times 0 - 3 \times 0)\, j + [2 \times (-1) - 3 \times 3]\, k$$
$$= 0 - 0 - 11\,k = -11\,k$$

(*b*) From the definition of dot product : $\vec{A} \cdot \vec{B} = AB \cos \theta$

$$\cos \theta = \frac{\vec{A} \cdot \vec{B}}{AB} = \frac{3}{\sqrt{2^2 + 3^2} \times \sqrt{3^2 + (-1)^2}}$$

$$= \frac{3}{\sqrt{13} \times \sqrt{10}} = 0.263$$

$$\theta = \cos^{-1}(0.263) = \mathbf{74.75^\circ}$$

Check : From the definition of cross product : $|\vec{A} \times \vec{B}| = AB \sin \theta$

$$\sin \theta = \frac{\vec{A} \times \vec{B}}{AB} = \frac{11}{\sqrt{13} \times \sqrt{10}} = 0.965$$

$$\theta = \sin^{-1}(0.965) = \mathbf{74.79^\circ}$$

$$(c)\ \vec{A} \times (\vec{A} \times \vec{B}) = \begin{vmatrix} i & j & k \\ 2 & 3 & 0 \\ 0 & 0 & -11 \end{vmatrix} = -33\,i + 22\,j$$

The above vector has a magnitude of $\sqrt{(-33)^2 + (-22)^2} = 39.66$

$$\sin \theta = \frac{|\vec{A} \times (\vec{A} \times \vec{B})|}{|\vec{A}| \times |\vec{A} \times \vec{B}|} = \frac{39.6}{\sqrt{13} \times 11} = 1.00$$

Thus the angle between $\vec{A}$ and $(\vec{A} \times \vec{B})$ is 90°. The included angle between vector $\vec{B}$ and the vector $(\vec{A} \times \vec{B})$ would also workout to be 90°.

Remarks : The included angle between the vector resulting from the cross product and either of the constituent vectors must be 90°.

EXAMPLE 1.5

Determine the work done when a force $\vec{F} = 2i + 3j - k$ moves form point P (3, 1, – 1) to point Q (2, – 1, 1).

Solution : Distance moved = position vector of point P – position vector of point Q

$$\vec{S} = (3i + j - k) - (2i - j + k) = i + 2j - 2k$$

Work done by the force is given by the dot product of force applied and distance moved

$$\text{Work done} = \vec{F}\cdot\vec{S} = (2i + 3j - k)\cdot(i + 2j - 2k)$$

$$= 2 \times 1 + 3 \times 2 + (-1) \times (-2) = \mathbf{10\ units}$$

1.10.7. Component of a force

Force is a vector quantity having both magnitude and direction

Let a force $\vec{F}$ be designated by vector OP. If through point P, planes are drawn parallel to the coordinate planes, then a rectangular box or parallelopiped gets formed (Fig. 1.22). The force $\vec{F}$ is then represented by the diagonal OP of the box, and the three components F_x, F_y and F_z are represented by the edges of the box.

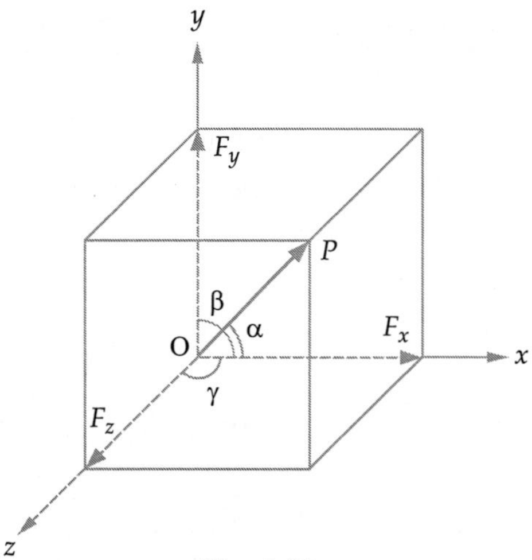

Fig. 1.22

Salient aspects :

(*i*) If ***i***, ***j*** and ***k*** are the unit vectors in the positive x, y and z directions respectively, then force $\vec{F}$ is designated as

$$\vec{F} = F_x i + F_y j + F_z k$$

$$\text{Magnitude } F = |\vec{F}| = \sqrt{F_x^2 + F_y^2 + F_z^2}$$

The components F_x, F_y and F_z are called the scalar components of the force vector $\vec{F}$.

(*ii*) If α, β and γ are the angles which the force vector $\vec{F}$ makes with the coordinate axes x, y and z axes respectively, then

$$F_x = F\cos\alpha; \quad F_y = F\cos\beta \quad \text{and} \quad F_z = F\cos\gamma$$

That gives

$$\vec{F} = (F\cos\alpha)\,i + (F\cos\beta)\,j + (F\cos\gamma)\,k$$

(*iii*) The cosines of angles α, β and γ are called direction cosines and are denoted as

$$l = \cos\alpha; \quad m = \cos\beta \quad \text{and} \quad n = \cos\gamma$$

The direction cosines are not independent and are related by the expression

$$l^2 + m^2 + n^2 = 1$$

EXAMPLE 1.6

Consider the $\vec{F} = (5i + 2j - 6k)$ kN that passes through the origin. Determine the magnitude of the force and the angles it makes with the x, y and z-axis.

Solution : Resultant of force, $R = \sqrt{5^2 + 2^2 + (-6)^2}$

$$= \sqrt{25 + 4 + 36} = 8.06 \text{ kN}$$

Then: $\cos\alpha = \dfrac{F_x}{R} = \dfrac{5}{8.06} = 0.62$; $\alpha = 51.68°$

$$\cos\beta = \frac{F_y}{R} = \frac{2}{8.06} = 0.248 \;; \quad \beta = 75.64°$$

$$\cos\gamma = \frac{F_z}{R} = \frac{-6}{8.06} = 0.744 \;; \quad \gamma = -\,138.07°$$

EXAMPLE 1.7

A point is acted upon by a set of three forces given by

$$(3\,i + 5\,j + 2\,k),\ (2\,i + 7\,j + 3\,k) \text{ and } (i + 2\,j + 5\,k).$$

Find the magnitude of the resultant force and its direction cosines.

Solution : Resultant force $\vec{R}$

$$= (3\,i + 5\,j + 2\,k) + (2\,i + 7\,j + 3\,k) + (i + 2\,j + 5\,k)$$
$$= 6\,i + 14\,j + 10\,k$$

Magnitude of resultant $= \sqrt{6^2 + 14^2 + 10^2}$ = **18.22** units

Direction cosines of the resultant are :

$$l = \cos\alpha = \frac{F_x}{F} = \frac{6}{18.22} = 0.329\,; \quad \alpha = 70.79°$$

$$m = \cos\beta = \frac{F_y}{F} = \frac{14}{18.22} = 0.768; \quad \beta = 39.82°$$

$$n = \cos\gamma = \frac{F_z}{F} = \frac{10}{18.22} = 0.5488; \quad \gamma = 56.71°$$

Check : $l^2 + m^2 + n^2 = (0.329)^2 + (0.768)^2 + (0.5488)^2 \simeq 1$

EXAMPLE 1.8

A force 25 kN starts from a point $P(1, 3, -2)$ and passes through the point $Q\,(-8, 5, 3)$. Express the force in terms of unit vectors i, j and k.

Solution : Vector joining the points $P(1, 3, -2)$ and $Q(-8, 5, 3)$ is

$$\overrightarrow{PQ} = \vec{r} = (-8 - 1)\,i + (5 - 3)\,j + [3 - (-2)]\,k$$
$$= -9\,i + 2\,j + 5\,k$$

$$\text{Unit vector along } PQ = \frac{-9i + 2j + 5k}{\sqrt{(-9)^2 + (2)^2 + (5)^2}} = \frac{-9i + 2j + 5k}{\sqrt{110}}$$
$$= -\,0.858\,i + 0.1907\,j + 0.4767\,k$$

Force vector $\vec{F}$ $= |\vec{F}| \times \vec{n}$

= magnitude of force × unit vector in the direction of force

$= 25\,(-\,0.858\,i + 0.1907\,j + 0.4767\,k)$

$= \mathbf{-\,21.45\,i + 4.767\,j + 11.917\,k}$

EXAMPLE 1.9

A space force has components of 10 N, 20 N and – 50 N along the *x*, *y* and *z* axes respectively. Write the vector equation for the force and calculate : (*a*) the magnitude of the force, (*b*) the direction cosines of the force and (*c*) the angles made by the force along the coordinate axes.

Solution : Refer Fig. 1.23 for the coordinate axes and the orientation of the given forces.

Vector equation of force

$$\vec{F} = F_x i + F_y j + F_z k = 10\,i + 20\,j - 50\,k$$

(*a*) Magnitude of force

$$F = |\vec{F}| = \sqrt{F_x^2 + F_y^2 + F_z^2}$$

$$= \sqrt{10^2 + 20^2 + (-50)^2} = \mathbf{54.77\ N}$$

(*b*) Direction cosines of the force

$$l = \cos\alpha = \frac{F_x}{F} = \frac{10}{54.77} = 0.1826$$

$$m = \cos\beta = \frac{F_y}{F} = \frac{20}{54.77} = 0.3652$$

$$n = \cos\gamma = \frac{F_z}{F} = \frac{-50}{54.77} = -0.9129$$

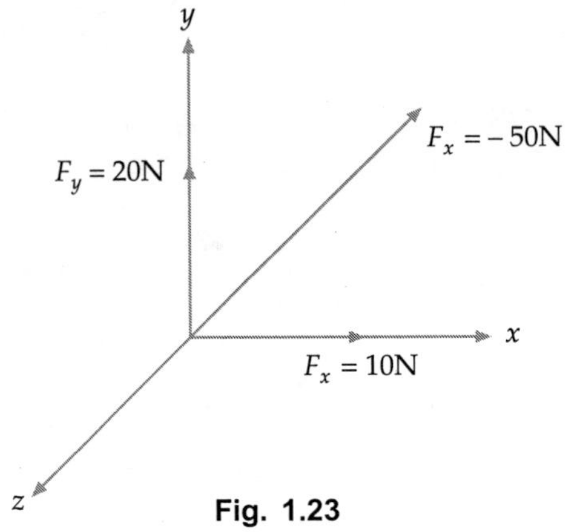

Fig. 1.23

Check : $l^2 + m^2 + n^2 = (0.1826)^2 + (0.3652)^2 + (-0.9129)^2$

$= 0.0333 + 0.1334 + 0.8334 \simeq 1.0$

(*c*) Angles made by the forces along the coordinate axes

$\cos\alpha = 0.1826;\quad \alpha = \mathbf{79.48^\circ}$

$\cos\beta = 0.3652;\quad \beta = \mathbf{68.58^\circ}$

$\cos\gamma = -0.929;\quad \gamma = \mathbf{155.91^\circ}$

EXAMPLE 1.10

A space force of magnitude 125 N makes angles of 30° and 65° with the *x* and *y* axes respectively.

Make calculations for the scalar components of the force along the *x*, *y* and *z* axes and write the vector equation for the force.

Solution : $l = \cos\alpha = \cos 30^\circ = 0.866$

$m = \cos\beta = \cos 65^\circ = 0.423$

From the relation; $l^2 + m^2 + n^2 = 1$, we have

$$n = \cos\gamma = \sqrt{1 - l^2 - m^2} = \sqrt{1 - (0.866)^2 - (0.423)^2}$$

$$= \sqrt{1 - 0.75 - 0.179} = 0.266$$

That gives

$$\gamma = \cos^{-1}(0.266) = 74.57^\circ$$

Scalar components of the force are :

$F_x = F\cos\alpha = 125\cos 30^\circ = 108.25\ N$

$F_y = F\cos\beta = 125\cos 65^\circ = 52.83\ N$

$F_z = F\cos\gamma = 125\cos 74.57^\circ = 33.26\ N$

Vector equation of the force

$$= F_x i + F_y j + F_z k = 108.25\, i + 52.83\, j + 33.26\, k$$

EXAMPLE 1.11

The x, y, z components of a force are 30 kN, – 20 kN and 16 kN respectively. Determine the component of this force along the line joining $A(1, 2, -3)$ and $B(-1, -3, 4)$.

Solution: Force $\vec{F} = 30i - 20j + 16k$

$$|\vec{F}| = F = \sqrt{30^2 + (-20)^2 + 16^2} = \sqrt{1556} = 39.446 \text{ kN}$$

$$\text{Unit vector of force} = \frac{30i - 20j + 16k}{39.446} = 0.760\, i - 0.507\, j + 0.406\, k$$

Then in terms of unit vector $\vec{F} = 39.446\,(0.760\, i - 0.507\, j + 0.406\, k)$.

Vector joining the points $A(1, 2, -3)$ and $B(-1, -3, 4)$ is

$$\overrightarrow{AB} = \vec{r} = (-1 - 1)\, i + (-3 - 2)\, j + [4 - (-3)]\, k$$
$$= -2\, i - 5\, j + 7\, k$$

$$\text{Unit vector along } AB = \frac{-2i - 5j + 7k}{\sqrt{(-2)^2 + (-5)^2 + (7)^2}} = \frac{-2i - 5j + 7k}{\sqrt{78}}$$
$$= -0.226\, i - 0.566\, j + 0.792\, k$$

The component of force F along AB is

$$= (\vec{F} \text{ in terms of unit vectors}) \cdot (\text{unit vector along } AB)$$
$$= 39.446\,(0.760\, i - 0.507\, j + 0.406\, k) \cdot (-0.226\, i - 0.566\, j - 0.792\, k)$$
$$= 39.446\,[0.760 \times (-0.226) - 0.507 \times (-0.566) + 0.406 \times 0.792]$$
$$= 39.446\,(-0.1718 + 0.2869 + 0.3215) = \mathbf{17.224 \text{ kN}}$$

EXAMPLE 1.12

A force $\vec{F}$ acts at the origin of a coordinate system in a direction defined by the angles $\theta_x = 68°$ and $\theta_z = 55°$. The component of force $\vec{F}$ along y-direction is – 125 N. Determine:

(*a*) angle θ_y **(*b*) magnitude of force $\vec{F}$**

(*c*) components of force along x and z directions, and

(*d*) component of force on a line that passes through the origin and point (1, 1, 1)

Solution: The direction cosines along the x and z directions are:

$$l = \cos\theta_x = \cos 68° = 0.3746$$
$$n = \cos\theta_z = \cos 55° = 0.5736$$

From the relation: $l^2 + m^2 + n^2 = 1$, we have

$$m = \cos\theta_y = \sqrt{1 - l^2 - n^2} = \sqrt{1 - (0.3746)^2 - (0.5736)^2} = \pm 0.7332$$

That gives: $\theta_y = \mathbf{42.85°}$ and $\mathbf{137.15°}$

(*b*) From the relation: $F_y = F\cos\theta_y$ we have

$$\text{Magnitude of force } F = \frac{F_y}{\cos\theta_y} = \frac{-125}{\cos 137.15} = \mathbf{170.53 \text{ N}}$$

(*c*) Then scalar components of force in the *x* and *z* directions are :

$$F_x = F\cos\theta_x = 170.53 \times \cos 68^\circ = \mathbf{63.88\ N}$$
$$F_z = F\cos\theta_z = 170.53 \times \cos 45^\circ = \mathbf{97.81\ N}$$

Vector equation of force

$$\vec{F} = F_x\,i + F_y\,j + F_z\,k = 63.88\,i - 125\,j + 97.81\,k$$

(*d*) Coordinates of the given point *A*(1, 1, 1)

$$\text{Vector } \overrightarrow{OA} = (1-0)\,i + (1-0)\,j + (1-0)\,k = i + j + k$$

$$\text{Unit vector along } OA = \frac{i+j+k}{\sqrt{1^2+1^2+1^2}} = \frac{i+j+k}{1.732}$$

$$\text{Force vector } \vec{F} = \text{magnitude of force} \times \text{unit vector along the direction } OA$$

$$= 170.53 \times \frac{i+j+k}{1.732} = 98.46\,(i + j + k)$$

Thus the components of force on a line that passes through the origin and point (1, 1, 1) are

$$F_x = 98.46\text{ N}\,; \quad F_y = 98.46\text{ N} \quad \text{and} \quad F_z = 98.96\text{ N}$$

1.11. DIMENSIONS AND UNITS

Dimensions refer to the qualitative characteristics for physical quantities, while units are standards of comparison for the quantitative measure of dimensions. For elaborating the difference, let us consider the distance between two points. The term length when applied to it gives the qualitative concept of this physical quantity. The term unit would, however, indicate the magnitude of distance. The distance may be quantitatively expressed as a metre or a mile. The metre and mile will then be identical dimensions. However, they will be different units as each contains a different amount of quantity of length.

Based on the units used to represent the three fundamental quantities namely mass (*m*), length (*l*) and time (*t*), the commonly used system of units are

- *FPS (Foot-Pound-Second) System of units:* In this system the length is expressed in foot, mass in pound and time in second. The unit of force is poundal which is defined as the force acting on a mass of one pound and producing an acceleration of 32.2 ft/s^2

$$1 \text{ poundal} = 32.2\ \frac{\text{pound ft}}{\text{s}^2}$$

- *CGS (Centimetre-Gram-Second) System of units:* In this system the length is expressed in centimetre, mass in gram and time in second. The unit of force is dyne which is defined as the force acting on a mass of one gram and producing an acceleration of 1 cm/s^2.

$$1 \text{ dyne} = 1\ \frac{\text{gm cm}}{\text{s}^2}$$

- *MKS (Metre-kilogram-Second) System of units* : In this system, the fundamental units are metre for length, kilogram for mass and second for time. The unit of force is kilogram force (Kg_f) which represents the force required to move a mass of one kilogram with an acceleration equal to 9.81 m/s^2 (gravitational acceleration near the earth's surface)

$$1 \text{ kg}_f = 9.81\ \frac{\text{kg m}}{\text{s}^2}$$

Almost all major countries of the world use SI units (System International d' Unit or International system of units) in which the fundamental units are metre (m) for measuring length, kilogram (kg) for measuring mass, and second (s) for measuring time. The force is expressed in newton (N); Newton is the amount of force which causes one kilogram mass to move with an acceleration of 1 m/s^2

$$1\,N = 1\,\frac{kg\,m}{s^2}$$

The relation between different units of force are:

$$1\,kg_f = 9.81\,N = 9.81 \times 10^5\,dyne$$

Some derived units pertinent to mechanics are:

Quantity	Unit	Symbol
Area	square metre	m^2
velocity	metre per second	m/s
Acceleration	metre per second square	m/s^2
Density	kilogram per cubic metre	kg/m^3
Force	newton	N
Moment	newton-metre	Nm
Stress	newton per square meter	N/m^2 = pascal
Work and energy	joule	J = Nm
Power	watt	W = J/s = Nm/s

- **Joule** is the work done when the point of application of a force of one newton is displaced through one meter along its line of action.

  Work = force × distance moved
  = newton × metre = Nm = J or joule

- **Watt** represents a work equivalent of one joule done per second

  Power = work done per unit time
  = J/s = W or watt ; W = 1 J/s

- **Stress** is force per unit area. A stress equivalent to one newton per square metre is called Pascal

$$1\,Pa = 1\,N/m^2$$

To facilitate working with magnitudes which may be multiples or submultiples of SI units, we use a prefix before the unit to suit a particular situation. The commonly used prefixes are listed in Table 1.1.

Table 1.1. SI System : Standard Multipliers

Factor	Prefix	Factor	Prefix
10^{12}	tera, T	10^{-3}	milli, m
10^{9}	giga, G	10^{-6}	micro, μ
10^{6}	mega, M	10^{-9}	nano, n
10^{3}	kilo, k	10^{-12}	pico, p
10^{-1}	deci, d	10^{-15}	femto, f
10^{-2}	centi, c	10^{-18}	atto, a

The use of tera, mega, kilo, milli, micro and pico is preferred.

The SI system of units is comprehensive, coherent, rational, brief and the easiest for the numerical and practical purposes, when used all by themselves.

Some of the additional comments on SI units are :

- While writing plurals, do not add s to the symbols. Thus write kg and not kgs.
- Units with name of scientists are not to be written with capital letters. Thus write newton and not Newton as unit of force.
- There are to be no dots and no dashes. For example work is in Nm, not N.m, not N-m etc.
- After every three digits, there is to be a gap but no comma, *e.g.*, 2 000 and not 2,000
- For practical purposes, use of units like tonne, degree celcius, litre, speed in km/hr is permitted but should be restricted as far as possible.

1.12. DIMENSIONAL HOMOGENEITY

The fundamental theory of dimensional analysis is based on the following axiom :

"Equations describing a physical phenomenon must be dimensionally homogeneous and the units therein must be consistent."

An equation is said to be dimensionally homogeneous if the dimensions of various terms on the two sides of the equation are identical. A dimensionally homogeneous equation is independent of the fundamental units of measurement if the units therein are consistent. This implies that a length dimension can be added to or subtracted from only a length dimension. Addition of area to volume is meaningless since area and volume are dimensionally incompatible. Consistency of units demands that if one term of an equation is measured in a particular unit of velocity say m/s, then all other terms in the equation must be measured in the same units of velocity.

The concept of dimensional homogeneity can be elaborated by considering the time period of oscillation T of a simple pendulum of length l and mass m :

$$T = 2\pi\sqrt{l/g}$$

Substituting the dimensions on both sides of the equation :

$$[T] = [1]\,[L/LT^{-2}]^{1/2} = [T]$$

The dimension of 2π, being constant, is taken as unity. The equation is thus dimensionally homogeneous because dimensions on both sides of the equation are same. Substituting the value of gravitational acceleration g = 9.807 m/s^2, equation is modified to

$$T = 2\pi\sqrt{l/9.807} = 2.006\sqrt{l}$$

This modified equation is not dimensionally homogeneous because the right hand side does not have the dimensions of time. Further, it is valid only when the time is measured in seconds and the length is measured in metres.

Some well known examples of dimensionally homogeneous relationships used in mechanics are:

- Force = rate of change of momentum

Now : force = mass × acceleration $\equiv M \times LT^{-2} = MLT^{-2}$

$$\text{rate of change of momentum} = \frac{\text{mass} \times \text{velocity}}{\text{time}} = \frac{M \times (LT^{-1})}{T} = MLT^{-2}$$

Obviously the dimensions of force are the same as that of the rate of change of momentum.

- Work done = change in kinetic energy

Now : work done = force × displacement $\equiv MLT^{-2} \times L = ML^2T^{-2}$

change in kinetic energy $= \frac{1}{2}$ mass × (velocity)$^2 \equiv \frac{1}{2} M \times (LT^{-1})^2 \equiv \frac{1}{2} ML^2T^{-2}$

Obviously the dimensions of work done are the same as those of change in kinetic energy.

It is worthwhile to point out that two dimensionally homogeneous equations can be multiplied or divided without affecting the homogeneity. However, an equation resulting from the addition or subtraction of two dimensionally homogeneous equations may not be dimensionally homogeneous. The students may verify this fact by considering the following two equations which are dimensionally homogeneous :

$$v = u + at\,; \qquad s = ut + \frac{1}{2}at^2$$

where s is the distance travelled in time t by a particle having velocity u and acceleration a; v represents the final velocity.

Dimensional homogeneity helps to convert the units from one system to another. Let it be required to convert pressure from kg_f/cm^2 to N/m^2. Recalling that : 1 kg_f = 9.807 N and 1 cm = 0.01 m, the conversion factor becomes

$$= \frac{kg_f/cm^2}{N/m^2} = \left[\frac{kg_f}{N}\right] \times \left[\frac{m}{cm}\right]^2 = 9.807 \times \left(\frac{1}{0.01}\right)^2 = 9.807 \times 10^4$$

Thus any pressure in kg_f/cm^2 can be converted to N/m^2 by multiplying it with 9.807×10^4.

It also helps to determine the dimensions of a physical quantity. Consider the hydrostatic equation $p = wh$, where p is the pressure intensity, h is the height of fluid column and w is the specific weight of the fluid.

Dimensionally, $p = [ML^{-2}T^{-2}]\,[L] = ML^{-1}T^{-2}$

Further, if w is taken in N/m^3 and h in m of fluid column, then units of pressure intensity would be

$$p = \left(\frac{N}{m^2}\right) \times m = \frac{N}{m^2}$$

REVIEW AND SUMMARY

1. Mechanics is the science that describes the conditions when a body is said to be at rest or in motion under the effect of forces.
2. Statics deals with bodies at rest, while dynamics concerns the bodies which are in motion. Dynamics is further divided into kinematics and kinetics.

   Kinematics deals with the relationship between displacement, velocity, acceleration and time of a given motion, without considering the forces that cause the motion.

   Kinetics is the study of relationship between forces acting on the body, and the mass and motion of the body.
3. A rigid body has definite amount of matter and the distance between the particles forming it remains constant before and after the application of force.
4. A quantity is said to be scalar when it is completely defined by its magnitude alone. Mass, volume, temperature and energy etc. are scalar quantities.

   Any physical quantity that possesses magnitude as well as direction is called a vector quantity. Displacement, velocity, acceleration and momentum etc. are vector quantities.
5. Force is an external agency that tends to change the state of rest or of uniform motion of a body.

Force is a vector quantity, and is completely defined by its magnitude, direction and point of application. The application of force on a body may

- change its state of rest or motion
- accelerate or retard its motion
- change its shape and size
- turn or rotate it, and
- keep it in equilibrium

6. Depending upon the position of line of action, forces may be collinear (along the same line), non-coplanar (not lying in the same plane), non-concurrent (not meeting at a point) and co-plnar concurrent (lying in the same plane and meeting at a point), etc.
7. According to the principle of transmissibility, the effect of forces on a body is same at every point in its line of action.
8. Law of dimensional homogeneity states that a basic equation representing a phenomenon phenoment must be valid for all system of units.
9. A vector is a quantity possessing magnitude and direction, and the two vector quantities are added by parallelogram law or triangle law. Depending upon orientation, a vector $\vec{A}$ may have component $\vec{A_x}$, $\vec{A_y}$ and $\vec{A_z}$ along the x, y and z-directions. Then

$$\vec{A} = A_x i + A_y j + A_z k$$

where i, j and k represent the unit vectors along the x, y and z-axes respectively.

- The unit vector is a vector of unit length and it is determined by dividing a vector by its magnitude

$$\text{Unit vector} = \frac{\vec{A}}{|\vec{A}|} = \frac{\vec{A}}{\sqrt{A_x^2 + A_y^2 + A_z^2}}$$

- When a vector is subtracted from itself, we get a zero vector or null vector.

The –ve of a vector $\vec{A}$ is $-\vec{A}$. The vector $-\vec{A}$ will have the same magnitude and inclination as $\vec{A}$ but its sense (direction) will be opposite to that of $\vec{A}$.

- When a line starts from the origin (0, 0, 0) and ends with a point $(x, -y, z)$ in space, its position vector will be

$= (x - 0)\, i + (-y - 0)\, j + (z - 0)\, k$

$= x\, i - y\, j + z\, k$

10. The dot product of vectors $\vec{A}$ and $\vec{B}$ is a scalar quantity, and is defined as the product of the magnitude of the vectors and cosine of their included angle.

$$\vec{A}.\vec{B} = |\vec{A}|.|\vec{B}|\cos\theta = AB\cos\theta$$

The dot product defines the term work in the field of dynamics.

- The cross product of vector $\vec{A}$ and $\vec{B}$ is a vector quantity and is defined as the product of the magnitude of vectors and the sine of the smaller angle between the vectors.

$$\vec{A}\times\vec{B} = AB\,\sin\theta\,\vec{n}$$

If $\vec{A} = A_x i + A_y j + A_z k$ and $\vec{B} = B_x i + B_y j + B_z k$, then the cross product in terms of rectangular components is given by

$$\vec{A} \times \vec{B} = \begin{vmatrix} i & j & k \\ A_x & A_y & A_z \\ B_x & B_y & B_z \end{vmatrix}$$

$$= (A_y B_z - B_y A_z)\, i - (A_x B_z - B_x A_z)\, j + (A_x B_y - B_x A_y)\, k$$

The cross product defines the term moment of force.

REVIEW QUESTIONS

A. Conceptual and conventional questions :

1. Differentiate between statics, kinetics and kinematics.
2. Define matter, particle and body. How does a rigid body differ from an elastic body?
3. Define space, time, motion and trajectory.
4. When is the motion of a particle called rectilinear, and when curvilinear ?
5. State Newton's law of motion.
6. What is Newton's law of gravitation?
7. Differentiate between scalar and vector quantities. How a vector quantity is represented?
8. Given below is a list of some of the quantities used in mechanics

   mass, displacement, momentum, area, volume, velocity, weight, speed, temperature and force.

   Classify these into scalar and vector quantities.
9. Make a clear distinction between mass, force and weight.
10. What are the two ways for the representation of forces ?
11. Represent graphically a force of 50 N acting upwards at 60° to the horizontal and towards the right.
12. State the effects which a force may produce when it acts on a body.
13. Enumerate the characteristics of a force.
14. What is a force system ? What is its point of concurrency ?
15. How forces are specified depending upon the position of their line of action?
16. Sketch a nonconcurrent-non parallel-coplanar force system.
17. State the difference between equilibrium, resultant and equilibrant.
18. Explain the principle of transmissibility of forces.
19. How does the term dimension differ from the term unit?
20. State the unit of force into CGS, MKS and SI system of unit. How are they related to each other ?
21. Name various system of units. List the salient features of the system which is being used these days.
22. What is meant by dimensional homogeneity ? Point out its utility.
23. Define the terms unit vector and position vector.
24. Explain the concept of dot and cross products in the context of vectors.
25. A line originates at the points (4, 1, – 2) and passes through the point (2, 2, 6). What will be the unit vector along this line?
26. Two vectors $\vec{A}$ and $\vec{B}$ are defined by the relations.

$$\vec{A} = 3i + 4j + 6k \quad \text{and} \quad \vec{B} = 4i + 5j - 9k$$

Determine the sum, difference and dot product of these vectors. Also find the angle between $\vec{A}$ and $\vec{B}$. **($7\,i + 9\,j - 3\,k$; $-i - j + 15\,k$; $-2\,z$; 109.2°)**

27. Two vectors $\vec{A}$ and $\vec{B}$ with magnitude 10 units and 15 units respectively have a dot product equivalent to 50 units. Find the magnitude of their cross product. **(144.4 units)**
28. The coordinates of the initial and terminal points of a vector are (3, 1, – 2) and (4, – 7, 10). Specify the vector, evaluate its magnitude and direction cosines. **($i - 8\,j + 12\,k$, 14.46, 0.069, – 0.553, 0.830)**
29. The line of action of a 100 N force $\vec{F}$ passes through points *A* (2, 5, 8) and *B* (7, 2, 6). The direction of force is from A to B. Express $\vec{F}$ in terms of the unit vectors *i*, *j* and *k*. **($81.11\,i - 48.67\,j - 32.44\,k$)**
30. A constant force $4\,i + 3\,j$ moves a particle along straight line from position (6, 9, 5) to position (4, 6, 8). What work will be done by the force?
Take force in newtons and distances in metres. **(– 17 Nm)**

B. Complete the following statements with most appropriate word/words :

1. Mechanics is a science which describes and predicts the state of .... or .... under the action of forces.
2. .... relates to bodies in motion without any reference to forces.
3. A body which does not deform under the action of applied forces is called the ....
4. A .... is a body of infinitely small volume and is considered to be concentrated at a point
5. A .... quantity is one that has magnitude as well as direction
6. Magnitude, direction, sense and .... are the characteristics of a force
7. According to .... of forces, the effect of a force upon a body is the same at every point in its line of action
8. Fundamental principles of mechanics are ....
9. Forces whose lines of action pass through a common point are called .... forces.
10. A .... is a single force which can replace two or more forces and produce the same effect as the forces.
11. Equlibrant is equal and opposite to the .... of several forces acting or a body.
12. .... is a method of designating a force by writing two capital letters one or either side of the force.
13. The units which are used for the measurement of fundamental quantities are called .... units.
14. The unit vector is obtained by dividing the vector by ....
15. A zero vector is also called a .... vector.
16. A vector $\vec{F} = 2\,i + 3\,j - 4\,k$ passes through the origin. The magnitude of its vector component along *y*-axis is ....
17. The dot product of two vector quanities is defined as the product of the magnitude of the vectors and .... of their included angle.
18. Two vectors will be .... to each other if their cross product is zero.

Answers :

1. rest, motion 2. kinematics 3. rigid body 4. particle 5. vector 6. point of application 7. principle of transmissibility 8. Newton's of motion and gravitation, parallelogram law, principle of transmissibility etc. 9. concurrent 10. resultant 11. resultant 12. Bow's notation 13. basic 14. its magnitude 15. null 16. three 17. cosine 18. parallel

C. Multiple choice questions :

1. Identify the wrong statement about matter, particle and body
   (*a*) matter is any thing that occupies space and possesses mass
   (*b*) a body has definite shape and consists of number of particles
   (*c*) a particle has infinitely small volume and its mass is considered to be concentrated at a point
   (*d*) a body exhibits no relative deformation between two parts when it is acted upon by a force system
2. For analysis purposes, a body can be modelled as rigid if
   (*a*) it has continuous distribution of mass
   (*b*) it is characterized by some preferred configuration of its own
   (*c*) relative motion between its parts is negligible
   (*d*) its dimensions are small compared with the co-ordinates describing its motion
3. Which of the following quantities is not a vector quantity?
   (*a*) energy (*b*) momentum (*c*) acceleration (*d*) force
4. Newtons first law of motion gives the concept of
   (*a*) work (*b*) force (*c*) inertia (*d*) energy
5. Newton's second law of motion connects
   (*a*) change of momentum and velocity
   (*b*) momentum and acceleration
   (*c*) momentum and rate of change of force
   (*d*) rate of change of momentum and external force
6. A force which combines with two or more forces to produce equilibrium is called
   (*a*) resultant (*b*) equilibrant (*c*) couple (*d*) moment
7. Which of the following products is not equal to zero ?
   (*a*) micro × mega (*b*) nano × giga (*c*) pico × tera (*d*) hecto × deci
8. According to the principle of transmissibility of forces, the effect of force on a body is
   (*a*) same at every point in its line of action
   (*b*) different at different points in its line of action
   (*c*) minimum when it acts at the centre of gravity of the body
   (*d*) maximum when its acts at the centre of gravity of the body
9. The unit of k in the relation $U = \frac{ky}{y^2 + a^2}$ is
   (*a*) m/s (*b*) ms (*c*) Jm (*d*) J/s
   In the given expression U represents the potential energy, y represents the displacement and a represents the maximum displacement, *i.e.*, amplitude.

Answers :

1. (*d*) 2. (*c*) 3. (*a*) 4. (*c*) 5. (*d*) 6. (*b*)
7. (*d*) 8. (*a*) 9. (*c*)

CHAPTER

2

TWO-DIMENSIONAL CON-CURRENT FORCE SYSTEMS

Learning objectives : The subject matter included in this chapter will enable the reader to

- ❑ *understand the concept of resultant and equilibrant, composition and resolution of forces*
- ❑ *state and explain the parallelogram law, triangle law and polygon law for addition of forces*
- ❑ *deal with the analytical and graphical methods for determining the resultant of coplanar force system*
- ❑ *draw the free body diagram; consideration of internal and external forces*
- ❑ *state the law of sines and Lami's theorem*

When a number of concurrent forces act on a body to keep the body in equilibrium, then resultant of these forces will be the vector sum of all these forces. The process of determination of resultant in known as *composition* of forces. Also from a given force, two forces in two directions can be obtained, which will produce the same effect as produced by single given force. This process is known as *resolution* of a force.

2.1. PARALLELOGRAM LAW OF FORCES

The parallelogram law of forces is used to determine the resultant of two forces acting at a point in a plane and inclined to each other at an angle.

It states that:

"*If two forces, acting at a point be represented in magnitude and direction by the two adjacent sides of a parallelogram, then their resultant is represented in magnitude and direction by the diagonal of the parallelogram passing through that point.*"

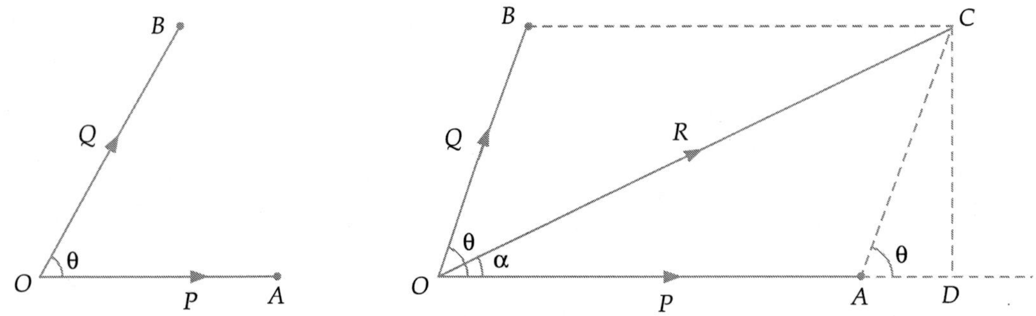

Fig. 2.1.

Consider two forces P and Q acting on a body at O as shown in Fig 2.1. The force P is represented in magnitude and direction by $\overrightarrow{OA}$ whereas the force Q is represented in magnitude and direction by $\overrightarrow{OB}$. Let the angle between the two forces be θ. The resultant of these two forces is obtained by the diagonal $\overrightarrow{OC}$ of the parallelogram $OACB$, as shown in Fig. 2.1.

The relationship between P, Q and R can be derived as follows:

Drop perpendicular from C and let it meet OA extend at point D. In $\Delta\,CAD$, side CA is parallel and equal to OB, *i.e.*, it represents force Q.

The resultant R of P and Q is given by

$$R = OC = \sqrt{OD^2 + CD^2} = \sqrt{(OA + AD)^2 + CD^2}$$

But
$$OA = P$$
$$AD = AC\cos\theta = Q\cos\theta$$
$$CD = AC\sin\theta = Q\sin\theta$$

$$\therefore \quad R = \sqrt{(P + Q\cos\theta)^2 + (Q\sin\theta)^2} = \sqrt{P^2 + Q^2\cos^2\theta + 2PQ\cos\theta + Q}$$

$$= \sqrt{P^2 + Q^2(\sin^2\theta + \cos^2\theta) + 2PQ\,c} = \sqrt{P^2 + Q^2 + 2PQ\cos\theta} \qquad ...(2.1)$$

The inclination of the resultant R to the direction of force P is given by

$$\tan\alpha = \frac{CD}{OD} = \frac{CD}{OA + AD} = \frac{Q\sin\theta}{P + Q\cos\theta}$$

$$\alpha = \tan^{-}\left[\frac{Q\sin\theta}{P + Q\cos\theta}\right] \qquad ...(2.2)$$

Note: It is not necessary that for the law of parallelogram for forces to be valid, one of the two forces should be along the x-axis. The forces P and Q may be in any direction as shown in Fig. 2.2.

If the angle between the forces is θ, then

$$R = \sqrt{P^2 + Q^2 + 2PQ\cos\theta}$$

The direction of resultant will be

$$\alpha = \tan^{-1}\frac{Q\sin\theta}{P + Q\cos\theta}$$

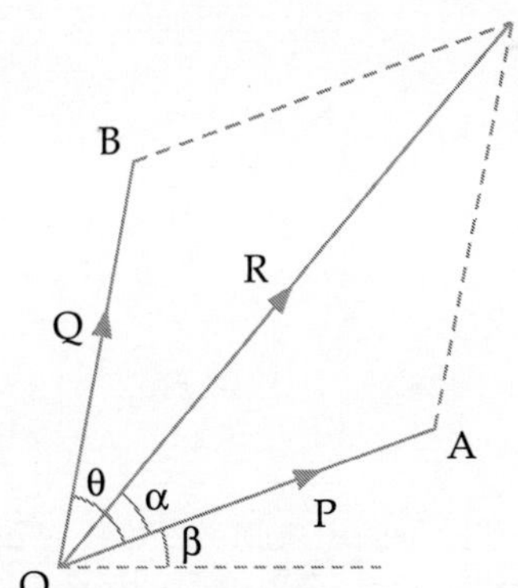

Fig. 2.2

α is the angle which the resultant makes with the direction of P. If β is the angle which the force P makes with the x-axis, the angle made by the resultant with x-axis will be $(\alpha + \beta)$.

Special cases:

(*i*) When the two forces are equal and θ is the angle between them:

$$R = -\sqrt{P^2 + Q^2 + 2PQ\cos\theta} = \sqrt{P^2 + P^2 + 2P \times P\cos\theta}$$

$$= \sqrt{2P^2(1 + \cos\theta)} = \sqrt{2P^2 \times 2\cos^2\frac{\theta}{2}} = 2P\cos\frac{\theta}{2} \qquad ...(2.3)$$

and $$\alpha = \tan^{-1}\left(\frac{Q\cos\theta}{P + Q\cos\theta}\right) = \tan^{-1}\left(\frac{P\sin\theta}{P + P\cos\theta}\right) = \tan^{-1}\left(\frac{\sin\theta}{1 + \cos\theta}\right)$$

$$= \tan^{-1}\left[\frac{2\sin\frac{\theta}{2}\cos\frac{\theta}{2}}{2\cos^2\frac{\theta}{2}}\right] = \tan^{-1}\left(\tan\frac{\theta}{2}\right)$$

$$\therefore \qquad \alpha = \frac{\theta}{2} \qquad ...(2.4)$$

i.e., the resultant bisects the angle between the forces.

(*ii*) When the two forces act at right angles; *i.e*, $\theta = 90°$

$$R = \sqrt{P^2 + Q^2 + 2PQ\cos 90°} = \sqrt{P^2 + Q^2} \quad (\because \cos 90° = 0) \qquad ...(2.5)$$

and $$\alpha = \tan^{-1}\frac{Q\sin 90°}{P + Q\cos 90°} = \tan^{-1}\left(\frac{Q}{P}\right) \qquad (\because \sin 90° = 1)$$

(*iii*) When the two forces act in the same line and same sense, *i.e.*, $\theta = 0°$

$$R = \sqrt{P^2 + Q^2 + 2PQ\cos 0°} = \sqrt{P^2 + Q^2 + 2PQ} \qquad (\because \cos 0° = 1)$$

$$= P + Q \qquad ...(2.6)$$

Apparently the resultant is maximum when the forces are collinear and act in the same direction.

(*iv*) When the two forces have the same line of action but opposite senses, *i.e.*, $\theta = 180°$

$$R = \sqrt{P^2 + Q^2 + 2PQ\cos 180°} = \sqrt{P^2 + Q^2 - 2PQ} \qquad (\because \cos 180° = -1)$$

$$= P - Q \qquad ...(2.7)$$

Obviously the resultant is minimum when the two forces are collinear but act in opposite direction.

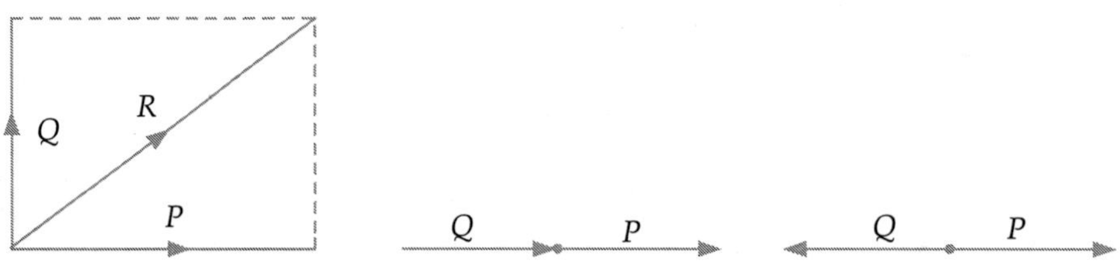

Fig. 2.3

The results obtained vide equations (2.5), (2.6) and (2.7) have been represented in Fig 2.3. These results lead us to conclude that when the forces acting on a body are collinear, their resultant is equal to the algebraic sum of the forces.

EXAMPLE 2.1

(*i*) Two forces of equal mangnitude *P* act at an angle θ to each other. What will be their resultant ?

(*ii*) Resultant of two equal forces is equal to either of them. Determine the angle between the forces.

(*iii*) The resultant of two forces (P + Q) and (P – Q) equals $\sqrt{3P^2 + Q^2}$. Show that the forces are then inclined to each other at an angle of 60°.

(*iv*) The resultant of two forces P and Q is R. If one of the forces is reversed in direction, The resultant is S. Then for the identity

$$R^2 + S^2 = 2(P^2 + Q^2)$$

to hold good, show that the forces can have any angle of inclination between them.

Solution : Solution to these forces stems from the parallelogram law of forces.

(i) $$R^2 = P^2 + Q^2 + 2PQ\cos\theta = P^2 + P^2 + 2P^2\cos\theta$$

$$= 4P^2\left(\frac{1+\cos\theta}{2}\right) = 4P^2\cos^2\frac{\theta}{2}$$

$$\therefore \quad R = 2P\cos\frac{\theta}{2}$$

(ii) $R = 2P\cos\frac{\theta}{2}$ as outlined above.

Since the resultant equals either of the two equal forces,

$$P = 2P\cos\frac{\theta}{2}; \quad \cos\frac{\theta}{2} = \frac{1}{2}$$

$$\frac{\theta}{2} = 60^\circ$$

$$\therefore \quad \theta = 120^\circ \text{ or } \frac{2\pi}{3}$$

(iii) $$3P^2 + Q^2 = (P+Q)^2 + (P-Q)^2 + 2(P+Q)(P-Q)\cos\theta$$

$$= 2P^2 + 2Q^2 + 2(P^2 - Q^2)\cos\theta$$

or $$(P^2 - Q^2) = 2(P^2 - Q^2)\cos\theta$$

or $$\cos\theta = \frac{1}{2}; \quad \theta = 60^\circ \text{ or } \frac{\pi}{3}$$

(iv) $$R^2 = P^2 + Q^2 + 2PQ\cos\theta$$

$$S^2 = P^2 + (-Q)^2 + 2P(-Q)\cos\theta = P^2 + Q^2 - 2PQ\cos\theta$$

upon summation

$$R^2 + S^2 = 2(P^2 + Q^2)$$

Obviously for the given identify to hold good, the forces can have any angle of inclination between them.

EXAMPLE 2.2

The resultant of two forces P and Q is at right angles to P. Show that the angle between the forces is $\cos^{-1}(-P/Q)$.

Solution : Let θ be the angle between the two given forces P and Q. Then the inclination α which the resultant makes with force P is

$$\tan\alpha = \frac{Q\sin\theta}{P + Q\cos\theta}$$

Since $\alpha = 90^\circ$ (given), we get

$$\tan 90^\circ = \frac{Q\sin\theta}{P + Q\cos\theta}$$

$$\infty = \frac{Q\sin\theta}{P + Q\cos\theta}$$

This is possible only when the denominator is zero

$$P + Q\cos\theta = 0; \quad \cos\theta = (-P/Q)$$

$$\therefore \quad \theta = \cos^{-1}(-P/Q)$$

EXAMPLE 2.3

Two locomotives on opposite banks of a canal pull a vessel moving parallel to the banks by means of two horizontal ropes. The tensions in these ropes have been measured to be 20 kN and 24 kN while the angle between them is 60°. Find the resultant pull on the vessel and the angle between each of the ropes and the sides of the canal.

Solution : The vessel is acted upon at point *O* by the force system as depicted in the figure given below:

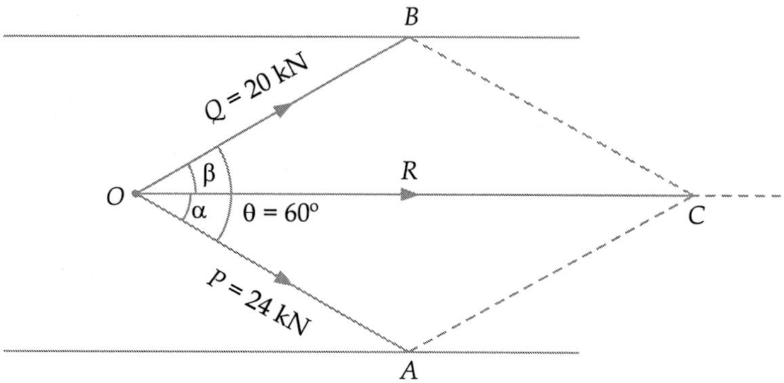

Fig. 2.4

The resultant force exerted on the vessel can be worked out by applying the parallelogram law of forces. That is

$$R = \sqrt{P^2 + Q^2 + 2PQ\cos\theta} = \sqrt{24^2 + 20^2 + 2\times 24\times 20\times \cos 60°}$$

$$= \sqrt{576 + 400 + 480} = \sqrt{1456} = \mathbf{38.16\ N}$$

The inclination of resultant with direction of force *P* is

$$\tan\alpha = \frac{Q\sin\theta}{P + Q\cos\theta} = \frac{20\sin 60°}{24 + 20\cos 60°} = \frac{20\times 0.866}{24 + 20\times 0.5} = 0.5094$$

$$\alpha = \mathbf{27°}$$

$$\therefore \quad \beta = 60 - 27 = \mathbf{33°}$$

EXAMPLE 2.4

The magnitude of two forces is such that when acting at right angles produce a resultant force of and when acting at 60 degree produce a resultant equal to $\sqrt{28}$. Work out the magnitude of the two forces.

Solution : Let *P* and *Q* be the magnitude of the two forces. From parallelogram law of forces,

$$R^2 = P^2 + Q^2 + 2PQ\cos\theta$$

where θ is the angle between the two forces.

Then according to the given conditions,

$$20 = P^2 + Q^2 + 2PQ\cos 90° = P^2 + Q^2 \qquad (\because \cos 90° = 0)$$

and

$$28 = P^2 + Q^2 + 2PQ\cos 60° = P^2 + Q^2 + PQ \qquad \left(\because \cos 60° = \frac{1}{2}\right)$$

From the above identities $PQ = 8$; $Q^2 = 64/P^2$

That gives: $20 = P^2 + 64/P^2$

or $P^4 - 20P^2 + 64 = 0$

or $(P^2 - 16)(P^2 - 4) = 0$

∴ $P = 2$ or 4 units

The corresponding values of force Q will be 4 or 2 units.

EXAMPLE 2.5

Find the magnitude of two concurrent forces which conform to the following data:

angle between the forces = 75°

angle between the resultant and one of forces = 45°

magnitude of the resultant = 2000 N

Solution : Let P and Q be the two forces and θ be the angle between them. Further, let α be the inclination which the resultant makes with force P. Then from parallelogram law of forces

(*i*) $\tan\alpha = \dfrac{Q\sin\theta}{P+Q\sin\theta}$; $\tan 45° = \dfrac{Q\sin 75°}{P+Q\cos 75°}$

$$1 = \frac{0.996Q}{P+0.259Q} \qquad \therefore \quad P = 0.707Q$$

(*ii*) $R^2 = P^2 + Q^2 + 2PQ\cos\theta$

$$(2000)^2 = (0.707Q)^2 + Q^2 + 2 \times (0.707Q) \times Q\cos 75°$$

$$= 0.5Q^2 + Q^2 + 0.366Q^2 = 1.866Q^2$$

∴ $$Q = \sqrt{\frac{(2000)^2}{1.866}} = \mathbf{1464\ N}$$

and $$P = 0.707Q = 0.707 \times 1464 = \mathbf{1035\ N}$$

EXAMPLE 2.6

The greatest and least resultant of two forces acting on a body are 35 kN and 5 kN respectively. Determine the magnitude of the forces. What would be the angle between these forces if the magnitude of the resultant is stated to be 25 kN?

Solution : Let P and Q be the two forces and θ be the angle of inclination between them. Then from the parallelogram law of forces, the resultant R is

$$R^2 = P^2 + Q^2 + 2PQ\cos\theta$$

The resultant will be maximum when the forces are collinear and act in the same direction, *i.e.*, θ = 0°. That gives

$$R^2 = \sqrt{P^2+Q^2+2PQ\cos 0°} = \sqrt{P^2+Q^2+2PQ} = P + Q$$

∴ $$35 = P + Q \qquad ...(i)$$

The resultant will be minimum when the forces are collinear and act in the opposite direction, *i.e.*, θ = 180°. That gives

$$R = \sqrt{P^2+Q^2+2PQ\cos 180°} = \sqrt{P^2+Q^2-2PQ} = P - Q$$

∴ $$5 = P - Q \qquad ...(ii)$$

From identities (*i*) and (*ii*)

$$P = \mathbf{20\ kN} \text{ and } Q = \mathbf{15\ kN}$$

(*b*) Let θ be the angle between the force P = 20 kN and Q = 15 kN when their resultant is 25 kN. Then

$$25^2 = 20^2 + 15^2 + 2 \times 20 \times 15\cos\theta$$

$$625 = 400 + 225 + 600 \cos \theta$$

$$\cos \theta = 0; \quad \theta = 90°$$

Thus the given system of forces are at right angles to each other when the resultant is 25 kN.

EXAMPLE 2.7

Two forces, one of which is double the other has resultant of 260 N. If the direction of the larger force is reversed and the other remains unaltered, the resultant reduces to 180 N. Determine the magnitude of the force and the angle between the forces.

Solution : Let P and Q (= $2P$) be the two forces and θ be the angle of inclination between them. From parallelogram law of forces

$$R^2 = P^2 + Q^2 + 2PQ \cos \theta$$

Then according to the given conditions,

$$260^2 = P^2 + (2P)^2 + 2P\,(2P) \cos \theta$$

$$= P^2 + 4P^2 + 4P^2 \cos \theta = 5P^2 + 4P^2 \cos \theta \qquad ...(i)$$

and

$$180^2 = P^2 + (-2P)^2 + 2P\,(-2P) \cos \theta$$

$$= P^2 + 4P^2 - 4P^2 \cos \theta = 5P^2 - 4P^2 \cos \theta \qquad ...(ii)$$

Adding identities (i) and (ii) we get

$$10P^2 = 260^2 + 180^2 = 67600 + 32400 = 100000$$

$$\therefore \quad P = \sqrt{\frac{100000}{10}} = \mathbf{100\ N}$$

$$Q = 2P = 2 \times 100 = \mathbf{200\ N}$$

Substituting $P = 100$ N in identity (*i*), we get

$$260^2 = 5 \times 100^2 + 4 \times 100^2 \cos \theta$$

$$\cos \theta = \frac{260^2 - 5 \times 100^2}{4 \times 100^2} = 0.44$$

$$\therefore \quad \theta = \mathbf{63.896°}$$

EXAMPLE 2.8

Two forces equal to P and 2 P respectively act on a particle. When the first force is increased by 120 N and the second force is doubled, the direction of the resultant remains the same. Determine the value of force P.

Solution : Let α be the inclination which the resultant makes with force P. Then for the Ist case

$$\tan \alpha = \frac{2P \sin \theta}{P + 2P \cos \theta} \qquad ...(i)$$

where θ is the angle between the two given forces.

Subsequently when the forces get changed

$$\tan \alpha = \frac{4P \sin \theta}{(P + 120) + 4P \cos \theta} \qquad ...(ii)$$

From identities (*i*) and (*ii*)

$$\frac{2P \sin \theta}{P + 2P \cos \theta} = \frac{4P \sin \theta}{(P + 120) + 4P \cos \theta}$$

or $$\frac{1}{P + 2P\cos\theta} = \frac{2}{(P+120) + 4P\cos\theta}$$

or $$(P + 120) + 4P\cos\theta = 2P + 4P\cos\theta$$

That gives: $\mathbf{P = 120\ N}$

EXAMPLE 2.9

The resultant of two forces P and Q acting at a point is R. The resultant R gets doubled when Q is either doubled or its direction is reversed. Show that P, Q and R conform to the ratio

$$P : Q : R = \sqrt{2} : \sqrt{3} : \sqrt{2}$$

Solution : From parallelogram law of forces, the resultant of forces P and Q with θ angle of inclination between them is given by

$$R^2 = P^2 + Q^2 + 2PQ\cos\theta \qquad ...(i)$$

When Q is doubled

$$(2R)^2 = P^2 + (2Q)^2 + 2P \times (2Q)\cos\theta$$

$$4R^2 = P^2 + 4Q^2 + 4PQ\cos\theta \qquad ...(ii)$$

When Q is reversed in direction,

$$(2R)^2 = P^2 + (-Q)^2 + 2P \times (-Q)\cos\theta$$

$$4R^2 = P^2 + Q^2 - 2PQ\cos\theta \qquad ...(iii)$$

Adding identities (*i*) and (*iii*)

$$5R^2 = 2P^2 + 2Q^2 \qquad ...(iv)$$

Multiplying identity (*iii*) by 2 and adding the result to identity (*ii*)

$$12R^2 = 3P^2 + 6Q^2$$

or $$4R^2 = P^2 + 2Q^2 \qquad ...(v)$$

From identities (*iv*) and (*v*)

$$5R^2 = 2P^2 + (4R^2 - P^2)$$

or $$R^2 = P^2; \qquad R = P$$

Then from identity (*v*)

$$4P^2 = P^2 + 2Q^2$$

$$3P^2 = 2Q^2 \ ; \ Q = \sqrt{3/2}\,P$$

Therefore, $$P : Q : R \equiv P : \sqrt{3/2}\,P : P \equiv \sqrt{2} : \sqrt{3} : \sqrt{2}$$

EXAMPLE 2.10

Determine the horizontal force and a force inclined at an angle of 60° with the vertical whose resultant equals a vertical force F.

Solution : Let horizontal force be F_1 and the inclined force be F_2. The angle of inclination θ between these forces is 150° and the resultant $R = F$ is at right angle to force F_1. Then from parallelogram law of forces

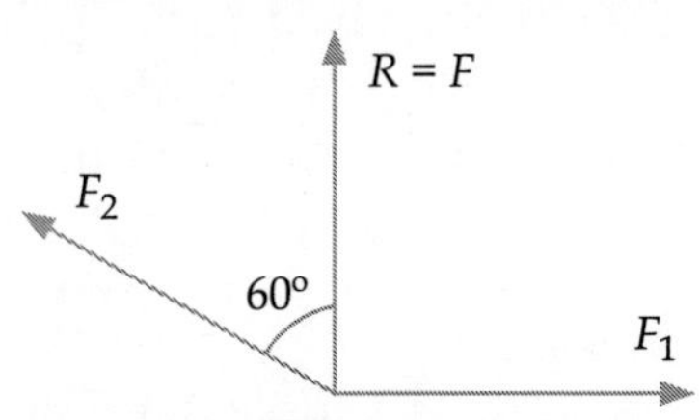

Fig. 2.5

$$R^2 = F^2 = F_1^2 + F_2^2 + 2F_1F_2\cos 150^\circ \qquad ...(i)$$

$$\tan\alpha = \tan 90^\circ = \frac{F_2\sin 150^\circ}{F_1 + F_2\cos 150^\circ}$$

Since tan 90° = ∞, the above identity will be true if

$$F_1 + F_2 \cos 150° = 0 \qquad ...(ii)$$

Upon squaring:

$$F_1^2 + F_2^2 \cos^2 150° + 2\, F_1 F_2 \cos 150° = 0$$

or

$$F_1^2 + 2F_1\, F_2 \cos 150° = -\, F_2^2 \cos^2 150°$$

Substituting the above result in identity (i), we get

$$F^2 = F_2^2 - F_2^2 \cos^2 150° = F_2^2\,(1 - \cos^2 150°) = F_2^2 \sin^2 150° = 0.25 F_2^2$$

That gives: $F_2 = \sqrt{\dfrac{F^2}{0.25}} = \mathbf{2\,F}$

Then from identity (*ii*),

$$F_1 + 2F \cos 150° = 0$$

or

$$F_1 = -\,2F \cos 150° = -2\,F\left(-\sqrt{3}/2\right) = \mathbf{\sqrt{3}}\,F$$

2.2. RESOLUTION OF FORCES

Finding the components of a given force in two given directions is called *resolution*. These component forces will have the same effect on the body as the given single force.

Let the given force be R and let it be required to find its components in directions making angles α and β with its line of action.

With reference to parallelogram $OACB$, the sides OA and OB represent the components of the given force R along OX and OY respectively. That is

$$OA = P \text{ and } OB = Q$$

Further $\angle OCA = \angle BOC$ (alternate angles)

$$= \beta$$

$$\therefore \quad \angle OAC = 180° - (\alpha + \beta)$$

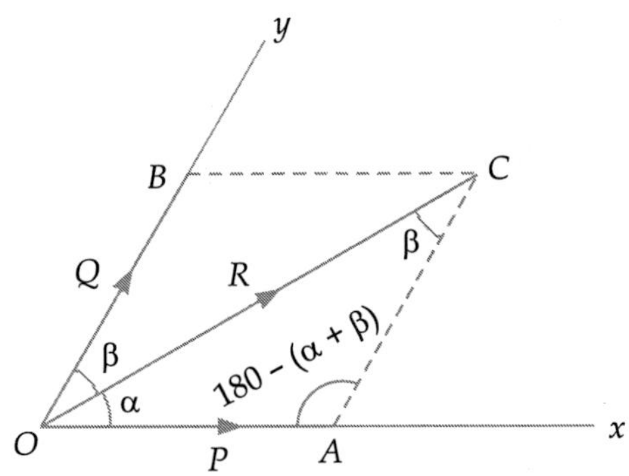

Fig. 2.6

Applying sine rule to Δ OAC, we get

$$\frac{OA}{\sin\beta} = \frac{AC}{\sin\alpha} = \frac{OC}{\sin\{180-(\alpha+\beta)\}}$$

AC is parallel and equal to OB which represents Q

$$\therefore \quad \frac{P}{\sin\beta} = \frac{Q}{\sin\alpha} = \frac{R}{\sin(\alpha+\beta)} \qquad ...(2.8)$$

Thus the resolved parts of the given force R are

$$P = R\,\frac{\sin\beta}{\sin(\alpha+\beta)} \quad \text{and} \quad Q = R\,\frac{\sin\alpha}{\sin(\alpha+\beta)} \qquad ...(2.9)$$

When the force R is to be resolved along perpendicular directions, then

OX and OY are at right angles and $OACB$ becomes a rectangle

$$\alpha + \beta = 90°$$

$$\beta = (90° - \alpha)$$

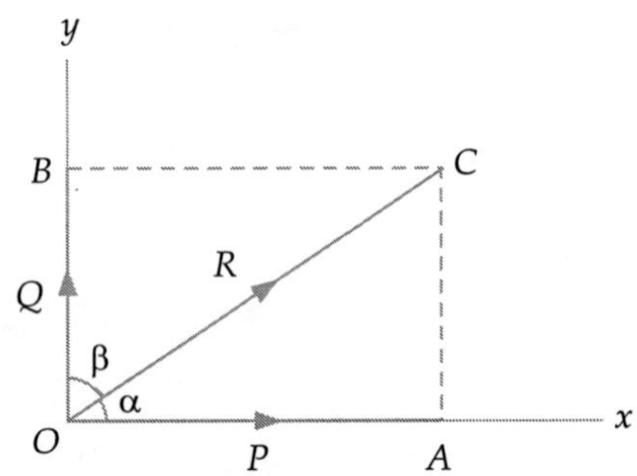

Fig. 2.7

$$\therefore \qquad P = R\frac{\sin\beta}{\sin(\alpha+\beta)} = R\frac{\sin(90^\circ-\alpha)}{\sin 90^\circ} = R\cos\alpha \qquad ...(2.10)$$

i.e., the projection of OC (representing R) on OX is a measure of resolved component along the direction OX

$$Q = R\frac{\sin\alpha}{\sin(\alpha+\beta)} = R\frac{\sin\alpha}{\sin 90^\circ} = R\sin\alpha \qquad ...(2.11)$$

i.e., the projection of OC (representing R) on OY is a measure of resolved component along the direction of OY

When the components P and Q are at right angles to each other, they are called the rectangular components of force R.

2.3. THEOREM OF RESOLVED PARTS

"The algebraic sum of the resolved parts of two forces in a given direction is equal to the resolved part of their resultant in the same direction."

Let OC represent the resultant of two forces OA and OB acting at point O as shown in Fig. 2.8. Further, let OX be the direction along which resolved parts of OA, OB and OC are to be worked out.

Drop perpendiculars form B, A and C to meet OX at B', A' and C' respectively.

The triangles OBB' and ACD are congruant and that gives $OB' = AD = A'C'$.

From the geometrical configuration of the system of forces, we have

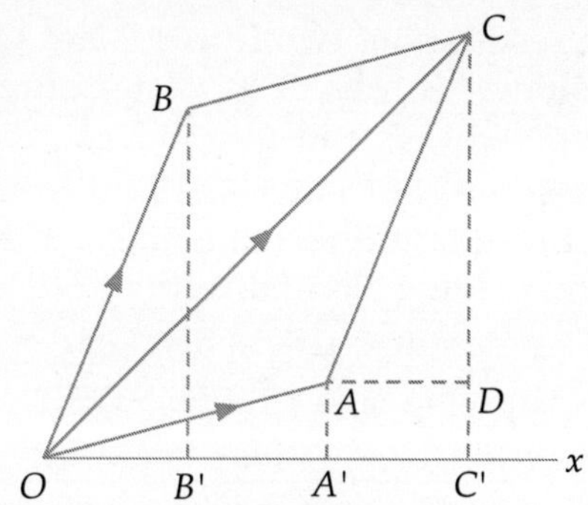

Fig. 2.8

$$OC' = OA' + A'C' = OA' + AD = OA' + OB'$$

That is: Resolved part of force OC on OX

= resolved part of force OA on OX + resolved part of force OB on OX

This theorem can be extended to any number of coplanar-concurrent forces.

EXAMPLE 2.11

The resultant of two forces is 10 N and it is inclined at 60° to one of the forces whose magnitude is 5 N. Determine the magnitude and direction of the other force.

Solution : The given force system may be depicted as shown in Fig. 2.9.

$$P = OA = 5\text{ N}$$

$$R = OC = 10\text{ N}$$

$$\alpha = 60^\circ$$

Let the other force Q be oriented along OB making an angle β with the resultant.

From the method of resolved parts,

$$P = R\frac{\sin\beta}{\sin(\alpha+\beta)} \qquad ...(i)$$

$$Q = R\frac{\sin\alpha}{\sin(\alpha+\beta)} \qquad ...(ii)$$

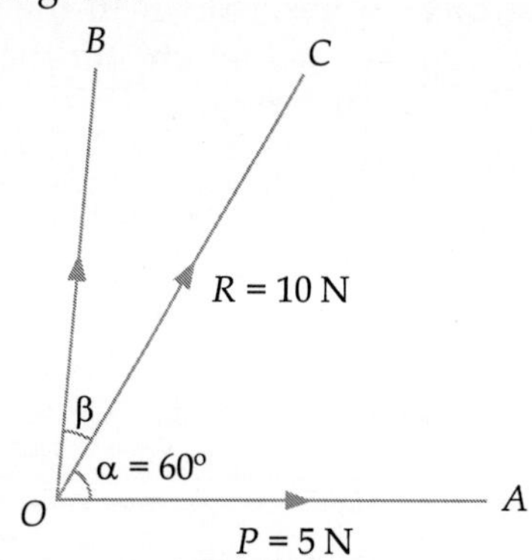

Fig. 2.9

Substituting P = 5 N and R = 10 N in expression (i), we get

$$5 = 10\,\frac{\sin\beta}{\sin(\alpha+\beta)}\,; \qquad 2\sin\beta = \sin(\alpha+\beta)$$

or
$$2\sin\beta = \sin\alpha\cos\beta + \cos\alpha\sin\beta = \sin 60^\circ\cos\beta + \cos 60^\circ\sin\beta$$
$$= 0.866\cos\beta + 0.5\sin\beta$$

That gives:
$$1.5\sin\beta = 0.866\cos\beta$$
$$\tan\beta = \frac{0.866}{1.5} = 0.5773\,; \qquad \beta = \mathbf{30^\circ}$$

Now, considering identity (ii) and substituting $\beta = 30^\circ$,

$$Q = 10\,\frac{\sin 60^\circ}{\sin 90^\circ} = 10 \times \frac{0.866}{1} = \mathbf{8.66\ N}$$

EXAMPLE 2.12.

The resultant of two forces *P* and *Q* is 200 N and it makes an angle of 30° with the horizontal as shown in Fig. 2.10. Determine the magnitude of components *P* and *Q*.

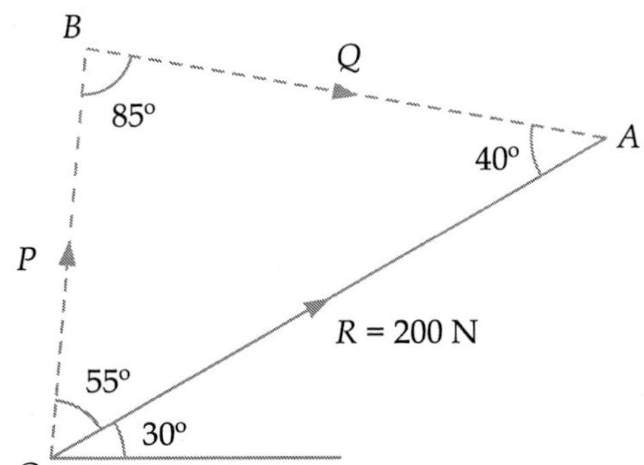

Fig. 2.10

Solution : The triangle OAB constitutes the force triangle with resultant directed along OA, and the components P and Q lying along directions OB and BA respectively.

Angle OBA = 180 − 55° − 40° = 85°

Applying sine rule to ΔOAB, we get

$$\frac{OA}{\sin 85^\circ} = \frac{OB}{\sin 40^\circ} = \frac{BA}{\sin 55^\circ}$$

or
$$\frac{200}{\sin 85^\circ} = \frac{P}{\sin 40^\circ} = \frac{Q}{\sin 55^\circ}$$

$$\therefore \quad P = 200 \times \frac{\sin 40^\circ}{\sin 85^\circ} = 200 \times \frac{0.643}{0.996} = \mathbf{129.1\ N}\ \text{from } O \text{ to } B$$

$$Q = 200 \times \frac{\sin 55^\circ}{\sin 85^\circ} = 200 \times \frac{0.819}{0.996} = \mathbf{164.4\ N}\ \text{from } B \text{ to } A$$

EXAMPLE 2.13.

The force system shown in the adjoining figure has a resultant of 2 kN acting up along the Y-axis. Make calculations for the force F both in magnitude and direction.

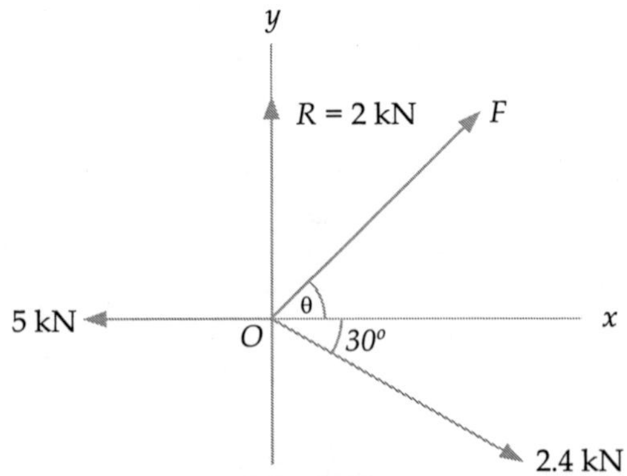

Fig. 2.11

Solution : The resultant force R = 2 kN acts upwards along O-Y axis. Resolving all the forces in vertical direction, we get

$$F\sin\theta - 2.4\sin 30^\circ = 2$$

or
$$F\sin\theta = 2 + 2.4\sin 30^\circ = 2 + 2.4 \times \frac{1}{2}$$
$$= 3.2\ \text{kN} \qquad \ldots(i)$$

Resolving all the forces in horizontal direction, we get

$$F \cos\theta + 2.4 \cos 30° - 5 = 0$$

$$F \cos\theta = 5 - 2.4 \cos 30° = 5 - 2.4 \times 0.866 = 2.922 \quad ...(ii)$$

Squaring identities (*i*) and (*ii*) and then adding, we get

$$F^2 (\sin^2\theta + \cos^2\theta) = 3.2^2 + 2.922^2 = 18.778$$

$$\therefore \quad F = \sqrt{18.778} = \mathbf{4.33\ kN}$$

The orientation of force can be obtained by dividing identity (*i*) by identity (*ii*). That gives

$$\tan\theta = \frac{3.2}{2.922} = 1.095$$

$$\therefore \quad \theta = \mathbf{47.6°}$$

EXAMPLE 2.14

Two cables which have known tensions of 40 N and 60 N are attached to the top of a tower *AB* as shown in Fig. 2.12. What tension will be induced in the guy wire *AC* if the resultant of the forces exerted at the top *A* by the cables acts vertically downwards?

Solution : Since the resultant of the forces acts vertically downwards, we have $\Sigma F_x = 0$ and that gives

$$40 \cos 15° - 60 \cos 30° + T \cos\theta = 0 \quad ...(i)$$

In the right angled triangle *ABC*

$$\theta = \tan^{-1}\left(\frac{AB}{BC}\right) = \tan^{-1}\left(\frac{15}{10}\right) = 56.31°$$

Substituting $\theta = 56.31°$ in expression (*i*), we get

$$40 \cos 15° - 60 \cos 30° + T \cos 56.31° = 0$$

$$\text{or} \quad 38.64 - 51.96 + 0.555\,T = 0$$

$$\therefore \quad \text{Tension in the guy wire } T = \frac{51.96 - 38.64}{0.555} = \mathbf{24\ N}$$

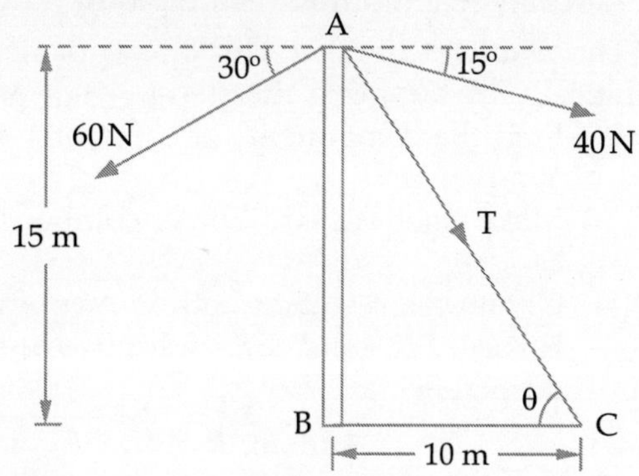

Fig. 2.12

EXAMPLE 2.15

A block of weight 300 N is acted upon by a horizontal force *F* = 500 N and a pressure *P* exerted by the inclined as shown in Fig. 2.13. If the resultant of the force system lies parallel to the plane, work out the magnitude of pressure *P* and the resultant force.

Solution : Let the co-ordinates be chosen parallel and perpendicular to the plane.

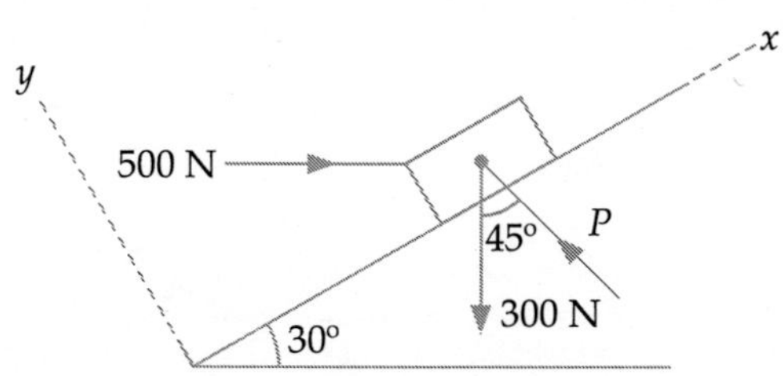

Fig. 2.13

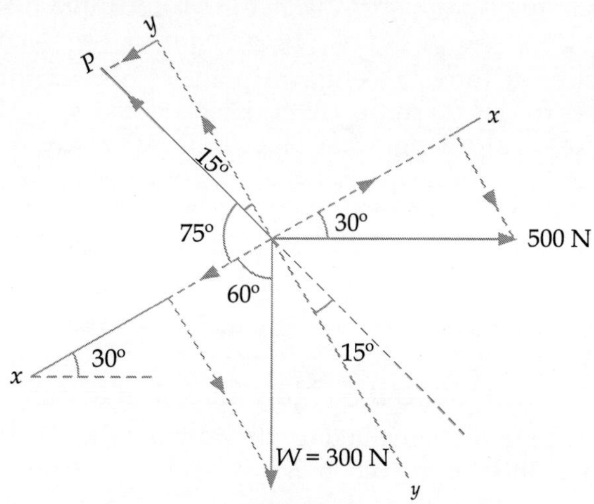

Fig. 2.14

As the resultant R lies along the plane, its component perpendicular to the plane is zero. Then with reference to Fig. 2.14.

$$R = -P \sin 15^\circ + 500 \cos 30^\circ - 300 \cos 60^\circ$$
$$= -0.259P + 433 - 150$$
$$= -0.259P + 283$$

Also
$$P \cos 15^\circ - 500 \sin 30^\circ - 300 \cos 60^\circ = 0$$
$$P \cos 15^\circ = 500 \sin 30^\circ + 300 \sin 60^\circ$$
$$P \times 0.966 = 250 + 259.8 = 509.8$$
$$P = \frac{509.8}{0.966} = 527.74 \text{ N}$$
$$\therefore \quad R = -0.259 \times 527.74 + 283 = \mathbf{146.31 \text{ N}}$$

2.4. RESULTANT OF COPLANAR-CONCURRENT FORCES

2.4.1. Analytical Method (Principle of resolved parts)

The resultant of a number of a coplanar-concurrent forces acting on a body is worked out analytically by adopting the step-by-step procedure outlined blew:

(*i*) Find the components of each force in the system in two mutually perpendicular X and Y directions.

(*ii*) Make algebraic addition of components in each direction to get two components ΣF_x and ΣF_y.

(*iii*) Obtain the resultant both in magnitude and direction by combining the two component forces ΣF_x and ΣF_y which are mutnally prependicular.

$$\text{Resultant, } R = \sqrt{(\Sigma F_x)^2 + (\Sigma F_y)^2}$$

and its inclination θ to X-axis is given by

$$\tan \theta = \frac{\Sigma F_y}{\Sigma F_x}$$

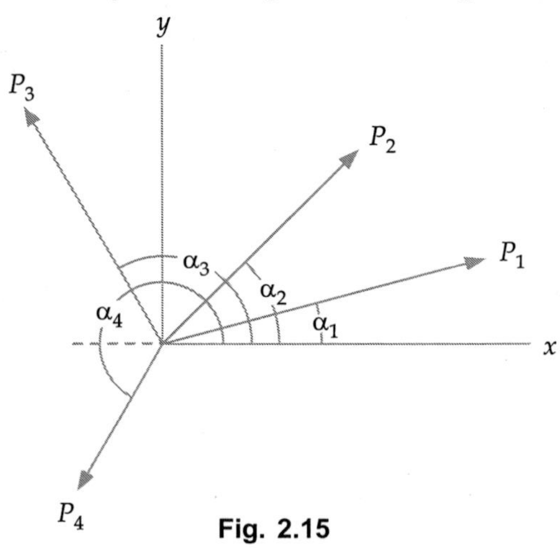

Fig. 2.15

This analytical method is based on theorem of resolved parts which states that "*the algebraic sum of the resolved parts of two forces in a given direction is equal to the resolved part of their resultant in the same direction*".

With reference to Fig. 2.15, P_1, P_2, P_3 and P_4 are the concurrent forces meeting at point O and making angles α_1, α_2, α_3 and α_4 with OX.

Resolving along x-axis and y-axis, we get

$$\Sigma F_x = P_1 \cos \alpha_1 + P_2 \cos \alpha_2 + P_3 \cos \alpha_3 + P_4 \cos \alpha_4$$
$$\Sigma F_y = P_1 \sin \alpha_1 + P_2 \sin \alpha_2 + P_3 \sin \alpha_3 + P_4 \sin \alpha_4$$
$$\text{Resultant, } R = \sqrt{(\Sigma F_x)^2 + (\Sigma F_y)^2}$$

and its inclination θ to x-axis is $\tan \theta = \dfrac{\Sigma F_y}{\Sigma F_x}$

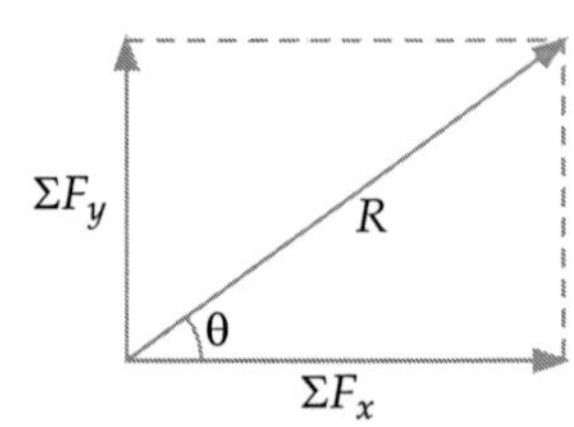

2.4.2. Graphical Method (Polygon law of forces)

The graphical method for determining the resultant of coplanar concurrent force systems has been dealt later in section 2.6.

2.4.3. Equilibrium of a Particle

A particle will be in equilibrium when resultant of all the forces acting on it is zero

$$\text{Resultant} \quad R = \sqrt{(\Sigma F_x)^2 + (\Sigma F_y)^2}$$

For a particle to be in equilibrium:

$$R = 0; \quad (\Sigma F_x)^2 + (\Sigma F_y)^2 = 0$$

Now $(\Sigma F_x)^2$ and $(\Sigma F_y)^2$ are positive quantities and their sum cannot be zero unless each of then is zero.

$$\therefore \qquad \Sigma F_x = 0 \quad \text{and} \quad \Sigma F_y = 0 \qquad \text{...(2.11)}$$

Hence, if any number of forces acting at a particle are in equilibrium, then the algebraic sum of their resolved parts in any two perpendicular directions are separately zero.

Equations 2.11 are called the equations of equilibrium of a particle in plane.

Conversely, if the sum of the resolved parts in any two perpendicular directions are separately zero, then the forces acting at any point are in equilibrium.

Equilibrium of a two-force and three-force body.

A two-force body is subjected to only two forces. For its equilibrium, the resultant force must be zero. Accordingly the two forces must be collinear and must have the same magnitude but opposite direction. These aspects have been shown below in Fig. 2.16 (*a*) and (*b*).

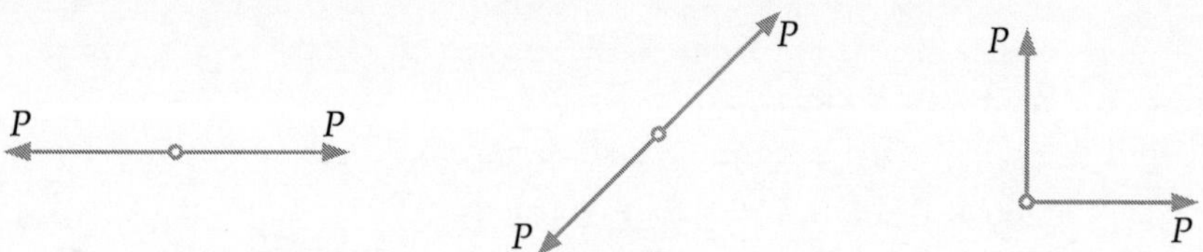

(*a*) (*b*) (*c*)

Fig. 2.16

The particle subjected to two forces of equal magnitude (Fig. 2.16 (*c*)) shall not be in equilibrium because the forces have different lines of action.

A three-force body is subjected to only three forces. For the equilibrium, the acting forces must be concurrent and must form a closed triangle. This aspect has been illustrated in Fig. 2.17.

Law of sines can then be used to obtain solution to the problem

$$\frac{F_1}{\sin\alpha} = \frac{F_2}{\sin\beta} = \frac{F_3}{\sin\gamma}$$

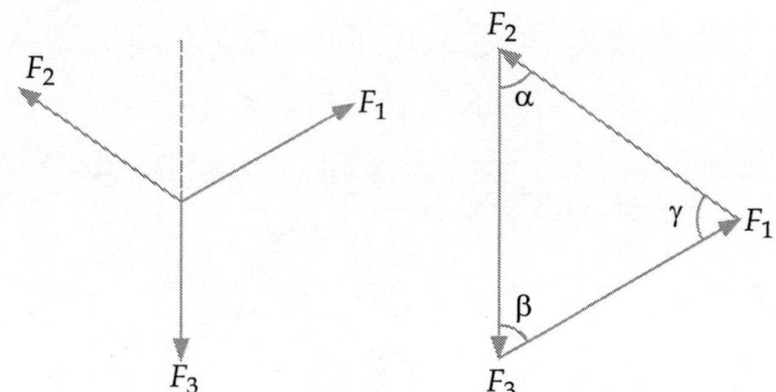

Fig. 2.17

EXAMPLE 2.16

Determine the magnitude and direction of the resultant of the following set of forces acting on a body

(*i*) 200 N inclined 30° with east towards north,

(*ii*) 250 N towards the north,

(*iii*) 300 N towards north west, and

(*iv*) 350 N inclined at 40° with west towards south.

What will be the equilibrant of the given force system ?

Solution : The given force system has been depicted in Fig. 2.18.

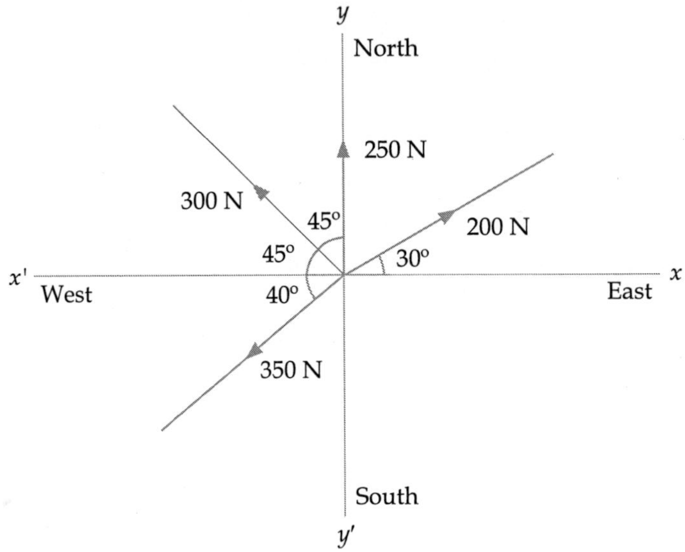

Fig. 2.18

Resolving all the forces along X-direction

$$\Sigma F_x = 200 \cos 30° + 250 \cos 90° + 300 \cos 135° + 350 \cos 220°$$
$$= 200 \times 0.866 + 250 \times 0 + 300 \times (-0.707) + 350 \times (-0.766)$$
$$= 173.2 + 0 - 212.1 - 268.1 = -307 \text{ N (along } OX')$$

Resolving all the forces along Y-direction

$$\Sigma F_y = 200 \sin 30° + 250 \sin 90° + 300 \sin 135° + 350 \sin 220°$$
$$= 200 \times 0.5 + 250 \times 1 + 300 \times 0.707 + 350 \times (-0.642)$$
$$= 100 + 250 + 212.1 - 224.7 = 337.4 \text{ N (along OY)}$$

The magnitude of resultant force is

$$R = \sqrt{(\Sigma F_x)^2 + (\Sigma F_y)^2} = \sqrt{(-307)^2 + (337.4)^2} = \mathbf{456\ N}$$

Since ΣF_x is – ve and ΣF_y is positive, the resultant lies in the second quadrant and its inclination θ′ with *OX*′ is

$$\theta' = \tan^{-1}\left(\frac{\Sigma F_y}{\Sigma F_x}\right) = \tan^{-1}\left(\frac{337.4}{307}\right) = 47.7°$$

Obviously the inclination of the resultant with *OX* (positive direction) will be

$$\theta = 180 - 47.7 = 132.3°$$

An equal and opposite force will be required to h old the body in position. The equilibrant will thus have a magnitude of 456 N, inclined at 47.7° to the *x*-axis (fourth quadrant) and is opposite to the resultant. Both the resultant and equilibrant are shown in Fig. 2.19.

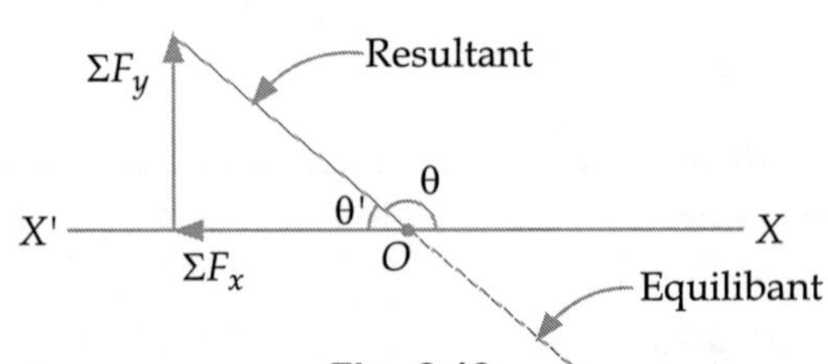

Fig. 2.19

EXAMPLE 2.17

A particle positioned at point *O* is acted upon by three forces as shown in Fig. 2.20. Determine the value of force *P* such that the resultant of these three forces is horizontal. What should be the magnitude and direction of a fourth force which acting at point *O* along with the given three forces will keep the particle in equilibrium state?

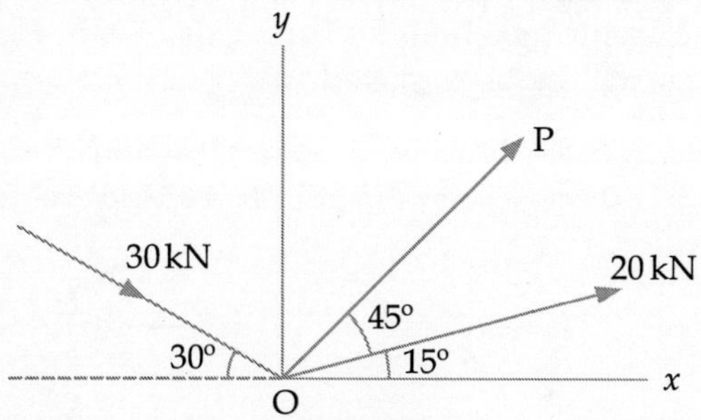

Fig. 2.20

Solution : Angle between force *P* and the axis 0 - *x* is (15 + 45) = 60°

Resolving all the forces along *y*-direction,

$\Sigma F_y = -30 \sin 30° + 20 \sin 15° + P \sin 50°$

The negative sign with force of 30 N is due to the fact that this force is acting towards the particle.

Since the resultant is horizontal, $\Sigma F_y = 0$

$$\therefore \quad 0 = -30 \times 0.5 + 20 \times 0.2588 + P \times 0.866 = -15 + 5.176 + 0.866\,P$$

That gives:

$$P = \frac{15 - 5.176}{0.866} = 11.344 \text{ N}$$

Resolving all forces along *X*-direction,

$$\Sigma F_x = 30 \cos 30° + 20 \cos 15° + 11.344 \cos 60° = 25.98 + 19.32 + 5.67 = 50.97 \text{ kN}$$

The equilibrant would be – 50.97 kN as shown in Fig. 2.21.

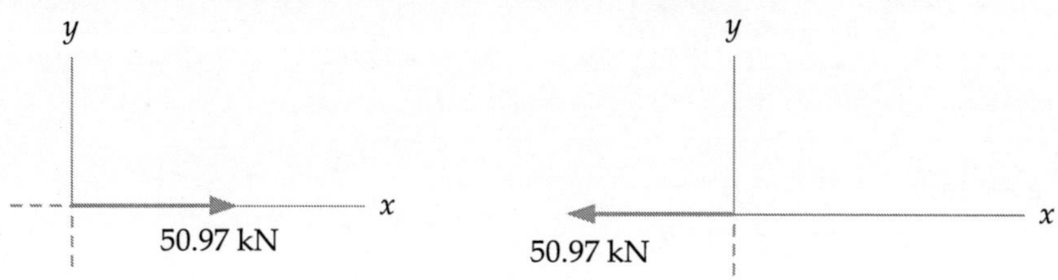

Fig. 2.21

EXAMPLE 2.18

Determine the resultant, both in magnitude and direction, of the four forces acting on the body as shown in the figure given below:

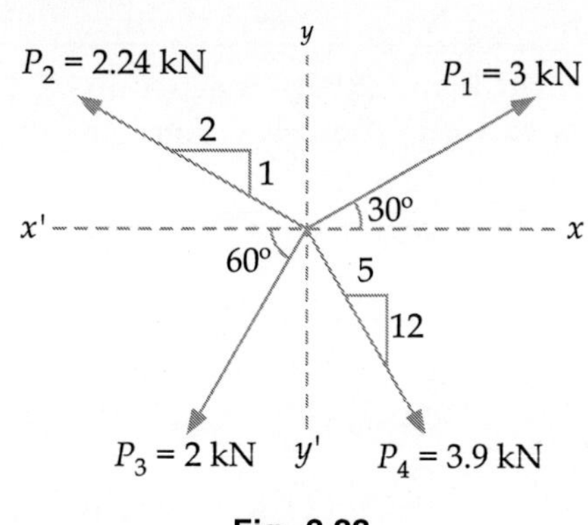

Fig. 2.22

Solution : Inclination of force 2.24 kN with OX′ = $\tan^{-1}(1/2)$ = 26.56°

Inclination of force 3.9 kN with OY′ = $\tan^{-1}(5/12)$ = 22.62°

Measuring all angles from axis O-X in the anticlockwise direction, the inclination of different forces will be as depicted in Fig. 2.23.

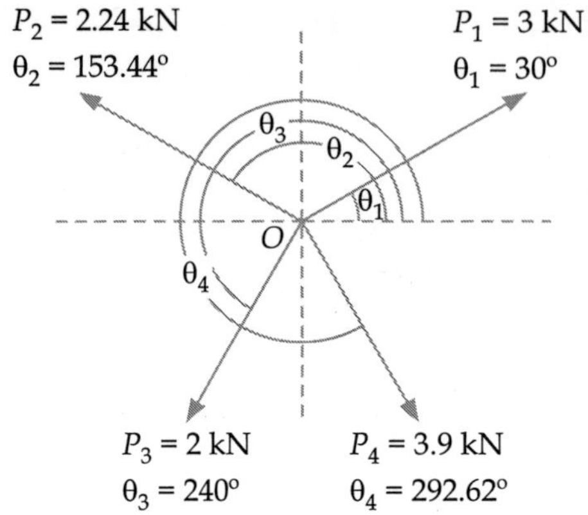

Fig. 2.23

Resolving all the forces in horizontal direction,

$$\Sigma F_x = P_1 \cos\theta_1 + P_2 \cos\theta_2 + P_3 \cos\theta_3 + P_4 \cos\theta_4$$
$$= 3 \cos 30° + 2.24 \cos 153.44° + 2 \cos 240° + 3.9 \cos 292.62°$$
$$= 2.598 - 2.004 - 1.0 + 1.50 = 1.094 \text{ kN}$$

Resolving all the forces in vertical direction,

$$\Sigma F_y = P_1 \sin\theta_1 + P_2 \sin\theta_2 + P_3 \sin\theta_3 + P_4 \sin\theta_4$$
$$= 3 \sin 30° + 2.24 \sin 153.44° + 2 \sin 240° + 3.9 \sin 292.62°$$
$$= 1.50 + 1.00 - 1.73 - 3.60 = -2.83 \text{ kN}$$

The magnitude of the resultant force is

$$R = \sqrt{(\Sigma F_x)^2 + (\Sigma F_y)^2} = \sqrt{(1.094)^2 + (-2.83)^2} = \sqrt{1.197 + 8.009} = \mathbf{3.034\ kN}$$

and the inclination of resultant with the horizontal is

$$\alpha = \tan^{-1}\left(\frac{\Sigma F_y}{\Sigma F_x}\right) = \tan^{-1}\left(\frac{-2.83}{1.094}\right) = \mathbf{68.86°}$$

EXAMPLE 2.19

Find the resultant of forces 2, $\sqrt{3}$, 5, $\sqrt{3}$ and 2 N that act at an angular point of a regular hexagon towards the other angular points taken in order.

Solution : Let *ABCDEF* be the regular hexagon with A as the angular point where forces are acting. The forces are:

2 N along *AB* ; $\sqrt{3}$ along *AC* ; 5 N along *AD* ;
$\sqrt{3}$ N along *AE* and 2 N along *AF*

Sum of the interior angles of a regular polygon
= (2n – 4) right angles

where *n* is the number of sides of the polygon. For a hexagon, *n* = 6 and therefore sum of interior angles
= (2 × 6 – 4) × 90 = 720°

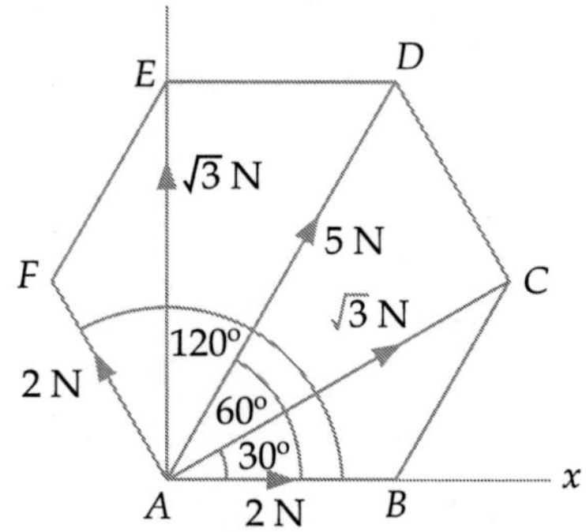

Fig. 2.24

$$\therefore \text{ Value of each interior angle } = \frac{720}{6} = 120°$$

Since included angle of a regular hexagon is 120°, the angular position of these forces will be as indicated in Fig. 2.25.

Resolving all the forces horizontally, *i.e.*, along *AX*

$$\Sigma F_x = 2 + \sqrt{3}\cos 30° + 5\cos 60° + \sqrt{3}\cos 90° + 2\cos 120°$$

$$= 2 + \sqrt{3} \times \frac{\sqrt{3}}{2} + 5 \times \frac{1}{2} + \sqrt{3} \times 0 + 2\left(-\frac{1}{2}\right) = 5 \text{ N}$$

Resolving all the forces vertically, i.e., along *AY*

$$\Sigma F_y = \sqrt{3}\sin 30° + 5\sin 60° + \sqrt{3}\sin 90° + 2\sin 120°$$

$$= \sqrt{3} \times \frac{1}{2} + 5 \times \frac{\sqrt{3}}{2} + \sqrt{3} \times 1 + 2 \times \frac{\sqrt{3}}{2} = 5\sqrt{3} \text{ N}$$

The magnitude of the resultant force is

$$R = \sqrt{(\Sigma F_x)^2 + (\Sigma F_y)^2} = \sqrt{5^2 + (5\sqrt{3})^2} = \sqrt{100} = \mathbf{10\ N}$$

and the inclination of the resultant with horizontal is

$$\alpha = \tan^{-1}\left(\frac{\Sigma F_y}{\Sigma F_x}\right) = \tan^{-1}\frac{5\sqrt{3}}{5} = \tan^{-1}(\sqrt{3}) = \mathbf{60°}$$

EXAMPLE 2.20

(*a*) State the conditions of equilibrium of a coplanar system of forces.

(*b*) Five forces of magnitude 2, *P*, *Q*, 4 and 3 kN act along *AB*, *CA*, *AD*, *AE* and *FA* respectively in a regular hexagon *ABCDEF* as shown in Fig. 2.25.

If the system is to be is equilibrium state, determine the value of forces *P* and *Q*.

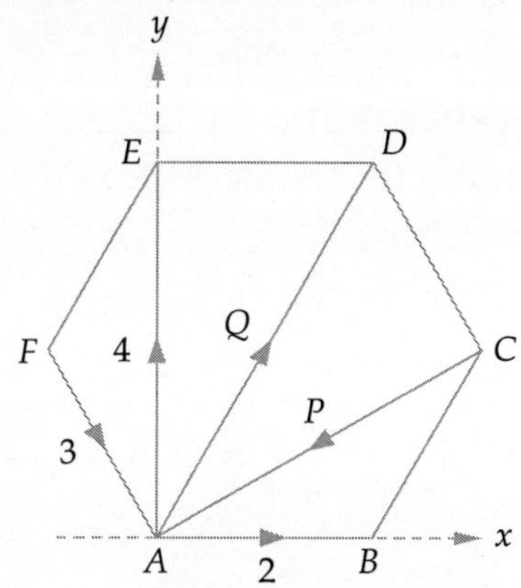

Fig. 2.25

Solution : The necessary conditions of equilibrium of a coplanar system of forces are:

(*i*) The algebraic sum of all the forces in a horizontal plane is equal to zero: $\Sigma F_x = 0$

(*ii*) The algebraic sum of all the forces in a vertical plane is equal to zero: $\Sigma F_y = 0$

(*iii*) The algebraic sum of all the moments about any point is equal to zero: $\Sigma M = 0$.

(*b*) Sum of the interior angles of a regular polygon

$$= (2n - 4) \text{ right angles}$$

where *n* is the number of sides of the polygon. For a hexagon, *n* = 6 and therefore sum of interior angles

$$= (2 \times 6 - 4) \times 90 = 720°$$

$$\therefore \quad \text{Value of each interior angle } = \frac{720}{6} = 120°$$

The angular position of the given forces at corner *A* will be as indicated in Fig. 2.26.

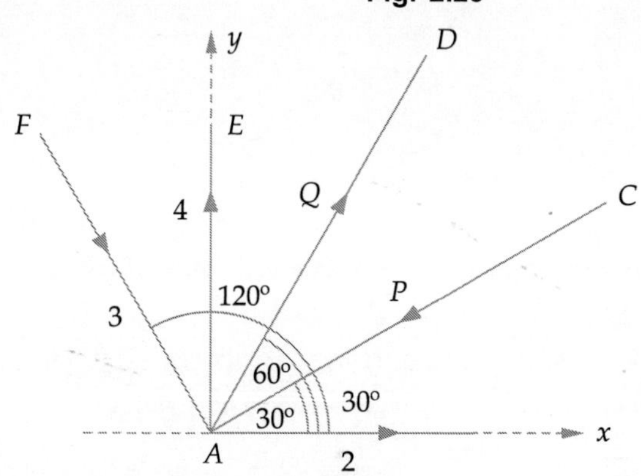

Fig. 2.26

Resolving all the forces horizontally, *i.e.*, along A-x

$$\Sigma F_x = 2 - P \cos 30° + Q \cos 60° - 3 \cos 120°$$
$$= 2 - 0.866\, P + 0.5\, Q + 1.5$$
$$= 3.5 - 0.866\, P + 0.5\, Q$$

Resolving all the forces vertically, *i.e.*, along A-y

$$\Sigma F_y = - P \sin 30° + Q \sin 60° + 4 - 3 \sin 120°$$
$$= - 0.5\, P + 0.866\, Q + 4 - 3 \times 0.866$$
$$= 1.402 - 0.5\, P + 0.866\, Q$$

Since the forces are in equilibrium: $\Sigma F_x = 0$ and $\Sigma F_y = 0$

$$\therefore \quad 3.5 - 0.866\, P + 0.5\, Q = 0 \quad ...(a)$$

$$\text{and} \quad 1.402 - 0.5\, P + 0.866\, Q = 0 \quad ...(b)$$

Multiplying identity (*a*) by 0.866 and identity (*b*) by 0.5, we get:

$$3.031 - 0.75\, P + 0.433\, Q = 0 \quad ...(c)$$

$$0.701 - 0.25\, P + 0.433\, Q = 0 \quad ...(d)$$

Subtracting identity (*d*) from identify (*c*) and solving for *P*, we get

$$\text{Force } P = \frac{3.031 - 0.701}{0.75 - 0.25} = \mathbf{4.66\ kN}$$

Substituting this value of *P* in identify (*c*), we get

$$\text{Force } Q = \frac{0.75 \times 4.66 - 3.031}{0.433} = \mathbf{1.071\ kN}$$

EXAMPLE 2.21

Calculate the tensile force in the cables AB and BC as shown in Fig. 2.27. Presume the pulleys to be frictionless.

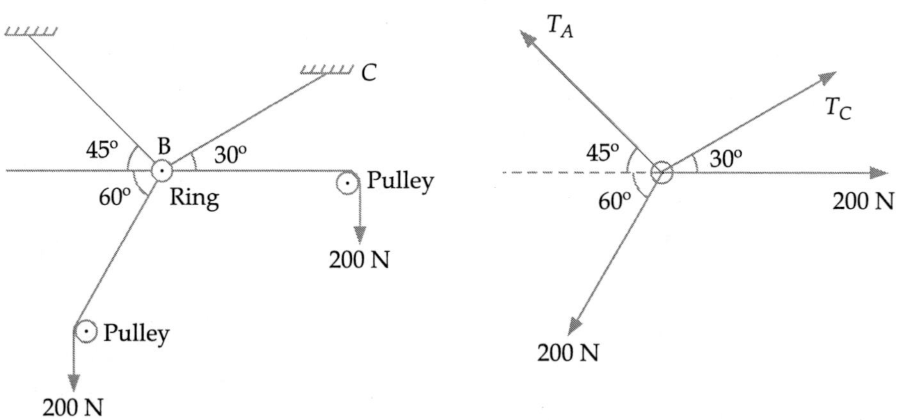

Fig. 2.27

Solution : Since the system is in equilibrium,

$$\Sigma F_x = 0\,; \quad 0 = - T_A \cos 45° + T_C \cos 30° + 200 - 200 \cos 60°$$
$$= - 0.707\, T_A + 0.866\, T_C + 200 - 100$$

$$\text{or} \quad - T_A + 1.22\, T_C + 141.0 = 0 \quad ...(i)$$

$$\Sigma F_y = 0\,; \quad 0 = T_A \sin 45° + T_C \sin 30° - 200 \sin 60° = 0.707\, T_A + 0.5\, T_C - 173$$

$$\text{or} \quad T_A + 0.707\, T_C - 244 = 0 \quad ...(ii)$$

Adding expressions (*i*) and (*ii*),

$$1.927\,T_C - 103 = 0;\quad T_C = \frac{103}{1.927} = \mathbf{53.45\ N}$$

Then from expression (*ii*)

$$T_A + 0.707 \times 53.45 - 244 = 0$$

$$\therefore\quad T_A = 244 - 0.707 \times 53.45 = 244 - 37.79 = \mathbf{206.21\ N}$$

EXAMPLE 2.22

A ball of weight W rests upon a smooth horizontal plane and has attached to its centre two strings AB and AC which pass over frictionless pulleys at B and C and carry loads P and Q as shown in Fig. 2.28. If the string AB is horizontal, find the angle α that the string AC makes with the horizontal when the ball is in a position of equilibrium. Also determine the pressure between the ball and the plane.

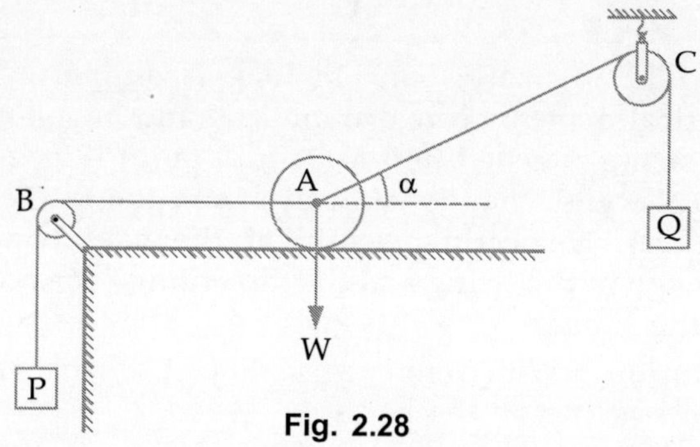

Fig. 2.28

Solution : Let T_1 and T_2 be the tensions in the strings AB and AC respectively. Since the pulleys are frictionless

$$T_1 = P \quad\text{and}\quad T_2 = Q$$

Resolving the forces horizontally we get

$$T_1 = T_2 \cos\alpha$$

$$\text{or}\quad P = Q\cos\alpha$$

$$\therefore\quad \alpha = \cos^{-1}\left(\frac{P}{Q}\right)$$

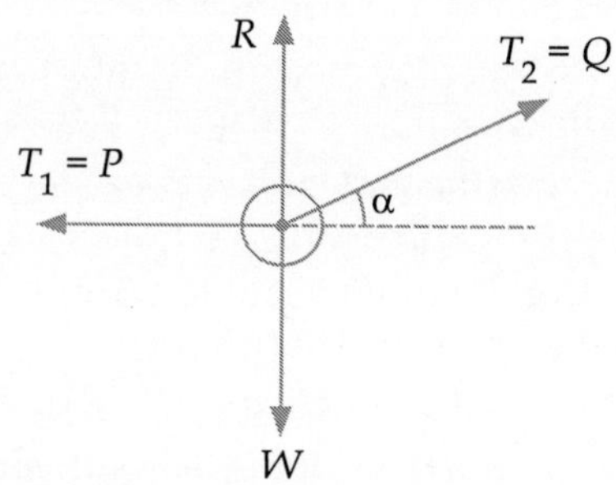

Resolving the forces vertically, the pressure R between the ball and the plane can be written as

$$R = W - T_2\sin\alpha = W - Q\sqrt{(1-\cos^2\alpha)}$$

$$= W - Q\sqrt{\left(1-\frac{P^2}{Q^2}\right)}$$

$$= W - \sqrt{Q^2 - P^2}$$

EXAMPLE 2.23

A right circular roller of weight W rests on a smooth horizontal plane and is held in position by string AC as shown in Fig. 2.29. The roller is pulled by a horizontal force P applied to its centre. Find the reaction at B and the tension of the string.

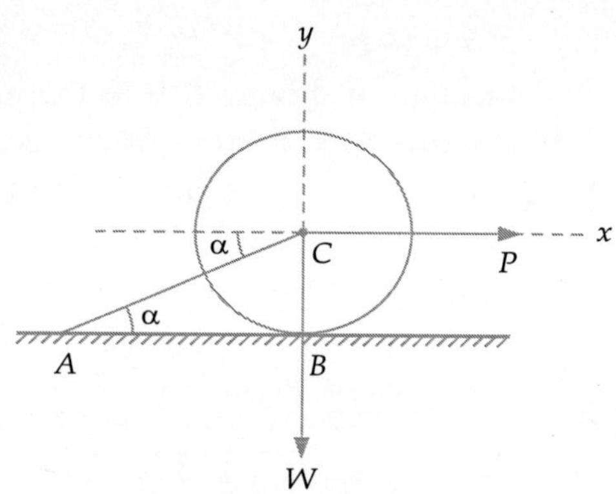

Fig. 2.29

Solution : The roller is in equilibrium under the action of following set of forces:

(*i*) Self weight acting vertically downward

(*ii*) Reaction R_b acting vertically upward

(*iii*) Force P acting horizontally towards right

(*iv*) Tension T of the string acting in the direction from C towards A

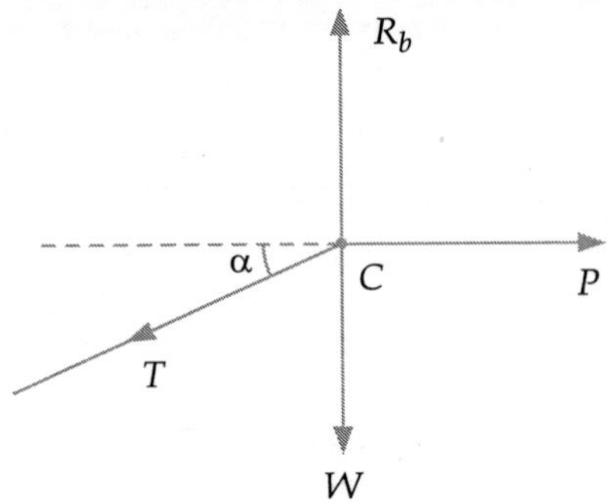

For these concurrent forces, the conditions of equilibrium are:

$\Sigma F_x = 0$; $\quad P - T\cos\alpha = 0$

$$\therefore \quad T = \frac{P}{\cos\alpha}$$

$\Sigma F_y = 0$; $\quad R_b = T\sin\alpha - W = 0$

$$\therefore \quad R_b = T\sin\alpha + W$$

$$= \frac{P}{\cos\alpha}\sin\alpha + W$$

$$= \mathbf{P\tan\alpha + W}$$

EXAMPLE 2.24

A rigid prismatic bar 5 m long is supported in a vertical plane by hinge at one end and by a horizontal string attached to it at its mid point (Fig. 2.30) A 150 N weight hangs vertically from the free end of the bar. Neglecting weight of the bar, determine tension in the string and the direction of the reaction at the hinge.

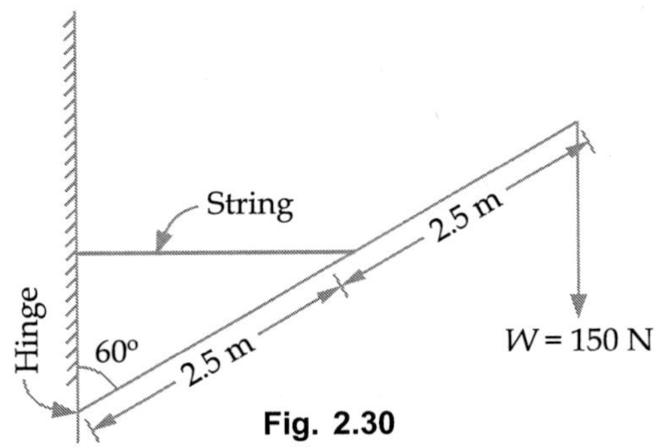

Fig. 2.30

Solution : With reference to Fig. 2.30, the various forces acting on the bar are:-

(*i*) load W = 150 N that acts at end B vertically downwards

(*ii*) tension T in the string acting horizontally

(*iii*) reaction R at the hinge point A.

Since the three forces keep the bar in equilibrium, they must be concurrent, i.e., meet at a point. Accordingly reaction R would act along AD and D would be the point of concurrency. Then

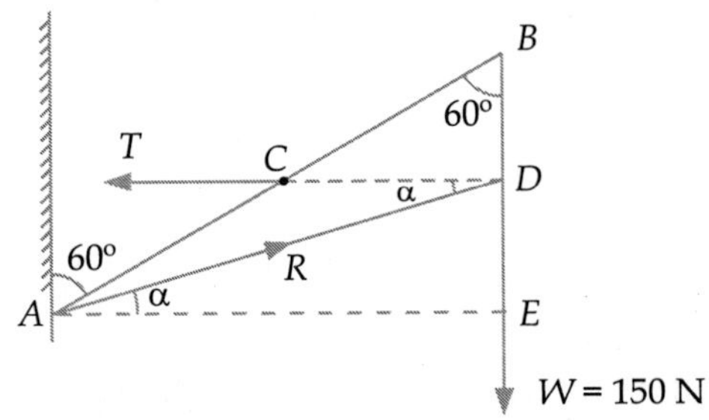

$$BD = BC\cos 60° = 2.5\cos 60° = 1.25 \text{ m} = DE$$

$$AE = AB\sin 60° = 5\sin 60° = 4.33 \text{ m}$$

$$\tan\alpha = \frac{DE}{AE} = \frac{1.25}{4.33} = 0.2887$$

$$\alpha = \mathbf{16.1°}$$

The direction of the reaction at the hinge is thus inclined at 16.1° with the horizontal.

Applying the equilibrium conditions

$\Sigma F_x = 0$; $\quad R\cos\alpha - T = 0$

$$R\cos\alpha = T \quad \text{...(}i\text{)}$$

$\Sigma F_y = 0$; $\quad R\sin\alpha - 150 = 0$

$$R\sin\alpha = 150° \quad \text{...(}ii\text{)}$$

From expressions (*i*) and (*ii*),

$$\tan\alpha = \frac{150}{T}; \quad T = \frac{150}{\tan\alpha} = \frac{150}{\tan 16.1°} = \mathbf{519.57\ N}$$

EXAMPLE 2.25

A string ABC of length 50 cm is tied to two points A and C at the same level. A smooth ring of weight 500 N which can slide along the string is at B, 30 cm away from A along the string and pulled by a horizontal force P as shown in Fig. 2.31. If point B is 15 cm below the level of AC, determine the magnitude of force P. It may be presumed that tensions in the string on both sides of B are same.

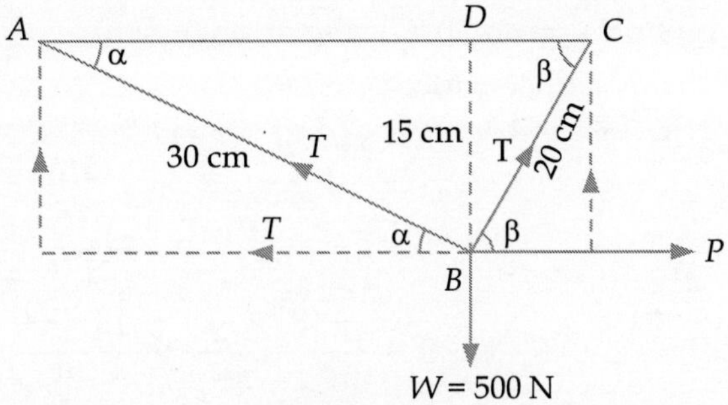

Fig. 2.31

Solution : From the geometrical configuration of the given system

$$\sin\alpha = \frac{15}{30}\,;\ \alpha = 30^\circ$$

$$\sin\beta = \frac{15}{30}\,;\ \beta = 48.59^\circ$$

Since the system of forces is in equilibrium,

$$\Sigma F_x = 0 \text{ and } \Sigma F_y = 0$$

Resolution of forces in the horizontal direction gives,

$$P + T\cos\beta - T\cos\alpha = 0$$

or $$P + T\cos 48.59^\circ - T\cos 30^\circ = 0$$

or $$P = T\cos 30^\circ - T\cos 48.59^\circ = 0.866\,T - 0.661\,T = 0.205\,T$$

Resolution of forces in the vertical direction gives,

$$T\sin\beta + T\sin\alpha - 500 = 0$$

or $$T\sin 48.59^\circ + T\sin 30^\circ = 500$$

or $$0.75\,T + 0.5\,T = 500$$

$\therefore$ $$T = \frac{500}{1.25} = \mathbf{400\ N}$$

and $$P = 0.205\,T = 0.205 \times 400 = \mathbf{82\ N}$$

EXAMPLE 2.26

A string of length l is fastened to two points A and B at the same level at a distance a apart. A ring of weight W can slide on the string and a horizontal force P is applied to it such that it is in equilibrium vertically below B. Show that $P = \frac{Wa}{l}$ and that the tension in the string is $\frac{W(l^2 + a^2)}{2l}$.

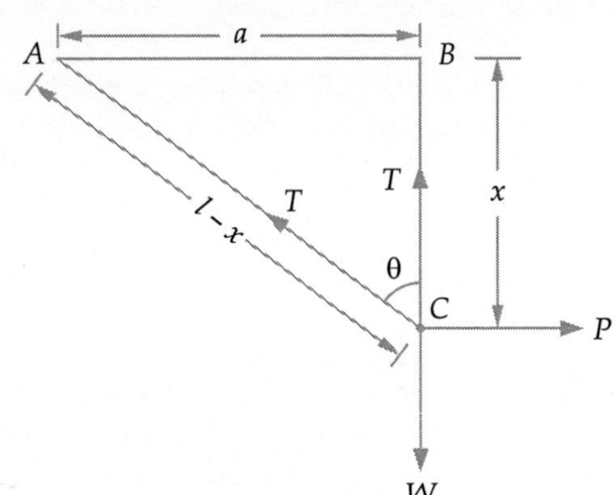

Fig. 2.32

Solution : Let C be the ring vertically below point B. Further, if $BC = x$ then $AB = (l - x)$.

The ring is acted upon by the following set of forces,

(*i*) tension along CB

(*ii*) tension along CA

(*iii*) Weight W of the ring acting vertically downwards

(*iv*) horizontal force P

Since the ring slides along the string

tension along CB = tension along CA = T (say)

For these concurrent forces, the conditions of equilibrium are:

$\Sigma F_y = 0;$ $\quad T + T\cos\theta = W$ or $T(1 + \cos\theta) = W$ ...(*i*)

$\Sigma F_x = 0;$ $\quad P = T\sin\theta$ ...(*ii*)

From the geometry of the arrangement

$$(l - x)^2 = x^2 + a^2$$

$$l^2 + x^2 - 2lx = x^2 + a^2; \quad x = \frac{l^2 - a^2}{2l}$$

$$l - x = l - \frac{l^2 - a^2}{2l} = \frac{l^2 + a^2}{2l}$$

$$\cos\theta = \frac{BC}{AC} = \frac{x}{l - x} = \frac{l^2 - a^2}{2l} \times \frac{2l}{l^2 + a^2} = \frac{l^2 - a^2}{l^2 + a^2}$$

$$\sin\theta = \frac{AB}{l - x} = a \times \frac{2l}{l^2 + a^2} = \frac{2al}{l^2 + a^2}$$

Substituting these values of $\sin\theta$ and $\cos\theta$ in expressions (*i*) and (*ii*), we get

$$T\left(1 + \frac{l^2 - a^2}{l^2 + a^2}\right) = W; \quad T\left(\frac{2l^2}{l^2 + a^2}\right) = W$$

$\therefore$ Tension of the string $\boldsymbol{T = \frac{W(l^2 + a^2)}{2l^2}}$

Horizontal force $P = T\sin\theta = \frac{W(l^2 + a^2)}{2l^2} \times \frac{2al}{l^2 + a^2} = \boldsymbol{\frac{Wa}{l}}$

2.5. TRIANGLE LAW OF FORCES

"If two forces acting simultaneously on a body are represented by the sides of a triangle taken in order, their resultant is represented by the closing side of the triangle taken in the opposite order."

Consider two forces *P* and *Q* acting on the body as shown in Fig 2.33 (*a*).

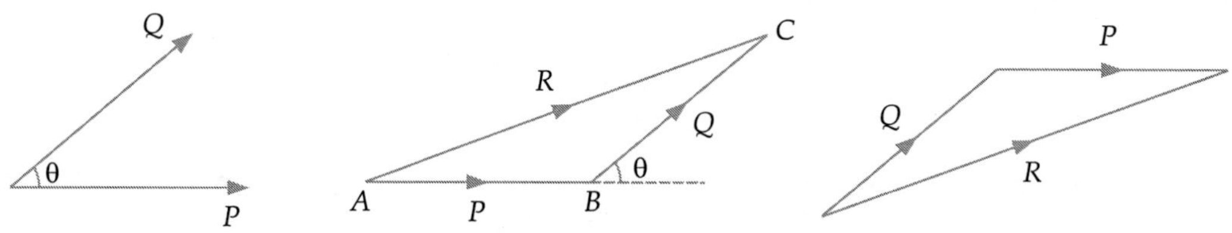

Fig. 2.33

Line *AB* be drawn to represent force *P* and *BC* to represent *Q*. The triangle *ABC* is completed by drawing the closing line *AC*. Line *AC* represents the resultant in magnitude, line of action and direction.

It is to be noted that addition of forces *P* and *Q* in any order gives the same resultant *R*.

The following trignometric relations can be applied while working out solutions by the triangle law of forces:

$$\frac{F_1}{\sin\alpha} = \frac{F_2}{\sin\beta} = \frac{F_3}{\sin\gamma}$$

and

$$F_1^2 = F_2^2 + F_3^2 - 2F_2F_3\cos\alpha$$
$$F_2^2 = F_1^2 + F_3^2 - 2F_1F_3\cos\beta$$
$$F_3^2 = F_1^2 + F_2^2 - 2F_1F_2\cos\gamma$$

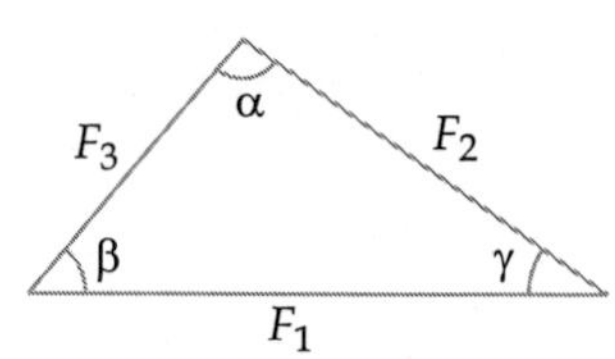

Fig. 2.34

The triangle law of forces is often stated as follows :

"If a system of three forces acting upon a body can be represented in magnitude and direction by the sides of a triangle, taken in order, then the system will be in equilibrium."

With reference to Fig 2.35, the three forces P, Q and R acting at point O have been repressed by the sides AB, BC and CA of the triangle ABC.

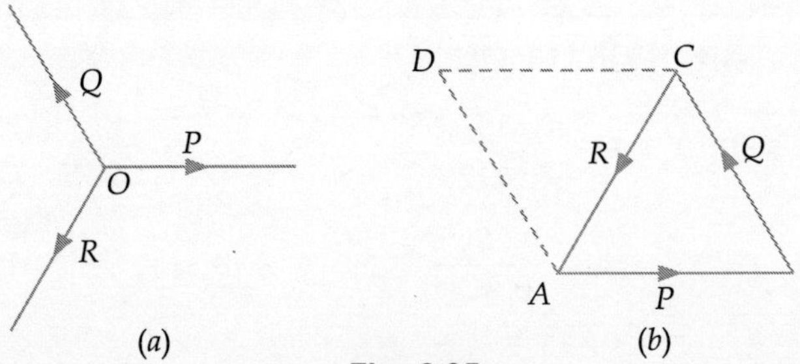

Fig. 2.35

Construct the parallelogram $ABCD$. The side AD, being equal and parallel to BC represents the force Q. By the parallelogram law of forces,

$$\vec{AB} + \vec{AD} = \vec{AC}$$

$$P + Q = -\vec{CA} = -R$$

$\therefore$ Resultant of P, Q and $R = -R + R = 0$

and hence the system is in equilibrium.

EXAMPLE 2.27

A uniform ladder weighing 80 N rests against a smooth vertical wall at a height of 12 m above the ground; the foot of ladder being 10 m from the wall. Determine the pressure due to wall.

Solution : The three foces acting in the system are:

(*i*) reaction R_1 due to wall (acts at rights angles to wall as the wall is smooth)

(*ii*) reaction R_2 of the ground

(*iii*) weight W of the ladder

Under equilibrium conditions, the reaction R_2 passes through the point of intersection of R_1 and W. The three forces R_1, R_2 and W are then represented respectively by the sides AN, OA and ON of the triangle OAN.

$$\frac{R_1}{AN} = \frac{R_2}{OA} = \frac{W}{ON};$$

$$AN = 5 \text{ m}$$

$$OA = \sqrt{5^2 + 12^2} = 13 \text{ m}$$

$$\therefore \quad \frac{R_1}{5} = \frac{R_2}{13} = \frac{80}{12}$$

Pressure R_1 due to wall $= \dfrac{80 \times 5}{12} =$ **33.3 N**

Alternatively :

$$\sin\theta = \frac{12}{13} = 0.923$$

$$\theta = 67.38°$$

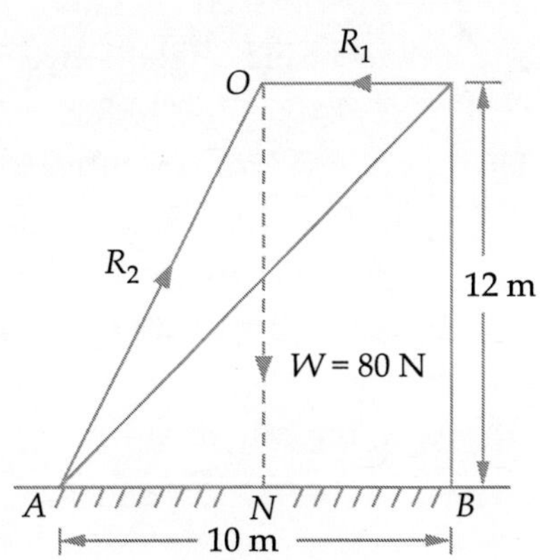

Fig. 2.36. (a)

Angle between R_1 and R_2 = 112.62°
Angle between R_1 and W = 90°
Angle between R_2 and W = 157.38°
Applying Lami's theorem,

$$\frac{R_1}{\sin 157.38°} = \frac{R_2}{\sin 90°} = \frac{W}{\sin 112.62°}$$

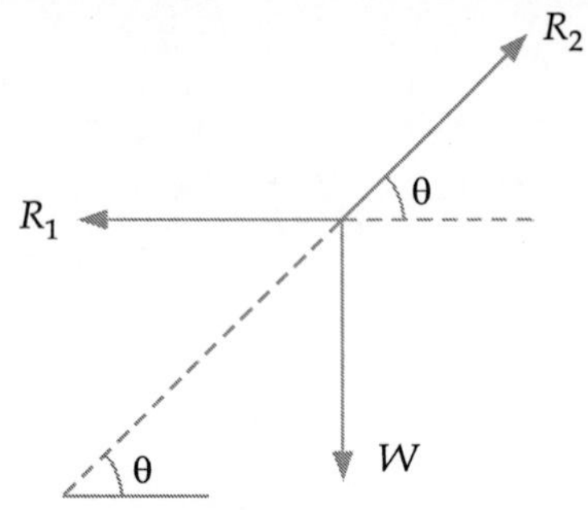

Fig. 2.36. (b)

$\therefore$ Pressure due to wall, $R_1 = W \dfrac{\sin 157.38°}{\sin 112.62°}$

$$= 80 \times \frac{0.3846}{0.923}$$

$$= \mathbf{33.33\ N}$$

EXAMPLE 2.28

A stud is acted upon by two forces P = 2 kN and Q = 3 kN as shown in Fig. 2.37. Determine the resultant of these concurrent forces by using

(*a*) Parallelogram law of forces
(*b*) Triangle law of forces.

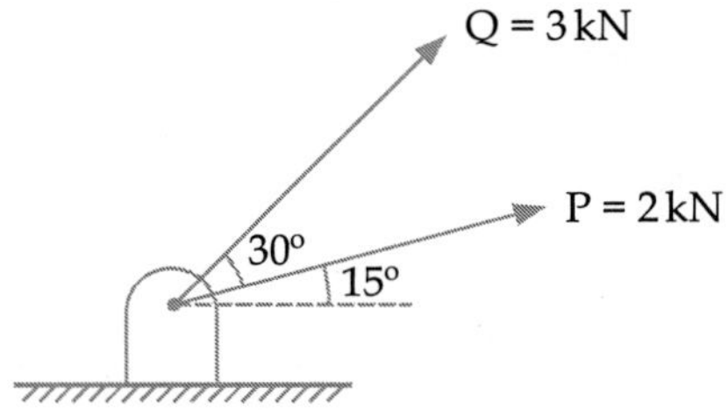

Fig. 2.37

Solution : (*a*) Resultant $R = \sqrt{P^2 + Q^2 + 2PQ\cos\theta}$

where θ is the angle between the forces P and Q; θ = 30°

$$R = \sqrt{2^2 + 3^2 + 2 \times 2 \times 3\cos 30°}$$

$$= \sqrt{4 + 9 + 10.39} = \mathbf{4.836\ N}$$

If α is the direction of resultant with respect to direction of force P, then

$$\tan\alpha = \frac{Q\sin\theta}{P + Q\cos\theta} = \frac{3\sin 30°}{2 + 3\cos 30°} = \frac{3 \times 0.5}{2 + 3 \times 0.866} = 0.326$$

$$\alpha = \tan^{-1} 0.326 = 18.06°$$

$\therefore$ Angle made by resultant with x-axis = 18.06 + 15 = **33.06°**

(*b*) The force diagram of these concurrent forces is as shown below:

$$P = OA = 2\ \text{kN}$$
$$Q = AB = 3\ \text{kN}$$

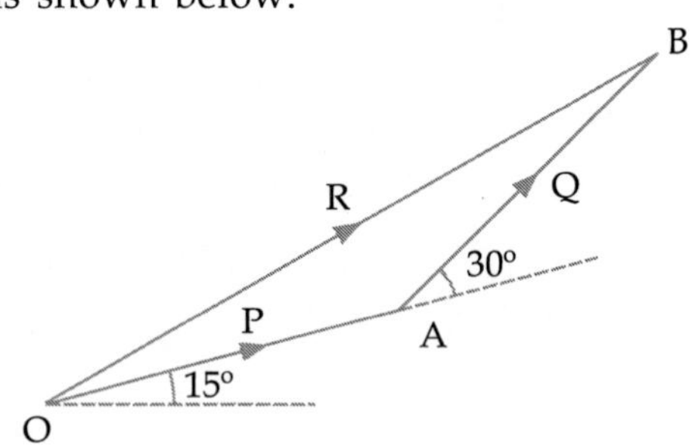

and the included angle between these forces is

$$\angle OAB = 180° - 30° = 150°$$

Then applying the law of cosines

$$R^2 = P^2 + Q^2 - 2PQ\cos(OAB)$$
$$= 2^2 + 3^2 - 2 \times 2 \times 3\cos 150°$$
$$= 4 + 9 + 10.39 = 23.39$$

$\therefore$ Resultant R = $\sqrt{23.39}$ = **4.836 N**

Applying the law of sines

$$\frac{R}{\sin OAB} = \frac{Q}{\sin AOB}$$

$$\sin AOB = \frac{Q}{R}\sin OAB = \frac{3}{4.836}\sin 150 = 0.310$$

$$\therefore \quad \angle AOB = \sin^{-1}(0.31) = 18.06°$$

This is the angle which the resultant makes with direction of force P.

$\therefore$ Angle of inclination of the resultant with the horizontal = 15 + 18.06 = **33.06°**

EXAMPLE 2.29

The resultant of two ropes pulling a cart, as shown in Fig. 2.38, acts along the axis of the cart and has a magnitude of 6 kN. What would be the tension in the ropes when θ = 30°? Proceed to workout the value of θ so that the tension in the rope *AB* is minimum. Also find out this minimum tension.

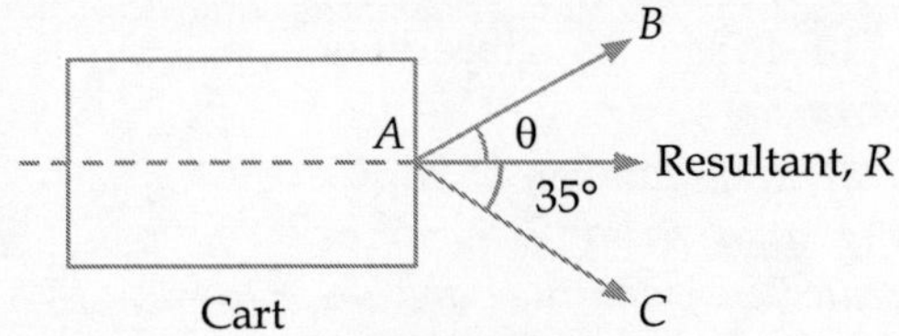

Fig. 2.38

Solution : With reference to the force diagram shown below, the law of sines gives

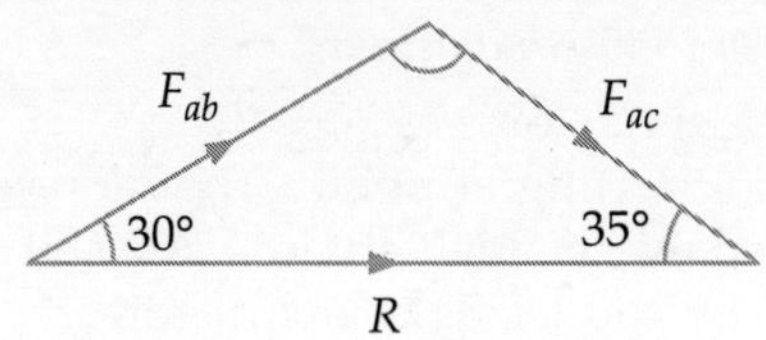

$$\frac{F_{ab}}{\sin 35°} = \frac{F_{ac}}{\sin 30°} = \frac{R}{\sin(180 - 30 - 35)}$$

$$\therefore \quad F_{ab} = R\frac{\sin 35°}{\sin 115°} = 6 \times \frac{0.573}{0.906} = \mathbf{3.795\ kN}$$

$$F_{ac} = R\frac{\sin 30°}{\sin 115°} = 6 \times \frac{0.5}{0.906} = \mathbf{3.311\ kN}$$

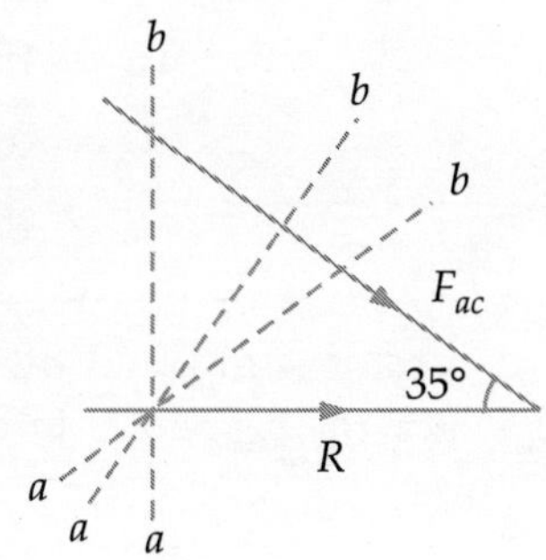

With fixed directions for the resultant and the line of action of tension in rope AC let a-b represent the line of action of tension in rope AB. It is observed that tension in rope AB will be minimum when the line of action of tensions in AB and AC are perpendicular. Hence for minimum tension in rope AB,

$$\theta = 90 - 35 = 55°$$

Minimum value of tension in rope AB will be then given by

$$F_{ab} = R\frac{\sin 35°}{\sin(180 - 35 - 55)} = 6 \times \frac{0.573}{1} = \mathbf{3.438\ N}$$

2.6. POLYGON LAW OF FORCES

"If a number of concurrent forces acting simultaneously on a body are represented in magnitude and direction by the sides of a polygon, taken in order, then the resultant is represented in magnitude and direction by the closing side of the polygon, taken in opposite order."

Consider force P_1, P_2, P_3, and P_4 acting on the body at point O as shown Fig. 2.39 (*a*). Line AB is drawn to represent force P_1, line BC to represent P_2, line CD to represent P_3 and line DE to represent P_4. The polygon is completed by drawing the closing line AE. This closing line AE represents the resultant of the given system in magnitude, line of action and direction.

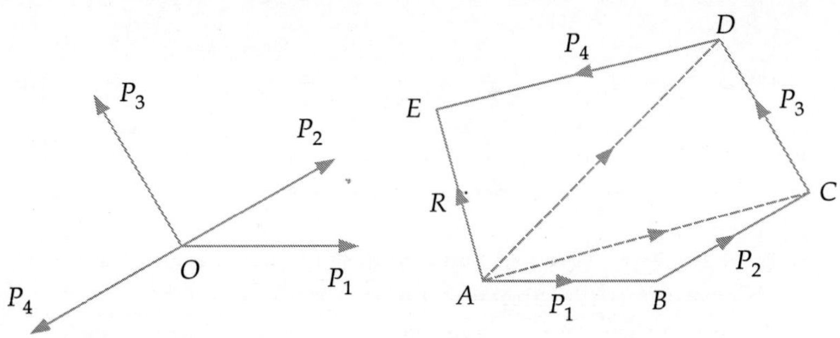

(*a*) Fig. 2.39 (*b*)

The polygon law of forces follows from the triangle law of forces as is evident from the steps outline below:

$$\vec{AC} = \vec{AB} + \vec{BC}$$

$$\vec{AD} = \vec{AC} + \vec{CD} = \vec{AB} + \vec{BC} + \vec{CD}$$

$$\vec{AE} = \vec{AD} + \vec{DE} = \vec{AB} + \vec{BC} + \vec{CD} + \vec{DE}$$

or $$R = P_1 + P_2 + P_3 + P_4.$$

Sometimes the polygon law of forces is stated as follows:

"If any number of forces acting at a point can be represented in magnitude and direction by the sides of a polygon taken in order, then the system of forces is in equilibrium."

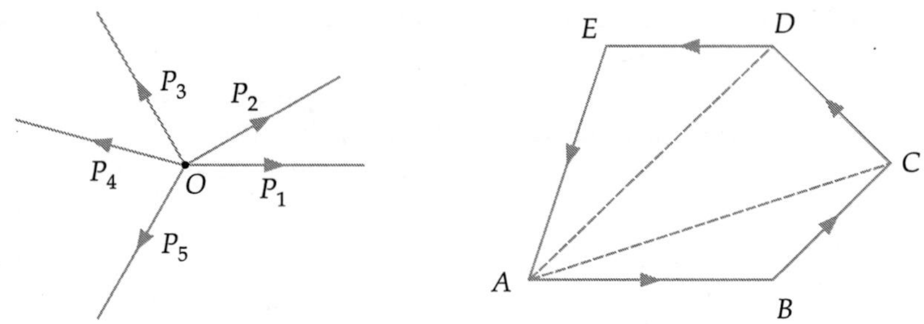

Fig. 2.40

Let the forces P_1, P_2, P_3, P_4 and P_5 acting at point O be represented in magnitude by the sides AB, BC, CD, DE and EA of the polygon $ABCDE$. Join AC and BD.

By the triangle law of forces.

$$\vec{AC} = \vec{AB} + \vec{BC}$$

$$\vec{AD} = \vec{AC} + \vec{CD} = \vec{AB} + \vec{BC} + \vec{CD}$$

$$\vec{AE} = \vec{AD} + \vec{DE} = \vec{AB} + \vec{BC} + \vec{CD} + \vec{DE}$$

i.e., $\vec{AE}$ represents the resultant of forces P_1, P_2, P_3, and P_4.

$\therefore$ Vector sum of all the forces $= \vec{AE} + \vec{EA} = 0$

and hence the system is in equilibrium.

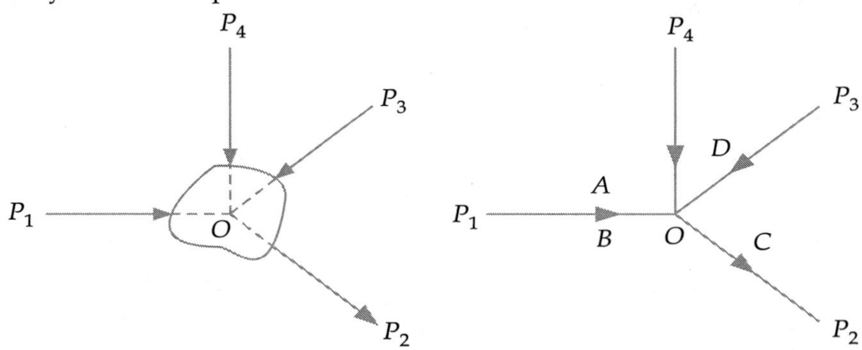

Fig. 2.41.

With reference to Fig. 2.41, let P_1, P_2, P_3 and P_4 be the coplanar concurrent forces acting simultaneously on a body, and O be the point of concurrency of three forces.

The polygon law of forces provides a convenient method for determining graphically the resultant of these coplanar concurrent forces. The steps involved in the method are outlined below:

(*a*) Construct the position (space) diagram showing the various forces (or loads) along with their magnitude and line of action (Fig. 2.41).

(*b*) Name the forces in the space diagram by using Bow's notation by which each force is named by two capital letters placed on its either side.

(*c*) Construct the force diagram by starting from a convenient point. Go on adding all the forces vectorially one by one to some suitable scale.

In the context of the force system indicated, draw vector *ab* equal to force P_1 to some suitable scale and parallel to the line of action of P_1. From point *b*, draw vector *bc* to represent P_2 in magnitude and direction. From point *c*, draw vector *cd* to represent force P_3 in magnitude and direction. Then from point draw vector *de* equal and parallel to force P_4.

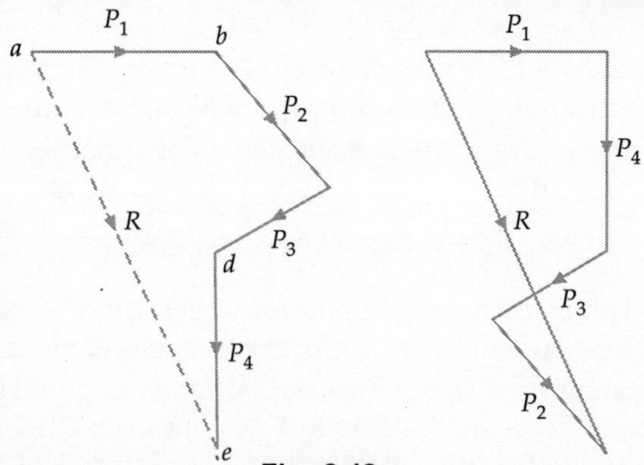

Fig. 2.42

The closing side *ae* of the polygon (Fig. 2.42) gives the required resultant in magnitude and direction. The direction is from *a* to *e* as shown in the vector diagram.

It is to be noted that the order of taking the forces does change the shape of the polygon. However, the resultant remains the same in magnitude, direction and sense.

EXAMPLE 2.30

Find graphically the resultant of the concurrent force system shown in Fig. 2.43.

Solution : Refer to Fig. 2.43 for the space diagram and name the various forces by Bow's notation.

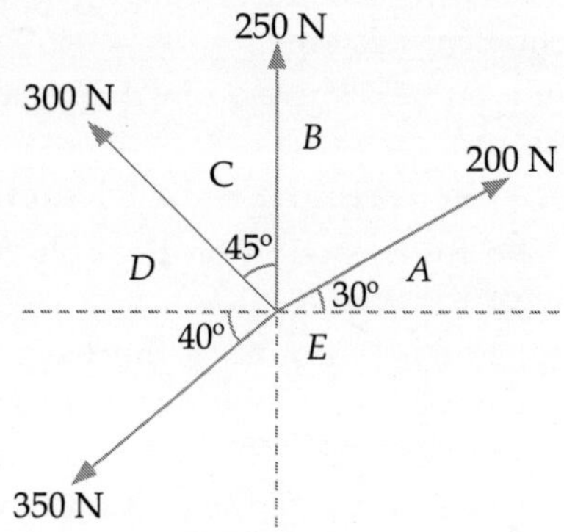

Fig. 2.43

Proceed to draw the vector diagram by adopting the following steps:

(*i*) Select some suitable point *a*, draw vector *ab* equal to force 200 N to some suitable scale and parallel to the line of action of this force. (Let 1 cm = 50 N).

(*ii*) From point *b*, draw vector *bc* to represent force 250 N in magnitude and direction.

(*iii*) From point *c*, draw vector *c d* to represent force 300 N in magnitude and direction.

(*iv*) Then from point *d*, draw vector *de* equal and parallel to force 350 N.

The closing side *ae* of the polygon gives the required resultant in magnitude and direction. The direction is from *a* to *e* as shown in the vector diagram.

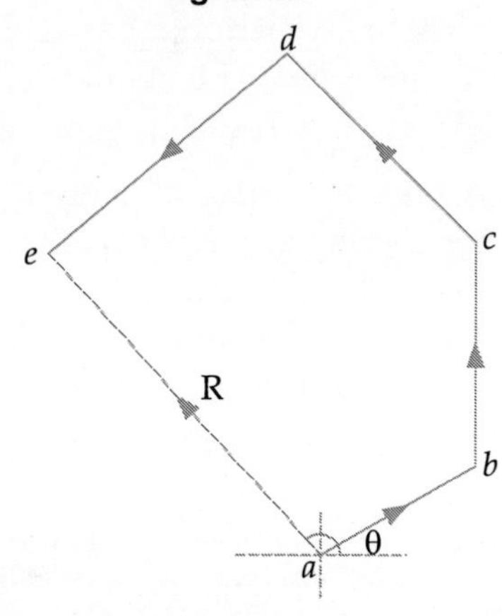

By measurement:

$$ae = 9.1 \text{ cm}$$

$\therefore$ Magnitude of resultant = 9.1 × 50 = 455 N

Angle θ (with the horizontal) = **133°**

2.7. FREE BODY DIAGRAM

The force analysis of a structure is made in a simplified way by considering the equilibrium of a portion of the structure. For that, the portion is drawn separately showing applied forces, self weight and reactions at the point of contact with other bodies. The resulting diagram is known as *free body diagram* (FBD). In a free body diagram, all the supports (like walls, floors, hinges etc) are removed and replaced by the reactions which these supports exert on the body.

A free body diagram can be drawn for any single body of a system, for any subsystem or for the entire system irrespective of whether the system is in equilibrium, *i.e.,* at rest or in uniform motion or in a dynamic state of motion.

Further, while drawing the free diagrams, one must have consideration of internal forces and external forces.

Internal forces: the forces which hold togather the particles of the body and help it to be rigid, *i.e.,* not deform. If more than one body is involved, internal forces hold the bodies together.

Imagine a bar being pulled by two equal and opposite forces applied at the ends. The internal forces will come into play to keep the bar undeformed. These forces cause stresses and strains distributed throughout the material of the body.

External forces: the forces which act on a body or system of bodies externally, *i.e.,* applied from outside. The forces essentially denote the action of other bodies (walls, floors, hinges etc.) on the rigid body being analysed.

Consider a sphere suspended by a string. The weight of the sphere and tension induced in the string represent the external forces. The reactions developed at the contact points, if any, also constitute the external forces.

A complete isolation and systematic representation of all external (applied and reactive) forces acting on a body is an important and effective tool for the solution of problems in mechanical systems.

Given below are a few examples of systems and mechanisms together with their free body diagrams:

(*i*) A sphere resting on a frictionless plane surface (Fig. 2.44)

The forces acting on the sphere when isolated from the surface are:

(*a*) Force *W* equal to the weight of the sphere. This weight acts downward through the centroid of the sphere.

(*b*) Reaction *R* at the point of contact with the surface. This reaction acts upwards normal to the surface as it is frictionless.

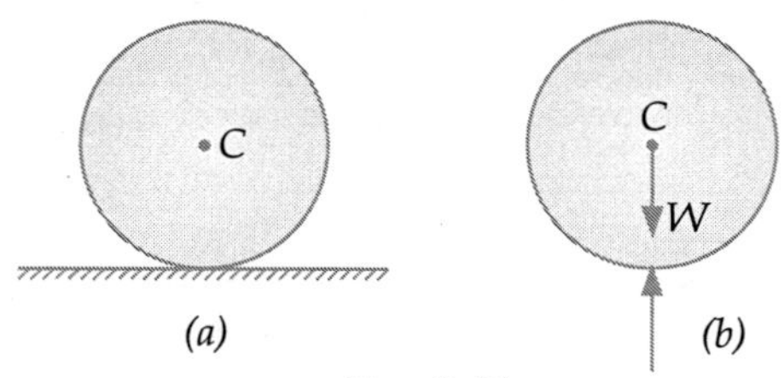

Fig. 2.44

As the sphere is in equilibrium, *R* and *W* are equal and collinear, and the free body diagram will be as shown in Fig. 2.44

(*ii*) A circular roller of weight *W* hangs by a string and rests against a smooth vertical wall (Fig. 2.45)

The force acting on the roller when isolated from the supports are:

(*a*) Force *W* equal to weight of the roller

(*b*) Wall reaction R_c at the point of contact C with the wall. The reaction will be normal to the wall as it is smooth

(*c*) Tension *T* in the string along *BA*.

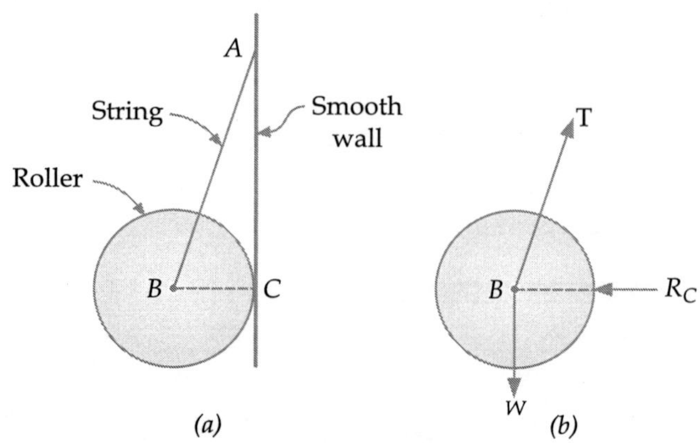

Fig. 2.45

As the roller is in equilibrium, all the forces will be concurrent and the free body diagram will be as shown in Fig. 2.45

(*iii*) A sphere resting in a *V*-shaped groove (Fig. 2.46)

The sphere is isolated from the inclined surface forming the groove, and is considered to be acted upon by the following set of forces:

(*a*) Weight of the sphere acting vertically down-ward through its centre *C*

(*b*) Reaction R_a acting nor-mal to the in-clined plane at the contact point *A*.

(*c*) Reaction R_b acting normal to the in-clined plane at the contact point *B*.

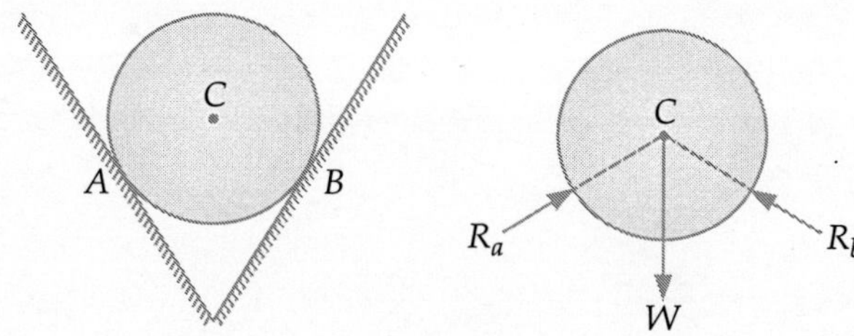

(*a*) (*b*)

Fig. 2.46

Since the sphere is in equilibrium, all the forces meet at point C and the free body diagram will be as shown in Fig. 2.46

(*iv*) A drum being rolled along the horizontal comes across a stepped obstacle (Fig. 2.47)

At the instant of obstacle being overcome, the forces acting on the drum are:

(*a*) Weight *W* of the drum acting vertically downwards through its centre *C*

(*b*) Pull *P* required to be applied at the position shown to overcome the obstacle

(*c*) Reaction at *A* that passes through point *B*, the intersection of *P* and *W*.

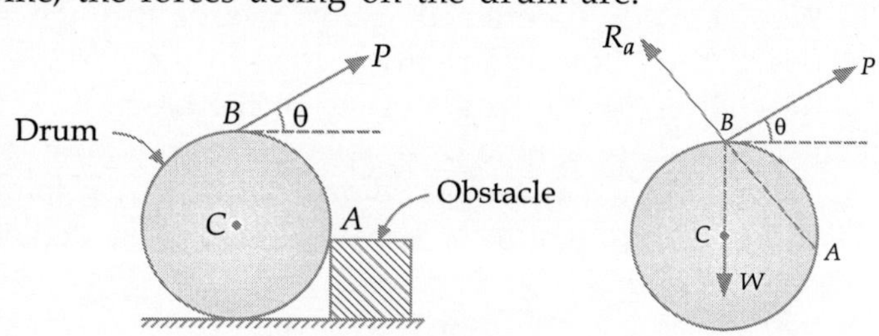

(*a*) (*b*)

Fig. 2.47

For free body diagram of the drum, refer to Fig. 2.47.

(*v*) A uniform ladder of weight *W* leans a against a smooth wall and rests on a rough floor (Fig. 2.48)

The various forces acting on the ladder when it is isolated from the wall and floor are:

(*a*) Force *W* equal to the weight of the ladder and acting vertically downward from the mid of the ladder.

(*b*) Reaction R_b of the wall. This acts right angles to the wall as the wall is smooth.

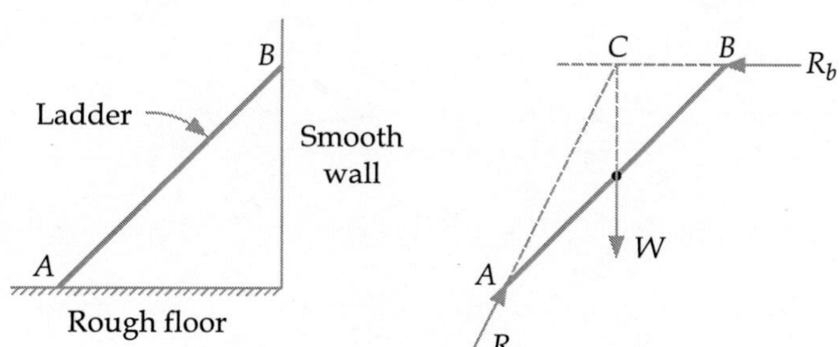

(*a*) (*b*)

Fig. 2.48.

(*c*) Reaction R_a at the ground.

Since the ladder is in equilibrium, all the forces must pass though a common point. This aspect then fixes the unknown direction of reaction R_a, and the free body diagram of the ladder will be as shown in Fig. 2.48

(*vi*) A beam loaded and supported (Fig. 2.49)

When the beam is detached from the supports, it subjected to following set of forces:

(*a*) Weight *W* acting vertically downwards through mass centre of the beam

(*b*) Reaction R_b normal to the beam at its smooth contact with the corner
(*c*) Horizontal applied force *P* and the couple *M*
(*d*) vertical and horizontal reactions (R_{av} and R_{ah}) exerted at the pin connection at *B*.

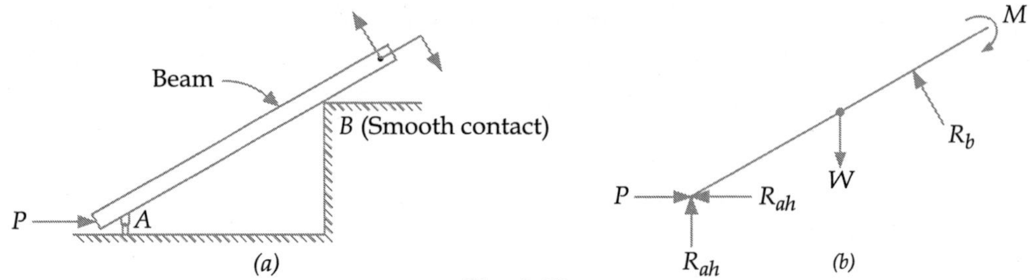

Fig. 2.49

The beam will be in equilibrium under the action of five forces and one couple, and its free body diagram will be as shown in Fig. 2.49.

(*vii*) Two spheres *P* and *Q* placed in a vessel (Fig. 2.50)

Forces acting on sphere P

(*a*) Weight W_1 of sphere acting downwards through its mass centre C_1
(*b*) Reaction R_a (towards right) normal to the vertical wall surface
(*c*) Reaction R_b (up wards) normal to the base
(*d*) Reaction R_d of sphere *Q* on sphere *P* at the point of contact *D*. This acts in a direction normal to the surface, *i.e.*, along the line C_1C_2 joining the mass centres of the spheres

The free body diagram for sphere *P* will be as depicted in Fig. 2.50 (*b*)

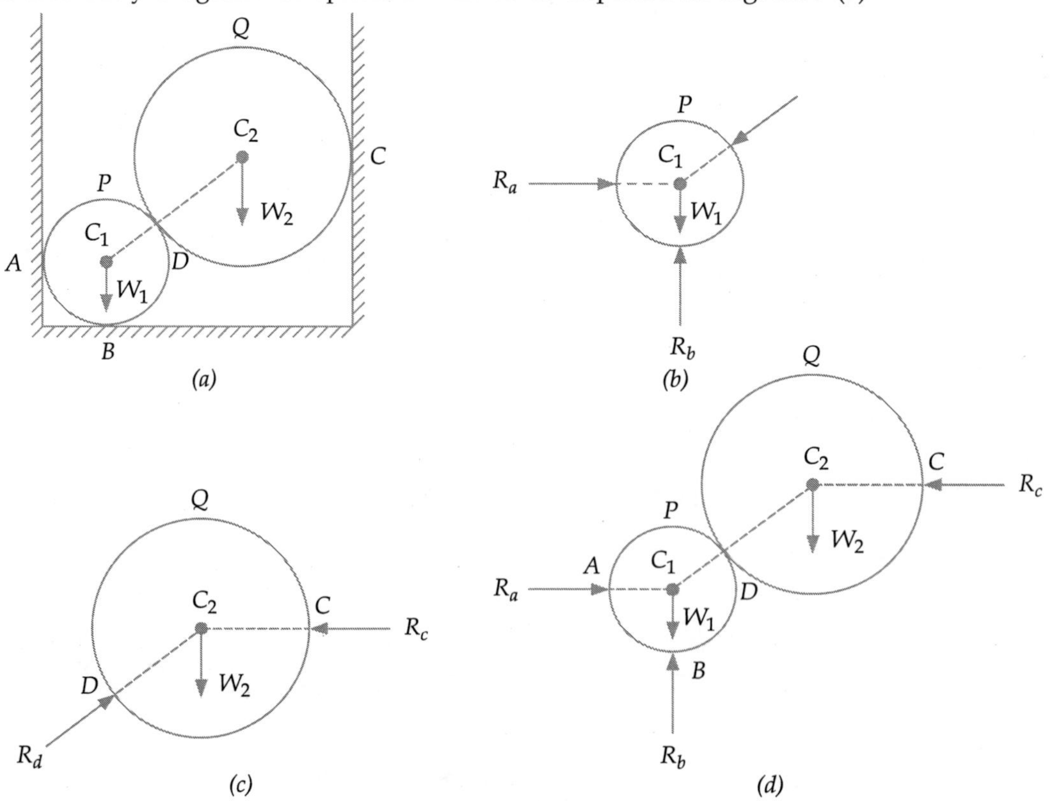

Fig. 2.50

Forces acting on sphere Q

(*a*) Weight of sphere acting downwards through its mass centre C_2

(*b*) Reaction R_c (towards left) and normal to the vertical wall surface.

(*c*) Reaction R_d of sphere *P* on sphere *Q* at the point of contact *D*. This acts in a direction normal to the surface, *i.e.*, along the line C_1C_2 joining the mass centres of the spheres.

The free body diagram for roller *Q* will be as shown in Fig. 2.50 *(c)*.

It is to be noted that

$$\begin{Bmatrix}\text{reaction of sphere } P \\ \text{on sphere } Q\end{Bmatrix} \text{ and } \begin{Bmatrix}\text{reaction of sphere } Q \\ \text{on sphere } P\end{Bmatrix}$$

are equal in magnitude, opposite in direction and are collinear. Obviously when both the spheres are taken together, these reactions cancel and do not appear in the free body diagram (Fig. 2.50 (*d*)).

Lami's Theorem

"If a body is in equilibrium under the action of three forces, then each force is proportional to the sine of the angle between the other two forces."

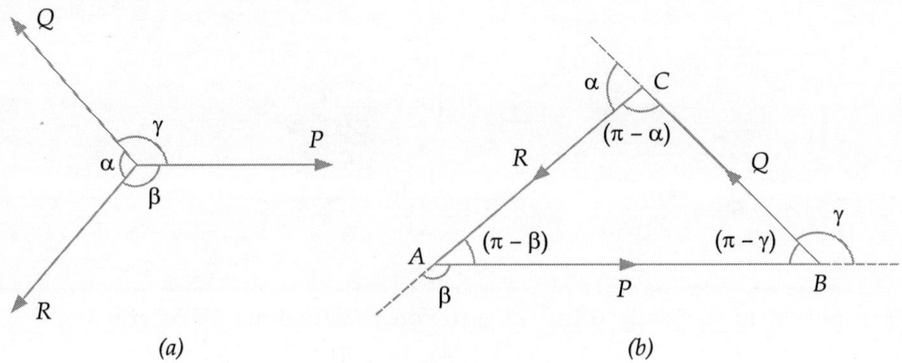

Fig. 2.51

Let *P*, *Q* and *R* be the three forces acting on a body along the directions as indicated in Fig 2.51 (*a*). Since these forces are in equilibrium, they can be represented by the sides of a triangle *ABC* [Fig 2.57 (*b*)] which is drawn so as to have its sides respectively parallel to the direction of forces. Thus

$$\vec{AB} = P; \quad \vec{BC} = Q; \quad \text{and} \quad \vec{CA} = R$$

The exterior and interior angles of the triangle *ABC* of forces will be as shown in Fig 2.51 (*b*).

Applying sine rule for the triangle *ABC*

$$\frac{AB}{\sin(\pi - \alpha)} = \frac{BC}{\sin(\pi - \beta)} = \frac{CA}{\sin(\pi - \gamma)}$$

$$\therefore \quad \frac{P}{\sin\alpha} = \frac{Q}{\sin\beta} = \frac{R}{\sin\gamma}$$

EXAMPLE 2.31

A weight of 2000 N is supported by two chains *AC* and *BC* as shown in Fig. 2.52. Determine the tension in each chain.

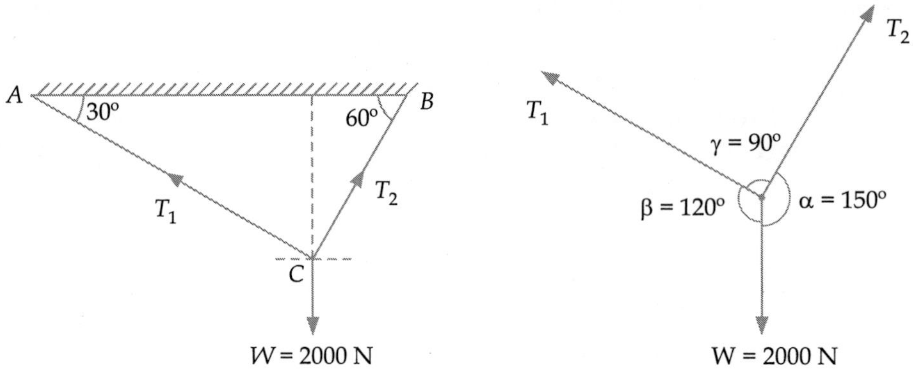

Fig. 2.52

Solution : Let T_1 and T_2 be the tensions in chains *AC* and *BC* respectively. Since the lines of action of these tensions and the weight *W* meet at a point, Lami's theorem can be applied. That is

$$\frac{T_1}{\sin \alpha} = \frac{T_2}{\sin \beta} = \frac{W}{\sin \gamma}$$

or
$$\frac{T_1}{\sin 150^\circ} = \frac{T_2}{\sin 120^\circ} = \frac{2000}{\sin 90^\circ}$$

∴
$$T_1 = 2000 \times \frac{\sin 150^\circ}{\sin 90^\circ} = 2000 \times \frac{0.5}{1} = 1000 \text{ N}$$

$$T_2 = 2000 \times \frac{\sin 120^\circ}{\sin 90^\circ} = 2000 \times \frac{0.866}{1} = 1732 \text{ N}$$

EXAMPLE 2.32

A string 2 m long is tied to the ends of a uniform rod that weighs 60 N and is 1.6 m long. The string passes over a nail, so that the rod hangs horizontally. Make calculations for the tension in the string.

Solution :

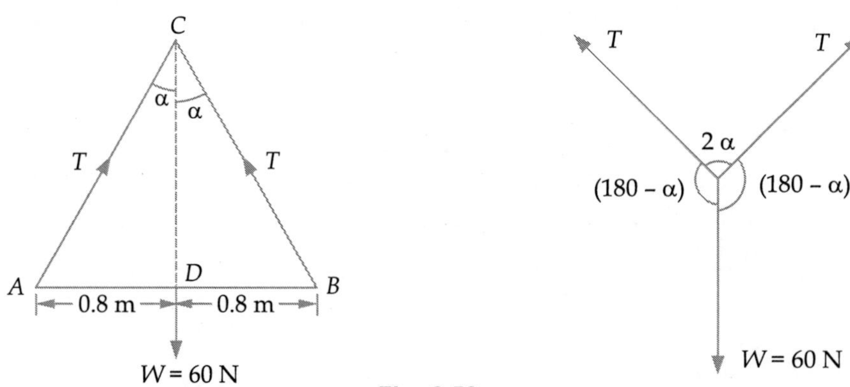

Fig. 2.53.

The line of actions of tensions *T*, *T* and weight *W* meet at point *C*, and therefore Lami's theorem is applicable. That is

$$\frac{W}{\sin 2\alpha} = \frac{T}{\sin (180 - \alpha)} = \frac{T}{\sin (180 - \alpha)} \qquad ...(i)$$

From the geometry of Fig. 2.53,

$$AC = BC = 1 \text{ m}; \quad CD = \sqrt{1^2 - 0.8^2} = 0.6 \text{ m}$$

$$\sin \alpha = 0.8 \quad \text{and} \quad \cos \alpha = 0.6$$

From identity (*i*), we get

$$T = W \frac{\sin (180 - \alpha)}{\sin 2\alpha} = W \frac{\sin \alpha}{2 \sin \alpha \cos \alpha} = 60 \times \frac{1}{2 \times 0.6} = \mathbf{50 \text{ N}}$$

EXAMPLE 2.33

In a jib crane, the jib and the tie rod are 5 m and 4 m long respectively. The height of crane post is 3 m and the tie rod remains horizontal. Determine the forces produced in the jib and tie rod when a load of 2 kN is suspended at the crane head.

Solution : Refer Fig. 2.54 for the arrangement of the system.

$$\sin \theta = \frac{3}{5} = 0.6; \quad \theta = 36.87^\circ$$

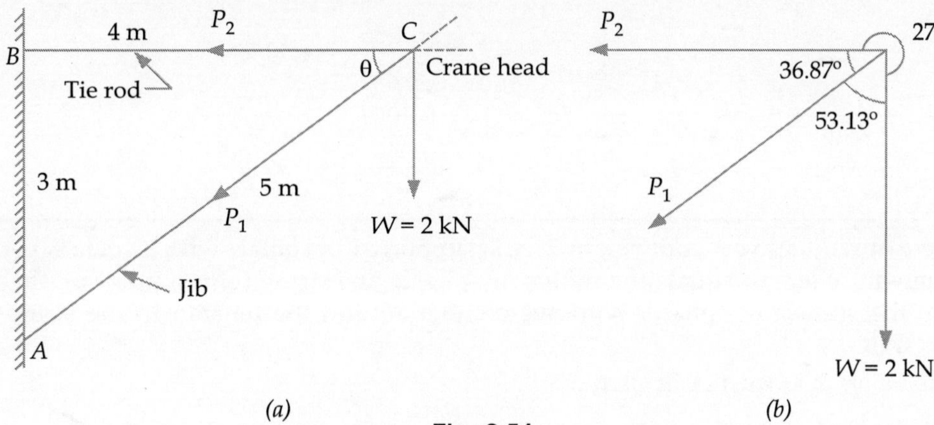

Fig. 2.54

Let P_1 and P_2 be the forces developed in the jib and tie rod respectively. The three forces P_1, P_2 and W are shown in Fig. 2.54 (*b*) with the angles between the forces calculated from the given directions. The lines of action of forces P_1, P_2 and weight W meet at the point C, and therefore Lami's theorem is applicable. That gives

$$\frac{P_1}{\sin 270^\circ} = \frac{P_2}{\sin 53.13^\circ} = \frac{2}{\sin 36.87^\circ}$$

$$\therefore \quad P_1 = 2 \times \frac{\sin 270^\circ}{\sin 36.87^\circ} = 2 \times \frac{1}{0.6} = \mathbf{-\,3.33 \text{ kN}}$$

$$P_2 = 2 \times \frac{\sin 53.13^\circ}{\sin 36.87^\circ} = 2 \times \frac{0.8}{0.6} = \mathbf{2.667 \text{ kN}}$$

The – ve sign indicates that the direction of force P_1 is opposite to that shown in Fig. 2.54 (*a*). Obviously the tie rod will be under tension and the jib will in compression.

EXAMPLE 2.34

A machine weighing 5 kN is supported by two chains attached to some point on the machine. One chain goes to the hook in the ceiling and has an indication of 45° with the horizontal. The other chain goes to the eye bolt in the wall and is inclined at 30° to the horizontal. Make calculations for the tensions induced in the chain.

Solution : Refer Fig. 2.55 for the arrangement. The machine is acted upon by the following set of forces:

(*i*) weight of machine $W = 5$ kN acting vertically downwards,

(*ii*) tension T_1 in the chain OA which goes to hook in the ceiling,

(*iii*) tension T_2 in the chain OB which goes to the eye bolt.

These forces are concurrent and meet at point O. Applying Lami's theorem

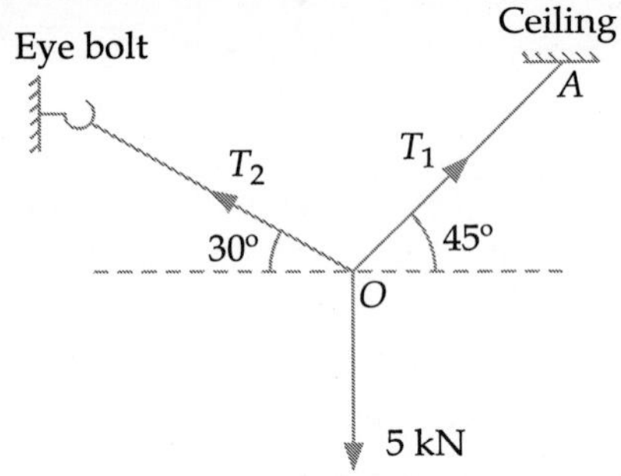

Fig. 2.55.

$$\frac{T_1}{\sin(90+30)} = \frac{T_2}{\sin(90+45)} = \frac{W}{\sin(180-30-45)}$$

or
$$\frac{T_1}{\sin 120°} = \frac{T_2}{\sin 135°} = \frac{W}{\sin 105°}$$

$$\therefore \quad T_1 = W \times \frac{\sin 120°}{\sin 105°} = 5 \times \frac{0.866}{0.966} = \mathbf{4.48\ kN}$$

$$T_2 = W \times \frac{\sin 135}{\sin 105} = 5 \times \frac{0.707}{0.966} = \mathbf{3.66\ N}$$

EXAMPLE 2.35

A smooth sphere of radius 15 cm and weight 2 N is supported in contact with a smooth vertical wall by a string whose length equals the radius of sphere. The string joins a point on the wall and a point on the surface of sphere. Workout inclination and the tension in the string and reaction of the wall.

Solution : Refer. Fig. 2.56 for the arrangement.

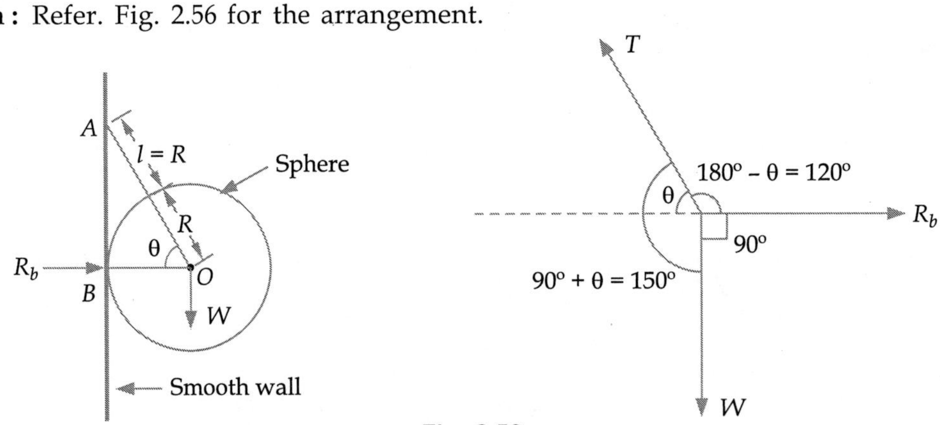

Fig. 2.56

In triangle AOB,

$$\cos\theta = \frac{OB}{OA} = \frac{R}{2R} = \frac{1}{2}; \quad \theta = \mathbf{60°}$$

The sphere is in equilibrium under the action of following forces:

(*i*) Weight $W = 2$ N of the sphere which acts vertically downwards through the centre O.

(*ii*) Tension T in the string

(*iii*) Reaction R_b of the wall at the point of contact B. Since the wall is smooth, this reaction acts perpendicular to the wall.

These three forces are concurrent, *i.e.*, meet at point O and as such the Lami's theorem is applicable.

$$\frac{T}{\sin 90^\circ} = \frac{W}{\sin 120^\circ} = \frac{R_b}{\sin 150^\circ}$$

$$\therefore \quad T = W\frac{\sin 90^\circ}{\sin 120^\circ} = 2 \times \frac{1}{0.866} = \mathbf{2.31\ N}$$

$$R_b = W\frac{\sin 150^\circ}{\sin 120^\circ} = 2 \times \frac{0.5}{0.866} = \mathbf{1.15\ N}$$

EXAMPLE 2.36

A roller of weight 500 N rests on a smooth inclined plane and is kept free from rolling down by a string as shown in Fig. 2.57. Work out tension in the string and reaction at the point of contact *B*.

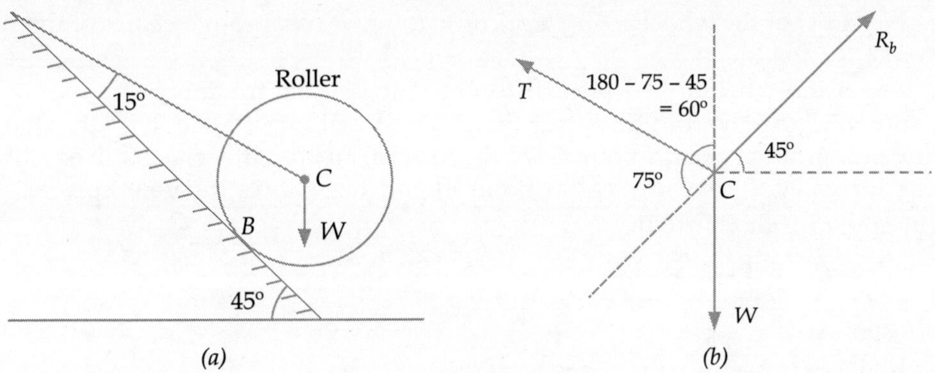

Fig. 2.57

Solution : The lines of action for tension T, weight W and reaction R_b at the contact point meet at C (the centre of the roller) and as such Lami's theorem is applicable. The angles between various segments around point C are as indicated in Fig. 2.57 (*b*).

Invoking Lami's theorem,

$$\frac{W}{\sin(60+45)} = \frac{R_b}{\sin(75+45)} = \frac{T}{\sin(90+45)}$$

$$\therefore \quad R_b = W \times \frac{\sin 120^\circ}{\sin 105^\circ} = 500 \times \frac{0.866}{0.966} = \mathbf{448.24\ N}$$

$$T = W \times \frac{\sin 135^\circ}{\sin 105^\circ} = 500 \times \frac{0.707}{0.966} = \mathbf{365.94\ N}$$

EXAMPLE 2.37

A uniform wheel of 50 cm diameter and 1 kN weight rests against a rigid rectangular block of thickness 20 cm (Fig. 2.58). Considering all surfaces smooth, determine

(*a*) least pull to be applied through the centre of wheel to just turn it over the corner of the block,

(*b*) reaction of the block

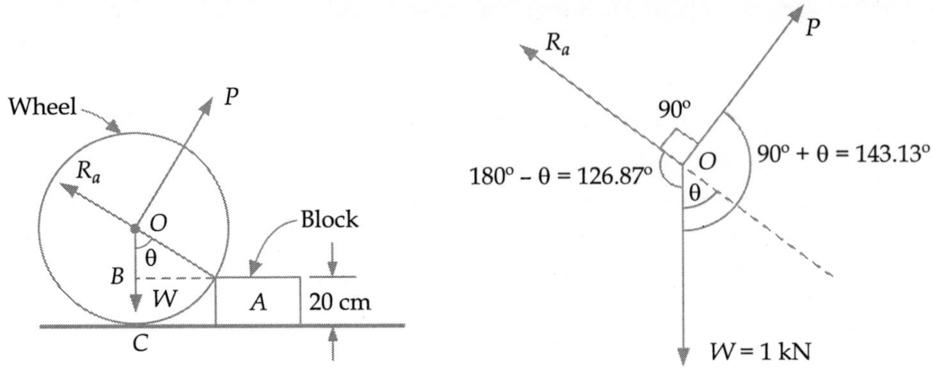

Fig. 2.58

Solution : The wheel is acted upon by the following forces when it is just about to turn over the block.

(*i*) weight $W = 1$ kN of the wheel acting vertically downwards through centre of the sphere

(*ii*) Reaction R_a of the block

(*iii*) Pull P which must be applied normal to OA if it is to be minimum.

When the sphere is just about to turn over the block, it will lose contact with the floor and apparently, the reaction at the contact point C would be zero. Further the wheel is in equilibrium and as such the forces W, R_a and P meet at point O and the Lami's theorem applies.

In the right angled triangle AOB,

$$\cos\theta = \frac{OB}{OA} = \frac{50-20}{50} = 0.6\ ; \quad \theta = 53.13^\circ$$

$$\frac{W}{\sin 90^\circ} = \frac{P}{\sin 126.87^\circ} = \frac{R_a}{143.13^\circ}$$

$$\therefore \quad P = W \times \frac{\sin 126.87^\circ}{\sin 90^\circ} = 1 \times \frac{0.8}{1} = \mathbf{0.8\ kN}$$

$$R_a = W \times \frac{\sin 143.13^\circ}{\sin 90^\circ} = 1 \times \frac{0.6}{1} = \mathbf{0.6\ kN}$$

EXAMPLE 2.38

Figure 2.59 shows a weight W tied to the end of a cord of length l. Determine the magnitude of force F to pull the weight at an angle α as indicated in the figure. Proceed to find the tension in the cord.

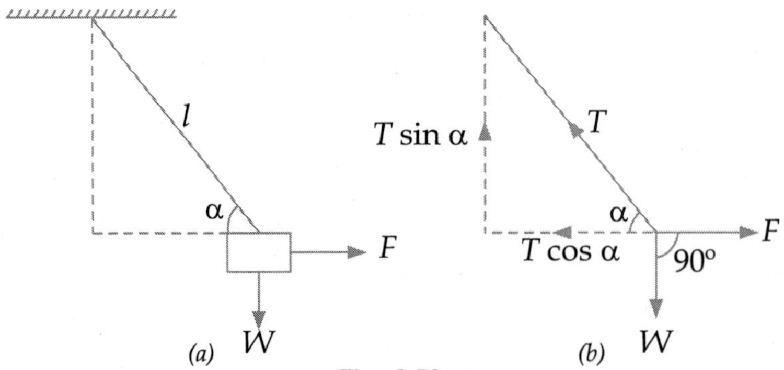

Fig. 2.59

Solution : The three forces F, T and W meet at a point and as such the Lami's theorem is applicable. Therefore

$$\frac{T}{\sin 90^\circ} = \frac{W}{\sin(180^\circ - \alpha)} = \frac{F}{\sin(90^\circ + \alpha)}$$

$$\therefore \quad F = W \times \frac{\sin(90 + \alpha)}{\sin(180 - \alpha)} = W \times \frac{\sin 90^\circ \cos\alpha + \cos 90^\circ \sin\alpha}{\sin\alpha} = W \times \frac{\cos\alpha}{\sin\alpha} = \mathbf{W \cot\alpha}$$

$$T = W \times \frac{\sin 90^\circ}{\sin(180^\circ - \alpha)} = W \times \frac{1}{\sin\alpha} = \mathbf{W \operatorname{cosec}\alpha}$$

Alternatively :

Since the system is in equilibrium

$$\Sigma F_x = 0; \quad F = T\cos\alpha \qquad ...(i)$$

$$\Sigma F_y = 0; \quad W = T\sin\alpha \qquad ...(ii)$$

Upon division: $\dfrac{F}{W} = \dfrac{\cos\alpha}{\sin\alpha}$ or $F = \mathbf{W \cot\alpha}$

Substituting $F = W\cot\alpha = W\dfrac{\cos\alpha}{\sin\alpha}$ in identity (i), we get

$$W\frac{\cos\alpha}{\sin\alpha} = T\cos\alpha$$

or $$T = \frac{W}{\sin\alpha} = \mathbf{W \operatorname{cosec}\alpha}$$

EXAMPLE 2.39

An electric light fixture weighing 50 N hangs from point C by two strings AC and BC as shown in Fig. 2.60. The string AC is inclined at 60° to the horizontal and string BC is 45° to the vertical. Using Lami's theorem or otherwise determine the forces in the strings AC and BC.

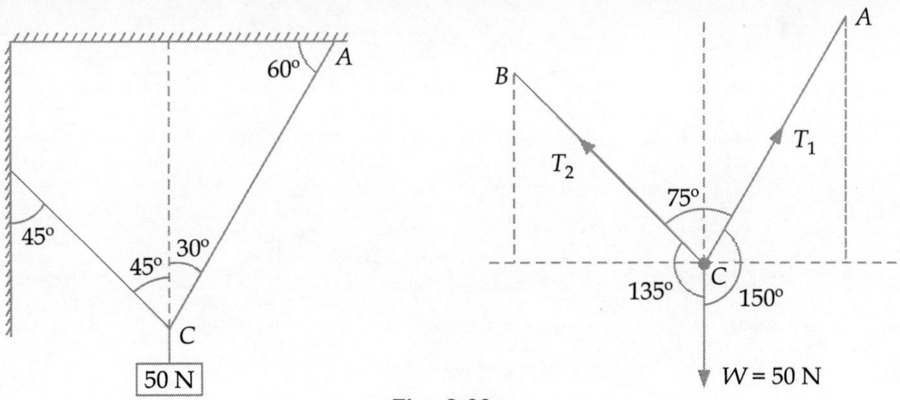

Fig. 2.60

Solution : Let T_1 and T_2 be the tensions in strings AC and BC respectively. The line of action of tensions in AC and BC and weight W of the light fixture meet at point C, and therefore Lami's theorem is applicable. That gives

$$\frac{T_1}{\sin 135^\circ} = \frac{T_2}{\sin 150^\circ} = \frac{50}{\sin 75^\circ}$$

$$\therefore \quad T_1 = 50 \times \frac{\sin 135^\circ}{\sin 75^\circ} = 50 \times \frac{0.707}{0.966} = \mathbf{36.594\ N}$$

$$T_2 = 50 \times \frac{\sin 150^\circ}{\sin 75^\circ} = 50 \times \frac{0.50}{0.966} = \mathbf{25.88\ N}$$

Alternatively :

Since the light fixture is in equilibrium, $\Sigma F_x = 0$ and $\Sigma F_y = 0$. Resolving all the forces horizontally, we get

$$T_1 \sin 30^\circ - T_2 \sin 45^\circ = 0$$

$$T_1 \times 0.5 - T_2 \times 0.707 = 0$$

$$T_2 = \frac{0.5}{0.707} T_1 = 0.707\ T_1$$

Resolving all the forces vertically, we get

$$T_1 \cos 30^\circ + T_2 \cos 45^\circ - 50 = 0$$

$$T_1 \times 0.866 + T_2 \times 0.707 = 50$$

Substituting $T_2 = 0.707\ T_1$ in the above identity, we get

$$0.866\ T_1 + 0.707 \times 0.707\ T_1 = 50$$

or $$0.866\ T_1 + 0.5\ T_1 = 50$$

$\therefore$ $$T_1 = \frac{50}{1.366} = \mathbf{36.6\ N}$$

$$T_2 = 0.707\ T_1 = 0.707 \times 36.6 = \mathbf{25.88\ N}$$

EXAMPLE 2.40

A body acted upon by three forces P_1, P_2 and P_3, as shown in Fig. 2.61, is in equilibrium.

If P_2 = 300 N, make calculations for the forces P_1 and P_3.

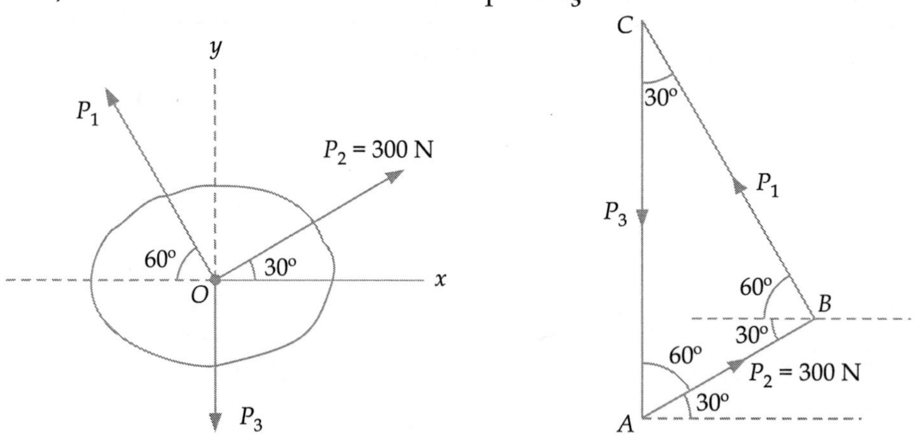

Fig. 2.61

Solution : The given forces can be represented by the sides *AB*, *BC* and *CA* of the triangle *ABC* (Fig. 2.61)

$$AB = P_2 = 300\text{ N}; \quad BC = P_1 \text{ and } CA = P_3$$

Then applying the law of sines,

$$\frac{P_1}{\sin 60^\circ} = \frac{P_2}{\sin 30^\circ} = \frac{P_3}{\sin 90^\circ}$$

That gives: $$P_1 = P_2 \frac{\sin 60^\circ}{\sin 30^\circ} = 300 \times \frac{0.866}{0.5} = \mathbf{519.6\ N}$$

$$P_3 = P_2 \frac{\sin 90^\circ}{\sin 30^\circ} = 300 \times \frac{1}{0.5} = \mathbf{600\ N}$$

Alternatively :

Since the system of forces is in equilibrium

$$\Sigma F_x = 0 \quad \text{and} \quad \Sigma F_y = 0$$

Resolution of forces along X-axis gives :

$$P_2 \cos 30^\circ + P_1 \cos 120^\circ + P_3 \cos 270^\circ = 0$$

All angles have been measured anticlockwise from axis O-X

$$0.866\, P_2 - 0.5\, P_1 + 0 = 0$$

$$\therefore \quad P_1 = \frac{0.866}{0.5} P_2 = \frac{0.866}{0.5} \times 300 = \mathbf{519.6\ N}$$

Resolution of all forces along Y-axis gives :

$$P_2 \sin 30^\circ + P_1 \sin 120^\circ + P_3 \sin 270^\circ = 0$$

$$0.5\, P_2 + 0.866\, P_1 - P_3 = 0$$

$$\therefore \quad P_3 = 0.5\, P_2 + 0.866\, P_1$$

$$= 0.5 \times 300 + 0.866 \times 519.6 = 150 + 450 = \mathbf{600\ N}$$

Alternatively :

Since the three forces meet at a point, solution to this problem can be obtained by invoking the Lami's theorem.

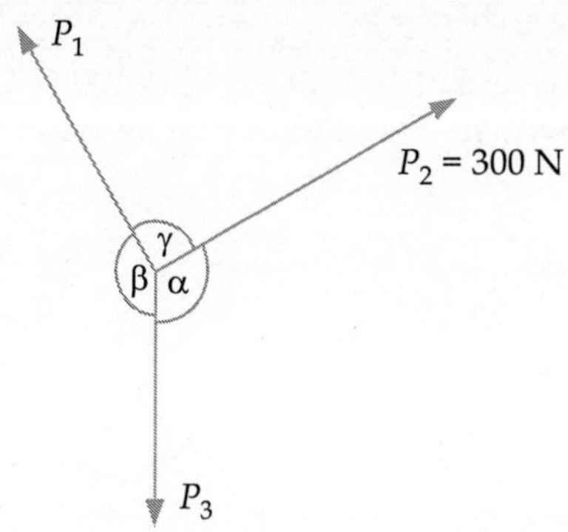

$$\alpha = 120^\circ;\ \beta = 150^\circ \text{ and } \gamma = 90^\circ$$

$$\frac{P_1}{\sin \alpha} = \frac{P_2}{\sin \beta} = \frac{P_3}{\sin \gamma}$$

$$\text{or} \quad \frac{P_1}{\sin 120^\circ} = \frac{300}{\sin 150^\circ} = \frac{P_3}{\sin 90^\circ}$$

$$\therefore \quad P_1 = 300 \times \frac{\sin 120^\circ}{\sin 150^\circ}$$

$$= 300 \times \frac{0.866}{0.5} = \mathbf{519.6\ N}$$

$$P_3 = 300 \times \frac{\sin 90^\circ}{\sin 150^\circ} = 300 \times \frac{1}{0.5} = \mathbf{600\ N}$$

EXAMPLE 2.41

A string *ABCDE* whose extremity *A* is fixed has weights W_1 and W_2 attached to it at *B* and *C*, and passes round a smooth peg at *D* carrying a weight of 800 N at the free end *E* (Fig. 2.62). If in a state of equilibrium, *BC* is horizontal and *AB* and *CD* make angles of 150° and 120° respectively with *BC*, make calculations for

(*a*) the tensions in portions *AB*, *BC*, *CD* and *DE* of the string

(*b*) the value of weights W_1 and W_2

(*c*) the pressure on the peg *D*.

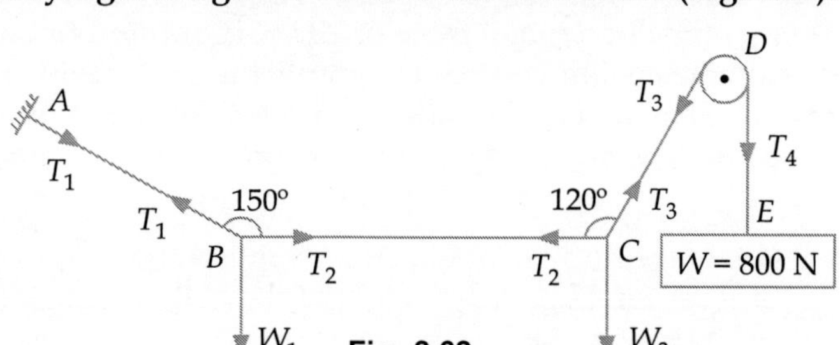

Fig. 2.62

Solution : Let T_1, T_2, T_3, T_4 be the tensions in segments *AB*, *BC*, *CD* and *DE* of the string.

Under equilibrium conditions, $T_3 = T_4 = 800$ N

Applying Lami's theorem at point C,

$$\frac{T_2}{\sin 150^\circ} = \frac{T_3}{\sin 90^\circ} = \frac{W_2}{\sin 120^\circ}$$

$$\therefore \quad T_2 = T_3 \frac{\sin 150^\circ}{\sin 90^\circ} = 800 \times \frac{0.5}{1} = \mathbf{400\ N}$$

$$W_2 = T_3 \frac{\sin 120^\circ}{\sin 90^\circ} = 800 \times \frac{0.866}{1} = \mathbf{692.8\ N}$$

Applying Lami's theorem at point B,

$$\frac{T_1}{\sin 90^\circ} = \frac{T_2}{\sin 120^\circ} = \frac{W_1}{\sin 150^\circ}$$

$$\therefore \quad T_1 = T_2 \frac{\sin 90^\circ}{\sin 120^\circ} = 400 \times \frac{1}{0.866} = \mathbf{461.89\ N}$$

$$W_1 = T_2 \frac{\sin 150^\circ}{\sin 120^\circ} = 400 \times \frac{0.5}{0.866} = \mathbf{230.95\ N}$$

(*c*) Pressure on the peg at $D = T_3 \sin 60^\circ + W$

$= 800 \sin 60^\circ + 800 = 692.82 + 800 =$ **1492.82 N**

EXAMPLE 2.42

A spherical ball of weight 100 N is attached to a string and is suspended from the ceiling as shown in Fig. 2.63. What tension would be induced in the string ?

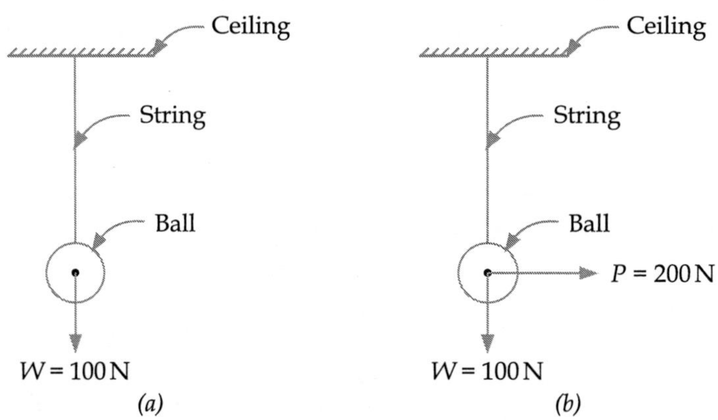

Fig. 2.63

Subsequently a horizontal force of 200 N is applied to the ball as shown in Fig. 2.70. Calculate the resultant tension in the string and the angle which the string makes with the vertical.

Solution : The ball is in equilibrium under the action of two forces namely the weight W of the ball and the tension T induced in the string

$$\therefore \quad T = W = 80 \text{ N}$$

(*ii*) When the horizontal force P is applied the ball, it will be in equilibrium under the action of three forces namely weight of the ball, horizontal force applied and the tension induced in the string.

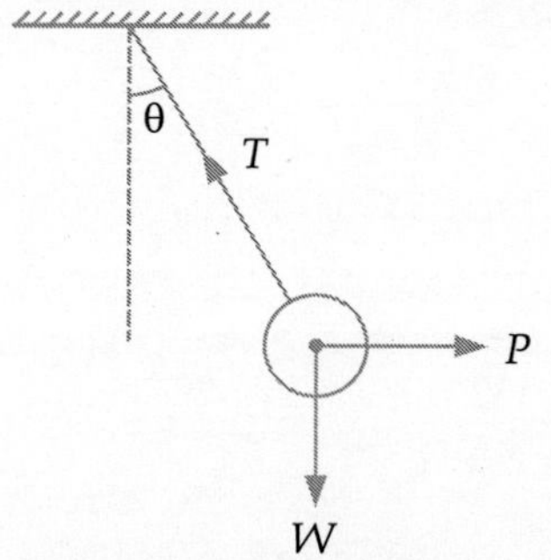

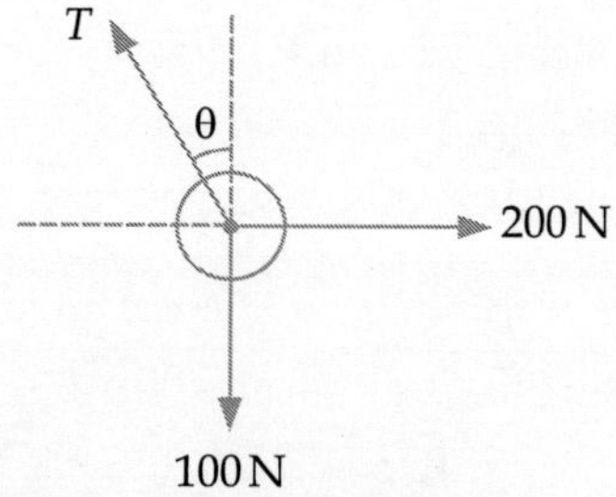

With reference to free body diagram, we have under equilibrium state

$$\Sigma F_x = 0: \quad -T \sin\theta + 200 = 0 \qquad ...(i)$$

$$\Sigma F_y = 0: \quad T \cos\theta - 100 = 0 \qquad ...(ii)$$

From expressions (*i*) and (*ii*)

$$\tan\theta = \frac{200}{100} = 2\,; \quad \theta = \mathbf{63.55°}$$

$$T = \frac{200}{\sin 63.55} = \mathbf{223.46\ N}$$

Alternatively

Invoking Lami's theorem

$$\frac{T}{\sin 90°} = \frac{100}{\sin(90+\theta)} = \frac{200}{\sin(180-\theta)}$$

or
$$\frac{T}{\sin 90} = \frac{100}{\cos\theta} = \frac{200}{\sin\theta}$$

and hence
$$\tan\theta = \frac{200}{100} = 2\,; \quad \theta = \mathbf{63.55°}$$

$$T = \frac{\sin 90}{\cos 63.55} \times 100 = \frac{1}{0.4454} \times 100 = \mathbf{224.52\ N}$$

EXAMPLE 2.43

A body of weight 20 N is suspended by two strings 5 m and 12 m long and other ends being fastened to the extremities of a rod of length 13 m. If the rod is to remain horizontal, determine the tensions in the strings.

Solution: Refer Fig. 2.64 for the arrangement of the system. In triangle *ABC*,

$$AC = 5\text{ m};\ BC = 12\text{ m and } AB = 13\text{ m}$$

$$5^2 + 12^2 = 13^2; \quad AC^2 + BC^2 = AB^2$$

and accordingly ΔABC is a right angled triangle with $\angle ACB = 90°$

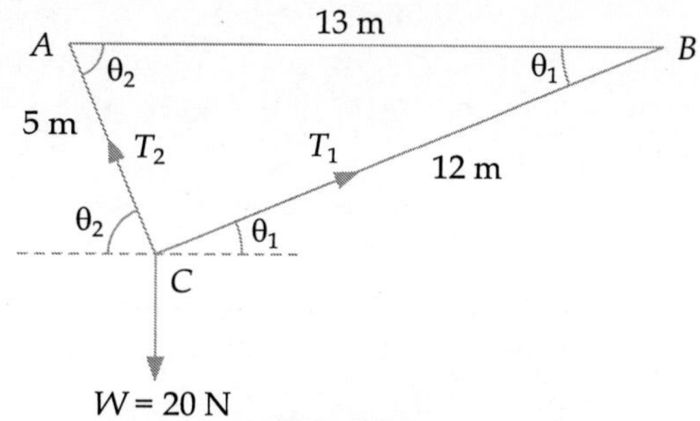

Fig. 2.64

Further:
$$\sin\theta_1 = \frac{5}{13} \text{ and } \cos\theta_1 = \frac{12}{13}$$

$$\sin\theta_2 = \frac{12}{13} \text{ and } \cos\theta_2 = \frac{5}{13}$$

The weight is acted upon by the following set of forces :

(i) weight of body $W = 20$ N acting vertically downwards,

(ii) tension T_1 in string CB, and

(iii) tension T_2 in string CA

Since the lines of action of the tensions and weight meet at point C, Lami's theorem holds good. That is

$$\frac{T_1}{\sin(90+\theta_2)} = \frac{T_2}{\sin(90+\theta_1)} = \frac{W}{\sin\angle ACB}$$

or $$\frac{T_1}{\cos\theta_2} = \frac{T_2}{\cos\theta_1} = \frac{W}{\sin 90°}$$

$$\therefore \quad T_1 = \frac{W}{\sin 90°} \times \cos\theta_2 = \frac{20}{1}\times\frac{5}{12} = \mathbf{7.69\ N}$$

$$T_2 = \frac{W}{\sin 90°} \times \cos\theta_1 = \frac{20}{1}\times\frac{12}{13} = \mathbf{18.46\ N}$$

EXAMPLE 2.44

A body of weight 20 N is suspended by two strings 5 m and 12 m long and other ends being fastened to the extremities of a rod of length 13 m. If the rod be so held that the body hangs immediately below the middle point, find out the tensions in the strings.

Solution: Refer Fig. 2.65 for the arrangement of the system. In triangle ABC,

$$AC = 5\text{ m};\ BC = 12\text{ m and } AB = 13\text{ m}$$

$$5^2 + 12^2 = 13^2;\quad (AC)^2 + (BC)^2 = AB^2$$

and accordingly ΔABC is a right angled triangle with $\angle ACB = 90°$

Applying sine rule to triangles ACD and BCD,

$$\frac{6.5}{\sin(90-\alpha)} = \frac{CD}{\sin(90-\beta)};\quad \frac{6.5}{\cos\alpha} = \frac{CD}{\cos\beta} \quad ...(i)$$

and $$\frac{6.5}{\sin\alpha} = \frac{CD}{\sin\beta} \quad ...(ii)$$

From expressions (i) and (ii),

$$\frac{6.5}{\cos\alpha}\times\cos\beta = \frac{6.5}{\sin\alpha}\times\sin\beta$$

or $$\tan\alpha = \tan\beta \quad \therefore\ \alpha = \beta$$

The system is acted upon by the following set of forces :

(i) weight $W = 20$ N of the body acting vertically downwards

(ii) tension T_1 in the string CB, and

(iii) tension T_2 in the string CA

These forces are concurrent and meet at point C, and therefore Lami's theorem is applicable. That gives

$$\frac{W}{\sin 90°} = \frac{T_1}{\sin(90+\alpha)} = \frac{T_2}{\sin(180-\alpha)}$$

or $$\frac{20}{1} = \frac{T_1}{\cos\alpha} = \frac{T_2}{\sin\alpha}$$

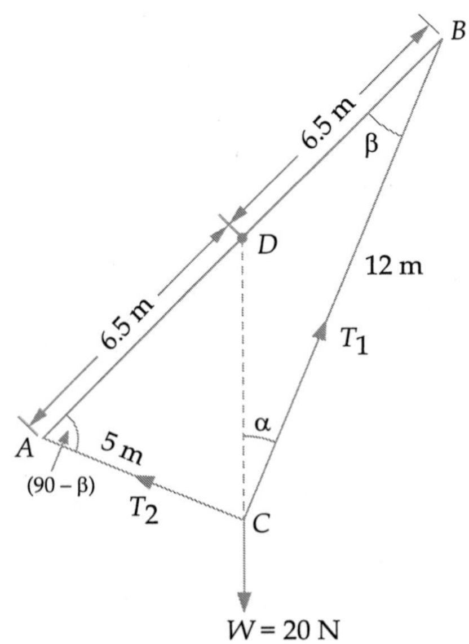

Fig. 2.65

Since $\alpha = \beta$, we can write

$$20 = \frac{T_1}{\cos\beta} = \frac{T_2}{\sin\beta}$$

From the given geometry $\cos\beta = \frac{12}{13}$ and $\sin\beta = \frac{5}{13}$. Therefore

$$T_1 = 20\cos\beta = 20\times\frac{12}{13} = \mathbf{18.46\ N}$$

and $$T_2 = 20\sin\beta = 20\times\frac{5}{13} = \mathbf{7.7\ N}$$

EXAMPLE 2.45

A small ball of weight *W* is held by two wires as shown in Fig. 2.66. Estimate the tension induced in the wire *AB* (*i*) before the wire *BC* is cut, and (*ii*) after the wire *BC* is cut.

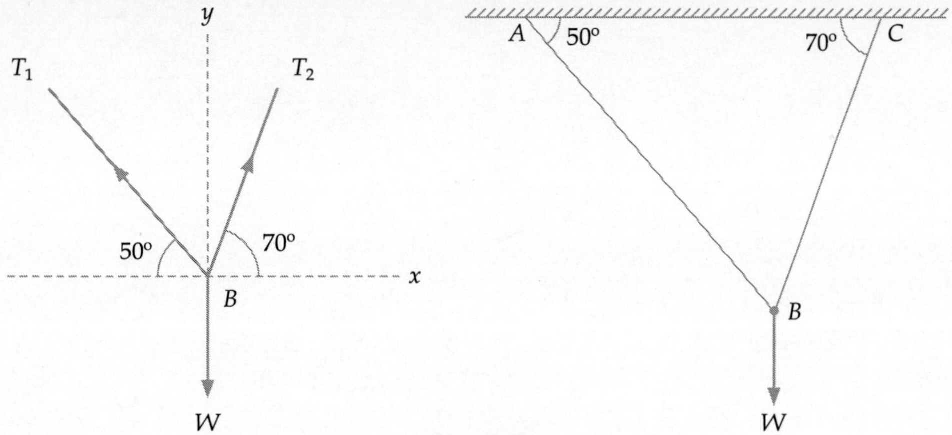

Fig. 2.66

Solution: Before the wire *CB* is cut, the ball is in static equilibrium.

Let T_1 and T_2 be the tensions in wires *BA* and *BC* respectively. Since the system is in equilibrium,

$\Sigma F_x = 0$; $\quad T_2\cos 70° - T_1\cos 50° = 0$

$$T_2 = \frac{T_1\cos 50°}{\cos 70°} = \frac{0.643\,T_1}{0.342} = 1.88\,T_1$$

$\Sigma F_y = 0$; $\quad T_2\sin 70° + T_1\sin 50° - W = 0$

or $\quad 0.94\,T_2 + 0.766\,T_1 - W = 0$

Substituting for $T_2 = 1.88T$, we get

$$0.94\times 1.88\,T_1 + 0.766\,T_1 = W$$

$\therefore$ Tension in wire *AB*, $T_1 = \dfrac{W}{0.94\times 1.88 + 0.766}$

$= \mathbf{0.395\ W}$

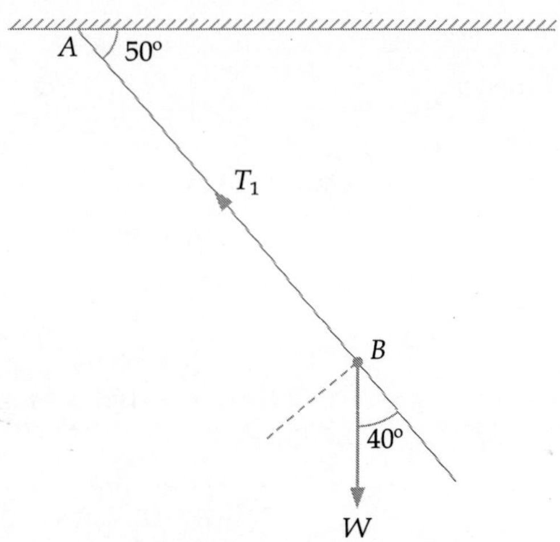

(*ii*) When the string *BC* is cut, the ball is no longer in static equilibrium. It is acted upon by an unbalanced force that acts normal to *AB* and accelerates the ball. Resolution of forces along the wire *AB* gives

$$T_1 - W\cos 40° = 0$$

or $$T_1 = W\cos 40° = \mathbf{0.766\ W}$$

EXAMPLE 2.46

A joist of length 4 m and weighing 200 N is raised by pulling a rope as shown in Fig. 2.67. Determine the tension *T* induced in the rope and reaction at end *A* of the joist

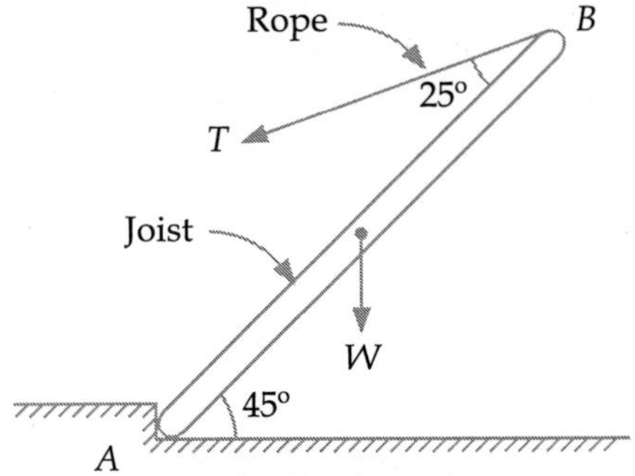

Fig. 2.67

Solution : The three forces acting in the system are:

(*i*) weight W of the joist

(*ii*) force T exerted by the rope

(*iii*) reaction R of the ground at A

Under equilibrium conditions, the reaction R passes through the point of intersection of W and T

The three forces R, T and W are then represented by the sides LM, MN and NL of the triangle LMN (Fig. 2.68).

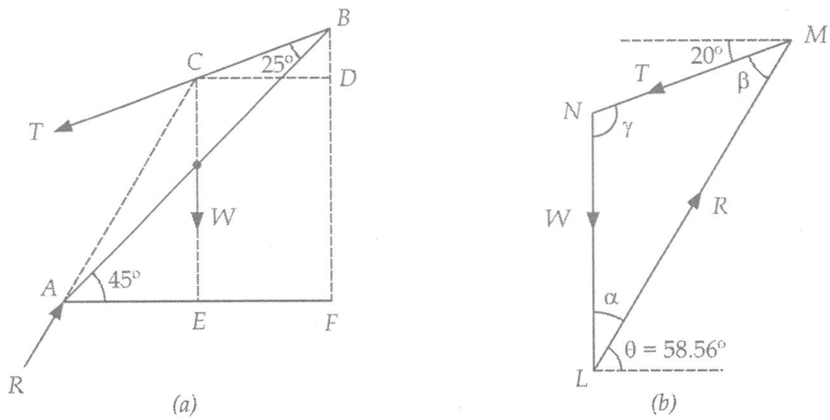

Fig. 2.68.

With reference to Fig. 2.68 (*a*)

$$AF = BF = AB \cos 45° = 4 \times 0.707 = 2.828 \text{ m}$$

$$EF = AE = \frac{1}{2} \times AF = 1.414 \text{ m}$$

In triangle BCD, angle $CBD = 25 + 45 = 70°$

$$BD = \frac{CD}{\tan 70°} = \frac{EF}{\tan 70°} = \frac{1.414}{2.447} = 0.515 \text{ m}$$

Further

$$CE = DF = BF - BD = 2.828 - 0.515 = 2.313 \text{ m}$$

$$\tan(\angle CAE) = \tan\theta = \frac{CE}{AE} = \frac{2.313}{1.414} = 1.6358$$

$$\theta = 58.56°$$

Then in the force triangle LMN,

$$\alpha = 90 - 58.56° = 31.44°$$

$$\beta = 58.56 - 20 = 38.56°$$

$$\gamma = 180 - 31.44 - 38.56 = 110°$$

From the law of sines

$$\frac{T}{\sin\alpha} = \frac{W}{\sin\beta} = \frac{R}{\sin\gamma}$$

$$\therefore \quad T = W\frac{\sin\alpha}{\sin\beta} = 200\frac{\sin 31.44°}{\sin 38.56°} = \mathbf{167.45\ N}$$

$$R = W\frac{\sin\gamma}{\sin\beta} = 200\frac{\sin 110°}{\sin 38.56°} = \mathbf{301.67\ N}$$

EXAMPLE 2.47

A heavy spherical ball of weight W rests in a V-shaped trough whose sides are inclined at α and β to the horizontal. Determine the pressure exterted on each side. Neglect friction.

Subsequently a similar spherical ball is placed on the side of inclination α and it is made to rest on the first ball. Work out the force exerted by the lower ball on the side inclined at β.

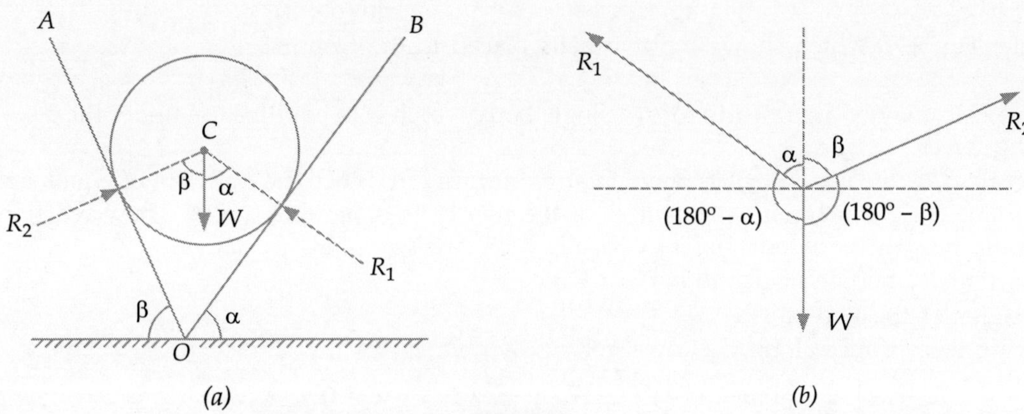

Fig. 2.69

Solution : Reference Fig. 2.69, the trough is formed by two planes OA and OB and these planes are inclined at α and β respectively with the horizontal. The spherical ball with centre C lies in this trough.

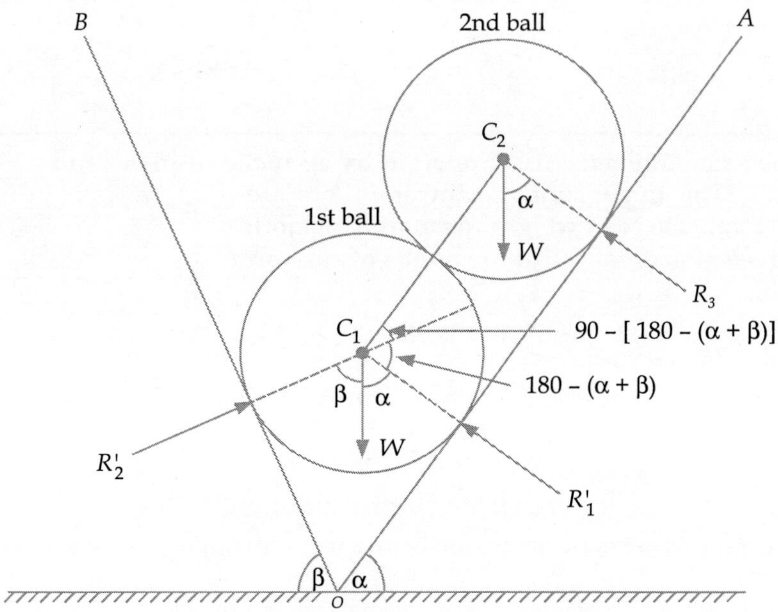

Fig. 2.70

Let R_1 and R_2 be the reactions at surfaces *OA* and *OB* respectively. Since there is no friction, these reactions will be normal to the respective surfaces.

Since the system is in equilibrium condition, the forces R_1, R_2 and W meet at point *C*, the centre of the ball. The system of forces is as shown in Fig. 2.69 (*b*).

Applying Lami's theorem at point *C*,

$$\frac{R_1}{\sin(180-\beta)} = \frac{R_2}{\sin(180-\alpha)} = \frac{W}{\sin\{180-(\alpha+\beta)\}}$$

or
$$\frac{R_1}{\sin\beta} = \frac{R_2}{\sin\alpha} = \frac{W}{\sin(\alpha+\beta)}$$

∴
$$R_1 = W\frac{\sin\beta}{\sin(\alpha+\beta)} \text{ and } R_2 = W\frac{\sin\alpha}{\sin(\alpha+\beta)}$$

(*b*) Refer Fig. 2.70 which shows the two balls placed in the trough.

The two balls may be assumed to form a single body which is in equilibrium under the action of following forces :

(*i*) reaction R_1' and R_3 acting perpendicular to surface *OA*. Since the balls are of equal size, the line C_1C_2 which joins the centres of the two balls is parallel to *OA*. These reactions would be then perpendicular to C_1C_2 also.

(*ii*) reaction R_2' perpendicular to surface OB.

(*iii*) weights of the two balls.

Resolving these forces along C_1C_2, we get

$$R_2' \cos\left[90-\{180-(\alpha+\beta\}\right] - W\sin \quad = 0$$

It is to be noted that components of R_1' and R_3 along C_1C_2 are zero.

$$R_2' \sin\{180-(\alpha+\beta\} = 2W\sin\alpha$$

or
$$R_2' \sin(\alpha+\beta) = 2\,W\sin\alpha$$

∴
$$R_2' = \frac{2W\sin\theta}{\sin(\alpha+\beta)}$$

EXAMPLE 2.48

Two rollers of the same diameter are supported by an inclined plane and a vertical wall as shown in Fig. 2.71. The upper and the lower rollers are respectively 200 N and 250 N in weight. Assuming smooth surfaces, find the reactions induced at the points of support *A*, *B*, *C* and *D*.

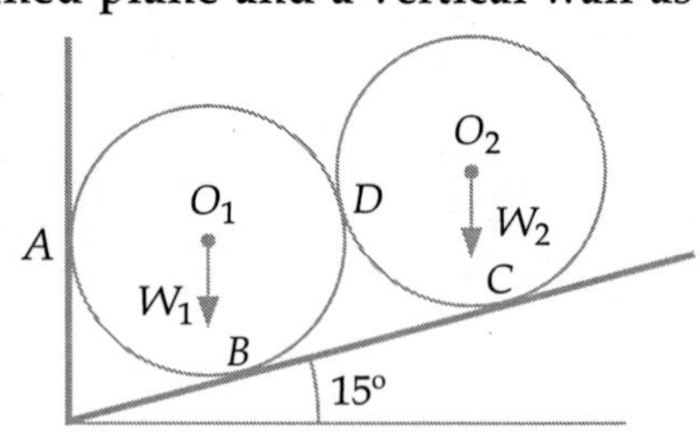

Fig. 2.71

Solution : The upper cylinder is kept in equilibrium by the following set of concurrent forces:

(*i*) weight 200 N acting vertically downward through its centre O_2

(*ii*) reaction R_c acting perpendicular to the inclined plane

(*iii*) pressure R_d from the lower cylinder in the direction O_1O_2

Since these three forces are concurrent, the Lami's theorem applies. That gives

$$\frac{R_d}{\sin(180-15)} = \frac{R_c}{\sin(90+15)} = \frac{W}{\sin 90}$$

$$\therefore \quad R_d = W \times \frac{\sin 165^\circ}{\sin 90^\circ} = 200 \times \frac{0.2588}{1} = \mathbf{51.76\ N}$$

$$R_c = W \times \frac{\sin 105^\circ}{\sin 90^\circ} = 200 \times \frac{0.9659}{1} = \mathbf{193.18\ N}$$

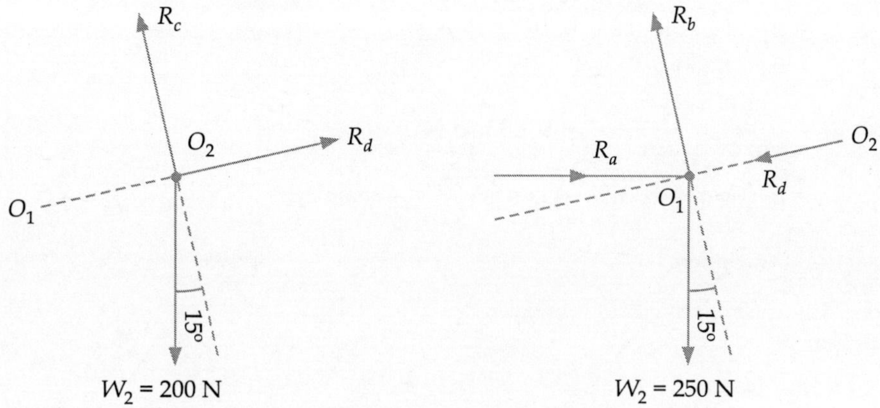

The following set of forces keep the lower cylinder in equilibrium :

(*i*) weight 250 N acting vertically downward through its centre O_1

(*ii*) reaction R_a acting perpendicular to the vertical wall

(*iii*) reaction R_b acting perpendicular to the inclined plane

(*iv*) pressure R_d from the upper cylinder in the direction O_2O_1

Resolving the forces along O_1O_2, we get

$$R_a \cos 15^\circ - W_1 \sin 15^\circ - R_d = 0$$

$$R_a \times 0.966 - 250 \times 0.259 - 51.76 = 0$$

$$\therefore \quad R_a = \frac{(250 \times 0.259) + 51.76}{0.966} = \frac{116.51}{0.966} = \mathbf{120.61\ N}$$

Resolving the forces perpendicular to O_1O_2, we get

$$R_b - R_a \sin 15^\circ - W_1 \cos 15^\circ = 0$$

$$R_b = R_a \sin 15^\circ + W_1 \cos 15^\circ$$

$$= (120.61 \times 0.2588) + (250 \times 0.9659)$$

$$= 31.21 + 241.47 = \mathbf{272.68\ N}$$

EXAMPLE 2.49

Two smooth spheres P, Q each of radius 25 cm and weighing 500 N, rest in a horizontal channel having vertical walls (Fig. 2.72). If the distance between the walls is 90 cm, make calculations for the pressure exerted on the wall and floor at points of contact A, B and C.

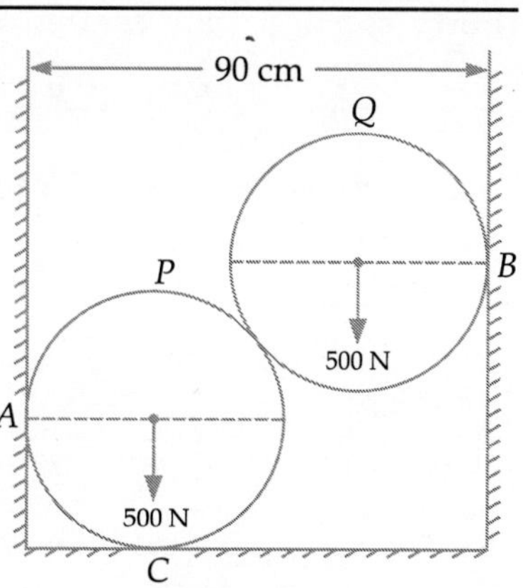

Fig. 2.72

Solution : The following points need consideration

(*i*) the spheres are smooth and as such the pressures at various points of contact would be normal to the surface.

(*ii*) at the point of contact between the two spheres, the reactions would act along the line joining their centres.

With reference to Fig. 2.73, the line $C_1 C_2$ makes an angle a with the horizontal line passing through centre C_1 of sphere P.

$$\cos\alpha = \frac{b-r-r}{2r} = \frac{90-25-25}{50} = \frac{40}{50}$$

$$\therefore \qquad \alpha = 36.87^\circ$$

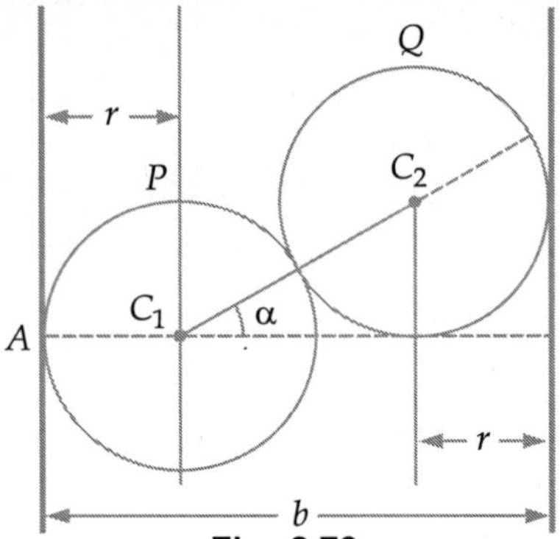

Fig. 2.73

Considering the equilibrium of sphere Q

$$\Sigma F_x = 0; \quad R_b - R\cos\alpha = 0$$

$$\Sigma F_y = 0; \quad R\sin\alpha - 500 = 0$$

$$\therefore \qquad R = \frac{500}{\sin\alpha} = \frac{500}{\sin 36.87} = \mathbf{833.33\ N}$$

$$R_b = R\cos\alpha = 833.33 \times \cos 36.87 = \mathbf{666.66\ N}$$

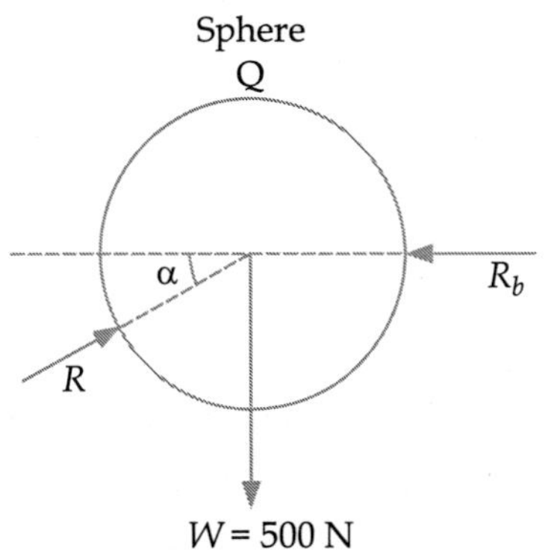

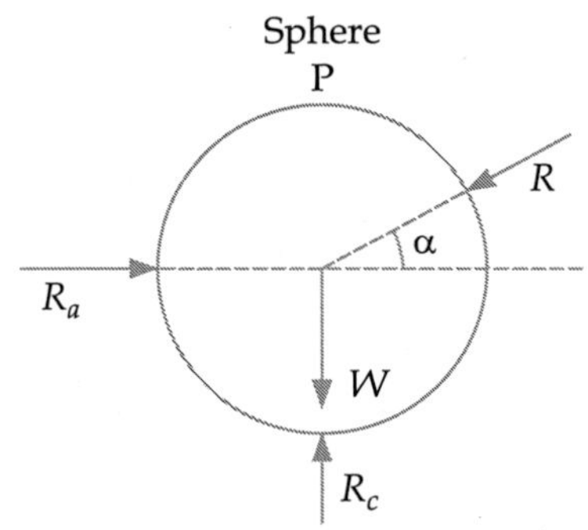

Considering equilibrium of sphere P,

$$\Sigma F_x = 0; \quad R_a - R\cos\alpha = 0$$

$$R_a = R\cos\alpha = 833.33 \times \cos 36.87 = \mathbf{666.67\ N}$$

$$\Sigma F_y = 0; \quad R_c - W - R\sin\alpha = 0$$

$$R_c = W + R\sin\alpha = 500 + 833.33 \times \sin 36.87 = \mathbf{1000\ N}$$

EXAMPLE 2.50

Refer to the system of cylinders arranged as depicted in Fig. 2.74. The cylinders A and B weigh 1000 N each and the weight of cylinder C is 2000 N. Determine the forces exerted at the contact points.

Solution : a, b and c are centres of spheres.

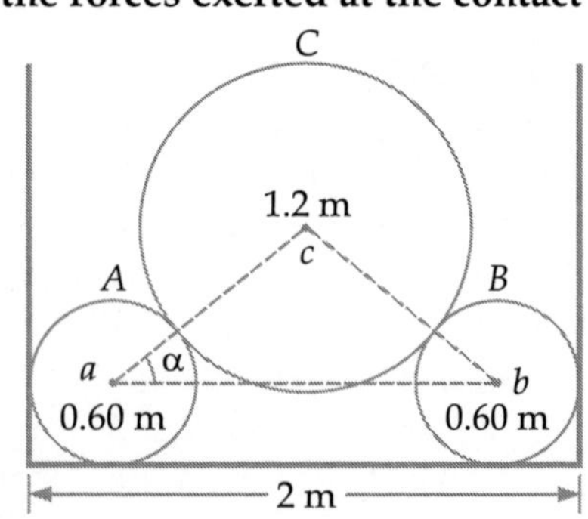

Fig. 2.74

$$ab = 2 - \frac{0.6}{2} - \frac{0.6}{2} = 1.4\text{ m}$$

$$ac = 0.3 + 0.6 = 0.9\text{ m}$$

$$\cos\alpha = \frac{1.4/2}{0.9} = 0.7777\ ; \quad \alpha = 38.94^\circ$$

Applying Lami's theorem to the forces acting on sphere C,

$$\frac{R_1}{\sin(90+\alpha)} = \frac{R_2}{\sin(90+\alpha)} = \frac{2000}{\sin(180-2\alpha)}$$

$$R_1 = R_2 = 2000 \times \frac{\sin(90+\alpha)}{\sin(180-2\alpha)} = 2000 \times \frac{\sin(90+38.94)}{\sin(180-2\times 38.94)}$$

$$= 2000 \times \frac{0.7777}{0.9777} = \mathbf{1590.87\ N}$$

Considering the free body diagram of cylinder A,

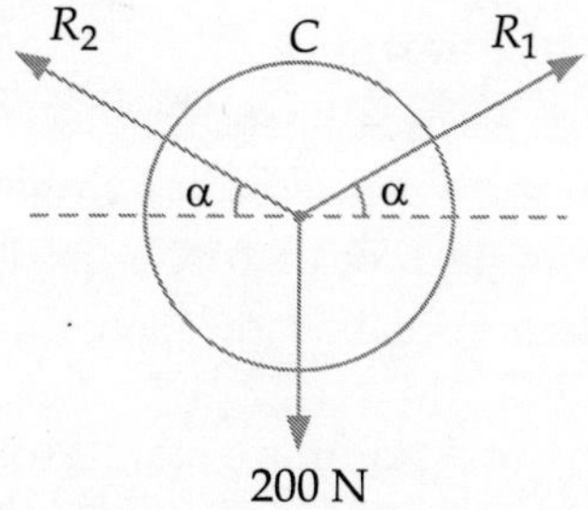

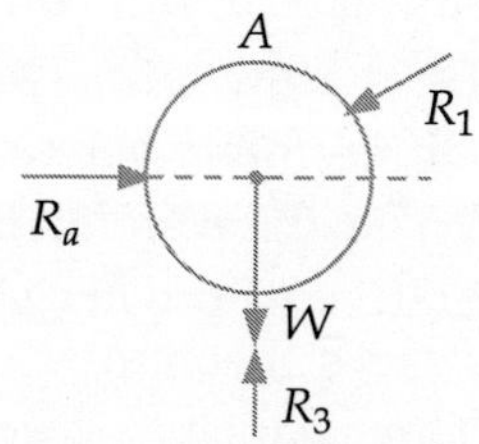

$$\Sigma F_x = 0; \quad R_a - R_1 \cos \alpha = 0$$

$$R_a = R_1 \cos \alpha = 1590.87 \times \cos 38.94$$

$$= \mathbf{1237.38\ N}$$

$$\Sigma F_y = 0; \quad R_1 \sin \alpha + W - R_3 = 0$$

$$R_3 = R_1 \sin \alpha + W$$

$$= 1590.87 \sin 38.94 + 1000$$

$$= \mathbf{1999.87\ N}$$

REVIEW AND SUMMARY

1. When a body is acted upon by a system of forces, then vectorial sum of all the forces is known as resultant. The resultant produces the same affect as is done by the combined effect of several forces.

   A body moving in the direction of resultant can be brought to rest by applying a force in opposite direction, equal in magnitude and collinear with resultant force. This single force which keeps the body at rest or in equilibrium is called the equilibrant force.

2. The process of determination of resultant of a force system is known as composition of forces. Conversely, the method of splitting a force into its components in two directions, without changing effect on the body, is called resolution of force.

3. For a body acting upon several concurrent, coplanar forces to be in equilibrium, these forces when added geometrically must form a closed polygon.

4. **Triangle law of forces:** If two forces acting on a body are represented in magnitude and direction by the two sides of a triangle taken in order, then their resultant is given by the third side of the triangle taken in the opposite order.

5. **Parallelogram law of forces:** If two forces acting at a point of a body are represented in magnitude and direction by the two adjacent sides of a parallelogram, then their resultant is given both in magnitude and direction by the diagonal of the parallelogram passing through that point.

6. **Lami's theorem:** If three forces acting on a particle keep it in equilibrium, then each force is proportional to the sine of the angle between the other two forces.

7. Free body diagram of a system is its portion that is drawn separately showing applied forces, self weight and reactions at the points of with other systems.

   In a free body diagram, all the supports (like walls, floors, hinges, etc.) are removed and replaced by the reactions which these supports exert on the body.

8. Internal forces are those forces which hold the particles of the body together and help it to be rigid. These forces cause internal stresses and strains distributed throughout the material of the body.

   External forces represent those forces which act on a body or a system of bodies externally, *i.e.*, applied from outside. These forces essentially denote the action of other bodies (walls, floors, hinges, etc.) on the rigid body being analysed.

REVIEW QUESTIONS

A. Conceptual and conventional questions :

1. Define the term resultant of a force system.
2. Differentiate between resolution and composition of forces.
3. Write the statement for triangle law of forces and parallelogram law of forces.
4. State and prove: (*a*) parallelogram law of forces, (*b*) theorem of resolution parts, and (*c*) Lami's theorem.
5. Three coplanar forces acting at a point are stated to be in equilibrium. Show that each force is proportional to the sine of the angle between the other two forces.
6. Two forces act at a point. State the range of the values of their resultant.
7. Is it possible for a force to have a component larger in magnitude than the force itself?
8. Comment on the validity of following statements:
   (*i*) Rectangular components of a force *P* are forces acting at right angles to the force *P*.
   (*ii*) Lines of action of any two components of a force intersect on the line of action of the force.
   (*iii*) There are many possible solutions to the determination of the components of a given force.
9. Outline the steps involved in the analytical method for the determination of resultant of coplanar force system.
10. What is a force polygon? How a force polygon is drawn to determine the magnitude of the resultant of a concurrent force system?
11. Define the term free body diagram and state the importance of drawing such a diagram.
12. Show that if a body is in equilibrium under the action of three forces (concurrent), those forces will always lie along the sides of a closed triangle.
13. The resultant of two equal forces acting at a point also equals to *P*. Determine the angle between the two forces.
14. A force of 200 N is resolved into two components. If one of the components is equal to 120 N and makes an angle of 30° with the 200 N force, find the other component and the angle between the components.
15. Refer Fig. 2.75 for a system of concurrent coplanar forces. If the resultant is zero, find the magnitude and direction of force *P*.
16. A 1 kN force has been resolved into components along *AB* and *AC* directions in the *x*-y plane specified by the angles α and β as shown in Fig. 2.76. If the component along *AC* is 2 kN and the component along *AB* is 1.6 N, determine the angles α and β.
17. A body is free to slide on a smooth vertical circular wire and is connected by a string, equal in length to the radius of the circle, to the highest point of the circle. Find the tension developed in the string, and the reaction of the circle.

(Each is equal to weight of the body)

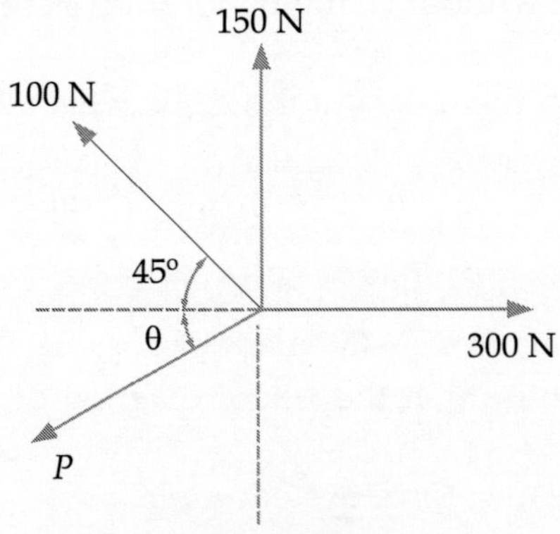

Fig. 2.75

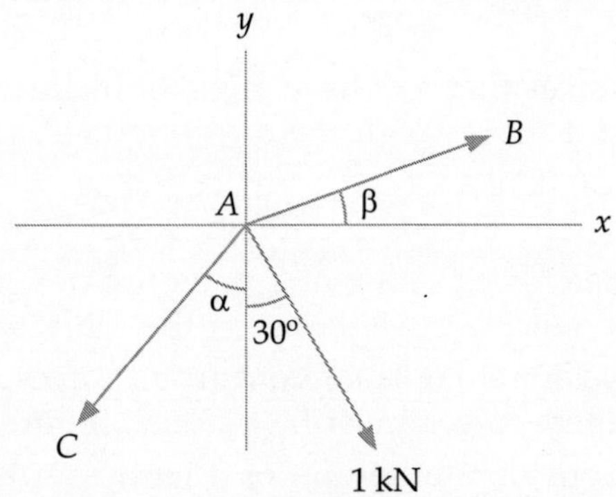

Fig. 2.76

18. A particle is acted upon by the force system shown in Fig. 2.77. Determine the magnitude and direction of the resultant.

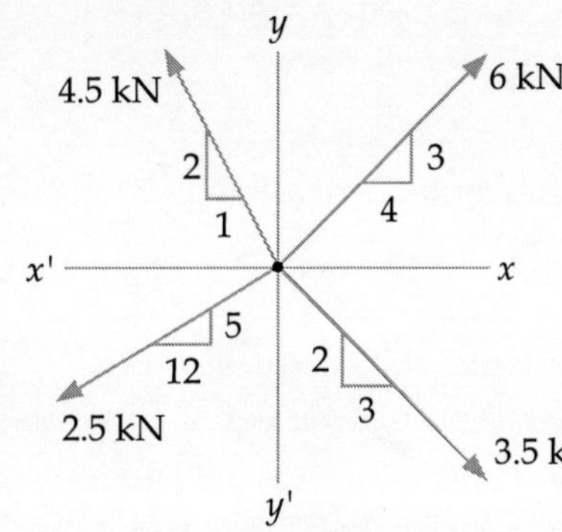

Fig. 2.77

19. Three forces $2P$, $3P$ and $4P$ act along three sides of an equilateral triangle taken in order. Find the magnitude and line of action of the resultant force.

(1.732 *P*, 210°)

20. A small block of weight 100 N is placed on an inclined plane which makes an angle $\theta = 30°$ with the horizontal. What is the component of this weight (*i*) parallel to the inclined plane and (*ii*) perpendicular to the inclined plane.

(8.66 N, 50 N)

21. A lamp weighing 5 N is suspended from the ceiling by a chain. It is pulled aside by a horizontal chord until the chain makes an angle of 60° with the ceiling. Determine the tensions in the chain and the chord.

(5.77 N, 2.89 N)

22. A weight *W* is supported by the two strings at right angles to one another and attached to two points in the same horizontal line. Prove that the tensions induced in the strings are inversely proportional to their lengths.
23. The top of an electric pole is connected with a stay wire which makes an angle of 60° with the horizontal. If a horizontal force of 100 N is essential for keeping the pole vertical, determine tension induced in the stay wire. Also workout vertical component of this tension and state the purpose served by it.

(200 N, 173 N, resistance against lifting)

24. A body of 50 N weight is suspended by two strings 5 m and 12 m long. The other ends of the strings are fastened to the extremities of a rod which is 13 m long. The rod is so held that the body hangs vertically below its mid point. Workout the tensions induced in the strings for the given arrangement.

(46.15 N, 19.23 N)

B. Complete the following statements with most appropriate word/words :

1. Forces whose lines of action pass through a common point are called ...... forces
2. A .... is a single force which can replace two or more forces and produce the same effect as the forces.

3. The splitting of a force into two perpendicular direction without changing its effect is called ....
4. Two forces P and Q have angle of inclination θ between them. If α is the angle which their resultant makes with the direction of P, then $\tan \alpha$ = ...
5. The resultant of two forces acting on a particle is equal to either of them. The angle of inclination between the forces is then equal to ....
6. "If three forces acting at a point are in equilibrium, then each force will be proportional to the sine of the angle between the other two forces". This is the statement of ....
7. If a system of coplanar concurrent forces is in equilibrium then the vector diagram drawn with these forces must be a .... figure.
8. Equations of equilibrium of a particle subjected to coplanar force system are ....
9. A body isolated from all other members which are connected to it is called the .... body.
10. .... is equal and opposite to the resultant of several forces acting on the body.

Answers :

1. concurrent 2. resultant 3. resolution of forces 4. $\dfrac{Q\sin\theta}{P+Q\cos\theta}$ 5. 120° 6. Lami's theorem 7. closed 8. $\Sigma F_x = 0$ and $\Sigma F_y = 0$ 9. free body 10. equilibriant

C. Multiple choice questions :

1. Resultant of two equal forces is equal to either of them. The angle between the forces is
   (*a*) 0° (*b*) 60° (*c*) 90° (*d*) 120°
2. Four forces P, $2P$, $3P$ and $4P$ act along the sides, taken in order, of a square. The resultant force is
   (*a*) zero (*b*) $12\sqrt{2}\,P$ (*c*) $2P$ (*d*) $\sqrt{5}\,P$
3. Given that $\vec{P}+\vec{Q}=\vec{R}$ and also $P = Q = R$. Then the angle between forces $\vec{P}$ and $\vec{Q}$ are
   (*a*) 0 (*b*) $\dfrac{\pi}{3}$ (*c*) $\dfrac{2\pi}{3}$ (*d*) π
4. The sum of two forces acting at a point is 16 N. If the resultant force is 8 N and its direction is perpendicular to the minimum force, then the forces are :
   (*a*) 6 N and 10 N (*b*) 8 N and 8 N
   (*c*) 4 N and 12 N (*d*) 2 N and 14 N
5. A system of three forces acts on a body and keeps it in equilibrium. The forces need to be
   (*a*) coplanar only (*b*) concurrent only
   (*c*) coplanar as well as concurrent (*d*) coplanat but may or may not be concurrent
6. Two forces can be in equilibrium if they are
   1. equal in magnitude
   2. opposite in direction
   3. collinear in action

   Which of these are the essential conditions for equilibrium?
   (*a*) 1 and 3 (*b*) 2 and 3 (*c*) 1 and 3 (*d*) 1, 2 and 3
7. A number of forces acting at a point will lie in equilibrium if
   (*a*) sum of the forces is zero
   (*b*) sum of the resolved parts is zero

(*c*) algebric sum of the forces is zero
(*d*) sum of the resolved parts in any two mutually perependicular direction is zero

8. A circular roller of weight *W* hangs by a tie rod and rests against a smooth vertical wall as shown in Fig. 2.78. The tension in the tie rod *AB* will be
(*a*) less than weight *W*
(*b*) greter then *W*
(*c*) equal to *W*
(*d*) data insufficient for making force analysis

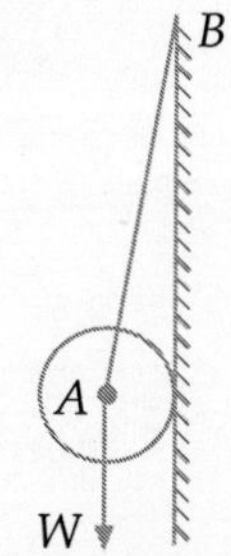

Fig. 2.78

9. The freee body diagram of a body shows the body
(*a*) with its surroundings and external forces acting on it
(*b*) isolated from all external forces
(*c*) isolated from all surroundings
(*d*) isolated from its surroundings and all external effects acting on it

Answers :

1. (*d*)	2. (*b*)	3. (*c*)	4. (*a*)	5. (*c*)	6. (*d*)
7. (*d*)	8. (*b*)	9. (*d*)			

❑❑❑

UNIT - II

CHAPTER

3

TWO-DIMENSIONAL NON-CONCURRENT FORCE SYSTEMS

Learning objectives : After having made a study of this chapter, the reader will develop a concept of

- *moment of force and principle of moments; Varignon's theorem*
- *like and unlike parallel forces and their resultant*
- *couple and its salient aspects ; force-couple system*
- *general conditions of equilibrium*

A coplanar non-concurrent force system consists of a set of forces that lie in the same plane but the line of action of all the forces do not meet at a single point.

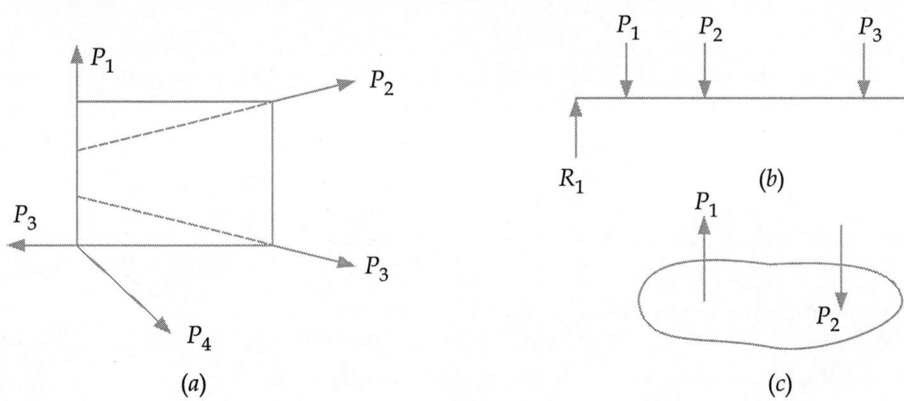

Fig. 3.1

Attention has been focussed in this chapter to study the effect of the resultant of these forces at a point which is away from the lines of action of the system of forces. Reference has also been made to the parallel forces and the effect they produce on a body. The general conditions for the equilibrium of a body have also been stated.

3.1. MOMENT OF A FORCE

Moment of force about a point is defined as the turning tendency of a force about that point. It is measured by the product of force and the perpendicular distance of the line of action of the force from that point.

With reference to Fig. 3.2,

P = force acting on a body

l = perpendicular distance between the point O and line of action of force P

Then moment of force P about point O

$= P \times l$

The point O is called the *moment centre* and the distance l is the *moment arm* or *arm of the force*.

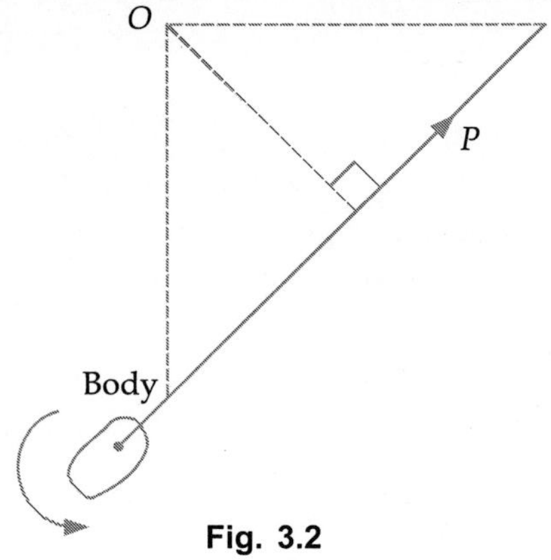

Fig. 3.2

- The moment of force about a point is a vector which is directed perpendicular to the plane containing the moment centre and the force.
- If force is measured in newton (N) and length in metres (m), then the units of force will be Nm.
- When a force acts on a body, it causes or tends to cause a change of state of rest or of uniform rectilinear motion of the body. The action of moment tends to cause a rotational motion to the body.

$M = -Pl$ Anticlockwise moment – ve

$M = Pl$ Clockwise moment + ve

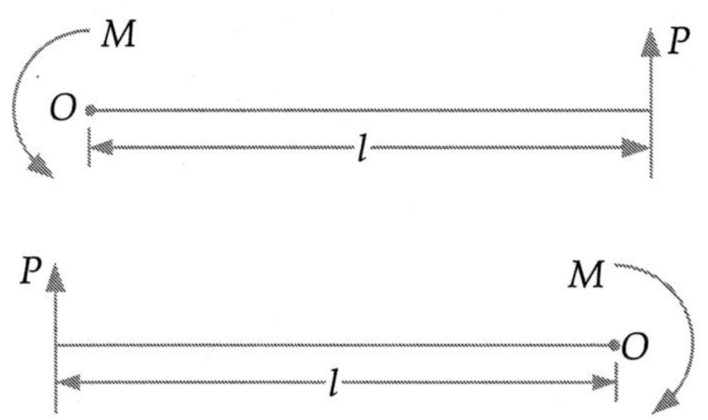

Fig. 3.3

- The tendency of rotation or turning of the body due to moment of force may be clockwise (in the same direction in which hands of the clock move) or anticlockwise (in a direction opposite to the movement of hands of a clock). The corresponding moments are referred to as clockwise moment and anticlockwise moment.

  The general convention is to take the clockwise moments as positive and anticlockwise moments as negative (Fig. 3.3).
- If a number of moments act on a body (Fig. 3.4), then the resultant moment will be the algebraic sum of these moments, and the effect of resultant moment on the body will be governed by its turning tendency.

With reference to Fig. 3.4, the resultant moment M about the point O is given by

$$M = -F_1 l_1 + F_2 l_2 - F_3 l_3 + F_4 l_4$$

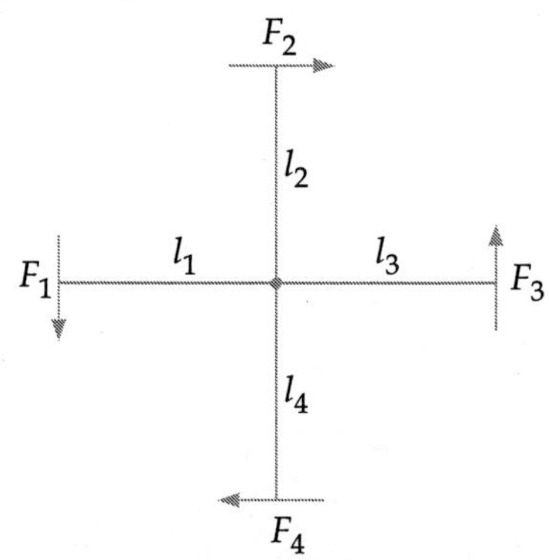

Fig. 3.4

3.2. GRAPHICAL REPRESENTATION OF MOMENT

Let a force P be represented in magnitude and direction by the vector AB. Further, let O be the point about which the moment is to be determined.

Moment of force P about point O,

$= P \times OM$

where OM is the perpendicular dropped from point O on line AB

$$= AB \times OM = 2\left[\frac{1}{2} \times AB \times OM\right]$$
$$= 2 \text{ (area of triangle } OAB).$$

Hence the moment of force about any point is geometrically equal to twice the area of the triangle whose base is the line that represents the force and whose vertex is the point about which the moment is required to be found out.

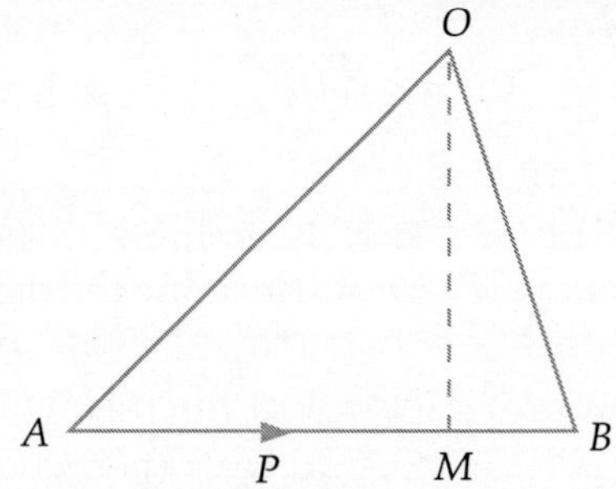

Fig. 3.5

3.3. VARIGNON'S THEOREM : LAW OF MOMENTS

"Moment of a resultant of two forces, about a point lying in the plane of the forces, is equal to the algebraic sum of moments of these two forces about the same point."

Consider two concurrent forces P and Q represented in magnitude and direction by AB and AC respectively. Let O be the point about which moment is to be taken. Through O, draw a line parallel to the direction of force P and let this line meet the line of action of force Q at point C. With AB and AC as adjacent sides, complete the parallelogram $ABCD$. The diagram AD of this parallelogram represents in magnitude and direction the resultant of forces P and Q.

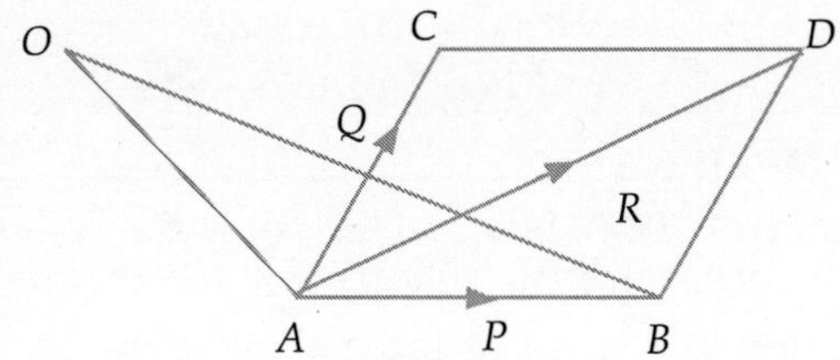

Fig. 3.6

Join O with points A and B.

Recalling that moment of force about a point is equal to twice of the area of the triangle so formed, we have

Moment of force P about O = 2 × area of triangle AOB = 2 × ΔAOB ...(*i*)

Likewise,

moment of force Q about O = 2 × ΔAOC ...(*ii*)

moment of force R about O = 2 × ΔAOD

From geometrical configuration of Fig. 3.6,

$$\Delta AOD = \Delta AOC + \Delta ACD = \Delta AOC + \Delta ABD$$

Further, the Δ's AOB and ABD are on the same base AB and between the same lines and as such are equal in area. Then

$$\Delta AOD = \Delta AOC + \Delta AOB$$

The moment of force R about O may then be re-written as

$$= 2 \times (\Delta AOC + \Delta AOB) \quad ...(iii)$$

From the identities (*i*), (*ii*) and (*iii*), it is evident that

Moment of forces P and Q about point O

= moment of resultant R about O

This principle can be extended for any number of forces and the generalised law of moments may be stated as

"moment of resultant of a number of forces about a point lying in the plane of forces is equal to the algebraic sum of the moments of these forces about the some point."

3.4. PRINCIPLE OF MOMENTS

A body acted upon by a number of coplanar forces will be in equilibrium, if the algebraic sum of moments of all the forces about a point lying in the same plane is zero. Mathematically

$$\Sigma M = 0$$

i.e., clockwise moments = anticlockwise moments

EXAMPLE 3.1

A force of 200 N is acting at a point *B* as shown in the adjoining figure. Determine the moment of this force about *O*.

Solution : Moment of force about point O

= force × (perpendicular distance between point O and the line of action of force)

$= P \times OM = P \times OB \cos 60°$

$= 200 \times (3 \times 0.866)$

= **519.6 Nm** (clockwise)

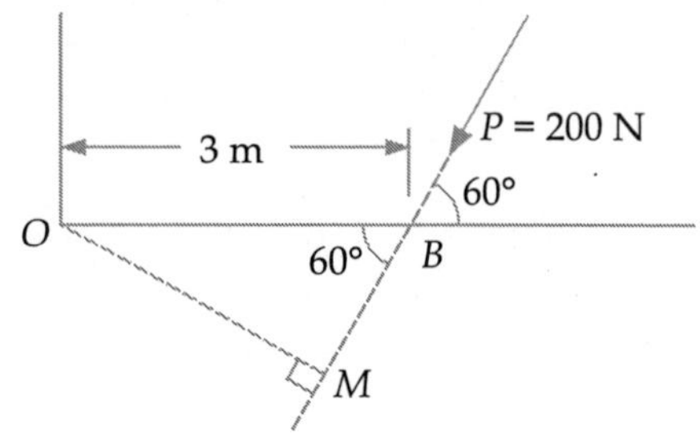

Fig. 3.7

EXAMPLE 3.2

A force of 200 N is acting at point *B* of a lever *AB* which is hinged at its lower end as shown in Fig. 3.8. Find the moment of force about the hinged end.

Solution : Moment of force about point A

= force × (perpendicular distance between point A and the line of action of force)

$= P \times AM$

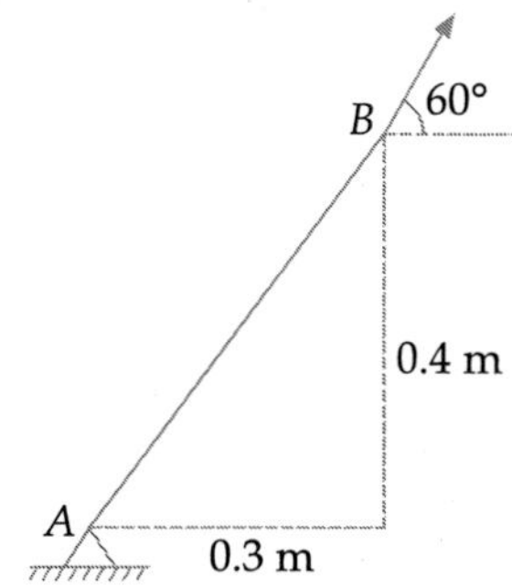

Fig. 3.8

Now, $AB = \sqrt{0.4^2 + 0.3^2} = 0.5$ m

$\alpha = \tan^{-1}\left(\dfrac{0.4}{0.3}\right) = 53.13°$

$\gamma = 90 - 53.13 = 36.87°$

$\beta = 36.87 - 30 = 6.87°$

$AM = AB \sin 6.87° = 0.5 \times 0.1196 = 0.0598$ M

∴ $M_a = 200 \times 0.0598 =$ **11.96 Nm** (anticlockwise)

Alternatively :

The horizontal and vertical components of the given force are:

$P_x = 200 \cos 60° = 100$ N

$P_y = 200 \sin 60° = 173.2$ N

From Varignon's theorem, moment of a given force about a specified point is equal to the sum of the moments of its components about that point.

∴ $M_a = 100 \times 0.4 - 173.2 \times 0.3$

$= 40 - 51.96 = -$ **11.96 Nm** (anticlockwise)

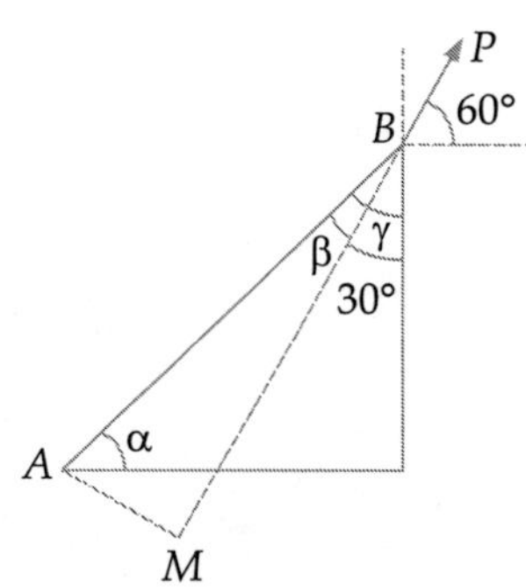

Fig. 3.9

EXAMPLE 3.3

A uniform wooden plank AB of length 3 m has a weight of 40 N. It is supported at end A and at a point D which is 1m from the other end B. Determine the maximum weight W that can be placed at end B so that the plank does not topple.

Solution : Refer to Fig. 3.10 for the given force system. The weight of the plank acts at its centre C.

When the plank is at the state of just being toppled, the reaction R_a at point A is zero.

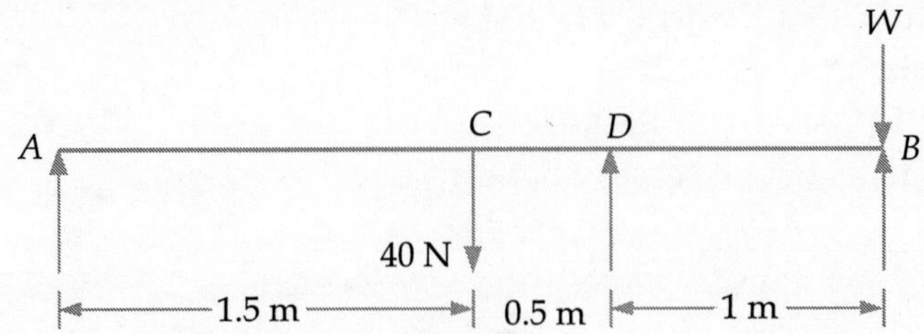

Fig. 3.10

Taking moments about point D,

$$W \times 1 = 40 \times 0.5$$

$$\therefore \quad W = \mathbf{20\ N}$$

EXAMPLE 3.4

A uniform rodof 10 m length has a self weight of 5 N. The rod carries a weight of 30 N hung from one of its ends. From what point the rod be suspended so that it remains horizontal?

Solution: The self weight of the rod 5 N acts at the mid point C of the rod.

Let the rod be suspended at point F which is at distance x from its end B where a load of 30 N is hung.

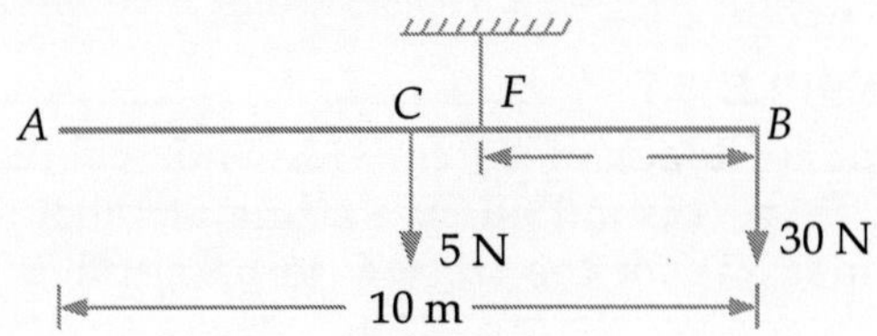

Fig. 3.11

Taking moments about point F, we get

$$5 \times CF - 30\,x = 0$$

$$5\,(BC - x) = 30\,x$$

or $$5\,(5 - x) = 30\,x$$

or $$25 - 5\,x = 30\,x$$

$$\therefore \quad x = \frac{25}{35} = \mathbf{0.7143\ m}$$

Hence for the rod to remain horizontal, it needs to be suspended from point F such that $BF = 0.7143$ m.

EXAMPLE 3.5

A uniform rod $ABCD$ weighs 8 N and is supported at B and C as shown in the figure given below. Determine the minimum force at A which will just overturn the rod.

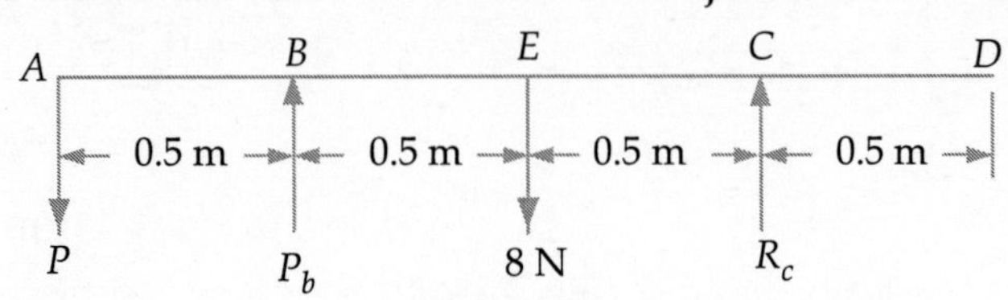

Fig. 3.12

Solution: Let P be the minimum force to be applied at end A which will just overturn the rod. At that instant reaction R_c will be zero.

Taking moments about point B,

$$8 \times 0.5 - P \times 0.5 = 0$$

$$\therefore \quad P = \frac{8 \times 0.5}{0.5} = \mathbf{8\ N}$$

EXAMPLE 3.6

A uniform rod AB of weight 100 N is hinged at A so as to be able to rotate about A in a vertical plane. It is held in horizontal position by a string attached to B and inclined at 60°

to the horizontal as shown in Fig. 3.13. Find the tension induced in the string.

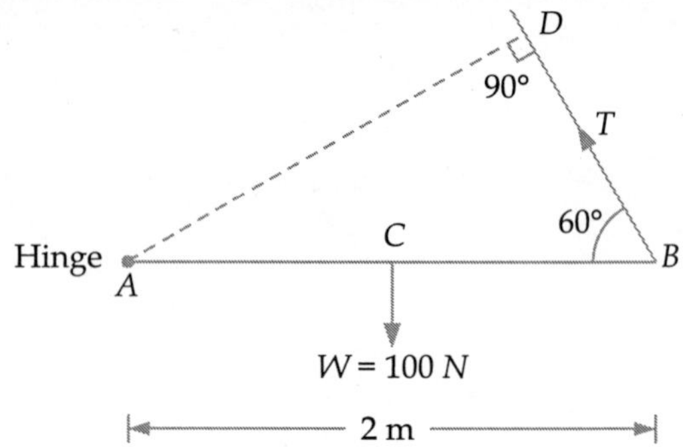

Fig. 3.13

Solution : For the equilibrium of the system, the algebraic sum of the moments of all forces about a point lying in the same plane in zero. Then taking moments about hinge point A and equating it to zero, we get

$$W \times AC - T \times AD = 0$$

The weight of rod acts at its mid point C and therefore $AC = 1$ m.

AD = perpendicular distance between hinge point A and line of action of tension T

$= AB \sin 60^\circ = 2 \times 0.866 = 1.732$ m

$$\therefore \quad 100 \times 1 - T \times 1.732 = 0$$

$$T = \frac{100}{1.732} = \mathbf{57.74\ N}$$

EXAMPLE 3.7

A square block of 25 cm side and weighing 20 N is hinged at A and rests on rollers at B (Fig. 3.14) It is pulled by a string attached at C and inclined at 30° with the horizontal. Make calculations for the force P to be applied so that the block gets just lifted off the roller.

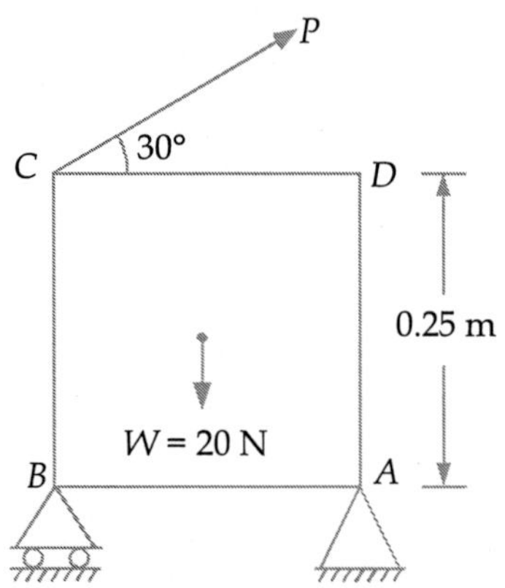

Fig. 3.14

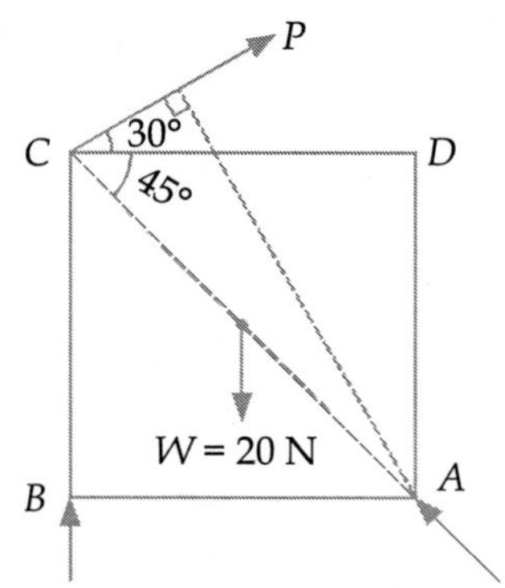

Fig. 3.14 (a)

Solution : With reference to free body diagram (Fig. 3.14 (*a*)), the block is subjected to the following set of forces :

(*i*) force P,

(*ii*) weight of block W = 20 N acting vertically downward,

(*iii*) reaction R_a at the hinge point A

(*iv*) when the block is at the state of just being lifted off the roller, reaction $R_b = 0$.

Taking moments about point A,

$$M_a = 20 \times \frac{0.25}{2} - P \times AE$$

Now, $AE = AC \sin 45° = \sqrt{0.25^2 + 0.25^2} \times \sin 45° = 0.3414$ m

$$\therefore \quad M_a = \frac{20 \times 0.25}{2} - P \times 0.3414$$

Under equilibrium state, this moment is zero. Therefore

$$0 = 20 \times \frac{0.25}{2} - P \times 0.3414$$

That gives : $P = \frac{20}{0.3414} \times \frac{0.25}{2}$ = **7.323 N**

EXAMPLE 3.8.

A vertical force 500 N is applied to the end *A* of a lever which is attached to a shaft at point *O* (Fig. 3.15). Make calculations for

(*a*) moment of force 500 N about the point *O*;

(*b*) horizontal force to be applied at *A* which would create the same moment about *O*;

(*c*) smallest force to be applied at *A* which would create the same moment about *O*;

(*d*) distance from the shaft where a 1000 N vertical force must act to create the same moment about *O*.

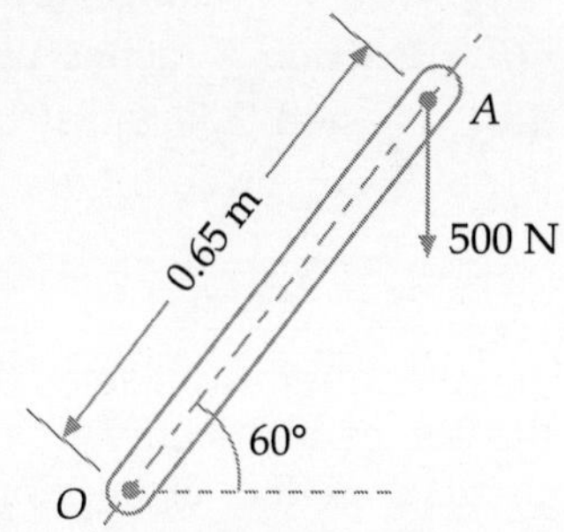

Fig. 3.15

Solution: Recall that moment of force about a point is measured by the product of force and the perpendicular distance of the line of action of the force from that point, and refer to the figures given below :

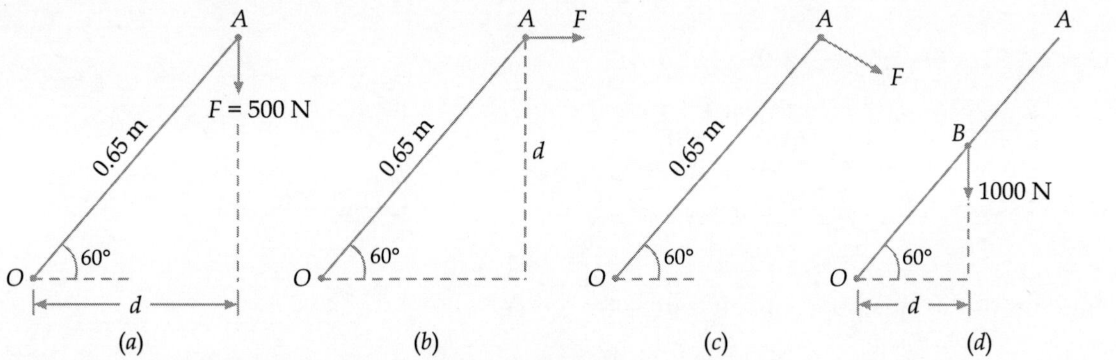

(*a*) Moment of vertical force F = 500 N about O is

$$M_o = F \times d = 500 \times (0.65 \cos 60°) = \mathbf{162.5\ Nm}$$

(*b*) Moment of horizontal force F about O is

$$M_o = F \times d = F \times (0.65 \text{ sins } 60°) = 0.563\ F$$

Then according to the given condition,

$$0.563\ F = 162.5 \text{ and that gives : } F = \frac{162.5}{0.563} = \mathbf{287.74\ N}$$

(*c*) The smallest force would occur when d is maximum and maximum value of d is 0.65 m. Accordingly

$$F \times 0.65 = 162.5 \quad \text{or} \quad F = \frac{162.5}{0.65} = \mathbf{250\ N}$$

(*d*) As per the given condition :

$$1000 \times OB \cos 60° = 162.5$$

where OB is the distance from the shaft where 1000 N vertical force has to be applied

$$OB = \frac{162.5}{1000 \times \cos 60°} = \frac{162.5}{1000 \times 0.5} = \mathbf{0.325\ m}$$

EXAMPLE 3.9.

A uniform rod *AB* of negligible weight and length *l* is hinged at the end *A* and supported by a string as shown in Fig. 3.16. What should be the inclination θ of the rod if it is in a state of equilibrium? Take *P* = 2*Q*.

Solution: The various forces acting on the rod are :

(*i*) Weight P acting downwards at end B

(*ii*) Reaction R_a acting at the hinge point A in a direction which is unknown

(*iii*) Tension T in the string. Presuming the pulley to be smooth and frictionless, $T = Q$

From the given configuration; $AB = AC$ and as such the triangle ABC is isosceles. Accordingly

$$\angle ABC = \angle ACB = \frac{180 - \theta}{2} = 90 - \frac{\theta}{2}$$

Under equilibrium state, the moment of all the forces about any point is zero. To eliminate the effect of reaction R_a, we take moments about point A

$$Pl \cos \theta - Ql \sin\left(90 - \frac{\theta}{2}\right) = 0$$

or $$P \cos \theta - Q \cos\frac{\theta}{2} = 0$$

or $$P\left(2\cos^2\frac{\theta}{2} - 1\right) - Q\cos\frac{\theta}{2} = 0$$

or $$2P \cos^2\frac{\theta}{2} - Q\cos\frac{\theta}{2} - P = 0$$

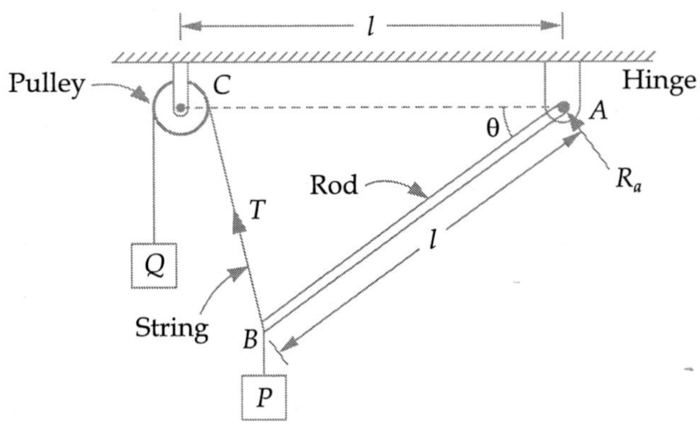

Fig. 3.16

Solution of this quadratic equation in $\cos\frac{\theta}{2}$ gives

$$\cos\frac{\theta}{2} = \frac{Q \pm \sqrt{Q^2 + 8P^2}}{4P} = \frac{1}{4}\left[\frac{Q}{P} \pm \sqrt{\frac{Q^2}{P^2} + 8}\right]$$

Substituting $\frac{Q}{P} = \frac{1}{2}$ (given) and taking the positive value

$$\cos\frac{\theta}{2} = \frac{1}{4}\left[\frac{1}{2} + \sqrt{\frac{1}{4} + 8}\right] = 0.843$$

$$\frac{\theta}{2} = \cos^{-1}(0.843) = 32.5°$$

∴ Inclination of the rod θ = **65°**

EXAMPLE 3.10

A uniform wheel weighing 20 kN and of 600 mm diameter rests against 150 mm thick rigid block as shown in Fig. 3.17.

Considering all surfaces to be smooth, determine (a) the least pull through the centre of wheel to just turn the wheel over the corner of the block, (b) the reaction of the block.

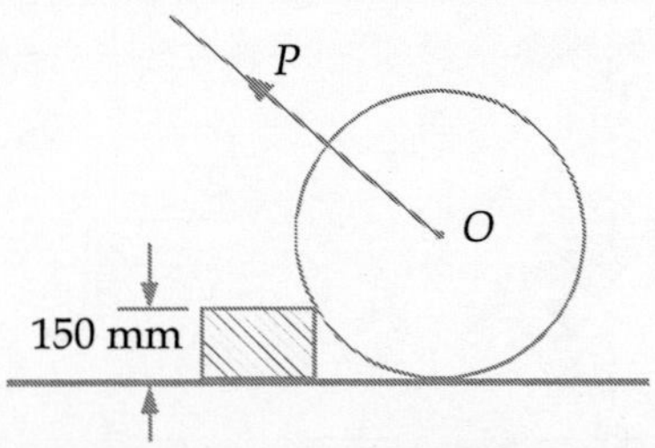

Fig. 3.17

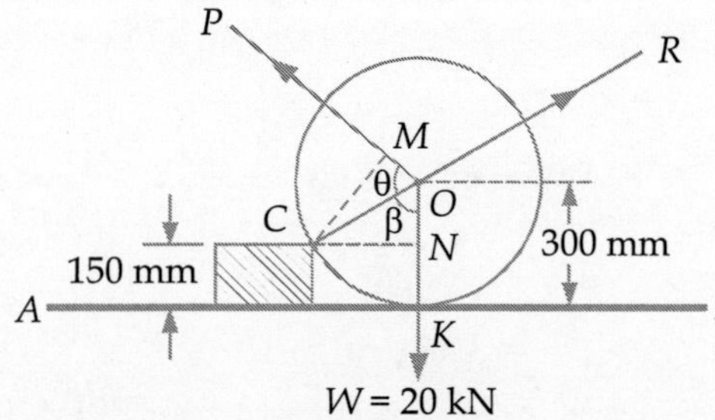

Fig. 3.18

Solution : When the wheel is about to turn, its contact with the ground will be lost. Hence the wheel has to be in equilibrium under the action of its weight *W* and the force *P*.

With reference to Fig. 3.18, *CM* and *CN* are the perpendiculars dropped on the lines of action of forces *P* and *W* respectively.

$$OC = OK = 300 \text{ mm}$$

$$ON = OK - NK = 300 - 150 = 150 \text{ mm}$$

$$CN = \sqrt{OC^2 - ON^2} = \sqrt{300^2 - 150^2} = 259.81 \text{ mm}$$

Taking moments about point *C*,

$$P \times CM - W \times CN = 0$$

$$P \times OC \sin\theta - 20 \times 259.81 = 0$$

$$\therefore \quad P = \frac{20 \times 259.81}{OC \sin\theta} = \frac{20 \times 259.81}{300 \sin\theta} = \frac{17.32}{\sin\theta}$$

The force *P* will be minimum when $\sin\theta$ is maximum. For that

$$\sin\theta = 1 \quad \text{or} \quad \theta = 90^\circ$$

Hence $\quad P_{min} =$ **17.32 kN** when pull is applied perpendicular to *OC*.

(*b*) The reaction *R* can be determined by resolving *W* along *OC*

$$R = W \cos\beta \qquad \text{where} \quad \beta = \angle CON$$

$$\cos\beta = \frac{ON}{OC} = \frac{150}{300} = \frac{1}{2}$$

$$\therefore \quad R = 20 \times \frac{1}{2} = \mathbf{10\ kN}$$

The force *P* acts perpendicular to *OC* and as such its resolved part along *OC* is zero.

EXAMPLE 3.11

A regular hexagon *ABCDEF* of side 2 m in length is subjected to forces 1, 2, 3, 4, 5 and 6 N along sides *AB*, *CB*, *DC*, *DE*, *EF* and *FA* respectively as shown in the Fig. 3.19 given below. Make calculations for the sum of moments at point *A*.

Solution : Starting from force represented by side *AB*, we take moments about point *A* and get

$$M_a = -1 \times 0 + 2 \times AN + 4 \times AC - 3 \times AE - 5 \times AM - 6 \times 0$$

The clockwise moments have been taken positive and anticlockwise moments as negative.

From the geometrical configuration of the system,

$$\angle BAN = \angle FAM = 30^\circ$$

$$\angle ACB = \angle BAC = 30^\circ$$

$$AN = AM = AB \cos 30^\circ = 2 \times \cos 30^\circ = 1.732 \text{ m}$$

Also $AN = AC \cos 60^\circ$

$$AC = \frac{1.732}{\cos 60^\circ} = 3.464 \text{ m} = AE$$

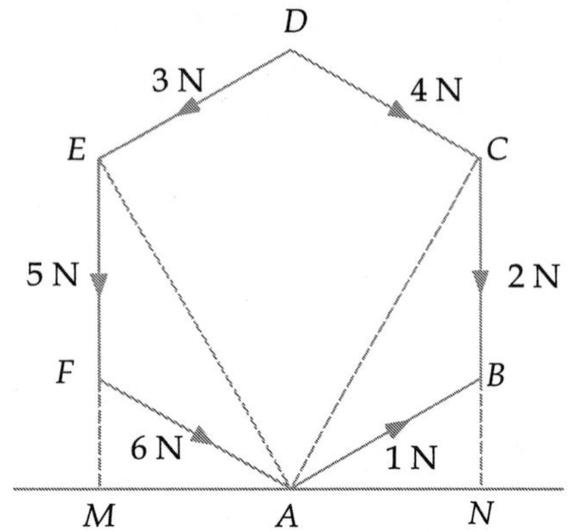

Fig. 3.19

$$\therefore \quad M_a = -(1 \times 0) + (2 \times 1.732) + (4 \times 3.464) - (3 \times 3.464) - (5 \times 1.732) - (6 \times 0)$$

$$= 0 + 3.464 + 13.586 - 10.392 - 8.66 - 0 = -\mathbf{1.728\ Nm}$$

The negative sign suggests that the resultant moment is anticlockwise.

EXAMPLE 3.12

Forces of 20, 30, 40, 50, 60 and 70 N act along the sides *AB, CB, CD, ED, EF* and *FA* respectively of a regular hexagon *ABCDEF* (Fig. 3.20) whose each side measures 10 cm. Make calculations for the algebraic sum of the moments of the forces about the centre of the hexagon and one of the vertices.

Solution : Let O be the centre of the hexagon and OP be the perpendicular drawn from the centre onto the side AB.

The triange OAB is an equilateral triangle and that gives

$$OP = \sqrt{(OA)^2 - (AP)^2} = \sqrt{10^2 - 5^2} = 8.66 \text{ cm}$$

Obviously the distance of each side from the centre O is 8.66 cm.

Algebraic sum of the moments about O,

$$M_O = -20 \times 8.66 + 30 \times 8.66 - 40 \times 8.66 + 50 \times 8.66 - 60 \times 8.66 - 70 \times 8.66$$

The clockwise moments have been taken as positive and anticlockwise moments negative.

$$M_O = 8.66\,(-20 + 30 - 40 + 50 - 60 - 70) = -952.60 \text{ N cm}$$

The negative sign suggests that the resultant moment about the centre of the hexagon is anticlockwise.

(*b*) From the geometrical configuration of the system

$$\angle BAN = \angle FAM = 30^\circ$$

$$\angle ACB = \angle BAC = 30^\circ$$

$$AN = AM = AB \cos 30^\circ = 10 \times 0.866 = 8.66 \text{ cm}$$

Also $AN = AC \cos 60^\circ$

$$AC = \frac{8.66}{\cos 60^\circ} = 7.32 \text{ cm} = AE$$

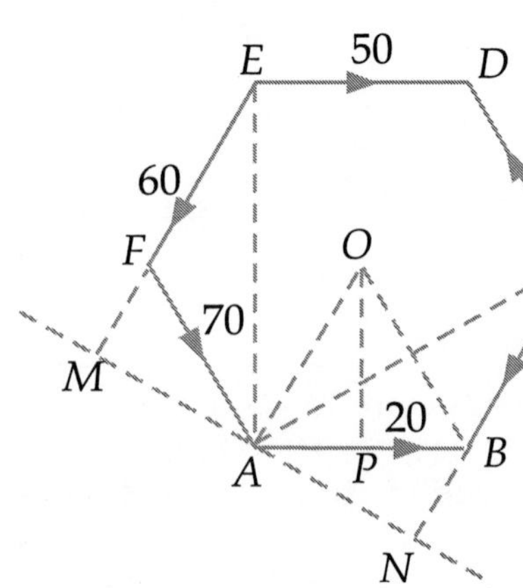

Fig. 3.20

∴ Moments about vertex A,

$$M_a = -(20 \times 0) + (30 \times 8.66) - (40 \times 17.32) + (50 \times 17.32) - (60 \times 8.66) - (70 \times 0)$$
$$= 8.66\,(30 - 80 + 100 - 60) = 8.66 \times (-10)$$
$$= \mathbf{-86.60\ N\ cm}$$

The negative sign suggests that the resultant moment is anticlockwise about the vertex A.

EXAMPLE 3.13

A beam *AB* of length 2 m, hinged at *A* and supported at *B* by a cord which passes over two frictionless pulleys, carries a 50 kN load as shown in Fig. 3.21. Determine the position of 100 kN load on the beam if it is to remain in equilibrium in horizontal position. Also determine the reaction at the hinged end.

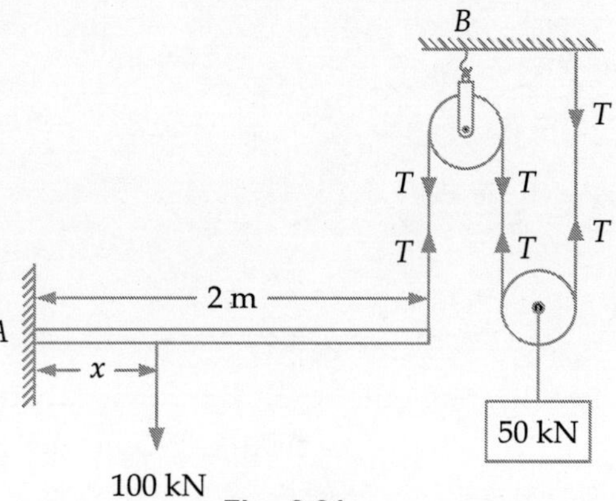

Fig. 3.21

Solution : Since the pulleys are weightless and frictionless, the tension T in the entire length of cord will be same.

For the equilibrium of pulley supporting 50 kN load (Fig. 3.21)

$$\Sigma F_y = 0\,;\quad T + T - 50 = 0\,;\quad T = 25\text{ kN}$$

The free body diagram of the beam is shown in Fig. 3.22, and for its equilibrium

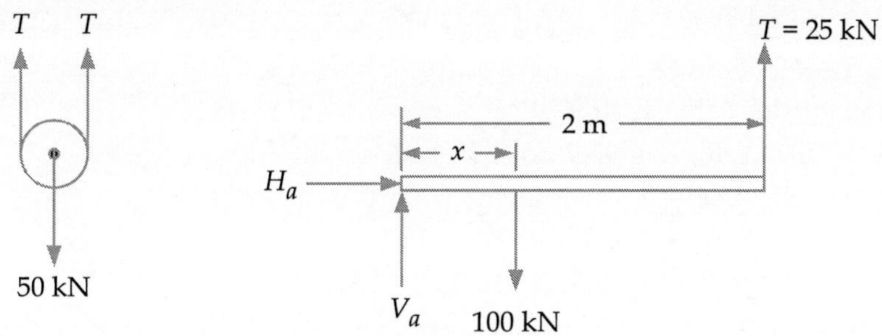

Fig. 3.22

$$\Sigma F_x = 0\,;\quad H_a = 0$$
$$\Sigma F_y = 0\,;\quad V_a - 100 + 25 = 0\,;\quad V_a = 75\text{ N}$$

Therefore the reaction at the hinged end is 75 N in the vertical direction.

$$\Sigma M_a = 0\,;\quad 100 \times x - 25 \times 2 = 0,\quad x = \mathbf{0.5\ m}$$

EXAMPLE 3.14

A load *P* of 3500 N is acting on the boom which is held by a cable *BC* as shown in Fig. 3.23. Neglecting weight of the boom,

(*a*) draw the free body diagram of the boom,

(*b*) find the tension in cable *BC*, and

(*c*) determine the reaction at *A*

Solution : (*b*) The free body diagram of the beam will be as shown in Fig. 3.24. Equations of equilibrium of a coplanar system of forces are:

$$\Sigma F_x = 0,\ \Sigma F_y = 0 \text{ and } \Sigma M = 0$$

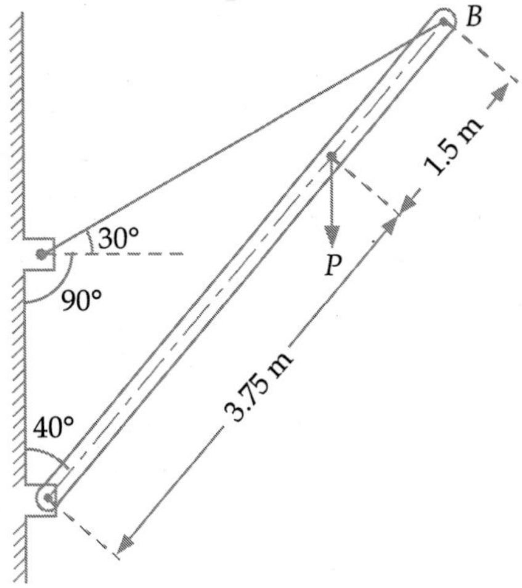

Fig. 3.23

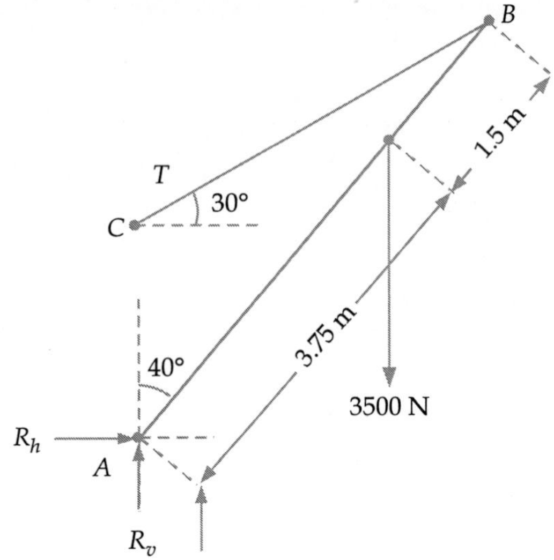

Fig. 3.24

Taking moments of all the forces about point A, we get

$$-3500 \times 3.75 \sin 40^\circ + T \cos 30^\circ \times (3.75 + 1.5) \cos 40^\circ - T \sin 30^\circ \times (3.75 + 1.5) \sin 40^\circ = 0$$

or $$-8436.6 + 3.483\,T - 1.687\,T = 0$$

That gives : $$T = \frac{8436.6}{3.483 - 1.687} = \mathbf{4697.4\ N}$$

Thus the tension in the cable is 4697.4 N

(*c*) $$\Sigma F_x = R_h - 4697.4 \cos 30^\circ = 0$$

$$\therefore \quad R_h = 4697.4 \times \cos 30^\circ = 4068.1 \text{ N}$$

$$\Sigma F_y = R_v - 3500 - 4697.4 \sin 30^\circ = 0$$

$$\therefore \quad R_v = 3500 + 4697.4 \sin 30^\circ = 5848.7 \text{ N}$$

Total reaction at point A,

$$R = \sqrt{R_h^2 + R_v^2}$$

$$= \sqrt{(4068.1)^2 + (5848.7)^2} = \mathbf{7124.5\ N}$$

If θ is the angle which the reaction at A makes with the x-axis, then

$$\theta = \tan^{-1}\left(\frac{R_v}{R_h}\right) = \tan^{-1}\left(\frac{5848.7}{4068.1}\right) = \mathbf{55.2^\circ}$$

EXAMPLE 3.15

A horizontal line *ABCD* measuring 9 m is acted upon by forces of magnitude 400, 600, 400 and 200 N at points *A*, *B*, *C* and *D* respectively with downward direction. These points are so located that *AB* = *BC* = *CD* = 3 m. The lines of action of the forces are inclined at 90°, 60°, 45° and 30° respectively with *AB*. Make calculations for the magnitude, position and direction of the resultant.

Solution : Refer to Fig. 3.25 for the given force system.

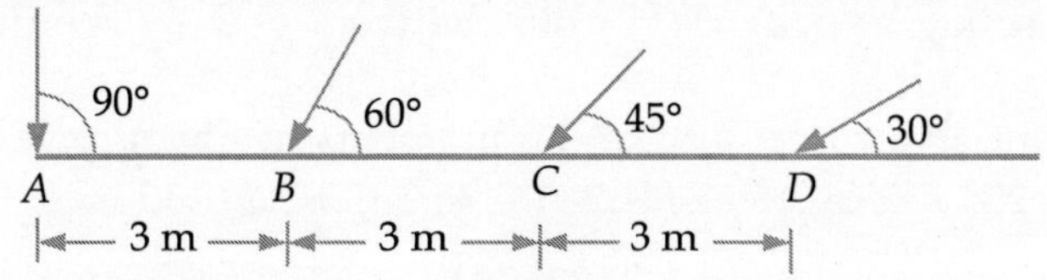

Fig. 3.25

Resolving all the forces horizontally,

$$\Sigma F_x = 200 \cos 30^\circ + 400 \cos 45^\circ + 600 \cos 60^\circ + 400 \cos 90^\circ$$
$$= 173.20 + 282.8 + 300 + 0 = 756 \text{ N}$$

Resolving all the forces vertically,

$$\Sigma F_y = 200 \sin 30^\circ + 400 \sin 45^\circ + 600 \sin 60^\circ + 400 \sin 90^\circ$$
$$= 100 + 282.8 + 519.6 + 400 = 1302.4 \text{ N}$$

Magnitude of the resultant force,

$$F = \sqrt{(\Sigma F_x)^2 + (\Sigma F_y)^2} = \sqrt{756^2 + 1302.4^2} = \mathbf{1505.9\ N}$$

Let α be the angle which the resultant makes with the horizontal direction, *i.e.*, with line *AD*. Then

$$\tan \alpha = \frac{\Sigma F_y}{\Sigma F_x} = \frac{1302.4}{756} = 1.723$$

$$\therefore \quad \alpha = \mathbf{59.87^\circ}$$

The distance *x* of the line of application of the resultant force from end *A* can be worked out by equating moments of the vertical components of the given forces and the resultant force about end *A*. That gives

$$1302.4 \times x = (600 \sin 60^\circ) \times 3 + (400 \sin 45^\circ \times 6 + (200 \sin 30^\circ \times 9)$$
$$= 1558.84 + 1697.06 + 900 = 4155.9$$

$$\therefore \quad x = \frac{4155.9}{1302.4} = \mathbf{3.191\ m}$$

EXAMPLE 3.16

Determine the magnitude, direction and position of the resultant of the system of forces shown in Fig. 3.26 given below. Use vector method.

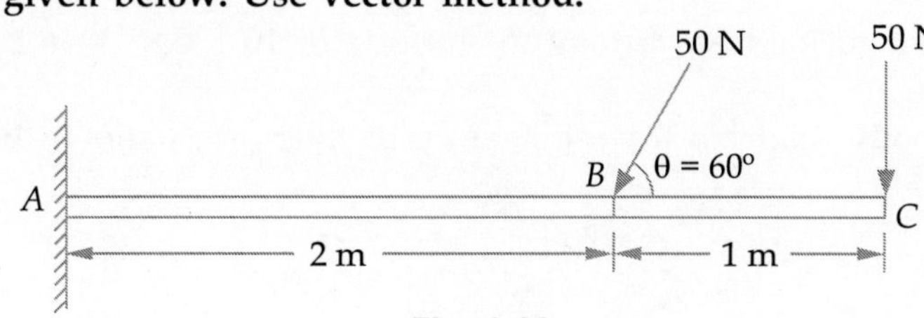

Fig. 3.26

Solution: Resolving the given forces horizontally and vertically, we get

$$\text{Resultant force } \bar{R} = -50\,j - 50 \cos 60^\circ\, i - 50 \sin 60^\circ\, j$$
$$= -50\,j - 25\,i - 43.3\,j = -25\,i - 93.3\,j$$

$$\therefore \quad \text{Magnitude of resultant } |R| = \sqrt{(-25)^2 + (-93.3)^2} = \mathbf{96.59\ N}$$

$$\text{Direction } \alpha = \tan^{-1}\left(\frac{-93.3}{-25}\right) = \mathbf{75^\circ}$$

Let the resultant R be acting at distance x from the fixed end A. Then under equilibrium conditions,

moment of resultant about A = sum of the moments of given forces about A

$$\therefore \quad x\,i\,(-25\,i - 93.3\,j) = 2\,i\,(-25\,i - 43.3\,j) + 3\,i\,(-50\,j)$$

$$\text{or} \quad -93.3\,x\,k = -86.6\,k - 150\,k$$

$$\therefore \quad x = \frac{86.6 + 150}{93.3} = \mathbf{2.535\ m}$$

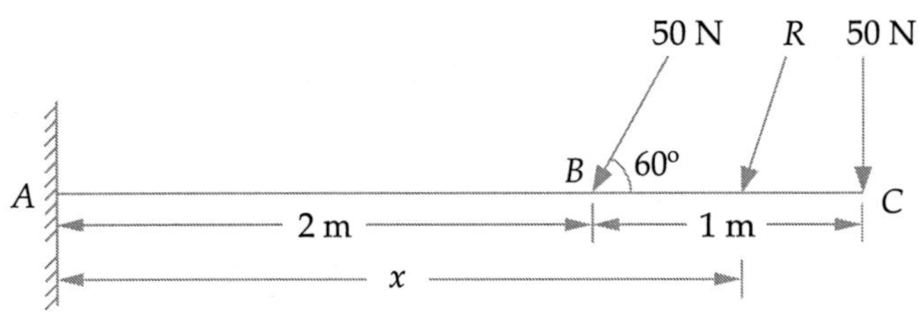

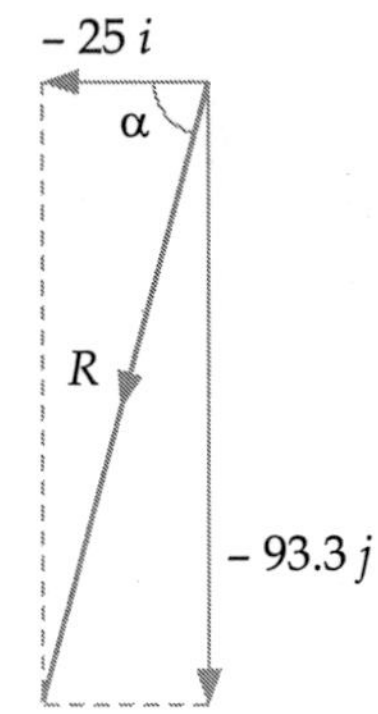

EXAMPLE 3.17

Determine reactions at A and B (Fig. 3.27)

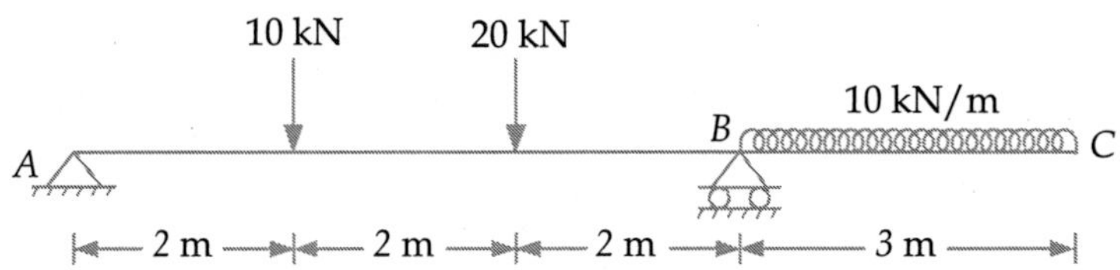

Fig. 3.27

Solution : At the hinged support, the reaction R_a can be in any direction. R_{av} and R_{ah} are its components in the vertical and horizontal directions. At the roller support, the reaction R_b will be in vertical direction only

The uniformly distributed load has a total magnitude of 10 × 3 = 30 kN, and acts at the middle of the 3 m length.

The equivalent free body diagram for the beam will then be as shown below:

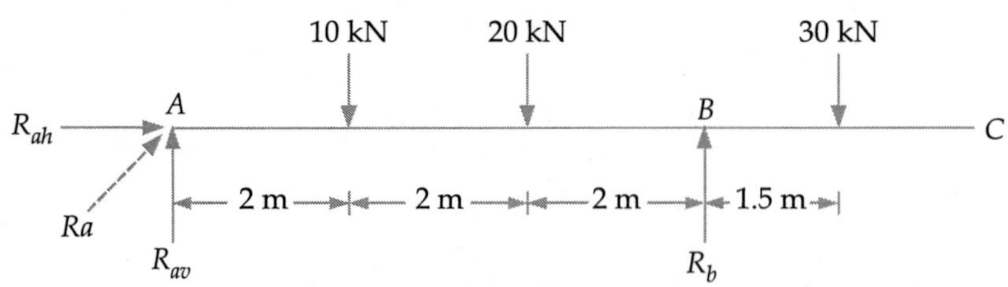

Fig. 3.28

The reactions R_{ah}, R_{av} and R_b can be evaluated by applying the conditions of equilibrium. That is

$$\Sigma M = 0; \quad \Sigma F_x = 0; \quad \Sigma F_y = 0$$

(*i*) Taking moments about A and equating it to zero;

$$(10 \times 2) + (20 \times 4) - (R_b \times 6) + (30 \times 7.5) = 0$$

$$20 + 80 - 6\,R_b + 225 = 0$$

$$\therefore \quad R_b = \frac{225 + 20 + 80}{6} = \frac{325}{6} = \mathbf{54.17\ kN}$$

(*ii*) $\Sigma F_x = 0$ for the whole system.

$$\therefore \quad R_{ah} = 0$$

This is because no other horizontal force, active or reactive, is acting on the beam.

Further, $\Sigma F_y = 0$. That gives $R_{av} - 10 - 20 + R_b - 30 = 0$

$$\therefore \quad R_{av} = 10 + 20 - R_b + 30 = 10 + 20 - 54.17 + 30 = \mathbf{5.83\ kN}$$

EXAMPLE 3.18

Determine reactions at *A* and *B* for the beam loaded as shown in Fig. 3.29.

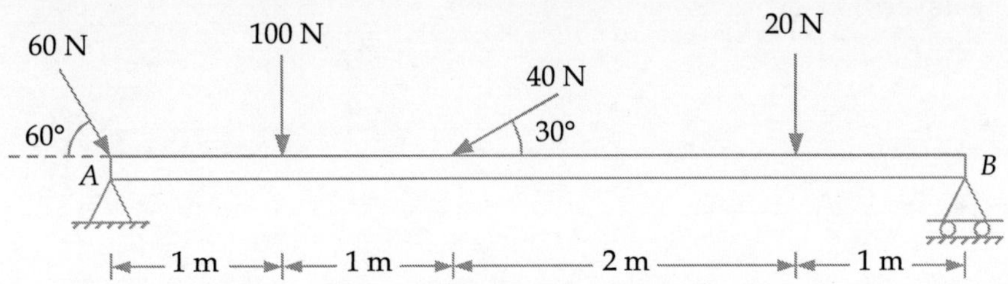

Fig. 3.29

Solution : At the hinged support the reaction R_a can be in any direction. R_{av} and R_{ah} are its components in the vertical and horizontal directions. At the roller support, the reaction R_b will be in vertical direction only.

The equivalent free body diagram for the beam will be then as shown below.

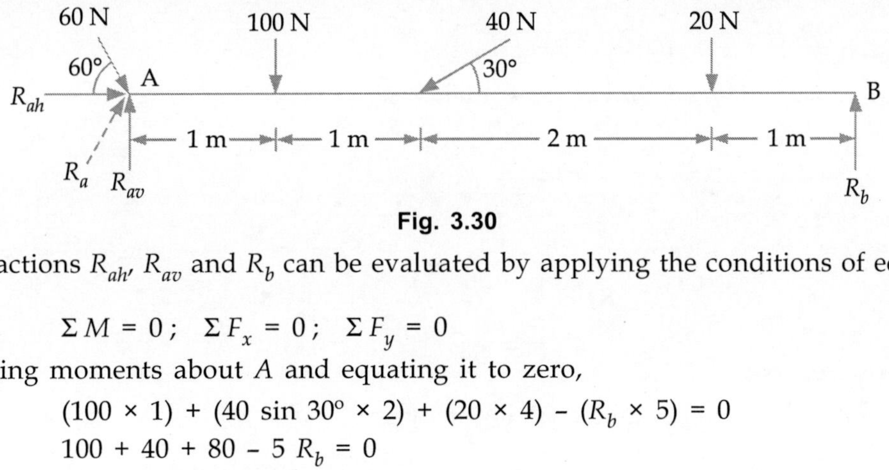

Fig. 3.30

The reactions R_{ah}, R_{av} and R_b can be evaluated by applying the conditions of equilibrium. That is

$$\Sigma M = 0; \quad \Sigma F_x = 0; \quad \Sigma F_y = 0$$

(*i*) Taking moments about *A* and equating it to zero,

$$(100 \times 1) + (40 \sin 30^\circ \times 2) + (20 \times 4) - (R_b \times 5) = 0$$

$$100 + 40 + 80 - 5\,R_b = 0$$

$$\therefore \quad R_b = \frac{100 + 40 + 80}{5} = \frac{220}{5} = \mathbf{44\ N}$$

(*ii*) $\Sigma F_x = 0$ for the whole system.

$$R_{ah} + 60 \cos 60 - 40 \cos 30^\circ = 0.$$

$$\therefore \quad R_{ah} = 40 \cos 30^\circ - 60 \cos 60^\circ = 34.64 - 30 = \mathbf{4.64\ N}$$

(*iii*) Further, $\Sigma F_y = 0$. That gives

$$R_{av} - 60 \sin 60° - 100 - 40 \sin 30° - 20 + Rb = 0$$
$$R_{av} = 60 \sin 60° + 100 + 40 \sin 30° + 20 - R_b$$
$$= 51.96 + 100 + 20 + 20 - 44 = \mathbf{147.96\ N}$$

Total reaction at hinge support,

$$R_a = \sqrt{R_{ah}^2 + R_{av}^2} = \sqrt{4.64^2 + 147.96^2} = 148 \text{ N}$$
$$\theta = \tan^{-1} \frac{R_{av}}{R_{ah}} = \tan^{-1} \frac{147.96}{4.64} = 88.20°$$

where θ is the angle which the reaction at A makes with the x-axis.

EXAMPLE 3.19

Figure 3.31 shows a system of levers supporting a load of 100 N. Determine the reactions at *A* and *B* on the lever.

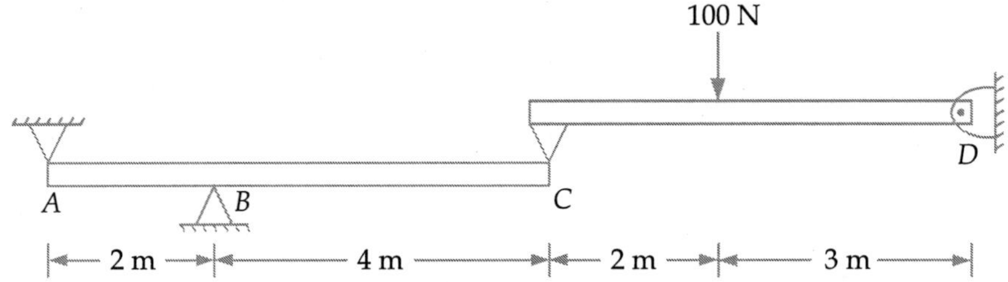

Fig. 3.31

Solution : Refer Fig. 3.32 for the free body diagrams for the upper and lower beams.

Taking moments about end C and equating it to zero,

$$100 \times 2 - R_d \times 5 = 0; \quad R_d = \frac{100 \times 2}{5} = 40 \text{ N}$$
$$\Sigma F_y = 0; \quad R_c - 100 + R_d = 0$$
$$R_c = 100 - R_d = 100 - 40 = 60 \text{ N}$$

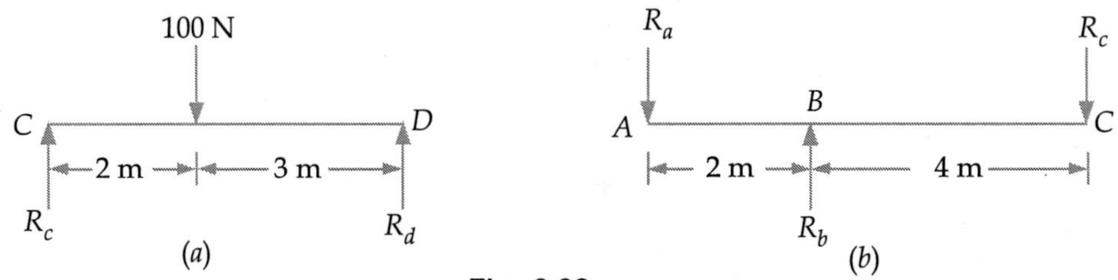

Fig. 3.32

From Fig. 3.32 (*b*),

Taking moments about end A and equating it to zero,

$$R_c \times 6 - R_b \times 2 = 0; \quad R_b = \frac{6R_c}{2} = \frac{6 \times 60}{2} = 180 \text{ N}$$
$$\Sigma F_y = 0; \quad R_a - R_b + R_c = 0$$
$$R_a = R_b - R_c$$
$$= 180 - 60 = 120 \text{ N}$$

Thus the reactions at points A and B on the lever are:

$$R_b = 180 \text{ N (upwards)}$$
$$R_a = 120 \text{ N (downwards)}$$

EXAMPLE 3.20

A weightless bar *AB* is placed in horizontal position on smooth inclines as shown in Fig. 3.33. Calculate the distance *x* at which the 100 N load be placed from end *B* so that the bar remains horizontal.

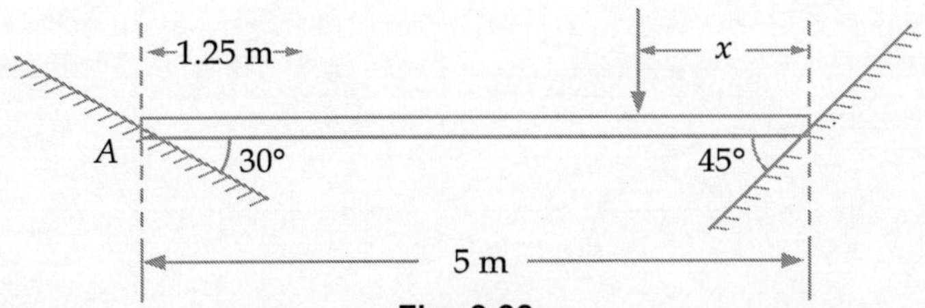

Fig. 3.33

Solution : Reactions at ends *A* and *B* of the bar will be normal to the smooth inclines and can be resolved horizontally and vertically as shown below

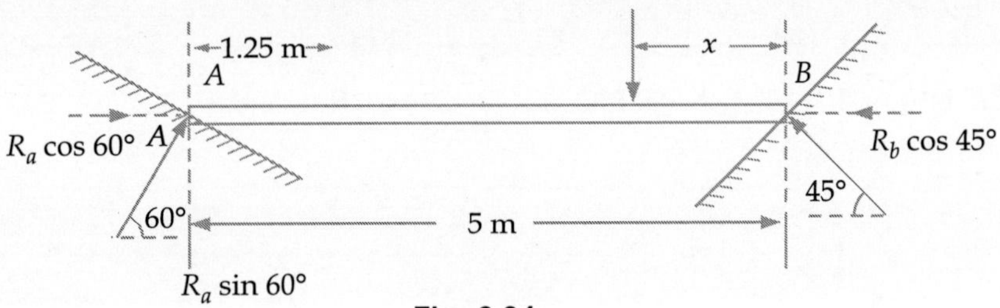

Fig. 3.34

Under equilibrium conditions : $\Sigma F_x = 0$, $\Sigma F_y = 0$ and $\Sigma M = 0$

$\Sigma F_x = 0$: $R_a \cos 60° - R_b \cos 45° = 0$

$$R_a = \frac{R_b \cos 45°}{\cos 60°} = 1.414\, R_b \qquad ...(i)$$

$\Sigma F_y = 0$: $R_a \sin 60° - 200 - 100 + R_b \sin 45° = 0$

$$0.866\, R_a - 300 + 0.707\, R_b = 0 \qquad ...(ii)$$

From expressions (*i*) and (*ii*),

$0.866 \times 1.414\, R_b - 300 + 0.707\, R_b = 0$ or $1.224 R_b - 300 + 0.707\, R_b = 0$

$$\therefore \quad R_b = \frac{300}{1.224 + 0.707} = \mathbf{155.36\ N}$$

$$R_a = 1.414 \times 155.36 = \mathbf{219.68}\ \text{N}$$

Taking moments of all forces about end *A* (clockwise positive), we get

$$-100x - 200 \times 3.75 + R_a \sin 60° \times 5 = 0$$

It is to be noted that component $R_a \cos 60°$ of R_a passes through *B*, and hence its moment about *B* is zero

$$-100\, x - 750 + 219.68 \sin 60° \times 5 = 0 \qquad (\because R_a = 219.68\ \text{N})$$

or

$$-100x - 750 + 951.24 = 0$$

$$\therefore \quad x = \frac{951.24 - 750}{100} = \mathbf{2.01\ m}$$

EXAMPLE 3.21

A beam has been loaded and supported as shown in Fig. 3.35 given below. Determine the reactions at the support points *A* and *B*.

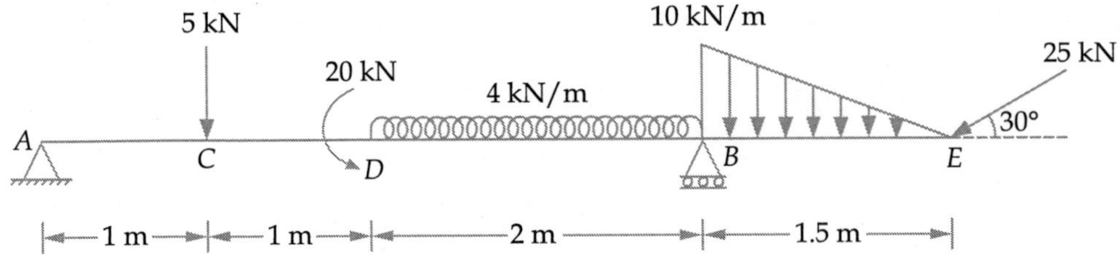

Fig. 3.35

Solution : Equivalent loads and their position of acting are as shown below :

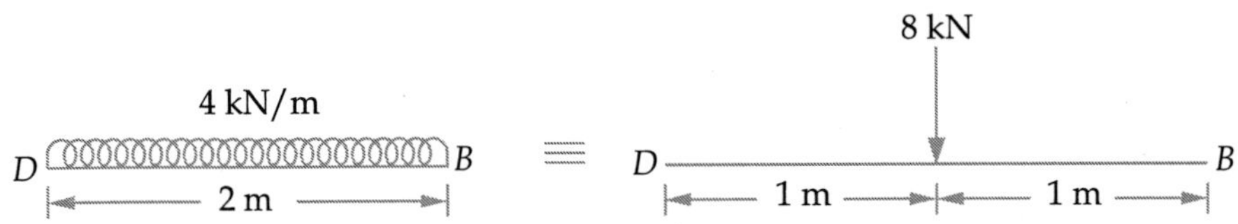

The total load is equal to 4 × 2 = 8 kN and it acts at the centre.

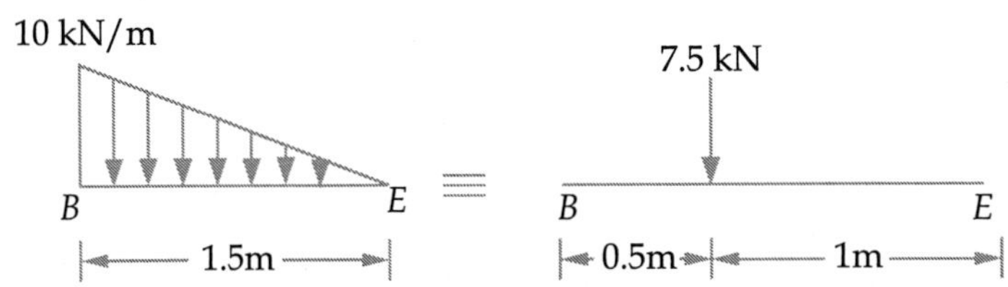

The total load is $\frac{1}{2} \times 1.5 \times 10 = 7.5$ kN and it acts at $\frac{1}{3} \times 1.5 = 0.5$ m from the base, *i.e.*, from point *B*.

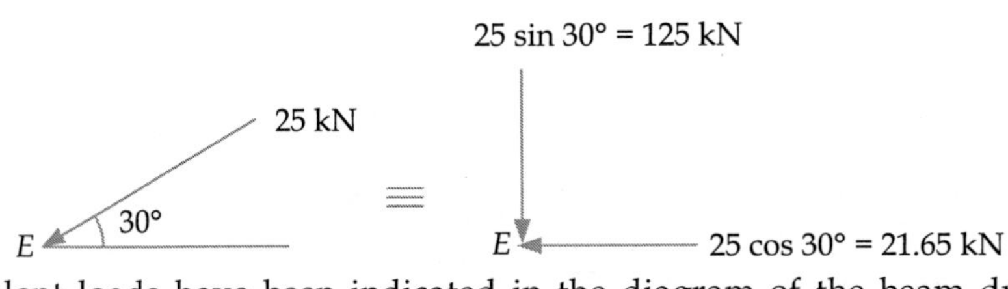

These equivalent loads have been indicated in the diagram of the beam drawn below.

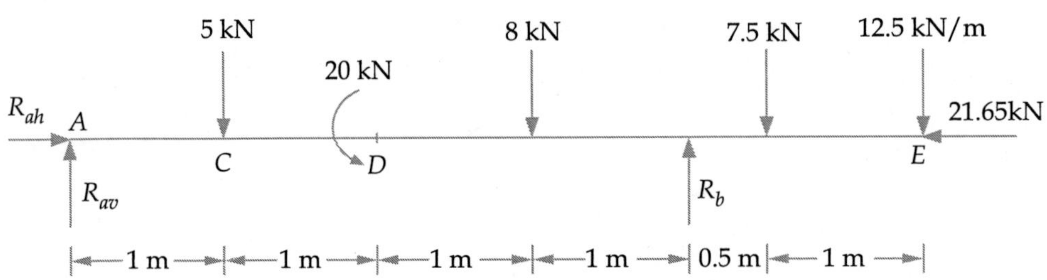

Fig. 3.36

For equilibrium of the beam : $\Sigma F_x = 0$; $\Sigma F_y = 0$ and $\Sigma M = 0$

$\Sigma F_x = 0$; $R_{ah} - 21.65 = 0$; $R_{ah} = 21.65$ kN

$\Sigma F_y = 0\ ;\ R_{av} - 5 - 8 + R_b - 7.5 - 12.5 = 0$

$\therefore \qquad R_{av} + R_b = 33$

Taking moments about the end A (clockwise positive)

$\Sigma M_a = 0\ ;\ 5 \times 1 - 20 + 8 \times 3 - R_b \times 4 + 7.5 \times 4.5 + 12.5 \times 5.5 = 0$

$5 - 20 + 24 - 4R_b + 33.75 + 68.75 = 0\ ;\ R_b = \mathbf{27.875\ kN}$

Then $\qquad R_{av} = 33 - R_b = 33 - 27.875 = \mathbf{5.125\ kN}$

Thus the reactions at support points A and B are

$$R_{ah} = 21.65\text{ kN}\ ;\ R_{av} = 5.125 \text{ and } R_b = 27.875\text{ kN}$$

EXAMPLE 3.22

A cantilever beam has been loaded as shown in Fig. 3.37 given below. Determine the reaction at cantilever end *A*.

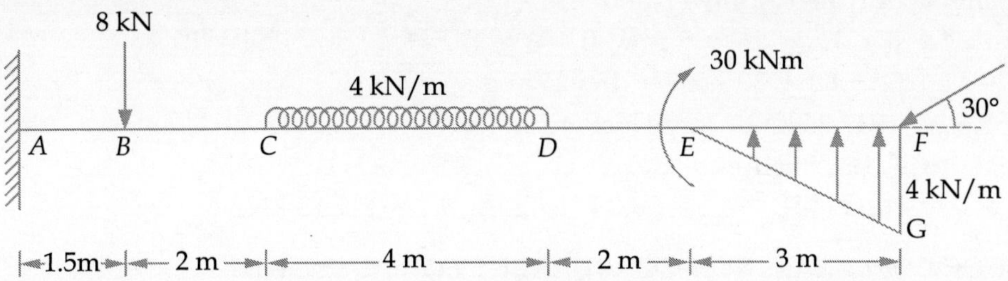

Fig. 3.37

Solution : Equivalent loads and their positions of acting are as shown below :

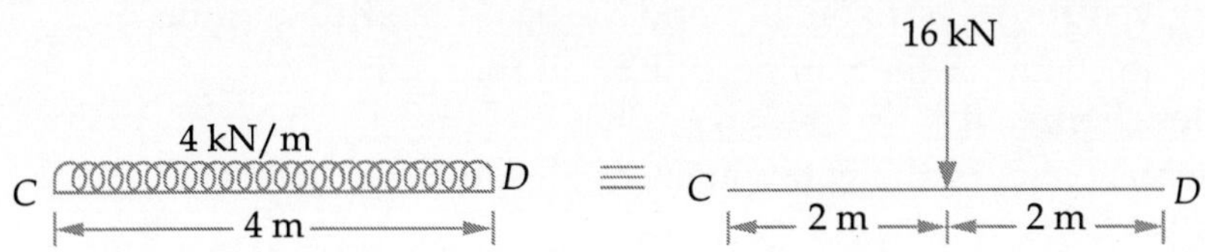

The total load is equal to 4 × 4 = 16 kN and it is taken to be acting at the centre.

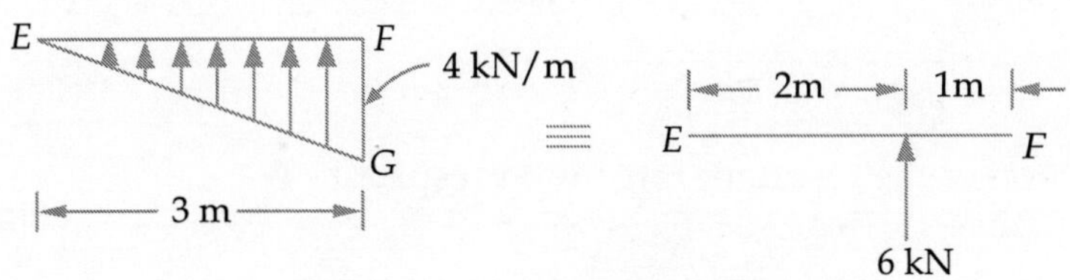

The total load is $\frac{1}{2} \times 3 \times 4 = 6$ kN and it is taken to be acting at $\frac{2}{3} \times 3 = 2$ m distance from the apex point E of the triangle EFG.

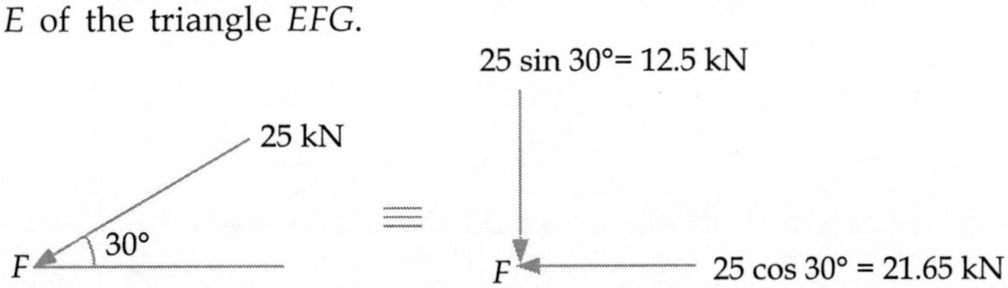

These equivalent loads have been indicated in the diagram of the beam drawn below.

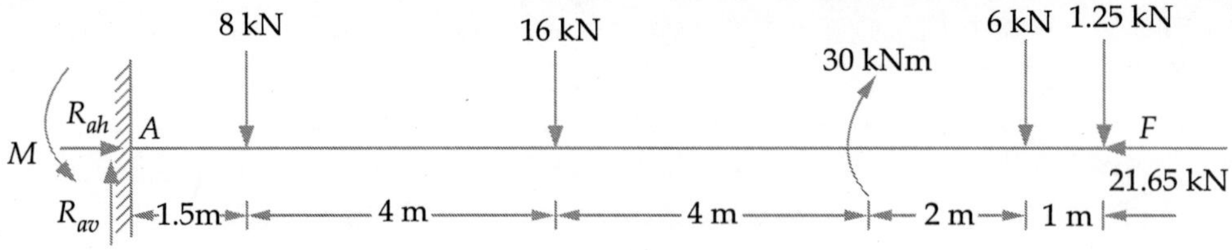

Fig. 3.38

For equilibrium of the beam : $\Sigma F_x = 0$; $\Sigma F_y = 0$ and $\Sigma M = 0$

$$\Sigma F_x = 0; \quad R_{ah} - 21.65 = 0 \quad ; \quad R_{ah} = \mathbf{21.65\ kN}$$

$$\Sigma F_y = 0; \quad R_{av} - 8 - 16 + 6 - 12.5 = 0 \quad ; \quad R_{av} = \mathbf{30.5\ kN}$$

Since the end A of the beam is fixed and end F is free, the beam AF is a cantilever and hence a moment of couple M is produced at A as shown in Fig. 3.45 .

Taking moments about the cantilever end A,

$$\Sigma M_a = 0 \quad ; \quad -M + 8 \times 1.5 + 16 \times 5.5 + 30 - 6 \times 11.5 + 12.5 \times 12.5 = 0$$

$$-M + 12 + 88 + 30 - 69 + 156.25 = 0$$

$$\therefore \qquad M = \mathbf{217.25}\ \text{kNm (anticlockwise).}$$

Thus the reactions at the fixed end are :

$$R_{ah} = 21.65\ \text{kN} \ ; \quad R_{av} = 30.5\ \text{kN} \quad \text{and} \quad M = 217.25\ \text{kNm}$$

3.5. RESULTANT OF COPLANAR, NON-CONCURRENT FORCE SYSTEM

Below are outlined the different steps which are followed to determine the resultant of coplanar, non-current force system in magnitude, direction and position :

1. Resolve the given forces horizontally and vertically and determine

ΣF_x = algebraic sum of the component of all forces in x-direction, and

ΣF_y = algebraic sum of the component of all forces in y-direction

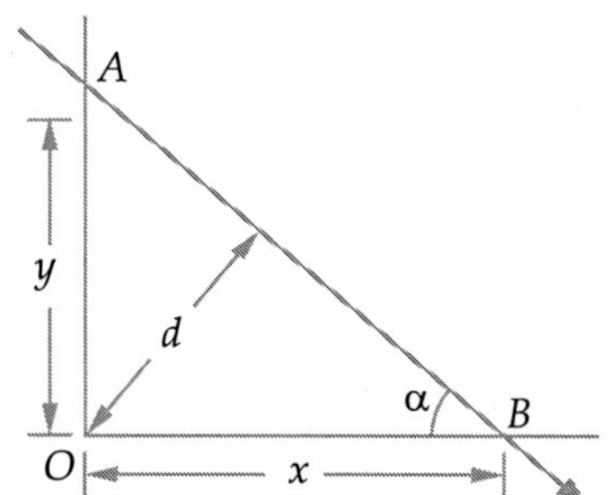

Fig. 3.39

2. Determine the magnitude of the resultant force by using the relation

$$R = \sqrt{(\Sigma F_x)^2 + (\Sigma F_y)^2}$$

3. Find the direction of the resultant force with respect to x-axis by using the relation

$$\alpha = \tan^{-1}\left(\frac{\Sigma F_y}{\Sigma F_x}\right)$$

Depending upon the sign of ΣF_x and ΣF_y , decide in which of the quadrant the resultant lies.

4. Obtain the algebraic sum of the moments of all forces about any given point say O.

5. Mark the position of resultant such that it produces the same direction of moment about point O.

6. Apply Varignon's theorem of moment to find the exact position of resultant. That is

$$\Sigma M_O = R \times d$$

where ΣM_O = moment of given forces about point O

d = perpendicular distance of line of action of resultant R from point O.

The x and y intercepts of the line of action of resultant can then be worked out as

$$x = \frac{d}{\sin \alpha} \quad \text{and} \quad y = \frac{d}{\cos \alpha}$$

Alternatively, the intercepts can be determined by considering the moment of R about O as the sum of moments of its components about O (Varignon's theorem)

$$R \times d = \Sigma M_O$$

$$\Sigma F_x \times 0 + \Sigma F_y \times x = \Sigma M_O$$

$$\therefore \quad x = \frac{\Sigma M_O}{\Sigma F_y}$$

Similarly by resolving the resultant into its components at A, it can be calculated that

$$y = \frac{\Sigma M_O}{\Sigma F_x}$$

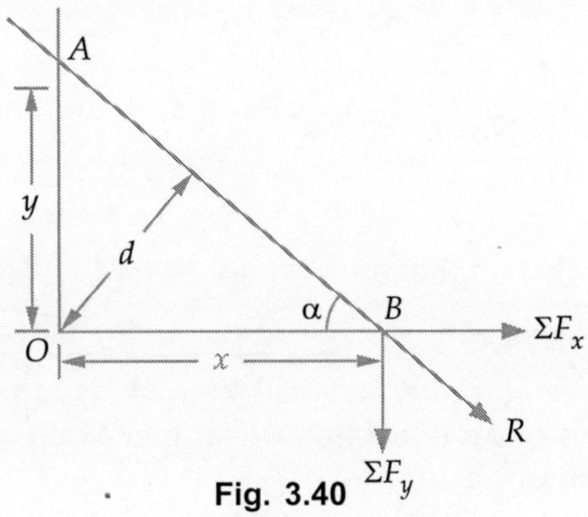

Fig. 3.40

EXAMPLE 3.27

Forces equal to P, 2P, 3P and 4P act along the sides AB, BC, CD and DA of a square ABCD. Find the magnitude, direction and line of action of the resultant.

Solution : Refer to Fig. 3.41 for the given force system

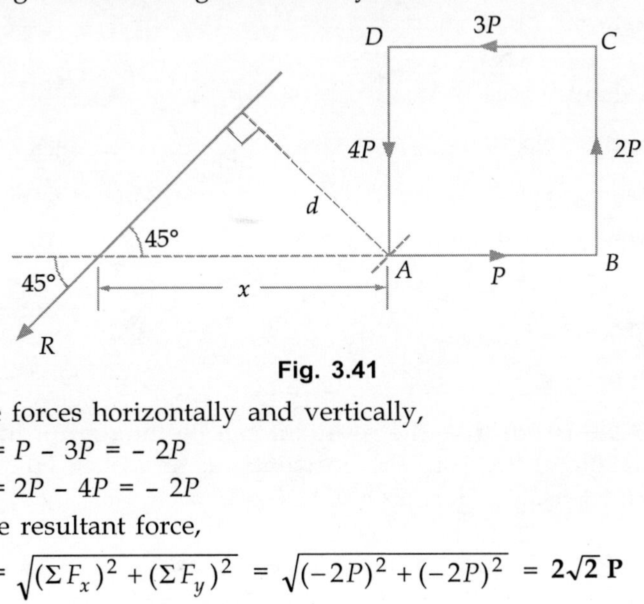

Fig. 3.41

Resolving all the forces horizontally and vertically,

$$\Sigma F_x = P - 3P = -2P$$

$$\Sigma F_y = 2P - 4P = -2P$$

Magnitude of the resultant force,

$$R = \sqrt{(\Sigma F_x)^2 + (\Sigma F_y)^2} = \sqrt{(-2P)^2 + (-2P)^2} = \mathbf{2\sqrt{2}\,P}$$

Let α be the angle which the resultant makes with the horizontal. Then

$$\tan \alpha = \frac{\Sigma F_y}{\Sigma F_x} = \frac{-2P}{-2P} = 1; \qquad \alpha = 45^\circ$$

Since both ΣF_x and ΣF_y are negative, resultant lies in the 3rd quadrant making an angle of 45° with horizontal as shown in Fig. 3.41.

Taking moments of all forces about point A (clockwise + ve), we get

$$\Sigma M_a = -2P \times a - 3P \times a = -5Pa$$

The resultant R should then lie as shown in Fig. 3.43 so that it can produce anti-clockwise (negative) moment about point A.

If d is the perpendicular distance of resultant from A, then

$$R \times d = \Sigma M_a$$

$$\therefore \quad d = \frac{\Sigma M_a}{R} = \frac{5Pa}{2\sqrt{2}P} = \frac{5a}{2\sqrt{2}}$$

The horizontal distance x of the line of action of resultant from point A is

$$x = \frac{d}{\sin\alpha} = \frac{5a}{2\sqrt{2}} \div \sin 45^\circ = 2.5a$$

In the above identities a is a measure of the side of each side of the square block.

EXAMPLE 3.23

Three forces equal to 1 kN, 2 kN and 3 kN are respectively acting in order along the three sides of an equilateral triangle. Make calculations for the magnitude, direction and position of their resultant.

Solution : Refer Fig. 3.42 for the given force system. Resolving all the forces horizontally and vertically.

$$\Sigma F_x = -1\cos 60^\circ + 2 - 3\cos 60^\circ$$
$$= -0.5 + 2 - 1.5 = 0$$
$$\Sigma F_y = -1\sin 60^\circ + 3\sin 60^\circ$$
$$= -0.866 + 2.598 = 1.732 \text{ kN}$$

Magnitude of the resultant force,

$$R = \sqrt{(\Sigma F_x)^2 + (\Sigma F_y)^2}$$
$$= \sqrt{0^2 + 1.732^2} = 1.732 \text{ N}$$

Let α be the angle which the resultant makes with the horizontal.

$$\tan\alpha = \frac{\Sigma F_y}{\Sigma F_x} = \frac{1.732}{0} = \infty$$

$$\therefore \quad \alpha = 90^\circ$$

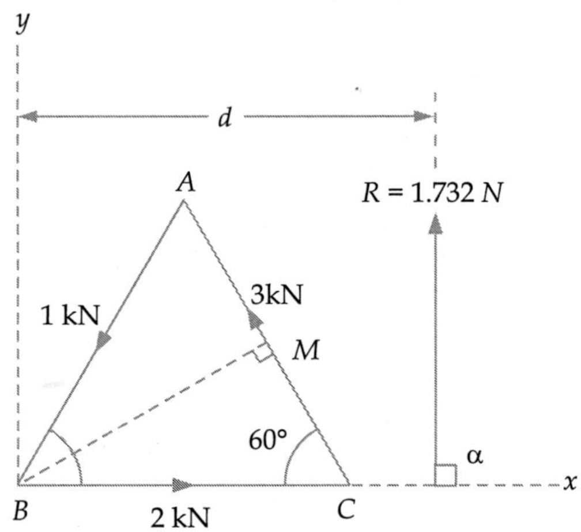

Fig. 3.42

Since ΣF_x is zero and ΣF_y is positive, the resultant can be thought of to lie in the 1st or 3rd quadrant making an angle of 90° with the horizontal as shown in Fig. 3.42.

Taking moments of all forces about point B (clockwise + ve), we get

$$\Sigma M_b = -3 \times BM$$
$$= -3 \times a\sin 60^\circ$$

The resultant R should then lie as shown in Fig. 3.42 so that it gives rise to anti-clockwise (negative) moment about point B.

If d is the perpendicular distance of resultant from B, then

$$R \times d = \Sigma M_b$$

$$\therefore\ d = \frac{\Sigma M_b}{R} = \frac{3a\sin 60^\circ}{1.732} = 3a$$

where a is a measure of each side of the equilateral triangle.

EXAMPLE 3.24

Compute the simplest resultant force for the loads acting on a cantilever beam as shown in the figure given below. What force and moment is transmitted by this force to the supporting wall at *A*?

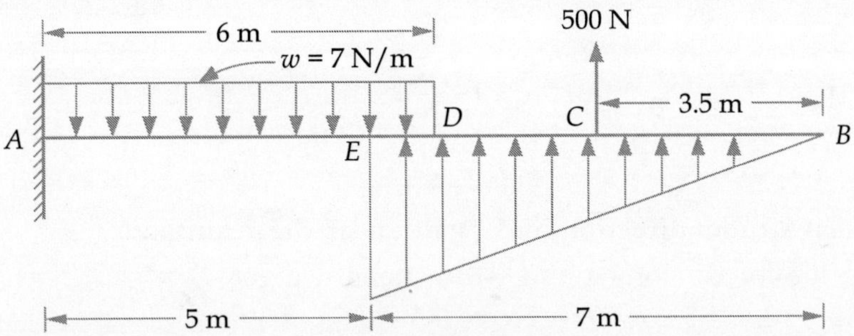

Fig. 3.43

Solution : Equivalent loads and their position of acting are as shown below:

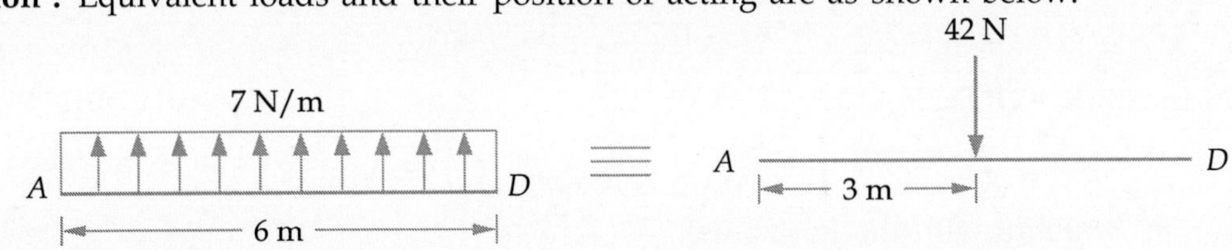

The load is equal to 7 × 6 = 42 N and its acts at the centre of portion *AD*

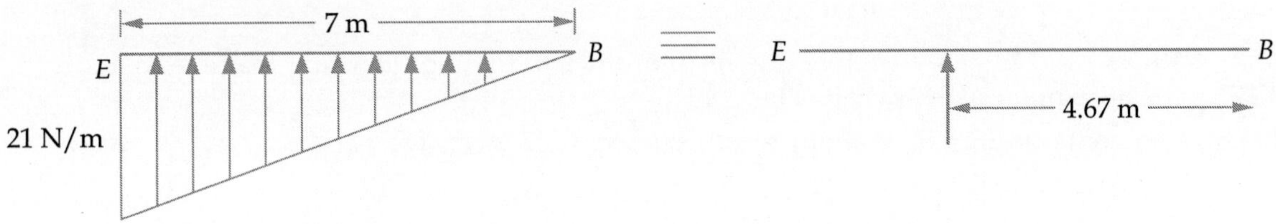

The total load is $\frac{1}{2} \times 7 \times 21 = 73.5$ N and it acts at $\frac{2}{3} \times 7 = 4.67$ from the apex, *i.e.*, from point *B*.

These equivalent loads have been indicated in the diagram of the beam drawn below:

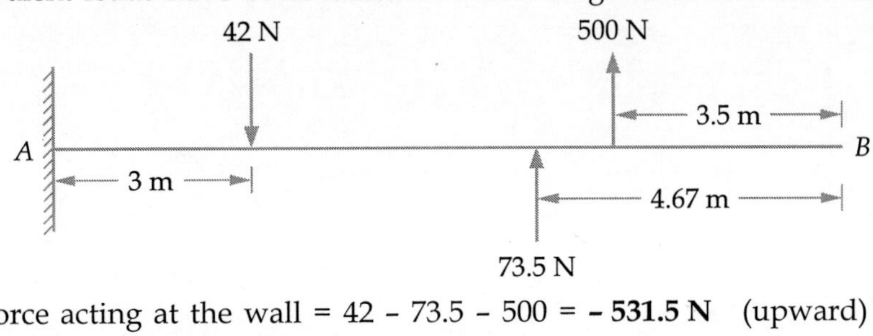

Resultant force acting at the wall = 42 – 73.5 – 500 = **– 531.5 N** (upward)

Moment acting at the wall,

$$M_A = 42 \times 3 - 500(12 - 3.5) - 73.5\,(12 - 4.67)$$
$$= 126 - 4250 - 538.75$$
$$= \mathbf{-\,4662.75\ N} \quad \text{(anti clockwise)}$$

EXAMPLE 3.25

A horizontal beam *AD* of length 12 m is acted upon by a set of forces as shown in Fig. 3.44

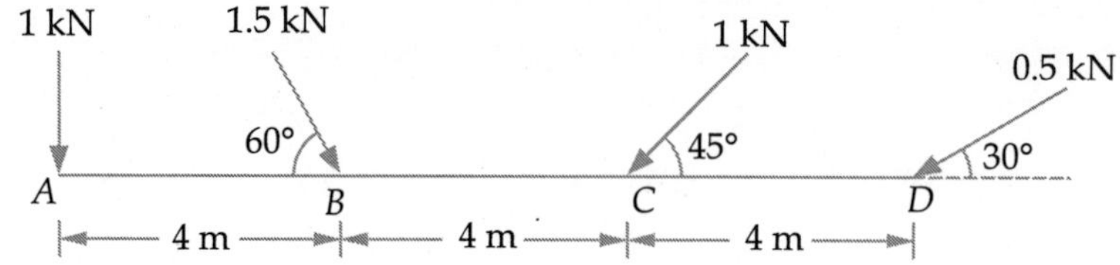

Fig. 3.44

Determine the magnitude, direction and position of the resultant.

Solution : Resolving the forces along x-and y-directions, we get

$$\Sigma F_x = 0 + 1.5 \cos 60° - 1 \cos 45° - 0.5 \cos 30°$$
$$= + 0.75 - 0.707 - 0.433 = - 0.39 \text{ kN}$$
$$\Sigma F_y = -1 - 1.5 \sin 60° - 1 \sin 45° - 0.5 \sin 30°$$
$$= -1 - 1.299 - 0.707 - 0.25 = - 3.256 \text{ kN}$$

Resultant force $R = \sqrt{(\Sigma F_x)^2 + (\Sigma F_y)^2}$

$$= \sqrt{(-0.39)^2 + (3.256)^2} = \sqrt{10.7536} = \mathbf{3.279} \text{ kN}$$

Inclination of resultant with the horizontal,

$$\tan \alpha = \frac{\Sigma F_y}{\Sigma F_x} = \frac{3.256}{0.39} = 8.333; \ \alpha = 83.17°$$

Since both ΣF_x and ΣF_y are negative, the resultant lies in the 3rd quadrant making an angle of 83.17° with horizontal as shown in Fig. 3.45

Taking moments of all forces about point A (clockwise $+ve$), we get

$$\Sigma M_a = (1.5 \sin 60°) \times 4 + (1 \sin 45°) \times 8 + (0.5 \sin 30°) \times 12$$
$$= 5.196 + 5.657 + 3 = 13.853 \text{ kNm}$$

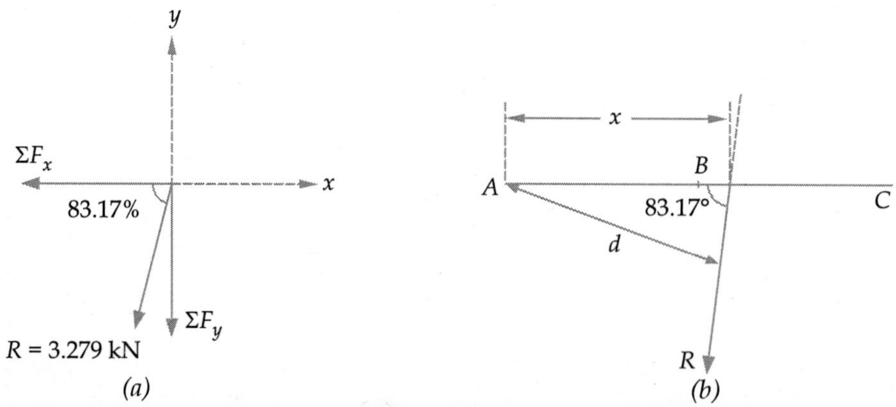

Fig. 3.45

The resultant R should then lie as shown in Fig. 3.45 so that it can produce clockwise (positive) moment about point A.

If d is the perpendicular distance of resultant from A, then

$$R \times d = \Sigma M_a$$

$$\therefore \quad d = \frac{\Sigma M_a}{R} = \frac{13.853}{3.279} = \mathbf{4.22\ m}$$

The horizontal distance x of the point of application of resultant from point A is

$$x = \frac{d}{\sin\alpha} = \frac{4.22}{\sin 83.17°} = \mathbf{4.25\ m}$$

EXAMPLE 3.26

Determine the resultant of the four forces acting tangentially to a circle of radius 3 m as shown in Fig. 3.46. What will be the location of the resultant with respect to centre of the circle?

Solution : For these coplanar and non-current forces

$$\Sigma F_x = 150 - 100 \cos 45° = 79.29 \text{ N}$$

$$\Sigma F_y = 50 - 80 - 100 \sin 45° = -100.71 \text{ N}$$

$$\text{Resultant } R = \sqrt{(\Sigma F_x)^2 + (\Sigma F_y)^2}$$

$$= \sqrt{(79.29)^2 + (-100.71)^2} = 128.18 \text{ N}$$

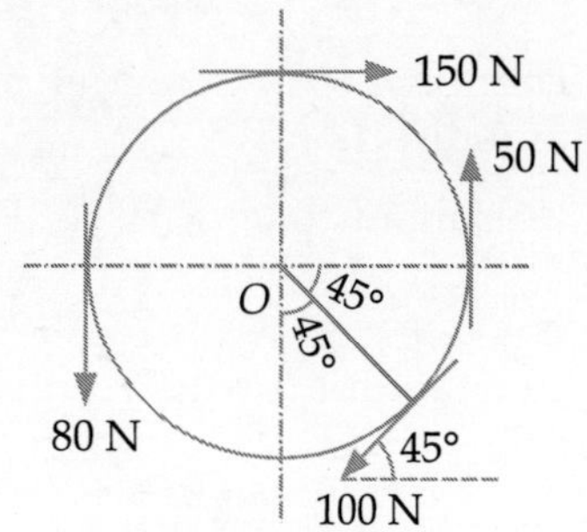

Fig. 3.46

Let α be the angle which the resultant makes with the horizontal direction as shown in Fig. 3.47. Then

$$\alpha = \tan^{-1}\left(\frac{\Sigma F_y}{\Sigma F_x}\right) = \tan^{-1}\left(\frac{100.71}{79.29}\right) = 51.8°$$

Since ΣF_x is positive and ΣF_y is negative, the resultant lies in the fourth quadrant.

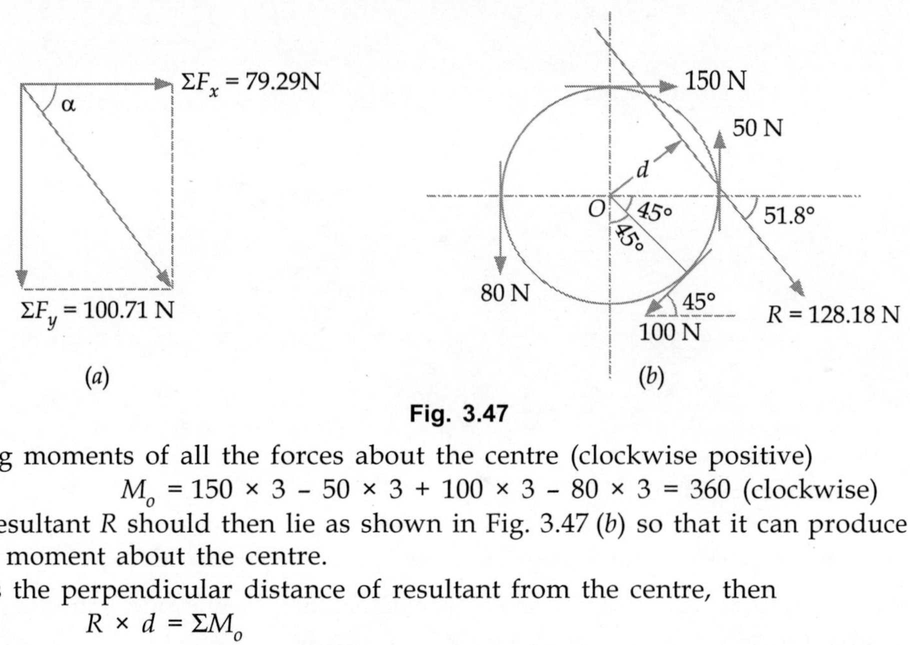

Fig. 3.47

Taking moments of all the forces about the centre (clockwise positive)

$$M_o = 150 \times 3 - 50 \times 3 + 100 \times 3 - 80 \times 3 = 360 \text{ (clockwise)}$$

The resultant R should then lie as shown in Fig. 3.47 (*b*) so that it can produce clockwise (positive) moment about the centre.

If d is the perpendicular distance of resultant from the centre, then

$$R \times d = \Sigma M_o$$

$$\therefore \quad d = \frac{\Sigma M_o}{R} = \frac{360}{128.18} = \mathbf{2.795\ m}$$

The resultant will act at a perpendicular distance of 2.795 m from the centre.

EXAMPLE 3.27

A plate measuring 6 m × 4 m is acted upon a set of forces in its plane as depicted in Fig. 3.48. Determine the magnitude, direction and position of the resultant force.

Solution : Resolving the forces in the horizontal and vertical directions, we get

$$\Sigma F_x = 80 - 100 \cos 30^\circ + 120 \cos 45^\circ - 60 \cos 30^\circ$$
$$= 80 - 86.6 + 84.85 - 51.96$$
$$= +26.29 \text{ N} \rightarrow$$

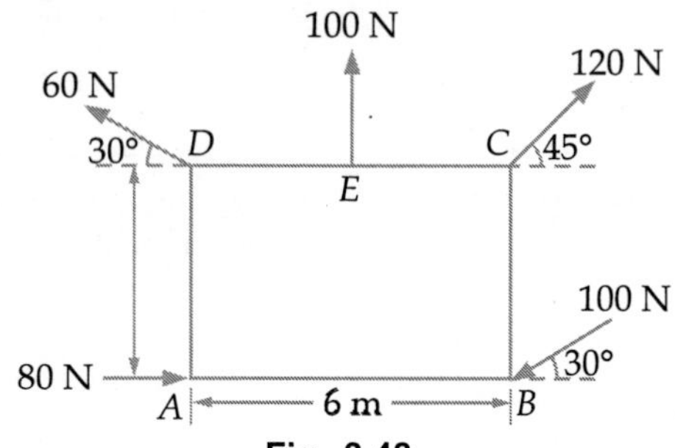

Fig. 3.48

The force 100 N acting at point E has no component in the x-direction.

$$\Sigma F_y = -100 \sin 30^\circ + 120 \sin 45^\circ + 100 + 60 \sin 30^\circ$$
$$= -50 + 84.85 + 100 + 30 = 164.85 \text{ N} \uparrow$$

The force 80 N acting at point A has no component in the y-direction.

$$\text{Resultant force } R = \sqrt{(\Sigma F_x)^2 + (\Sigma F_y)^2}$$
$$= \sqrt{(26.29)^2 + (164.85)^2} = \sqrt{27866.68} = \mathbf{166.93 \text{ N}}$$

Inclination α of the resultant with the horizontal,

$$\tan \alpha = \frac{\Sigma F_y}{\Sigma F_x} = \frac{164.85}{26.29} = 6.27; \quad \alpha = 80.94^\circ$$

Since both ΣF_x and ΣF_y are positive, the resultant lies in first quadrant at an angle 80.94° with horizontal, *i.e.*, with x-axis as shown in Fig. 3.49 (*a*).

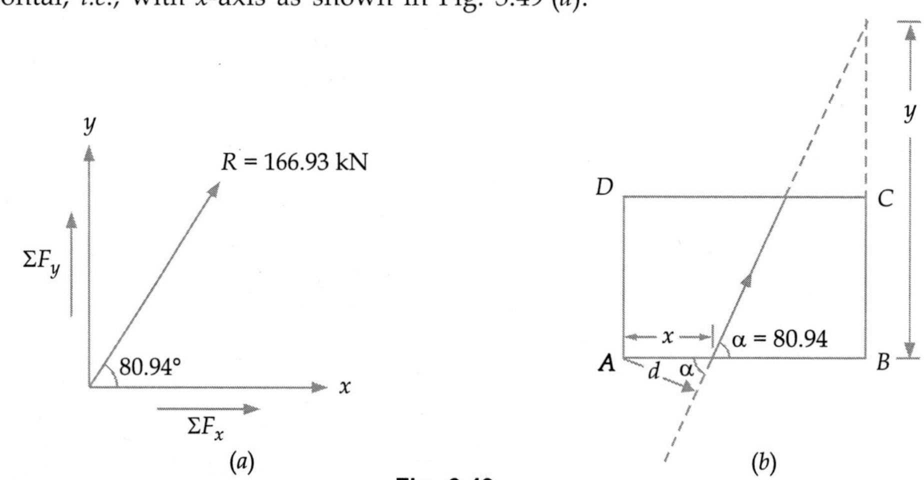

Fig. 3.49

(*b*) **Position of resultant force:**

Taking moments about point A (clockwise + *ve*), we get

$$\Sigma M_a = (100 \sin 30^\circ) \times 6 - (120 \sin 45^\circ) \times 6 + (120 \cos 45^\circ) \times 4 - 100 \times 3 - (60 \cos 30^\circ) \times 4$$
$$= 300 - 509.1 + 339.4 - 300 - 207.8 = -377.5 \text{ Nm (anticlockwise)}$$

The resultant R should then lie as shown in Fig. 3.49 (*b*) so that it can produce anticlockwise (negative) moment about point A.

If d is the perpendicular distance of the resultant from A, then

$$R \times d = \Sigma M_a$$

$$\therefore \qquad d = \frac{\Sigma M_a}{R} = \frac{377.5}{166.93} = 2.261 \text{ m}$$

The intercepts x on x-axis and y on y-axis are then given by

$$x = \frac{d}{\sin\alpha} = \frac{2.261}{\sin 80.94} = 2.29 \text{ m}$$

$$y = \frac{d}{\cos\alpha} = \frac{2.261}{\cos 80.94} = 14.36 \text{ m}$$

The intercepts on x-axis and y-axis can also be worked out as

$$x = \frac{\Sigma M_a}{\Sigma F_y} = \frac{377.5}{164.85} = 2.29 \text{ m}$$

$$y = \frac{\Sigma M_a}{\Sigma F_x} = \frac{377.5}{26.29} = 14.36 \text{ m}$$

Since intercept y is greater than BC, the resultant meets the arm when produced.

EXAMPLE 3.28

Determine the resultant of four forces and one couple that act on the plate as shown in the figure given below:

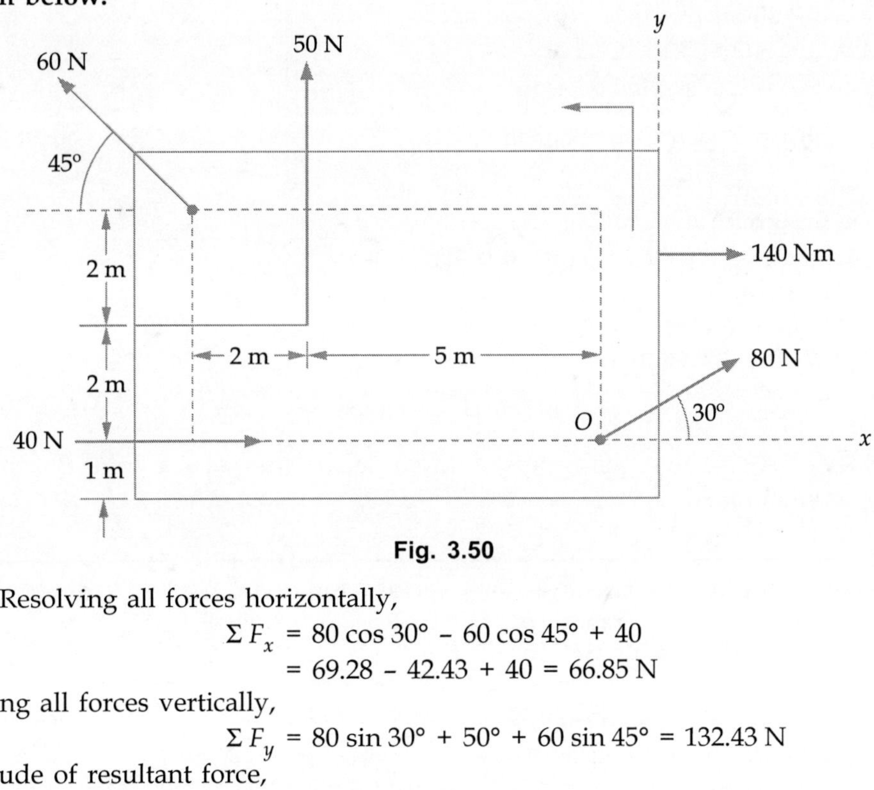

Fig. 3.50

Solution : Resolving all forces horizontally,

$$\Sigma F_x = 80 \cos 30° - 60 \cos 45° + 40$$
$$= 69.28 - 42.43 + 40 = 66.85 \text{ N}$$

Resolving all forces vertically,

$$\Sigma F_y = 80 \sin 30° + 50° + 60 \sin 45° = 132.43 \text{ N}$$

Magnitude of resultant force,

$$F = \sqrt{(\Sigma F_x)^2 + (\Sigma F_y)^2} = \mathbf{148.34 \text{ N}}$$

Let α be the angle which the resultant force makes with the horizontal direction, Then

$$\tan \alpha = \frac{\Sigma F_y}{\Sigma F_x} = \frac{132.43}{66.85} = 1.981$$

That gives: $\alpha =$ **63.21°**

Taking moments about point O (considering clockwise moments positive),

$$M_0 = -140 + 50 \times 5 + 60 \sin 45° \times 7 - 60 \cos 45° \times 4$$

The lines of application of 40 N and 80 N forces pass through O and as such they will have no contribution towards moments.

$$M_0 = -140 + 250 + 296.98 - 169.70 = \mathbf{237.28\ Nm}$$

Thus the system will have a clockwise moment equivalent to 237.28 Nm at point O.

EXAMPLE 3.29

One end of a rope of length l is fixed to a vertical telegraph pole. A man standing on the ground pulls at the other end of the rope with a given force. Find the point at which the rope should be tied to the pole in order that the man will have the best chance of over turning the pole about its end fixed to the ground.

Solution : With reference to Fig. 3.51, AB is the vertical pole grounded at point A.

Let P be the force applied through rope BC (length l) towards C where the man is standing.

Moment of force P about point A,

$$M = P \times AD = P \times AC \sin \theta$$
$$= P \times (BC \cos \theta) \sin \theta$$
$$= Pl \sin \theta \cos \theta = \frac{1}{2} Pl \sin 2\theta$$

The force P and length l are fixed and as such the moment M will be maximum when

$\sin 2\theta = 1$, *i.e.*, when $2\theta = 90°$ and $\theta = 45°$

Obviously the man will have the best chance of over-turning the vertical pole when he pulls it through the rope at an angle of 45° with the horizontal.

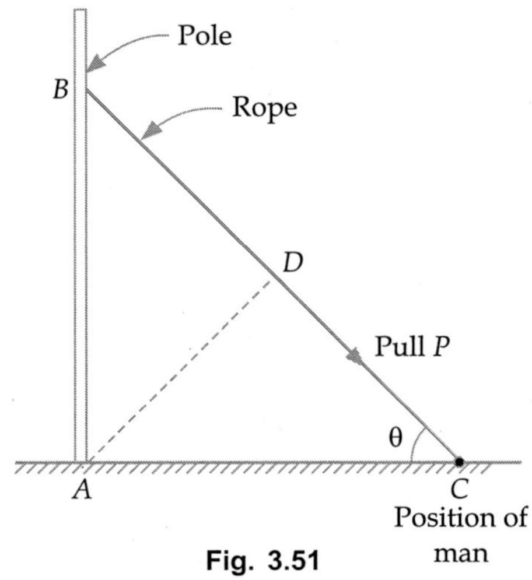

Fig. 3.51

$$AB = BC \sin \theta = l \sin 45° = \frac{l}{\sqrt{2}}$$

Hence to achieve the objective, the rope should be tied to the pole at point B which is at height $l/\sqrt{2}$ above the point A.

EXAMPLE 3.30

For the force system applied to an angle bracket as shown in Fig. 3.52, determine the magnitude and direction of the resultant. What would be the distance of line of action of the resultant from point O.

Solution : The component of forces in the horizontal and vertical directions are determined as below :

$$\Sigma F_x = 150 \cos 35° - 80$$
$$= 122.87 - 80 = 42.87 \text{ kN}$$
$$\Sigma F_y = 150 \sin 35° + 20 - 40$$
$$= 66 \text{ kN}$$

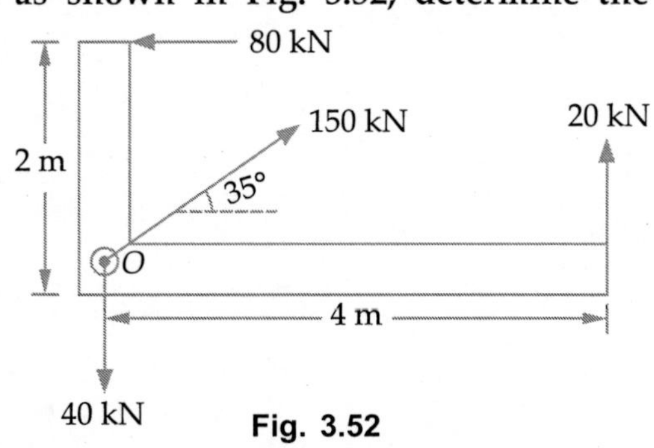

Fig. 3.52

Resultant $R = \sqrt{(\Sigma F_x)^2 + (\Sigma F_y)^2}$

$$= \sqrt{42.87^2 + 66^2} = \sqrt{6193.84} = 78.7 \text{ kN}$$

Inclination α of the resultant with horizontal,

$$\alpha = \tan^{-1}\left(\frac{\Sigma F_y}{\Sigma F_x}\right) \tan^{-1}\left(\frac{66}{42.87}\right) = \tan^{-1}(1.5395) = 57^\circ$$

Since both ΣF_x and ΣF_y are positive, the resultant lies in the first quadrant making an angle 57° with horizontal, *i.e.*, with x-axis as shown in Fig. 3.53 (*b*).

Position of resultant force

Taking moments of all forces about point A (clockwise + *ve*) we get

$$\Sigma M_o = -80 \times 2 - 20 \times 4 = -240 \text{ kN} \qquad \text{(anticlockwise)}$$

The resultant R should then lie as shown in Fig. 3.53 so that it can produce anticlockwise (negative) moment about point O.

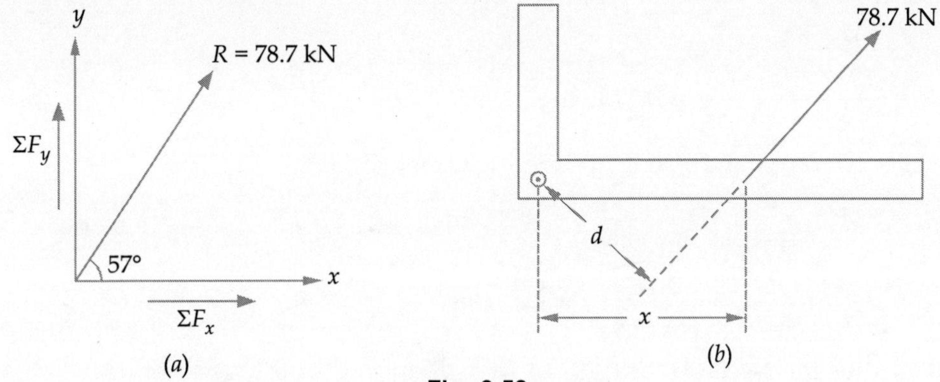

Fig. 3.53

If d is the perpendicular distance of resultant from O, then

$$R \times d = \Sigma M_o$$

$$\therefore \quad d = \frac{\Sigma M_o}{R} = \frac{240}{78.7} = 3.049 \text{ m}$$

Intercept on x-axis is given by

$$x = \frac{d}{\sin\alpha} = \frac{3.049}{\sin 57} = 3.625 \text{ m}$$

EXAMPLE 3.31

An angle bracket has been subjected to three forces and a couple as shown in Fig. 3.54. Determine the resultant of this system of forces. Proceed to locate the points where the line of action of the resultant intersects line AB and the line BC.

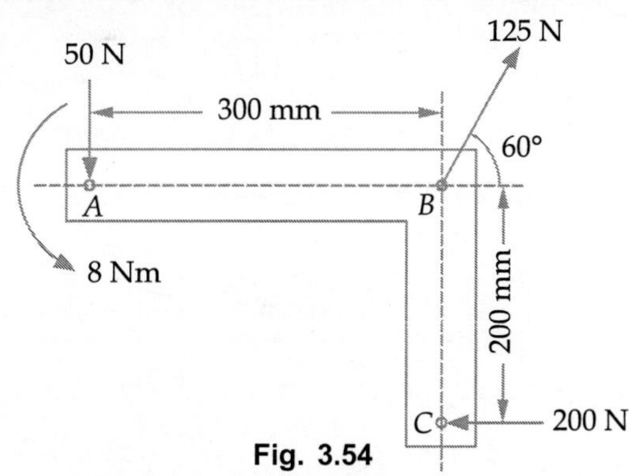

Fig. 3.54

Solution : Resolving all the forces horizontally and vertically

$$\Sigma F_x = 125 \cos 60^\circ - 200 = -137.5 \text{ N}$$

$$\Sigma F_y = 125 \sin 60^\circ - 50 = 58.25 \text{ N}$$

Magnitude of the resultant force,

$$R = \sqrt{(\Sigma F_x)^2 + (\Sigma F_y)^2}$$
$$= \sqrt{(-137.5)^2 + (58.25)^2}$$
$$= 149.33 \text{ N}$$

If α be the angle which the resultant makes with the horizontal, then

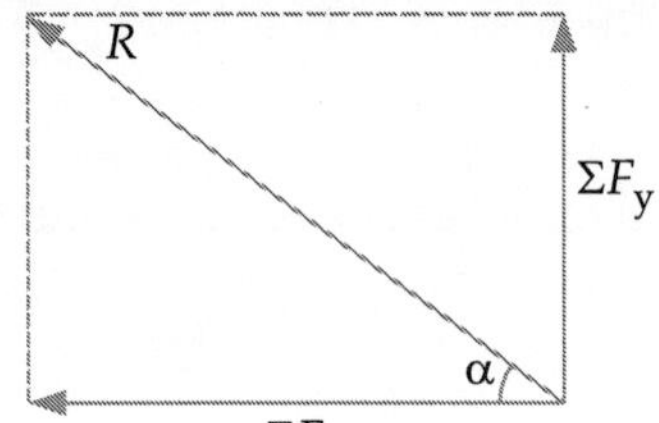

Fig. 3.55 (a)

$$\tan \alpha = \frac{\Sigma F_y}{\Sigma F_x} = \frac{58.25}{137.5} = 0.4236 \quad ; \quad \alpha = \mathbf{22.96^\circ}$$

(*b*) Taking moments of all forces about point *B* (clockwise moment + *ve*)

$\Sigma M_b = -8 - 50 \times 0.3 + 200 \times 0.2 = 17$ Nm (clockwise)

The resultant should lie below *B* and act as shown in Fig. 3.55 (*b*) so that it can produce a clockwise (positive) moment about *B*.

Let *d* be the distance of resultant from point *B*. Then,

$$R \times d = \Sigma M_b$$

$$d = \frac{\Sigma M_b}{R} = \frac{17}{149.33} = 0.1138 \text{ m}$$

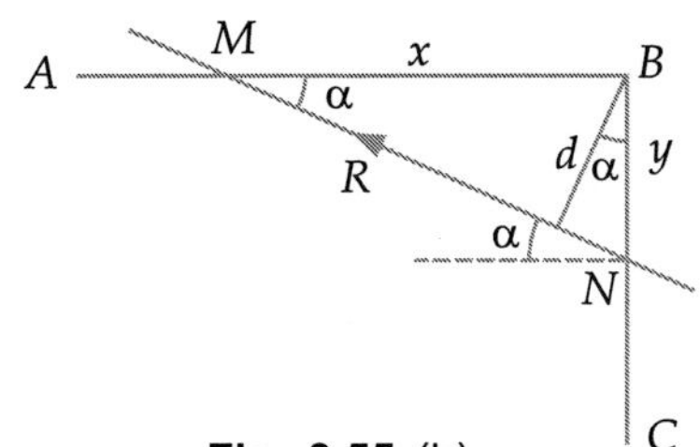

Fig. 3.55 (b)

The intercepts on *x*-axis and *y*-axis are then worked out as

$$x = BM = \frac{d}{\sin \alpha} = \frac{0.1138}{\sin 22.96^\circ} = 0.292 \text{ m}$$

$$y = BN = \frac{d}{\cos \alpha} = \frac{0.1138}{\cos 22.96^\circ} = 0.124 \text{ m}$$

The resultant thus intersects *AB* at 0.292 m from *B* and *BC* at 0.124 m from *B*.

EXAMPLE 3.32

Determine the resultant of the system of forces and a couple acting on a lamina shown in Fig. 3.56. Also locate the points of intersection of the resultant with the arms *AB*, *BC* and *CD* of the lamina.

Solution : Resolving all the forces horizontally and vertically, we get

$\Sigma F_x = 50 + 75 \sin 30^\circ = 87.5 \text{ N } (\rightarrow)$

$\Sigma F_y = -100 + 75 \cos 30^\circ = -35.05 \text{ N } (\downarrow)$

Magnitude of the resultant force,

$$R = \sqrt{(\Sigma F_x)^2 + (\Sigma F_y)^2}$$
$$= \sqrt{(87.5)^2 + (35.05)^2} = 94.26 \text{ N}$$

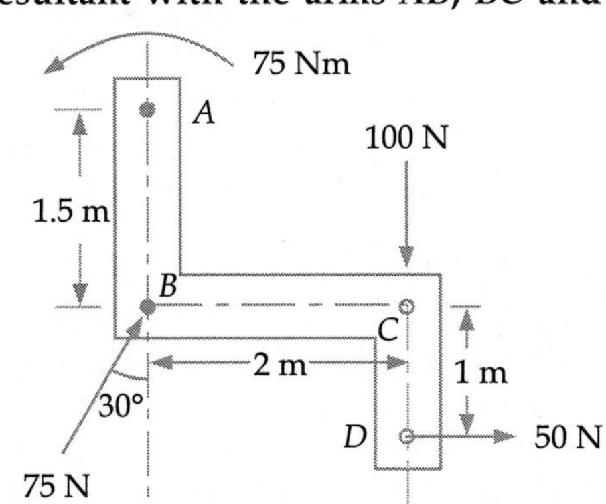

Fig. 3.56

Let α be the angle which the resultant makes with the horizontal. Then

$$\alpha = \tan^{-1}\left(\frac{\Sigma F_y}{\Sigma F_x}\right) = \tan^{-1}\left(\frac{35.05}{87.5}\right) = 21.83^\circ$$

Since ΣF_x is positive and ΣF_y is negative the resultant lies in the fourth quadrant.

(*b*) Taking moments of all forces about point *B* (clockwise + *ve*)

$$\Sigma M_b = 100 \times 2 - 50 \times 1 - 75 = 75 \text{ Nm (clockwise)}$$

To produce a clockwise moment about point *B*, the orientation of the resultant should be as shown in Fig. 3.57.

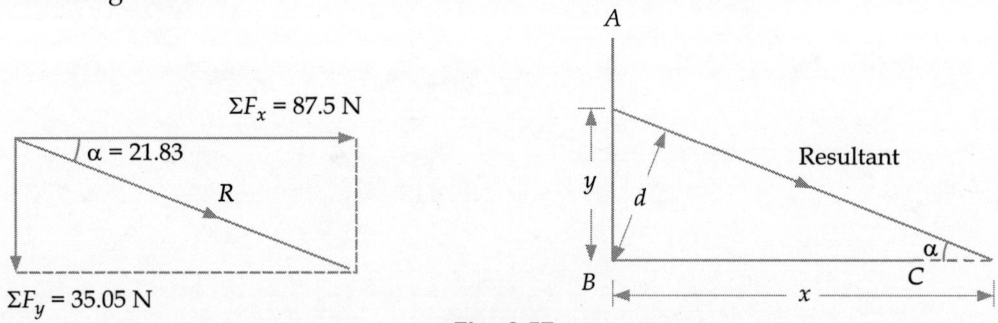

Fig. 3.57

If *d* is the perpendicular distance of line of action of resultant from *B*, then

$$R \times d = \Sigma M_b \quad ; \quad d = \frac{\Sigma M_b}{R} = \frac{75}{94.26} = \mathbf{0.796 \text{ m}}$$

The intercepts on *x*-axis and *y*-axis are then worked out as

$$x = \frac{d}{\sin\alpha} = \frac{0.796}{\sin 21.83} = \mathbf{2.141 \text{ m}}$$

Since the intercept *x* is greater than *BC*, the resultant meets the arm *BC* when produced.

$$y = \frac{d}{\cos\alpha} = \frac{0.796}{\cos 21.83} = \mathbf{0.857 \text{ m}}$$

EXAMPLE 3.33

Determine the magnitude and position of the resultant of set of forces acting on the bell crank lever as shown in Fig. 3.58.

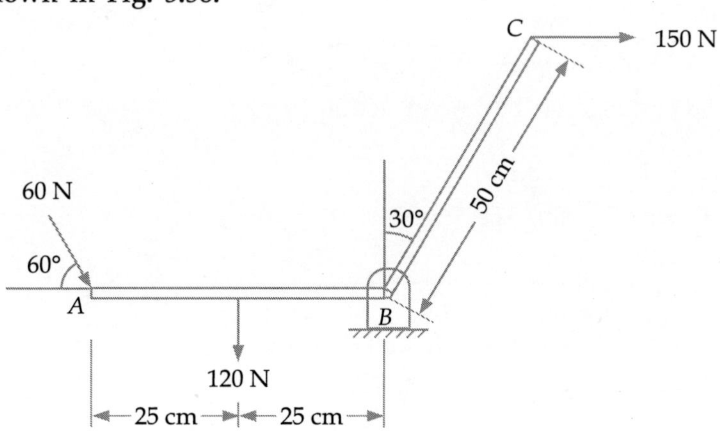

Fig. 3.58

Solution : Resolving the given forces horizontally and vertically, we get

$$\Sigma F_x = 150 + 60 \cos 60^\circ = 180 \text{ N}$$

$$\Sigma F_y = -60 \sin 60^\circ - 120 = -171.96 \text{ N}$$

$$\therefore \quad \text{Resultant } R = \sqrt{(\Sigma F_x)^2 + (\Sigma F_y)^2} = \sqrt{180^2 + (-171.96)^2} = 248.94 \text{ N}$$

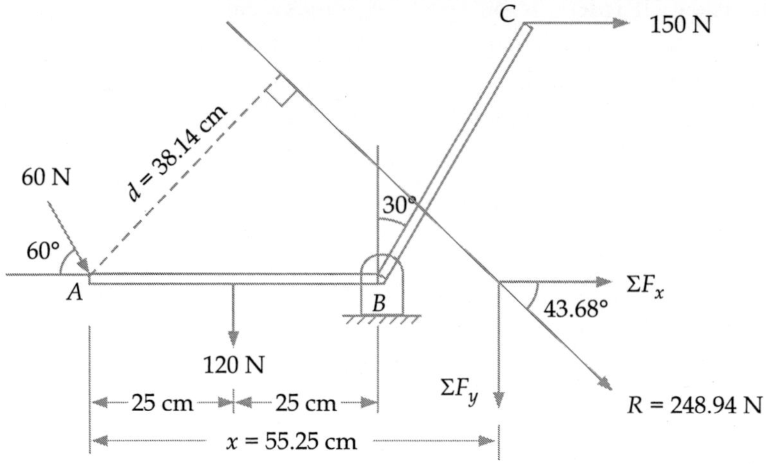

Fig. 3.59

Let α be the angle which the resultant makes with the horizontal. Then

$$\tan\alpha = \frac{\Sigma F_y}{\Sigma F_x} = \frac{-171.96}{180} = -0.955 \;; \quad \alpha = -43.68^\circ$$

Since ΣF_x is positive and ΣF_y is negative, the resultant lies in the fourth quadrant.

Taking moments of all forces about point A (clockwise positive)

$$M_a = 120 \times 25 + 150 \times (50 \cos 30^\circ)$$
$$= 3000 + 6495.2 = 9495.2 \text{ N cm}$$

The resultant R should lie as shown in Fig. 3.67, so that it can produce clockwise (positive) moment about the point A.

If d is the perpendicular distance of resultant from point A, then

$$R \times d = \Sigma M_a$$

$$\therefore \qquad d = \frac{\Sigma M_a}{R} = \frac{9495.2}{248.94} = 38.14 \text{ cm}$$

If x is the horizontal distance of the line of action from point A, then from the geometry of Fig. 3.69,

$$x = \frac{d}{\sin\alpha} = \frac{38.14}{\sin 43.68} = \mathbf{55.23\ cm}$$

EXAMPLE 3.34

For the force system acting on a body OABC as shown in Fig. 3.60, make calculations for the magnitude and direction of resultant force. Proceed to determine (*i*) the distance of resultant from point O, and (*ii*) the points where the resultant meets the x-axis and y-axis.

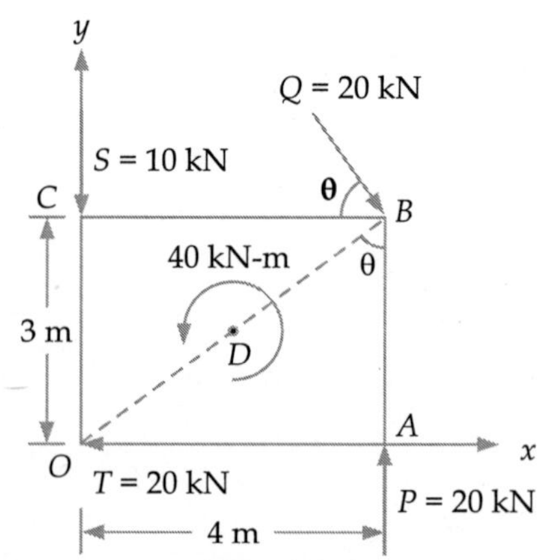

Fig. 3.60

Solution : From the geometry of the given force system

$$\theta = \tan^{-1}\left(\frac{4}{3}\right) = 53.12^\circ$$

Resolving the given forces horizontally and vertically, we get

$$\Sigma F_x = 20 \cos 53.12^\circ - 20 = -8 \text{ kN}$$

$$\Sigma F_y = -10 - 20 \sin 53.12^\circ + 20 = -6 \text{ kN}$$

$$\text{Resultant } R = \sqrt{(\Sigma F_x)^2 + (\Sigma F_y)^2} = \sqrt{(-8)^2 + (-6)^2} = \mathbf{10 \text{ kN}}$$

If α is the angle which the resultant makes with the horizontal, then

$$\tan \alpha = \frac{\Sigma F_y}{\Sigma F_x} = \frac{6}{8} = 0.75 \quad ; \quad \alpha = \mathbf{36.87^\circ}$$

Since ΣF_x and ΣF_y are both negative, the resultant lies in the third quadrant making an angle of 36.87° with the horizontal or x-axis as shown in Fig. 3.61.

Taking moment of the given system of forces about point O (clockwise positive) we have,

$$\Sigma M_o = (20 \cos 53.12^\circ) \times 3 + (20 \sin 53.12^\circ) \times 4 - 20 \times 4 - 40$$

$$= 36 + 64 - 80 - 40 = -20 \text{ kN m}$$

The resultant R should lie as shown in Fig. 3.72, so that it can produce anticlockwise (negative) moment about the point O.

If d is the perpendicular distance of line of action of resultant from point O, then

$$R \times d = \Sigma M_o ; \quad d = \frac{\Sigma M_o}{R} = \Sigma \frac{20}{10} = 2 \text{ m}$$

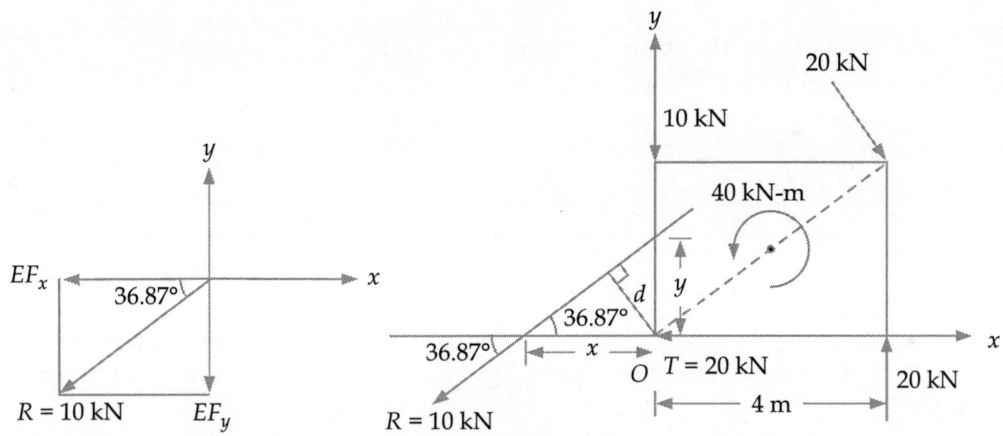

Fig. 3.61

The intercepts x on x-axis and y on y-axis are then given by (from the geometry of Fig. 3.61),

$$x = \frac{d}{\sin \alpha} = \frac{2}{\sin 36.87} = \frac{2}{0.6} = \mathbf{3.33 \text{ m}}$$

$$y = \frac{d}{\cos \alpha} = \frac{2}{\cos 36.87} = \frac{2}{0.8} = \mathbf{2.5 \text{ m}}$$

3.6. PARALLEL FORCES

Forces which donot meet at a point but lie in the same plane are called coplanar non-concurrent forces. The non-concurrent forces may be parallel or non-parallel. The lines of action of parallel forces are parallel to each other and those of non-parallel forces are not parallel. The parallel forces are classified as

- *Like parallel forces.* These forces have their lines of action parallel to each other and all of them act in the same direction [Fig. 3.62 (*a*)]. These forces may be equal or unequal in magnitude
- *Unlike Parallel forces.* These forces have their lines of action parallel to each other and all of them do not act in the same direction [Fig. 3.62 (*b*)]. These forces may be equal or unequal in magnitude.

Fig. 3.62

3.6.1. Resultant of Like Parallel Forces

With reference to Fig. 3.63, two like parallel forces act at points A and B, and their resultant R meets AB at C. The resultant R will be equal to $(P + Q)$ and parallel to P and Q.

Through C, draw MN perpendicular to P and Q. Then

$$CM = AC \cos\theta$$

and $$CN = BC \cos\theta$$

Taking moments of forces P, Q and R about C,

$$P \times CM + 0 - Q \times CN = 0$$

The resultant R passes through C, and accordingly its moment about C has been taken zero. Therefore

$$P \times CM = Q \times CN$$

or $$\frac{P}{Q} = \frac{CN}{CM} = \frac{BC\cos\theta}{AC\cos\theta}$$

$$\therefore \quad \frac{P}{Q} = \frac{BC}{AC}$$

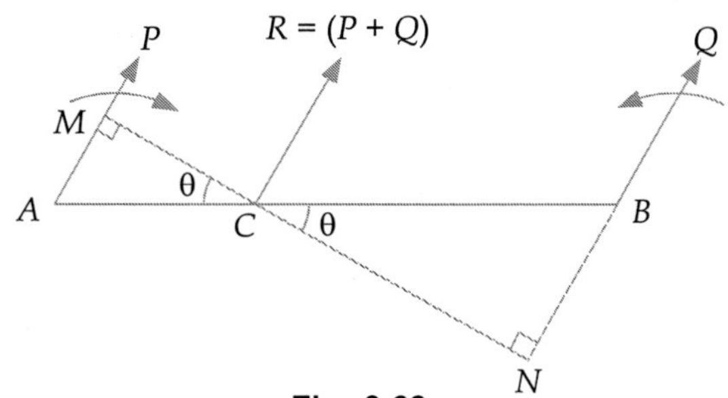

Fig. 3.63

i.e., the point of application of the resultant of two like parallel forces lies between the two forces. Further, it divides the distance between these forces internally in the inverse ratio of the magnitude of forces.

EXAMPLE 3.35

Two like parallel forces are acting at a distance of 200 mm from each other. These forces are equivalent to a single force of 250 N and its line of action is at a distance of 50 mm from one of the forces. Make calculations for the magnitude of the two forces.

Solution : With reference to Fig. 3.63, let P and Q be the two like parallel forces which act at points A and B, and let the resultant meet AB at C. Then

$$AB = 200 \text{ mm} \; ; \quad AC = 50 \text{ mm} \quad \text{and} \quad BC = 200 - 50 = 150 \text{ mm}$$

$$P + Q = R = 250 \text{ N} \qquad ...(i)$$

Applying the principle of moments and taking moments about point C,

$$P \times AC = Q \times BC$$

$$P = Q \times \frac{BC}{AC} = Q \times \frac{150}{50} = 3\,Q \qquad ...(ii)$$

From expressions (*i*) and (*ii*),

$$3Q + Q = 250\ ; \quad Q = \frac{250}{4} = \mathbf{62.5\ N}$$

$$P = 3 \times 62.5 = \mathbf{187.5\ N}$$

3.6.2. Resultant of Unlike Parallel Forces

With reference to Fig. 3.64, two unlike parallel forces act at points *A* and *B* and their resultant *R* meets *AB* at *C*. The point *C* will lie outside of *B* and on the same side as the larger force *P*. The resultant *R* will be equal to $(P - Q)$ and parallel to *P* and *Q*.

The moment of resultant *R* about *C* is zero, and as such the algebraic sum of the moments of *P* and *Q* about *C* must be zero. That implies that these moments must be equal and opposite and that is possible only if *C* lies outside *AB*.

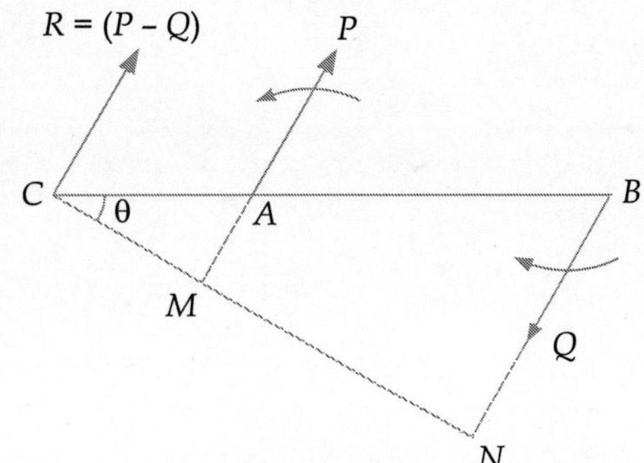

Fig. 3.64

Through *C*, draw *MN* perpendicular to *P* and *Q*. Then

$$CM = AC \cos\theta \quad \text{and} \quad CN = BC \cos\theta$$

Taking moments of forces *P*, *Q* and *R* about *C*

$$P \times CM - Q \times CN = 0$$

or $$P \times AC \cos\theta = Q \times BC \cos\theta$$

$$\therefore \quad \frac{P}{Q} = \frac{BC}{AC}$$

i.e., the point of application of the resultant of two unlike parallel forces lies outside the two forces nearer to the larger of the two forces. Further, it divides the distance between these forces externally in the inverse ratio of the magnitude of forces.

EXAMPLE 3.36

Determine the magnitude of two unlike parallel forces acting at a distance of 400 mm from each other. The resultant of these forces is 80 N and it acts at a distance of 160 mm from the greater of the two forces.

Solution : With reference to Fig. 3.64, let *P* and *Q* be the two unlike parallel forces which act at *A* and *B*, and let the resultant *R* meet *AB* externally at *C*. Then

$$AB = 400\text{ mm}\ ; \quad AC = 160\text{ mm} \quad \text{and} \quad BC = 560\text{ mm}$$

$$P - Q = R = 80\text{ N} \qquad \ldots(i)$$

Applying the principle of moments and taking moments about point C,

$$P \times AC = Q \times BC$$

$$P = Q \times \frac{BC}{AC} = Q \times \frac{560}{160} = 3.5\,Q \qquad \ldots(ii)$$

From expressions (*i*) and (*ii*),

$$3.5\,Q - Q = 80; \quad Q = \frac{80}{2.5} = \mathbf{32\ N}$$

$$P = 3.5\,Q = 3.5 \times 32 = \mathbf{112\ N}$$

3.6.3. Centre of Parallel Forces

When a number of parallel forces act at different points in a body, the magnitude of their resultant is obtained by the algebraic sum of all these parallel forces. The centre of the forces is worked out by taking moments.

Let A (x_1, y_1), B (x_2, y_2) and C (x_3, y_3) be the points on a body which are acted upon by forces P_1, P_2 and P_3 respectively.

The resultant of these forces is

$$R = P_1 + P_2 + P_3 + \ldots$$

and its point of application $(\bar{x}, \bar{y})$ is worked out by taking moments about the two axes.

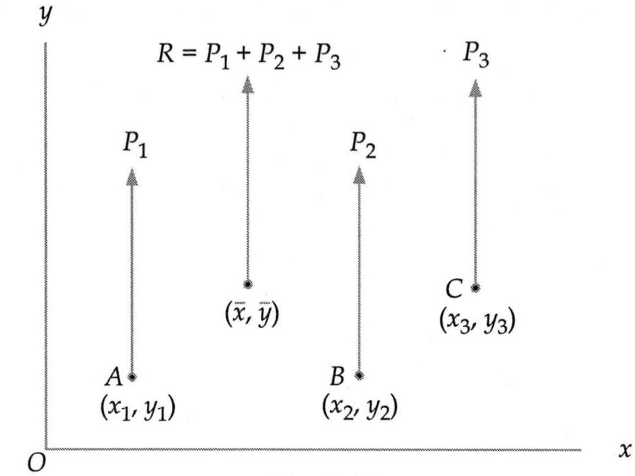

Fig. 3.65.

Talking moments about y-axis, we get

$$(P_1 + P_2 + P_3 + \ldots)\,\bar{x} = P_1 x_1 + P_2 x_2 + P_3 x_3 + \ldots$$

$$\therefore \qquad \bar{x} = \frac{P_1 x_1 + P_2 x_2 + P_3 x_3 + \ldots}{P_1 + P_2 + P_3 + \ldots} = \frac{\Sigma Px}{\Sigma P}$$

Similarly when moments are taken about x-axis, we get

$$\bar{y} = \frac{P_1 y_1 + P_2 y_2 + P_3 y_3 + \ldots}{P_1 + P_2 + P_3} = \frac{\Sigma Py}{\Sigma P}$$

EXAMPLE 3.37

A system of parallel forces acting on a lever is as shown in Fig. 3.66. Determine the magnitude, direction and position of resultant.

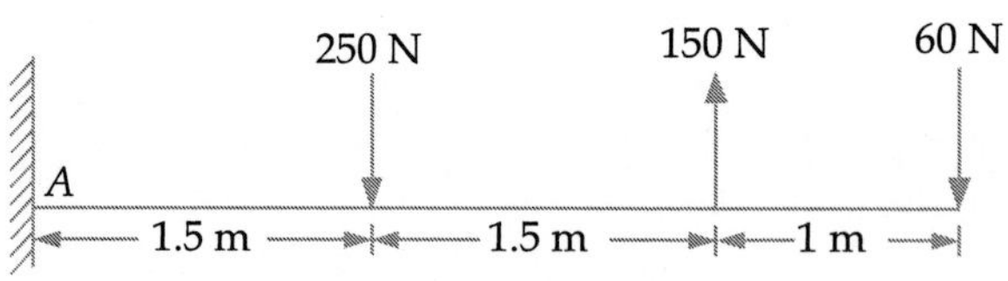

Fig. 3.66

Solution : Taking downward forces as positive, the resultant of the given force system is

$$R = 250 - 150 + 60 = \mathbf{160\ N}$$

Thus the resultant has a magnitude of 160 N and it acts vertically downwards.

Let this resultant act at a distance x from the fixed end A. Taking moments about point A (clockwise moments positive) and applying the principle of moments

moment of resultant = sum of the moments of its components

$$160x = (250 \times 1.5) - (150 \times 3) + (60 \times 4)$$

$$160x = 375 - 450 + 240 = 165$$

$$\therefore \quad x = \frac{165}{160} = \mathbf{1.031\ m}$$

Thus the resultant lies at a distance of 1.031 m from the fixed end A.

EXAMPLE 3.38

For the system of parallel forces depicted in Fig. 3.67, the resultant has a magnitude of 600 N vertically upwards and it acts through a point which is 4 m to the right of force 1500 N. Workout the values of forces *P* and *Q*.

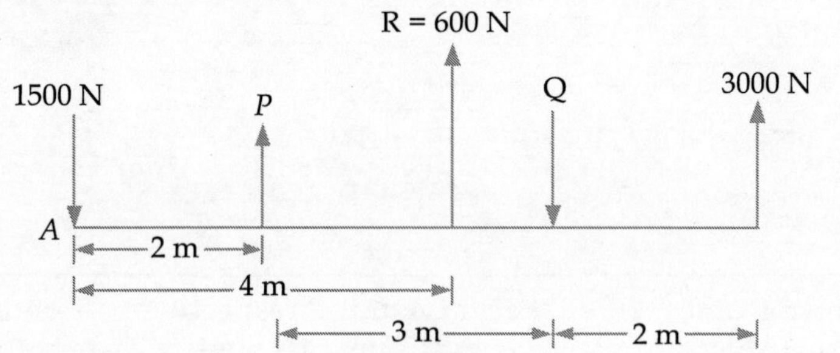

Fig. 3.67

Solution : Taking downward forces as positive, we have

$$-600 = 1500 - P + Q - 3000 = -P + Q - 1500$$

$$P - Q = -900 \quad \text{...}(i)$$

Taking moments about point A (clockwise moments positive) and applying the principle of moments,

Sum of moments of components = moment of resultant

$$-P \times 2 + Q \times 5 - 3000 \times 7 = -600 \times 4$$

$$-2P + 5Q = 21000 - 2400 = 18600 \quad \text{...}(ii)$$

Substituting the value of $Q = P + 900$ from expression (i)

$$-2P + 5(P + 9800) = 18600$$

$$3P = 18600 - 4500 = 14100$$

$$\therefore \quad P = \frac{14100}{3} = \mathbf{4700\ N}$$

$$900 = P + 900 = 4700 + 900 = \mathbf{5600\ N}$$

EXAMPLE 3.39

A coplanar parallel force system consisting of three forces acts on a rigid bar *AB* as shown in Fig. 3.68. Determine the simplest equivalent action for the force system.

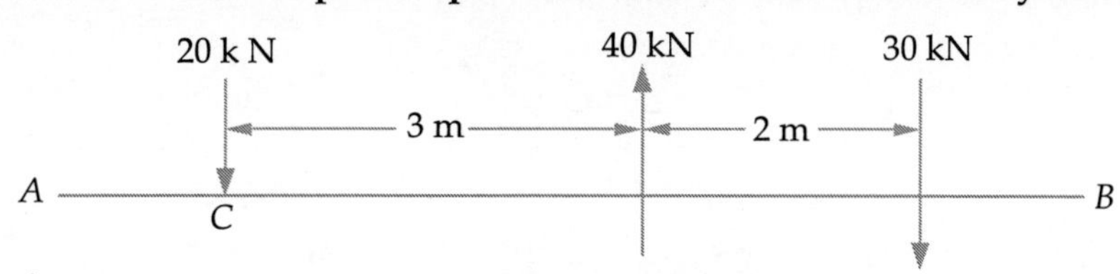

Fig. 3.68

If an additional force of 10 kN acts along the bar from *A* to *B*, what would be the simplest equivalent action?

Solution : Let R be the simplest equivalent action. Then

$$R = \Sigma F = -20 + 40 - 30 = -10 \text{ kN} \quad (\downarrow)$$

The location of R can be worked out by taking moments about point C. That gives

$$10 \times x - 40 \times 3 + 30 \times 5 = 0; \quad x = 3 \text{ m}$$

Thus the equivalent action R acts downwards along the line of action of 40 kN force.

(*b*) The resultant of 10 kN acting downwards and 10 kN force acting horizontally from A to B is

$$R_1 = \sqrt{10^2 + 10^2} = 14.14 \text{ kN}$$

This resultant acts at $\alpha = \tan^{-1}(10/10) = 45°$ below x-axis as shown in figure.

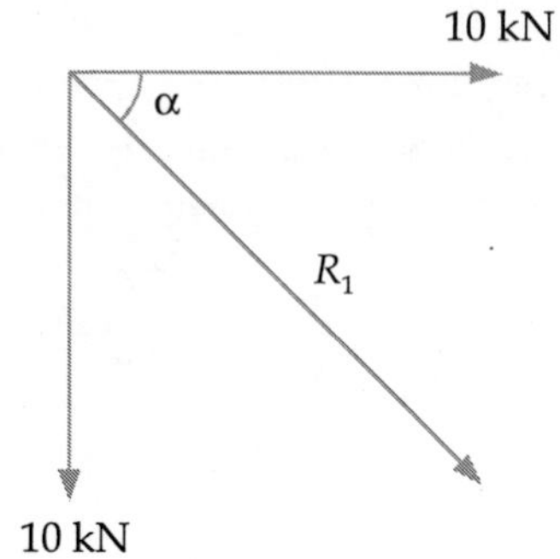

EXAMPLE 3.40

Two unlike parallel force P and Q ($P > Q$) act at A and B respectively. If both of these forces are increased by x, show that the resultant will move through a distance d given by

$$d = \frac{x}{(P-Q)} AB$$

Solution : Let R be the resultant of unlike parallel forces P and Q acting respectively at points A and B.

$$R = (P - Q)$$

Let this resultant act at distance y from point A.

Taking moments about point A, we get

$$R \times y = Q \times AB \quad ; \quad y = \frac{Q}{R} \times AB = \frac{Q}{P-Q} \times AB$$

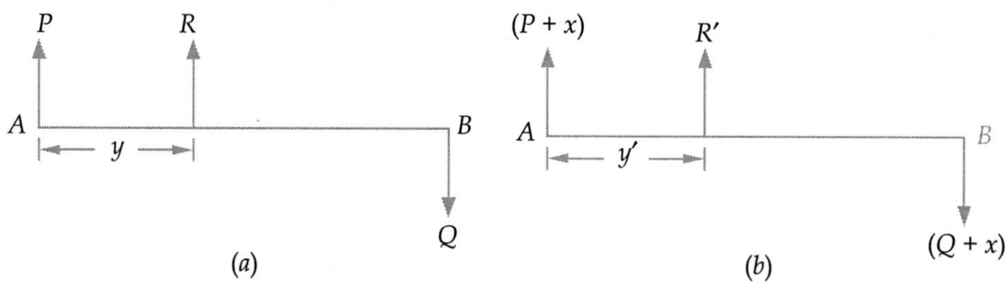

Fig. 3.69

When forces P and Q are increased by amount x, then

$$R' = (P + x) - (Q + x) = P - Q$$

and $$R' \times y' = (Q + x) \times AB \quad ; \quad y' = \frac{Q+x}{P-Q} AB$$

$\therefore$ Shift of resultant $d = y' - y$

$$= \frac{Q+x}{P-Q} AB - \frac{Q}{P-Q} AB = \frac{x}{P-Q} AB$$

EXAMPLE 3.41

The equivalent of a system of four forces is a couple of moment of magnitude 4 kNm acting in anticlockwise direction. What should be the magnitude and location of the fourth force if three of the four forces are as shown in Fig. 3.70?

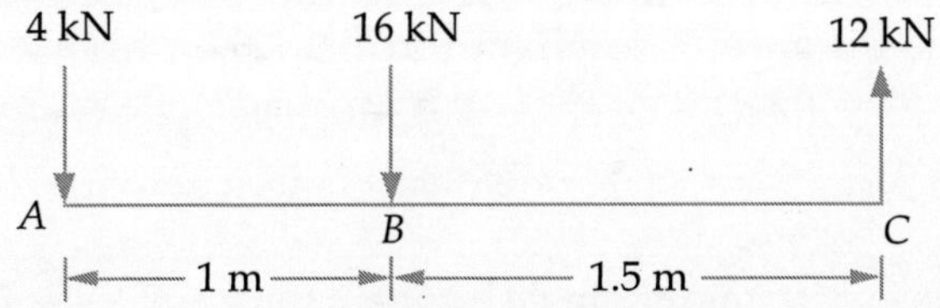

Fig. 3.70

Solution : The equivalent of the given system of forces is a couple and accordingly

(*i*) $\Sigma F_h = 0$, *i.e.*, the algebraic sum of the horizontal components of different forces is zero.

Since the given three forces are all vertical, they do not have any horizontal component. Obviously the horizontal component of the fourth force would also be zero. That implies that even the fourth force would be a vertical force.

(*ii*) $\Sigma F_v = 0$, *i.e.*, the algebraic sum of the vertical components of different forces is zero.

Let P be the fourth force acting vertically upwards. Then

$$P - 4 - 16 + 12; \quad P = 8 \text{ kN } (\uparrow)$$

Algebraic sum of the moments of given three forces about point A

$$= 12 \times 2.5 - 16 \times 1 = 14 \text{ kNm (anticlockwise)}$$

Moment of couple due to equivalent force system

$$= 4 \text{ kNm (anticlockwise)}$$

$\therefore$ Moment to be developed by the fourth force

$$= 14 - 4 = 10 \text{ kNm}$$

This moment of 10 kNm has to be clockwise so that net moment about point A is equal to 4 kNm (anticlockwise). For providing clockwise moment, the vertical force P = 8 kN should lie on left of point A at distance

$$x = \frac{10}{8} = \mathbf{1.25\ m}$$

EXAMPLE 3.42

A cantilever beam is acted upon by two vertical forces and a couple of moment 1250 Nm as shown in the figure given below:

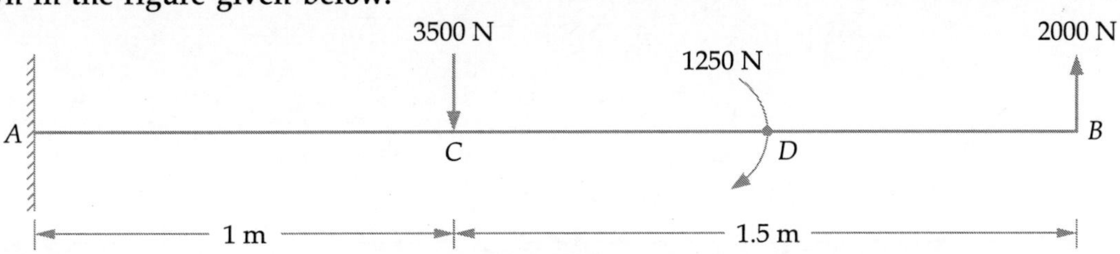

Fig. 3.71

Determine: (*i*) resultant of the system and (*ii*) and equivalent system through the fixed and *A*.

Solution : Resultant force R = 3500 – 2000 = 1500 N acting downward

Resultant moment about A (clockwise moment positive)

$= 3500 \times 1 + 1250 - 2000 \times 2.5$

$= 3500 + 1250 - 5000$

$= -250$ Nm (anticlockwise)

The resultant force acts downward and it will give the anti-clockwise moment if it acts towards left of *A*. Let *x* be the distance of resultant force (1500 N) from *A*. Then

$$1500x = 250; \quad x = \mathbf{0.166\ m}$$

Hence the resultant of the system is 1500 N downward and it acts at a distance of 0.166 m left of fixed and *A*.

(*ii*) This means that we have to determine a single resultant force and a single moment through *A*.

Single resultant force = 1500 N (downward)

Single resultant moment = 250 N (anticlockwise)

EXAMPLE 3.43

A rectangle *ABCD* has forces *P* and *Q* acting along its sides as shown in Fig. 3.72. Show that the perpendicular distance between the resultants of *P* and *Q* at *A* and that of *P* and *Q* at *C* is given by

$$= \frac{Pa - Qb}{\sqrt{P^2 + Q^2}}$$

where $a = AB = CD$ and $b = BC = DA$

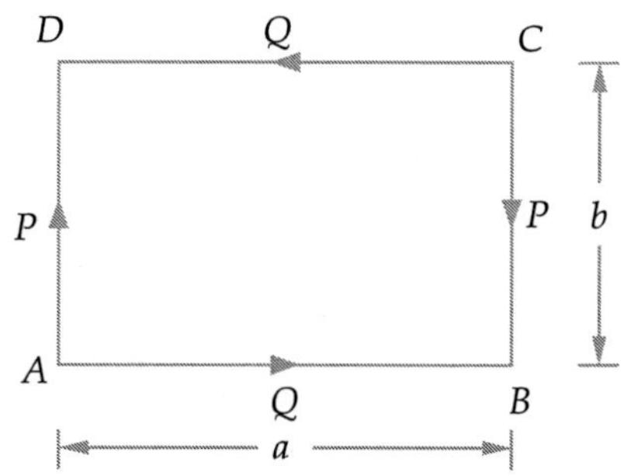

Fig. 3.72

Solution : Refer Fig. 3.72 for the system of forces and their resultants.

Net moment of the couples formed by forces *P* and *Q*

$$= P \times a - Q \times b \qquad ...(i)$$

Resultant of the forces *P* and *Q* at point *A*

$$R_1 = \sqrt{P^2 + Q^2}$$

Resultant of the forces *P* and *Q* at point *C*

$$R_2 = \sqrt{P^2 + Q^2}$$

The resultants R_1 and R_2 are equal and constitute a couple with perpendicular distance *x* between them.

Moment of the couple formed by the resultants

$$= \sqrt{P^2 + Q^2} \times x \qquad ...(ii)$$

Equating the moments given by expressions (*i*) and (*ii*),

$$\sqrt{P^2 + Q^2} \times x = Pa - Qb$$

$$\therefore \quad x = \frac{Pa - Qb}{\sqrt{P^2 + Q^2}}$$

which is the required expression.

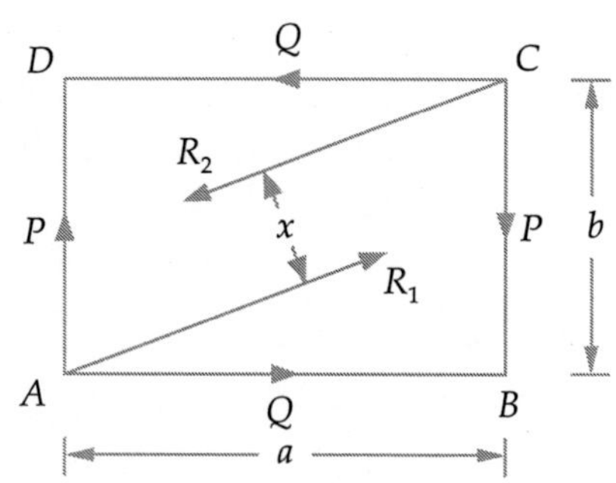

EXAMPLE 3.44

Check whether the parallel force system depicted in Fig. 3.73 is in equilibrium. If not, can this system be brought to equilibrium by applying a single force.

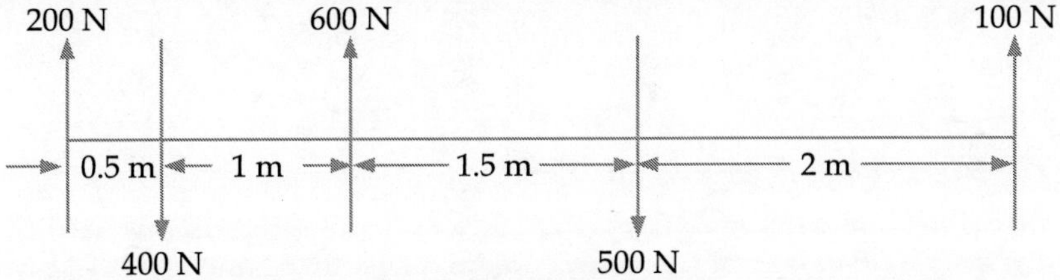

Fig. 3.73

Solution: Considering downward forces as positive, the resultant of the given force system is

$$R = -200 + 400 - 600 + 500 - 100 = 0$$

Thus the resultant force is zero.

Taking moments (clockwise moments positive) about a point on the line of action of force 200 N at the left end,

$$M = (400 \times 0.5) - (600 \times 1.5) + (500 \times 3) - (100 \times 5)$$
$$= 200 - 900 + 1500 - 500 = 300 \text{ Nm } (\circlearrowright)$$

Obviously the given force system of parallel forces reduces to a clockwise couple of 300 Nm. It may be recalled that the conditions of equilibrium of coplanar parallel system are

(*a*) $\Sigma F = 0$ and (*b*) $\Sigma M = 0$

The given system is thus not in equilibrium.

It may be further recalled that a couple cannot be balanced by a single force. Therefore, the given force system cannot be brought to equilibrium by applying a single force.

EXAMPLE 3.45

A rigid bar is subjected to a system of parallel forces as shown in Fig. 3.74.

Reduce this system to

(*a*) a single force

(*b*) a single force-moment system at *A*

(*c*) a single force moment system at *B*

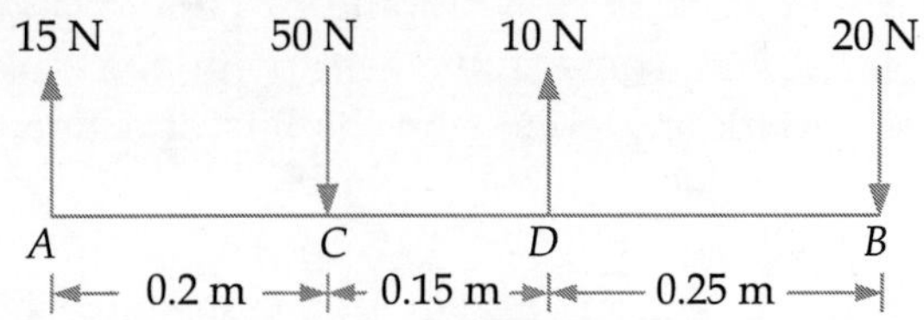

Fig. 3.74

Solution : (*a*) **A single force or resultant**

Taking downward forces as positive, the resultant of the given force system is

$$R = -15 + 50 - 10 + 20 = \mathbf{45\ N}$$

Thus the resultant has a magnitude of 45 N, its line of action is parallel to that of given forces and it acts vertically downwards.

Let this resultant act at a distance x from the end A.

Taking moments about A (clockwise moments positive) and applying the principle of moments,

Moment of the resultant = sum of the moments of its components

$$45x = (50 \times 0.2) - (10 \times 0.35) + (20 \times 0.6) = 10 - 3.5 + 12 = 18.5$$

$$\therefore \quad x = \frac{18.5}{45} = \mathbf{0.411\ m}$$

Thus the resultant lies at a distance of 0.411 from A.

(*b*) **Single force-moment at *A***

When a force of 45 N acting at E is shifted to A, it is a accompanied by a moment

$$M_a = (45 \times 0.411) = \mathbf{18.5\ Nm}$$

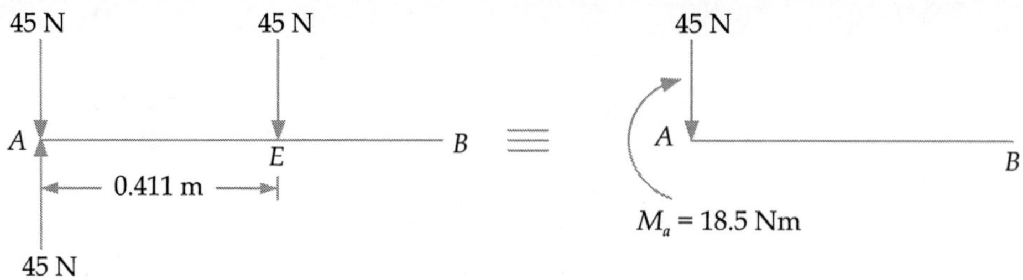

(*c*) **Single force - moment at *B***

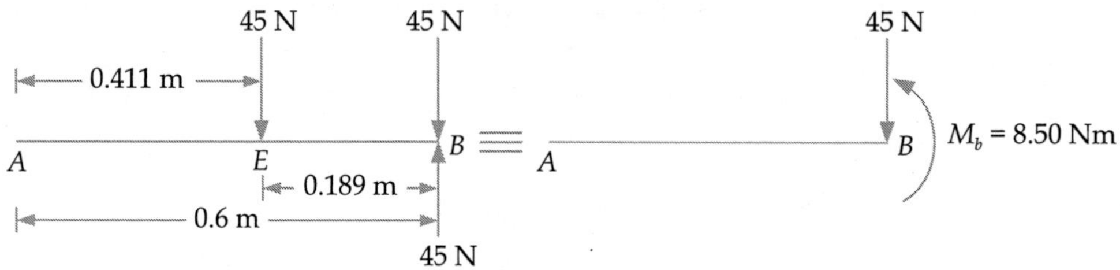

When the force of 45 N acting at *E* is moved to *B*, it is accompanied by a anticlock wise moment

$$M_b = -(45 \times 0.189) = \mathbf{-8.50\ Nm}$$

3.6.4. Graphical Method for the Resultant of Parallel Forces

Consider a body acted upon by like parallel forces P_1, P_2, P_3. Draw the relevant space diagram and let the forces be named according to Bow's notation.

The vector diagram for the force system is then drawn by following the steps outlined below:

1. Take any suitable point *a*, draw *ab* parallel and equal to force P_1 as per some suitable scale.
2. Proceed to draw *bc* and *cd* parallel and equal to force P_2 and P_3 respectively.
3. Taking some convenient point *o*, join *oa*, *ob*, *oc* and *od*.
4. Mark any point *l* on the line of action of force P_1 and through this point draw line *Rl* parallel to line *ao*.

   Subsequently,

   lm is drawn || to *ob* meeting line of action of force P_2,
   mn is drawn || to *oc* meeting line of action of force P_3
   ns is drawn || to *od*
5. Extend *Rl* and *Sn* to meet at point *k*.

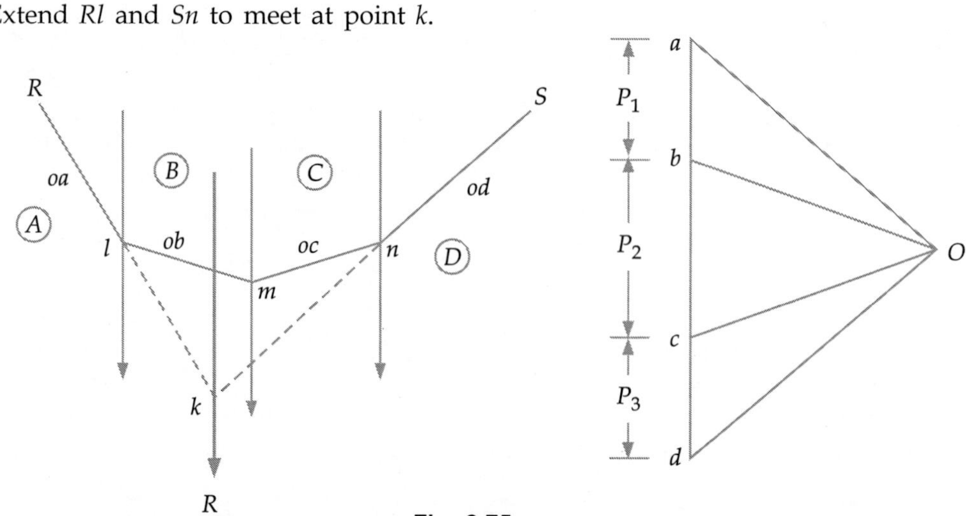

Fig. 3.75

Result : The magnitude of the resultant force is given by *ad* (as per the chosen scale) and its position is indicated by line drawn through *k* and parallel to *ad*.

While finding the resultant of unlike parallel forces, the same approach is adopted but due consideration is given to the direction of the forces while drawing the vector diagram. This is illustrated in the figure given below :

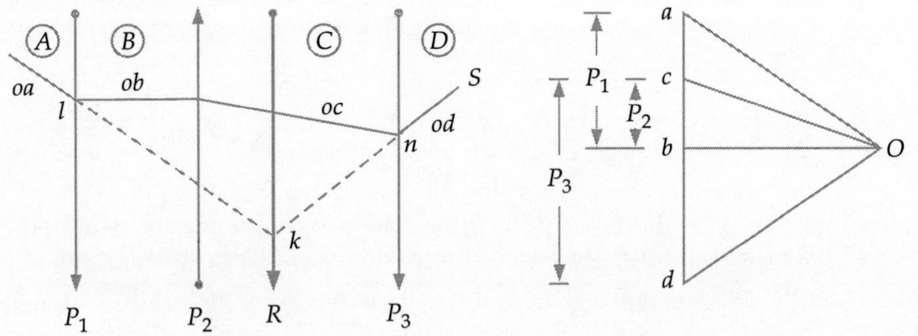

$ab = P_1$ (drawn downward), $bc = P_2$ (drawn upward)
$cd = P_3$ (drawn downward)

Fig. 3.76

3.7. COUPLE

Two parallel forces equal in magnitude but opposite in direction, and separated by a finite distance are said to form a couple.

The rotational effect of a couple is measured by its moment which is defined as the product of either of the forces and the perpendicular distance between the forces. The perpendicular distance separating the two forces is called *arm of the couple.*

With reference to Fig. 3.77, the moment of couple M is

$$M = P \times l$$

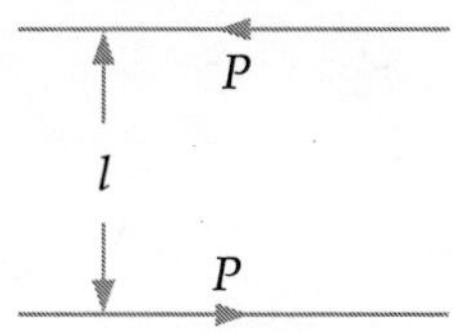

Fig. 3.77

Examples of a couple :

(*i*) Winding of a watch or a clock
(*ii*) Opening or closing a water tap
(*iii*) Unscrewing the cap of an ink bottle
(*iv*) Locking/unlocking of a lock with a key
(*v*) Turn of the cap of a pen

Properties of a couple :

(*i*) If moment is taken about any point lying in the plane of couple, then moment of the couple remains the same.

Taking moments about point O, (Fig. 3.78)

$= P \times AO - P \times BO$
$= P\ (AO - BP)$
$= P \times AB$ clockwise
$=$ moment of couple

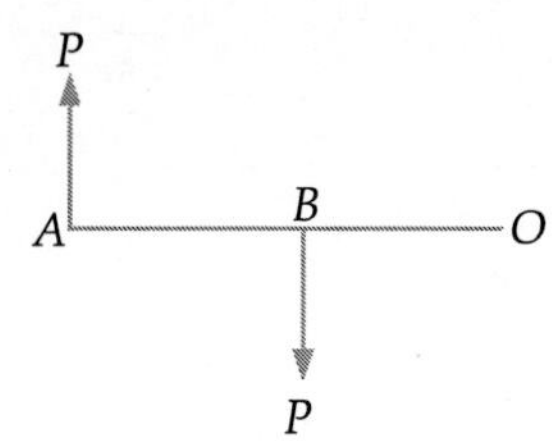

Fig. 3.78

Obviously a deduction can be made that

"the algebraic sum of the moments of two forces forming a couple about any point in their plane is constant and equal to the moment of the couple."

(*ii*) Two caplanar couples, whose moments are equal and opposite, balance each other.

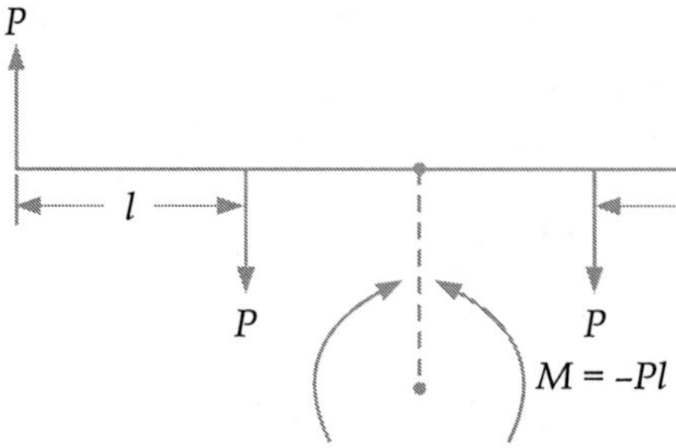

Fig. 3.79

With reference to Fig. 3.79, there act two couples whose moments are equal and opposite. These couples will balance each other and accordingly there will be no turning effect at point *O*.

(*iii*) Any two couples will be equivalent if their moments are equal, both in magnitude and direction.

Consider the system of forces depicted in Fig. 3.80

In Fig. 3.80 (*a*) : $M = P \times l = P\,l$ (anticlockwise)

In Fig. 3.80 (*b*) : $M = \dfrac{P}{2} \times 2\,l = P\,l$ (anticlockwise)

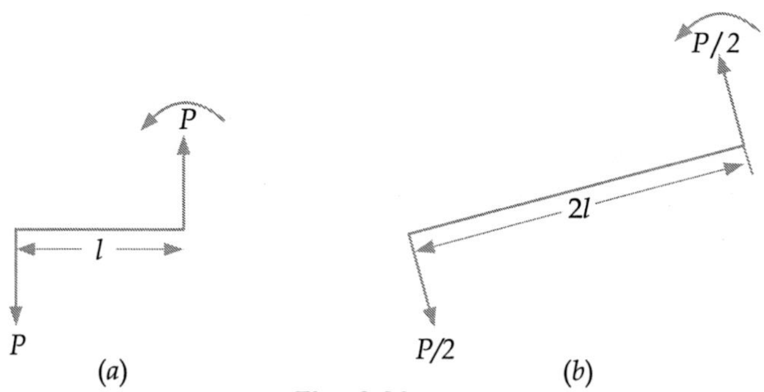

Fig. 3.80

(*iv*) Algebraic sum of moments of a number of couples is equal to the moment of a single couple.

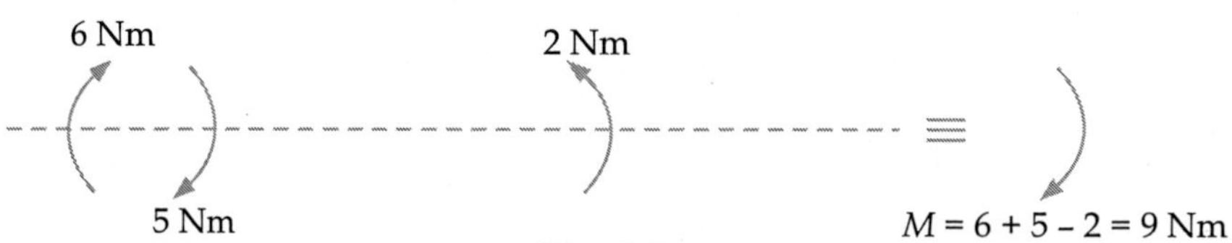

Fig. 3.81

(*v*) A single force *P* and a couple *M* acting in the same plans on a body cannot balance each other. However, they are together equivalent to a single force at a distance $e = M/P$ from its original line of action.

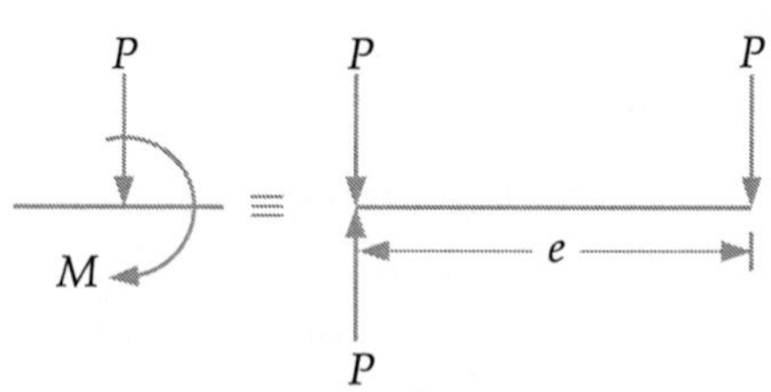

Fig. 3.82

The salient aspects of a couple may then be summarised as:

- a couple consists of a pair of equal and opposite forces separated by a definite distance.

- the algebraic sum of the vertical and horizontal components of the forces forming a couple is zero.
- a couple cannot be balanced by a single force, but can be balanced only by a couple of opposite sense.
- any number of coplanar couples can be reduced to a single couple of moment equal to the algebraic sum of the moments of all the couples.
- the translatory effect of a couple on a body is zero.
- the effect of couple on a body remains unchanged if the couple is
  (*a*) rotated through an angle,
  (*b*) shifted to any other position,
  (*c*) replaced by another pair of forces whose rotational effect is same

EXAMPLE 3.46

A rectangle *ABCD* has sides *AB* = *CD* = 80 mm and *BC* = *DA* = 60 mm. Forces of 150 N each act along *AB* and *CD*, and forces of 100 N each act along BC and DA. Make calculations for the resultant moment of the force system.

Solution : Refer to Fig. 3.83 for the arrangement of the force system.

The forces of 150 N each acting along *AB* and *CD* generate a couple of moment

$$= 150 \times 60 = 9000 \text{ Nmm}$$

The forces of 100 N each acting along *BC* and *DA* produce a couple of moment,

$$= 100 \times 80 = 8000 \text{ Nm}$$

$\therefore$ Resultant moment $= 9000 + 8000$
$= \mathbf{17000}$ Nmm (anticlockwise).

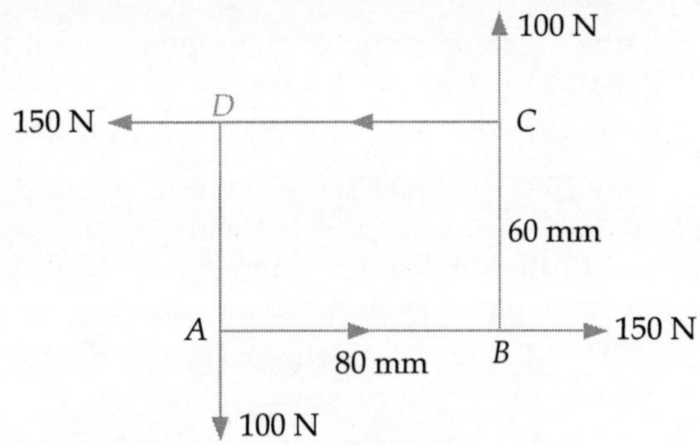

Fig. 3.83

Alternatively, the above result can be obtained by taking moments of all forces about any point, say *A*. That gives

$$M = (100 \times 80) + (150 \times 60) + (100 \times 0) - (150 \times 0)$$
$$= 8000 + 9000 = \mathbf{17000\ Nmm} \text{ (anticlockwise)}$$

EXAMPLE 3.47

A square block of each side 1.5 m is acted upon by a system of forces along its sides as shown in the adjoining figure. If the system reduces to a couple, determine the magnitude of the forces *P* and *Q*, and the couple

Solution : The system is said to reduce to a couple and as such the algebraic sum of the horizontal and vertical components of different forces is zero.

That gives :

$$\Sigma F_x = 150 - 150 \cos 45° - P = 0$$
$$\therefore \quad P = 150 - 150 \times 0.707 = \mathbf{43.95\ N}$$

Also, $\Sigma F_y = 300 - 150 \sin 45° - Q = 0$

$$\therefore \quad Q = 300 - 150 \times 0.707 = \mathbf{193.95\ N}$$

Moment of couple = algebraic sum of the moments of forces about any corner, say *A*

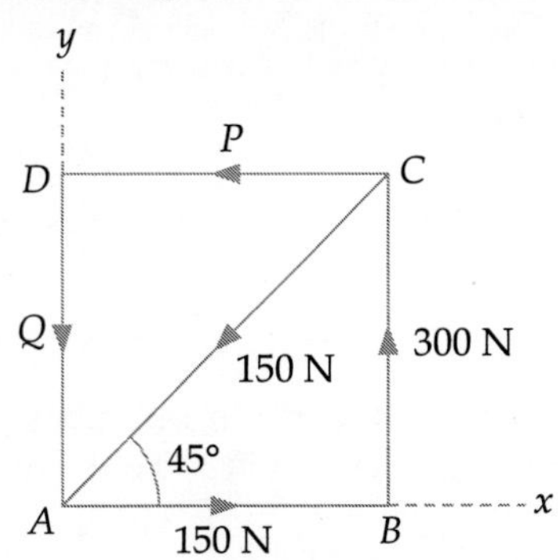

Fig. 3.84

$= -300 \times 1.5 - 43.95 \times 1.5 =$ **– 515.925 Nm**

The negative sign implies that the moment is anticlockwise.

3.8. GENERAL CONDITIONS FOR EQUILIBRIUM

A system of forces acting upon a body changes or tends to change the state of rest or of uniform motion of the body. The body may also undergo rotation if a couple also acts on the body.

Obviously, the body is likely to experience

- displacement parallel to x-axis due to component F_x of the resultant force
- displacement parallel to y-axis due to component F_y of the resultant force
- rotation due to resultant couple.

When acted upon by a system of **coplanar concurrent** forces, the body undergoes only displacement and the rotation is zero. The system is then said to be in equilibrium if there is no unbalanced force acting on it, *i.e.*, the resultant of all forces acting on the body is zero. Analytically then

(*i*) the algebraic sum of components of all forces in horizontal direction in their plane is zero, *i.e.*, $\Sigma F_x = 0$.

(*ii*) the algebraic sum of components of all forces in vertical direction in their plane is zero, *i.e.*, $\Sigma F_y = 0$.

Graphically, the force polygon, *i.e.*, the force (vector) diagram must close.

When acted upon by a system of **coplanar non-concurrent** forces, the body undergoes displacement (due to resultant force) as well as rotation due to resultant couple. The system is then said to in equilibrium if the algebraic sum of all the forces and their moments about any point in their plane is zero. Mathematically,

- $\Sigma F_x = 0$, *i.e.*, the algebraic sum of components of all the forces in horizontal direction is zero.
- $\Sigma F_y = 0$, *i.e.*, the algebraic sum of components of all the forces in vertical direction is zero.
- $\Sigma M = 0$, i.e., the algebraic sum of all the moments is zero.

The conditions $\Sigma F_x = 0$ and $\Sigma F_y = 0$ ensure that the system does not reduce to a single force and condition $\Sigma M = 0$ ensures that it does not reduce to a couple.

EXAMPLE 3.48

A 30 cm long rod rests on two pegs whose distance apart is 15 cm. Weights of 3*W* and 5*W* respectively are suspended from its ends. Determine the position of pegs if its reactions are to be equal.

Solution : Let the pegs be so located at points *E* and *F* such that

$$AE = x \text{ cm}$$

and $$BF = (15 - x) \text{ cm}$$

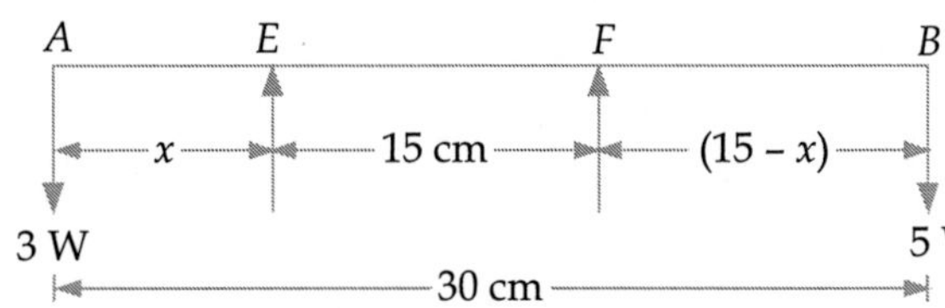

Fig. 3.85

Since the reactions at ends *E* and *F* are equal, we have

$$R_e = R_f = \frac{1}{2}(3\,W + 5\,W) = 4\,W$$

For the rod to be in a state of equilibrium, the algebraic sum of the moments of the forces about *E* must be zero. That gives:

$$-3\,Wx - 4\,W \times 15 + 5\,W(30 - x) = 0$$

The clockwise moments have been taken positive and anticlock moments negative

$$-3Wx - 60W + 150W - 5Wx = 0$$

Solving: $x = \dfrac{90}{8} = \mathbf{11.25\ cm}$

∴ $AE = 11.25$ cm

and $BF = 15 - 11.25 = \mathbf{3.75\ cm}$

EXAMPLE 3.49

A weight of 800 N is hanging from a wooden beam which is carried by two persons as indicated in the adjoining figure. Neglecting weight of the beam, determine the load shared by each person.

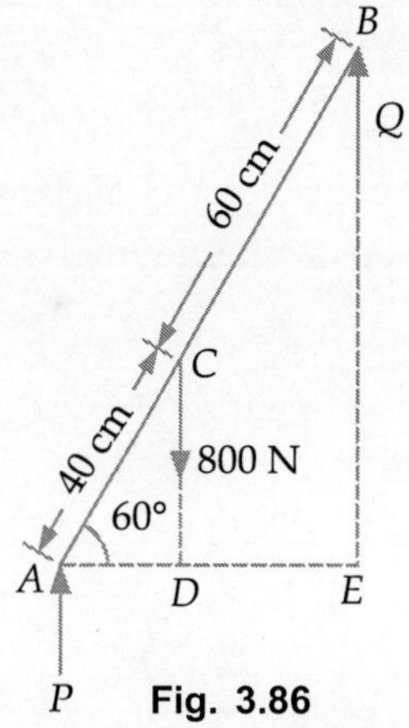

Fig. 3.86

Solution : Considering equilibrium of the beam,

$\Sigma F_y = 0;$ $P + Q - 800 = 0$

$P + Q = 800$

Taking moments about lower end A of the beam,

$$Q \times AE - W \times AD = 0$$

or $Q \times AB \cos 60° - W \times AC \cos 60° = 0$

or $Q \times (1 \times 0.5) - 800 \times (0.4 \times 0.5) = 0$

or $0.5\,Q = 160\,;\quad Q = \dfrac{160}{0.5} = \mathbf{320\ N}$

$P = 800 - Q$

$= 800 - 320 = \mathbf{480\ N}$

EXAMPLE 3.50

The lever of a lever safety valve is 100 cm long, weighs 4 N and this weight acts at its CG which is 12 cm away from the valve. The valve weighs 3 N, is 10 cm in diameter and is located at a distance of 8 cm from the fulcrum. Make calculations for the weight to be suspended at the end of the lever which will just release the steam at a pressure of 1×10^5 N/m².

Solution : Refer Fig. 3.87

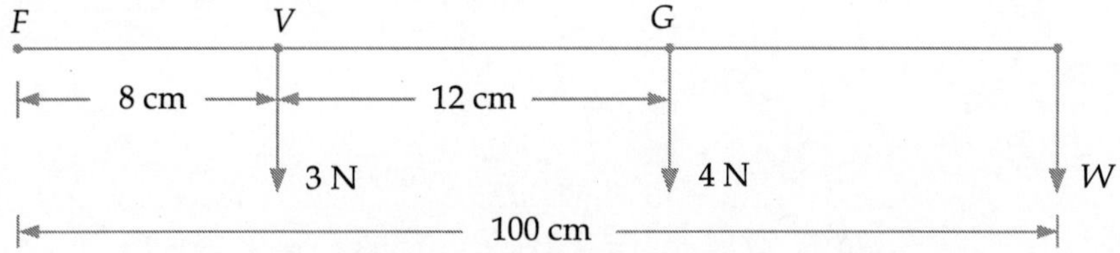

F = fulcrum; V = valve; G is position of CG

Fig. 3.87

Force on the valve, P = steam pressure × area of valve

$$= 10^5 \times \left\{\frac{\pi}{4} \times (0.1)^2\right\} = 785 \text{ N}$$

Taking moments about the fulcrum,

$$(W \times 100) + (4 \times 20) + (3 \times 8) - (785 \times 8) = 0$$

or $100\,W = (785 \times 8) - (4 \times 20) - (3 \times 8)$

$= 6280 - 80 - 24 = 6176$

∴ $W = \dfrac{6176}{100} = \mathbf{61.76}$ N

EXAMPLE 3.51

Two rods *AB* and *BC* of equal length, are freely jointed at *B*, are placed in a vertical plane with *A* and *C* joined by an inextensible string and rest on a smooth horizontal plane as shown in Fig. 3.88. The weight of rod *AB* is 2 *W* and that of *BC* is 4 *W*. If the inclination of rods at *B* is 90°, make calculations for tension in the string.

Solution : The weight of rod *AB* acts at its mid point *D* and that of rod *BC* acts through its mid point *E*.

Considering the equilibrium of the entire system

$\Sigma F_y = 0; \quad R_a + R_c = 2\,W + 4\,W = 6\,W$

$\Sigma M = 0;$ Taking moments of all forces about point *A*,

$$R_c \times AC = 2\,W \times AF + 4\,W \times AH$$

From the geometry of the arrangement

$$AC = 2\,AG = 2 \times l \cos 45^\circ = 2\,l \times 0.707 = 1.414\,l$$

$$AF = AD \cos 45^\circ = \frac{l}{2} \times 0.707 = 0.3535\,l$$

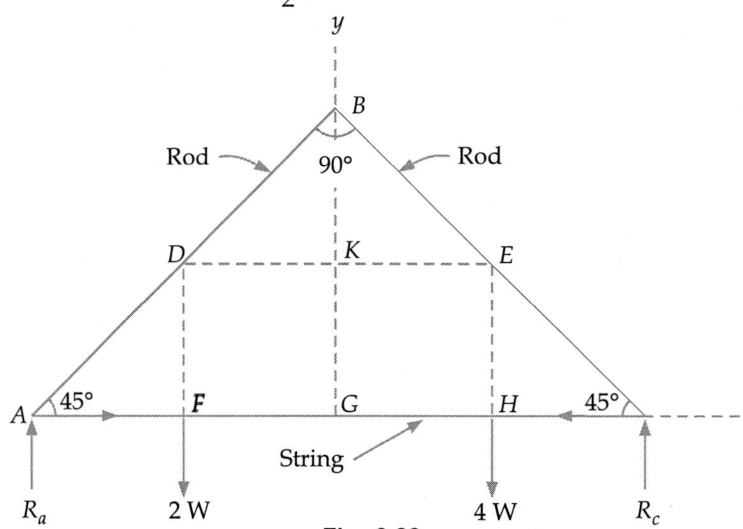

Fig. 3.88

$$AH = AC - HC = AC - AF = 1.414\,l - 0.3535\,l = 1.0605\,l$$

$\therefore \quad R_c \times 1.414\,l = 2\,W \times 0.3535\,l + 4\,W \times 1.0605\,l$

or $\quad 1.414\,R_c = 0.707\,W + 4.242\,W$

$\therefore \quad R_c = \dfrac{(0.707 + 4.242)\,W}{1.414} = \mathbf{3.5\ W}$

$$R_a = 6\,W - 3.5\,W = \mathbf{2.5\ W}$$

(*b*) Considering equilibrium of rod *BC*, take moments about end *B*

$$T \times BG + 4W \times EK - R_C \times GC = 0$$

or $\quad T \times AB \sin 45^\circ + 4W \times BE \sin 45^\circ - R_C \times BC \sin 45^\circ = 0$

or $\quad T \times AB + 4W \times BE - RC \times BC = 0$

or $\quad T \times l + 4W \times \dfrac{l}{2} - 3.5W \times l = 0$

$\therefore \quad T = 3.5W - 2W = \mathbf{1.5\ W}$

EXAMPLE 3.52

A smooth uniform ladder of length 2 a and weight W rests on the ground and against a vertical wall. To prevent its slipping, a string of length l is tied to its lower end and to the junction of the wall with the ground.

How the value of this tension would be affected if a man whose weight is twice that of ladder has ascended the ladder three fourth of the way up.

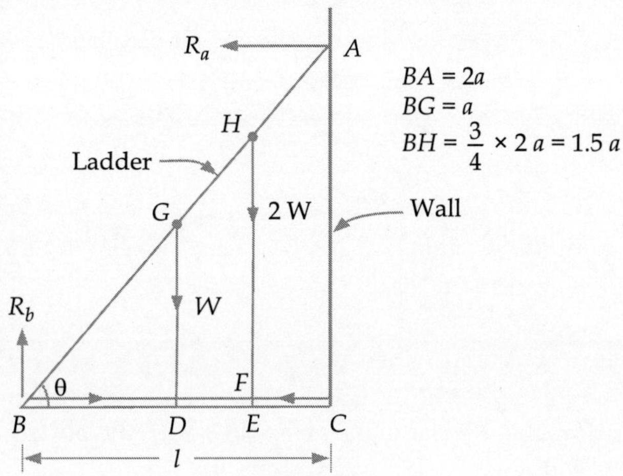

Fig. 3.89

Solution : Refer Fig. 3.89 for the ladder leaning against a wall. Let R_a and R_b be the reactions at ends A and B, and T be the tension in the string tied at B and C.

For equilibrium of the ladder,

$$\Sigma F_x = 0; \quad R_a = T$$
$$\Sigma F_y = 0; \quad R_b = W$$
$$\Sigma M = 0$$

Taking moments about point B,

$$R_a \times AC = W \times BD$$

or $$R_a \times 2a \sin\theta = W \times a \cos\theta \qquad \Bigg| \quad \cos\theta = \frac{l}{2a}$$

or $$T \times 2a \times \frac{\sqrt{4a^2 - l^2}}{2a} = Wa\,\frac{l}{2a} \qquad \Bigg| \quad \sin\theta = \sqrt{1 - \cos^2\theta}$$

∴ $$T = \frac{Wl}{2\sqrt{4a^2 - l^2}} \qquad \Bigg| \quad = \sqrt{1 - \frac{l^2}{4a^2}} = \frac{\sqrt{4a^2 - l^2}}{2a}$$

(b) When a man with weight 2 W stands at H located at distance

$$AH = \frac{3}{4} \times 2a = 1.5a$$

For equilibrium of the system

$$\Sigma F_x = 0; \quad R_a = T$$
$$\Sigma F_y = 0; \quad R_b = W + 2W = 3W$$
$$\Sigma M = 0; \quad \text{Taking moments about point } B$$
$$R_a \times AC = W \times BD + 2W \times BE$$

or $$R_a \times 2a \sin\theta = W \times a\cos\theta + 2W \times 1.5a\cos\theta = 4Wa\cos\theta$$

or $$T \times 2a\,\frac{\sqrt{4a^2 - l^2}}{2a} = 4Wa \times \frac{l}{2a} = 2Wl$$

∴ $$T = \frac{2Wl}{\sqrt{4a^2 - l^2}}$$

EXAMPLE 3.53

A heavy uniform bar AB rests in equilibrium with end A in contact with a smooth vertical wall and a point D in contact with another smooth vertical as shown in Fig. 3.90. Determine the inclination of the bar to the horizontal.

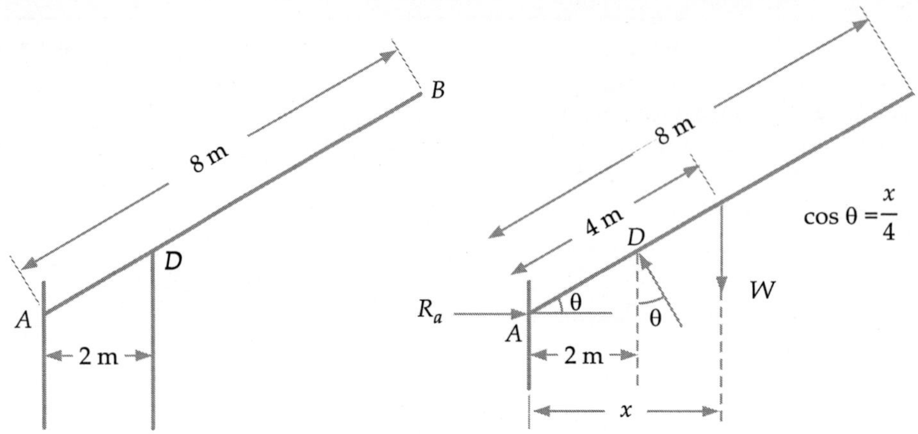

Fig. 3.90

Solution : Let the bar make an angle θ with the horizontal. At D, a reaction R_d acts perpendicular to the bar

Considering equilibrium of the bar,

$$\Sigma F_x = 0; \quad R_a - R_d \sin\theta = 0; \quad R_a = R_d \sin\theta \qquad ...(i)$$

$$\Sigma F_y = 0; \quad R_d \cos\theta - W = 0; \quad W = R_d \cos\theta \qquad ...(ii)$$

$\Sigma M = 0$, Taking moments about point A

$$Wx = R_d \times AD = \frac{W}{\cos\theta} \times \frac{2}{\cos\theta} = \frac{2W}{\cos^2\theta} = \frac{2W \times 16}{x^2}$$

$$\therefore \quad x^3 = 32; \quad x = 3.174 \text{ m}$$

$$\cos\theta = \frac{3.174}{4} = 0.7935$$

$$\theta = \mathbf{37° \, 28'}$$

EXAMPLE 3.54

Two identical uniform bars AC and BC each of length l and weight W are hinged together at C and supported in a vertical plane by two legs at D and E as shown in Fig. 3.91. Neglecting friction, find the angle θ that each bar will make with the horizontal in the condition of equlibrium.

Solution : Let θ be angle which each bar should make with the horizontal.

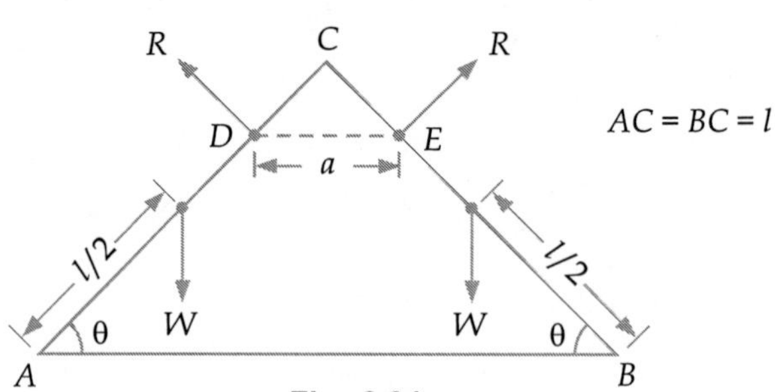

Fig. 3.91

For equilibrium of the whole system,

$$\Sigma F_y = 0; \qquad 2W = 2R\cos\theta$$

$$R = \frac{W}{\cos\theta} \qquad ...(i)$$

For equilibrium of bar BC, $\Sigma M = 0$

Taking moments about point C,

$$R \times EC = W \times \frac{l}{2}\cos\theta$$

or

$$R \times \frac{a}{2\cos\theta} = W \times \frac{l}{2}\cos\theta$$

or $$R = \frac{Wl}{a}\cos^2\theta \qquad ...(ii)$$

Equating identities (*i*) and (*ii*), we get

$$\frac{Wl}{a}\cos^2\theta = \frac{W}{\cos\theta}$$

$$\cos^3\theta = \frac{a}{l}; \quad \cos\theta = (a/l)^{1/3}$$

$$\therefore \qquad \theta = \cos^{-1}(a/l)^{1/3}$$

EXAMPLE 3.55

The cable and the rigid bar shown in Fig. 3.92 support a load of 1000 N. Determine the tensions in the cable and the reaction at the hinge *A*.

Solution : Refer Fig. 3.93 for the free body diagram. The various forces acting in the rigid bar are:

(*i*) weight W = 1000 N and the tension in the cable at *B*.

(*ii*) reactive forces R_{ah} and R_{av} at the hinged end *A*.

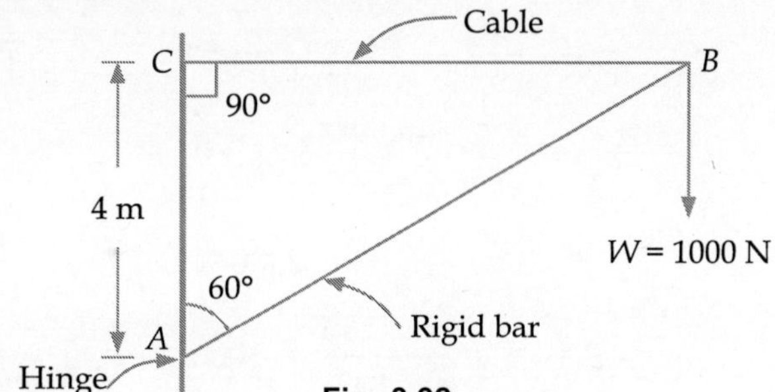

Fig. 3.92

Since the bar is in equilibrium, the force system (T, W, R_{ah} and R_{av}) acting on the bar satisfy the three conditions of equilibrium. That is

$$\Sigma M = 0; \quad \Sigma F_x = 0 \quad \text{and} \quad \Sigma F_y = 0$$

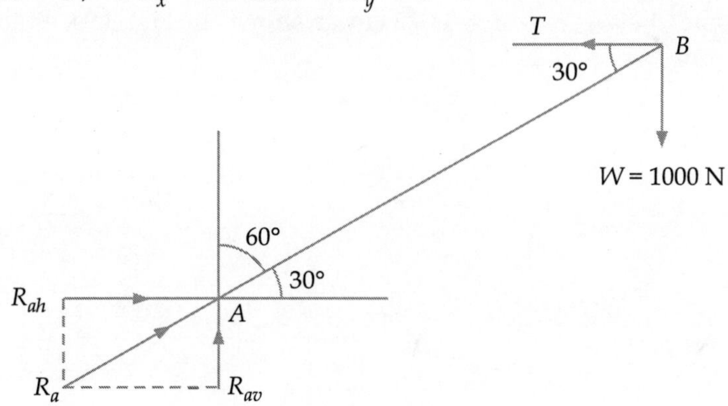

Fig. 3.93

Taking moment about the hinge A and equating to zero,

$$1000 \times BC - T \times AC = 0$$

$$T = 1000 \times \frac{BC}{AC}$$

From the geometrical configuration,

$$BC = AC \tan 60^\circ = 4 \tan 60^\circ = 6.928 \text{ m}$$

$\therefore$ Tension in the cable, $T = 1000 \times \dfrac{6.928}{4} = \mathbf{1732\ N}$

$$\Sigma F_x = 0; \quad R_{ah} - T = 0; \quad R_{ah} = T$$

$$\Sigma F_y = 0; \quad W - R_{av} = 0; \quad R_{av} = W = 1000 \text{ N}$$

∴ Total reaction at the hinge,

$$R_a = \sqrt{R_{ah}^2 + R_{av}^2} = \sqrt{1732^2 + 1000^2} = \mathbf{2000\ N}$$

$$\theta = \tan^{-1}\frac{R_{av}}{R_{ah}} = \tan^{-1}\frac{1000}{1732} = \mathbf{30°}$$

where θ is the angle which the reaction at A makes with the x-axis.

Alternatively :

Since the bar is in equilibrium, the three forces W, T and R_a would be concurrent and Lami's theorem would apply.

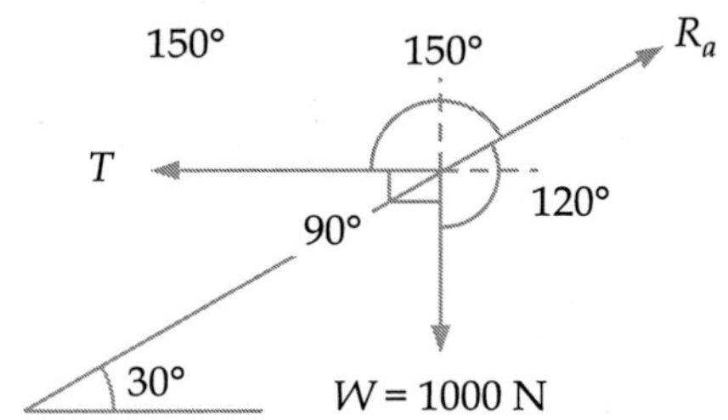

That gives

$$\frac{W}{\sin 150°} = \frac{R_a}{\sin 90°} = \frac{T}{\sin 120°}$$

$$\therefore \quad R_a = W \times \frac{\sin 90°}{\sin 150°}$$

$$= 1000 \times \frac{1}{0.5} = \mathbf{2000\ N}$$

$$T = W \times \frac{\sin 120°}{\sin 150°} = 1000 \times \frac{0.866}{0.5} = \mathbf{1732\ N}$$

EXAMPLE 3.56

Two smooth spheres, each of radius 20 cm and weight 200 N, rest in a horizontal channel having vertical walls, the distance between which is 72 cm as shown in Fig. 3.94. Find the pressures at contact points *E*, *F* and *G*.

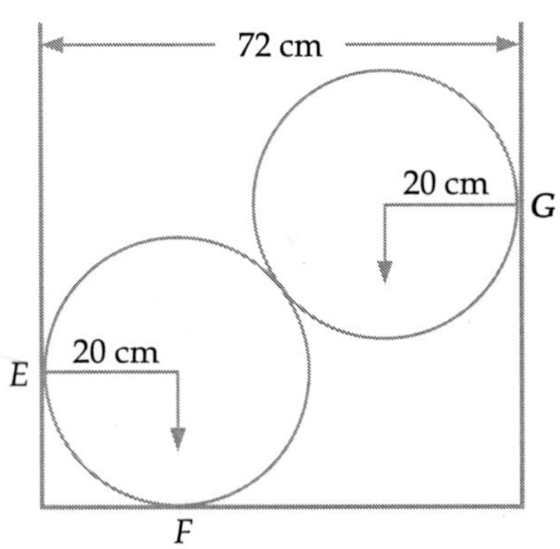

Fig. 3.94

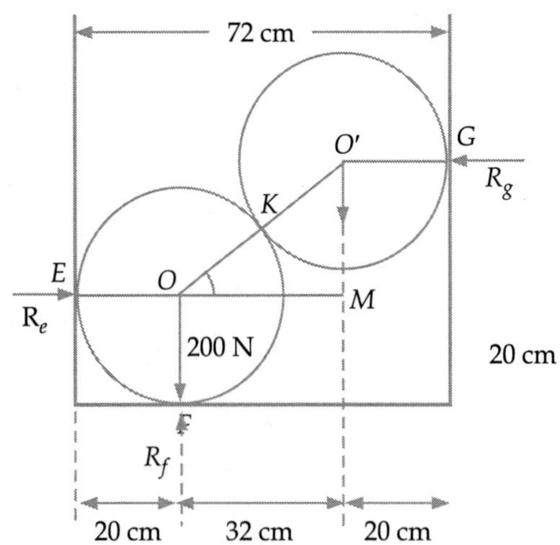

Fig. 3.95

Solution : Let R_e, R_f and R_g be the reactions at surface points E, F and G respectively.

From the geometrical configuration (Fig. 3.95)

$$OO' = 20 + 20 = 40 \text{ cm}$$

$$OM = 32 \text{ cm}$$

$$\cos\theta = \frac{32}{40} = 0.8\ ;\ \theta = 36.87^\circ\ ;\ \sin\theta = 0.6$$

The upper sphere is in equilibrium under the action of forces R_g, R_k and its weight: For that

$$\Sigma F_x = 0;\quad R_k \cos\theta - R_g = 0$$

$$\Sigma F_y = 0;\quad R_k \sin\theta - 200 = 0$$

$$\therefore\quad R_k = \frac{200}{\sin\theta} = \frac{200}{0.6} = \mathbf{333.33\ N}$$

$$R_g = R_k \cos\theta = 333.33 \times 0.8 = \mathbf{266.66\ N}$$

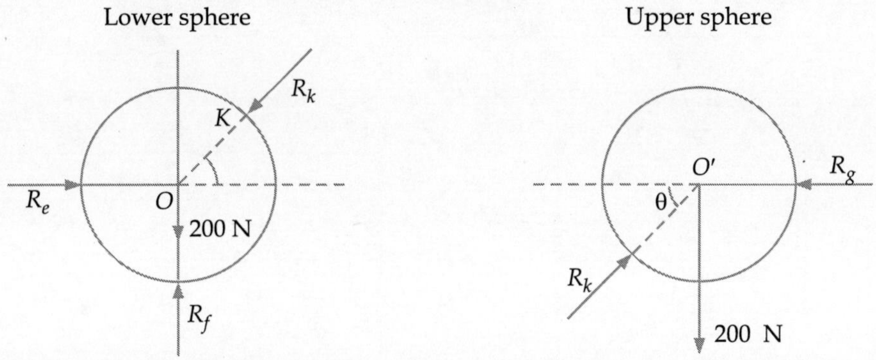

The lower sphere is in equilibrium under the action of R_e, R_f, R_k and its weight. For that

$$\Sigma F_x = 0;\quad R_k \cos\theta - R_e = 0$$

$$\Sigma F_y = 0;\quad R_k \sin\theta + 200 - R_f = 0$$

$$\therefore\quad R_e = R_k \cos\theta = 333.33 \times 0.8 = \mathbf{266.66\ N}$$

$$R_f = R_k \sin\theta + 200 = 333.33 \times 0.6 + 200 = 200 + 200 = \mathbf{400\ N}$$

EXAMPLE 3.57

Three cylinders are piled in a rectangular ditch as shown in Fig. 3.96. Neglecting friction, determine the reaction between cylinder A and the vertical wall.

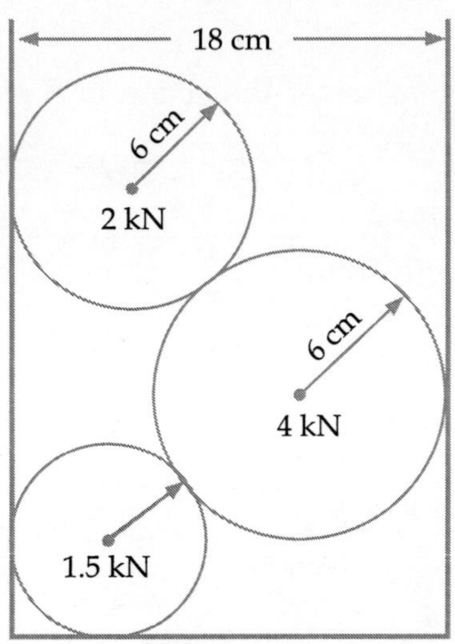

Fig. 3.96

Solution : (*i*) The cylinder *C* is in equilibrium under the action of reactions R_1 and R_2 and its weight (Fig. 3.97).

In Δ *CBD*:

$$BC = 6 + 5 = 11 \text{ cm}$$

$$BD = 18 - 5 - 6 = 7 \text{ cm}$$

$$\cos\theta = \frac{BD}{BC} = \frac{7}{11} = 0.636\ ; \quad \theta = 50.48^\circ$$

For equilibrium of cylinder C,

$$\Sigma F_y = 0; \quad R_2 \sin\theta - 2 = 0$$

$$\therefore \qquad R_2 = \frac{2}{\sin\theta} = \frac{2}{\sin 50.48^\circ}$$

$$= \frac{2}{0.7714} = \mathbf{2.593\ kN}$$

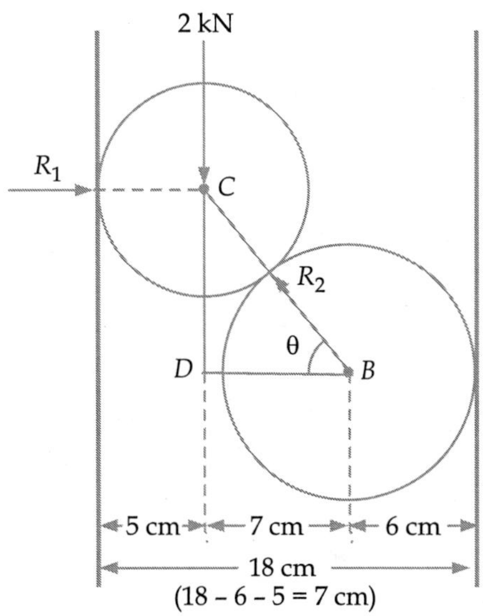

Fig. 3.97

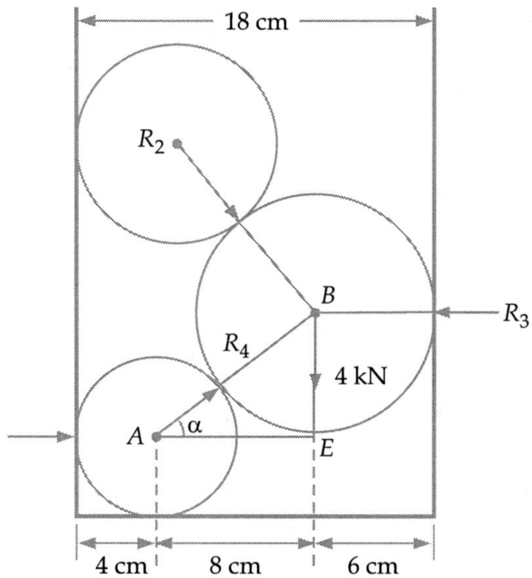

Fig. 3.98

(*ii*) The cylinder *B* is in equilibrium under the action of forces R_2, R_3, R_4 and its weight (Fig. 3.98)

In Δ *ABE* :

$$AB = 6 + 4 = 10 \text{ cm}$$

$$AE = 18 - 6 - 4 = 8 \text{ cm}$$

$$\cos\alpha = \frac{AE}{AB} = \frac{8}{10} = 0.8\ ; \alpha = 36.86^\circ$$

For equlibrium of cylinder *B*,

$$\Sigma F_y = 0; \ R_2 \sin\theta + 4 - R_4 \sin\alpha = 0$$

or $$2.593 \sin 50.48^\circ + 4 - R_4 \sin 36.86^\circ = 0$$

or $$2.593 \times 0.7714 + 4 - R_4 \times 0.6 = 0$$

$$\therefore \qquad R_4 = \frac{2.593 \times 0.7714 + 4}{0.6} = \frac{2 + 4}{0.6} = \mathbf{10\ kN}$$

(*iii*) The cylinder A is in equilibrium under the action of forces R_4, R_5, R_6 and its weight Applying the condition $\Sigma F_x = 0$, we have

$$R_6 - R_4 \cos \alpha = 0$$
$$R_6 = R_4 \cos \alpha = 10 \times \cos 36.86 = \mathbf{8\ kN}$$

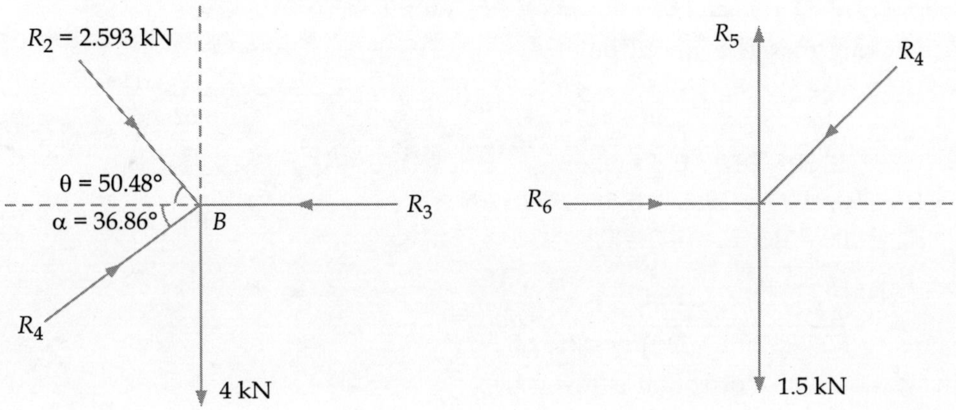

EXAMPLE 3.58

A chain of weight 200 N is suspended from its two ends on a wall in such a way that its ends make angles of 30° and 45° with the horizontal. Determine the tensions produced in the ends of the chain.

Solution : Refer to Fig. 3.99 which shows the chain AB suspended from its ends A and B. The tangents drawn to the curve at A and B have an inclination of 30° and 45° with the horizontal. Since the system is in equilibrium, these tangents (representing tensions T_1 and T_2) meet at point O where weight of the chain also acts.

Resolving all forces horizontally at O,

$$\Sigma F_x = T_1 \cos 30^\circ - T_2 \cos 45^\circ = 0$$
$$T_1 \times 0.866 - T_2 \times 0.707 = 0$$
$$\therefore \quad T_2 = T_1 \times \frac{0.866}{0.707} = 1.225\ T_1 \quad ...(i)$$

Resolving all forces vertically at O,

$$\Sigma F_y = T_1 \sin 30^\circ + T_2 \sin 45^\circ - 200 = 0$$
$$T_1 \times 0.5 + T_2 \times 0.707 = 200 \quad ...(ii)$$

From (*i*) and (*ii*),

$$0.5\ T_1 + 0.707 \times 1.225\ T_1 = 200$$
$$0.5\ T_1 + 0.866\ T_1 = 200$$
$$\therefore \quad T_1 = \frac{200}{0.5 + 0.866} = \mathbf{146.41\ N}$$
$$T_2 = 1.225\ T_1 = 1.225 \times 146.41 = \mathbf{179.35\ N}$$

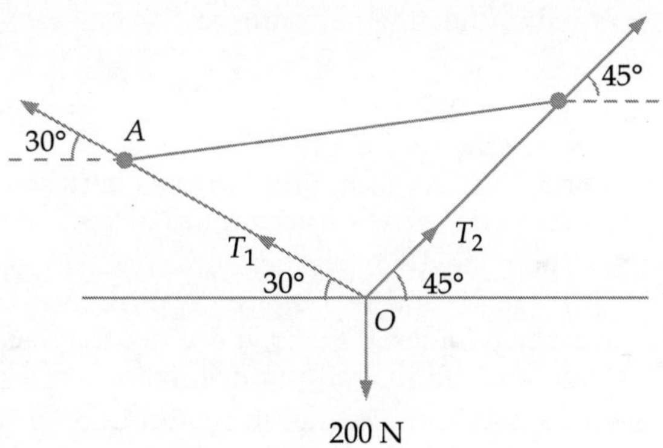

Fig. 3.99

EXAMPLE 3.59

The rail AB of a foundry crane is horizontal and is 20 m long. End A is hinged to vertical wall and end B is tied to a tie rod BC making an inclination of 30° with the rail. The other end C of tie rod is connected to wall. If a load of 400 N is placed on the rail at a distance of 8 m from end A, find the tension produced in tie rod and reaction developed at end A.

Solution : Let T be tension produced in the tie rod, and R_x and R_y be the horizontal and vertical components of reactions at end A. (Fig. 3.100)

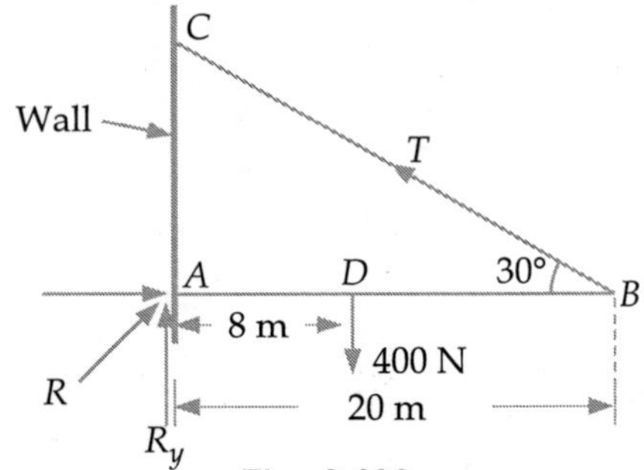

Fig. 3.100

Under equilibrium conditions,

$$\Sigma F_x = 0; \quad R_x - T\cos 30^\circ = 0; \quad R_x = 0.866\,T$$

$$\Sigma F_y = 0; \quad R_y - 400 + T\sin 30^\circ = 0; \quad R_y = 400 - 0.5\,T$$

$\Sigma M_a = 0$; Taking moments about point A,

$$T\sin 30^\circ \times 20 = 400 \times 8$$

$$\therefore \quad T = \frac{400\times 8}{\sin 30\times 20} = \frac{400\times 8}{0.5\times 20} = \mathbf{320\ N}$$

$$R_x = 0.866\,T = 0.866 \times 320 = \mathbf{277.12\ N}$$

$$R_y = 400 - 0.5\,T = 400 - 0.5 \times 320 = \mathbf{240\ N}$$

$$\therefore \text{ Reaction, } R = \sqrt{R_x^2 + R_y^2} = \sqrt{(277.12)^2 + (240)^2} = \mathbf{366.60\ N}$$

and its inclination with horizontal is given by

$$\tan\theta = \frac{R_y}{R_x} = \frac{240}{277.12} = 0.86\,; \quad \theta = \mathbf{40.69^\circ}$$

REVIEW AND SUMMARY

1. The moment of force about a point is defined as the product of force and the perpendicular distance of the line of action of the force from that point. The moment of force measures the tendency of the force to rotate the rigid body about an axis.
2. **Varignon's theorem :** The moment of the resultant of a number of forces about any point is equal to the algebraic sum of the moments of all the forces of the system about the same point.
3. The lines of action of parallel forces are parallel to each other.
   (*a*) When the two parallel forces act in the same direction, their resultant equals their sum and its point of application lies between the two forces. Further, it divides the distance between the two forces internally in the inverse ratio of the magnitude of forces.
   (*b*) When the two parallel forces act in opposite directions, their resultant equals their difference and its point of application lies outside the two forces nearer to the larger of the two forces. Further, it divides the distance between the forces externally in the inverse ratio of the magnitude of forces.
4. Two parallel forces equal in magnitude but opposite in direction, and separated by a finite distance are said to form a couple.
   The rotational effect of a couple is measured by its moment which equals the product of either of the two forces and the perpendicular distance between the forces.
5. The salient aspects of a couple are :
   - a couple consists of a pair of equal and opposite forces separated by a finite distance
   - the algebraic sum of the vertical and horizontal components of forces forming a couple is zero
   - a couple can not be balanced by a single force, but can be balanced only by a couple of opposite sense
   - a system of forces acting on a rigid body can be replaced by a single force acting at a given point and a couple; this force and couple is known as equivalent force-couple system

- two system of forces are said to be equivalent when they can be replaced by the same force-couple system at a given point.
- the translatory effect of a couple on a body is zero.
- the effect of couple on a body remains unchanged if the couple is
  (*a*) rotated through an angle
  (*b*) shifted to any other position
  (*c*) replaced by another pair of forces whose rotational effect is same

6. When acted upon by a system of coplanar non-current forces, the body undergoes displacement (due to resultant force) as well as rotation due to resultant couple. The system would then be in equilibrium if

$$\Sigma F_x = 0\ ;\quad \Sigma F_y = 0 \quad \text{and} \quad \Sigma M = 0$$

The condtions $\Sigma F_x = 0$ and $\Sigma F_y = 0$ ensure that the system does not reduce to a single force, and the condition $\Sigma M = 0$ ensures that it does not reduce to a couple.

REVIEW QUESTIONS

A. Conceptual and conventional questions :

1. Define moment and moment of force.
2. List some engineering applications of moment.
3. State the possible displacements of a body subjected to a general force system.
4. State and prove Varignon's theorem.
5. What is a couple? Enumerate its various characteristics.
6. Can a couple be balanced by a single force applied to a body?
7. Is it true that the line of action of the resultant of the parallel forces always passes between the lines of action of these forces?
8. Outline the steps involved in the graphical method for the determination of the resultant of parallel forces.
9. State the equilibrium conditions for bodies acted upon by a system of (*i*) coplanar concurrent forces, (*ii*) coplanar non-concurrent forces.
10. Find the magnitude of the two like parallel forces acting at a distance of 24 cm, whose resultant is 200 N and its line of action is at a distance of 6 cm from one of the forces. **(50 N, 150 N)**
11. Three like parallel force 20 N, 40 N and 60 N are acting at points *A*, *B* and *C* respectively on a straight line *ABC*. The distances are *AB* = 3 m, *BC* = 4 m. Find the resultant and also the distance of the resultant from point *A* on line *ABC*. **(120 N, 4.5 m)**
12. Four parallel forces of magnitudes 100 N, 200 N, 50 N and 400 N are shown in Fig. 3.101. Determine the magnitude of the resultant and also the distance of the resultant from point *A*.

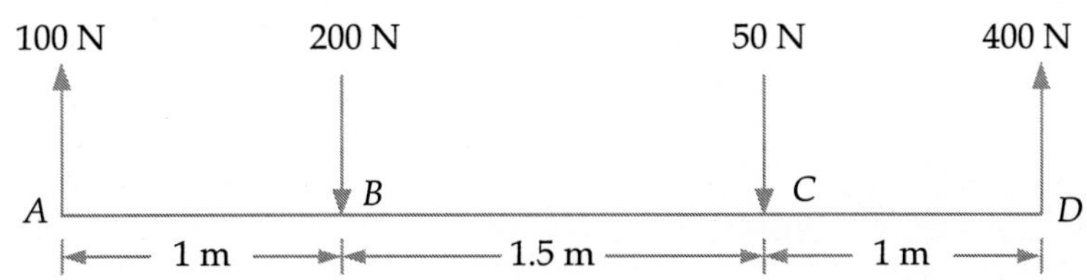

Fig. 3.101

(*R* = 250 N & 4.3 m)

13. Two like parallel forces P and Q act on a rigid body at A and B respectively. If P and Q be inter changed in position, show that the point of application of the resultant will be displaced through a distance d along AB, where

$$d = \frac{P - Q}{P + Q} \times AB$$

14. $ABCD$ is a rectangle in which $AB = CD = 100$ mm and $BC = DA = 80$ mm. Forces of 100 N each act along AB and CD and forces of 50 N each act along BC and DA. Find the resultant moment of two couples. **(13 Nmm)**

15. Three forces, acting on a rigid body, are represented in magnitude, direction and line of action by the three sides of a triangle taken in order. Prove that the forces are equivalent to a couple whose moment is equal to twice the area of the triangles.

16. A beam AB of span 8 m is hinged at A and is on rollers at B. It carries a uniformly distributed load, a concentrated load and an externally applied moment as shown in Fig 3.102. Determine the reactions at A and B for the loading shown.

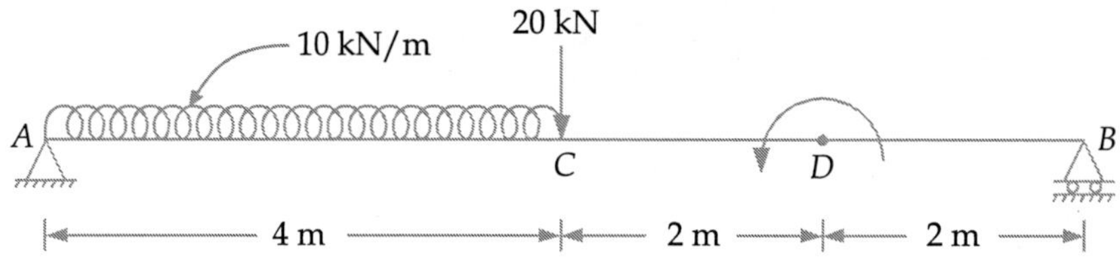

Fig. 3.102

(R_b = 10 kN down and, R_a = 70 kN)

17. For the lever system depicted in Fig. 3.103, determine the force P required to keep it in equilibrium.

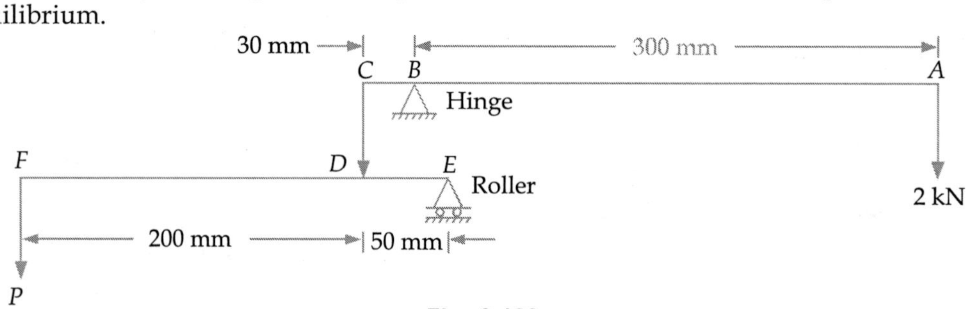

Fig. 3.103

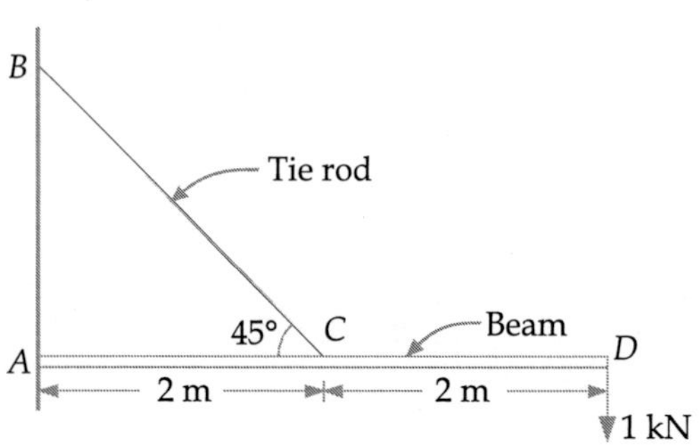

Fig. 3.104

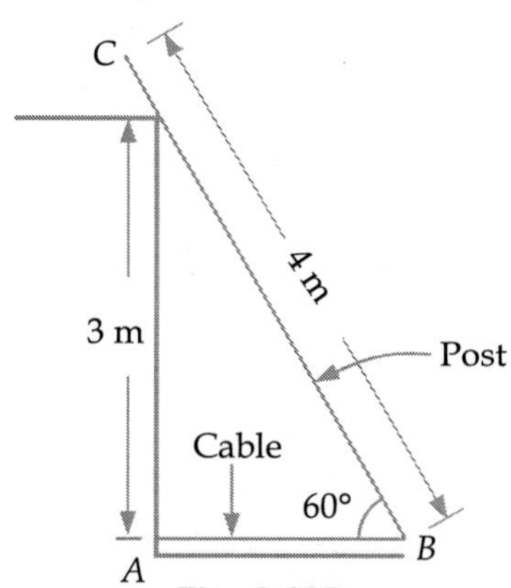

Fig. 3.105

18. Figure 3.104 shows a horizontal beam *AD* of a jib-crane supported horizontally at the point *A* and by a tie rod *BC*. Calculate the tension in the tie rod when a load of 1 kN is acting at point *D* on the beam.
19. A post *BC* of weight 18 N is prevented from sliding with the help of cable *AB* as shown in Fig. 3.105. Assuming all surfaces to be smooth, work out the tension developed in the cable. **(5.19 N)**
20. A uniford rod *AB* of 1 m length and 1 kg mass is suspended from its ends as shown in Fig. 3.106. What amount of tensions would be induced in the strings when a mass of 5 kg is suspended from a point *C* on the rod? **(T_1 = 16.85 N, T_2 = 41.85 N)**

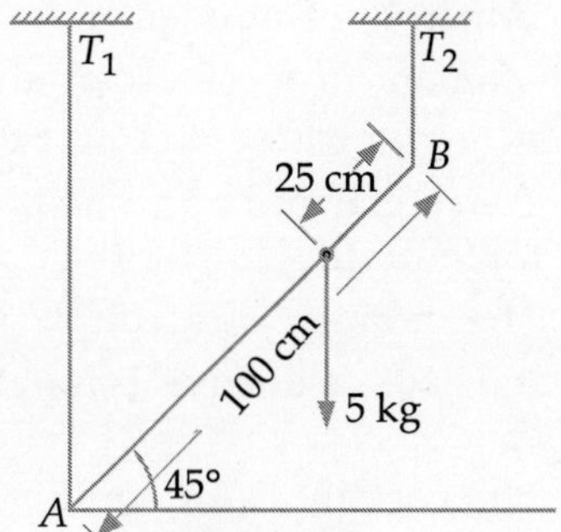

Fig. 3.106

21. Determine the position and magnitude of the resultant force acting on the frame work shown in Fig. 3.107.
What is the moment of the force system.
22. A plate measuring 800 × 500 mm is acted upon by a set of forces as shown in Fig. 3.108.
Determine the resultant of these forces and also locate the points where the line of action of resultant intersets the edges of the plate. **(1706.76 N, 54,1°, 193.19 mm, 238.48 mm, 329.48 mm)**

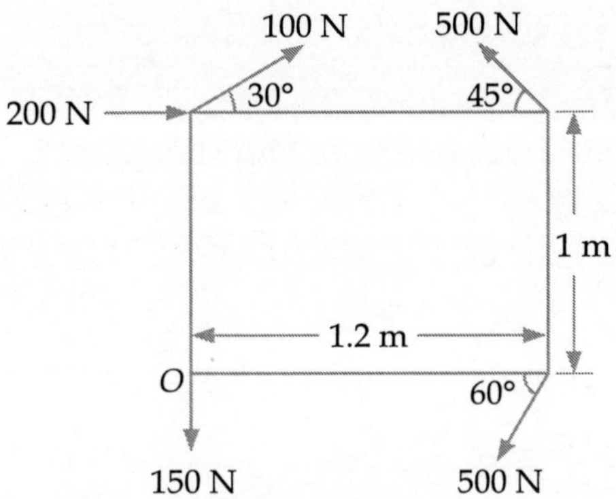

Fig. 3.107

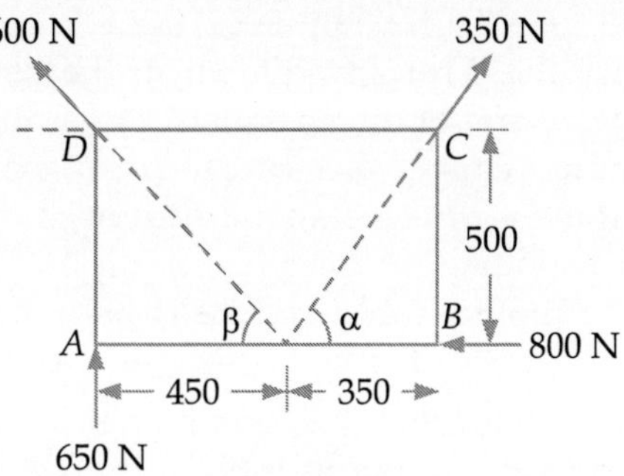

Fig. 3.108

23. A circular disc of 8 cm radius is acted upon by set of two forces as shown in Fig. 3.109.
The resultant moment of these forces about a point A on the disc is stated to be zero. Make calculations for :
(*a*) the magnitude of force P
(*b*) the magnitude of the resultant of two forces
(*c*) the point on the y-axis through which the line of action of the resultant passes through
(61.6 N, 141.25 N, 2.20 cm)

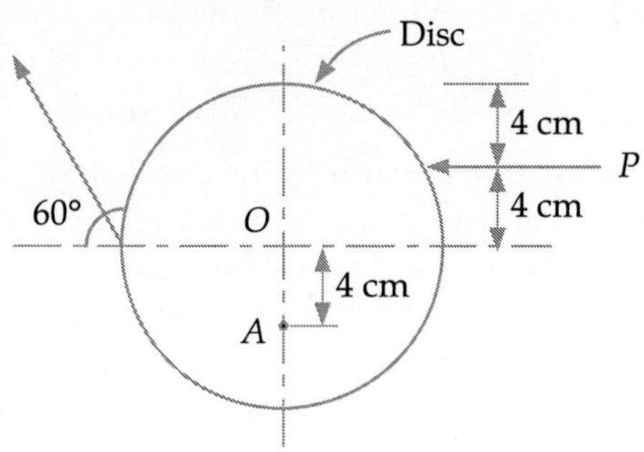

Fig. 3.109

B. Complete the following statements with most appropriate word/words :

1. The rotational tendency of a force is called ....
2. The moment of force about any point is the .... of force and .... distance between the point and line of action of force.

3. The moment of force about any pivot point is a .... quantity and it varies .... with its distance from the point.
4. "The algebraic sum of the moments of a system of forces about any point in their plane is equal to the moment of their resultant about that point."
   This is the generalised statement of ....
5. Force causes .... displacement while moment causes .... displacement.
6. Principle of moment is based on .... theorem.
7. Moment of force about an axis indicates .... of the body about that axis.
8. A couple is formed by two .... and .... parallel forces.
9. Two equal and opposite parallel forces produce a .... whose moment is equal to either force multiplied by their .... distance.
10. The moment of a couple is known as ....
11. If a body is in equilibrium under the action of three forces, then the forces must be .... or ....
12. The conditions of equilibrium for coplanar and non-concurrent forces are ....

Answers :

1. moment 2. product, perpendicular 3. vector, directly 4. Varignon's theorem 5. linear, angular 6. Varignon's 7. rotational tendency 8. equal, opposite 9. couple, perpendicular 10. torque 11. concurrent, parallel 12. $\Sigma F_x = 0$, $\Sigma F_y = 0$ and $\Sigma M = 0$

C. Multiple choice questions :

1. Identify the wrong statement in the context of moment, couple and torque
   (*a*) Two parallel forces equal in magnitude but opposite in direction, and separated by a finite distance are said to form a couple.
   (*b*) Moment of force about a point is defined as the turning tendency of the force about that point.
   (*c*) The moment of a couple is known as torque
   (*d*) The sum of forces forming a couple in any direction has a definite non-zero value.
2. Consider the following statements :
   The effect of couple in a body remains unchanged if the couple is
   1. rotated through an angle 2. shifted to any other position
   3. replaced by another pair of forces whose rotational effect is same
   Which of these statements are correct ?
   (*a*) 1 and 2 (*b*) 1 and 3 (*c*) 2 and 3 (*d*) 1, 2, and 3
3. Consider the following statements
   1. The translatory effect of a couple on a body is zero.
   2. A couple can be balanced by a couple of opposite sense
   3. The moment of a couple about any point is the same.
   Which of these statements is/are correct?
   (*a*) 2 and 3 (*b*) 3 only (*c*) 1 and 2 (*d*) 1, 2, and 3
4. A rectangle *ABCD* has sides *AB* = *CD* = 80 mm and *BC* = *DA* = 60 mm. Forces of 150 N each act along *AB* and *CD*, and forces of 100 N each act along *BC* and *DA*. The resultant moment of the force system is
   (*a*) 8000 Nmm (*b*) 9000 Nmm (*c*) 1000 Nmm (*d*) 17000 Nmm

Answers :

1. (*d*) 2. (*d*) 3. (*d*) 4. (*d*)

CHAPTER

4

FRICTION

Learning objectives : Attention has been focussed in this chapter to make the reader conversant with:

- *friction and its characteristics; dry and fluid friction; static and dynamic friction*
- *laws of solid friction*
- *terms related to friction such as coefficient of friction, angle of friction, angle of respose and cone of friction*
- *ladder friction, rope and belt friction and wedge friction*

When a body slides over another body, a force is exerted at the surface of contact by the stationary body on the moving body. This resisting force is called the force of friction and acts in a direction opposite to the direction of motion. The frictional forces are present throughout in nature and exist to a considerable extent in all machines; no matter how accurately they are made. Friction is quite undesirable and needs to be mitigated in some machines and processes such as

- power screws
- bearings and gears
- flow of fluids in pipes

Its presence would cause loss of power, wearing out of parts and huge economic losses. However, the working of many devices such as

- friction brakes and clutches
- belt and rope drives
- holding and fastening devices

depends on friction and there the presence of friction forces is advantageous. Obviously friction is both a liability and a necessity, and is often referred to as necessary evil.

4.1. FRICTION AND ITS CHARACTERISTICS

Consider a block resting on a horizontal plane. Experience and observations indicate that the mating surfaces are generally rough and have irregularities. The irregularities are somewhat like hills and valleys. The valleys of one surface mesh with the hills of the other surface. Obviously, there is some degree of interlocking and wedging action between the contact surfaces.

When a force P is applied to the block in a direction parallel to the surface of contact, a force F automatically gets induced, acts in a direction opposite to that in which motion is sought and offers resistance to the movement of block. This resisting force acts tangentially to the contact surface and is known as friction force. Such forces always develop whenever there exists a tendency for the sliding of contact surfaces. The friction is attributed to the interlocking of minute projections on the surface which oppose the relative motion.

Refer Fig. 4.2 for the idealised plot of friction force F and the tractive (applied) force P. Its salient features are:

(*i*) When static conditions prevail, *i.e.*, when there is no motion of the block, balance between applied force $\boldsymbol{P}$ and the friction force $\boldsymbol{F}$ is maintained. The frictional force increases in direct proportion to the applied force and this is depicted by line 0-1 which slopes at 45° with the axis of applied force. Line 0-1 represents the *zone of static friction* and point 1 corresponds to the maximum static friction.

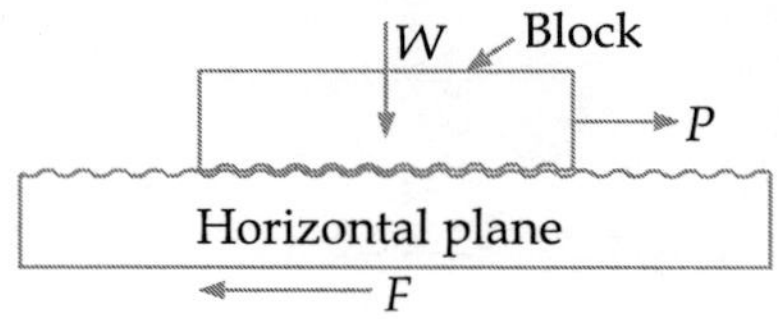

Fig. 4.1

(*ii*) There comes a limit beyond which frictional force cannot increase and the block is set into motion. The friction force at this instant is maximum and is called the limiting friction.

The *limiting friction* is the maximum friction force at the time of impending motion, i.e., when the motion is about to begin.

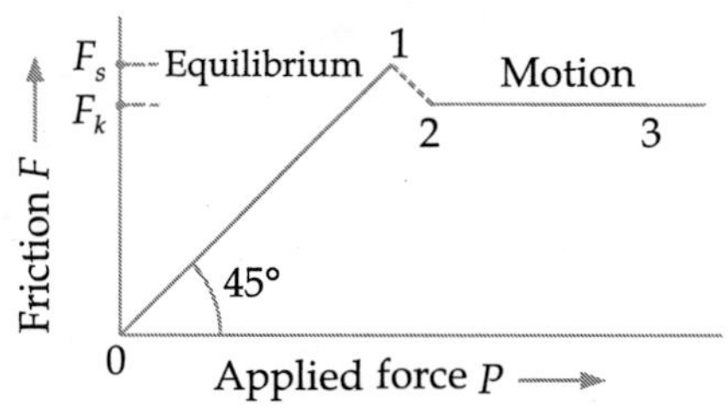

Fig. 4.2

(*iii*) When the motion of block commences, the friction force drops to somewhat low value called the kinetic friction. The decrease may be atribtuted to the fact that the chance of meshing of hills and valleys of the two surfaces becomes less. Such a stage is represented by straight line 2-3 of the graph and it is known as zone of *kinetic friction*:

Betwen points 1 and 2, the variation is uncertain and is shown by a dotted line.

The main characteristics of friction force are:

- It always acts in a direction opposite to that in which motion is intended; it always opposes the tractive force.
- It is a passive force; it exists as long as the tractive force acts.
- It is a self-adjusting force; only that much comes into play as is just sufficient to prevent motion.

4.2. NATURE OF FRICTION

There are essentially two types of friction, namely, dry friction and fluid friction.

4.2.1. Dry friction

Dry friction (also called *coulomb friction*) manifests when the contact sufaces are dry and there is tendency for relative motion.

Dry friction is further subdivided into:

- Sliding friction: friction between two surfaces when one surface slides over another.
- Rolling friction: friction between two surfaces which are separated by balls or rollers.

It may be pointed out that rolling friction is always less than sliding friction.

4.2.2. Fluid Friction

Fluid friction manifests when a lubricating fluid is introduced between the contact surfaces of two bodies.

If the thickness of the lubricant or oil between the mating surfaces is small, then the friction between the surfaces is called *greasy* or *non-viscous friction*. The oil gets absorbd by the surfaces and the contact between them is no more a metal to metal contact. Instead the contact is through thin layer of oil and that ultimately results is less friction.

When the two surfaces are separated by a thick film of lubricant, metallic contact is entirely non-existent. The friction is due to viscosity of the oil, or the shear resistance between the layers of the oil rubbing against each other. Obviously then there occurs a great reduction in friction. This frictional force is knwon as *viscous* or *fluid friction*.

4.3. STATIC AND DYNAMIC FRICTION

The *static friction* is the frictional force that develops between mating surfaces when subjected to external forces but there is no relative motion between them.

The *dynamic friction* is the frictional force that develops between mating surfaces when subjected to external forces and there is relative motion between them. The dynamic friction is also known as *kinetic friction*.

The kinetic friction is always somewhat less than the maximum static friction force.

4.4. LAWS OF SOLID FRICTION (STATIC OR DYNAMIC)

Based upon experimental evidence, the following laws of friction have been established for dry contact surfaces.

(*i*) Friction acts tangential to the surfaces in contact and is in a direction opposite to that in which motion is to impend, i.e., take place.

(*ii*) Friction force is maximum at the instant of impending motion. Its variation from zero to maximum value (limiting friction) depends upon the resultant force tending to cause motion.

(*iii*) The magnitude of limiting friction bears a constant ratio to the normal reaction between the mating surfaces.

This ratio drops to a slightly lower value when the motion starts.

(*iv*) Limiting friction is independent of the area and shape of contact surfaces.

(*v*) Limiting friction depends upon the nature (roughness or smoothness) of the surfaces in contact.

(*vi*) At low velocities between sliding surfaces, the friction force is practically independent of the velocity. However, slight reduction in friction occurs when the speeds are high.

4.5. TERMS RELATED TO FRICTION

4.5.1. Coefficient of Friction

The frictional force F depends upon the wedging action between irregularities of contact surfaces and becomes more pronounced with increase in normal reaction R between them. Mathematically $F \propto R$. The ratio F/R is called *coefficient of friction* and is denoted by μ.

The coefficient of friction is defined as the ratio of force of friction to the normal reaction between the contact surfaces.

The coefficient of friction would be high if the contact surfaces are rough. Obviously, the coefficient of friction measures the roughness between a pair of contact surfaces and incorporates the geometric property of the mating contours.

(*i*) When the system is in a state of impending motion, the frictional force has the limiting (maximum) value F_s and the ration F_s/R is called the coefficient of static friction μ_s. Thus $\mu_s = F_s/R$.

(*ii*) When motion starts, the maximum friction falls to a lower value F_k called kinetic friction. Experimental results indicate that the friction force developed during motions remain fairly constant. The ratio F_k/R is called the coefficient of kinetic friction μ_k. Thus $\mu_k = F_k/R$.
The value of μ_k is found to be less than μ_s for the same pair of contact surfaces.

(*iii*) The approximate values of static friction coefficients for pair of contacting surfaces are:
- wood on wood = 0.3 to 0.5
- metal on metals = 0.15 to 0.25
- rubber tyre on concrete = 0.7 to 0.9
- wood on metals = 0.2 to 0.6
- masonry on clay = 0.3 to 0.5.

4.5.2. Total Reaction

The block *A* resting on a rough surface is in equilibrium under the action of following forces:
(*i*) weight *W* of the block, acting downwards through its centre of gravity
(*ii*) normal reaction *R* between the block and surface
(*iii*) frictional force $F = \mu R$ where μ is the coefficient of friction
(*iv*) tractive force *P*

The forces *R* and *F* act at right angles to each other. Their resultant equals $\sqrt{R^2 + F^2}$ and is called the total reaction or resultant reaction.

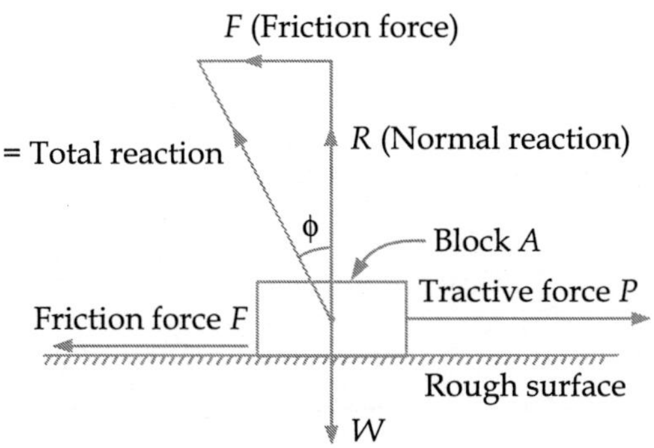

Fig. 4.3

4.5.3. Angle of Friction

It is defined as the angle which the resultant of normal reaction and limiting force of friction makes with the normal reaction.

With reference to Fig. 4.3

$$R = \text{normal reaction;}$$
$$F = \text{limiting force of friction}$$
$$S = \sqrt{R^2 + F^2} = \text{total or resultant reaction}$$
$$\tan\phi = \frac{F}{R} \quad \text{where } \phi \text{ is the angle of friction}$$

The ratio F/R is also called the coefficient of friction, μ.

$$\therefore \quad \mu = \tan\phi \qquad ...(4.1)$$

The angle of friction, ϕ, is a measure of the limiting position of total reaction between the two contacting surfaces.

4.5.4. Angle of Repose

Consider a block of weight W resting on an inclined plane OA making an angle α with the horizontal. Let the angle α be increased gradually till the block is just at the point of sliding. The block is then in equilibrium state under the influence of following set of forces:

(*i*) weight W of the block acting vertically downwards,

(*ii*) normal reaction R acting at right angles to the inclined plane, and

(*iii*) limiting forces of friction, $F = \mu R$; acting up the plane as the block is to slide downwards.

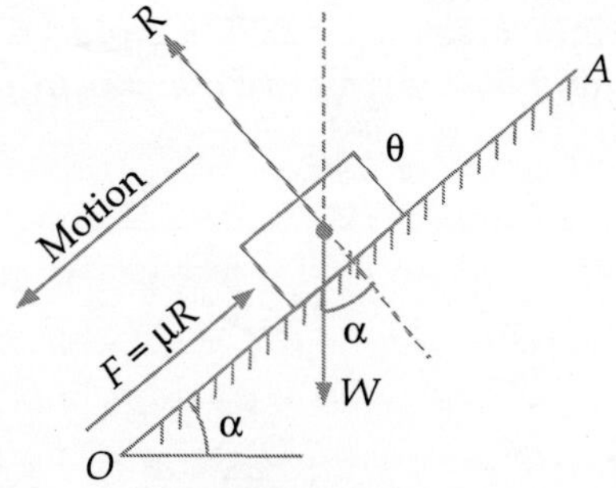

Fig. 4.4

Resolving these forces along and perpendicular to the plane

$$\mu R = W \sin \alpha$$

$$R = W \cos \alpha$$

These two expressions give : $\mu = \tan \alpha$

In terms of angle of friction ϕ, the coefficient of friction is given as: $\mu = \tan \phi$

$$\therefore \quad \tan \phi = \tan \alpha \quad \text{or} \quad \phi = \alpha \qquad ...(4.2)$$

The angle α of the inclined plane at which a block resting on it is about to slide down the plane is called the angle of repose, and it is equal to the angle of friction between the block and the inclined plane.

4.5.5. Cone of Friction

With reference to Fig. 4.5, the lines of action of normal reaction R and the frictional force F meet at point O. The magnitude of total reaction is $\sqrt{F^2 + R^2}$ and it makes an angle ϕ with the normal reaction. Obviously O is the vertex of a cone whose axis is R and semi-vertex angle is ϕ. Such a right circular cone is called the cone of friction.

It is worthwhile to point out that in case of impending motion, the friction force is maximum and the total reaction lies on the surface of cone. When the friction force is less than the limiting friction, the total reaction would lie within the cone. In that case, the forces acting on the body are not sufficient to cause motion. This aspect forms the working principle for self-locking mechanisms.

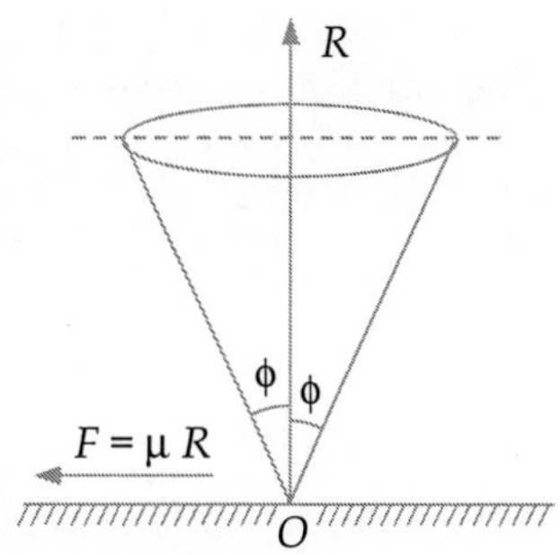

Fig. 4.5

EXAMPLE 4.1

(*a*) A block of weight 120 N is placed on a rough horizontal plane. If a horizontal force of 50 N just causes the body to slide over the horizontal plane, calculate the coefficient of friction between the block and the plane, and the angle of friction.

(*b*) A force of 10 N is required to start a uniform plank of wood slide over a bench for which the coefficient of friction is constant. The plank is cut into three equal pieces and these pieces are stacked one above the other. What force will be now needed to start movement of the pile?

(*c*) A block of mass 80 kg rests on a floor and the static coefficient of friction for the contact surface is 0.15. It is acted upon by a horizontal force of 50 N as shown in Fig. 4.6. What frictional force will be developed along the contact surface?

(*d*) Refer Fig. 4.7 which shows a mass m_1 = 0.5 kg resting on a frictionless surface and another mass m_2 = 0.25 kg placed above m_1; the coefficient of friction between m_1 and m_2 being 0.3. If m_2 is to slide on m_1, what amount of force is needed to pull mass m_1.

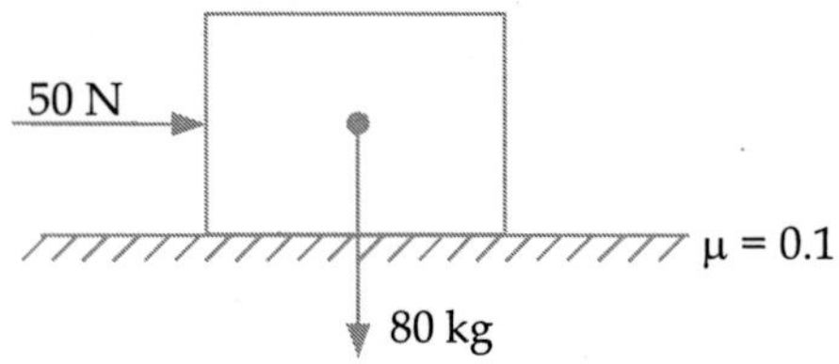

Fig. 4.6

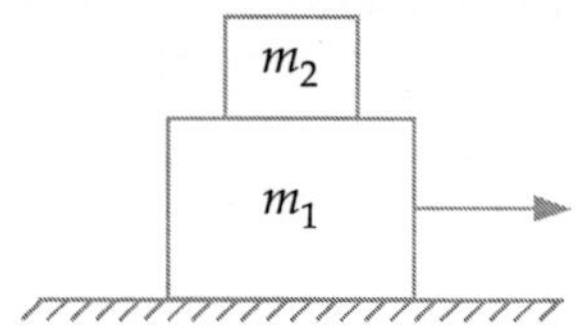

Fig. 4.7

Solution : (*a*) $\mu = \frac{F}{R} = \frac{50}{120} = \mathbf{0.4167}$

Also $\mu = \tan\phi$, where ϕ is the angle of friction

$\phi = \tan^{-1}(\mu) = \tan^{-1}(0.4167) = \mathbf{22.62^\circ}$

(*b*) $F = \mu R$. The coefficient of friction μ and the normal reaction $R = mg$ remain constant. As such the same force of 10 N would be needed to start the movement of pile system.

(*c*) Maximum value of static friction $= \mu R = \mu mg$

$= 0.15 \times (80 \times 9.81) = 117.72$ N

When the applied force is less than this maximum value, the friction force is obtained by the equations of equilibrium. $\Sigma F_y = 0$ and that gives $F = 50$ N.

(*d*) Frictional force between mass m_1 and $m_2 = \mu R = \mu\ (m_2 g)$

$= 0.3 \times (0.25 \times 9.81) = 0.736$ N

The maximum accelaration with which it can move without sliding

$$a = \frac{0.736}{0.25} = 2.944 \text{ N} \quad (F = m\,a)$$

The total mass which the applied force has to move

$= 0.25 + 0.5 = 0.75$ kg

Force to be applied $= 0.75 \times 2.944 = \mathbf{2.208}$ N

EXAMPLE 4.2

A block of mass 25 kg is at rest on a horizontal floor. To slide the block, a force *P* is applied along the horizontal direction. The coefficient of static friction and dynamic friction between the block and floor are stated to be 0.36 and 0.3 respectively. For each value of *P* given below, work-out the corresponding value of frictional force *F* and complete the table.

Applied force *P* (N)	20	40	60	80	88.3	90
Frictional Force *F* (N)						

Solution : Weight of the block $W = 25 \times 9.81 = 245.25$ N

Normal reaction $R = W = 245.25$ N

Limiting value of force of friction $F = \mu_s R = 0.36 \times 245.25 = \mathbf{88.3\ N}$

The block will remain at rest till the applied force is less than 88.3 N

When the applied force exceeds μR, the block starts sliding. The dynamic conditions then prevail and the frictional force reduces to

$$F = \mu_d R = 0.3 \times 245.25 = \mathbf{73.57\ N}$$

Applied force (N)	20	40	60	80	88.3	90
Frictional Force (N)	20	40	60	80	88.3	73.57

EXAMPLE 4.3

A body weighing 300 N is resting on a rough horizontal table. A pull of 100 N applied at an angle of 15 degree with the horizontal just causes the body to slide over the table. Make calculations for the normal reaction and the coefficient of friction.

Solution : In the limiting equilibrium, the forces are balanced. That is

$$\Sigma F_x = 0; \quad F - 100 \cos 15^\circ = 0$$
$$F = 100 \cos 15^\circ$$
$$= 96.59 \text{ N}$$

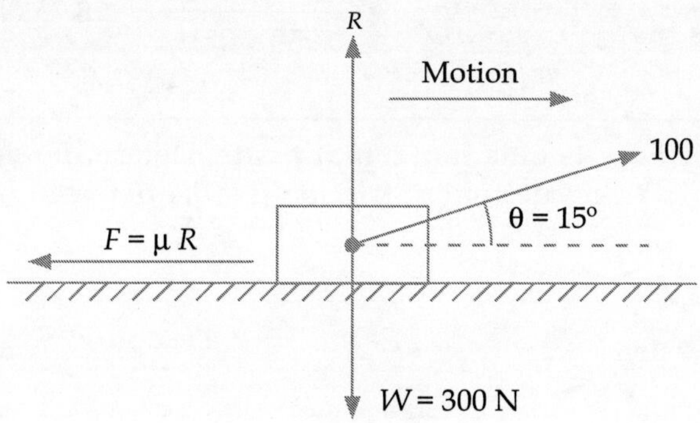

Fig. 4.8

$$\Sigma F_y = 0; \quad R + 100 \sin 15^\circ - 300 = 0$$
$$R = 300 - 100 \sin 15^\circ$$
$$= 300 - 25.88 = \mathbf{274.12\ N}$$
$$\mu = \frac{F}{R} = \frac{96.59}{274.12} = \mathbf{0.352}$$

EXAMPLE 4.4

A body of weight 100 N rests on a rough horizontal surface (μ = 0.3) and is acted upon by a force applied at an angle of 30 degree to the horizontal. What force is required to just cause the body to slide over the surface?

(*b*) Proceed to determine the inclination and magnitude of minimum force required to set the block into impending motion.

Solution : In the limiting equilibrium, the forces are balanced. That is

$$\Sigma F_x = 0; \quad F = P \cos \theta$$
$$\Sigma F_y = 0; \quad R = W - P \sin \theta$$
$$\text{Also} \quad F = \mu R$$
$$\therefore \quad \mu (W - P \sin \theta) = P \cos \theta$$

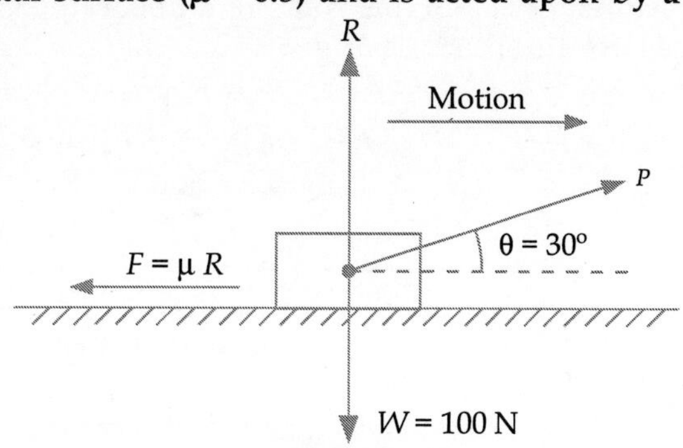

Fig. 4.9

$$\text{or} \quad P = \frac{\mu W}{\cos \theta + \mu \sin \theta} = \frac{0.3 \times 100}{\cos 30^\circ + 0.3 \sin 30^\circ} = \frac{0.3 \times 100}{0.866 + 0.15} = \mathbf{29.53\ N}$$

(*b*) Let α be the inclination of the minimum force required to just cause the body to slide over the surface. Then

$$P = \frac{\mu W}{\cos\alpha + \mu \sin\alpha} \qquad ...(i)$$

For force P to be minimum, the denominator must be maximum. For that

$$\frac{d}{dx}(\cos\alpha + \mu\sin\alpha) = 0$$

or $\quad -\sin\alpha + \mu\cos\alpha = 0;\quad \tan\alpha = \mu = 0.3$

∴ $\quad \alpha = 16.7°$

Substituting $\alpha = 16.7°$ in identity (*i*) above, we get

$$P_{min} = \frac{0.3 \times 100}{\cos 16.7° + 0.3 \sin 16.7°} = \frac{30}{0.958 + 0.086} = \mathbf{28.73\ N}$$

EXAMPLE 4.5

A wooden block of weight 50 N rests on a horizontal plane. Determine the force required to just (*a*) pull it, and (*b*) push it. Take coefficient friction μ = 0.4 between the mating surfaces. Comment on the result.

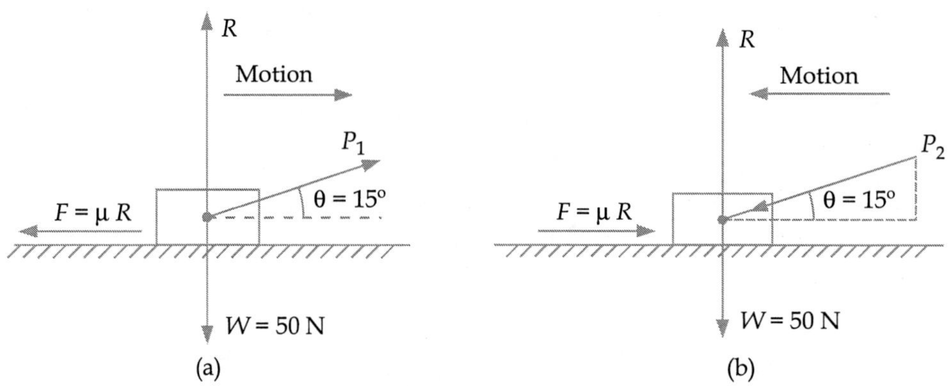

Fig. 4.10

Solution : Let P_1 be the force required to just pull the block. In the limiting equilibrium, the forces are balanced. That gives

$$\Sigma F_x = 0;\quad F = P_1 \cos\theta$$
$$\Sigma F_y = 0;\quad R = W - P_1 \sin\theta.$$

Also $\quad F = \mu R \quad ∴ \quad \mu(W - P_1 \sin\theta) = P_1 \cos\theta$

or $$P_1 = \frac{\mu W}{\cos\theta + \mu\sin\theta} = \frac{0.4 \times 50}{\cos 15° + 0.4 \sin 15°} = \frac{0.4 \times 50}{0.966 + 0.103} = 18.70\text{ N}$$

(*b*) Let P_2 be the force required to just push the block. With reference to the free body diagram [Fig. 4.10 (*b*)],

Let us write the equations of equilibrium,

$$\Sigma F_x = 0;\quad F = P_2 \cos\theta$$
$$\Sigma F_y = 0;\quad R = W + P_2 \sin\theta$$

Also $\quad F = \mu R$

∴ $\quad \mu(W + P_2 \sin\theta) = P_2 \cos\theta$

or $$P_2 = \frac{\mu W}{\cos\theta - \mu\sin\theta} = \frac{0.4 \times 50}{\cos 15^\circ - 0.4\sin 15^\circ} = \frac{0.4 \times 50}{0.966 - 0.103} = \mathbf{23.17\ N}$$

Comments. It is easier to pull the block than push it.

EXAMPLE 4.6

A body resting on a rough horizontal plane required a pull of 24 N inclined at 30° to the plane just to move it. It was also found that a push of 30 N at 30° to the plane was just enough to cause motion to impend. Make calculations for the weight of body and the coefficient of friction.

Solution : Refer to Fig. 4.11 for the system of forces when pull and push are applied to the body. Force of friction is always opposite to the direction of motion.

In the limiting equilibrium, the forces are balanced. With reference to the free body diagram (Fig. 4.11 (*a*) when pull is applied)

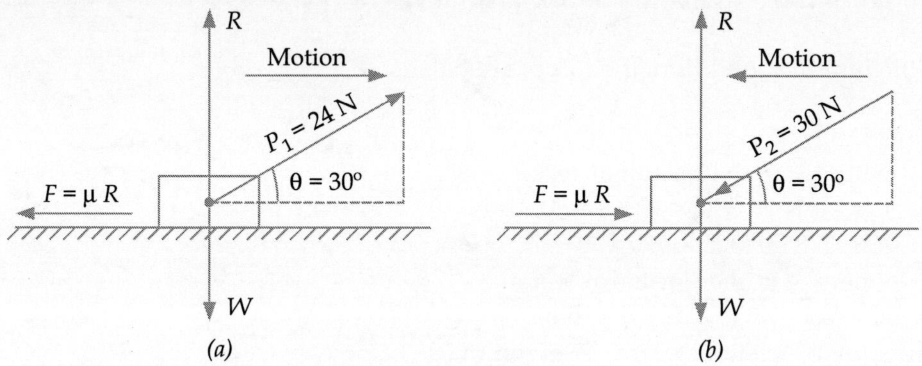

Fig. 4.11

$$\Sigma F_x = 0\ ; \quad F = P_1 \cos\theta$$
$$\Sigma F_y = 0; \quad R = W - P_1 \sin\theta$$

Also $$F = \mu R$$

$\therefore$ $\mu (W - P \sin\theta) = P_1 \cos\theta$

or $$P_1 = \frac{\mu W}{\cos\theta + \mu \sin\theta} \quad ...(i)$$

With reference to the free body diagram (Fig. 4.11 (*b*) when push is applied)

$$\Sigma F_x = 0\ ; \quad F = P_2 \cos\theta$$
$$\Sigma F_y = 0\ ; \quad R = W + P_2 \sin\theta$$

Also $F = \mu R \quad \therefore \quad \mu (W + P_2 \sin\theta) = P_2 \cos\theta$

or $$P_2 = \frac{\mu W}{\cos\theta - \mu \sin\theta} \quad ...(ii)$$

From expression (*i*) and (*ii*),

$$\frac{P_1}{P_2} = \frac{\cos\theta - \mu \sin\theta}{\cos\theta + \mu \sin\theta}$$

or $$\frac{24}{30} = \frac{\cos 30^\circ - \mu \sin 30^\circ}{\cos 30^\circ + \mu \sin 30^\circ} = \frac{0.866 - 0.5\,\mu}{0.866 + 0.5\,\mu}$$

or $0.6928 + 0.4\,\mu = 0.866 - 0.5\,\mu$

$$\therefore \quad \mu = \frac{0.866 - 0.6928}{0.9} = \mathbf{0.192}$$

For expression (*i*)

$$W = \frac{P_1 (\cos\theta + \mu \sin\theta)}{\mu} = \frac{P_1 (\cos 30° + \mu \sin 30°)}{\mu}$$

$$= \frac{24 (0.866 + 0.192 \times 0.5)}{0.192} = \mathbf{120.25\ N}$$

EXAMPLE 4.7

A block weighing 600 N has been subjected to the load system as shown in Fig. 4.12. Work-out the frictional force for the block and state whether it is in equilibrium or in motion. What additional force needs to be added to 120 N force so that the block just moves to the left?

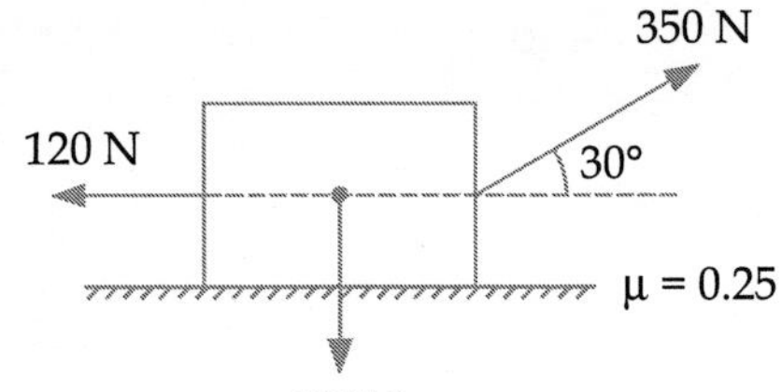

Fig. 4.12

Solution : In the limiting condition, the forces are balanced. That is

$\Sigma F_x = 0$; $-120 + 350 \cos 30 - F = 0$

where F is the frictional force

$F = -120 + 350 \times 0.866 = 183.1$ N

$\Sigma F_y = 0$; $R + 350 \sin 30° - 600 = 0$

where R is the normal reaction

$R = 600 - 350 \times 0.5 = 425$ N

Maximum value of frictional force is given by

$$F_{max} = \mu \times R = 0.25 \times 425 = 106.25 \text{ N}$$

Since $F_{max} < F$, the block moves towards right and friction force = 106.25 N acts towards left.

(*b*) Let P be the force which needs to be added to 120 N force to just move the block to the left. When the block moves towards left, the friction force will act towards right. Then from the relation $\Sigma F_x = 0$,

$$-(120 + P) + 350 \cos 30° + F_{max} = 0$$

$$-(120 + P) + 350 \times 0.866 + 106.25 = 0$$

$$\therefore \quad P = -120 + 350 \times 0.866 + 106.25 = \mathbf{289.35\ N}$$

EXAMPLE 4.8

Consider a system of three blocks resting upon one another as shown in Fig. 4.13. The block A weighs 150 N, B weighs 50 N and C weighs 100 N. The coefficients of friction are 0.3 between A and B, 0.2 between B and C, and 0.1 between C and the ground. Determine the least horizontal force P necessary to start motion of any part of the system.

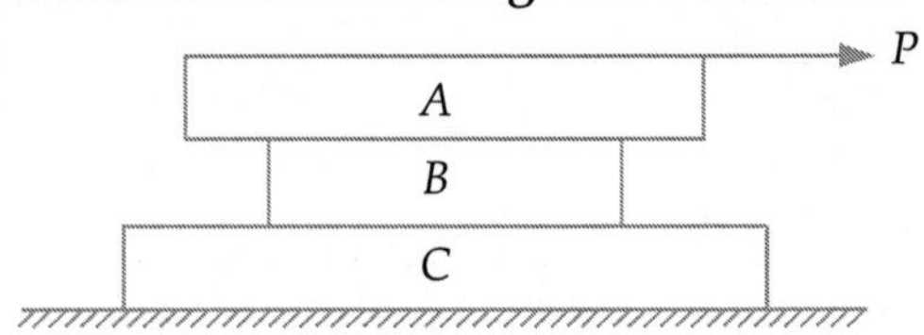

Fig. 4.13

Solution : Limiting force of friction between contacting surfaces,

$$F = \mu R = \mu W$$

$\therefore$ Limiting force of friction between A and B,

$$F_{ab} = \mu_{ab} \times W_a = 0.3 \times 150 = 45 \text{ N}$$

Limiting force of friction between B and C,

$$F_{bc} = \mu_{bc} (W_a + W_b) = 0.2\ (150 + 50) = 40\ \text{N}$$

Limiting force of friction between B and C,

$$F_{cg} = \mu_{cg} (W_a + W_b + W_c) = 0.1\ (150 + 50 + 100) = 30\ \text{N}$$

With gradual increase in applied force P, the frictional force increases till it attains the maximum (limiting) values. Any further increase in P sets the body in motion. Now

(*i*) When $P = 30$ N, the block A exerts a force of 30 N on block B. However this force is less than the limiting friction between A and B, and obviously there will be no relative motion betwen A and B.

(*ii*) The block B also exerts a force of 30 N on block C. This force is again less than limiting friction between B and C. Obviously no relative motion between B and C.

(*iii*) The block C too is subjected to a pull of 30 N. This equals the limiting friction between block C and the ground. As such the block C will be on the point of motion.

The above remarks tell us that when a horizontal P equal to 30 N is applied, all the three blocks would be in a state of impending motion as a single body.

EXAMPLE 4.9

Two blocks A and B of weight 4 kN and 2 kN respectively are in equilibrium position as shown in Fig. 4.14. Presuming that coefficient of friction between the two blocks as well as between block and floor is 0.25, make calculations for the force P required to move the block A.

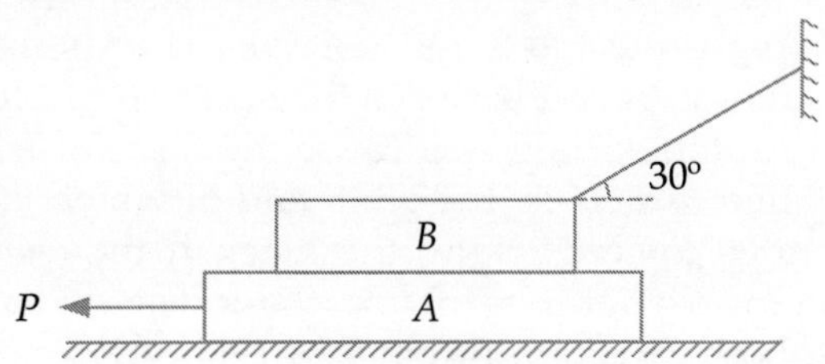

Fig. 4.14

Solution : Refer Fig. 4.15 for the free body diagrams for the two blocks.

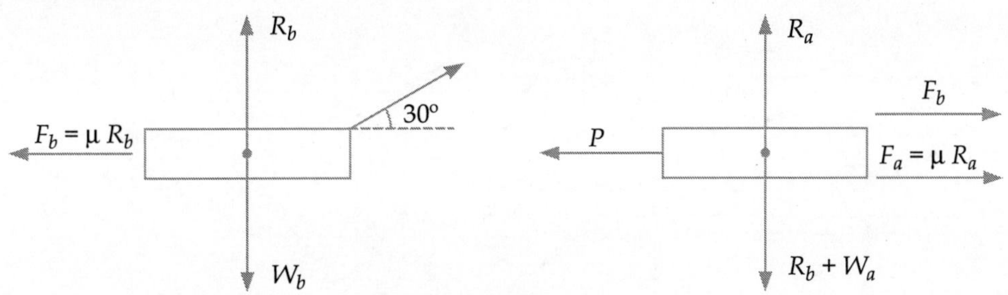

Fig. 4.15

Considering equilibrium of block B. Resolving the forces along the horizontal and vertical directions,

$$T \cos 30° - \mu R_b = 0\ ; \qquad T \cos 30° = \mu R_b$$

$$R_b + T \sin 30° - W_b = 0; \quad T \sin 30° = W_b - R_b$$

$$\therefore \quad \tan 30° = \frac{W_b - R_b}{\mu R_b} = \frac{2 - R_b}{0.25 R_b}$$

or

$$0.5773 = \frac{2 - R_b}{0.25 R_b}\ ; \quad 0.1443\ R_b = 2 - R_b$$

$$\therefore \quad R_b = \frac{2}{1.1443} = 1.748\ \text{N}$$

$$F_b = \mu R_b = 0.25 \times 1.748 = 0.437 \text{ N}$$

Considering equilibrium of block A. Resolving the forces along the horizontal and vertical directions,

$$F_b + \mu R_a - P = 0 ; \quad P = F_b + \mu R_a$$
$$R_a - R_b - W_a = 0 ; \quad R_a = R_b + W_a = 1.748 + 4 = 5.748$$
$$\therefore \quad P = 0.437 + 0.25 \times 5.748 = \mathbf{1.874 \text{ N}}$$

EXAMPLE 4.10

Two blocks *A* and *B* weighing 50 kg and 80 kg respectively are in equilibrium in the position shown in Fig. 4.16. Calculate the force *P* required to move the lower block *B* and tension in the cable. Take coefficient of friction at all contact surfaces to be 0.3.

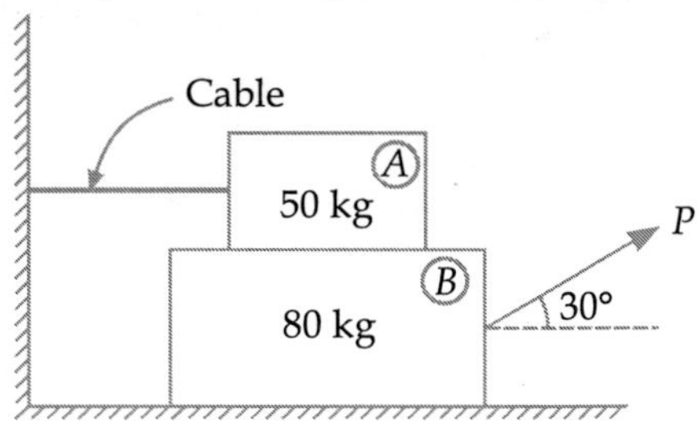

Fig. 4.16

Solution : Weight of block $A = 50 \times 9.81$
$= 490.5$ N

Weight of block $B = 80 \times 9.81$
$= 784.8$ N

As the block *B* tends to move towards right, there will be rubbing action between the surfaces of blocks *A* and *B*. The rubbing action will cause the force of friction to act between the two surfaces. The block *A* is tied to a cable, the other end of which is fixed to a vertical plane. When the block *B* tends to move towards right, the block *A* moves towards left with respect to block *B*. Hence the force of friction $F = \mu R_1$ on the lower surface of block *A* will act towards right as shown in free body diagram of block *A* (Fig. 4.17)

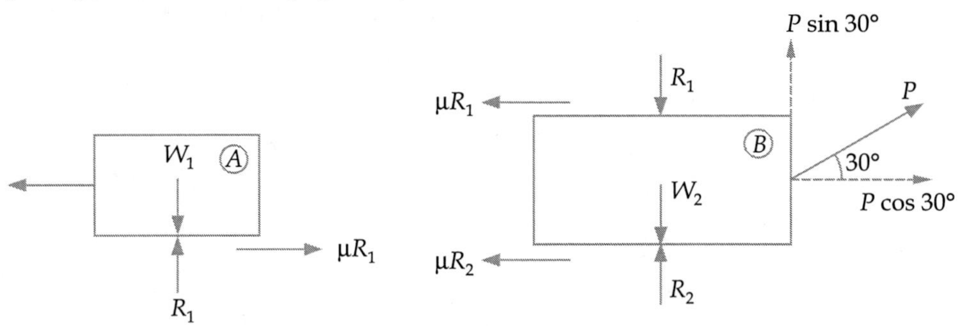

Fig. 4.17

For equilibrium of block *A*,

$$\Sigma F_x = 0 ; \quad -T + \mu R_1 = 0 ; \quad T = \mu R_1 = 0.3R_1$$
$$\Sigma F_y = 0 ; \quad R_1 - W_1 = 0 ; \; R_1 = W_1 = 490.5 \text{ N}$$

$\therefore$ Tension in the cable $T = 0.3R_1 = 0.3 \times 490.5 = 147.15$ N

Considering free body diagram of block *B* (Fig. 4.17),

$$\Sigma F_x = 0 ; \quad -\mu R_1 - \mu R_2 + P \cos 30° = 0$$

or $\quad -0.3 \times 490.5 - 0.3R_2 + 0.866\,P = 0 \quad \quad ...(i)$

$$\Sigma F_y = 0 ; \quad R_2 - W_2 - R_1 + P \sin 30° = 0$$

or $\quad R_2 - 784.8 - 490.5 + 0.5P = 0$

or $\quad R_2 = 1275.3 - 0.5P$

Then from identity (*i*)

$$-0.3 \times 490.5 - 0.3(1275.3 - 0.5P) + 0.866P = 0$$

or $$-147.15 - 382.59 + 0.15P + 0.866P = 0$$

∴ Force required to move the lower block,

$$P = \frac{382.59 + 147.15}{0.15 + 0.866} = \mathbf{521.4\ N}$$

EXAMPLE 4.11

Two blocks *A* and *B* weighing 50 N and 80 N respectively are positioned as shown in Fig. 4.18. The coefficient of friction between ground and block *B* is 0.1 and that between block *B* and block *A* is 0.28. State whether *B* is stationary with respect to ground and *A* moves or *B* is stationary with respect to *A*. Proceed to determine the minimum value of weight *W* in the pan so that motion starts.

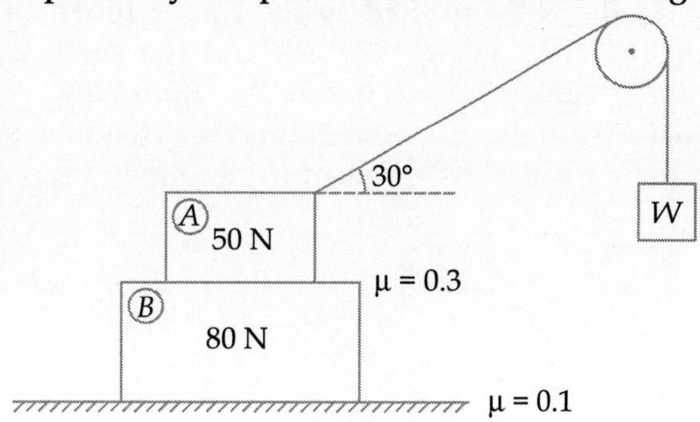

Fig. 4.18

Solution : The coefficient of friction between the block *B* and ground (μ = 0.1) is much less than that between the blocks *A* and *B* (μ = 0.28). Obviously when weight *W* is put in the pan, motion will start of both the blocks together. As such the block *B* will be stationary with respect to block *A*.

With reference to free body diagram of the whole system (Fig. 4.19), the conditions for limiting equilibrium are :

$$\Sigma F_x = 0 = -0.1\,R + W\cos 30°$$

$$\therefore \quad R = \frac{W\cos 30°}{0.1} = 8.66\ W$$

$$\Sigma F_y = 0 = R - (50 + 80) + W\sin 30°$$

$$= 8.66\,W - 130 + 0.5\,W$$

$$\therefore \quad W = \frac{130}{8.66 + 0.5} = \mathbf{14.21\ N}$$

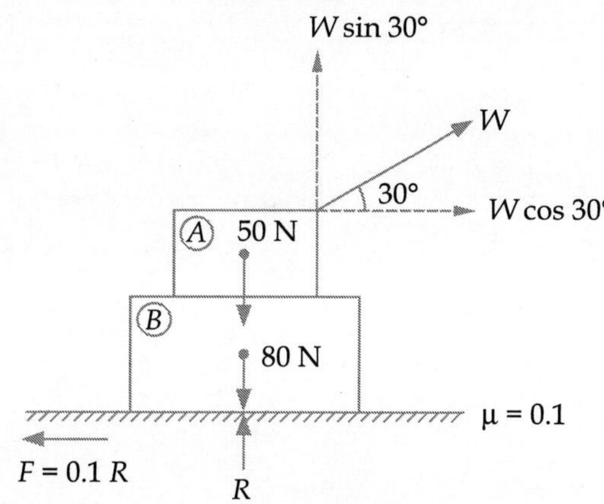

Fig. 4.19

This is the minimum weight in the pan required to start the motion.

(*b*) When the weight in the pan exceeds beyond a certain limit, the block *A* may move with respect to *B*. The block *B* will then remain stationary with respect to ground.

Considering free body diagram of block *A*, the conditions of limiting equilibrium are :

$$\Sigma F_x = 0 = W\cos 30° - 0.3R$$

$$R = \frac{W\cos 30°}{0.3} = 2.887\,W$$

$$\Sigma F_y = 0 = R - 50 + W\sin 30°$$

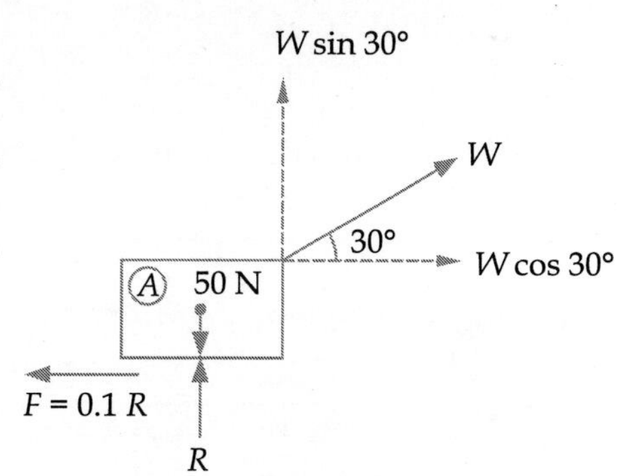

Fig. 4.20

$$= 2.887\,W - 50 + 0.5\,W$$

$$\therefore \quad W = \frac{50}{2.887 + 0.5} = \mathbf{14.76\ N}$$

EXAMPLE 4.12

Two blocks of weight W_1 and W_2 are connected by a string and rest on a rough horizontal plane as shown in Fig. 4.21. Set up an expression for the magnitude and direction of the least force P that needs to be applied to the upper block to induce sliding. For each block and the surface, the coefficient of friction may be taken as μ.

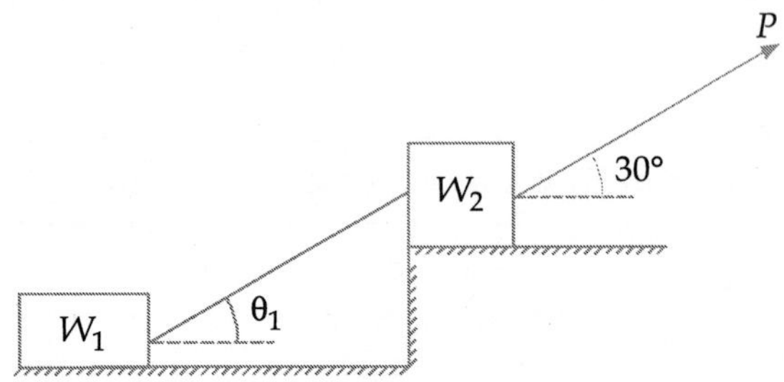

Fig. 4.21

Solution : Refer Fig. 4.22 for the free body diagrams for the two blocks for the impending motion.

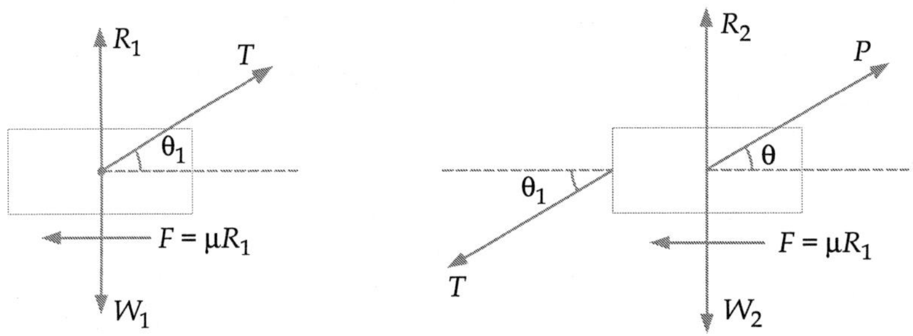

Fig. 4.22

Writing the equations of equilibrium for block W_1, we get

$$T\cos\theta_1 - \mu R_1 = 0\,; \quad T\cos\theta_1 = \mu R_1$$

$$T\sin\theta_1 + R_1 - W_1 = 0; \quad R_1 = W_1 - T\sin\theta_1$$

$$\therefore \quad T\cos\theta_1 = \mu(W_1 - T\sin\theta_1) = \mu W_1 - \mu T\sin\theta_1$$

$$T(\cos\theta_1 + \mu\sin\theta) = \mu W_1$$

$$T = \frac{\mu W_1}{\cos\theta_1 + \mu\sin\theta_1} \qquad ...(i)$$

Writing the equations of equilibrium for block W_2, we get

$$P\cos\theta - T\cos\theta_1 - \mu R_2 = 0\,; \quad T\cos\theta_1 = P\cos\theta - \mu R_2$$

$$R_2 - W_2 + P\sin\theta - T\sin\theta_1 = 0\,; \quad R_2 = W_2 - P\sin\theta + T\sin\theta_1$$

$$\therefore \quad T\cos\theta_1 = P\cos\theta - \mu(W_2 - P\sin\theta + T\sin\theta_1)$$

$$= P\cos\theta - \mu W_2 + \mu P\sin\theta - \mu T\sin\theta_1$$

or

$$T(\cos\theta_1 + \mu\sin\theta_1) = P(\cos\theta + \mu\sin\theta) - \mu W_2$$

$$T = \frac{P(\cos\theta + \mu\sin\theta) - \mu W_2}{\cos\theta_1 + \mu\sin\theta_1} \qquad ...(ii)$$

Equating expression (*i*) and (*ii*),

$$\frac{\mu W_1}{\cos\theta_1 + \mu\sin\theta_1} = \frac{P(\cos\theta + \mu\sin\theta) - \mu W_2}{\cos\theta_1 + \mu\sin\theta_1}$$

or

$$\mu W_1 = P(\cos\theta + \mu\sin\theta) - \mu W_2$$

$$\therefore \qquad P = \frac{\mu (W_1 + W_2)}{\cos\theta + \mu \sin\theta}$$

Substituting $\mu = \tan\phi = \dfrac{\sin\phi}{\cos\phi}$ where ϕ is the angle of friction,

$$P = \frac{\dfrac{\sin\phi}{\cos\phi}(W_1 + W_2)}{\cos\theta + \dfrac{\sin\phi}{\cos\phi}\sin\theta}$$

$$= \frac{\sin\phi (W_1 + W_2)}{\cos\theta\cos\phi + \sin\phi\sin\theta} = \frac{(W_1 + W_2)\sin\phi}{\cos(\theta - \phi)} \qquad ...(4.3)$$

The pull P will be least if the demoninator is maximum. The parameter $\cos(\theta - \phi)$ has a maximum value equal to 1.

$$\cos(\theta - \phi) = 1 = \cos 0$$

$$\theta - \phi = 0 ; \quad \theta = \phi \qquad ...(4.4)$$

Thus the tractive force P will be minimum if its inclination with the horizontal is equal to the angle of friction. Further

$$P_{min} = (W_1 + W_2)\sin\phi \qquad ...(4.5)$$

EXAMPLE 4.13

A homogeneous block of weight *W* and height *b* rests on a horizontal plane on its square base of side *a*. A force *P* is applied horizontally to the block at height *h*. If the coefficient of friction is μ, determine the maximum value of *h* so that the block slides without toppling.

Solution : Refer to Fig. 4.23 for the force system and the free body diagram.

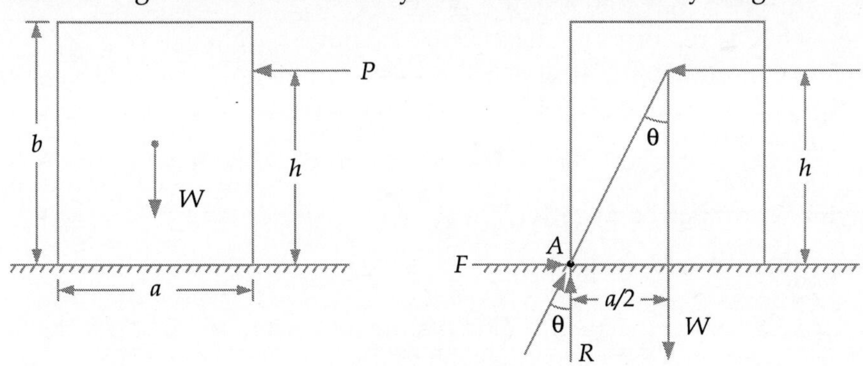

Fig. 4.23

At the instant of toppling, the entire reaction between the plane and the block will be at point A. The body is acted upon by the following set of forces:

(*i*) Weight W of the block

(*ii*) Force applied P

(*iii*) Resultant force due to the frictional force F and the normal reaction

Since the system is in equilibrium, these forces would be concurrent and pass through a point. Then

$$\tan\theta = \frac{a/2}{h} = \frac{a}{2h} ; \quad h = \frac{a}{2\tan\theta}$$

Also $\mu = \tan\theta$

$$\therefore \quad h = \frac{a}{2\mu}$$

This is the maximum height at which a horizontal force can be applied to slide the block without toppling.

EXAMPLE 4.14

A steel shelf 1.5 m high × 1.0 m wide and weighing 400 N is mounted on bushes *A* and *B* (Fig. 4.24). These bushes donot rotate when the shelf is moved along the floor. Assuming that coefficient of friction between the bushes and floor is 0.75, work-out the force required just to cause the shelf to move. If the shelf is not to tip over, determine the maximum height at which the force can be applied.

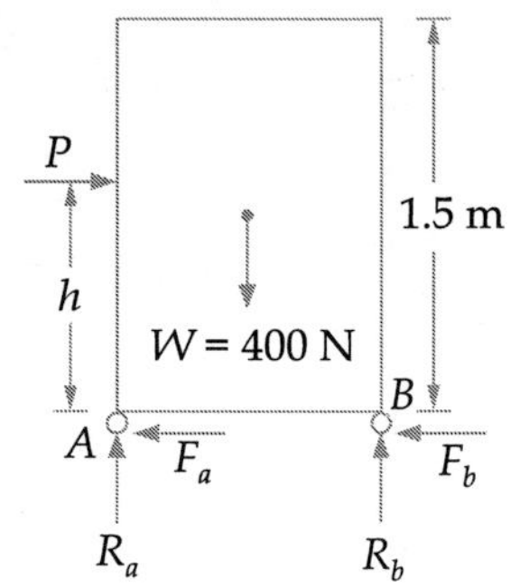

Fig. 4.24

Solution : Due to symmetry of the system, the reactions R_a and R_b at the rollers are equal

$$R_a = R_b = \frac{400}{2} = 200 \text{ N}$$

Let *P* be the push that needs to be applied to the shelf to just set it into motion towards right. Force of friction is always opposite to the direction of motion. Accordingly frictional force at the rollers would act towards left and it equals

$$F_a = F_b = \mu R = 0.75 \times 200 = 150 \text{ N}$$

In the limiting equilibrium (impending motion), the forces are balanced. That is

$$\Sigma F_x = 0; \quad P - F_a - F_b = 0$$

$$P = F_a + F_b = 150 + 150 = 300 \text{ N}$$

Thus a force of 300 N is required for impending motion to right.

(*b*) Let *h* be the maximum height at which the push can be applied without causing tip over of the shelf. Taking moments about the point B (clockwise positive)

$$300\,h - 400 \times 0.5 + R_a \times 1 = 0$$

$$R_a = 200 - 300\,h$$

Since the reaction R_a cannot be negative

$$h_{\max} = \frac{200}{300} = 0.667 \text{ m}$$

Obviously when $h \le 0.667$ m the shelf will just slide and when $h > 0.667$ m the shelf will tip over.

EXAMPLE 4.15

How high can a particle rest inside a hollow sphere of radius *r* if the coefficient of friction be $\frac{1}{\sqrt{3}}$?

Solution : With reference to Fig. 4.25, let the particle be at rest in limiting equilibrium at point *A* which is at height *h* = *AD* from the lowest point of the sphere. Further, the particle is in equilibrium under the following set of forces:

(*i*) Weight *W* of particle acting vertically downwards,

(*ii*) Normal reaction *R* which is at right angle to the tangent *AE* and passes through the centre *O* of the sphere, (*iii*) Friction force $F = \mu R$ which acts along the tangent *AE*.

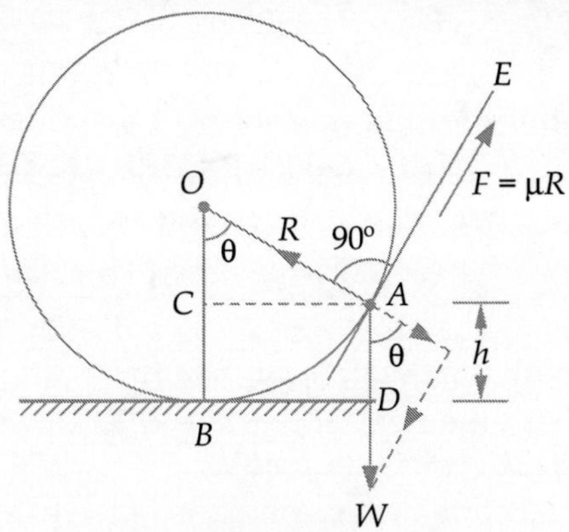

Fig. 4.25

Resolving the forces along and perpendicular to *AE*, we have

$$\mu R - W \sin\theta \quad \text{and} \quad R - W\cos\theta$$

For the system to be in equilibrium, the resolved parts of the forces along any two mutually perpendicular directions are separately equal to zero. Accordingly

$$\mu R - W\sin\theta = 0 \quad \text{and} \quad R - W\cos\theta = 0$$

From these two relations,

$$\mu W\cos\theta - W\sin\theta = 0$$

or $$\tan\theta = \mu = \frac{1}{\sqrt{3}}; \quad \theta = 30°$$

From the figure,

Height or particle, $h = DA = BC = BO - CO$

$$= BO - \sqrt{(AO^2 - AC^2)}$$

$$= r - \sqrt{r^2 - (r\sin 30°)^2}$$

$$= r - \sqrt{r^2 - \frac{r^2}{4}} = r - r\sqrt{1 - \frac{1}{4}}$$

$$= r\left(1 - \frac{\sqrt{3}}{2}\right) = \mathbf{0.135\ r}$$

EXAMPLE 4.16

A rectangular block, 1.25 m high × 1 m wide, weighing 200 N rests on a rough horizontal plane (μ = 0.25). A pull *P* is applied to the block at its left hand top corner at an angle of 25° with the horizontal. Make calculations for the value of *P* if the block is to

(*i*) slide over the horizontal surface, and

(*ii*) tip about the left hand bottom edge.

Comment on the results.

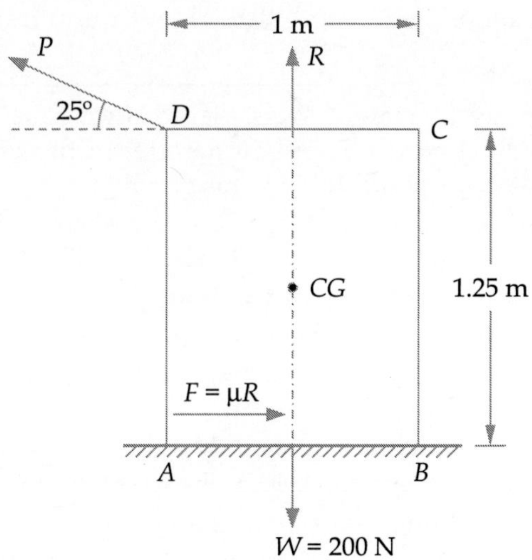

Fig. 4.26

Solution : When the block slides, it is in equilibrium under the following set of forces:

(*i*) Weight *W* of block acting vertically downwards through its centre of gravity *CG*.

(*ii*) the normal reaction *R* acting upward along the line of action of weight.

(*iii*) Friction force $F = \mu R$ acting along the bottom side towards right (opposite to the direction in which the block slides).

Under equilibrium, the components of all forces along horizontal and vertical directions are zero. That is

$$F - P\cos 25° = 0; \quad \mu R = P\cos 25° = 0.906\,P \quad \text{...}(i)$$

and $R - W + P \sin 25° = 0$; $R = W - P \sin 25°$

$$= 200 - 0.423\,P$$

Substituting this value of R in identity (*i*), we get

$$0.25(200 - 0.423P) = 0.906P$$

or

$$50 - 0.106P = 0.906P$$

$\therefore$ Magnitude of pull, $P = \dfrac{50}{0.906 + 0.106} = \mathbf{49.41\ N}$

(*ii*) When the block is to tip about the corner point A, both the normal reaction R and frictional force $F = \mu R$ will act at this point.

At equilibrium condition of the block, the moments of all forces about any point is zero. Then taking moments of forces about point A at which the block is to tip, we get

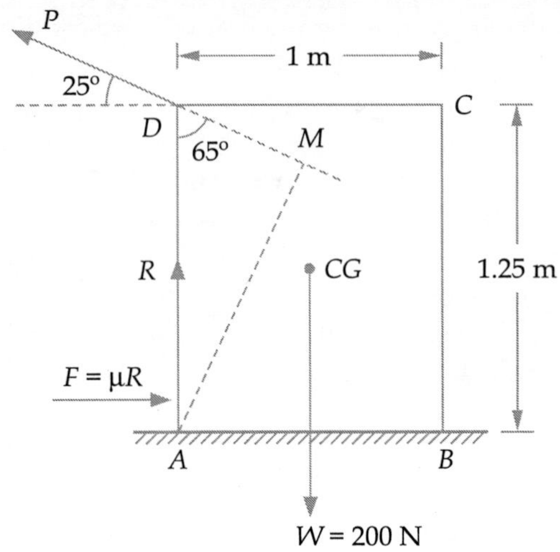

Fig. 4.27

$$P \times AM - W \times \frac{AB}{2} = 0$$

(moments of R and F are each equal to zero)

where AM is the perpendicular drawn on the line of action of the required force.

$$P \times 1.25 \sin 65° - 200 \times \frac{1}{2} = 0$$

$\therefore$ Magnitude of pull, $P = \dfrac{100}{1.25 \sin 65°} = \mathbf{88.3\ N}$

Comments: When the pull P is applied gradually, the block will slide before tipping. This stems from the fact that when P reaches a value of 49.41 N, sliding starts and it continues till P reaches 88.3 N. At P = 88.3 N, the block tips about the corner point A.

4.6. INCLINED PLANES

An inclined plane is a plane that makes a certain angle to the horizontal. Such planes are quite common, e.g., steep hill road and ramp in a building.

EXAMPLE 4.17

A body of weight W is placed on a rough inclined plane of inclination α. Find the force which should be applied parallel to the plane so that the body is on the point of motion (*i*) down the plane (*ii*) up the plane.

Solution : Refer to Fig. 4.28 which shows a body of weight W resting on a rough inclined plane of inclination α with the horizontal. The body is subjected to a force P acting along the plane

(*i*) When tendency of the body is to move downwards (Fig. 4.29):

The friction force on the body acts up the plane, since the body has impending motion down the plane. The various forces which keep the body in equilibrium and as shown in the free body diagram are:

(*i*) weight W of the body, (*ii*) reaction R of the plane,

(*iii*) tractive force P, and (*iv*) frictional force $F = \mu R$

Resolving these forces along and perpendicular to the plane, we get

$$P + \mu R - W \sin \alpha = 0\ ; \quad P = W \sin \alpha - \mu R$$

$$R - W\cos\alpha = 0\ ;\quad R = W\cos\alpha$$

$$\therefore\quad P = W\sin\alpha - \mu W\cos\alpha$$

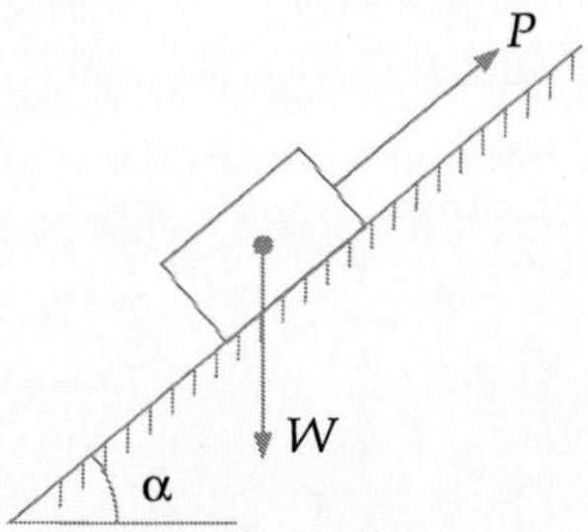

Fig. 4.28

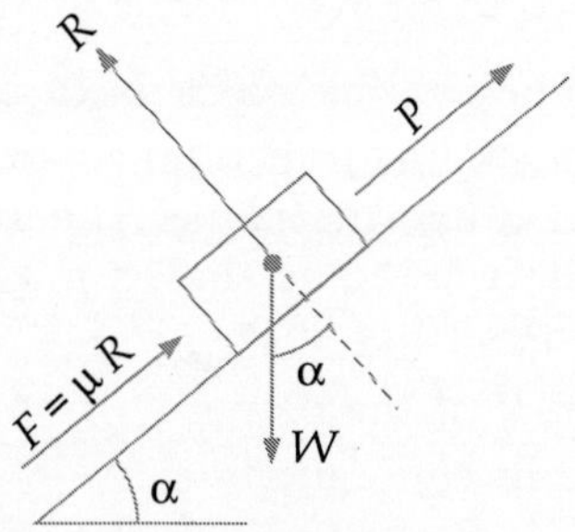

Fig. 4.29

Substituting $\mu = \tan\phi$ where ϕ is the angle of friction

$$P = W\sin\alpha - \tan\phi\, W\cos\alpha = W\sin\alpha - \frac{\sin\phi}{\cos\phi} W\cos\alpha$$

$$= \frac{W}{\cos\phi}(\sin\alpha\cos\phi - \sin\phi\cos\alpha) = \frac{W}{\cos\phi}\sin(\alpha - \phi) \qquad ...(4.6)$$

(*ii*) **When tendency of the body is to move upwards (Fig. 4.30):**

The friction force on the body acts down the plane, since the body has impending motion up the plane. The various forces which keep the body in equilibrium and as shown in the free body diagram are:

(*i*) weight W of the body,

(*ii*) reaction R of the plane,

(*iii*) tractive force P, and

(*iv*) frictional force $F = \mu R$

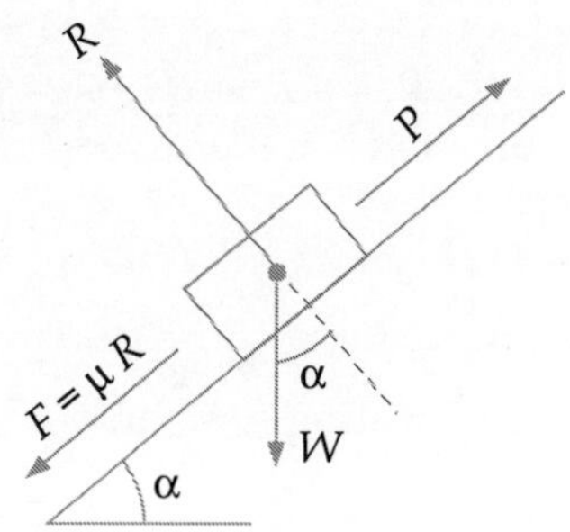

Fig. 4.30

Resolving these forces along and perpendicular to the plane, we get

$$P - \mu R - W\sin\alpha = 0$$

$$P = W\sin\alpha + \mu R$$

$$R - W\cos\alpha = 0\ ;\quad R = W\cos\alpha$$

$$\therefore\quad P = W\sin\alpha + \mu W\cos\alpha$$

Substituting $\mu = \tan\phi$ where ϕ is the angle of friction

$$P = W\sin\alpha + \tan\phi\, W\cos\alpha = W\sin\alpha + \frac{\sin\phi}{\cos\phi} W\cos\alpha$$

$$= \frac{W}{\cos\phi}(\sin\alpha\cos\phi + \sin\phi\cos\alpha) = \frac{W}{\cos\phi}\sin(\alpha + \phi) \qquad ...(4.7)$$

EXAMPLE 4.18

Find the least force required to drag a block of weight W placed on a rough inclined plane having inclination α with the horizontal. The tractive force applied to the block makes an angle θ to the inclined plane. Consider the following the cases: (*i*) the block is to move up the plane, (*ii*) the block is to move down the plane.

Solution : Refer to Fig. 4.31 which shows a block of weight W resting on a rough inclined plane of inclination α with the horizontal. The block is subjected to a force P acting at angle θ with the inclined plane.

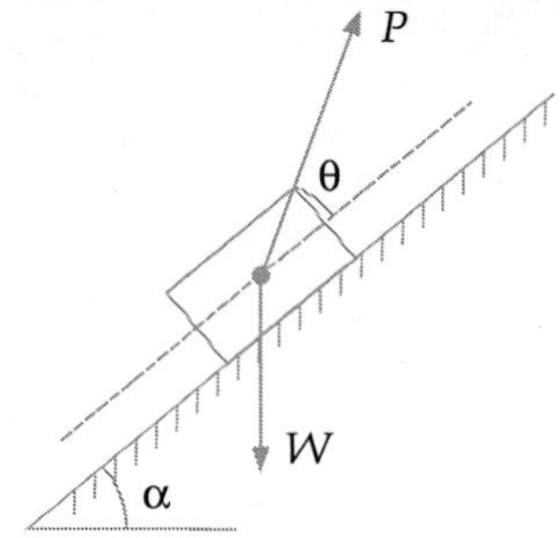

Fig. 4.31

(*i*) **When tendency of the block is to move up the plane (Fig. 4.31)**

Since the block has impending motion up the plane, the friction force would act down the plane. The various forces which keep the block in equilibrium and as shown in the free body diagram are:

(*i*) weight W of the block,

(*ii*) reaction R of the plane,

(*iii*) tractive force P, and

(*iv*) friction force $F = \mu R$

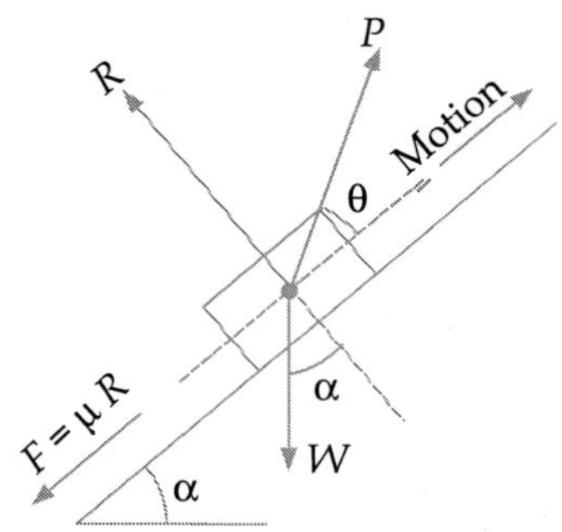

Fig. 4.32

Resolving these forces along and perpendicular to the plane, we get

$$P \cos\theta - \mu R - W \sin\alpha = 0$$

or $$P \cos\theta = W \sin\alpha + \mu R$$

and $$R - W \cos\alpha + P \sin\theta = 0$$

or $$R = W \cos\alpha - P \sin\theta$$

$$\therefore \quad P \cos\theta = W \sin\alpha + \mu (W \cos\alpha - P \sin\theta)$$

$$= W \sin\alpha + \mu W \cos\alpha - \mu P \sin\theta$$

or $$P \cos\theta + \mu P \sin\theta = W \sin\alpha + \mu W \cos\alpha$$

or $$P (\cos\theta + \mu \sin\theta) = W (\sin\alpha + \mu \cos\alpha)$$

$$\therefore \quad P = W \frac{\sin\alpha + \mu \cos\alpha}{\cos\theta + \mu \sin\theta}$$

Substituting $\mu = \tan\phi = \dfrac{\sin\phi}{\cos\phi}$, where ϕ is the angle of friction.

$$P = W \frac{\sin\alpha + \dfrac{\sin\phi}{\cos\phi}\cos\alpha}{\cos\theta + \dfrac{\sin\phi}{\cos\phi}\sin\theta} = W \frac{\sin\alpha \cos\phi + \sin\phi \cos\alpha}{\cos\theta \cos\phi + \sin\theta \sin\phi}$$

$$= W \frac{\sin(\alpha + \phi)}{\cos(\theta - \phi)} \qquad \text{...(4.8)}$$

The pull P will be least if the denominator $\cos(\theta - \phi)$ is maximum, and the maximum value of $\cos(\theta - \phi)$ is equal to 1

$$\cos(\theta - \phi) = \phi\ ; \quad \cos(\theta - \phi) = \cos 0$$

$$\theta - \phi = 0\ ; \quad \theta = \phi \qquad \text{...(4.9)}$$

Thus the tractive force P will be minimum if its inclination with the inclined plane is equal to the angle of friction. Further

$$P_{min} = W \sin(\alpha + \phi)$$

(*ii*) **When tendency of the block is to move down the plane**

Since the block has impending motion down the plane, the friction force would act up the plane. The various forces which keep the block in equilibrioum and as shown in the free body diagram are:

(*i*) weight W of the block, (*ii*) reaction R of the plane,
(*iii*) tractive force P, and (*iv*) frictional force $F = \mu R$

Resolving these forces along and perpendicular to the plane, we get

$$P \cos\theta + \mu R - W \sin\alpha = 0$$

or $\quad P \cos\theta = W \sin\alpha - \mu R$

and $\quad R - W \cos\alpha + P \sin\theta = 0\,;$

or $\quad R = W \cos\alpha - P \sin\theta$

$\therefore \quad P \cos\theta = W \sin\alpha - \mu (W \cos\alpha - P \sin\theta)$

$\quad = W \sin\alpha - \mu W \cos\alpha + \mu P \sin\theta.$

or $\quad P(\cos\theta - \mu \sin\theta) = W(\sin\alpha - \mu\cos\alpha)$

$$\therefore \quad P = W\,\frac{\sin\alpha - \mu\cos\alpha}{\cos\theta - \mu\sin\theta}$$

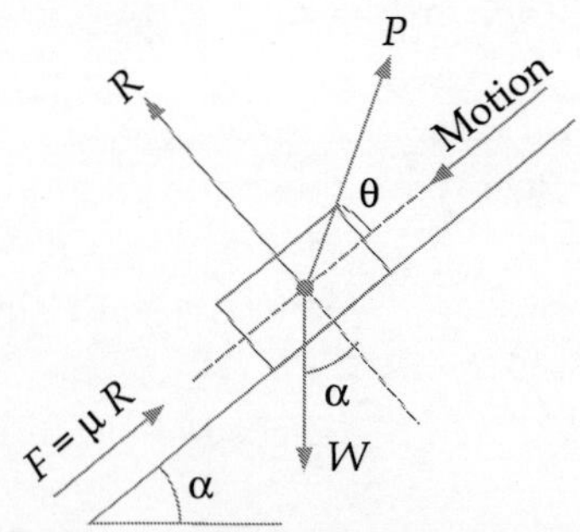

Fig. 4.33

Substituting $\mu = \tan\phi = \dfrac{\sin\phi}{\cos\phi}$ where ϕ is the angle of friction

$$P = W\,\frac{\sin\alpha - \dfrac{\sin\phi}{\cos\phi}\cos\alpha}{\cos\theta - \dfrac{\sin\phi}{\cos\phi}\sin\theta}$$

$$= W\,\frac{\sin\alpha\cos\phi - \sin\phi\cos\alpha}{\cos\theta\cos\phi - \sin\theta\sin\phi} = W\,\frac{\sin(\alpha - \phi)}{\cos(\theta + \phi)} \qquad ...(4.10)$$

EXAMPLE 4.19

A block of weight *W* rests on a rough inclined plane of inclination α with the horizontal. The block is subjected to a force *P* acting horizontal. Set up the following relations for the equilibrium of the block:

$P = W \tan(\alpha - \phi)$ when the block is at the point of sliding down

$\quad = W \tan(\alpha + \phi)$ when the block is at the point of sliding up

Here ϕ is the angle of friction.

Solution : Refer to Fig. 4.34 which shows a block of weight W resting on a rough inclined plane of inclination α with the horizontal. The block is subjected to a force P that acts horizontally.

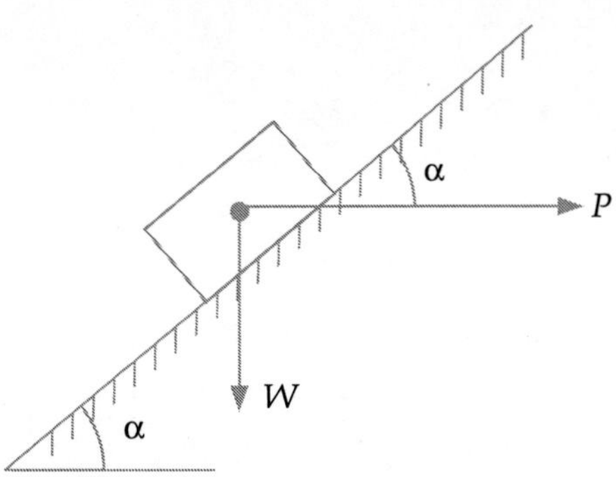

Fig. 4.34

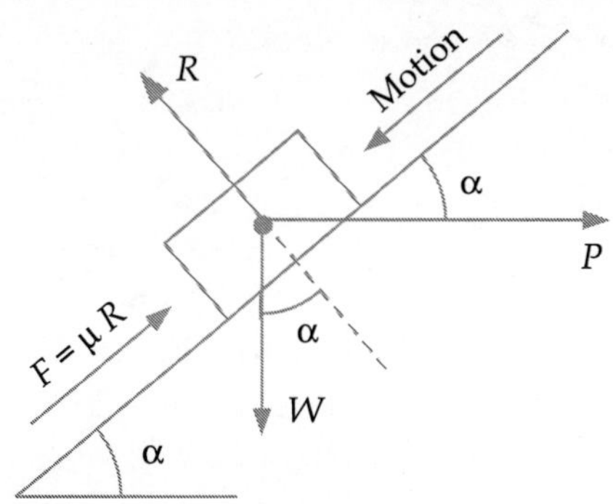

Fig. 4.35

(*i*) **When tendency of the block is to move down the plane. (Fig. 4.35)**

Since the block has impending motion down the plane, the friction force would act up the plane. The various forces which keep the block in equilibrium and as shown in the free body diagram are:

(*i*) weight W of the block, (*ii*) reaction R of the plane,
(*iii*) tractive force P, and (*iv*) frictional force $F = \mu R$

Resolving these forces along and perpendicular to the plane, we get

$$P \cos \alpha + \mu R - W \sin \alpha = 0$$

or $$P \cos \alpha = W \sin \alpha - \mu R$$

and $$R - W \cos \alpha - P \sin \alpha = 0 \; ; \quad R = W \cos \alpha + P \sin \alpha$$

$\therefore$ $$P \cos \alpha = W \sin \alpha - \mu (W \cos \alpha + P \sin \alpha)$$
$$= W \sin \alpha - \mu W \cos \alpha - \mu P \sin \alpha$$

or $$P (\cos \alpha + \mu \sin \alpha) = W (\sin \alpha - \mu \cos \alpha)$$

$\therefore$ $$P = W \frac{\sin \alpha - \mu \cos \alpha}{\cos \alpha + \mu \sin \alpha}$$

Substituting $\mu = \tan \phi = \dfrac{\sin \phi}{\cos \phi}$ where ϕ is the angle of friction

$$P = W \frac{\sin \alpha - \dfrac{\sin \phi}{\cos \phi} \cos \alpha}{\cos \alpha + \dfrac{\sin \phi}{\cos \phi} \sin \alpha} = W \frac{\sin \alpha \cos \phi - \sin \phi \cos \alpha}{\cos \alpha \cos \phi + \sin \alpha \sin \phi}$$

$$= W \frac{\sin (\alpha - \phi)}{\cos (\alpha - \phi)} = W \tan (\alpha - \phi) \qquad \text{...(4.11)}$$

Note : The above relation is essentially valid when $\alpha > \phi$. In case $\phi > \alpha$, this relation takes the form

$$P = W \tan (\phi - \alpha)$$

(*ii*) **When tendency of the block is to move up the plane (Fig. 4.36)**

Since the block has impending motion up the plane, the friction force would act down the plane. The various forces which keep the block in equilibrium and as shown in the free body diagram are:

(*i*) weight W of the block,
(*ii*) reaction R of the plane
(*iii*) tractive force P, and
(*iv*) frictional force $F = \mu R$

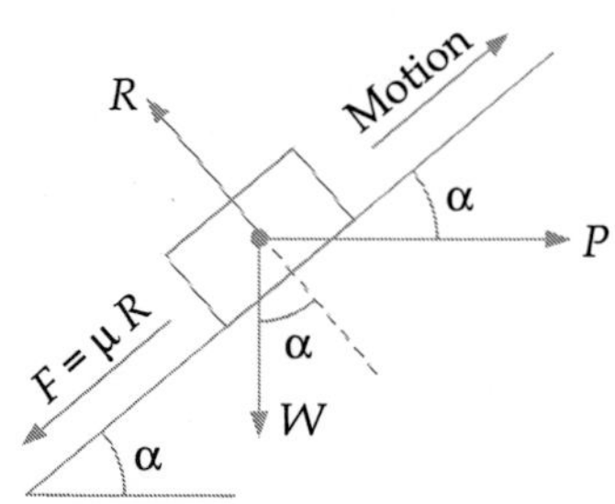

Fig. 4.36

Resolving these forces along and perpendicular to the plane, we get

$$P \cos \alpha - \mu R - W \sin \alpha = 0 \; ; \quad P \cos \alpha = W \sin \alpha + \mu R$$
$$R - W \cos \alpha - P \sin \alpha = 0 \; ; \quad R = W \cos \alpha + P \sin \alpha$$

$\therefore$ $$P \cos \alpha = W \sin \alpha + \mu (W \cos \alpha + P \sin \alpha)$$
$$= W \sin \alpha + \mu W \cos \alpha + \mu P \sin \alpha$$

or $$P (\cos \alpha - \mu \sin \alpha) = W (\sin \alpha + \mu \cos \alpha)$$

$\therefore$ $$P = W \frac{\sin \alpha + \mu \cos \alpha}{\cos \alpha - \mu \sin \alpha}$$

Substituting $\mu = \tan\phi = \dfrac{\sin\phi}{\cos\phi}$ where ϕ is the angle of friction

$$P = W\,\frac{\sin\alpha + \dfrac{\sin\phi}{\cos\phi}\cos\alpha}{\cos\alpha - \dfrac{\sin\phi}{\cos\phi}\sin\alpha} = W\,\frac{\sin\alpha\cos\phi + \sin\phi\cos\alpha}{\cos\alpha\cos\phi - \sin\phi\sin\alpha}$$

$$= W\,\frac{\sin(\alpha+\phi)}{\cos(\alpha+\phi)} = W\tan(\alpha+\phi) \quad ...(4.12)$$

EXAMPLE 4.20

A block, in the shape of a rectangular prism rests on a rough inclined plane, as shown in Fig. 4.37. The block is tied up by a horizontal string which has a tension of 10 N. If the block weighs 35 N, determine:

(*i*) the frictional force on the block,

(*ii*) the normal reaction of the inclined plane, and

(*iii*) the coefficient of friction between the contacting surfaces.

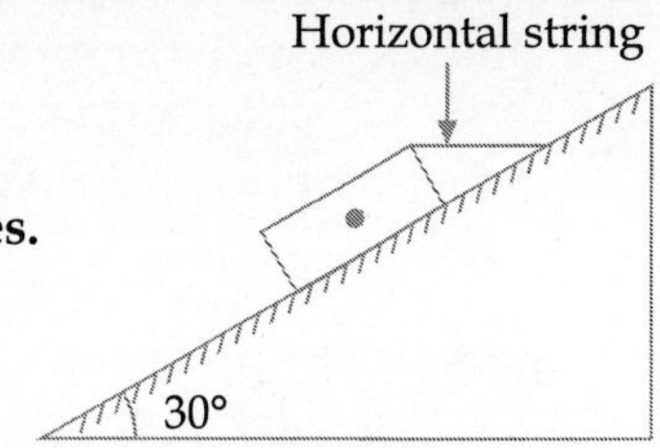

Fig. 4.37

Solution : The various forces acting on the block are:

(*i*) weight W = 35 N of the block,

(*ii*) tension T = 10 N in the string,

(*iii*) normal reaction R at the contact surface, and

(*iv*) friction force $F = \mu R$

Refer to Fig. 4.38 for the free body diagram of the force system under equilibrium conditions,

$$\Sigma F_x = 0 \text{ (along the plane)}$$

$$\mu R - W\sin 30^\circ + T\cos 30^\circ = 0$$

$$\mu R = W\sin 30^\circ - T\cos 30^\circ$$

$$= 35\sin 30^\circ - 10\cos 30^\circ = 35 \times 0.5 - 10 \times 0.866$$

$$= 17.5 - 8.66 = 8.84 \text{ N} \quad ...(i)$$

$$\Sigma F_y = 0 \text{ (perpendicular to the plane)}$$

$$R - W\cos 30^\circ - T\sin 30^\circ = 0$$

$$R = W\cos 30^\circ + T\sin 30^\circ = 35 \times 0.866 + 10 \times 0.5$$

$$= 30.31 + 5 = 35.31 \text{ N} \quad ...(ii)$$

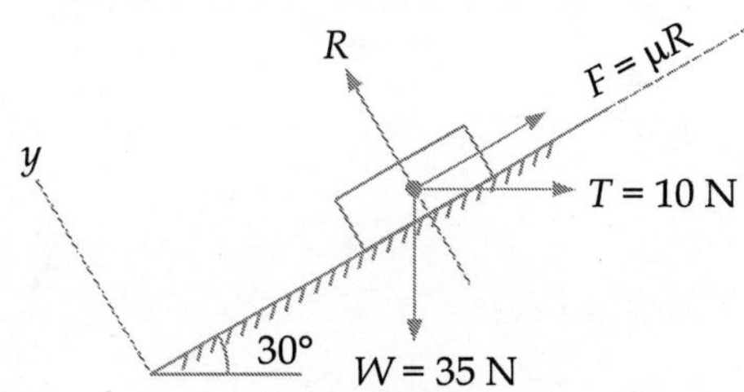

Fig. 4.38

From identities (*i*) and (*ii*)

$$\mu = \frac{8.84}{35.31} = \mathbf{0.25}$$

EXAMPLE 4.21

Two blocks of weight W_1 and W_2 rest on a rough inclined plane and are connected by a string. The coefficients of friction between these blocks and the plane are μ_1 and μ_2 respectively. Show that the blocks will be on the point of motion when the inclination of the plane with the horizontal is given by

$$\tan\theta = \frac{\mu_1 W_1 + \mu_2 W_2}{W_1 + W_2}$$

What will be this inclination if the blocks are of equal weight and the friction coefficients are $\mu_1 = 1/2$ and $\mu_2 = 1/3$?

Solution : Refer Fig. 4.39 for the arrangement and the free body diagrams for the two blocks.

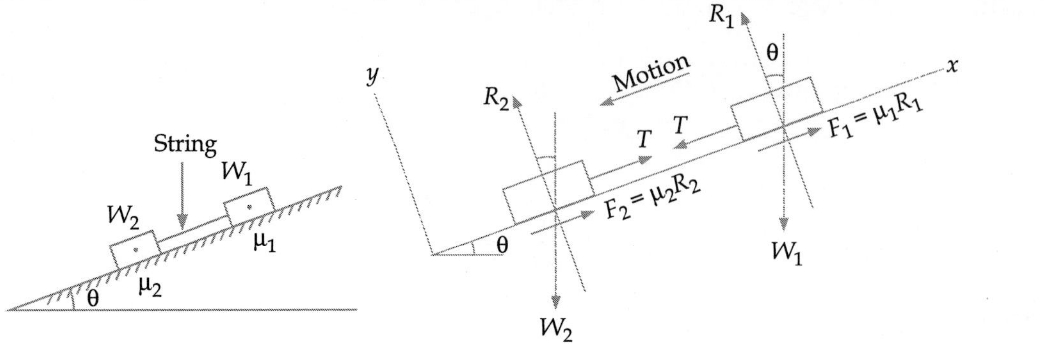

Fig. 4.39

Let T be the tension in the string and θ be the inclination of the plane with the horizontal.

Considering equilibrium for block of weight W_1

$\Sigma F_x = 0$ (along the plane)

$$\mu_1 R_1 - T - W_1 \sin\theta = 0 \quad ...(i)$$

$\Sigma F_y = 0$ (perpendicular to the plane)

$$R_1 - W_1 \cos\theta = 0; \quad R_1 = W_1 \cos\theta \quad ...(ii)$$

From identities (*i*) and (*ii*)

$$\mu_1 W_1 \cos\theta - W_1 \sin\theta - T = 0 \quad ...(a)$$

Similarly, considering equilibrium for block of weight W_2, we may write

$$\mu_2 R_2 - W_2 \sin\theta + T = 0$$

$$R_2 - W_2 \cos\theta = 0; \quad R_2 = W_2 \cos\theta$$

or

$$\mu_2 W_2 \cos\theta - W_2 \sin\theta + T = 0 \quad ...(b)$$

Adding expressions (*a*) and (*b*), we get

$$(\mu_1 W_1 + \mu_2 W_2)\cos\theta - (W_1 + W_2)\sin\theta = 0$$

$$\therefore \quad \tan\theta = \frac{\mu_1 W_1 + \mu_2 W_2}{W_1 + W_2}$$ which is the required expression.

(*b*) When $W_1 = W_2 = W$, the above expression reduces to

$$\tan\theta = \frac{\mu_1 + \mu_2}{2}$$

Substituting $\mu_1 = 1/2$ and $\mu_2 = 1/3$,

$$\tan\theta = \frac{1/2 + 1/3}{2} = \frac{5}{12}$$

∴ Inclination of plane, $\theta = \tan^{-1}(5/12) = 22.62°$

EXAMPLE 4.22

A block of weight W_1 = 100 N rests on an inclined plane and another weight W_2 is attached to the first weight through a string as shown in Fig. 4.40. If the coefficient of friction between the block and plane is 0.3, determine the maximum and minimum values of W_2 so that equilibrium can exist.

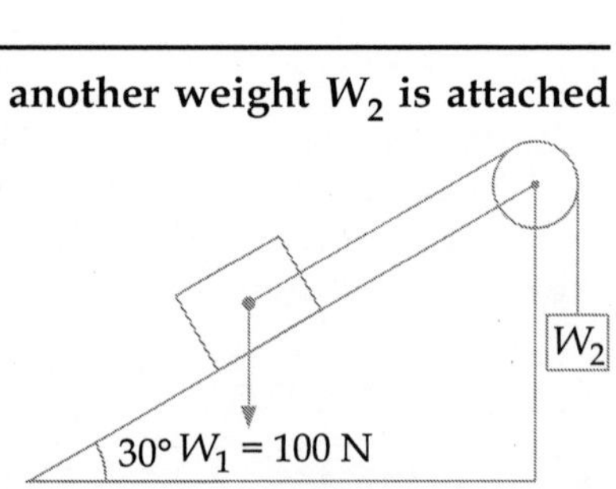

Fig. 4.40

Solution : There will be two values for W_2

(*i*) When weight W_1 = 100 N is about to slide down the plane

(*ii*) When weight W_1 = 100 N is about to slide up the plane

Case (*i*): Refer Fig. 4.41 (*a*) for the free body diagram for impending motion down the plane. Under equilibrium conditions,

$$\Sigma F \text{ (along the plane)} = 0$$

$$W_2 + \mu R - W_1 \sin 30° = 0$$

$$W_2 = W_1 \sin 30° - \mu R = 100 \sin 30° - 0.3 R = 50 - 0.3 R$$

$$\Sigma F \text{ (perpendicular to the plane)} = 0$$

$$R = 100 \cos 30° = 86.6 \text{ N}$$

$$\therefore \quad W_2 = 50 - 86.6 \times 0.3 = 24.02 \text{ N} \sim \mathbf{24} \text{ N}$$

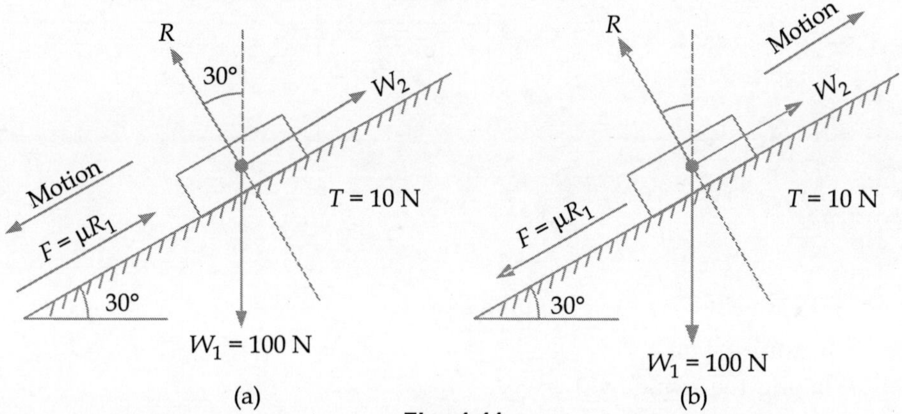

Fig. 4.41

Case (*ii*): Refer Fig. 4.41 (*b*) for the free body diagram for impending motion up the plane. Under equilibrium conditons,

$$\Sigma F \text{ (along the plane)} = 0$$

$$W_2 - W_1 \sin 30° - \mu R = 0$$

$$W_2 = \mu R + W_1 \sin 30° = 0.3 R + 100 \sin 30° = 0.3 R + 50$$

$$\Sigma F \text{ (perpendicular to plane)} = 0$$

$$R = 100 \cos 30° = 86.6$$

$$\therefore \quad W_2 = 0.3 \times 86.6 + 50 = 75.98 \text{ N} \sim \mathbf{76 \text{ N}}$$

Therefore the limiting values of weight W are 24 N and 76 N.

EXAMPLE 4.23

Two blocks of weight 200 N and 300 N and connected by a string passing over a frictionless pulley rest on rough surfaces; block of weight 200 N on horizontal surface and the other on an inclined surface as shown in Fig. 4.42.

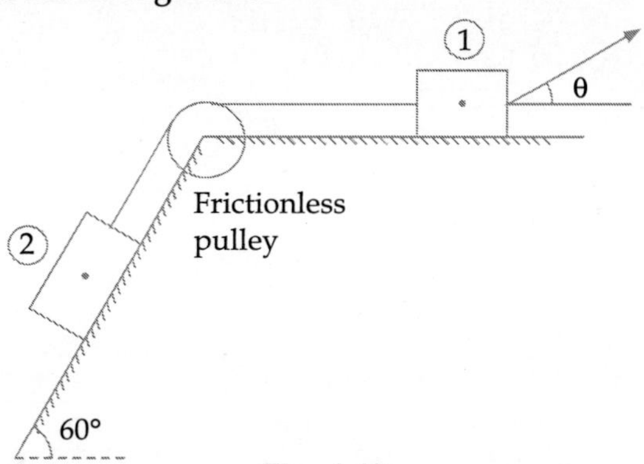

Fig. 4.42

For both the surfaces the coefficients of friction μ = 0.25. Find out the minimum value of force, both in magnitude and direction, for the motion to impend.

Solution : Refer to Fig. 4.43 for the various forces acting on the two blocks:

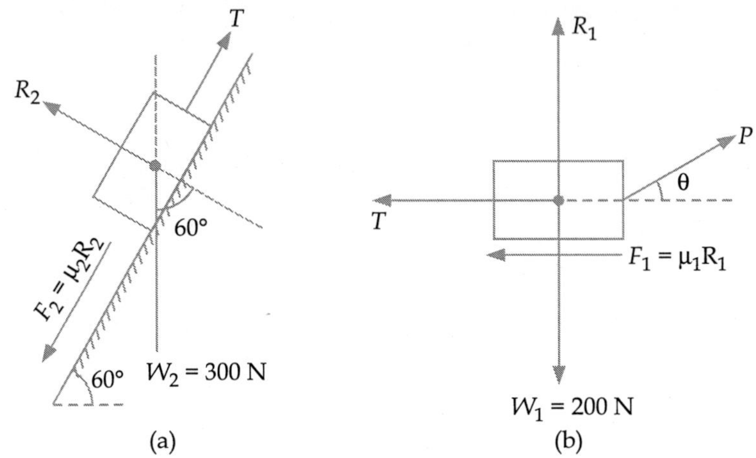

Fig. 4.43

Considering equilibrium of block 2

$$\Sigma F \text{ (along the plane)} = 0$$
$$T - W_2 \sin 60^\circ - \mu_2 R_2 = 0$$
$$T = W_2 \sin 60^\circ + \mu_2 R_2 = 300 \times 0.866 + 0.25\, R_2$$
$$\Sigma F \text{ (perpendicular to the plane)} = 0$$
$$R_2 - W_2 \cos 60^\circ = 0\ ;$$
$$R_2 = W_2 \cos 60^\circ = 300 \times 0.5 = 150 \text{ N}$$
$$\therefore \quad T = 300 \times 0.866 + 0.25 \times 150 = 259.8 + 37.5 = \mathbf{297.3\ N}$$

Considering equilibrium of block 1

$$\Sigma F \text{ (along the plane)} = 0$$
$$P \cos\theta - T - \mu_1 R_1 = 0$$
or $$P \cos\theta - 297.3 - 0.25\, R_1 = 0 \qquad ...(i)$$
$$\Sigma F \text{ (perpendicular to the plane)} = 0$$
$$W_1 - R_1 - P \sin\theta = 0$$
or $$R_1 = W_1 - P\sin\theta = 200 - P \sin\theta$$

Substituting this value of R_1 in expression (i), we get
$$P\cos\theta - 297.3 - 0.25\,(200 - P\sin\theta) = 0$$
or $$P\cos\theta - 297.3 - 50 + 0.25\, P \sin\theta = 0$$
or $$P = \frac{347.3}{\cos\theta + 0.25\sin\theta}$$

For minimum value of P, the denominator $(\cos\theta + 0.25\sin\theta)$ should be maximum, i.e., its differential should be zero.

$$\frac{d}{d\theta}(\cos\theta + 0.25\sin\theta) = 0$$
or $$-\sin\theta + 0.25\cos\theta = 0\ ; \quad \tan\theta = 0.25$$
$$\therefore \quad \theta = 14^\circ$$

$$\therefore \quad \text{Minimum value of } P = \frac{347.3}{\cos 14^\circ + 0.25 \sin 14^\circ} = \frac{347.3}{0.97 + 0.25 \times 0.24} = \mathbf{337.18\ N}$$

EXAMPLE 4.24

Two blocks of weight 50 N and 200 N and connected by a cord rest on two inclined planes as shown in Fig. 4.44. Determine the maximum tension in the cord when limiting friction conditions develop for both the blocks.

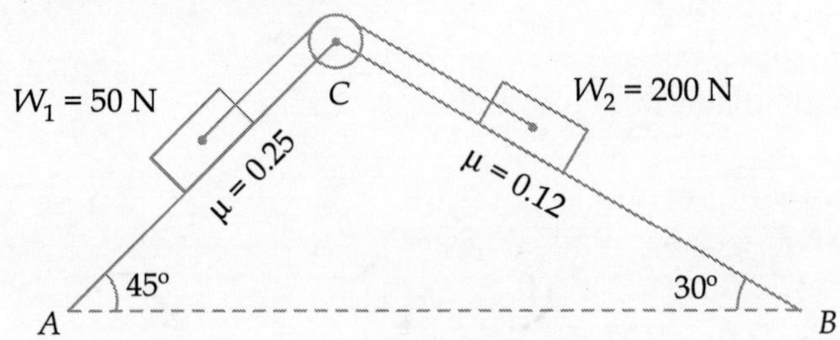

Fig. 4.44

Solution : Refer to Fig. 4.45 for the free body diagram of the two blocks

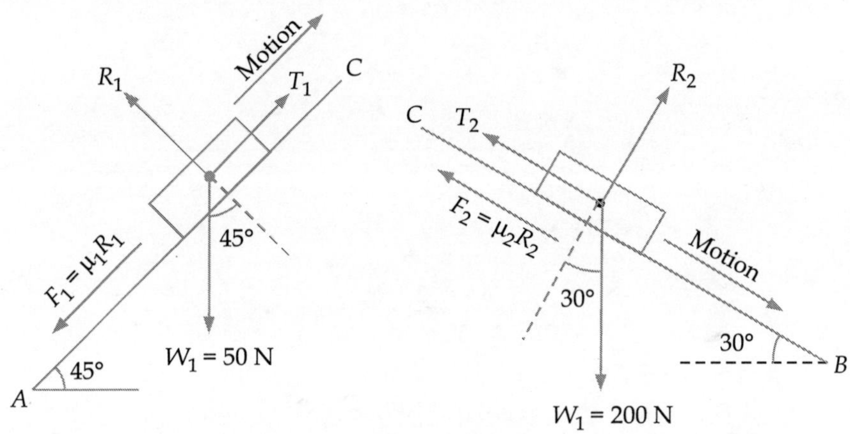

Fig. 4.45

Considering equilibrium of block resting on the plane of 45° inclination:

Resolving forces along and perpendicular to the plane AC

$$\mu_1 R_1 + W_1 \sin 45^\circ - T_1 = 0$$

$$R_1 - W_1 \cos 45^\circ = 0 \; ; \quad R_1 = 50 \cos 45^\circ = 35.35 \text{ N}$$

$$\therefore \quad T_1 = \mu R_1 + W_1 \sin 45^\circ = 0.25 \times 35.35 + 50 \sin 45^\circ$$

$$= 8.84 + 35.35 = \mathbf{44.19\ N}$$

Considering equilibrium of block resting on the plane of 30° inclination:

Resolving forces along and perpendicular to plane BC

$$T_1 + \mu_2 R_2 - W_2 \sin 30^\circ = 0$$

$$R_2 - W_2 \cos 30^\circ = 0 \; ; \quad R_2 = 200 \cos 30^\circ = 173.2 \text{ N}$$

$$\therefore \quad T_2 = W_2 \sin 30^\circ - \mu_2 R_2 = 200 \sin 30^\circ - 0.12 \times 173.2$$

$$= 100 - 20.78 = \mathbf{79.22} \text{ N}$$

Hence cord will be subjected to a maximum tension of **79.22 N**

EXAMPLE 4.25

Two masses m_1 = 22.5 kg and m_2 = 14 kg are tied together by a rope parallel to the inclined plane surface as shown in Fig. 4.46. The coefficient of friction between m_1 and the plane is 0.25 while that of mass m_2 and the plane is 0.5. Determine :

(*a*) the value of the inclination of the plane surface θ for which the masses will just start sliding downwards, and

(*b*) the tension is the rope.

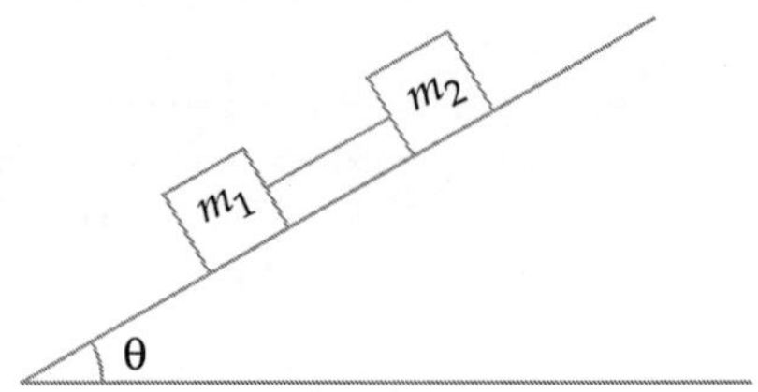

Fig. 4.46

Solution : Consider free body diagram of block of m_1 = 22.5 kg mass.

Applying conditions of equilibrium :

$\Sigma F_x = 0$; $\quad T + F - 22.5 \times 9.81 \sin\theta = 0$

or $\quad T + F_1 = 220.73 \sin\theta$

$\Sigma F_y = 0$; $\quad R_1 - 22.5 \times 9.81 \cos\theta$

or $\quad R_1 = 220.73 \cos\theta$

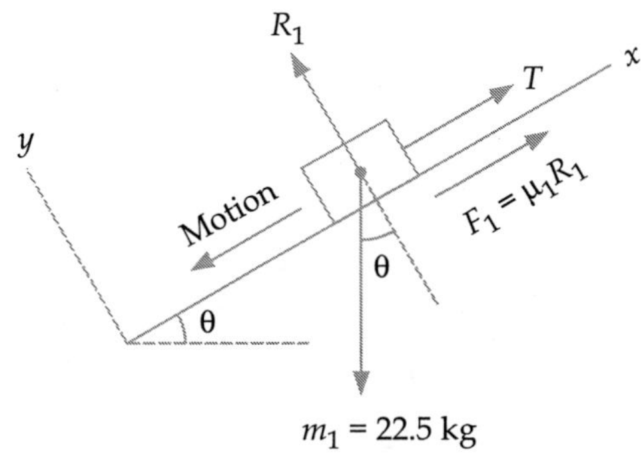

Fig. 4.47

But $\quad F_1 = \mu_1 R_1 = 0.25 \times 220.73 \cos\theta = 55.18 \cos\theta$

Accordingly $\quad T + 55.18 \cos\theta = 220.73 \sin\theta$

or $\quad T = 220.73 \sin\theta - 55.18 \cos\theta \quad$ ...(*a*)

Consider free body diagram of block of m_2 = 14 kg mass.

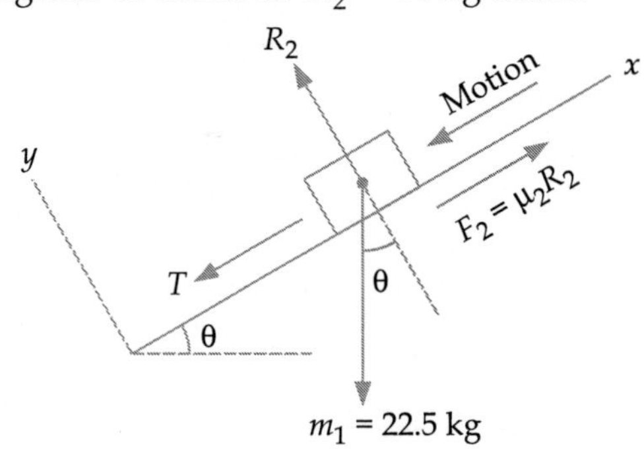

Fig. 4.48

Apply conditions of equilibrium :

$\Sigma F_x = 0$; $\quad -T + F - 14 \times 9.81 \sin\theta = 0$

or $\quad T = F_2 - 137.34 \sin\theta$

$\Sigma F_y = 0;$ $\quad R_2 - 14 \times 9.81 \cos\theta = 0$

or $\quad R_2 = 137.34 \cos\theta$

But $F_2 = \mu_2 R_2 = 0.5 \times 137.34 \cos\theta = 68.67 \cos\theta$

Accordingly $\quad T = 68.67 \cos\theta - 137.34 \sin\theta \qquad ...(b)$

From expressions (*a*) and (*b*).

$220.73 \sin\theta - 55.18 \cos\theta = 68.67 \cos\theta - 137.34 \sin\theta$

or $\quad 358.07 \sin\theta = 123.85 \cos\theta$

or $\quad \tan\theta = \dfrac{123.85}{358.07} = 0.3459; \quad \theta = 19.08°$

(*b*) Substituting $\theta = 19.08°$ in expression (*a*),

Tension in the rope $T = 220.73 \sin 19.08 - 55.18 \cos 19.08$

$= 72.15 - 52.15 = \mathbf{20\ N}$

EXAMPLE 4.26

The coefficient of static and kinetic friction between the 100 kg block and inclined plane (Fig. 4.49) are 0.3 and 0.2 respectively. Determine :

(*a*) the friction force *F* acting on the block when *P* is applied with a magnitude of 200 N to the block at rest,

(*b*) the force *P* required to initiate motion up the incline from rest, and

(*c*) the friction force *F* acting on the block if *P* = 600 N.

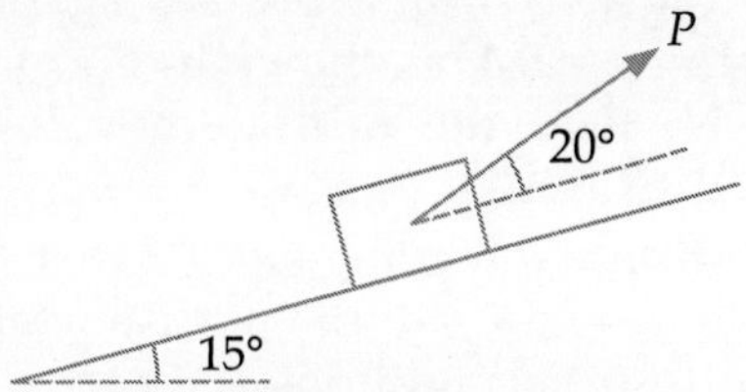

Fig. 4.49

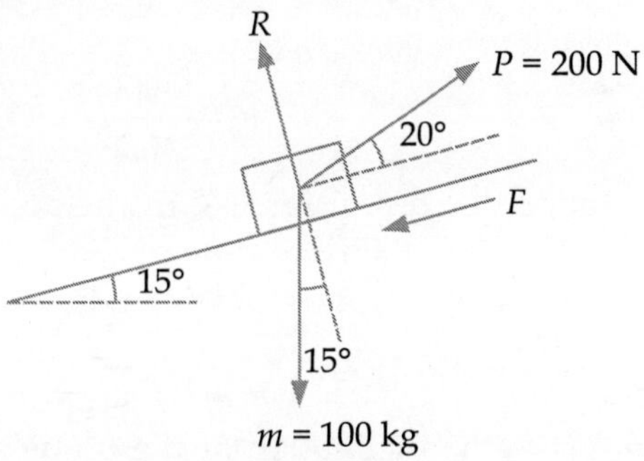

Fig. 4.50

Solution : Consider free body diagram of the block when $P = 200$ N is applied to the block at rest.

From condition of equilibrium,

$\Sigma F_x = 0;$

$200 \cos 20° - F - 100 \times 9.81 \sin 15° = 0$

$F = 200 \times 0.9397 - 100 \times 9.81 \times 0.2588$

$= 187.94 - 253.88 = \mathbf{-65.94\ N}$

The – *ve* sign indicates that when the block is at rest, the friction force is acting upwards along the incline.

(*b*) When the block just moves up the plane

The friction force would then act opposite to the direction of motion, i.e., downwards along the plane.

$F_x = 0;$ $\quad P \cos 20° - F - 100 \times 9.81 \sin 15° = 0$

$P \cos 20° = F + 100 \times 9.81 \times 0.259$

$0.9397P = 0.3\, R + 253.88 \qquad ...(i)$

$F_y = 0;$ $\quad R + P \sin 20° - 100 \times 9.81 \times \cos 15 = 0$

$R = 947.57 - 0.342P \qquad ...(ii)$

From expressions (*i*) and (*ii*)

$0.9397P = 0.3\,(947.57 - 0.342P) + 253.88$

$= 284.27 - 0.1026P + 253.88 = 538.15 - 0.1026P$

$$\therefore \qquad P = \frac{538.15}{0.9397 + 0.1026} = \mathbf{516.31\ N}$$

(*c*) Since the applied force P = 600 N is more than 516.31 N (a force which just initiates the motion), the block will move up along the plane and the friction force would act down the slope. Further, the kinetic friction would prevail with $\mu = 0.2$. Then from condition of equilibrium in the y-direction,

$$\Sigma F_y = 0; \qquad 600 \sin 20^\circ + R - 100 \times 9.81 \cos 15^\circ = 0$$

$$R = 100 \times 9.81 \cos 15^\circ - 600 \sin 20^\circ$$

$$= 947.57 - 205.21 = 742.36$$

Friction force $F = \mu R = 0.2 \times 742.36 =$ **148.47 N**

EXAMPLE 4.27

A block *A* weighing *W* newtons is placed on a rough inclined plane having μ = 0.2 and is held in position by means of a horizontal rod hinged to the block *B* which presses against a rough vertical wall having μ = 0.4, as shown in Fig. 4.51. If the block *B* weighs 500 N, find the minimum value of *W* consistent with equilibrium.

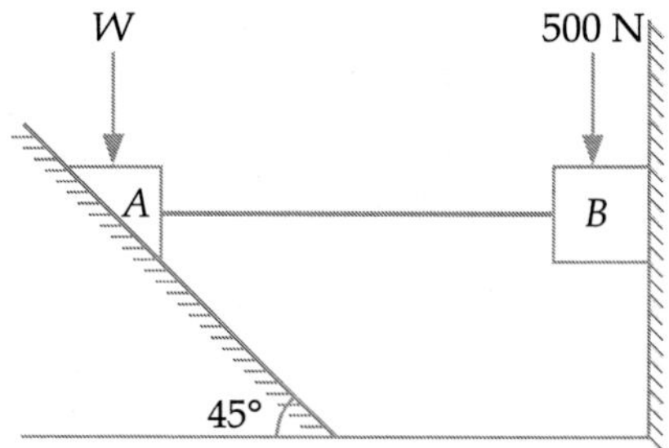

Fig. 4.51

Solution : When the block B just slides down, the frictional force $F = \mu R_b$ at the wall surface acts upwards. Under equilibrium conditions,

$$\Sigma F_x = 0; \qquad P - R_b = 0; \quad P = R_b$$

$$\Sigma F_y = 0; \qquad \mu R_b - 500 = 0$$

$$R_b = \frac{500}{0.4} = \mathbf{1250\ N}$$

Hence horizontal force P in the rod is 1250 N.

With block B sliding downwards, the block A has a tendency to move up the plane. The following relationship then holds good between W and P,

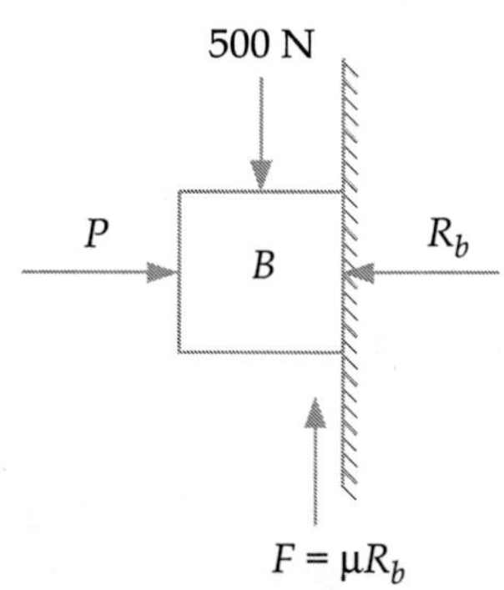

Fig. 4.52

$$P = W \tan(\alpha + \phi)$$

where α is the inclination of the plane and ϕ is the angle of friction for the block

$$\tan\phi = \mu = 0.2; \quad \phi = 11.30^\circ$$

$$\therefore \qquad 1250 = W \tan(45 + 11.30) = 1.5\,W$$

That gives :

$$W = \frac{1250}{1.5} = \mathbf{833.3\ N}$$

Thus the minimum value of W consistent with equilibrium is 833.3 N.

EXAMPLE 4.28

Two identical blocks *A* and *B* are connected by a rod and rest respectively against vertical wall and horizontal floor as shown in Fig. 4.53. The sliding motion of the blocks impends when rod makes an angle of 45° with the horizontal. Estimate the coefficient of friction assuming it to be same both at the floor and wall.

Solution : Let F_{ab} be the force in the rod. With reference to free body diagram of block A (Fig. 4.54*a*) and for its limiting equilibrium

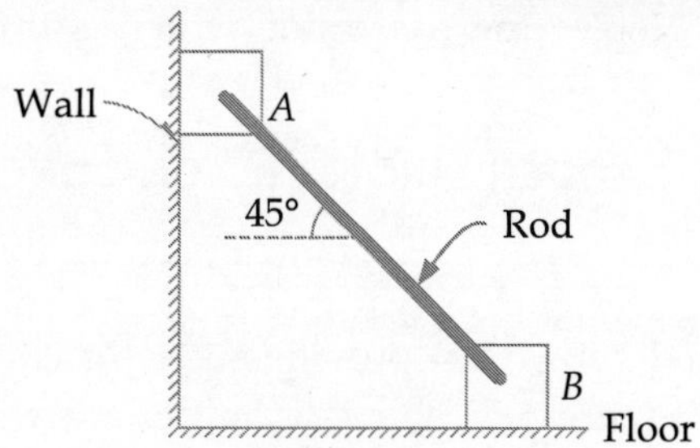

Fig. 4.53

$$\Sigma F_x = 0$$

and $$\Sigma F_y = 0$$

$$\therefore \quad \Sigma F_x = R - F_{ab} \cos 45° = 0$$

or $$R = 0.707\, F_{ab}$$

$$\Sigma F_y = F - W - F_{ab} \sin 45° = 0$$

Making substitution for $F = \mu R = \mu \times 0.707\, F_{ab}$, we have

$$\mu \times 0.707\, F_{ab} - W - 0.707\, F_{ab} = 0$$

$$F_{ab} = \frac{W}{0.707\,(1+\mu)} \qquad ...(i)$$

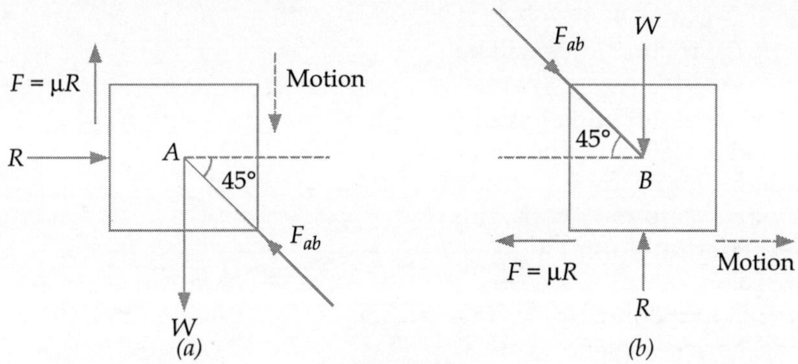

Fig. 4.54

From free body diagram of block B (Fig. 4.54 *b*), the equations for limiting equilibrium are

$$\Sigma F_x = -\mu R + F_{ab} \cos 45° = 0; \quad R = \frac{0.707\, F_{ab}}{\mu}$$

$$\Sigma F_y = -F_{ab} \sin 45° - W + R = 0$$

Making substitution for $R = 0.707\, F_{ab}/\mu$, we get

$$-0.707\, F_{ab} - W + \frac{0.707\, F_{ab}}{\mu} = 0$$

or $$0.707\, F_{ab}\left(\frac{1}{\mu} - 1\right) = W\,; \quad F_{ab} = \frac{W}{0.707}\,\frac{\mu}{1-\mu} \qquad ...(ii)$$

From expressions (*i*) and (*ii*)

$$\frac{W}{0.707\,(1+\mu)} = \frac{W}{0.707}\,\frac{\mu}{1-\mu}$$

or $$(1+\mu)\,\mu = 1 - \mu\,; \quad \mu^2 + 2\mu - 1 = 0$$

Solution of this quadratic equation gives :

$$\mu = \frac{-2 \pm \sqrt{2^2 - 4 \times (-1) \times 1}}{2 \times 1} = \frac{-2 \pm 2.828}{2}$$

Neglecting – *ve* value which is impossible, we obtain

$$\mu = \mathbf{0.414}$$

4.7. LADDER FRICTION: NON-CONCURRENT FORCE SYSTEM

A ladder is an arrangement used for climbing on the walls or roofs. It essentially consists of two long uprights of wood or iron and connected by a number of cross bars. These cross bars are called rungs and provide steps for climbing.

Figure 4.55 shows a ladder *AB* with its end *A* resting on the ground and end *B* leaning against a wall. The ladder is acted upon by the following set of forces:

(*i*) weight *W* acting down-wards at its mid point.

(*ii*) normal reaction R_b and friction force $F_b = \mu R_b$ at the end *B* leaning against the wall. Since the ladder has a tendency to slip downwards, the friction force will be acting upwards. If the wall is smooth ($\mu = 0$), the friction force will be zero.

(*iii*) normal reaction R_a and friction force $F_a = \mu R_a$ at the end *A* resting on the floor. Since the ladder, upon slipping, tends to move away from the wall, the direction of friction force will be towards the wall.

Applying equilibrium conditions, the algebraic sum of the horizontal and vertical component of forces would be zero.

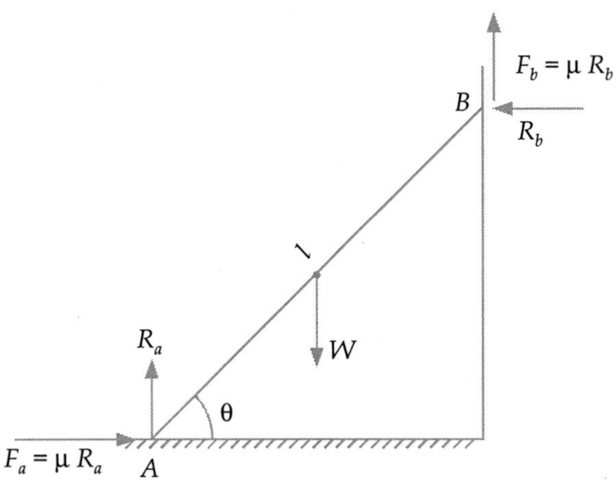

Fig. 4.55

EXAMPLE 4.29

Determine the minimum angle θ at which a uniform ladder can be placed against a wall without slipping under its own weight. The coefficient of friction for all surfaces is 0.2.

Solution : Refer Fig. 4.55 for the ladeder *AB* leaning against a wall and various forces acting on it.

Resolving all the forces along horizontal and vertical directions,

$$R_b = \mu R_a \text{ and } R_a + \mu R_b = W$$

$$\therefore \quad R_a + \mu(\mu R_a) = W \quad \text{or} \quad R_a = \frac{W}{(1+\mu^2)}$$

Taking moments about end *B*,

$$R_a \, l \cos\theta = W \times \frac{l}{2}\cos\theta + \mu R_a \, l \sin\theta$$

where *l* is the length of the uniform ladder

Substituting $R_a = \dfrac{W}{1+\mu^2}$ in the momentum equation, we get

$$\frac{Wl}{1+\mu^2}\cos\theta = \frac{Wl}{2}\cos\theta + \mu\frac{W}{1+\mu^2}\,l\sin\theta$$

or

$$\frac{\mu}{1+\mu^2}\sin\theta = \left(\frac{1}{1+\mu^2} - \frac{1}{2}\right)\cos\theta = \frac{1-\mu^2}{2(1+\mu^2)}\cos\theta$$

or $$\tan\theta = \frac{1-\mu^2}{2\mu}\ ; \quad \theta = \tan^{-1}\left(\frac{1-\mu^2}{2\mu}\right)$$

$$= \tan^{-1}\left(\frac{1-0.2^2}{2\times 0.2}\right) = \mathbf{67.38^\circ}$$

EXAMPLE 4.30

A uniform ladder of length 5 m and weighing 20 N is placed against a smooth vertical wall with its lower end 4 m away from the wall. If the ladder is just to slip, determine (*a*) the co-efficient of friction between the ladder and floor (*b*) the frictional force acting on the ladder at the point of contact between ladder and floor.

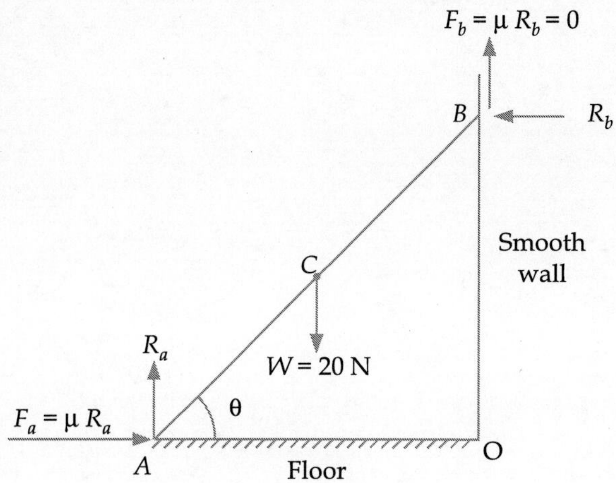

Fig. 4.56

Solution : Refer Fig. 4.56 for the ladder and various forces acting on it.

$$OA = 4\text{ m}\ ; \quad AB = 5\text{ m}$$

$$OB = \sqrt{5^2 - 4^2} = 3\text{ m}$$

$$\sin\theta = \frac{3}{5} \text{ and } \cos\theta = \frac{4}{5}$$

Resolving all the forces along horizontal and vertical directions,

$$R_a = W = 20\text{ N}\ ; \quad R_b = \mu R_a$$

Taking moments about end *A*,

$$R_b \times OB = 20 \times AC\cos\theta$$

$$R_b \times 3 = 20 \times \left(\frac{5}{2}\times\frac{4}{5}\right) = 40$$

$$\therefore \quad R_b = \frac{40}{3} = 13.33\text{ N}$$

$$\mu = \frac{R_b}{R_a} = \frac{13.33}{20} = \mathbf{0.667}$$

Friction force between the ladder and the floor,

$$F_a = \mu R_a = 0.667 \times 20 \simeq \mathbf{13.4\ N}$$

EXAMPLE 4.31

A uniform ladder of weight 250 N and length 5 m is placed against a vertical wall in a position where its inclination to the vertical is 30°. A man weighing 800 N climbs the ladder. At what position will he induce slipping? Take coeficient of friction μ = 0.2 at both the contact surfaces of the ladder.

Solution : Refer Fig. 4.57 for the ladder and the forces acting on it. Let *x* be the distance from end *A* along the ladder where the man will induce slipping.

Resolving all the forces along horizontal and vertical directions

$$R_b = \mu R_a; \quad R_a = \frac{R_b}{\mu} = \frac{R_b}{0.2} = 5 R_b$$

and $\quad R_a + \mu R_b = 250 + 800 = 1050$

$\therefore \quad 5R_b + 0.2 R_b = 1050$

$$R_b = \frac{1050}{5.2} = 201.92 \text{ N}$$

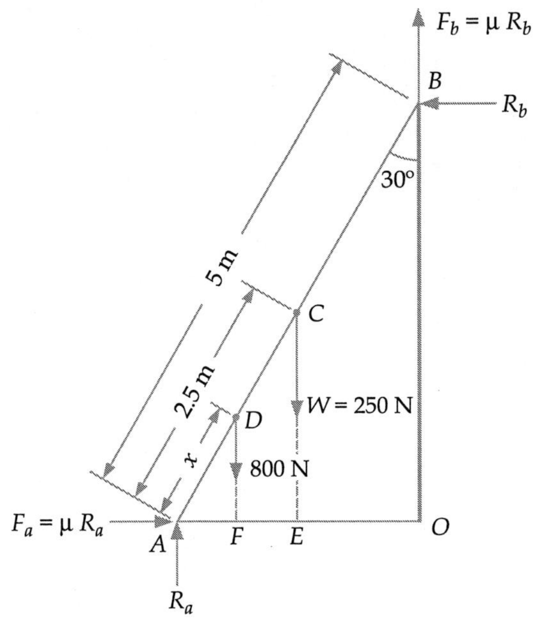

Fig. 4.57

Taking momenbts about end A,

$$R_b \times OB + \mu R_b \times OA = 250 \times EA + 800 \times FA$$

From the geometrical configuration,

$$OB = 5 \cos 30° = 4.33 \text{ m} ; \quad OA = 5 \sin 30° = 2.5 \text{ m}$$

$$EA = CA \sin 30° = 2.5 \sin 30° = 1.25 \text{ m}$$

$$FA = x \sin 30° = 0.5x$$

Substituting the above values in momentum equation, we get

$$(201.92 \times 4.33) + (0.2 \times 201.92 \times 2.5) = 250 \times 1.25 + (800 \times 0.5x)$$

or $\quad 874.31 + 100.96 = 312.5 + 400x$

$$\therefore \quad x = \frac{874.31 + 100.96 - 312.5}{400} = \mathbf{1.66 \text{ m}}$$

EXAMPLE 4.32

A uniform ladder 3 m long and weighing 200 N is placed against a wall making an angle of 60° with the ground. The ladder, in addition to its own weight has to support a 800 N weight of a man standing at its upper end. Presuming that coefficient of friction between the wall and ladder is 0.3 and that between the ground and ladder is 0.4, make calculations for the horizontal force *P* to be applied at the ground level to prevent slipping. If the force *P* is not

applied, what should be the minimum inclination of the ladder with the horizontal so that there is no slipping of it with the man standing its upper end.

Soluton : Refer Fig. 4.58 for the ladder *AB* placed against a wall and various forces acting on it. *P* is the horizontal force which has been applied at the ground level to prevent slipping.

Resolving all the forces along horizontal and vertical directions,

$$P + \mu_a R_a = R_b \qquad ...(i)$$

$$R_a + \mu_b R_b = W_1 + W_2 = 1000 \qquad ...(ii)$$

Taking moments about the end *A*,

$$W_2 \times OA + W_1 \times DA = R_b \times OB + \mu_b R_b \times OA$$

From the geometrical configuration,

$$OA = 3 \cos 60° = 1.5 \text{ m} ; \quad DA = 1.5 \cos 60° = 0.75 \text{ m}$$

$$OB = 3 \sin 60° = 2.6 \text{ m}$$

$$\therefore \quad 800 \times 1.5 + 200 \times 0.75 = R_b \times 2.6 + 0.3 R_b \times 1.5$$

or $$1200 + 150 = 3.05 R_b ; \quad R_b = \frac{1350}{3.05} = \mathbf{442.6 \ N}$$

Then from identities (*ii*) and (*i*), we have

$$R_a = 1000 - \mu_b R_b = 1000 - 0.3 \times 442.6 = 867.2 \text{ N}$$

$$P = R_b - \mu_a R_a = 442.6 - 0.4 \times 867.2 = \mathbf{95.72 \ N}$$

Thus a minimum horizontal force of 95.72 N needs to be applied at the ground level to prevent slipping of the ladder.

(*b*) If force *P* is not applied, let α be the angle that the ladder should make with the horizontal to prevent any chance of slipping.

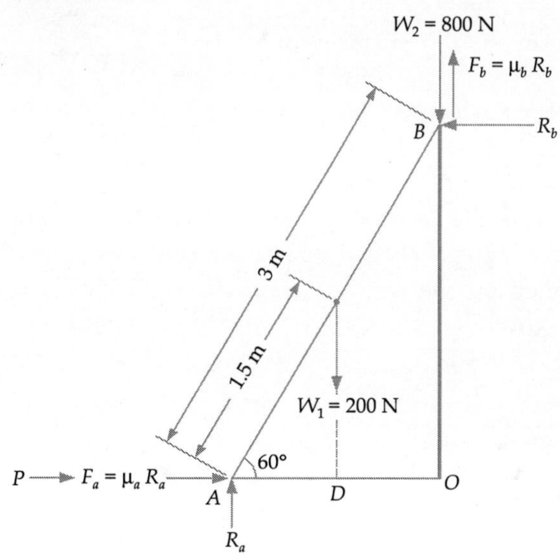

Fig. 4.58

Applying conditions for equilibrium, we have

$$R_b = \mu_a R_a = 0.4 R_a$$

$$R_a + \mu_b R_b = W_1 + W_2 = 1000$$

These identities give

$$R_a + 0.3 \times (0.4\ R_a) = 1000\ ; \quad R_a = \frac{1000}{1 + 0.3 \times 0.4} = 892.8 \text{ N}$$

and $\quad R_b = 0.4\,R_a = 0.4 \times 892.8 = 357.1 \text{ N}$

Taking moments about the end A,

$$W_2 \times OA + W_1 \times DA = R_b \times OB + \mu_b R_b \times OA$$

or $\quad (800 \times 3 \cos \alpha) + (200 \times 1.5 \cos \alpha) = 357.1 \times 3 \sin \alpha + (0.3 \times 357.1 \times 3 \cos \alpha)$

or $\quad 2400 \cos \alpha + 300 \cos \alpha = 1071.3 \sin \alpha + 321.4 \cos \alpha$

or $\quad 2378.6 \cos \alpha = 1071.3 \sin \alpha\ ; \quad \tan \alpha = \dfrac{2378.6}{1071.3} = 2.22$

$\therefore \quad \alpha = \mathbf{65.75°}$

4.8. WEDGE FRICTION

A wedge is a piece of metal or wood in the shape of a prism whose section is usually triangular or trapezoidal. It is used for lifting loads, for tightening fits or keys for shafts. When lifting loads, the wedge is placed below the load and a horizontal force P is applied. The wedge moves towards left and the load moves upwards. Obviously, there is sliding at the surfaces AB, CD and GH.

For the wedge-load system depicted in Fig. 4.59, let

ϕ_1 be the angle of friction for horizontal surface and bottom surface of the wedge,

ϕ_2 be the angle of friction for the surfaces of wedge and block, and

ϕ_3 the angle of friction for surfaces of block and wall.

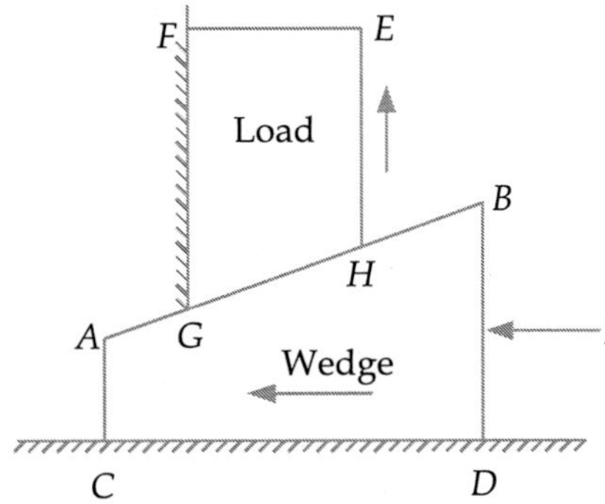

Fig. 4.59

The wedge is in equilibrium under the action of the following set of forces:

(*i*) horizontal force P applied on the wedge

(*ii*) total reaction S_1 which is the resultant of normal reaction R_1 and friction force $\mu_1 R_1$

(*iii*) total reaction S_2 which is the resultant of normal reaction R_2 and frictional force $\mu_2 R_2$.

When the force P is applied on the wedge, the surface CD moves towards left and hence frictional force acts towards right. Likewise, the frictional force on surface AB would act upwards along AB.

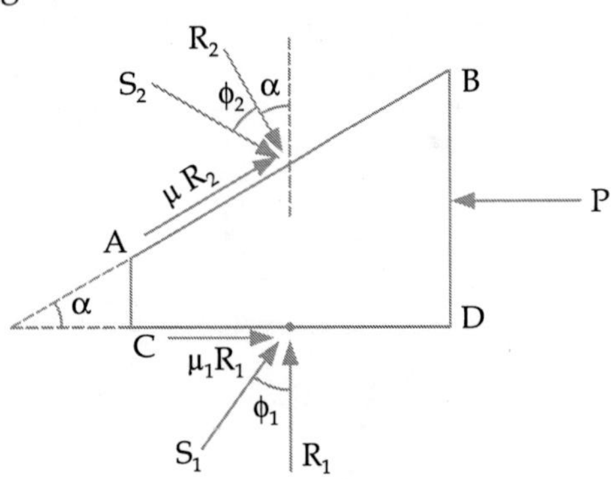

Fig. 4.60

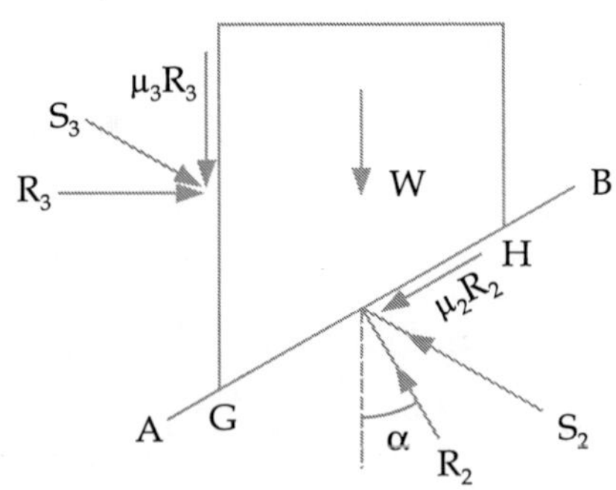

Fig. 4.61

Resolving all the forces horizontally and vertically, we get

$$P - S_1 \sin \phi_1 - S_2 \sin (\alpha + \phi_2) = 0 \quad ...(i)$$

$$S_1 \cos \phi_1 - S_2 \cos (\alpha + \phi_2) = 0 \quad ...(ii)$$

The **block** is in equilibrium under the action of following set of forces:

(*i*) W (the weight of block)

(*ii*) S_2 (a total reaction of normal reaction R_2 and frictional force $\mu_2 R_2$)

(*iii*) S_3 (a total reaction of normal reaction R_3 and frictional force $\mu_3 R_3$)

When wedge is pushed, the block is raised. The surface of the block is in contact with surface of wedge and therefore the total reaction S_2 will be equal and opposite to that of S_2 which was acting on wedge. Surface moves up and so the frictional force $\mu_3 R_3$ acts downward.

Resolving all the forces along horizontal and vertical directions, we get:

$$S_3 \cos \phi_3 - S_2 \sin (\alpha + \phi_2) = 0 \quad ...(iii)$$

$$S_3 \sin \phi_3 + W - S_2 \cos (\alpha + \phi_2) = 0 \quad ...(iv)$$

The solution of the identities (*i*) to (*iv*) helps to determine the values of unknown parameters.

As wedges are subjected to concurrent force system, solution can be obtained by applying Lami's theorem.

EXAMPLE 4.33

A block weighing 2500 N, overlying a 10-degree wedge on a horizontal floor and leaning against a vertical wall, is to be raised by applying a horizontal force to the wedge. The coefficient of friction between all surfaces in contact is 0.3. Make calculations for the minimum horizontal force to be applied to raise the block.

Solution : Refer to Fig. 4.59 for the wedge-load system

$$\mu = \tan \phi = 0.3 \quad \text{(given)}$$

$$\therefore \quad \phi = \tan^{-1} 0.3 = 16.7^\circ$$

Since the load is given, the equilibrium of the block would be considered first. Fig. 4.61 shows the various forces acting on the block. When these forces are resolved along horizontal and vertical directions, we have:

$$S_3 \cos \phi_3 - S_2 \sin (\alpha + \phi_2) \,; \quad \phi_2 = \phi_3 = 16.7^\circ$$

$$S_3 \cos 16.7^\circ = S_2 \sin (10 + 16.7)$$

$$\therefore \quad S_2 = S_3 \frac{\cos 16.7^\circ}{\sin 26.7^\circ} = S_3 \times \frac{0.9578}{0.4493} = 2.132\, S_3$$

Also:

$$S_3 \sin \phi_3 + W - S_2 \cos (\alpha + \phi_2) = 0$$

$$S_3 \times \sin 16.7^\circ + 2500 - 2.132\, S_3 \times \cos (10 + 16.7) = 0$$

$$0.2874\, S_3 + 2500 - 1.9047\, S_3 = 0$$

$$\therefore \quad S_3 = \frac{2500}{1.9047 - 0.2874} = \mathbf{1545.8}\ \text{N}$$

$$S_2 = 2.132\, S_3 = 2.132 \times 1545.8 = 3295.6\ \text{N}$$

For considering equilibrium of wedge, refer to Fig. 4.60 which shows the various forces acting on it. When these forces are resolved against horizontal and vertical directions, we obtain

$$P - S_1 \sin \phi_1 - S_2 \sin (\alpha + \phi_2) = 0$$

$$P - S_1 \times \sin 16.7^\circ - 3295.6 \sin (10 + 16.7) = 0$$

$$P - 0.2874\, S_1 - 1480.8 = 0$$

Also

$$S_1 \cos \phi_1 - S_2 \cos (\alpha + \phi_2) = 0$$

$$S_1 = S_2 \frac{\cos(\alpha + \phi_2)}{\cos \phi_1} = 3295.6 \times \frac{\cos 26.7^\circ}{\cos 16.7^\circ} = 3074 \text{ N}$$

$$\therefore \quad P = 0.2874\ S_1 + 1480.8 = 0.2874 \times 3074 + 1480.8 = \mathbf{2364.3} \text{ N}$$

Alternatively : As wedges are subjected to concurrent force system, solution can also be obtained by applying Lami's theorem.

The block is in equilibrium under the action of forces: weight W and total reactions S_2 and S_3 (Fig. 4.62)

$$\therefore \quad \frac{S_2}{\sin(90 - \phi)} = \frac{S_3}{\sin\{180 - (\alpha + \phi)\}} = \frac{W}{\sin\{\phi + 90 + (\alpha + \phi)\}}$$

or $$\frac{S_2}{\cos \phi} = \frac{S_3}{\sin(\alpha + \phi)} = \frac{W}{\sin(90 + \alpha + 2\phi)}$$

or $$\frac{S_2}{\cos 16.7^\circ} = \frac{S_3}{\sin 26.7^\circ} = \frac{W}{\sin 133.4^\circ}$$

$$\therefore \quad S_2 = W \frac{\cos 16.7^\circ}{\sin 133.4^\circ} = 2500 \times \frac{0.9578}{0.7265} = 3295.9 \text{ N}$$

$$S_3 = W \frac{\sin 26.7^\circ}{\sin 133.4^\circ} = 2500 \times \frac{0.4493}{0.7265} = 1546.1 \text{ N}$$

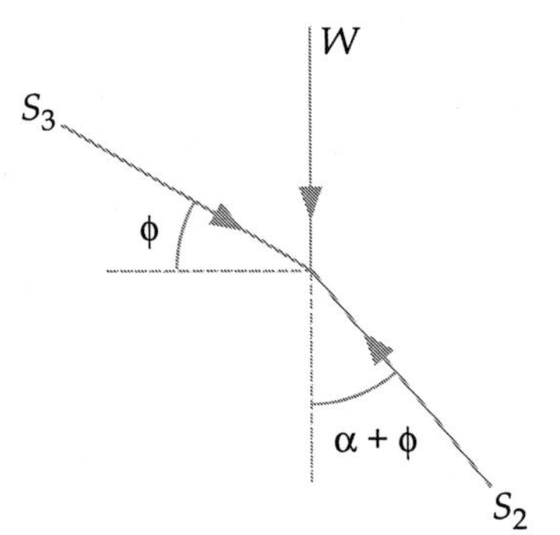

Fig. 4.62

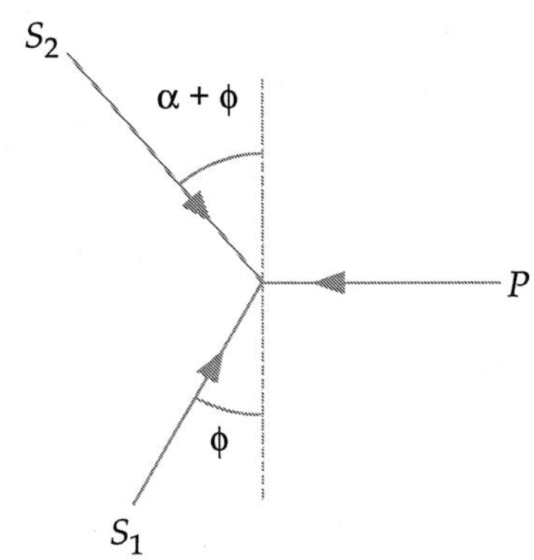

Fig. 4.63

The wedge is in equilibrium under the action of forces: P, reactions S_1 and S_2 (Fig. 4.63).

$$\therefore \frac{P}{\sin\{180 - \phi - (\alpha + \phi)\}} = \frac{S_2}{\sin(90 + \phi)} = \frac{S_1}{\sin\{90 + (\alpha + \phi)\}}$$

or $$\frac{P}{\sin[180 - 16.7 - (10 + 16.7)]} = \frac{S_2}{\cos \phi} = \frac{S_2}{\cos 16.7}$$

$$\therefore \quad P = S_2 \times \frac{\sin 136^\circ}{\cos 16.7^\circ}$$

$$= 3295.9 \times \frac{0.687}{0.9578} = \mathbf{2364\ N}$$

EXAMPLE 4.34

A block C of 60 kN is to be raised by means of two wedges *A* and *B* as shown in Fig. 4.64. The limiting coefficient of friction μ = 0.25 for all surfaces. Neglecting the weight of the wedge blocks, determine the value of force *P* for upward impending motion of block *C*.

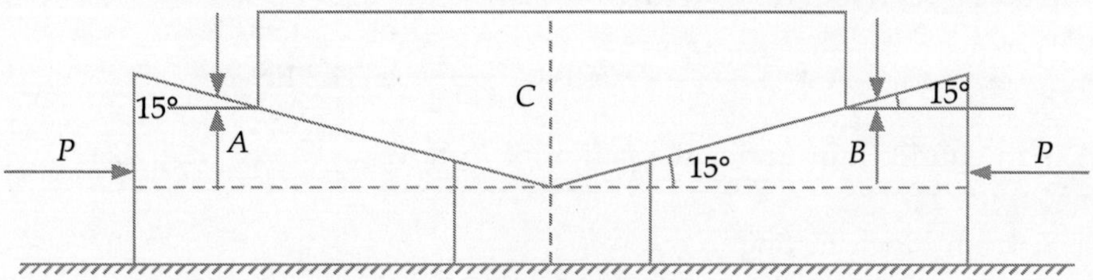

Fig. 4.64

Solution : $\mu = \tan\phi = 0.25$ where ϕ is the angle of friction

$\phi = \tan^{-1}(0.25) = 14°$

The **block** C is in equilibrium under the action of

(*i*) weight W = 60 kN of block C

(*ii*) total reaction S_1 which is the resultant of normal reaction R_1 and friction force $\mu_1 R_1$ at point E

(*iii*) total reaction S_2 which is the resultant of normal reaction R_2 and friction force $\mu_2 R_2$ at point F

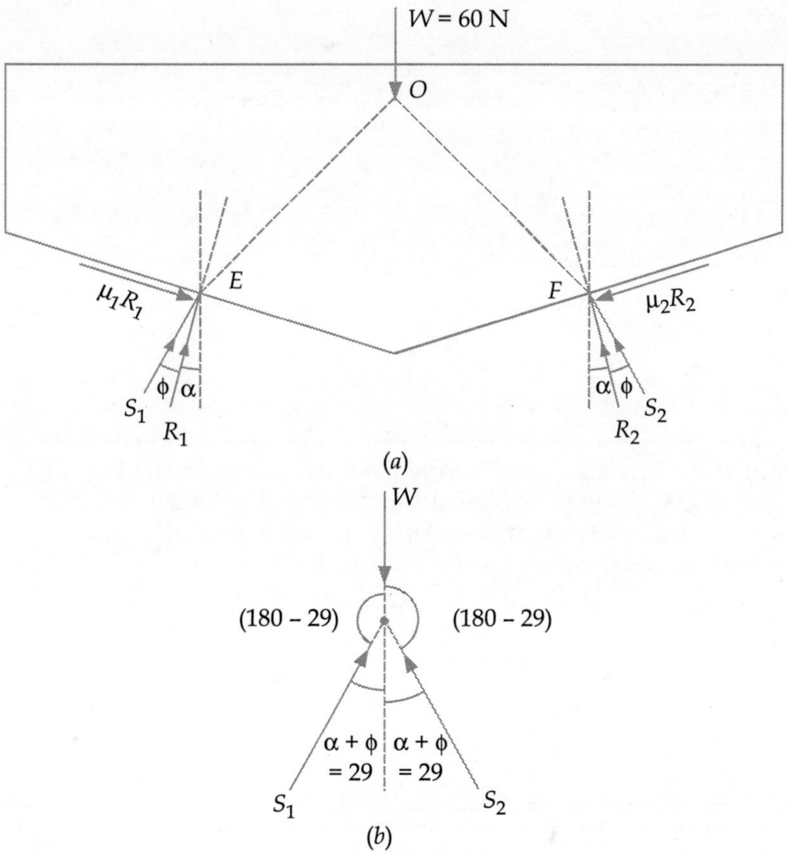

Fig. 4.65

These forces meet at point O

Applying Lami's theorem

$$\frac{W}{\sin 58^\circ} = \frac{S_1}{\sin(180-29)} = \frac{S_2}{\sin(180-29)}$$

$$\therefore \quad S_1 = S_2 = W \times \frac{\sin 151^\circ}{\sin 58^\circ} = 60 \times \frac{0.4849}{0.848} = \mathbf{34.3}\text{ kN}$$

The block A is in equilibrium under the action of following forces (Fig. 4.66)

(*i*) horizontal force P,

(*ii*) reaction S_1 from block C, and

(*iii*) total reaction S_3 which is the resultant of normal reaction R_3 and friction force $\mu_3 R_3$ at the horizontal surface.

Applying Lami's theorem,

$$\frac{P}{\sin(180-29-14)} = \frac{S_1}{\sin(90+14)} = \frac{S_3}{\sin(29+180+14)}$$

$$\therefore \quad P = S_1 \times \frac{\sin 137^\circ}{\sin 104^\circ} = 34.3 \times \frac{0.682}{0.97} = \mathbf{24.12\ kN}$$

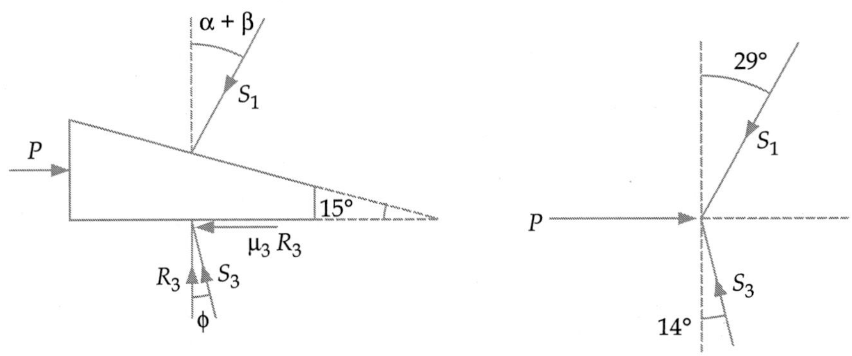

Fig. 4.66

EXAMPLE 4.35

Refer to the system of three blocks placed on the floor as shown in the Fig. 4.67 given below. Block C has a mass of 100 kg, and the identical blocks A and B have masses of 75 kg each. If the coefficient of friction μ = 0.2 for all mating surfaces, can this arrangement remain in equilibrium?

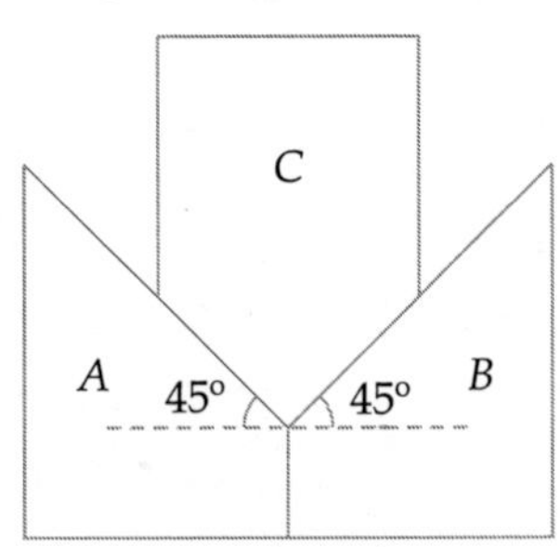

Fig. 4.67

Solution : Refer Fig. 4.67 given below for the force body diagram of the blocks taken separately.

Considering block C: $= \Sigma F_y = 0$

$$2R \sin 45^\circ + 2\,\mu R \sin 45^\circ = gM_c$$

or $$2R \times 0.707 + 2 \times 0.2\,R \times 0.707 = 9.81 \times 100$$

or $$R = \frac{9.81 \times 100}{2 \times 707 + 2 \times 0.2 \times 0.707} = 578.15\text{ N}$$

For the whole arrangement: $\Sigma F_y = 0$

$$(100 + 75 + 75) \times 9.81 = 2\,R_1$$

$$\text{or} \qquad R_1 = \frac{250 \times 9.81}{2} = 1226.25 \text{ N}$$

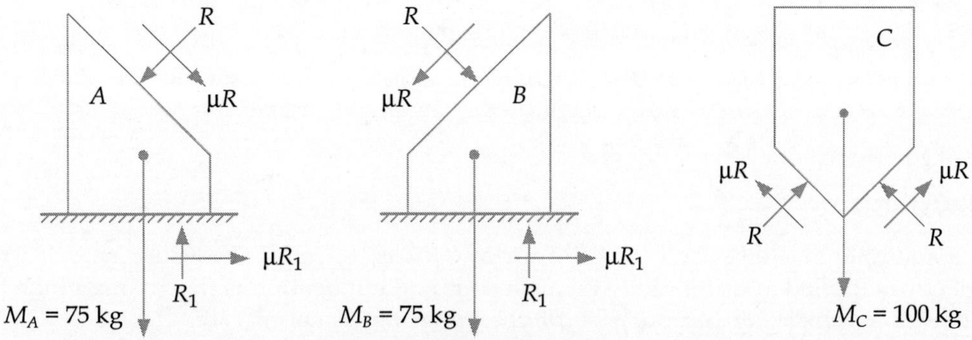

Fig. 4.68

Considering equilibrium of block A or B

$$\begin{aligned}\Sigma F_x &= R\cos 45° - \mu R\cos 45° - \mu R_1 \\ &= 578.15 \times 0.707 - 0.2 \times 578.15 \times 0.707 - 0.2 \times 1226.5 \\ &= 408.75 - 81.75 - 245.3 = 81.7 \text{ N}\end{aligned}$$

For equilibrium ΣF_x should be zero. Since it is not, the arrangement will not remain in equilibrium state.

4.9. INCLINED PLANE

Fig. 4.69 shows an inclined plane AB making an angle α with the horizontal. A body of weight W is placed at point C and it is connected to an inextensible string which passes over a pulley positioned at B. An effort is applied at the free end of the string.

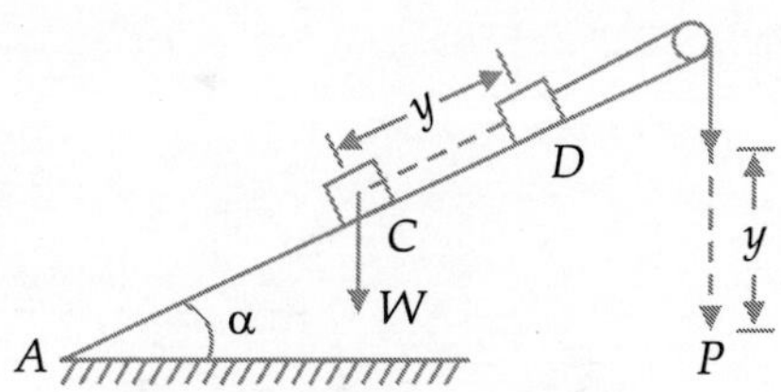

Fig. 4.69. Inclined plane

$$MA = \frac{W}{P}$$

Let the effort P move downward through a vertical distance y. Since the string is inextensible, the weight W would also move upward along the plane through the same distance y. The corresponding vertical movement of the load is $y \sin \alpha$

$$\therefore \qquad VR = \frac{\text{distance moved by the effort}}{\text{distance moved by the load}} = \frac{y}{y \sin \alpha} = \text{cosec } \alpha$$

EXAMPLE 4.36

An effort of 60 N is required to lift a load of 100 N lying on an inclined plane having an inclination of 30° with the horizontal. Determine the velocity ratio, mechanical advantage and efficiency. How this efficiency would be affected if coefficient of friction between the load and inclined surface is 0.3? Also calculate the effort lost in friction.

Solution : $VR = \text{cosec } \alpha = \text{cosec } 30° = 2$

$$MA = \frac{W}{P} = \frac{100}{60} = 1.67$$

$$\eta = \frac{MA}{VR} = \frac{1.67}{2} = 0.835 \text{ or } 83.5\%$$

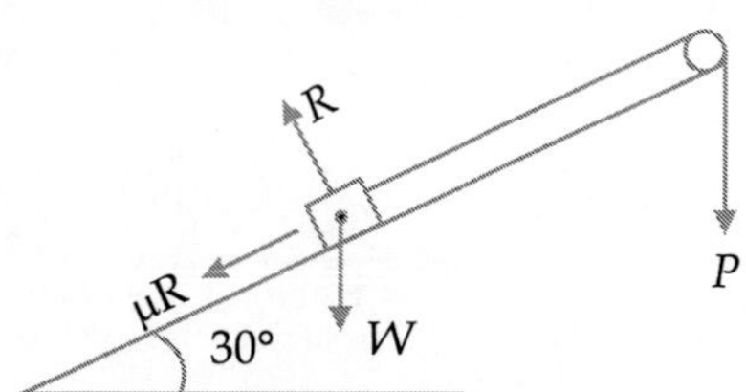

Fig. 4.70

(*b*) Refer Fig. 4.70. Resolving the forces perpendicular and parallel to the inclined plane,

$$R = 100 \cos 30° = 86.6 \text{ N}$$

$$\mu R + 100 \sin 30° = P' \qquad (P' = \text{actual effort})$$

$$\therefore \quad P' = 0.30 \times 86.6 + 100 \times 0.5 = 75.9 \text{ N}$$

$$\eta = \frac{MA}{VR} = \frac{W/P'}{VR} = \frac{100}{75.9 \times 2} = 0.658 \text{ or } \mathbf{65.8\%}$$

Effort lost in friction = 75.9 – 60 = **15.9 N**

4.10. SCREW JACK

Screw jack is a simple machine used for lifting heavy loads, through short distances, with the help of small effort applied at its handle. The most common application is the raising of the front or rear portion of a vehicle for the purpose of changing the wheel or tyre.

Construction: The unit consists of a base which is in the shape of the frustrum of a hollow cone. The frustrum has a hole which is threaded to receive a strong square threaded screw. A collar is fitted onto the top of the screw in such a manner that the screw can rotate freely in it. The load to be raised or lowered is placed on the top of the collar. A lever or handle, often called tommy bar, can be inserted into a hole provided in the collar. When an effort is applied at the end of the lever, the top of the screw goes either up or down depending on the direction in which the lever is turned.

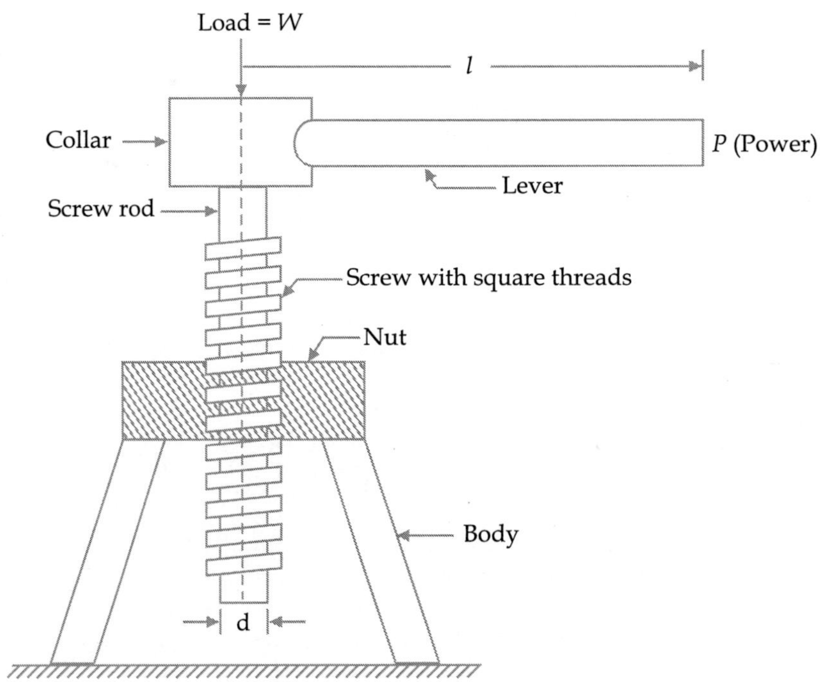

Fig. 4.71. Screw jack

When one rotation is given to the handle,

distance moved by the effort = $2\pi l$

where l is the length of lever arm from the centre of screw.

distance through which load is lifted = pitch of the thread = p

$$\therefore \text{ Velocity ratio } VR = \frac{\text{distance moved by effort}}{\text{distance moved by the load}} = \frac{2\pi l}{p}$$

It may be pointed out that for a multiple threaded screw, the pitch equals np where

n = number of starts, and

p = distance between two consective threads (apparent pitch)

Analysis : The working principle of a screw jack is the same as that of the inclined plane.

Refer Fig. 4.72 which shows development of one complete turn of a screw thread. Distance AB will be equal to the circumference and distance BC will be equal to pitch of the screw. The inclination of the thread is given by

$$\tan \alpha = \frac{p}{\pi d_m} \qquad ...(10.13)$$

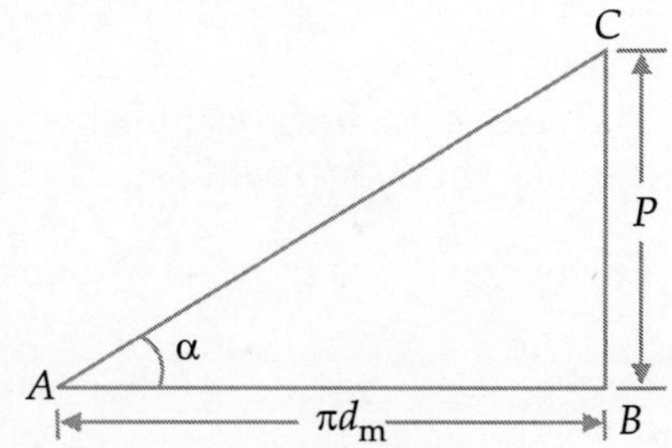

Fig. 4.72

where d_m is the mean diameter of the thread. The angle α is referred to as the helix angle.

***(a)* Effort required to lift the load:** The situation corresponds to that of lifting a load W up an inclined plane by a horizontal force P as shown in Fig. 4.73.

Since the block has impending motion up the plane, the friction force would act down the plane. The various forces which keep the block in equilibrium and as shown in the free body diagram are:

(*i*) weight W of the block,

(*ii*) reaction R of the plane

(*iii*) tractive force P, and

(*iv*) frictional force $F = \mu R$

Resolving these forces along and perpendicular to the plane, we get

$$P \cos \alpha - \mu R - W \sin \alpha = 0 \quad ; \quad P \cos \alpha = W \sin \alpha + \mu R$$

$$R - W \cos \alpha - P \sin \alpha = 0 \quad ; \quad R = W \cos \alpha + P \sin \alpha$$

$$\therefore \quad P \cos \alpha = W \sin \alpha + \mu (W \cos \alpha + P \sin \alpha)$$

$$= W \sin \alpha + \mu W \cos \alpha + \mu P \sin \alpha$$

or $$P (\cos \alpha - \mu \sin \alpha) = W (\sin \alpha + \mu \cos \alpha)$$

$$\therefore \quad P = W \frac{\sin \alpha + \mu \cos \alpha}{\cos \alpha - \mu \sin \alpha}$$

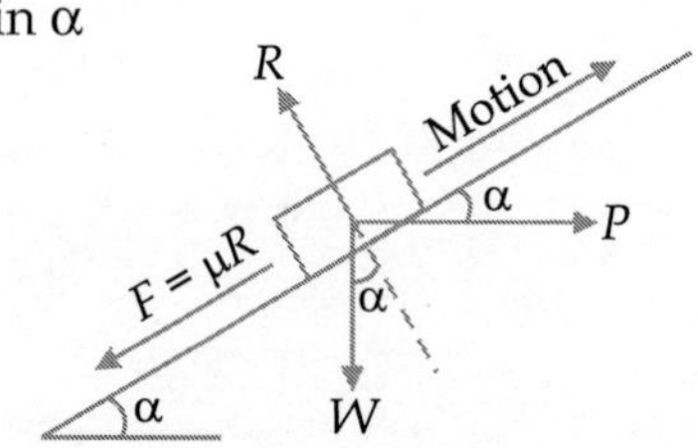

Fig. 4.73

Substituting $\mu = \tan \phi = \dfrac{\sin \phi}{\cos \phi}$ where φ is the angle of friction

$$P = W \frac{\sin \alpha + \dfrac{\sin \phi}{\cos \phi} \cos \alpha}{\cos \alpha - \dfrac{\sin \phi}{\cos \phi} \sin \alpha} = W \frac{\sin \alpha \cos \phi + \sin \phi \cos \alpha}{\cos \alpha \cos \phi - \sin \phi \sin \alpha}$$

$$= W \frac{\sin (\alpha + \phi)}{\cos (\alpha + \phi)} = W \tan (\alpha + \phi) \qquad ...(4.12)$$

Here P is the effort applied at the mean radius of screw jack. But in practice, the effort is applied at the end of the handle of the jack.

Let P_h = horizontal force applied at the end of the handle
l = length of the handle

Taking moments of P_h and P about the axis of screw jack, we get

$$P_h \times l = P \times \frac{d_m}{2} = W \tan(\alpha + \phi)\frac{d_m}{2}$$

$$\therefore \quad P_h = \frac{W d_m}{2l}\tan(\alpha + \phi) \qquad ...(4.13)$$

The above identity establishes the relation between the effort required at the end of the handle and the load lifted.

Further, $$P_h = \frac{W d_m}{2l}\tan(\alpha + \phi) = \frac{W d_m}{2l}\,\frac{\tan\alpha + \tan\phi}{1 - \tan\alpha\tan\phi}$$

Substituting $\tan\alpha = \dfrac{p}{\pi d_m}$ and $\tan\phi = \mu$, we get

$$P_h = \frac{W d_m}{2l}\left[\frac{\dfrac{p}{\pi d_m} + \mu}{1 - \dfrac{p}{\pi d_m}\mu}\right] = \frac{W d_m}{2l}\left[\frac{p + \mu\pi d_m}{\pi d_m - p\mu}\right] \qquad ...(4.14)$$

The above relation gives the value of P_h in terms of coefficient of friction and pitch of the screw.

(*b*) **Effort required to lower the load:** Since the block has impending motion down the plane, the friction force would act up the plane. The various forces which keep the block in equilibrium and as shown in the free body diagram are:

(*i*) weight W of the block,
(*ii*) reaction R of the plane,
(*iii*) tractive force P, and
(*iv*) frictional force $F = \mu R$

Resolving these forces along and perpendicular to the plane, we get

$$P\cos\alpha + \mu R - W\sin\alpha = 0$$

or $$P\cos\alpha = W\sin\alpha - \mu R$$

$$R - W\cos\alpha - P\sin\alpha = 0; \quad R = W\cos\alpha + P\sin\alpha$$

$$\therefore \quad P\cos\alpha = W\sin\alpha - \mu(W\cos\alpha + P\sin\alpha)$$
$$= W\sin\alpha - \mu W\cos\alpha - \mu P\sin\alpha$$

or $$P(\cos\alpha + \mu\sin\alpha) = W(\sin\alpha - \mu\cos\alpha)$$

$$\therefore \quad P = W\,\frac{\sin\alpha - \mu\cos\alpha}{\cos\alpha + \mu\sin\alpha}$$

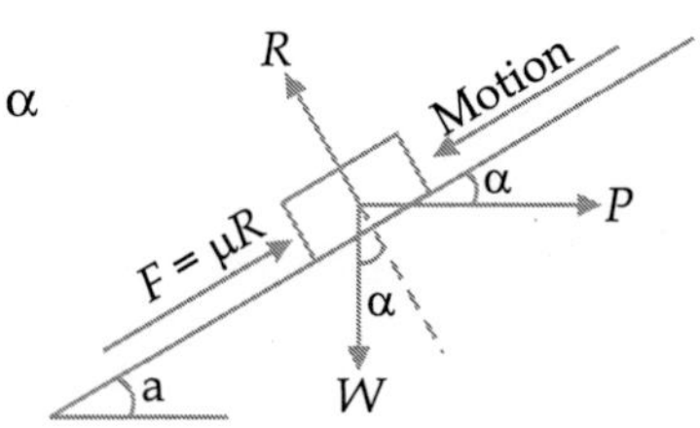

Fig. 4.74

Substituting $\mu = \tan\phi = \dfrac{\sin\phi}{\cos\phi}$ where ϕ is the angle of friction

$$P = W\,\frac{\sin\alpha - \dfrac{\sin\phi}{\cos\phi}\cos\alpha}{\cos\alpha + \dfrac{\sin\phi}{\cos\phi}\sin\alpha} = W\,\frac{\sin\alpha\cos\phi - \sin\phi\cos\alpha}{\cos\alpha\cos\phi + \sin\alpha\sin\phi} = W\,\frac{\sin(\alpha - \phi)}{\cos(\alpha - \phi)}$$

$$= W\tan(\alpha - \phi) \qquad ...(4.15)$$

Here P is the effort applied at the mean radius of the jack. But in actual practice, an effort P_h is applied at the end of the handle of the jack.

Equating the moments of P and P_h about the axis of the jack,

$$P_h \times l = P \times \frac{d_m}{2} = W \tan(\alpha - \phi) \times \frac{d_m}{2}$$

$$\therefore \quad P_h = W \frac{d_m}{2l} \tan(\alpha - \phi) \qquad ...(4.16)$$

The above identity establishes the relation between the effort required at the end of the handle to lower the load.

Further, $$P_h = W \frac{d_m}{2l} \frac{\tan\alpha - \tan\phi}{1 + \tan\alpha \tan\phi}$$

Substituting $\tan\alpha = \frac{p}{\pi d_m}$ and $\tan\phi = \mu$, we have

$$P_h = \frac{W d_m}{2l}\left[\frac{\frac{p}{\pi d_m} - \mu}{1 + \frac{p}{\pi d_m}\mu}\right] = \frac{W d_m}{2l}\left[\frac{p - \mu \pi d_m}{\pi d_m + \mu p}\right] \qquad ...(4.17)$$

The above relation gives the value of P_h in terms of coefficient of friction and pitch of the screw.

Efficiency of screw jack: The effort required at the mean radius of screw jack to lift the load is given by

$$P = W \tan(\alpha + \phi)$$

In the absence of friction, ϕ will be zero. In that case the ideal effort would equal

$$P_0 = W \tan\alpha$$

$$\therefore \quad \text{Efficiency } \eta = \frac{\text{ideal effort}}{\text{actual effort}} = \frac{W \tan\alpha}{W \tan(\alpha + \phi)} = \frac{\tan\alpha}{\tan(\alpha + \phi)} \qquad ...(4.18)$$

The condition for maximum efficiency can be worked out by differentiating the above expression with respect to α and setting the derivative equal to zero. That is

$$\frac{d\eta}{d\alpha} = \frac{d}{d\alpha}\left[\frac{\tan\alpha}{\tan(\alpha + \phi)}\right] = 0$$

or $$\frac{\sec^2\alpha \tan(\alpha + \phi) - \sec^2(\alpha + \phi) \tan\alpha}{\tan^2(\alpha + \phi)} = 0$$

or $$\sec^2\alpha \tan(\alpha + \phi) = \sec^2(\alpha + \phi) \tan\alpha$$

or $$\frac{1}{\cos^2\alpha} \times \frac{\sin(\alpha + \phi)}{\cos(\alpha + \phi)} = \frac{1}{\cos^2(\alpha + \phi)} \times \frac{\sin\alpha}{\cos\alpha}$$

or $$\sin(\alpha + \phi)\cos(\alpha + \phi) = \sin\alpha \cos\alpha$$

or $$\sin 2(\alpha + \phi) = \sin 2\alpha$$

$$\therefore \quad 2(\alpha + \phi) = \pi - 2\alpha \ ; \ \alpha = \frac{\pi}{4} - \frac{\phi}{2} = 45^\circ - \frac{\phi}{2}$$

Maximum efficiency can be worked out by substituting $\alpha = 45^\circ - \frac{\phi}{2}$ in the expression for efficiency

$$\eta_{max} = \frac{\tan(45° - \phi/2)}{\tan(45° + \phi/2)} = \frac{\dfrac{1 - \tan\phi/2}{1 + \tan\phi/2}}{\dfrac{1 + \tan\phi/2}{1 - \tan\phi/2}} = \frac{(1 - \tan\phi/2)^2}{(1 + \tan\phi/2)^2}$$

$$= \frac{\left[\dfrac{\cos\phi/2 - \sin\phi/2}{\cos\phi/2}\right]^2}{\left[\dfrac{\cos\phi/2 + \sin\phi/2}{\cos\phi/2}\right]^2} = \frac{(\cos\phi/2 - \sin\phi/2)^2}{(\cos\phi/2 + \sin\phi/2)^2}$$

$$= \frac{\cos^2\phi/2 + \sin^2\phi/2 - 2\cos\phi/2}{\cos^2\phi/2 + \sin^2\phi/2 + 2\cos\phi/2}$$

Since $\sin^2\phi/2 + \cos^2\phi/2 = 1$ and $2\sin\phi/2\cos\phi/2 = \sin\phi$, we can write

$$\eta_{max} = \frac{1 - \sin\phi}{1 + \sin\phi} \qquad ...(4.19)$$

A jack is self locking if its efficiency is less than 50 percent or $\alpha < \phi$. If the jack is not self-locking, the weight resting on the collar starts coming down. An effort equal to $P = W\tan(\alpha - \phi)$ has then to be applied at the mean radius of the jack to balance the load which is descending.

The efficiency of a screw jack is independent of load and depends only on the helix angle of screw and the angle of friction ϕ.

$$\eta = \frac{\tan\alpha}{\tan(\alpha + \phi)}$$

Further $\mu = \tan\phi$ and $\tan\alpha = \dfrac{P}{\pi d_m}$

The following observations may thus be made:

(*i*) efficiency increases with increase in helix angle, *i.e.*, with increase in pitch of screw thread.

(*ii*) higher the angle of friction ϕ, lower will be the efficiency. The efficiency drops with increase in coefficient of friction between the nut and screw.

(*iii*) in case of V-thread with included angle 2β, the coefficient of friction μ, as compared to that for square thread is higher and is given by

$$\mu_1 = \frac{\mu}{\cos\beta}$$

Apparently the efficiency of a V-threaded screw jack would be less.

(*iv*) the maximum efficiency of screw jack is

$$\eta_{max} = \frac{1 - \sin\phi}{1 + \sin\phi}$$

The maximum efficiency is independent of load, helix angle and depends only on the angle of friction. Higher the angle of friction, lower will the efficiency. Since $\mu = \tan\phi$, the efficiency drops with increase in coefficient of friction between the fixed nut and screw.

EXAMPLE 4.37

A power screw moves a nut against a force of 10 kN at a speed of 6 mm/s. The screw has a single square thread of 6 mm pitch on a major diameter of 45 mm. Calculate the power of the driving motor. The coefficient of friction at the screw thread is 0.15.

Solution: Thickness of threads $t = \frac{p}{2} = \frac{6}{2} = 3$ mm

Outer diameter of screw $d_o = 45$ mm (given)

Diameter at the base of screw $d_i = d_0 - 2t = 45 - 2 \times 3 = 39$ mm

$$\text{Mean diameter of screw} = \frac{d_0 + d_i}{2} = \frac{45+39}{2} = 42 \text{ mm}$$

Now $\tan\alpha = \frac{p}{\pi d_m} = \frac{6}{\pi \times 42} = 0.0455;\quad \alpha = 2.605°$

$\tan\phi = \mu = 0.15;\quad \phi = 8.53°$

The force needed to raise or lower the nut,

$$P = W\tan(\alpha + \phi) = 100 \tan(2.605 + 8.53) = 19.68 \text{ kN}$$

Power of driving motor $= P \times V$

$= 19.68 \times (6 \times 10^{-3}) = 0.1181$ kN m/s $=$ **0.118 kW**

EXAMPLE 4.38

A load of 500 N is required to be lifted by a screw jack whose spindle has the square threads with the following specifications:

outer diameter = 8 cm; threads in 1.6 cm = 2

thread thickness = half the pitch

coefficient of friction between nut and screw = 0.1

What force needs to be applied at the end of a handle of 50 cm length?

Solution: For 2 square threads in 1.6 cm length,

pitch of screw $p = \frac{1.6}{2} = 0.8$ cm

thickness of threads $t = \frac{1}{2} \times 0.8 = 0.4$ cm

outer diameter of screw $d_1 = 8$ cm

diameter at the base of screw $d_2 = d_1 - 2t = 8 - 2 \times 0.4 = 7.2$ cm

mean diameter of screw $d_m = \frac{d_1 + d_2}{2} = \frac{8+7.2}{2} = 7.6$ cm

Now $\tan\alpha = \frac{p}{\pi d_m} = \frac{0.8}{3.14 \times 7.6} = 0.0335$

$\alpha = 1.92°$

$\tan\phi = \mu = 0.1\ ;\quad \phi = 5.72°$

The effort P applied at the mean radius of the jack in lifting a given load is given by

$$P = W\tan(\alpha + \phi) = 500 \tan(1.92 + 5.72) = 67.07 \text{ N}$$

If P_h is the effort applied at the end of a handle of length l, then equating the moments of P and P_h about the axis of jack, we get

$$P_h \times l = P \times \frac{d_m}{2}\ ;\quad P_h = \frac{Pd_m}{2l} = 67.07 \times \frac{7.6}{2 \times 30} = \textbf{5.097 N}$$

EXAMPLE 4.39

A square threaded screw of 7.5 cm mean diameter and 1.5 cm pitch is utilized to lift a weight of 600 N. Determine

(*a*) the tangential force which needs to be applied to it at the end of a handle of length 40 cm.

(*b*) the efficiency of jack and state whether the jack is self locking or not.

If the jack is self locking, calculate the torque necessary to lower the load. If not, what amount of torque needs to be applied to keep the load from descending?

Assume coefficient of friction at the screw thread = 0.05

Solution: The tangential force applied at the mean radius of the screw is given by

$$P = W \tan(\alpha + \phi) = W\left[\frac{\tan\alpha + \tan\phi}{1 - \tan\alpha \tan\phi}\right]$$

Now, $$\tan\alpha = \frac{p}{\pi d_m} = \frac{1.5}{\pi \times 7.5} = 0.0637 \ ; \quad \alpha = 3.639^\circ$$

$$\tan\phi = \mu = 0.05 \ ; \quad \phi = 2.862^\circ$$

$$\therefore \quad P = 600\left[\frac{0.0637 + 0.05}{1 - 0.0637 \times 0.05}\right] = 68.44 \text{ N}$$

The effort P_h to be applied at the end of a handle of length can be worked out from the correlation

$$P_h \times l = P \times \frac{d_m}{2} \ ; \quad P_h = \frac{P d_m}{2l} = \frac{68.44 \times 7.5}{2 \times 40} = \mathbf{6.416 \text{ N}}$$

(*b*) Efficiency of the screw jack is given by

$$\eta = \frac{\tan\alpha}{\tan(\alpha + \phi)} = \frac{\tan 3.639}{\tan(3.639 + 2.862)} = \frac{0.0637}{0.11395} = 0.558 \text{ or } 55.8\%$$

Since the efficiency of the screw jack is more than 50%, it is not self locking. The weight W = 600 N will than start coming down. The tangential force required at the mean radius to prevent the load from descending is

$$P_1 = W \tan(\alpha - \phi) = 600 \tan(3.639 - 2.862) = 8.137 \text{ N}$$

$$\therefore \text{ Torque to be applied, } T = P_1 \times \frac{d_m}{2} = 8.137 \times \frac{7.5}{2}$$

$$= 30.51 \text{ N cm} = \mathbf{0.3051} \text{ Nm}$$

EXAMPLE 4.40

The spindle of a screw jack has single start square threads of 60 mm root diameter and 8 mm pitch. The jack supports a load of 500 N on its head and that revolves with the screw. If the coefficient of friction at the screw threads is 0.1, determine

(*a*) the effort to be applied at the end of a handle of 50 cm length to raise the load,

(*b*) work done in lifting the load through 10 mm,

(*c*) efficiency of the screw, and

(*d*) ratio between efforts required to raise and lower the load.

Solution: The effort P applied at the mean radius of the jack in lifting a given load is given by

$$P = W \tan(\alpha + \phi)$$

Now, $\tan \alpha = \dfrac{P}{\pi d_m} = \dfrac{8}{\pi \times 60} = 0.04246$; $\alpha = 2.43°$

$\tan \phi = 0.1$; $\phi = 5.72°$

$\therefore \quad P = 500 \tan (2.43 + 5.72) = 71.60$ N

If P_h is the effort applied at the end of a handle of length l, then equating the moments of P and P_h about the axis of the jack, we get

$$P_h \times l = P \times \frac{d_m}{2}; \; P_h = \frac{P d_m}{2l} = \frac{71.60 \times 60}{2 \times 50} = \mathbf{4.296\ N}$$

(*b*) Torque required at the mean radius of the jack,

$$T = P \times \frac{d_m}{2} = 71.60 \times \frac{60}{2} = 2148 \text{ N mm} = 2.148 \text{ Nm}$$

Number of turns required to lift the load through 100 mm,

$$= \frac{\text{distance moved}}{\text{pitch of thread}} = \frac{100}{8} = 12.5$$

Since one turn ≡ 2π radians, we have

total angular displacement = 12.5 × 2π = 78.5 radians

Work done = torque required × angular displacement

= 2.148 × 78.5 = **168.62** Nm

$$\text{(c) Efficiency of screw jack} = \frac{\text{work done by screw}}{\text{work done on screw}}$$

$$= \frac{\text{load lifed} \times \text{distance moved}}{\text{work done on screw}} = \frac{500 \times 0.1}{168.62}$$

= 0.2965 = **29.65%**

The efficiency of a screw jack can also be worked out from the correlation

$$\eta = \frac{\tan \alpha}{\tan (\alpha + \phi)} = \frac{\tan 2.43}{\tan (2.43 + 5.72)} = \frac{0.04246}{0.1432} = 0.2965 = 29.65\%$$

(*d*) The effort required at the mean radius to lower the load is given by

$$P = W \tan (\alpha - \phi) \quad \text{if } \alpha > \phi$$
$$= W \tan (\phi - \alpha) \quad \text{if } \alpha < \phi.$$

For the given case $\alpha = 2.43°$ and $\phi = 5.72°$

$\therefore \quad P = 500 \tan (5.72 - 2.43) = 28.74$ N

$$\therefore \quad \text{Required ratio} = \frac{71.60}{28.74} = \mathbf{2.49}$$

EXAMPLE 4.41

The threads used in a turnbuckle are specified to be single start square threads having a mean diameter of 50 mm and pitch 12 mm. If the coefficient of friction between the screw and the nut is 0.25, make calculations for the turning moment necessary to (*a*) tighten and (*b*) slacken the wire when the pull is 1200 N. It may be assumed that no twising occurs in the rope.

Solution: $\tan \alpha = \dfrac{P}{\pi d_m} = \dfrac{12}{\pi \times 50} = 0.0764$; $\alpha = 4.37°$

$\tan \phi = \mu = 0.15$; $\phi = 8.55°$

(*a*) Tightening of the rope is equivalent to moving the load up the inclined plane. For that, the effort applied at the mean radius of the jack is

$$P = W \tan(\alpha + \phi) = 1200 \tan(4.37 + 8.55) = 275.28 \text{ N}$$

Torque required at the mean radius of jack,

$$T = P \times \frac{d_m}{2} = 275.28 \times \frac{0.05}{2} = \mathbf{6.882} \text{ Nm}$$

(*b*) Slackening of the rope is equivalent to moving the load down the inclined plane. For that, the effort applied at the mean radius of the jack is

$$P = W \tan(\phi - \alpha) \qquad (\because \alpha < \phi)$$

$$= 1200 \tan(8.55 - 4.37) = 87.70 \text{ N}$$

Torque required at the mean radius of the jack,

$$T = P \times \frac{d_m}{2} = 87.70 \times \frac{0.05}{2} = \mathbf{2.192} \text{ Nm.}$$

4.11. DIFFERENTIAL SCREW JACK

Fig. 4.75 shows a differential screw jack which provides an increased velocity ratio compared to that of a simple screw jack.

The unit consists of two threaded elements *A* and *B* of different diameters and pitches, but having threads in the same direction. The element *A* is essentially a cylinder with threads both on its outer and inner surface. The threads on its outer surface mesh with the thread nut *C*; this nut also functions as body of the jack. The segment *B* has threads only on its outer surface and these engage with the internal threads of the element *A*. Obviously the element *A* behaves as a screw for the nut and also as a nut for the element *B*.

The weight *W* to be raised or lowered is placed on the top of a platform (collar). The platform does not rotate but moves in the vertical direction only. The effort *P* is applied at the end of a lever or handle which is inserted into a hole provided in the collar.

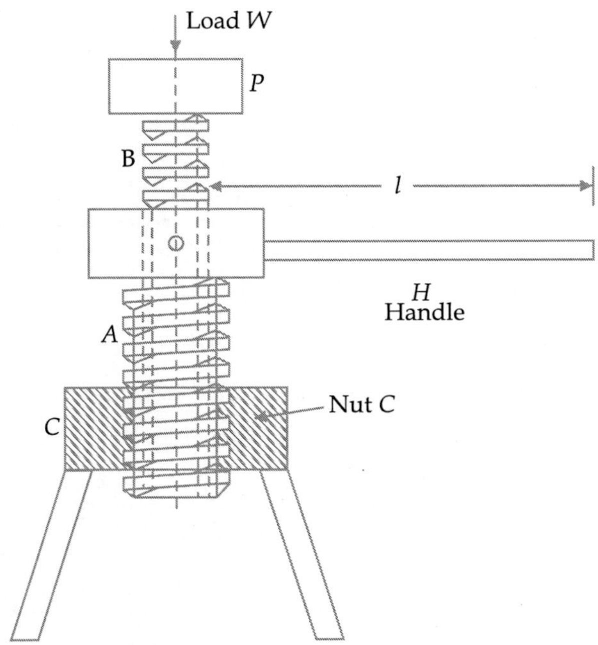

Fig. 4.75. Differential screw jack

Let p_a and p_b is the pitches of screws *A* and *B* respectively. When effort is applied and the lever is made to turn one complete rotation, the screw *A* lifts up by pitch length p_a and in the same time the screw *B* comes done by pitch length p_b. Then

Net distance through which load is lifted = $(p_a - p_b)$.

If l is the length of lever, then

distance moved by the effort = $2\pi l$

$$\therefore \quad \text{Velocity ratio } (VR) = \frac{\text{distance moved by the effort}}{\text{distance moved by the load}} = \frac{2\pi l}{p_a - p_b}$$

$$\text{Mechanical advantage } (MA) = \frac{W}{P}, \text{ and}$$

$$\text{Efficiency, } \eta = \frac{MA}{VR}$$

EXAMPLE 4.42

A differential screw jack has the following specifications :

pitch of larger screw = 12 mm

pitch of smaller screw = 8 mm

length of lever arm from centre of screw = 400 mm.

Find its velocity ratio. If an effort of 75 N applied at the end of the lever lifts a load of 10 KN, determine the efficiency of the jack.

Solution : For a differential screw jack,

$$VR = \frac{2\pi l}{p_a - p_b} = \frac{2\pi \times 400}{12 - 8} = \mathbf{628}$$

$$MA = \frac{W}{P} = \frac{10 \times 10^3}{75} = 133.3$$

$$\eta = \frac{MA}{VR} = \frac{133.3}{628} = 0.212 \text{ or } \mathbf{21.2\%}$$

4.12. ROPE AND BELT FRICTION

The transmission of power in factories from one rotating shaft to another that lies at a considerable distance is achieved through belts and ropes. The shafts are fitted with pulleys, the belt is wrapped round the pulleys and its ends are connected to form an endless connector. The belts and the pulley remain in contact by frictional grip.

With reference to Fig. 4.76, the pulley A which is connected to the rotating shaft is called the *driver*. The pulley B that needs to be driven is termed as *follower*. When the driver rotates, it carries the belt because of friction that exists between the pulley and the belt. The frictional

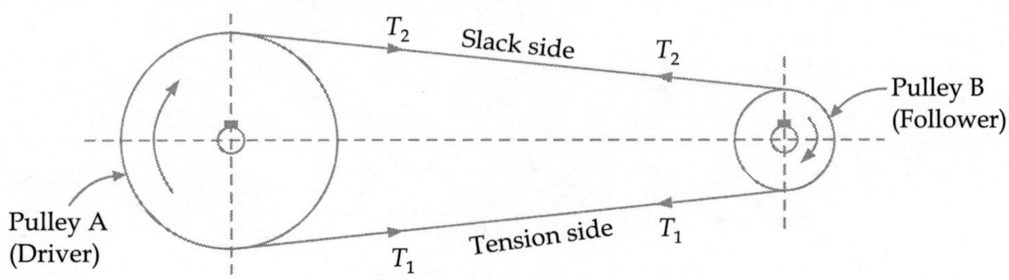

Fig. 4.76. Belt drive-open system

resistance develops all along the contact surfaces. That makes the belt carry the follower which too starts rotating. The driving pulley pulls the belt from one side (called tension side) and delivers it to the other side (called slack side). The tension T_1 in the belt on the tension side is more than tension T_2 on the slack side.

4.13. RATIO OF TENSIONS

Fig. 4.77 shows a flexible belt resting over the flat rim of a stationary pulley. The tensions T_1 and T_2 are such that the motion is impending (just to take place) between the belt and the pulley.

Considering the impending motion to be clockwise relative to the drum, the tension T_1 is more than T_2. It is to be noted that only a part of the belt is in contact with the pulley. The angle subtended at the centre of the pulley by the position of belt in contact with it is called the *angle of contact* or the *angle of lap*

Angle of contact $\theta = \angle AOB$

Let attention be focused on small element EF of the belt which subtends an angle $\delta\theta$ at the centre. The segment EF is acted upon by the following set of forces :

- tension T in the belt acting tangentially at F,
- tension $(T + \delta T)$ in the belt acting tangentially at E
- normal reaction R exerted by the pulley rim, and
- friction force $F = \mu R$ which acts against the tendency to slip and is perpendicular to normal reaction R.

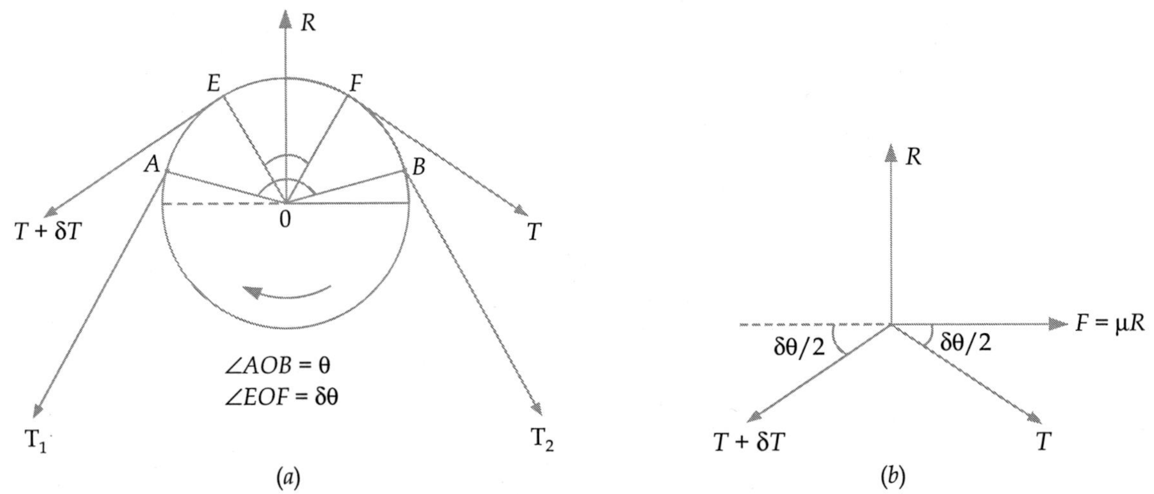

Fig. 4.77. Ratio of tension on tight and block side

Considering equilibrium of forces in the radial (vertical) direction,

$$R = (T + \delta T) \sin \frac{\delta\theta}{2} + T \sin \frac{\delta\theta}{2}$$

For small values of $\delta\theta$; $\quad \sin \frac{\delta\theta}{2} \to \frac{\delta\theta}{2}$

$$R = (T + \delta T)\frac{\delta\theta}{2} + T\frac{\delta\theta}{2} = T\frac{\delta\theta}{2} + \delta T\frac{\delta\theta}{2} + T\frac{\delta\theta}{2}$$

The term $\delta T \frac{\delta\theta}{2}$ is small in magnitude and can be neglected

$$R = T\frac{\delta\theta}{2} + T\frac{\delta\theta}{2} = T\,\delta\theta \qquad ..(i)$$

Considering equilibrium of forces in tangential (horizontal) direction,

$$\mu R = (T + \delta T)\cos\frac{\delta\theta}{2} - T\cos\frac{\delta\theta}{2}$$

For small values of $\delta\theta$; $\quad \cos\frac{\delta\theta}{2} \to 1$

$$\mu R = (T + \delta T) - T = \delta T$$

$$R = \frac{\delta T}{\mu} \quad \text{...}(ii)$$

From expressions (*i*) and (*ii*)

$$T\,\delta\theta = \frac{\delta T}{\mu}$$

Separating the variables and integrating between the limit $T = T_2$ at $\theta = 0$ and $T = T_1$ at $\theta = \theta$, we get,

$$\int_{T_2}^{T_1} \frac{\delta T}{T} = \mu \int_{o}^{\theta} \delta\theta \;;\quad \log_e \frac{T_1}{T_2} = \mu\theta \quad \therefore \quad \frac{T_1}{T_2} = e^{\mu\theta} \quad \text{...(4.20)}$$

When two pulley of unequal diameters are connected by open belt drive, the slip occurs first on the smaller pulley where the force of friction is less. Accordingly, the angle of contact on smaller pulley is taken into account while using the equation 4.20.

(*ii*) ***V-belt and rope drive :*** A V-belt or a rope runs into a grooved pulley.

Refer Fig. 4.78 (*a*) which shows a V-belt of trapezoidal section resting in a grooved pulley.

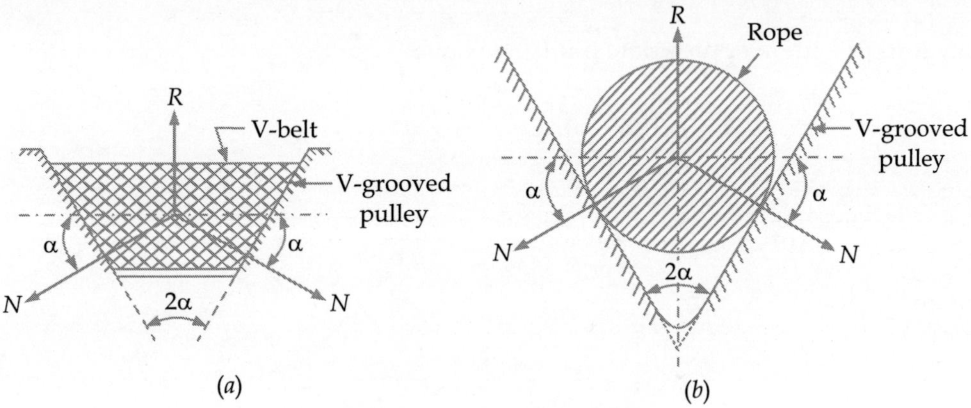

Fig. 4.78. V-belt and rope drive

Let 2α = angle of groove

N = normal reaction between belt and sides of the V-grooved pulley

R = total reaction in the plane of groove

μ = coefficient of friction between belt and groove

Considering equilibrium between R and N, we have

$$R = N \sin\alpha + N \sin\alpha = 2N \sin\alpha$$

$$\therefore \quad N = \frac{R}{2\sin\alpha} = \frac{R}{2}\operatorname{cosec}\alpha$$

$$\text{Friction resistance} = \mu N + \mu N = 2\,\mu N = 2\mu \times \frac{R}{2}\operatorname{cosec}\alpha = \mu R \operatorname{cosec}\alpha$$

Then with reference to Fig. 4.79

$$R = (T + \delta T)\sin\frac{\delta\theta}{2} + T \sin\frac{\delta\theta}{2}$$

$$= (T + \delta T)\frac{\delta\theta}{2} + T\frac{\delta\theta}{2} \quad (\because \delta\theta \text{ is small})$$

$$= T\frac{\delta\theta}{2} + \frac{\delta T\,\delta\theta}{2} + T\frac{\delta\theta}{2}$$

The term $\frac{\delta T\,\delta\theta}{2}$ is the product of two small quantities and hence neglected. That gives

$$R = T\frac{\delta\theta}{2} + T\frac{\delta\theta}{2} = T\,\delta\theta \qquad ...(a)$$

Also : $\mu R \operatorname{cosec}\alpha = (T + \delta\theta)\cos\frac{\delta\theta}{2} - T\cos\frac{\delta\theta}{2} = (T + \delta T) - T = \delta T \qquad ...(b)$

This is because $\frac{\delta\theta}{2}$ is small and $\cos\frac{\delta\theta}{2} \to 1$

From expressions (*a*) and (*b*),

$$\mu\,(T\,\delta\theta)\operatorname{cosec}\alpha = \delta T \qquad ...(b)$$

or $$(\mu \operatorname{cosec}\alpha)\,\delta\theta = \frac{\delta T}{T}$$

Upon integration with in appropriate limits, we have

$$\mu \operatorname{cosec}\alpha \int_{o}^{\theta} \delta\theta = \int_{T_2}^{T_1} \frac{\delta T}{T}$$

or $$(\mu \operatorname{cosec}\alpha) \times \theta = \log_e \frac{T_1}{T_2}$$

∴ Ratio of tensions for V-belt,

$$\frac{T_1}{T_2} = e^{(\mu \operatorname{cosec}\alpha)\,\theta} \qquad ...(4.21)$$

Equations 4.21 is equally valid for a rope drive (Fig 4.78 *b*).

(*iii*) ***Power transmitted :***

Let T_1 = tension on the tight side

T_2 = tension on the slack side

V = velocity of the belt

Then, effective turning force = $(T_1 - T_2)$

work done = force × distance moved = $(T_1 - T_2)\,V$ in Nm/s

and power = $(T_1 - T_2)\,V$ in watts ...(4.22)

where values of T_1 and T_2 are in N and V is the velocity of belt in m/s.

EXAMPLE 4.43

A belt is stretched over two identical pulleys of diameter *D* meter. The initial tension in the belt through out is 2.4 kN when the pulleys are at rest. When this belt-pulley system is used to transmit power, it is found that increase in tension on one side is equal to decrease in tension on the other side. Find the maximum torque that can be transmitted by the belt drive. Take coefficient of friction between belt and pulley as 0.3.

Solution : Since both the pulleys are of the same diameter, angle of embrace (contact) θ = 180° = π radian.

Let increase/decrease in tension be Δt. Then

$$T_1 = 2.4 + \Delta t \quad \text{and} \quad T_2 = 2.4 - \Delta t$$

Invoking the relation $\frac{T_1}{T_2} = e^{\mu\theta}$, we have

$$\frac{2.4 + \Delta T}{2.4 - \Delta T} = e^{0.3 \times \pi} = 2.57$$

or $$2.4 + \Delta t = 2.57 \times 2.4 - 2.57\Delta T; \qquad \Delta T = 1.06 \text{ kN}$$

$\therefore$ $$T_1 = 2.4 + 1.06 = 3.46 \text{ kN} \quad \text{and} \quad T_2 = 2.4 - 1.06 = 1.34 \text{ kN}$$

Maximum torque that can be transmitted by the belt

$$= (T_1 - T_2) \times \frac{D}{2} = (3.46 - 1.34)\frac{D}{2} = \mathbf{1.06\, D \text{ kNm}}$$

EXAMPLE 4.44

Find out the number of turns a hauling rope must be wound round a rotating capstan in order to haul a load of 3 MN up to a gradient of 1 in 30. Presume the following data :

Resistance due to rolling = 0.00375 per newton load

Friction coefficient between the rope and drum = 0.35

Pull on the free end of the rope = 250 N

Solution : The hauling is required up an inclined plane and for that T_1 has to overcome the frictional resistance as well as effort component of the incline. That is

$$T_1 = (3 \times 10^6 \times 0.00375) + \left(3 \times 10^6 \times \frac{1}{30}\right) = 111250 \text{ N}$$

$$T_2 = 250 \text{ N}$$

Invoking the relation $\frac{T_1}{T_2} = e^{\mu\theta}$, we have

$$e^{\mu\theta} = \frac{111250}{250} = 445$$

$$\mu\theta = \log_e 445 \text{ ; } \theta = \frac{\log_e 445}{0.35} = 41742 \text{ radians}$$

$$\text{Number of turns} = \frac{17.42}{2\pi} = 2.77$$

Hence three turns need to be given.

EXAMPLE 4.45

A horizontal drum of belt drive carries the belt over a semicircle around it. The drum has a diameter of 1 m and when it rotates, 400 Nm of torque is transmitted. The drum has a mass of 25 kg and the coefficient of friction between the belt and drum is 0.25. Neglecting mass of belt, determine tensions induced in the limbs of the belt, and the vertical reaction on the bearings.

Solution: Since the drum carries the belt over a semi-circle around it,

$$\text{angle of contact} = 180° = \pi \text{ radians}$$

If T_1 and T_2 are the tensions on the tight and slack side of the belt, then

$$\frac{T_1}{T_2} = e^{\mu\theta} = e^{0.25 \times \pi} = 2.19 \qquad ...(i)$$

From the relation for torque transmitted: $T = (T_1 - T_2)\, r$

$$T_1 - T_2 = \frac{T}{r} = \frac{400}{0.5} = 800 \qquad \text{...}(ii)$$

From expressions (*i*) and (*ii*)

$$2.19\, T_2 - T_2 = 800\,; \quad T_2 = \frac{800}{1.19} = \mathbf{672.27\ N}$$

$$T_1 = 2.19 \times 672.27 = \mathbf{1472.27\ N}$$

Vertical reaction on the drum bearings,

$$= (T_1 + T_2) + W \qquad \text{where } W \text{ is the weight of drum}$$

$$= (1472.27 + 672.27) + (25 \times 9.81) = \mathbf{2389.79\ N}$$

EXAMPLE 4.46

Determine the minimum value of weight W required to cause motion of a block which rests on a horizontal plane. The block weighs 300 N and the coefficient of friction between the block and plane is 0.6. Angle of wrap over the pulley is 90° and the coefficient of friction between the pulley and rope is 0.3.

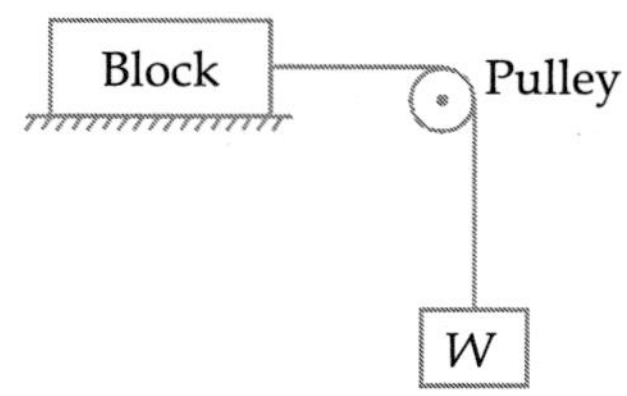

Fig. 4.79

Solution : Since the weight W impends vertical motion in the downward direction, the tension on the two sides of the pulley will be as shown in Fig. 4.70.

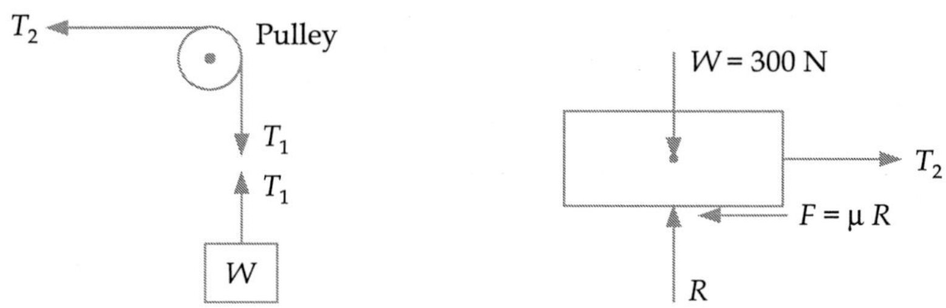

Fig. 4.80.

$$T_1 = W\,; \ \mu = 0.3\,; \ \theta = 90^\circ = \frac{\pi}{2} \text{ radians}$$

Invoking the relation

$$\frac{T_1}{T_2} = e^{\mu\theta}; \quad \frac{W}{T_2} = e^{0.3 \times \pi/2} = (2.718)^{0.471} = 1.60$$

Considering equilibrium of block,

$$\Sigma F_x = 0\,; \ T_2 = F = \mu R$$

$$\Sigma F_y = 0\,; \ R = 300 \text{ N}$$

$\therefore \qquad T_2 = 0.6 \times 300 = 180$ N

Hence $\qquad W = 1.6 \times 180 = \mathbf{288\ N}$

EXAMPLE 4.47

A flat belt running on a pulley 1m in diameter is to transmit 7.5 kW power at a speed of 200 rev/min. The angle of embrace of belt and pulley is 170 degree and the coefficient of friction between the belt and pulley is 0.25. If the pull in belt is not to exceed 200 N/cm , determine the width of belt.

Solution : Angle of embrace (lap)

$$= 170° = \frac{170}{180} \times \pi = 2.965 \text{ radians}$$

If T_1 and T_2 are the tensions on the tight and slack side of the belt, then

$$\frac{T_1}{T_2} = e^{\mu\theta} = e^{0.25 \times 2.965} = e^{0.741} = 2.1$$

$$\therefore \quad T_1 = 2.1\, T_2$$

Speed of belt, $V = \dfrac{\pi d_1 N}{60} = \dfrac{\pi \times 1 \times 200}{60} = 10.46 \text{ m/s}$

Power transmitted by the belt

$$= (T_1 - T_2) \times V$$

$$= (2.1\, T_2 - T_2) \times 10.46 = 11.506\, T_2 \text{ Nm/s}$$

$$\therefore \quad 7.5 \times 10^3 = 11.506\, T_2$$

$$T_2 = 651.83 \text{ N} \quad \text{and}$$

$$T_1 = 651.83 \times 2.1 = 1368.85 \text{ N}$$

Hence necessary width of belt = $\dfrac{1368.85}{200}$ = **6.84** cm

REVIEW AND SUMMARY

1. Experience indicates that there is always some degree of interlocking and wedge action between the contact surfaces. Obviously then, when a block is made to slide over a surface, a force gets automatically induced. This force acts in a direction opposite to that in which motion is sought and offers resistance to the movement of blocks. This resisting force acts tangentially to the contact surface and is called as *frictional resistance* or *force of friction.*
2. There are two types of friction
   (*a*) dry friction or solid friction or coulomb friction
   (*b*) fluid friction
   Dry friction develops when the contact surfaces are dry (no lubrication) and there is tendency for relative motion. Fluid friction manifests when a lubricating film is introduced between the contact surfaces and is due to the adjacent layers of fluid moving at different velocities.
3. The friction experienced by the block when at rest is called static friction. The limiting friction is the maximum friction force at the time of impending motion, *i.e.*, when the motion is about to commence. When the motion of block commences, the friction force drops to some what low value called the dynamic or kinetic friction.
4. There are two types of dynamic friction namely the sliding friction and the rolling friction. Sliding friction is the friction experienced by a body when it slides over another body. Rolling friction manifests when a body rolls over another body.
5. The ratio of the force of friction F to the normal reaction R between the contact surfaces is called the coefficient of friction μ. That is $\mu = F/R$.
   The angle of friction ϕ is the angle which the resulatant of normal reaction R and limiting force of friction F makes with the normal reaction. That is $\tan \phi = F/R = \mu$.

The angle α of an inclined plane at which a block resting on it is about to slide down is called the angle of repose, and it is equal to the angle of friction between the block and the inclined plane.

The cone with the point of contact as the vertex and the normal at the point of contact as the axis, and having semi-vertex angle ϕ is known as cone of friction.

6. Friction is involved in wedges, ladders, belts and ropes, square threaded screws (screw jack), journal and thrust bearings.

REVIEW QUESTIONS

A. Conceptual and conventional questions :

1. Define the term friction. How does it come into play?
2. Print out the characteristics of frictional force.
3. Cite some engineering applications where (*a*) friction is undesirable, (*b*) friction is quite necessary.
4. Friction is a necessary evil. Comment on the validity of this statement.
5. Explain, with the help of neat diagram, the concept of limiting friction.
6. Differentiate between (*a*) static and kinetic friction (*b*) sliding and rolling friction.
7. How will you distinguish between viscous friction and non-viscous friction?
8. State the main laws of dry (solid) friction.
9. Define the following terms:
   (*a*) coefficient of friction, (*b*) angle of friction,
   (*c*) angle of repose, and (*d*) cone of friction.
10. (*a*) Can we talk of friction coefficient for a single surface?
    (*b*) What is wrong with the following statement:
    "coefficient of static friction for concrete is 0.6."
11. What is cone of friction? State its significance.
12. A weight of 200 N is resting on a rough horizontal plane, and can be just moved by a force of 80 N applied horizontally. Determine the coefficient of friction, and the direction and magnitude of the resultant reaction of the plane.
    (0.4, 215.4 N at an angle of 68.2° with horizontal)
13. A body of weight 200 N is placed on a rough horizontal plane. Determine the coefficient of friction if a horizontal force of 130 N is just sufficient to cause the body slide over the plane. Also calculate the resultant reaction. **(0.65, 238 N)**
14. Find the least friction required to pull a body of weight W placed on a rough horizontal plane, when the force is applied at an angle θ with the horizontal. **($W \sin \theta$)**
15. A body of weight 450 N is pulled up an inclined plane, by a force of 300 N. The inclination of the plane is 30° to the horizontal and the force is applied parallel to the plane. Find the coefficient of friction. **(0.192)**
16. The force required to pull a body of weight 100 N on a rough horizontal plane is 30 N. Determine the coefficient of friction if the force applied is at an angle of 15° with the horizontal.
    (0.313)
17. A wooden block of 5 kg mass rests on a rough horizontal plane. What pull inclined at 15 degree to the horizontal is required to make the block just move? How this result will be affected if push is applied along the same line of action? Comment on the change. Take coefficient of friction between the block and the plane, $\mu = 0.4$.

18. A body resting on a rough horizontal plane can be just moved by a force inclined to the horizontal. If friction coefficient is 0.5, compare the force with the weight of the body.

(0.45)

19. A body of 400 N is pulled up along an inclined plane having inclination 30° to the horizontal at a steady speed. If the coefficient of friction between the body and the plane is 0.3 and force is applied parallel to the inclined plane, find the force required. Find also the work done on the body if the distance travelled by the body is 120 m along the plane.

20. A cord connects two bodies of weight 500 N and 1000 N. The two bodies are placed on an inclined plane and cord is parallel to the inclined plane. The coefficient of friction for the weight of 500 N is 0.2 and that of 1000 N is 0.4. Determine the inclination of the plane to the horizontal and tension in the cord when the motion is about to take place, down the inclined plane. The body weighing 500 N is below the body weighing 1000 N.

(18.417°, 63.086 N)

21. A block of weight 500 N rests on a surface for which the coefficient of friction is 0.2. Work out the range of angle α such that there is no movement of the block when a force of 100 N is applied as shown in Fig. 4.81. **(135°)**

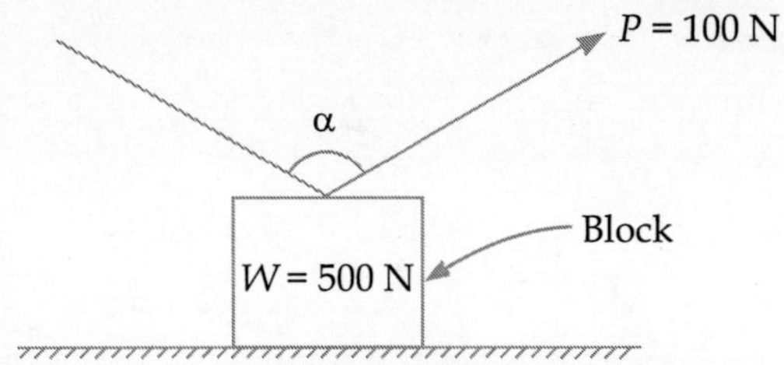

Fig. 4.81

22. A uniform ladder rests in limiting equilibrium with one end on a rough horizontal plane and the other against a smooth vertical wall. Show that a man ascending the ladder cannot go more than halfway up.

23. A uniform ladder of length 13 m and weighing 30 N is placed against a smooth vertical wall with its lower end 10 m from the wall. In this position the ladder is just to slip. Determine the

(*a*) coefficient of friction between ladder and the floor,

(*b*) frictional force acting on the ladder at the point of contact between ladder and floor.

((*a*) 0.602, (*b*) 18.06 N)

24. A uniform ladder of weight 800 N and of length 7 m rests on a horizontal ground and leans against a smooth vertical wall. The angle made by the ladder with the horizontal is 60°. When a man of weight 600 N stands on the ladder at a distance of 4 m from the top of ladder, the ladder is at the point of sliding. Determine the coefficient of friction between ladder and the floor. **(0.237)**

25. A uniform ladder of length 15 m rests against a vertical wall making an angle of 60° with the horizontal Coefficient of friction between wall and the ladder, and the ground and ladder are 0.3 and 0.25 respectively. A man weighing 500 N ascends the ladder. How high will he be able to go before the ladder slips? Find the weight necessary to be put at the bottom of the ladder so as to just sufficient to permit the man to go the top. Assume weight of ladder is 850 N.

(6.39 m, 607.6 m)

26. A uniform ladder 3 m long weighs 18 N. It is placed against a wall 60° with floor as shown in Fig. 4.82. The coefficient of friction between the wall and the ladder is 0.25 and that

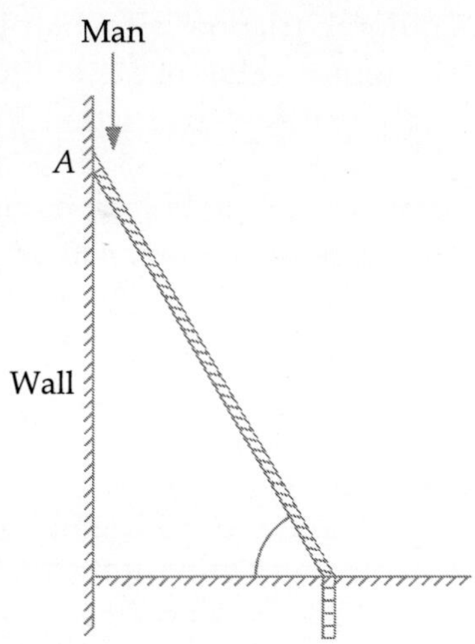

Fig. 4.82

between the floor and the ladder is 0.35. The ladder, in addition to its own weight has to support a man weighing 90 N at its top at A. Calculate

(*a*) The horizontal force F to be applied to the ladder at the floor level to prevent slipping.

(*b*) If the force F is not applied what would be the minimum inclination of the ladder with the horizontal so that there is no slipping of it with the man at the top?

((*a*) 16.6 N, (*b*) 69°)

27. A wedge A of 50 N weight is to be driven between an inclined support and a block B of 2000 N weight (Fig. 4.83). If coefficient of friction between all surfaces of contact is 0.3, make calculations for the magnitude of the force P required to start moving the wedge.

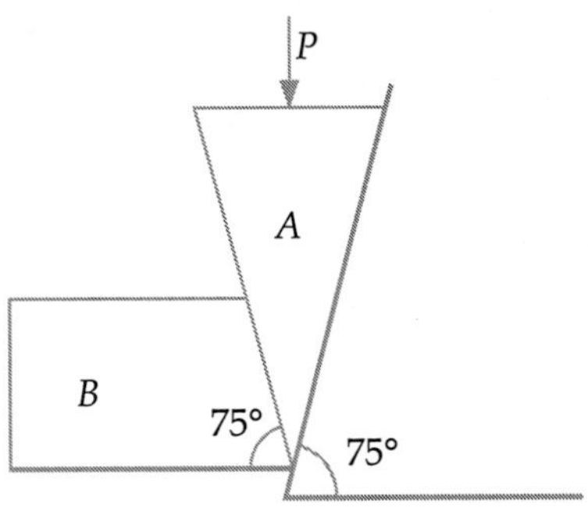

Fig. 4.83

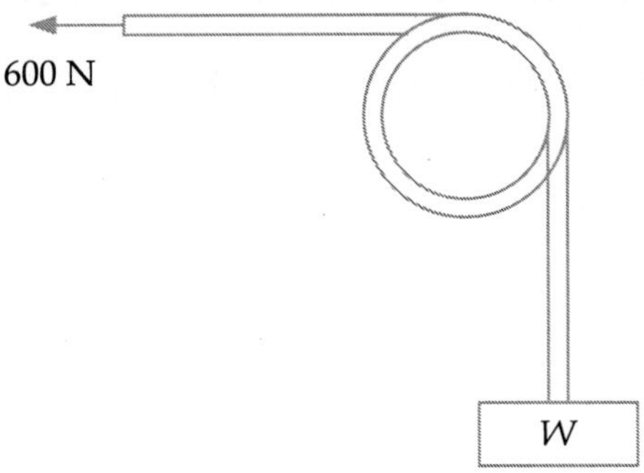

Fig. 4.84

28. A pull of 600 N is exerted on a rope which is wrapped 1.25 turns round a stationary horizontal drum as shown in Fig. 4.84 Presuming that coefficient of friction between the brake and drum is 0.3, make calculations for the weight that can be supported when

(*a*) impending motion is upward,

(*b*) impending motion is downward. **(56.85 N, 6330 N)**

B. Complete the following statements with most appropriate word/words :

1. Frictional force is .... of area of contact and acts in .... direction to the surfaces in contact.
2. .... friction is another name for dynamic friction.
3. Coulomb friction is friction between .... surfaces.
4. Maximum value of static friction is known as .... friction.
5. The coefficient of friction μ and the angle of friction ϕ are related by the expression ....
6. Semi-central angle in cone of friction is equal to ....
7. The force of friction, acting on a moving body is called the ....
8. The angle made by the resultant of normal reaction and the limiting force of friction with the normal reaction is called the ....
9. The ratio of limitating force of friction to the normal reaction is known as ....
10. If a ladder leans again a smooth vertical wall, then the force of friction between the ladder and wall is ....
11. The .... is defined as the maximum inclination of a plane at which a body remains in equilibrium over the inclined plane due to friction only.
12. Screw jack works on the principle of ....

Answers:

1. independent, tangential 2. kinetic 3. dry 4. limiting 5. $\mu = \tan\theta$ 6. angle of friction 7. dynamic friction 8. angle of friction 9. coefficient of friction 10. zero 11. angle of respose 12. inclined plane

C. Multiple choice questions :

1. The force of friction between two bodies in contact is
   (*a*) a function of the relative velocity between them
   (*b*) always normal to the surface of contact
   (*c*) never shown in the free body diagram of the system of these two bodies
   (*d*) dependent on the areas of contact

2. Mark the wrong statement:
   (*a*) The maximum force of friction is independent of the magnitude of area in contact between the surfaces
   (*b*) Coefficient of friction between two surfaces is a constant of proportionality between the limiting force of the friction and the normal reaction
   (*c*) The coefficient of static friction must be less than the coefficient of dynamic friction for a given pair of bodies
   (*d*) If a block slides over an inclined plane, the reaction by the plane on the block may not be in line with the weigth force

3. Which of the following statements about friction is true ?
   (*a*) friction can be reduced to zero (*b*) frictional force can accelerate a body
   (*c*) frictional force is proportional to the area of contact between the two surfaces
   (*d*) kinetic friction is always greater than rolling friction

4. The coefficient of friction between two surfaces is a constant of proportionality between the applied tangential force and the normal reaction
   (*a*) at the instant of the application of force
   (*b*) at any instant when the body is at rest
   (*c*) at the instant of impending motion (*d*) at an instant after motion takes place

5. The magnitude of coefficient of static friction between two bodies in contact
   (*a*) always lies between zero and unity
   (*b*) may exceed unity
   (*c*) is less than the coefficient of dynamic friction between the same pair of bodies
   (*d*) depends upon areas of contact

6. Coefficient of kinetic friction between a moving body and the surface on which it moves
   (*a*) is dependent on the velocity of body
   (*b*) depends only on the nature of the surfaces
   (*c*) depends on the nature of surfaces and the relative velocity between the surface and body
   (*d*) depends on the nature of the surfaces, weight of the body and its velocity

7. Identify the wrong statement:
   (*a*) The maximum inclination of the plane at which a body remains in equilibrium over the inclined plane by the frictional force alone is called the angle of repose

(*b*) Angle of friction is the angle between the normal reaction and the resultant of limiting friction and normal reaction
(*c*) Angle of friction equals the angle of repose
(*d*) The coefficient of friction μ and the angle of friction ϕ are related by the expression $\mu = \sin \phi$

8. The coefficient of friction between two surfaces is 0.2. Then the angle of friction as
(*a*) $\sin^{-1}$ (0.2) (*b*) $\cos^{-1}$ (0.2) (*c*) $\tan^{-1}$ (0.2) (*d*) $\cot^{-1}$ (5)

9. A zero angle of friction implies that
(*a*) frictional force is infinite (*b*) frictional force is zero
(*c*) frictional force acts normal to the plane
(*d*) frictional force acts along the direction of motion.

10. Brakes are applied to a car travelling at 90 km/hr and it comes to stop 5 seconds thereafter. The coefficient of friction between the wheels and road surface is
(*a*) 0.18 (*b*) 0.25 (*c*) 0.36 (*d*) 0.5
Take g = 10 m/s^2.

11. Consider the following statements
1. Friction acts in a direction opposite to that in which motion is intended
2. Frictioin comes only that much into play as is just sufficient to prevent motion
3. Friction exists as long as the tractive force acts
Which of these statements is/are correct?
(*a*) 1 only (*b*) 1 and 2 (*c*) 1 and 3 (*d*) 1, 2 and 3

12. All of the following statements are correct, except
(*a*) Coulomb friction manifests when the contact surfaces are dry and there is tendency for relative motion
(*b*) Friction force acts normal to the surfaces in contact and is maximum at the instant of impending motion
(*c*) A great reduction in friction occurs when lubricating film is introduced between the contact surfaces
(*d*) The kinetic friction is always somewhat less than the limiting static friction

13. Which of the following statements is not correct about the limiting friction?
(*a*) it is the maximum friction force at the time of impending motion
(*b*) it depends upon the roughness or smoothness of the surfaces in contact
(*c*) it is affected by the area and shape of the contact surfaces
(*d*) its magnitude bears a constant ratio to the normal reaction between the mating surfaces

14. A block resting on an inclined plane begins to slide down the plane when the angle of inclination is gradually increased to 30°. Then the coefficient of friction between the block and plane is closest to
(*a*) 0.50 (*b*) 0.58 (*c*) 0.72 (*d*) 0.87

Answers :

1. (*c*)	2. (*c*)	3. (*d*)	4. (*a*)	5. (*c*)	6. (*c*)
7. (*d*)	8. (*d*)	9. (*b*)	10. (*b*)	11. (*d*)	12. (*b*)
13. (*c*)	14. (*b*)				

CHAPTER

5

STRUCTURE: TRUSSES AND FRAMES

Learning objectives : The concepts of this chapter are so designed that its study makes the reader conversant with

- *framed structures and their application*
- *types of frames : plane frame and space frame ; statically determinate and statically indeterminate frames ; perfect, deficient and redundant beams*
- *reaction at supports of a frame*
- *analysis of a frame : method of joints, method of sections and the graphical method*

A framed structure is an assemblage of a number of bars or rods joined together in such a way to form a rigid framework; this structure is designed to resist geometrical distortion under any applied system of loading. The ends are joined by riveting, welding or by nuts and bolts. The bars used in the framed structure are slender (cross-sectional dimensions quite small compared to length) and are usually of angle section, T-section, I-section etc. The bars are called two-force members because they are either in tension or in compression. The framed structure finds application in the roofs of sheds at railway platforms, workshops and industrial buildings, and bridge trusses etc. Attention has been focussed in this chapter on the study of plane structures, that is, the structures whose members lie in one plane.

5.1. CLASSIFICATION FRAMES

The frames are generally classified as:

- **Plane frame and space frame:** In a plane frame, all the members lie in a single plane and the forces act along the plane of the frame. Bridge trusses and roof trusses are the plane frames.

  A frame in which all the members do not lie in the some plane is called a space frame. Tripod and suspention towers are the space frames.

- **Statically determinate and statically indeterminate frames:** The force analysis of the members of a statically determinate frame can be made by applying the equations of statics only. Equations of static equilibrium are not sufficient to determine the forces in statically indeterminate frames; there is need of considering their deformation also.

- **Perfect, deficient and redundant frames:** The structure is said to be **perfect** if the number of members is just sufficient to prevent its distortion of shape when subjected to external loads. For a perfect frame, the following correlation exists between the number of joints j and the number of members m,

$$m = 2j - 3$$

The number of joints and the number of members in the frames shown in Fig. 5.1 satisfy the identity $m = 2j - 3$ and as such these are perfect structures (frames). A perfect structure is statically determinate. Equations of static equilibrium are sufficient to determine the forces in its members without the need of considering their defor-mation.

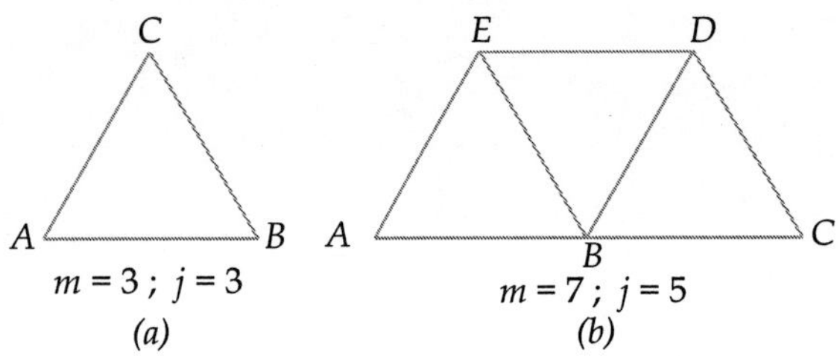

Fig. 5.1

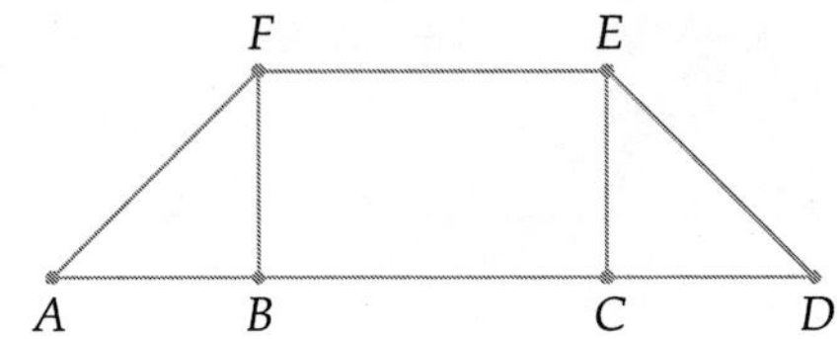

Fig. 5.2

A structure is termed **imperfect** or **deficient** frame if the number of memebrs in it is less than that required for a perfect frame. Refer Fig. 5.2 which shows a frame having 6 joints and 8 members. The number of members required for the frame to be perfect is

$$m = 2j - 3 = 2 \times 6 - 3 = 9$$

Since there are only 8 members (one less than that required for a perfect frame), the given frame is deficient or imperfect. An imperfect frame is under rigid or collapsible. Such a frame cannot prevent geometrical distortion when loaded.

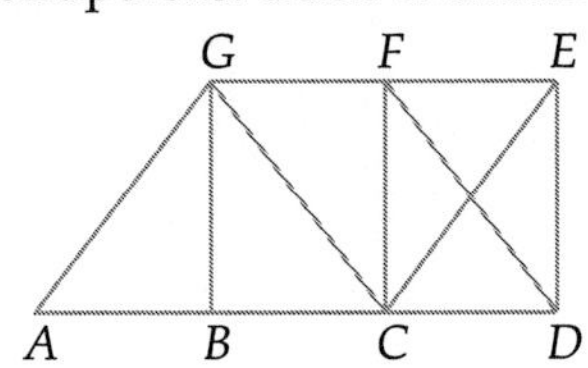

Fig. 5.3

A structure is termed **redundant** frame if the number of members in it is more than that required for a perfect frame. Refer Fig. 5.3 which shows a frame which has 7 joints and 12 members. The number of members required for the frame to be perfect is

$$m = 2j - 3 = 2 \times 7 - 3 = 11$$

Since the frame has 12 members (one more than that required for a perfect frame), the given frame is redundant. A redundant frame is over rigid and statically indeterminate. Further, each extra member adds one degree of indeterminancy. Thus the frame depicted in Fig. 5.3 is redundant to a single degree because one member is extra.

The frame depicted in Fig. 5.3 would remain stable even if one of the members CE or DF is removed.

5.2. REACTIONS AT SUPPORTS OF A FRAME

Statically determinate frames normally have two types of supports: one hinged and the other on rollers.

Roller supports (Fig. 5.4 *a*) are frictionless and provide a reaction at right angles to the roller base. At a hinged support (Fig. 5.4 *b*), the line of action of the reaction depends upon the load system on the frame. This reaction can be split into two components acting in the horizontal and vertical directions.

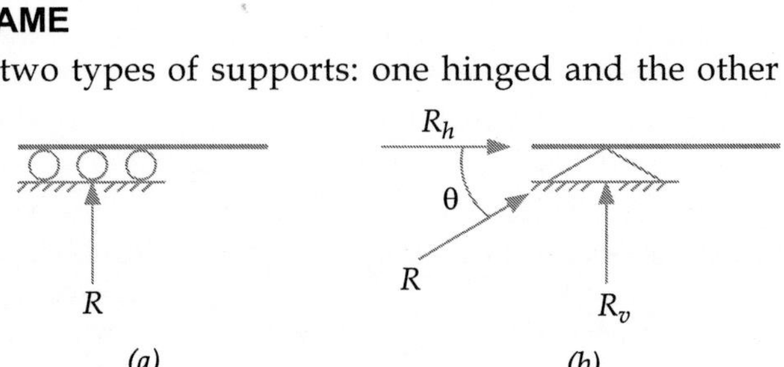

Fig. 5.4

$$R_v = R \sin \theta \text{ and } R_h = R \cos \theta$$

A roller support prevents motion only in the direction perpendicular to its seat. With pin (hinged) support, the beam has no translatory motion in any direction. In both types of supports, there is no prevention of rotation about the connections.

5.3. ANALYSIS OF A FRAME

The analysis of a truss involves the determination of:

(*i*) The reaction at the supports,

(*ii*) The internal stresses induced in the members due to external loads. That helps to workout the size and shape of the structural members forming the truss.

While making such an analysis, the following assumptions are made:

(*i*) The truss is a perfect one and statically determinate.

(*ii*) All the members comprising the truss are rigid and lie in the same plane.

(*iii*) The members are slender and of uniform cross-section.

(*iv*) The external loads and reactions act at the joints only.

(*v*) The self weight of the members is neglected because the self weight is small compared to the loads they carry.

(*vi*) The forces are transmitted from one member to another through smooth pins (no friction) fitting perfectly in the members.

The stresses in the members of a framed structure are generally worked out by adopting the following methods:

(1) Analytical method: method of joints and method of sections

(2) Graphical method

5.3.1. Method of Joints

Every joint is treated separately as a free body in equilibrium, i.e., the sum of all the vertical forces as well as the horizontal forces acting on the joint is equated to zero.

$$\Sigma F_v = 0 \quad \text{and} \quad \Sigma F_h = 0$$

The steps involved are:

- All the pin joints are labelled.
- A free body diagram of the entire frame is drawn and the reaction at the supports is determined using the conditions of equilibrium.
- Each joint is treated separately as a free body. A certain direction of forces acting on the joint is assumed and the magnitude of forces is worked out by applying the conditions of equilibrium. If the magnitude of a particular force comes out positive, the assumption in respect of its direction is correct. However if the magnitude of the force comes out negative, the actual direction of force is reversed, i.e., opposite to what has been assumed.
- Start is made from a joint where there are not more than two members in which the forces are unknown, and the process is repeated from one joint to another until all the unknowns have been determined.
- Assumption is made about the nature of forces in a member and the force is determined. If the answer is positive, the assumption is correct. A negative answer would imply that the nature of force is opposite to that assumed.

Finally, the force in the member will be tensile if the member pulls the joint to which it is connected (force is directed away from the pin) whereas the force in the member will be compressive if the member pushes the joint to which it is connected (force is towards the pin). These aspects have been illustrated in Fig. 5.5. A member under tension is called a *tie*, and a member under compression is called a **strut**.

(*a*) Tension: Pulling at each end (*b*) Compression: Pushing at each end

Fig. 5.5

The following principles help to identify the members not subjected to any force when the truss is loaded:

(*i*) A single force cannot form a system in equilibrium. This implies that if there is only one force acting at a joint, then for the equilirbium of this joint, this force equals zero.
With reference to Fig. 5.6 (*a*),

$$F_1 = 0.$$

(*ii*) When two members meeting at a joint are not collinear and there is no external force acting at the joint, then the forces in both the members are zero.
With reference to Fig. 5.6(*b*),

$$F_1 = F_2 = 0$$

(*iii*) When three members are meeting at an unloaded joint and out of them two are collinear, then the force in the third member will be zero.
With reference to Fig. 5.6(*c*),

$$F_3 = 0.$$

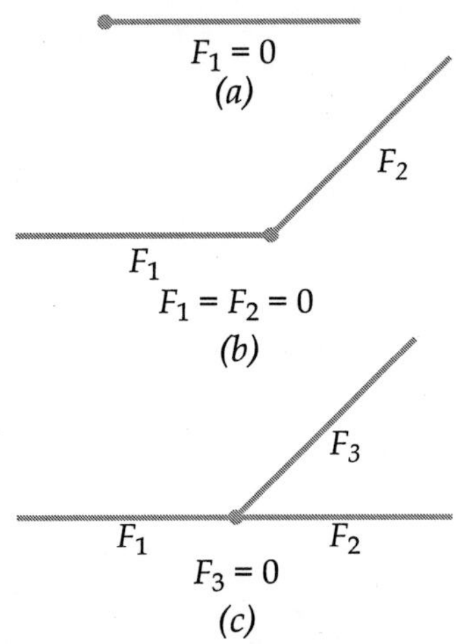

Fig. 5.6

EXAMPLE 5.1

With reference to trusses shown in Fig. 5.7, identify the members which are not subjected to any force.

Solution : (*i*) Consider joint *D* for the truss shown in Fig. 5.7 *a*. Forces at this joint are:

F_1 in member *DC* ; F_2 in member *DA* ; F_3 in member *DF*

The joint is unloaded and the forces F_2 and F_3 are collinear

$$\therefore \quad F_1 = 0$$

Likewise forces in the members *EF* and *HG* are zero.

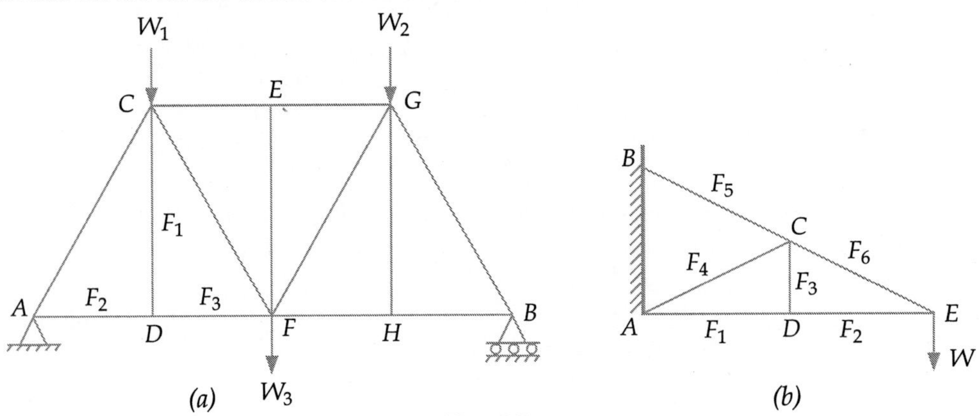

Fig. 5.7

(*ii*) Considering joint *D* of of the cantilever truss (Fig. 5.7 *b*), there are three forces namely F_1, F_2 and F_3.

Since the joint is unloaded and the forces F_1 and F_2 are collinear

$$\therefore \quad F_3 = 0$$

Considering joint *C*;

Since $F_3 = 0$, the forces acting at this joint are F_4, F_5 and F_6.

The joint is unloaded and out of these three forces, two forces namely F_5 and F_6 are collinear. Accordingly, the third force F_4 is zero.

Thus the members *AC* and *DC* will not be subjected to any load.

(*iii*) Considering loading and the support system for the truss depicted in Fig. 5.7 *c*, the reaction at roller support is vertical and $R_b = W$. No reaction will be experienced at the hinged support, *i.e.*, $R = 0$.

Obviously then, there will no force in the members other than *BD*, *DF* and *FH*.

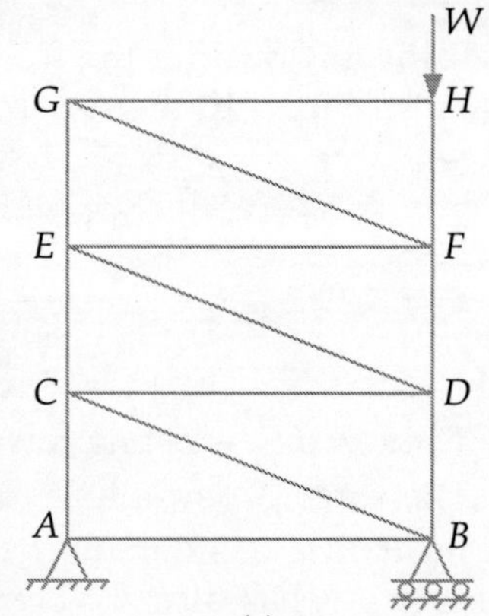

(*c*)

Fig. 5.7

EXAMPLE 5.2

Determine the forces in all the members of a truss with the loading and support system shown in Fig. 5.8.

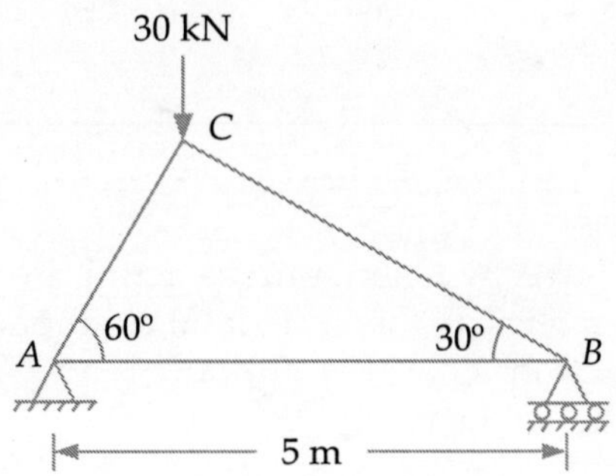

Fig. 5.8

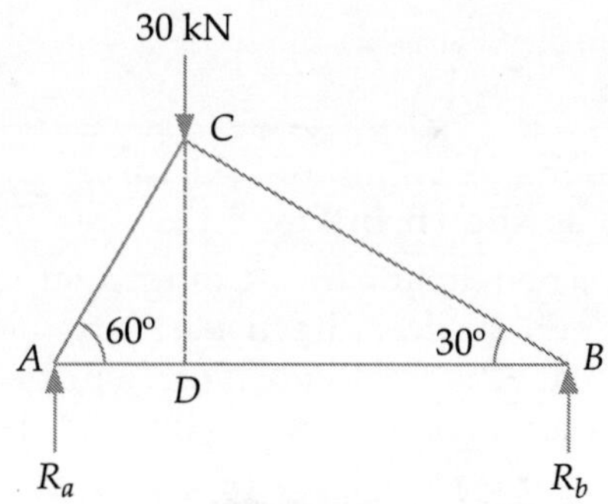

Fig. 5.9

Solution : Roller support (end B) is frictionless and provides a reaction R_b at right angles to the roller base. The reaction at the hinged support (end A) can have two components acting in the horizontal and vertical directions. Since the load of 30 kN acts vertically downward, the horizontal component of reaction at A is zero and there will be only vertical reaction R_a.

Considering the free-body diagram of the whole truss (Fig. 5.9)

The triangle *ABC* is a right angled triangle with angle $ACB = 90°$

$$AC = AB \cos 60°$$

$$= 5 \times 0.5 = 2.5 \text{ m}$$

Distance of line of action of 30 kN force from *A*,

$$AD = AC \cos 60°$$

$$= 2.5 \times 0.5 = 1.25 \text{ m}$$

Taking moments about A, we obtain

$$R_b \times 5 = 30 \times 1.25$$

$$R_b = \frac{30 \times 1.25}{5} = 7.5 \text{ kN}$$

$$R_a = 30 - 7.5 = 22.5 \text{ kN}$$

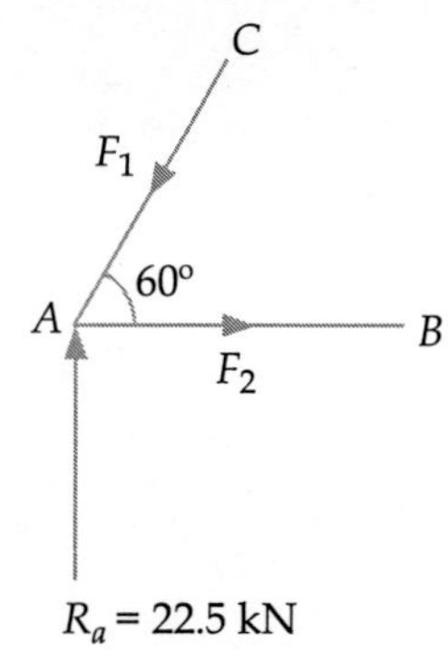

Fig. 5.10

Joint *A*: Consider the free body diagram of joint A with the direction of forces assumed as shown in Fig. 5.10. Equations of equilibrium can be written as:

$$\Sigma F_x = 0: \quad F_2 - F_1 \cos 60^\circ = 0$$

$$\Sigma F_y = 0: \quad F_1 \sin 60^\circ - R_a = 0$$

$$\therefore \quad F_1 = \frac{R_a}{\sin 60^\circ} = \frac{22.5}{0.866} = \mathbf{25.97\ kN} \text{ (compressive)}$$

and

$$F_2 = F_1 \cos 60^\circ = 25.97 \times 0.5 = 12.99 \text{ kN (tensile)}$$

The force F_1 is acting towards the pin which means that the member *AC* is in compression.

The force F_2 is pulling the joint *A* which means that the member *AB* is in tension.

Joint *B*: Consider the free body diagram of joint *B*. The force F_2 in member *AB* has already been calculated above and found to be tensile. Hence force F_2 will pull the joint *B* and will be directed away from it as shown in Fig. 5.11.

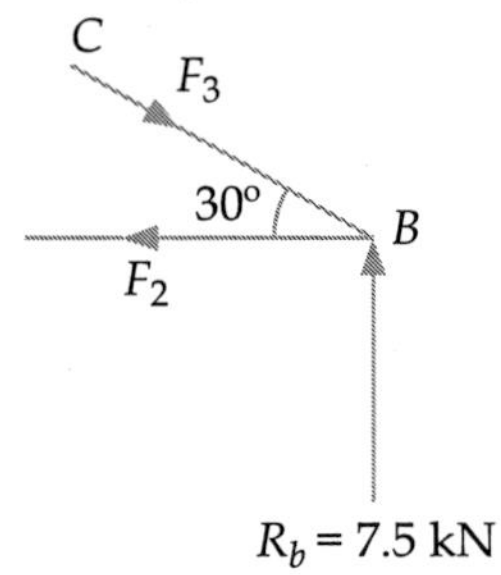

Fig. 5.11

From the equations of equilibrium,

$$\Sigma F_x = 0: \quad F_3 \cos 30^\circ - F_2 = 0$$

$$\therefore \quad F_3 = \frac{F_2}{\cos 30^\circ} = \frac{12.99}{0.866} = \mathbf{15\ kN} \text{ (compressive)}$$

The force F_3 is acting towards the joint *B* which means that the member *BC* is in compression.

EXAMPLE 5.3

Determine the reactions and the forces in each member of a simple triangle truss supporting two loads as shown in Fig. 5.12.

Solution : The reaction at the hinged support (end *A*) can have two components acting in the horizontal and vertical directions. Since the applied loads are vertical, the horizontal component of reaction at *A* is zero and there will be only vertical reaction R_a. Roller support (end *C*) is frictionless and provides a reaction R_c at right angles to the roller base.

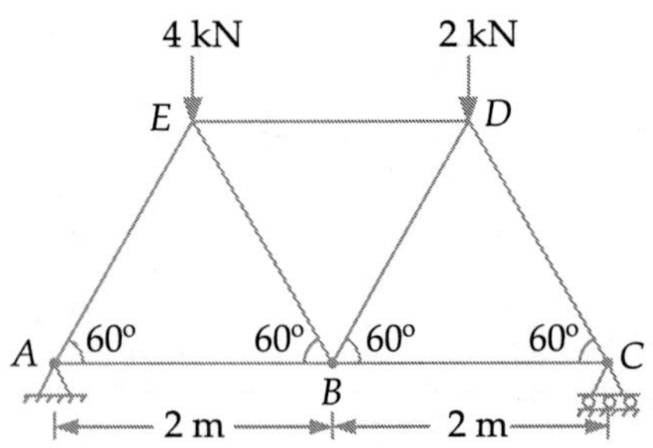

Fig. 5.12

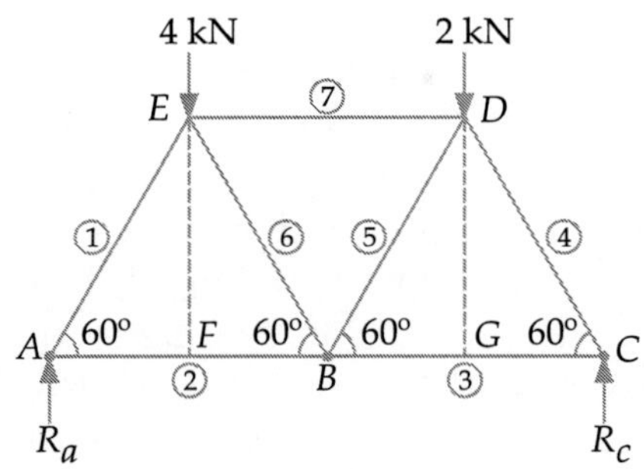

Fig. 5.13

With reference to free body diagram of the whole truss (Fig. 5.13), each of the triangular (element) constitutes an equilateral triangle of each side equal to 2 m. Then

Distance of line of action of 4 kN force from A,

$$AF = AE \cos 60^\circ = 2 \times 0.5 = 1 \text{ m}$$

Distance of line of action of 2 kN force from *A*,

$$AG = AB + BG = AB + BD \cos 60^\circ = 2 + 2 \times 0.5 = 3 \text{ m}$$

Taking moments about A, we get

$$R_c \times 4 = 2 \times 3 + 4 \times 1 = 10; \qquad R_c = \frac{10}{4} = 2.5 \text{ kN}$$

$$R_a = (4 + 2) - 2.5 = 3.5 \text{ kN}$$

Joint *A*: Consider the free body diagram of joint *A* with the direction of forces assumed as shown in Fig. 5.14. Start can be made with this joint because there are only two unknown forces acting at this joint.

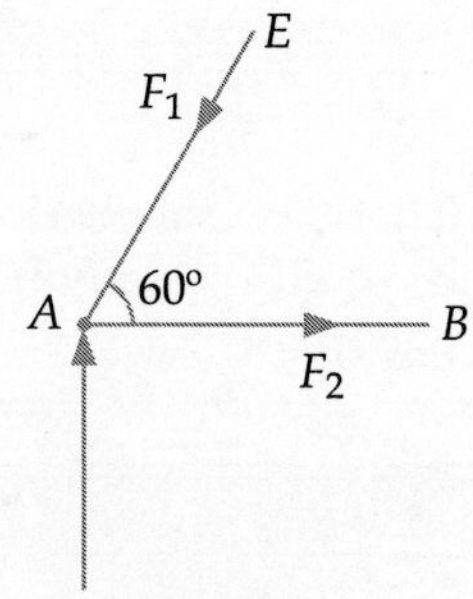

Fig. 5.14

Equations of equilibrium can be written as:

$$\Sigma F_x = 0; \qquad F_2 - F_1 \cos 60° = 0$$

$$\Sigma F_y = 0; \qquad R_a - F_1 \sin 60° = 0$$

$$\therefore \qquad F_1 = \frac{R_a}{\sin 60°} = \frac{3.5}{0.866}$$

$$= \mathbf{4.04} \text{ kN (compressive)}$$

$$\text{and} \qquad F_2 = F_1 \cos 60° = 4.04 \times 0.5$$

$$= \mathbf{2.02} \text{ kN (tensile)}$$

The magnitude of forces F_1 and F_2 are both coming out to be positive and therefore the assumed direction of forces are correct.

The force F_1 is acting towards the pin *A* which means that member *AE* is in compression. The force F_2 is pulling the pin *A* which means that member *AB* is in tension.

Joint *C*: Consider the free body diagram of joint C with the direction of forces assumed as shown in Fig. 5.15. Analysis of this joint is also possible because there are only two unknown forces F_3 and F_4 acting at this joint.

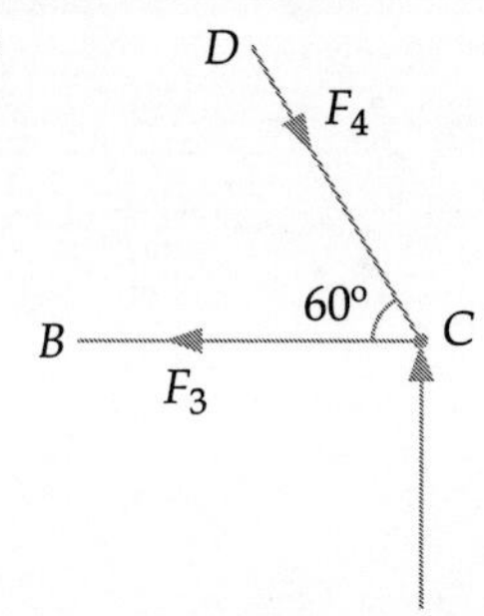

Fig. 5.15

Equations of equilibrium can be written as:

$$\Sigma F_x = 0: \qquad F_4 \cos 60° - F_3 = 0$$

$$\Sigma F_y = 0: \qquad R_C - F_4 \sin 60° = 0$$

$$\therefore \qquad F_4 = \frac{R_C}{\sin 60°} = \frac{2.5}{0.866}$$

$$= \mathbf{2.88 \text{ kN}} \text{ (compression)}$$

$$\text{and} \qquad F_3 = F_4 \cos 60° = 2.88 \times 0.5$$

$$= \mathbf{1.44} \text{ kN (tensile)}$$

Joint *B*: Consider the free body diagram of joint *B* with the two known values F_2 (tensile force in member *AB*) and F_3 (tensile force in member *BC*) inserted:

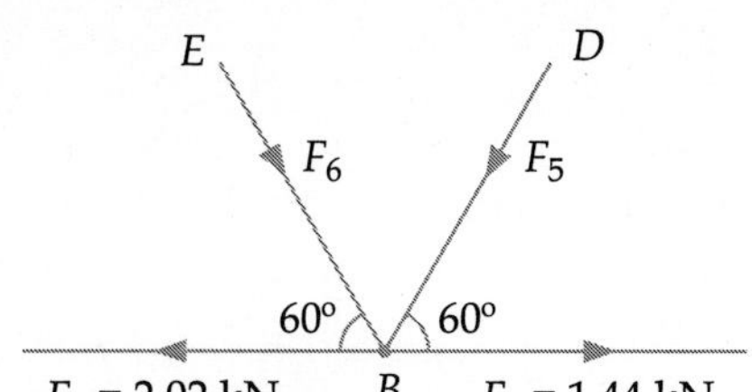

Fig. 5.16

Equations of equilibrium for this joint are:

$$\Sigma F_x = 0: \quad 1.44 - 2.02 + F_5 \cos 60° + F_6 \cos 60°$$

$$\text{or} \quad (F_5 + F_6) \times 0.5 = 0.58$$

$$\text{or} \quad F_5 + F_6 = 1.16 \qquad ...(i)$$

$$\Sigma F_y = 0: \quad F_5 \sin 60° - F_6 \sin 60° = 0$$

$$\text{or} \quad F_5 = F_6 \qquad ...(ii)$$

From expression (*i*) and (*ii*)

$$F_5 = F_6 = \frac{1.16}{2} = \mathbf{0.58} \text{ kN}$$

Force F_5 in member *BD* is 0.58 kN tensile and force F_6 in member *BE* is 0.58 kN compressive.

Joint D: Consider the free body diagram of joint D with the known values F_4 (compressive force in member DC) and F_5 (tensile force member DB) inserted.

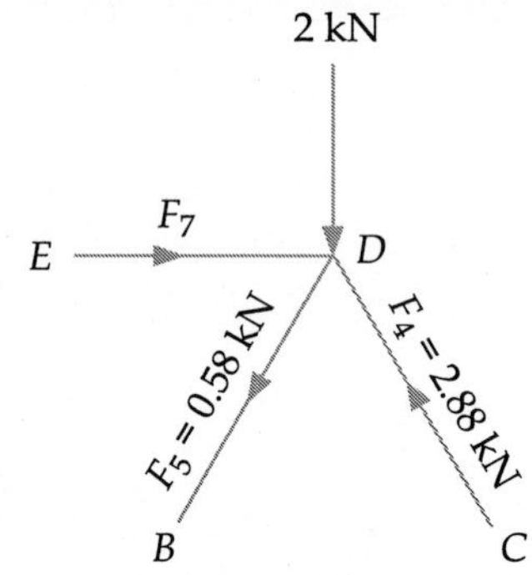

Fig. 5.17

Equations of equilibrium for this joint are:

$\Sigma F_x = 0:$ $\quad F_7 - 0.58 \cos 60° - 2.88 \cos 60°$

or $\quad F_7 = 0.58 \times 0.5 + 2.88 \times 0.5$

$= 0.29 + 1.44 = \mathbf{1.73}$ **kN** (compressive)

There is no need to consider the equilibrium of joint E as the forces in all the members have been determined. The magnitude and the nature of the forces in all the members of the given truss for the applied system of loading are tabulated below :

Member	Force	Member	Force
AE	F_1 = 4.04 kN (C)	BE	F_5 = 0.58 kN (T)
AB	F_2 = 2.02 kN (T)	BD	F_6 = 0.58 kN (C)
BC	F_3 = 1.44 kN (T)	DE	F_7 = 1.73 kN (C)
CD	F_4 = 2.88 kN (C)		

EXAMPLE 5.4

A cantilever truss Wassen type is loaded as shown in the Fig. 5.18. Use the method of sections or the method of joints to find the forces in all members of the truss.

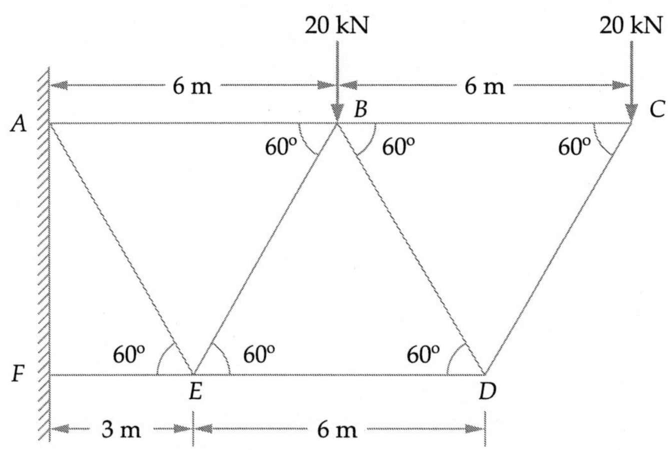

Fig. 5.18

Solution : Consider the free body diagram of ***joint* C** with the direction of forces assumed as shown

Equations of equilibrium are :

$\Sigma F_x = 0; \quad -F_1 + F_2 \cos 60° = 0$

$\Sigma F_y = 0; \quad -20 + F_2 \sin 60° = 0$

$\therefore \quad F_2 = \dfrac{20}{\sin 60°} = 23.1$ kN (compression)

and $F_1 = F_2 \cos 60° = 23.1 \cos 60°$

$= 11.55$ kN (tension)

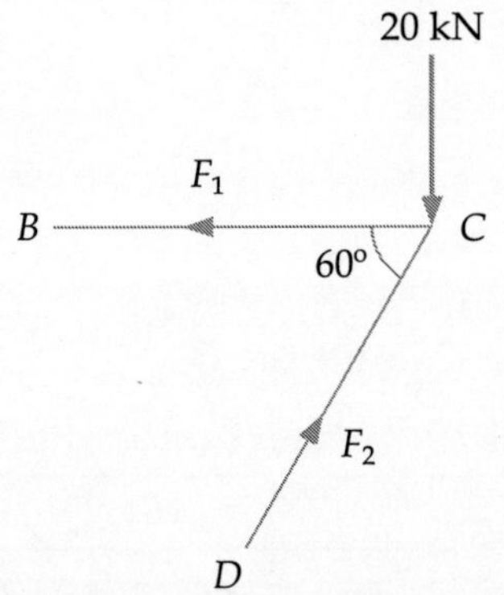

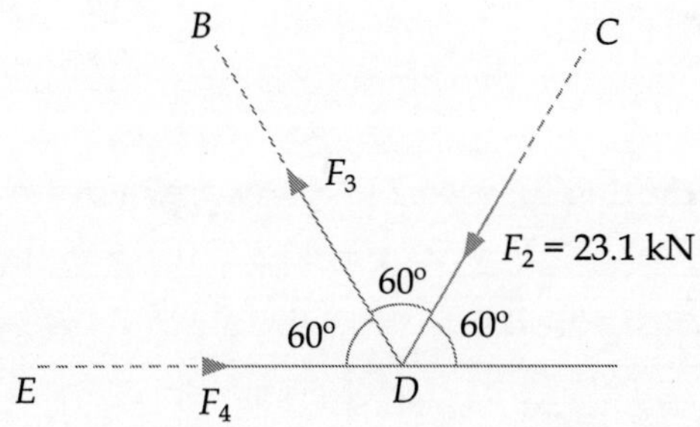

Joint D: From the equations of equilibrium,

$\Sigma F_x = 0;$ $\quad - F_2 \cos 60° - F_3 \cos 60° + F_4 = 0$

or $\quad 23.1 \times 0.5 + 0.5 F_3 = F_4$

$\Sigma F_y = 0;$ $\quad - F_2 \sin 60° + F_3 \sin 60° = 0$

or $\quad F_3 = F_2 = 23.1$ kN (tension)

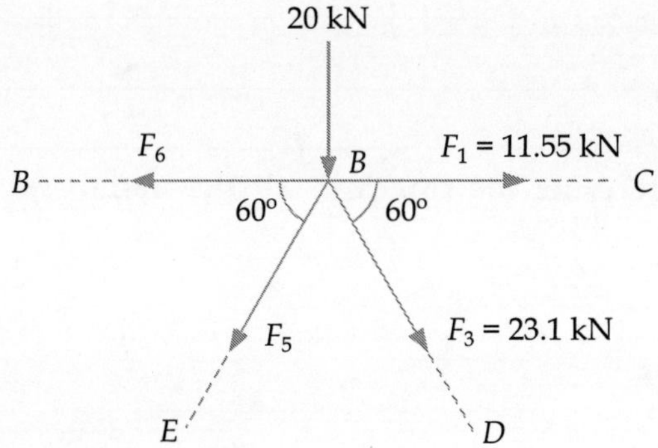

Substituting $F_3 = 23.1$ kN in expression for F_4, we get

$F_4 = 23.1 \times 0.5 + 0.5 \times 23.1 = 23.1$ kN (compression)

Joint B: From the equations of equilibrium,

$$\Sigma F_x = 0: \quad F_1 + F_3 \cos 60° - F_5 \cos 60 - F_6 = 0$$

$$\text{or} \quad 11.55 + 23.1 \times 0.5 - 0.5 F_5 = F_6$$

$$\Sigma F_y = 0: \quad - 20 - F_3 \sin 60° - F_5 \sin 60° = 0$$

$$\text{or} \quad - 20 - 23.1 \times 0.866 = 0.866 F_5$$

$$\therefore \quad F_5 = \frac{-20 - 23.1 \times 0.866}{0.866} = -46.2 \text{ kN}$$

The negative sign with the magnitude of force F_5 shows that a wrong choice has been made while assuming its direction. Obviously the assumed direction of force in member *BE* need to be reversed. Therefore $F_5 = 46.2$ kN (compression) and *BE* is a compressive member.

Substituting $F_5 = -46.2$ kN in expression for F_6, we get

$$F_6 = 11.55 + 23.1 \times 0.5 - 0.5 \times (-46.2) = 46.2 \text{ kN} \quad \text{(tension)}$$

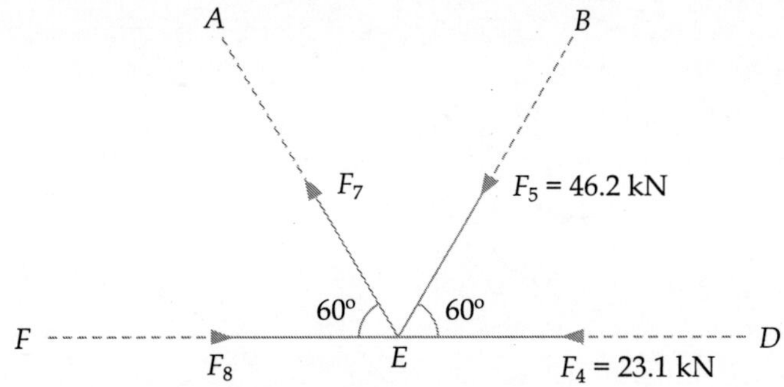

Joint E: From the equations of equilibrium,

$$\Sigma F_x = 0; \quad - F_4 - F_5 \cos 60° - F_7 \cos 60° + F_8 = 0$$

$$\text{or} \qquad F_8 = F_4 + F_5 \cos 60° + F_7 \cos 60°$$
$$= 23.1 + 46.2 \times 0.5 + 0.5\,F_7$$
$$\Sigma F_y = 0; \qquad - F_5 \sin 60° + F_7 \sin 60°$$
$$\therefore \qquad F_7 = F_5 = 46.2 \text{ kN} \quad \text{(tension)}$$

Substituting F_7 = 46.2 kN is expression for F_8, we get

$$F_8 = 23.1 + 46.2 \times 0.5 + 0.5 \times 46.2 = 69.3 \text{ kN} \quad \text{(compression)}$$

The forces in the various members have been tabulated below:

Member	*AB*	*BC*	*CD*	*DE*	*EF*	*AE*	*BE*	*BD*
Force (kN)	46.2	11.55	23.1	23.1	69.3	46.2	46.2	23.1
Nature	*C*	*T*	*C*	*C*	*C*	*T*	*C*	*T*

EXAMPLE 5.5

Determine the forces in all the members of the truss loaded and supported as shown in Fig. 5.19.

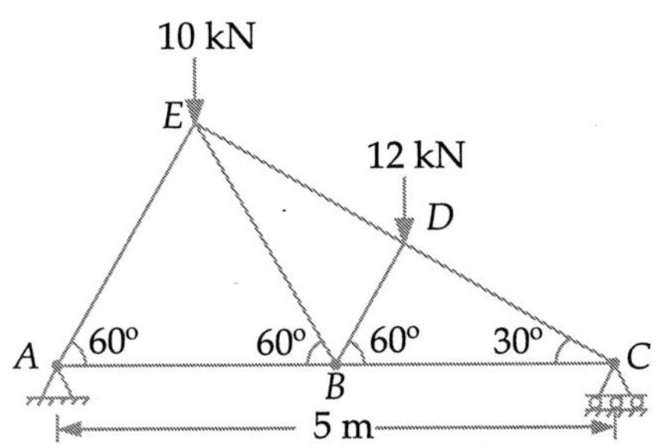

Fig. 5.19

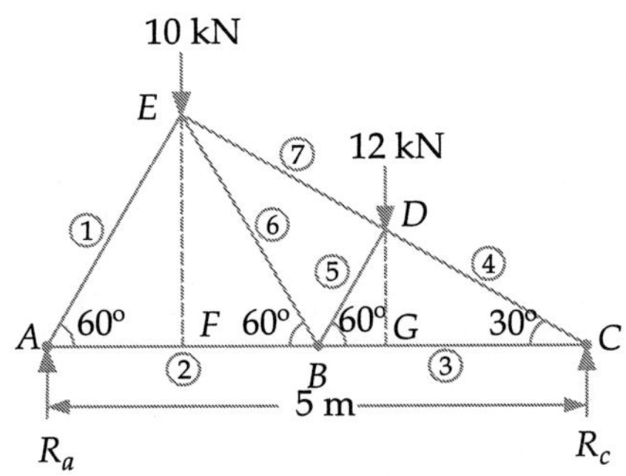

Fig. 5.20

Solution : The reactions at the supports can be determined by considering equilibrium of the entire truss. Since both the external loads are vertical, only the vertical component of the reaction at the hinged end *A* need to be considered.

With reference to free body diagram of the whole truss (Fig. 5.20), the triangle *AEC* is a right angled triangle with angle *AEC* = 90°. Then

$$AE = AC \cos 60° = 5 \times 0.5 = 2.5 \text{ m}$$
$$CE = AC \sin 60° = 5 \times 0.866 = 4.33 \text{ m}$$

Further, from the give geometry, Δ *ABE* is an equilateral triangle and therefore

$$AB = BC = AE = 2.5 \text{ m}$$

Distance of line of action of force 10 kN from joint *A*,

$$AF = AE \cos 60° = 2.5 \times 0.5 = 1.25 \text{ m}$$

Again, the traingle *BDC* is a right angled triangle with angle *BDC* = 90°. Also

$$BC = AC - AB = 5 - 2.5 = 2.5 \text{ m}$$
$$BD = BC \cos 60° = 2.5 \times 0.5 = 1.25 \text{ m}$$

Distance of line of action of force 12 kN from joint *A*,

$$AG = AB + BG = AB + BD \cos 60°$$
$$= 2.5 + 1.25 \times 0.5 = 3.125 \text{ m}$$

Taking moments about end *A*, we get

$$R_C \times 5 = 12 \times 3.125 + 10 \times 1.25 = 50$$

$$\therefore \quad R_C = \frac{50}{5} = 10 \text{ kN} \quad \text{and} \quad R_a = (10 + 12) - 10 = 12 \text{ kN}$$

Joint *A* : Consider the free body diagram of joint A with the direction of forces assumed as shown in Fig. 5.21. Start can be made with this joint because there are only two unknown forces F_1 (in member *AE*) and F_2 (in member *AB*) acting at this joint.

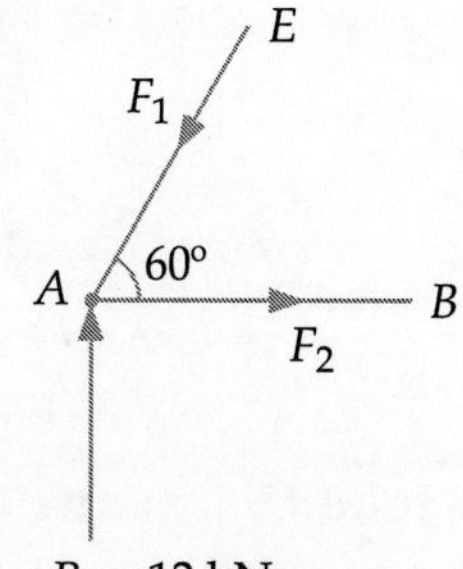

Fig. 5.21

Equations of equilibrium can be written as

$\Sigma F_x = 0 :\quad F_2 - F_1 \cos 60°$

$\Sigma F_y = 0 :\quad R_a - F_1 \sin 60° = 0$

$$\therefore \quad F_1 = \frac{R_a}{\sin 60°} = \frac{12}{0.866}$$

$$= \mathbf{13.85} \text{ kN (compression)}$$

$$F_2 = F_1 \cos 60°$$

$$= 13.85 \times 0.5 = \mathbf{6.92} \text{ kN (tension)}$$

Joint C: Consider the free body diagram of joint C with the direction of forces assumed as shown in Fig. 5.22. Analysis of this joint is also possible because there are only two unknown forces F_3 (in member *BC*) and F_4 (in member *CD*) acting at this joint.

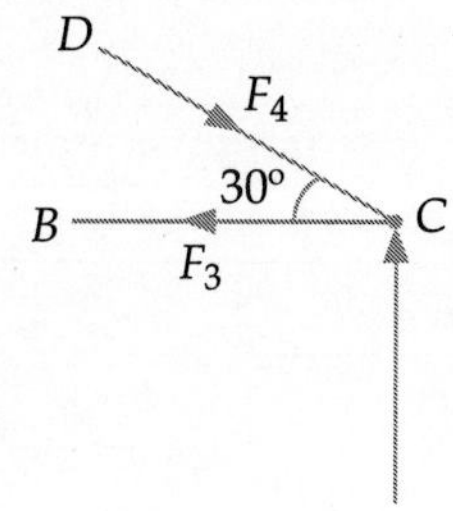

Fig. 5.22

Equations of equilibrium can be written as

$\Sigma F_x = 0 :\quad F_4 \cos 30° - F_3 = 0$

$\quad R_C - F_4 \sin 30° = 0$

$$\therefore \quad F_4 = \frac{R_C}{\sin 30°}$$

$$= \frac{10}{0.5} = \mathbf{20\ kN} \text{ (compression)}$$

and $\quad F_3 = F_4 \cos 30°$

$$= 20 \times 0.866$$

$$= \mathbf{17.32\ kN} \text{ (Tension).}$$

Joint *B*: Consider the free body diagram of joint *B* with the known values of forces F_2 (tensile force in member *AB*) and F_3 (tensile force in member *BC*) inserted.

Equations of equilibrium for this joint are:

$\Sigma F_x = 0 :\quad 17.32 - 6.92 - F_5 \cos 60° - F_6 \cos 60° = 0$

or $\quad (F_5 + F_6) \times 0.5 = 10.40$

or $\quad F_5 + F_6 = 20.80 \quad$ ...(*i*)

$\Sigma F_y = 0 :\quad - F_5 \sin 60° + F_6 \sin 60° = 0$

or $\quad F_5 = F_6 \quad$ ...(*ii*)

From expressions (*i*) and (*ii*)

$$F_5 = F_6 = 10.40 \text{ kN}$$

Force F_5 in member *BD* is 10.40 kN compressive and force F_6 in member *BE* is 10.40 kN tensile.

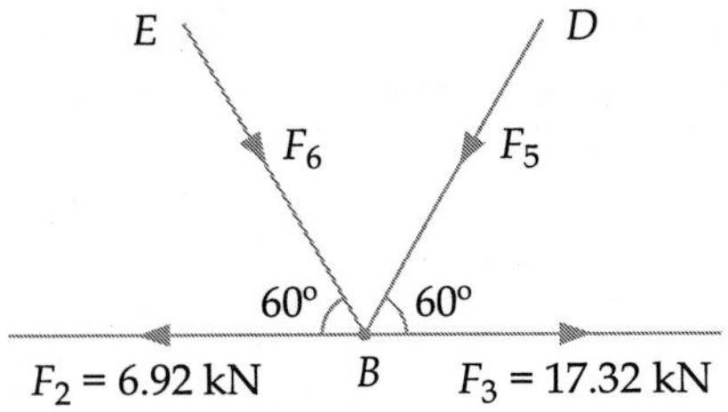

Fig. 5.23

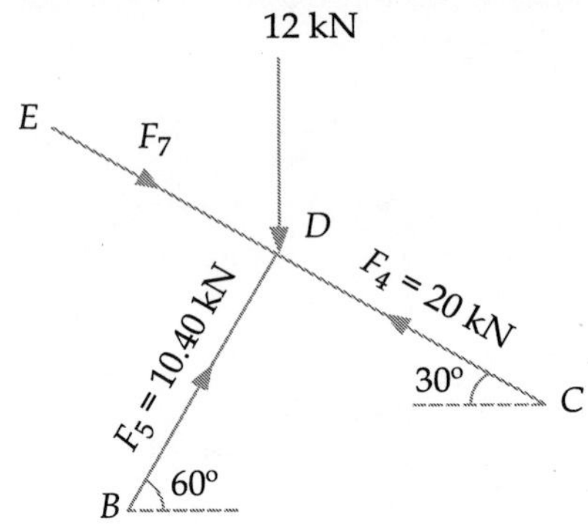

Fig. 5.24

Joint *D*: Consider the free body diagram of joint *D* with known values of forces F_4 (compressive force in member *DC*) and F_5 (compressive force in member *DB*) inserted.

Equations of equilibrium for this joint are:

$$\Sigma F_x = 0; \qquad F_7 \cos 30^\circ + F_5 \cos 60 - F_4 \cos 30$$

$$\text{or} \qquad F_7 \cos 30^\circ = -F_5 \cos 60^\circ + F_4 \cos 30^\circ$$

$$= -10.40 \times 0.5 + 20 \times 0.866 = 12.12$$

$$\therefore \qquad F_7 = \frac{12.12}{0.866} = 13.99 \text{ kN (compressive)}$$

The magnitude and the nature of the forces in all the members of the given truss for the applied system of loading are tabulated below:

Member	AE	AB	BC	CD	BD	BE	DE
Force	F_1 = 13.85 kN	F_2 = 6.92 kN	F_3 = 17.32 kN	F_4 = 20 kN	F_5 = 10.40 kN	F_6 = 10.40 kN	F_7 = 13.99 kN
Nature	C	T	T	C	C	T	C

EXAMPLE 5.6

Determine the reactions and the forces in each member of a truss supporting two loads as shown in Fig. 5.25.

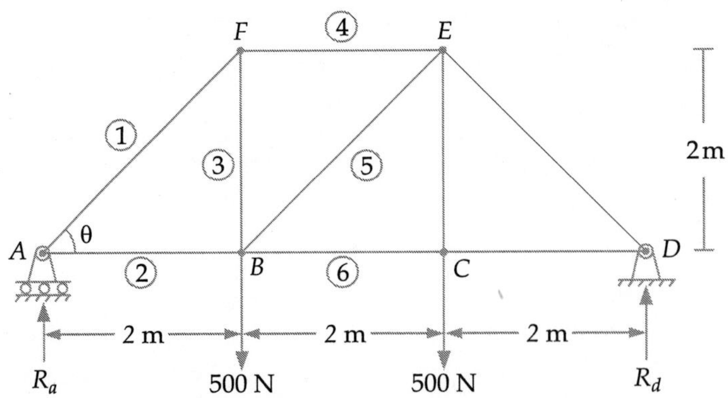

Fig. 5.25

Solution : The reactions at the supports can be determined by considering equilibrium of the entire truss. Since both the external loads are vertical, only the vertical component of reaction at the hinged end D need to be considered.

Taking moments about end A, we get

$$R_d \times 6 = 500 \times 4 + 500 \times 2 = 3000$$

$$\therefore \quad R_d = \frac{3000}{6} = \mathbf{500\ N}; \quad R_a = (500 + 500) - 500 = \mathbf{500\ N}$$

Joint *A*: Consider the free body diagram of joint A with the direction of forces assumed as shown in Fig. 5.26. Start can be made with this joint because there are only two unknown forces F_1 (in member AF) and F_2 (in member AB) acting at this joint.

$$\tan\theta = \frac{BF}{AB} = \frac{2}{2} = 1; \quad \theta = 45^\circ$$

Equations of equilibrium can be written as,

$\Sigma F_x = 0: \quad F_2 - F_1 \cos 45^\circ = 0$

$\Sigma F_y = 0: \quad R_a - F_1 \sin 45^\circ = 0$

$$\therefore \quad F_1 = \frac{R_a}{\sin 45^\circ} = \frac{500}{0.707} = \mathbf{707.21\ N}\ \text{(compression)}$$

$$F_2 = F_1 \cos 45^\circ$$
$$= 707.21 \times 0.707 = \mathbf{500\ N}\ \text{(tension).}$$

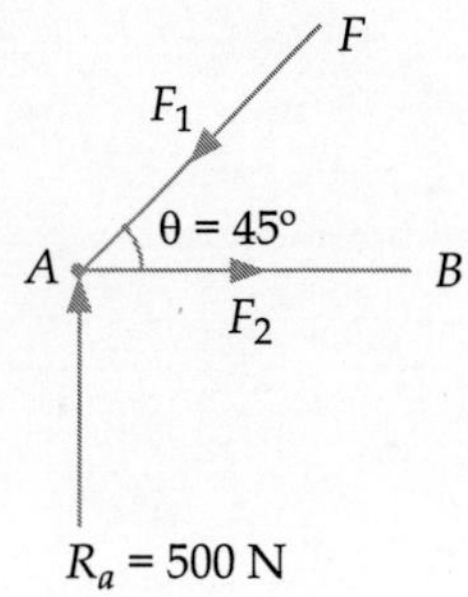

Fig. 5.26

Joint *F*: Equations of equilibrium for the joint F (Fig. 5.27) can be written as

$\Sigma F_x = 0: \quad F_4 + 707.21 \sin 45^\circ = 0$

$\therefore \quad F_4 = -707.21 \times 0.707 = -500$ N

$\Sigma F_y = 0: \quad F_3 + 707.21 \cos 45^\circ = 0$

$\therefore \quad F_3 = -707.21 \times 0.707 = -500$ N

The – ve sign with the magnitude of forces F_3 and F_4 shows that a wrong choice has been made while assuming their directions. Obviously the assumed direction of the forces in members FB and FE and need to be reversed.

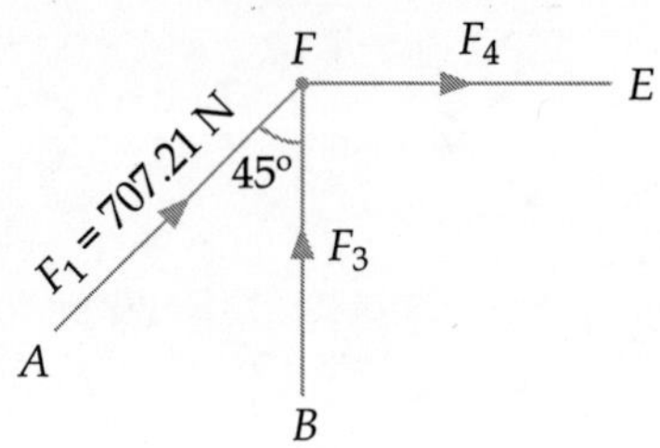

Fig. 5.27

Therefore the member FB is a tension member and the member FE is a compression member.

Joint *B*: Equations of equilibrium for the joint B (Fig. 5.28) can be written as

$\Sigma F_x = 0: \quad F_6 + F_5 \cos 45^\circ - 500 = 0$

$\Sigma F_y = 0: \quad 500 - 500 + F_5 \sin 45^\circ = 0$

$\therefore \quad F_5 = 0$

Putting $F_5 = 0$ in the above identity, we get

$F_6 = 500$ N (Tension)

The zero value for force F_5 signifies that no force is induced in the member BE.

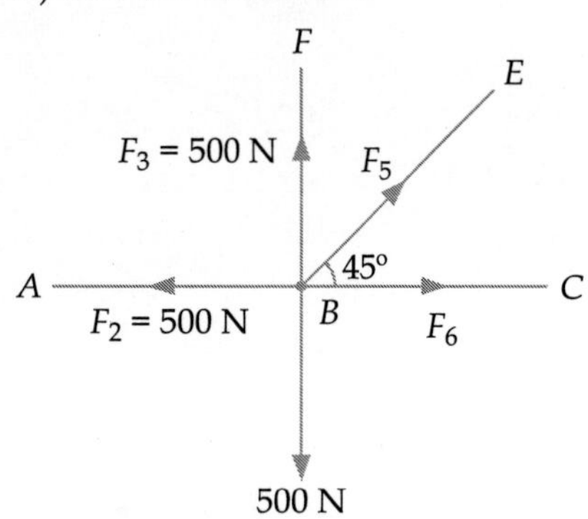

Fig. 5.28

Since the truss is symmetrical, we have

Force in member CE = force in member BF
= 500 N (Tension)

Force in member CD = force in member AB
= 500 N (Tension)

Force in member DE = force in member AF
= 707.21 (Compression)

The forces in all the members of the given truss for the applied system of loading are tabulated below:

Member	AF, DE	AB, CD	BF, CE	FE	BE	BC
Force	707.21 N	500 N	500 N	500 N	0	500 N
Nature	C	T	T	C	–	T

EXAMPLE 5.7

A truss has been loaded and supported as shown in Fig. 5.29. Make calculations for the reactions at the supports and the forces in the members of the truss.

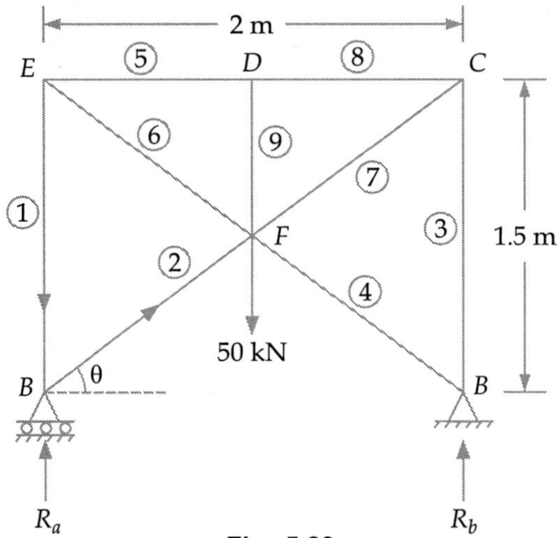

Fig. 5.29

Solution : Both the supports will have only vertical reactions as the external load of 50 kN at the joint F is vertical. Further, due to symmetry

$$R_a = R_b = \frac{50}{2} = 25 \text{ kN}$$

From the geometry of the truss, the inclination of the inclined members with the horizontal is

$$\tan\theta = \frac{0.75}{1} = \frac{3}{4}$$

That gives: $\sin\theta = \dfrac{3}{5}$ and $\cos\theta = \dfrac{4}{5}$

Joint *A*: From the equations of equilibrium,

$\Sigma F_x = 0 :\quad F_2 \cos\theta = 0\,;$

$\therefore\quad F_2 = 0$ (no force)

$\Sigma F_y = 0 :\quad R_a - F_1 = 0\,;$

$\therefore\quad F_1 = R_a =$ **25 kN** (compressive)

Joint *B*: From the equations of equilibrium,

$\Sigma F_x = 0 :\quad F_4 \cos\theta = 0\,;$

$\therefore\quad F_4 = 0$ (no force)

$\Sigma F_y = 0 :\quad R_b - F_3 = 0\,;$

$\therefore\quad F_3 = R_b =$ **25 kN** (compressive)

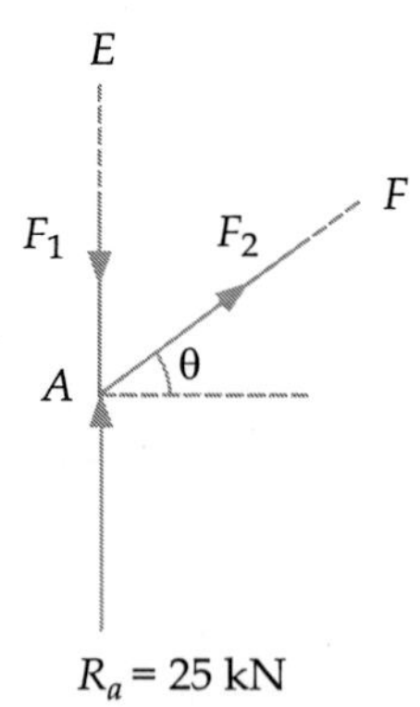

Fig. 5.30

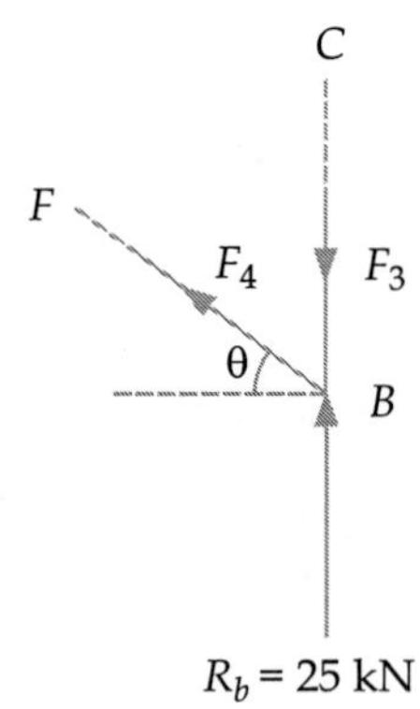

Fig. 5.31

Joint C: From the equations of equilibrium,

$$\Sigma F_x = 0: \quad F_6 \cos\theta - F_5 = 0$$

$$\Sigma F_y = 0: \quad F_1 - F_6 \sin\theta = 0$$

$$\therefore \quad F_6 = \frac{F_1}{\sin\theta} = \frac{25}{3/5} = \mathbf{41.67\ kN}\ \text{(tensile)}$$

and $F_5 = F_6 \cos\theta = 41.67 \times \frac{4}{5} = 33.34$ kN (compression)

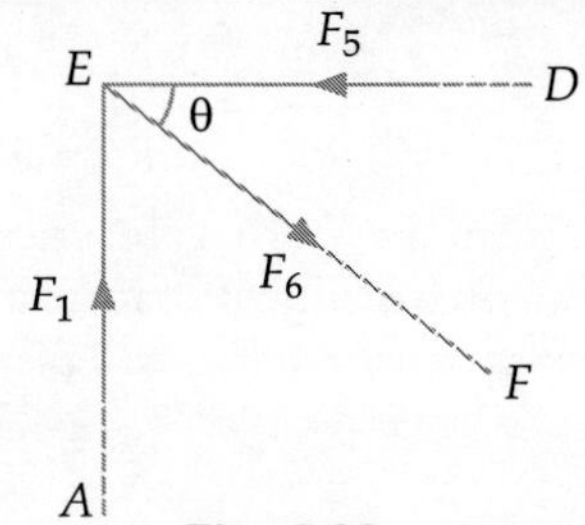

Fig. 5.32

Likewise considering equilibrium of **joint** C,

$$F_7 = 41.67 \text{ kN (tensile)}$$

and $\quad F_8 = \mathbf{33.34\ kN}$ (compression)

Joint D: Considering equilibrium of forces in vertical direction,

$$F_9 = 0 \text{ (no force)}$$

The forces in various members of the truss are tabulated below:

Members	AE, BC	DE, DC	FE, FC	FD, FA, FB
Force	25 kN	33.34 kN	41.67 kN	0 (no force)
Nature	compressive	compressive	tensile	---

EXAMPLE 5.8

Determine the forces in all the members of the truss loaded and supported as shown in Fig. 5.33.

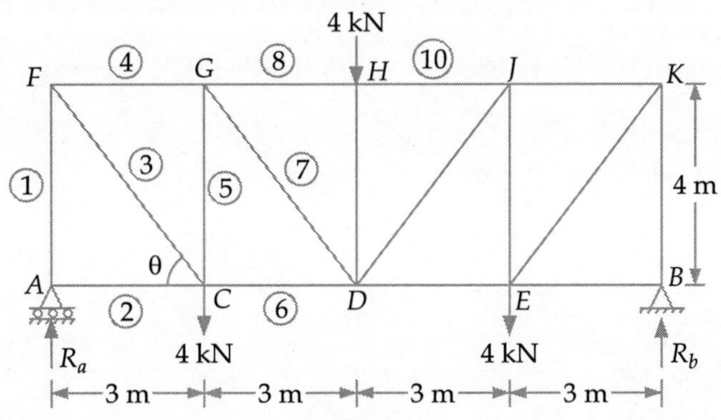

Fig. 5.33

Solution : The reactions at the supports can be determined by considering equilibrium of the entire truss. Since all the three external loads are vertical, only the vertical component of reaction at the hinged end B need to be considered. Further, due to symmetry of truss and loading

$$R_a = R_b = \frac{4+4+4}{2} = 6 \text{ kN}$$

In right angled triangle ACF, $CF = \sqrt{3^2 + 4^2} = 5$ m

$$\therefore \quad \sin\theta = \frac{4}{5} = 0.8 \text{ and } \cos\theta = \frac{3}{5} = 0.6$$

Joint A: With reference to free body diagram of joint A (Fig. 5.34) the equations of equilibrium are:

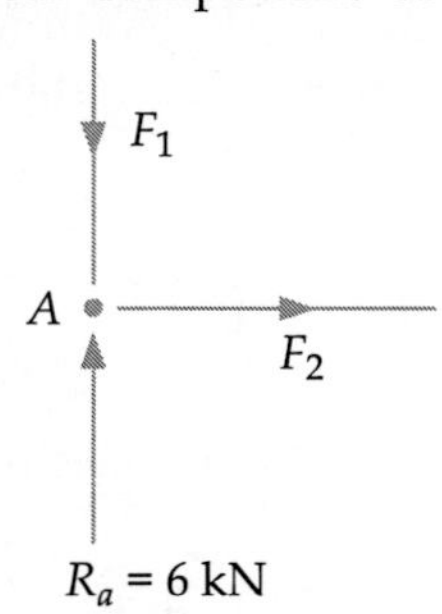

Fig. 5.34

$\Sigma F_x = 0:$ $\quad F_2 = 0$

$\Sigma F_y = 0:$ $\quad F_1 =$ **6 kN** (compression)

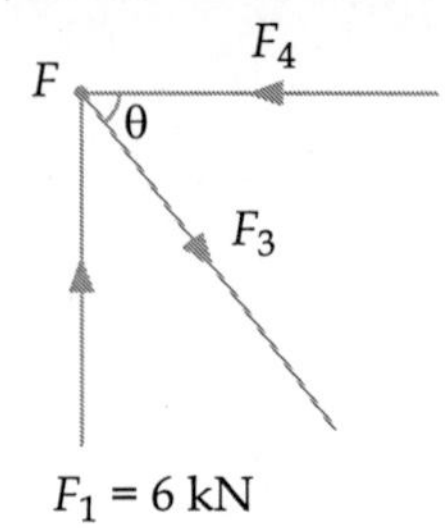

Fig. 5.35

Joint *F*: With reference to free body diagram of joint *F* (Fig. 5.35), the equations of equilibrium are :

$\Sigma F_x = 0:$ $\quad -F_4 + F_3 \cos\theta = 0$

$\Sigma F_y = 0:$ $\quad 6 - F_3 \sin\theta = 0$

$\therefore \quad F_3 = \dfrac{6}{\sin\theta} = \dfrac{6}{0.8} =$ **7.5 kN** (tension)

and $\quad F_4 = F_3 \cos\theta = 7.5 \times 0.6$

$= $ **4.5 kN** (compression)

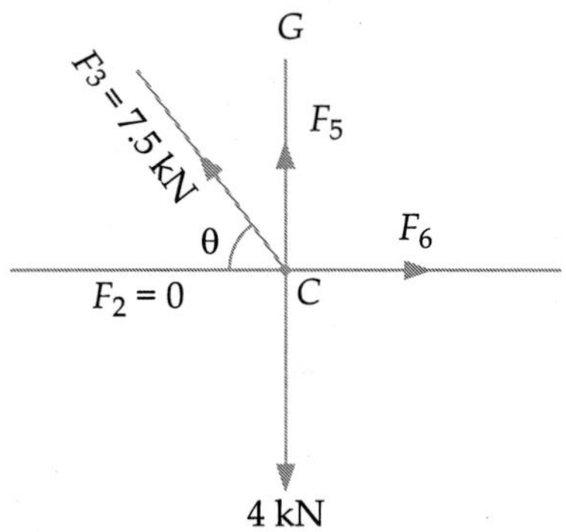

Fig. 5.36

Joint *C*: With reference to free body diagram of joint C (Fig. 5.36), the equations of equilibrium are:

$F_x = 0:$ $\quad F_6 - F_3 \cos\theta = 0$

$\therefore \quad F_6 = F_3 \cos\theta = 7.5 \times 0.6$

$=$ **4.5 kN** (tension)

$F_y = 0:$ $\quad F_5 + F_3 \sin\theta - 4 = 0$

$\therefore \quad F_5 = 4 - F_3 \sin\theta$

$= 4 - 7.5 \times 0.8 = -$ **2 kN**

The negative sign with the magnitude of force F_5 shows that a wrong choice has been made while assuming its direction. Obviously the assumed direction of force in member *CG* need to be reversed.

Therefore, F_5 = 2 kN (compression) and *CG* is a compression member.

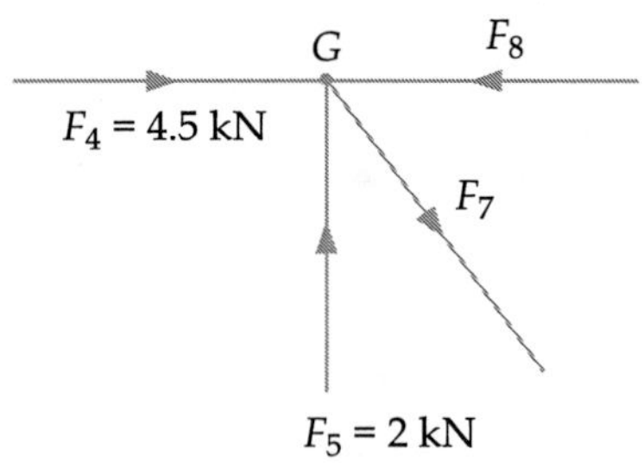

Fig. 5.37

Joint *G* : With reference to free body diagram of joint *G* (Fig. 5.37) the equations of equilibrium are:

$\Sigma F_x = 0:$ $\quad 4.5 - F_8 + F_7 \cos\theta = 0$

$\Sigma F_y = 0:$ $\quad 2 - F_7 \sin\theta = 0$

$\therefore \quad F_7 = \dfrac{2}{\sin\theta} = \dfrac{2}{0.8}$

$=$ **2.50 kN** (tension)

and $\quad F_8 = 4.5 + F_7 \cos\theta$

$= 4.5 + 2.50 \times 0.6$

$=$ **6 kN** (compression)

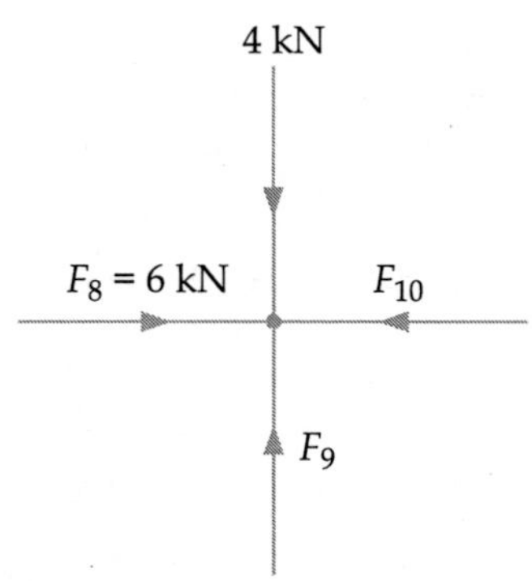

Fig. 5.38

Joint *H* : With reference to free body diagram of joint *H* (Fig. 5.38), the equations of equilibrium are:

$\Sigma F_x = 0:$ $\quad F_{10} - 6 = 0$

$\therefore \quad F_{10} =$ **6 kN** (compression)

$\Sigma F_y = 0:$ $\quad F_9 - 4 = 0$

$\therefore \quad F_9 =$ **4 kN** (compression)

The truss is symmetrical and as such, there is no need to consider equilibrium of the joints left. Forces in all the members of the given truss for the applied system of loading are tabulated below:

Member	Force	Nature
AF, BK	6 kN	compressive
AC, BE	0	–
CF, EK	7.5 kN	tensile
FG, JK	4.5 kN	compressive
GC, JE	2 kN	compressive
CD, DE	4.5 kN	tensile
DG, DJ	2.5 kN	tensile
GH, HJ	6 kN	compressive
HD	4 kN	compressive

EXAMPLE 5.9

Determine the forces in each member of the cantilever truss loaded as shown in Fig. 5.39.

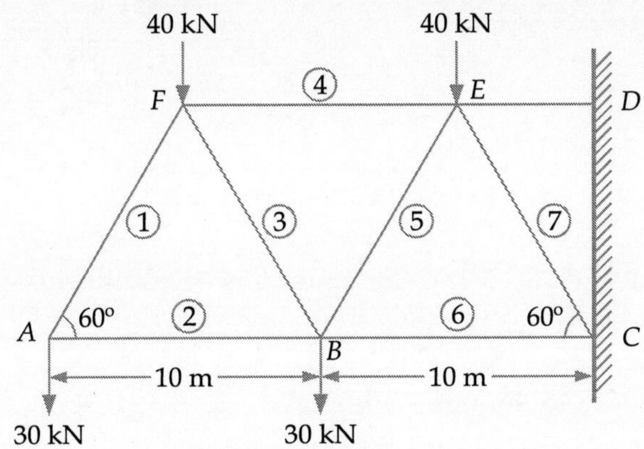

Fig. 5.39

Solution : Consider the free body diagram of **Joint A** with the direction of forces assumed as shown in Fig. 5.40.

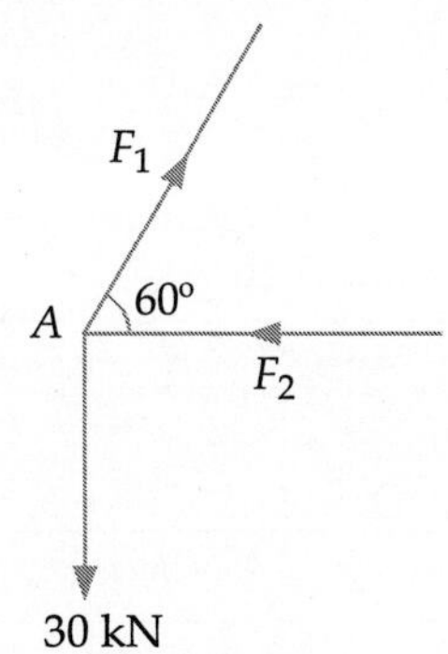

Fig. 5.40

Equations of equilibrium are:

$\Sigma F_x = 0: \quad F_1 \cos 60° - F_2 = 0$

$\Sigma F_y = 0: \quad F_1 \sin 60° - 30 = 0$

$$\therefore \quad F_1 = \frac{30}{\sin 60°} = \frac{30}{0.866} = \mathbf{34.64\ kN}\ \text{(tension)}$$

and $\quad F_2 = F_1 \cos 60° = 34.64 \times 0.5$

$= \mathbf{17.32\ kN}$ (compression)

Joint *F*: With reference to free body diagram of joint F (Fig. 5.41), the equations of equilibrium are:

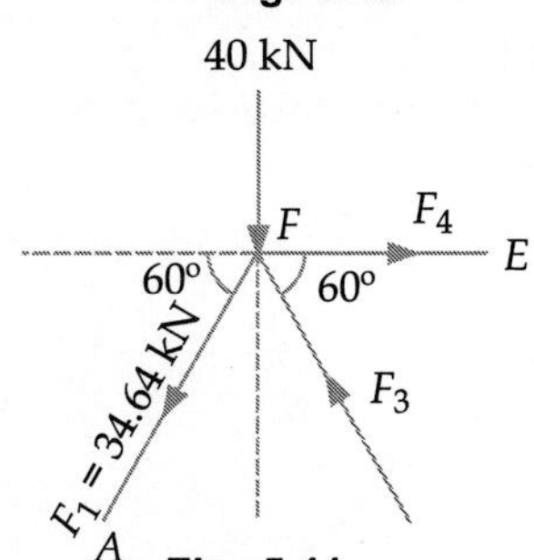

Fig. 5.41

$\Sigma F_x = 0: \quad F_4 - F_3 \cos 60° - 34.64 \cos 60°$

or $\quad F_4 = 0.5\ F_3 + 17.32$

$\Sigma F_y = 0: \quad F_3 \sin 60 - 34.64 \sin 60° - 40 = 0$

$$\therefore \quad F_3 = \frac{(34.64 \times 0.866) + 40}{0.866} = \mathbf{80.81\ kN}\ \text{(compression)}$$

and $\quad F_4 = 0.5 \times 80.81 + 17.32 = \mathbf{57.72\ kN}$ (tension)

Joint *B*: Considering joint *B* (Fig. 5.42), the equations of equilibrium are:

$\Sigma F_x = 0: \; F_5 \cos 60° + 80.81 \cos 60° + 17.32 - F_6 = 0$

$F_6 = 0.5\, F_5 + 80.81 \times 0.5 + 17.32 = 0.5\, F_5 + 57.72$

$\Sigma F_y = 0: \; F_5 \sin 60° - 80.81 \sin 60° - 30 = 0$

$$\therefore \quad F_5 = \frac{80.81 \times 0.866 + 30}{0.866} = \mathbf{115.45\ kN}\ \text{(tension)}$$

and $\quad F_6 = 0.5 \times 115.45 + 57.72 =$ **115.44** (compression)

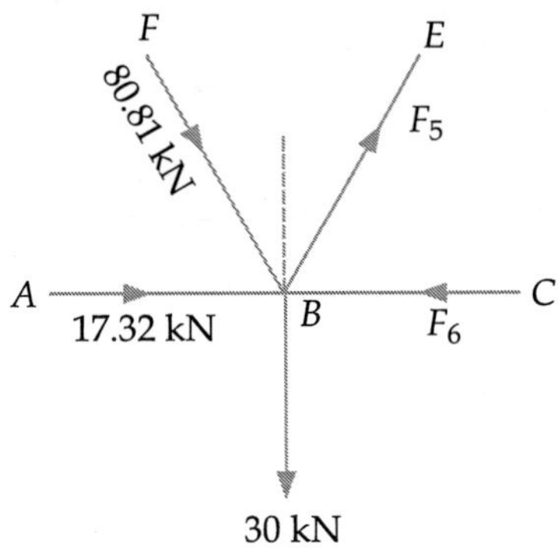

Fig. 5.42

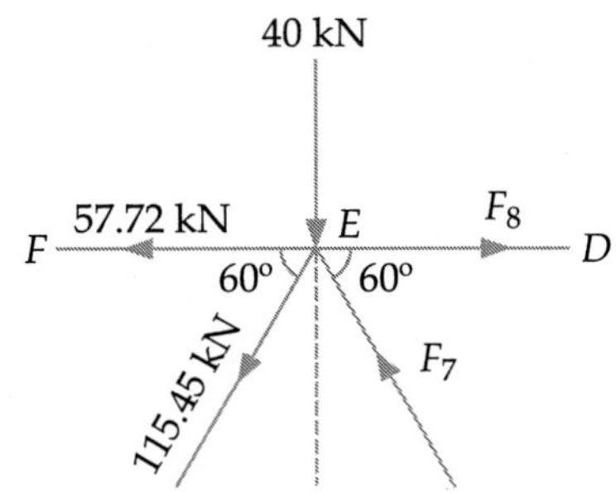

Fig. 5.43

Joint *E*: Considering joint *E* (Fig. 5.43), the equations of equilibrium are:

$\Sigma F_x = 0: \; F_8 - 57.72 - F_7 \cos 60° - 115.45 \cos 60°$

or $\quad F_8 = 57.72 + 0.5\, F_7 + 115.45 \times 0.5 = 0.5\, F_7 + 115.44$

$\Sigma F_y = 0: \quad F_7 \sin 60° - 115.45 \sin 60° - 40 = 0$

$$\therefore \quad F_7 = \frac{115.45 \times 0.866 + 40}{0.866} = \mathbf{161.62\ kN}\ \text{(compression)}$$

and $\quad F_8 = 0.5 \times 161.62 + 115.44 =$ **196.25 kN** (tension)

The magnitude and the nature of forces in the various members of the truss have been tabulated below:

Member	AF	AB	BF	FE	BE	BC	CE	ED
Force kN	34.64	17.32	80.81	57.72	115.45	115.44	161.62	196.25
Nature	tensile	compressive	compressive	tensile	tensile	compressive	compressive	tensile

EXAMPLE 5.10

Determine the forces in the various members of the cantilever truss loaded and supported as shown in Fig. 5.44.

Solution : From the geometrical configuration of the truss,

$$BC = \sqrt{2^2 + 1^2} = 2.236 \text{ m}$$

$$\sin\theta = \frac{1}{2.236} = 0.447$$

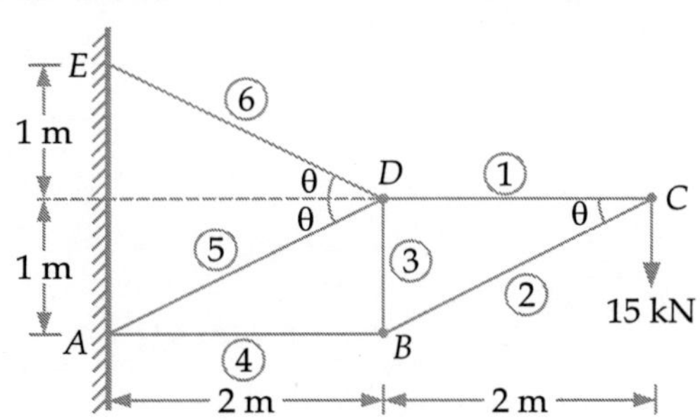

Fig. 5.44

$$\cos\theta = \frac{2}{2.236} = 0.894$$

Joint C: Consider the free body diagram of joint *C* with the direction of forces assumed as shown in Fig. 5.45.

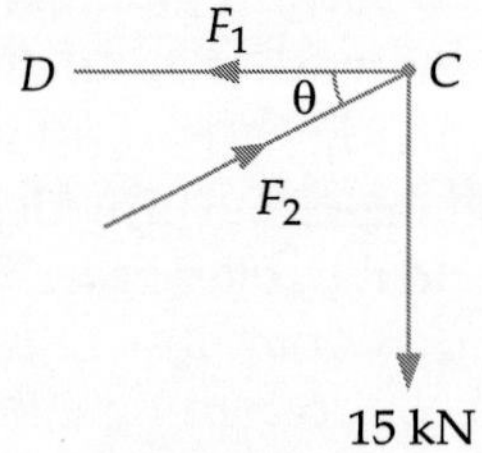

Fig. 5.45

Equation of equilibrium are:

$\Sigma F_x = 0:$ $\quad F_2 \cos\theta - F_1 = 0$

$\Sigma F_y = 0:$ $\quad F_2 \sin\theta - 15 = 0$

$$\therefore \quad F_2 = \frac{15}{\sin\theta} = \frac{15}{0.447}$$

$$= \mathbf{33.56\ kN}\ \text{(compression)}$$

and $\quad F_1 = F_2 \cos\theta = 33.56 \times 0.894 = \mathbf{30.00\ kN}$ (tension)

Joint *B*: With reference to free body diagram of joint *B* (Fig. 5.46), the equations of equilibrium are:

$\Sigma F_x = 0:$ $\quad F_4 - 33.56 \cos\theta$

$\therefore \quad F_4 = 33.56 \times 0.894 = \mathbf{30.00\ kN}$ (compression)

$\Sigma F_y = 0:$ $\quad F_3 - 33.56 \sin\theta = 0$

$\therefore \quad F_3 = 33.56 \times 0.447 = \mathbf{15.00\ kN}$ (tension)

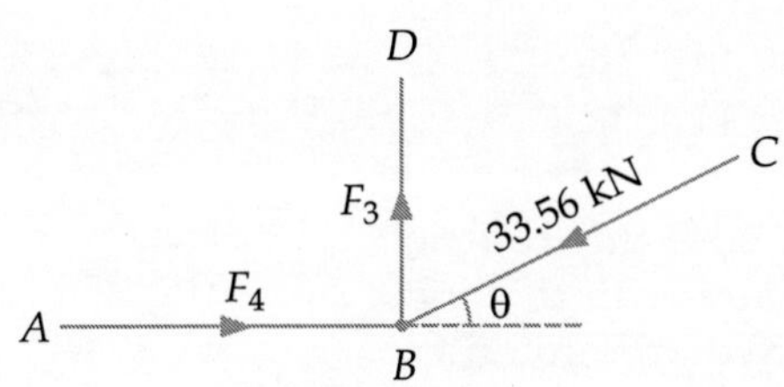

Fig. 5.46

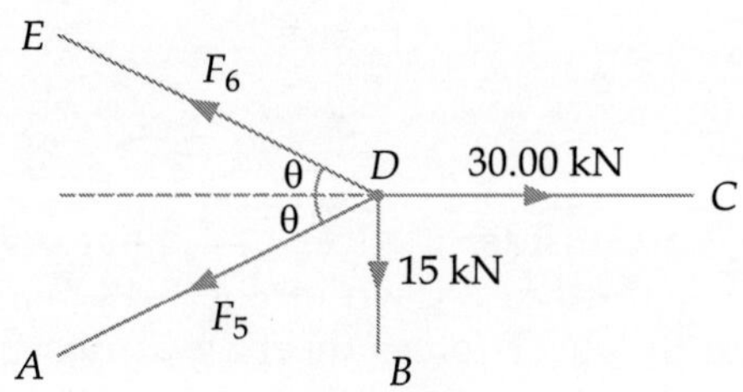

Fig. 5.47

Joint *D*: Considering joint *D* (Fig. 5.47), the equations of equilibrium are:

$\Sigma F_x = 0:$ $\quad 30 - F_5 \cos\theta - F_6 \cos\theta = 0$

or $$F_5 + F_6 = \frac{30}{0.894} = 33.56 \qquad ...(i)$$

$\Sigma F_y = 0:$ $\quad F_6 \sin\theta - F_5 \sin\theta - 15 = 0$

or $$F_6 - F_5 = \frac{15}{0.447} = 33.56 \qquad ...(ii)$$

From identities (*i*) and (*ii*),

$F_6 = \mathbf{33.56\ kN}$ (tension)

and $\quad F_5 = 0$ (no force)

The magnitude and the nature of forces in the various members of the truss have been tabulated below:

Member	CD	CB	BD	BA	AD	DE
Force	30 kN	33.56 kN	15 kN	30 kN	0	33.56 kN
Nature	tensile	compressive	tensile	compressive	–	tensile

EXAMPLE 5.11

Determine the magnitude and nature of forces in the various members of the triangular truss loaded and supported as shown in Fig. 5.48.

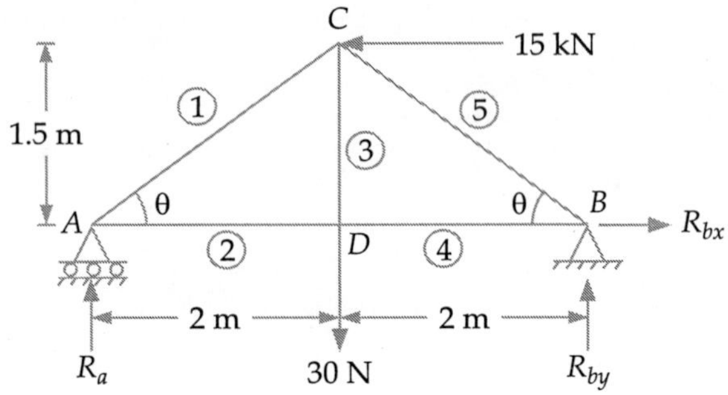

Fig. 5.48

Solution : The roller support (joint A) is frictionless and provides a reaction R_a at right angles to the roller base. The reaction at the hinged support (joint *B*) will have two components R_{bx} and R_{by} acting in the horizontal and vertical direction respectively.

Consider free body diagram of the entire truss and take moments about joint B;

$$R_a \times 4 = 30 \times 2 + 15 \times 1.5 = 82.50 \,; \quad R_a = 20.62 \text{ kN}$$

From $\Sigma F_y = 0$; $R_{by} = 130 - 20.62 = 9.38$ kN

From $\Sigma F_x = 0$; R_{bx} 15 kN

Further, from the geometrical configuration of the truss

$$AC = \sqrt{(AD)^2 + (CD)^2} = \sqrt{2^2 + 1.5^2} = 2.5 \text{ m}$$

$$\sin\theta = \frac{CD}{AC} = \frac{1.5}{2.5} = 0.6$$

$$\cos\theta = \frac{AD}{AC} = \frac{2}{2.5} = 0.8$$

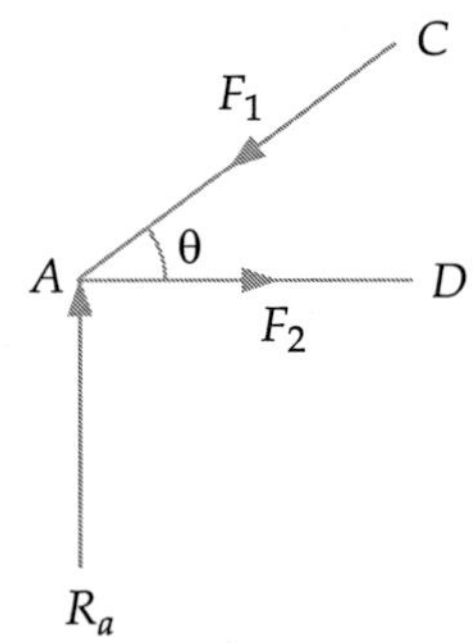

Fig. 5.49

Joint A : Consider the free body diagram of joint A with the direction of forces assumed as shown in Fig. 5.49. Start can be made with this joint because of the three forces meeting at this point, only two are unknown.

Equations of equilibrium can be written as

$\Sigma F_x = 0$: $F_2 - F_1 \cos\theta = 0$

$\Sigma F_y = 0$: $R_a - F_1 \sin\theta = 0$

$$\therefore \quad F_1 = \frac{R_a}{\sin\theta} = \frac{20.62}{0.6} = \mathbf{34.36\ kN} \text{ (compression)}$$

and $F_2 = F_1 \cos\theta = 34.36 \times 0.8 =$ **27.48 kN** (tensile)

Joint *D*: Considering joint *D* (Fig. 5.50), the equations of equilibrium would give

$F_4 - 27.48 = 0$; $F_4 =$ **27.48 kN** (tensile)

$F_3 - 30 = 0$; $F_3 =$ **30 kN** (tensile)

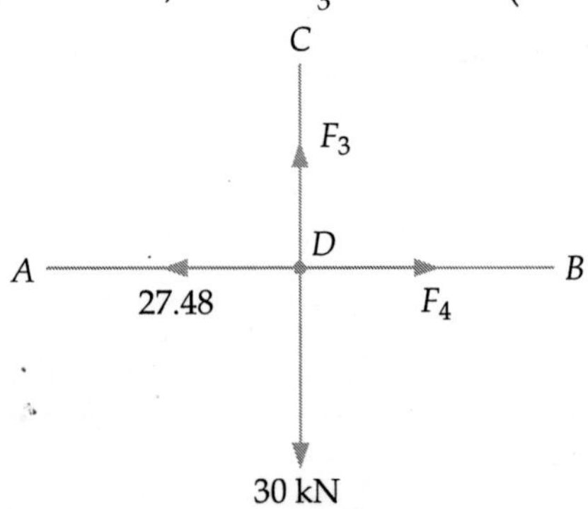

Fig. 5.50

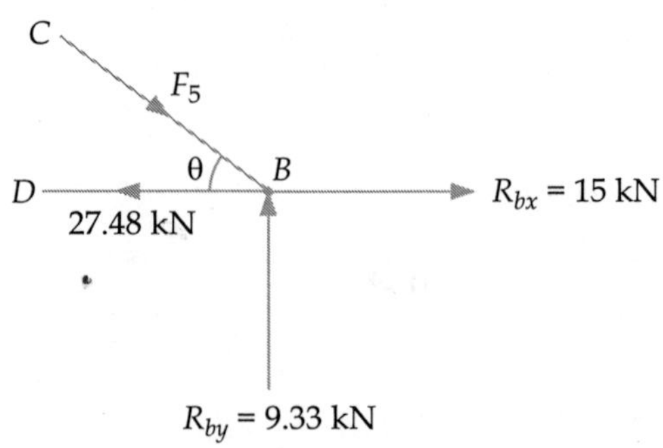

Fig. 5.51

Joint *B*: Considering joint *B* (Fig. 5.51), the equations of equilibrium are :

$\Sigma F_x = 0$: $\quad 15 - 27.48 + F_5 \cos\theta = 0$

$$\therefore \quad F_5 = \frac{27.48 - 15}{\cos\theta} = \frac{12.48}{0.8} = \mathbf{15.60\ kN} \text{ (compression)}$$

Check:

$\Sigma F_y = 0$: $\quad F_5 \sin\theta - R_{by} = 0$

$$F_5 = \frac{R_{by}}{\sin\theta} = \frac{9.38}{0.6} = \mathbf{15.63\ kN} \text{ (compression)}$$

The magnitude and the nature of forces in the various members of the truss have been tabulated below:

Member	AD	AC	DC	DB	BC
Force	27.48 kN	34.36 kN	30 kN	27.48 kN	15.60 kN
Nature	T	C	T	T	C

EXAMPLE 5.12

A truss has been loaded and supported as shown in Fig. 5.52. Make calculations for the reactions at the supports and the forces in the members of the truss.

Solution : The reaction at the hinged support (joint *A*) will hve two R_{ax} and R_{ay} components acting in the horizontal and vertical directions respectively. The roller support (joint *B*) is frictionless and provides a reaction R_b at right angles to the roller base.

Height of the truss = 2 tan 60° = 3.464 m

Consider free body diagram of the entire truss and take moments about joint *A*;

$$R_b \times 4 = (10 \times 2) + (15 \times 3.464)$$
$$= 20 + 51.96 = 71.96$$

$$\therefore \quad R_b = \frac{71.96}{4} = 17.99 \text{ kN}$$

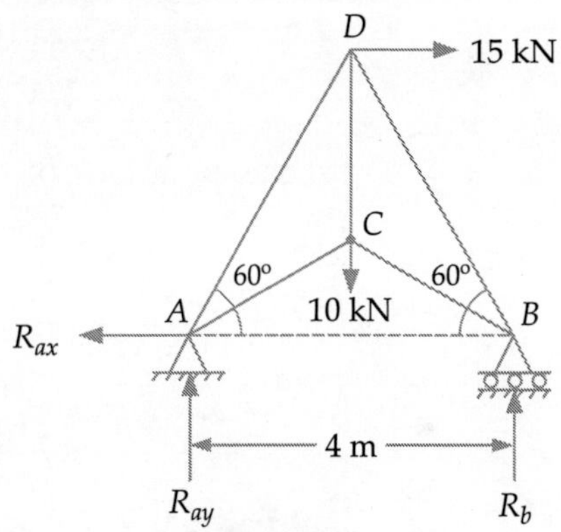

Fig. 5.52

From $\Sigma F_y = 0$;

$$R_{ay} = 10 - 17.99 = -7.99 \text{ kN}$$

The negative sign indicates that the reaction R_{ay} is acting downwards.

From $\Sigma F_x = 0$; $R_{ax} = 15$ kN

Joint *B* : Consider free body diagram of joint *B* with the direction of forces assumed as shown in Fig. 5.53. Equations of equilibrium can be written as :

$\Sigma F_x = 0$: $\quad F_2 \cos 60° - F_1 \cos 30° = 0$

or $\quad 0.5\,F_2 - 0.866\,F_1$; $F_2 = 1.732\,F_1$ ...(*i*)

$\Sigma F_y = 0$: $\quad -F_2 \sin 60° + F_1 \sin 30° - R_b = 0$

or $\quad -0.866\,F_2 + 0.5\,F_1 + 17.99 = 0$

or $\quad F_1 = 1.732\,F_2 - 35.98$ ...(*ii*)

From expressions (*i*) and (*ii*)

$$F_2 = 1.732\,(1.732\,F_2 - 35.98) = 3F_2 - 62.32$$

$$\therefore \quad F_2 = \frac{62.32}{2} = \mathbf{31.16\ kN}\ \text{(compressive)}$$

$$\text{and} \quad F_1 = \frac{F_2}{1.732} = \frac{31.16}{1.732} = \mathbf{17.99\ kN}\ \text{(tensile)}$$

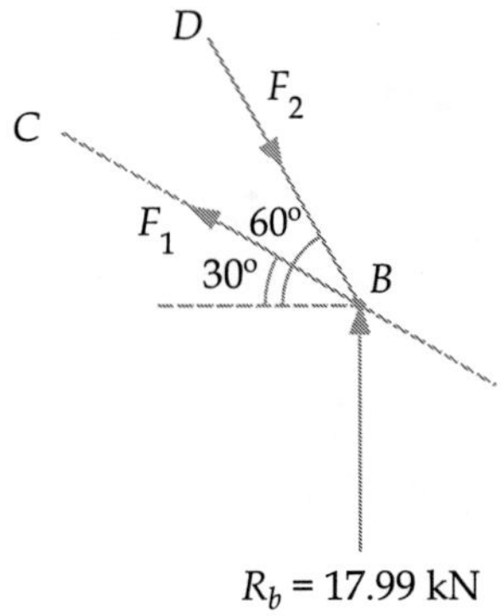

Fig. 5.53

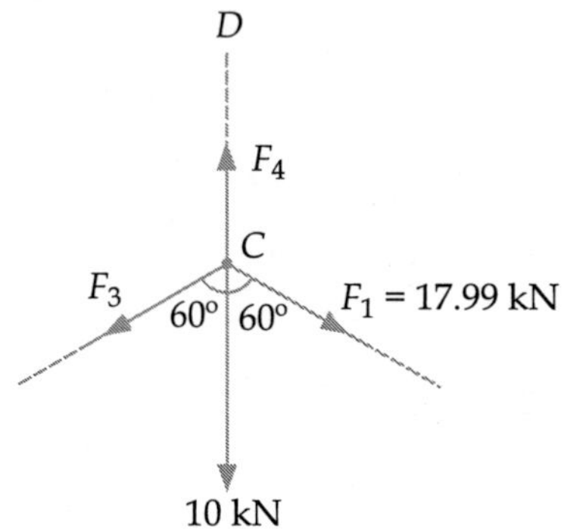

Fig. 5.54

Joint *C*: From the equations of equilibrium,

$\Sigma F_x = 0: \quad F_1 \sin 60° - F_3 \sin 60° = 0$

$\therefore \quad F_3 = F_1 = \mathbf{17.99\ kN}$ (tensile)

$\Sigma F_y = 0: \quad F_4 - 10 - F_3 \cos 60° - F_1 \cos 60° - 0$

$\therefore \quad F_4 = 10 + F_3 \cos 60° + F_1 \cos 60°$

$= 10 + 17.99 \times 0.5 + 17.99 \times 0.5$

$= \mathbf{27.99\ kN}$ (tensile)

Joint *A*: From the equations of equilibrium

$\Sigma F_x = 0: \quad F_3 \cos 30° - F_5 \cos 60° - 15 = 0$

$$\therefore \quad F_5 = \frac{F_3 \cos 30° - 15}{\cos 60°}$$

$$= \frac{17.99 \times 0.866 - 15}{0.5}$$

$= \mathbf{1.16\ kN}$ (compressive)

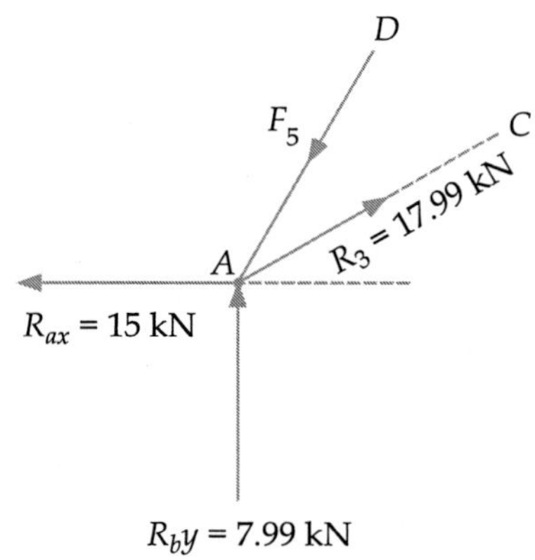

Fig. 5.55

The forces in the various members of the truss are tabulated below :

Member	AC	AD	CD	BC	BD
Force	17.99 kN	1.16 kN	27.99 kN	17.99 kN	31.16 kN
Nature	tensile	compressive	tensile	tensile	compressive

EXAMPLE 5.13

Make calculations for the forces in each member of the truss loaded and supported as shown in Fig. 5.56. All members of the truss are 2 m in length.

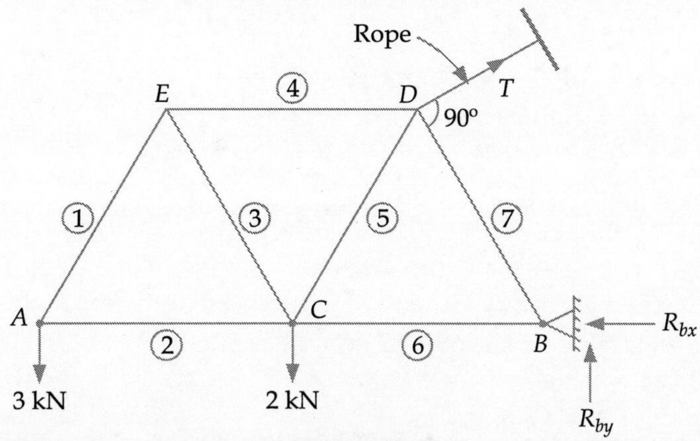

Fig. 5.56

Solution : Let T be the tension in rope ; R_{bx} and R_{by} be the reactions at the hinged support as shown in the figure.

Consider equilibrium of the entire truss

$M_b = 0$: $\quad T \times DB - 2 \times CB - 3 \times AB = 0$

$\therefore \quad T = \dfrac{2 \times CB + 3 \times AB}{DB} = \dfrac{2 \times 2 + 3 \times 4}{2} = 8$ kN

$\Sigma F_x = 0$: $\quad T \cos 30 - R_{bx} = 0$

$\therefore \quad R_{bx} = T \cos 30 = 8 \times 0.866 = 6.93$ kN

$\Sigma F_y = 0$: $\quad R_{by} + T \sin 30° - 2 - 3 = 0$

$\therefore \quad R_{by} = 2 + 3 - 8 \sin 30° = 2 + 3 - 8 \times 0.5 = 1$ kN

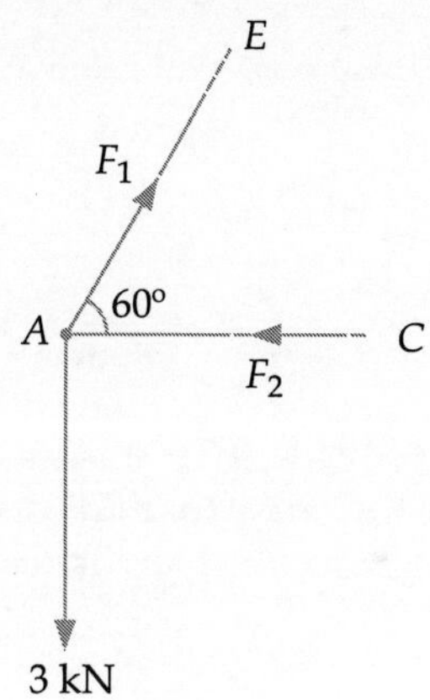

Fig. 5.57

Joint *A*: From the equations of equilibrium

$\Sigma F_x = 0$: $\quad F_1 \cos 60° - F_2 = 0$

$\Sigma F_y = 0$: $\quad F_1 \sin 60° - 3 = 0$

$\therefore \quad F_1 = \dfrac{3}{\sin 60°} =$ **3.464 kN** (tensile)

and $\quad F_2 = F_1 \cos 60° = 3.464 \times 0.5$

$=$ **1.732 kN** (compression)

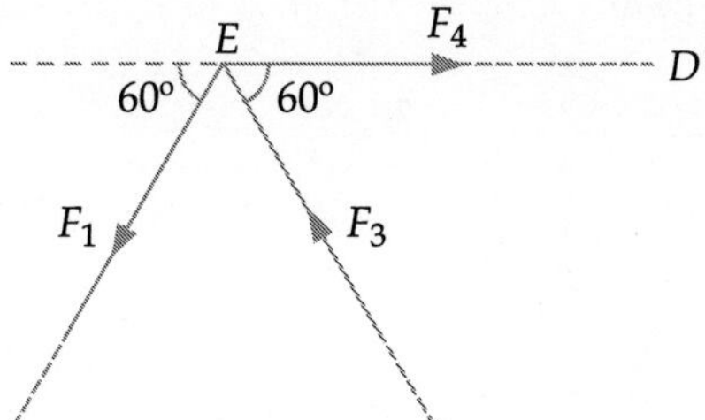

Fig. 5.58

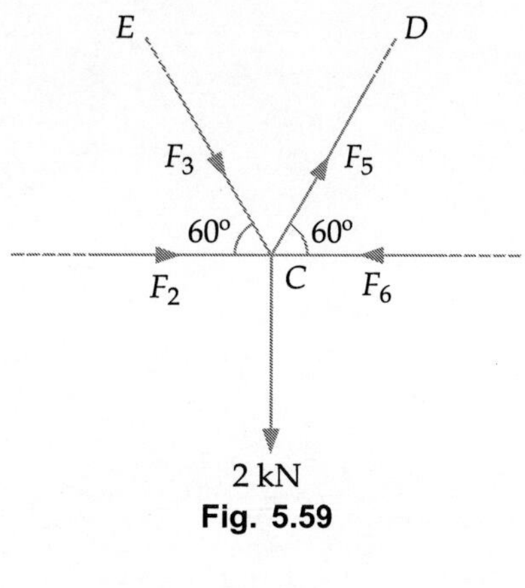

Fig. 5.59

Joint *E*: From the equations of equilibrium

$\Sigma F_x = 0$: $\quad F_4 - F_1 \cos 60° - F_3 \cos 60° = 0$

$\Sigma F_y = 0:$ $\quad F_1 \sin 60° + F_3 \sin 60° = 0$

$\therefore \quad F_3 = F_1 = \mathbf{3.464\ kN}$ (compression)

and $\quad F_4 = (F_1 + F_3) \cos 60°$

$= (3.464 + 3.464) \times 0.5 = \mathbf{3.464\ kN}$ (tensile)

Joint *C*: Writing the equations of equilibrium,

$\Sigma F_x = 0:$ $\quad F_2 - F_6 + F_3 \cos 60° + F_5 \cos 60° = 0$

or $\quad 1.732 - F_6 + 3.464 \times 0.5 + 0.5\ F_5 = 0$

$F_6 - 0.5\ F_5 = 3.464$ ...(*i*)

$\Sigma F_y = 0:$ $\quad - F_3 \sin 60 + F_5 \sin 60° - 2 = 0$

or $\quad - 3.464 \times 0.866 + F_5 \times 0.866 = 2$

$\therefore \quad F_5 = \dfrac{2 + 3.464 \times 0.866}{0.866} = \mathbf{5.77\ kN}$ (tensile)

Then from expression (*i*)

$F_6 = 3.464 + 0.5 \times 5.77$

$= \mathbf{6.35\ kN}$ (compression)

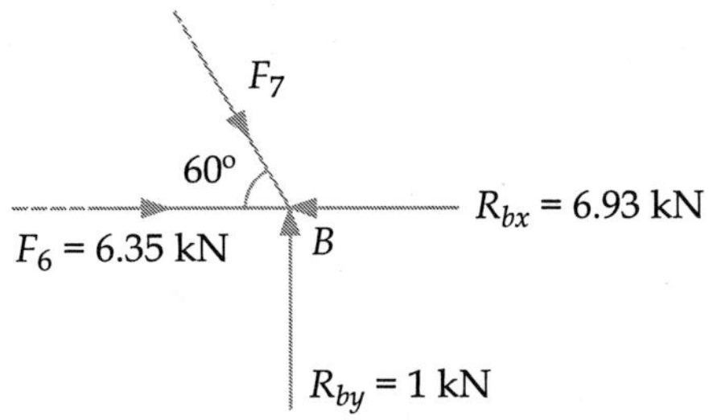

Fig. 5.60

Joint *B*: Writing the equilibrium equation for the vertical direction:

$\Sigma F_y = 0:$ $\quad F_7 \sin 60° - 1 = 0$

$\therefore \quad F_7 = \dfrac{1}{0.866} = \mathbf{1.155\ kN}$ (compression)

EXAMPLE 5.14

For the truss loaded as shown in Fig. 5.61, make calculations for the reactions at the supports and forces set up in each member of the truss.

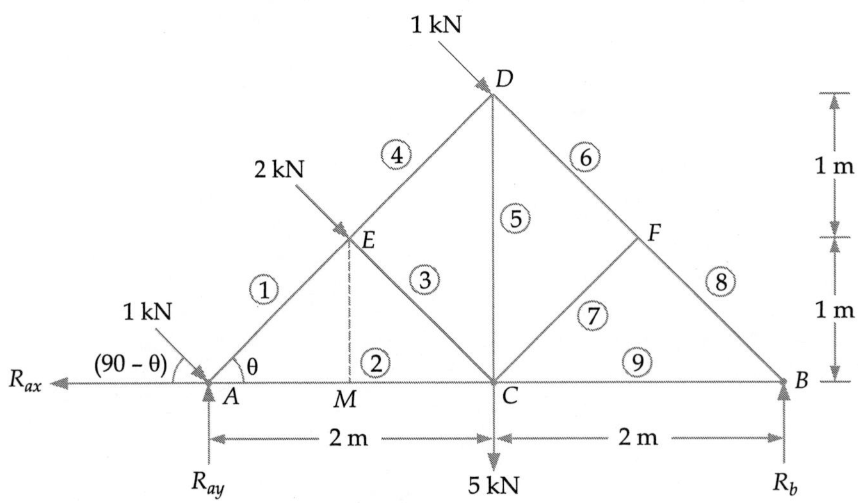

Fig. 5.61

Solution : $\quad AD = \sqrt{(AC)^2 + (CD)^2} = \sqrt{2^2 + 2^2} = \sqrt{8} = 2.828$ m

$AE = \dfrac{AD}{2} = 1.414$ m

$$\sin\theta = \frac{EM}{AE} = \frac{1}{1.414} = 0.707$$

$$\cos\theta = \frac{AM}{AE} = \frac{1}{1.414} = 0.707$$

Applying equilibrium conditions for the entire truss,

$\Sigma M_a = 0:$ $\quad -2 \times AE - 1 \times AD + R_b \times AB - 5 \times AC = 0$

or $\quad -2 \times 1.414 - 1 \times 2.828 + R_b \times 4 - 5 \times 2$

or $\quad R_b = \dfrac{2 \times 1.414 + 1 \times 2.828 + 10}{4} = 3.914$ kN

$\Sigma F_x = 0:$ $\quad 1\cos(90 - \theta) + 2\cos(90 - \theta) + 1\cos(90 - \theta) - R_{ax} = 0$

or $\quad R_{ax} = (1 + 2 + 1)\sin\theta = 4 \times 0.707 = 2.828$ kN

$\Sigma F_y = 0:$ $\quad R_{ay} - 5 + R_b - 1\sin(90 - \theta)$

$\quad - 2\sin(90 - \theta) - 1\sin(90 - \theta)$

or $\quad R_{ay} = 5 - R_b + (1 + 2 + 1)\cos\theta$

$\quad = 5 - 3.914 + 4 \times 0.707$

$\quad = 3.914$ kN

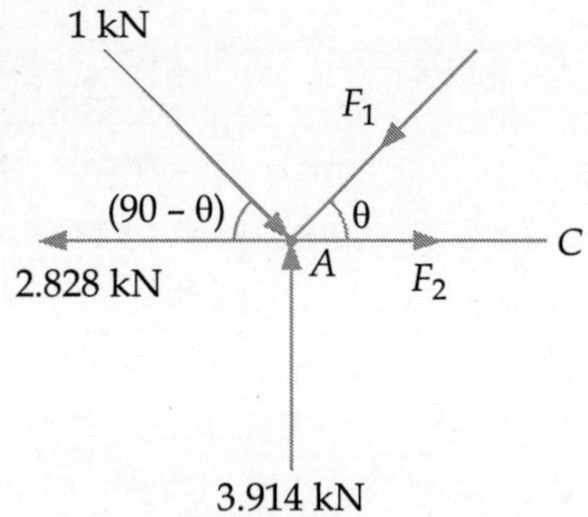

Fig. 5.62

Joint *A*: Consider the free body diagram of joint *A* with the direction of forces assumed as shown in Fig. 5.62.

Equations of equilibrium are:

$\Sigma F_x = 0:$ $\quad F_2 - F_1\cos\theta + 1\cos(90 - \theta) - 2.828 = 0$

or $\quad F_2 = F_1\cos\theta - \sin\theta + 2.828$

$\Sigma F_y = 0:$ $\quad -F_1\sin\theta - 1\sin(90 - \theta) + 3.914$

$\therefore \quad F_1 = \dfrac{3.914 - \cos\theta}{\sin\theta}$

$$= \frac{3.914 - 0.707}{0.707} = \mathbf{4.536\ kN}\ \text{(compression)}$$

and $\quad F_2 = 4.536 \times 0.707 - 0.707 + 2.828 = 5.328$ (tension)

Joint *E*: With reference to free body diagram of joint *E* (Fig. 5.63), the equations of equilibrium along the direction *AD* and perpendicular to it are:

$F_4 - 4.536 = 0;\quad F_4 =$ **4.536 kN** (compression)

$F_3 - 2 = 0;\quad F_2 =$ **2 kN** (compression)

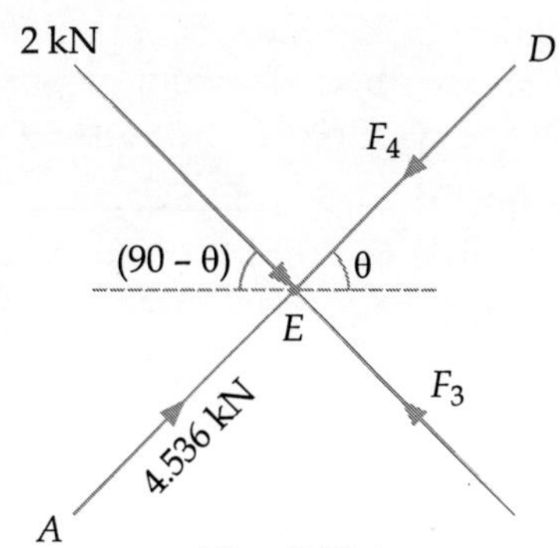

Fig. 5.63

Joint *D*: With reference to free body diagram of joint *D* (Fig. 5.64) the equations of equilibrium are:

$\Sigma F_x = 0:$ $\quad 1\cos(90 - \theta) + 4.536\cos\theta - F_6\cos\theta$

or $\quad F_6 = \dfrac{\sin\theta + 4.536\cos\theta}{\cos\theta}$

$$= \frac{0.707 + 4.536 \times 0.707}{0.707} = \mathbf{5.536\ kN}\ \text{(compression)}$$

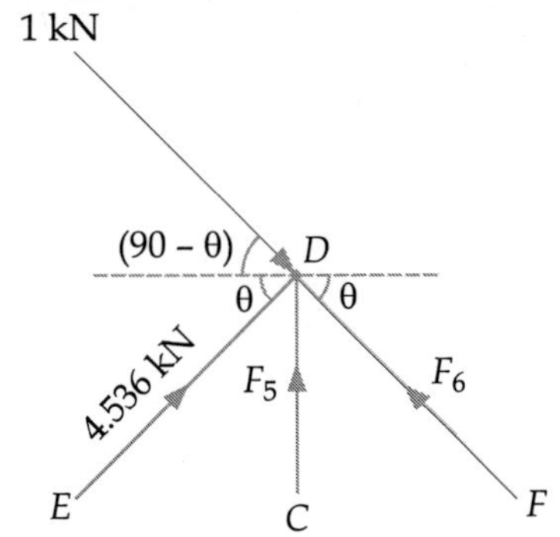

Fig. 5.64

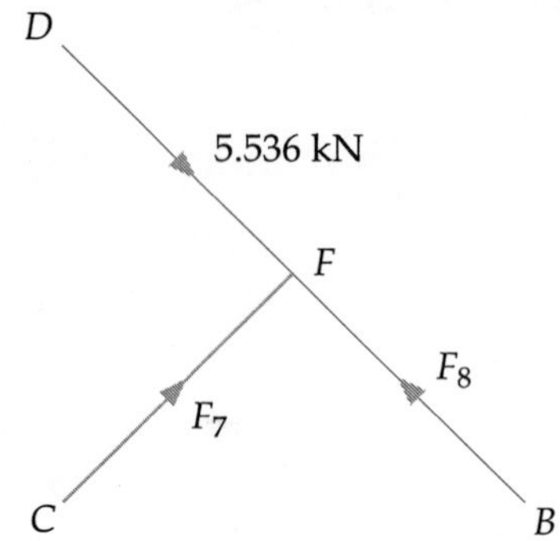

Fig. 5.65

It is apparent from Fig. 5.65 that for the static equilibrium of joint *F*,

$$F_8 = \mathbf{5.536\ kN}\ \text{(compression)}$$

$$\text{and} \quad F_7 = 0\ \text{(no force)}$$

Applying conditions of equilibrium to point *B* (Fig. 5.66), we have

$$\Sigma F_x = 0 : \qquad F_9 - 5.536 \cos \theta = 0$$

$$\therefore \qquad F_9 = 5.536 \times 0.707 = \mathbf{3.914\ kN}\ \text{(tension)}$$

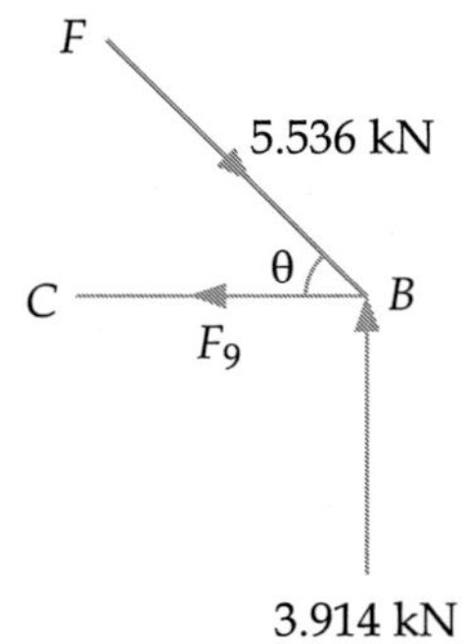

Fig. 5.66

5.3.2. Method of Sections

The various steps involved are:

(*i*) The truss is split into two parts by passing an imaginary section.

(*ii*) The imaginary section has to be such that it does not cut more than three members in which the forces are to be determined.

(*iii*) The conditions of equilibrium

$$\Sigma F_x = 0; \quad \Sigma F_y = 0 \text{ and } \Sigma M = 0$$

are applied for the one part of the truss and the unknown force in the member is determined.

(*iv*) While considering equilibrium, the nature of force in any member is chosen arbitrarily to be tensile or compressive.

If the magnitude of a particular force comes out positive, the assumption in respect of its direction is correct. However, if the magnitude of the forces comes out negative, the actual direction of the force is opposite to that what has been assumed.

The method of section is particularly convenient when the forces in a few members of the frame is required to be worked out.

EXAMPLE 5.15

Determine the forces in members *AC* and *AB* of a simple triangular truss with the loading and support system shown in adjoining figure. Use the method of section.

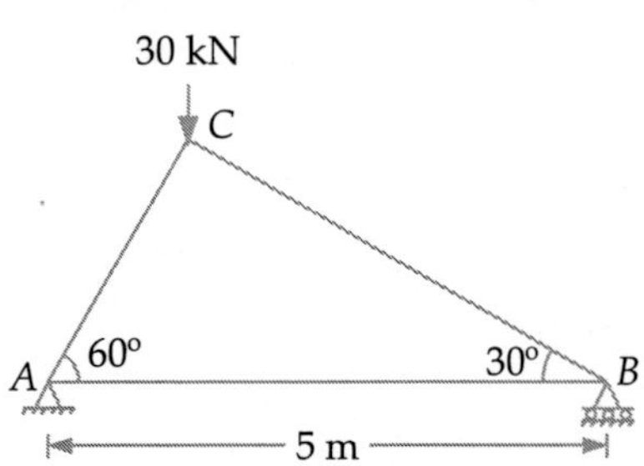

Fig. 5.67

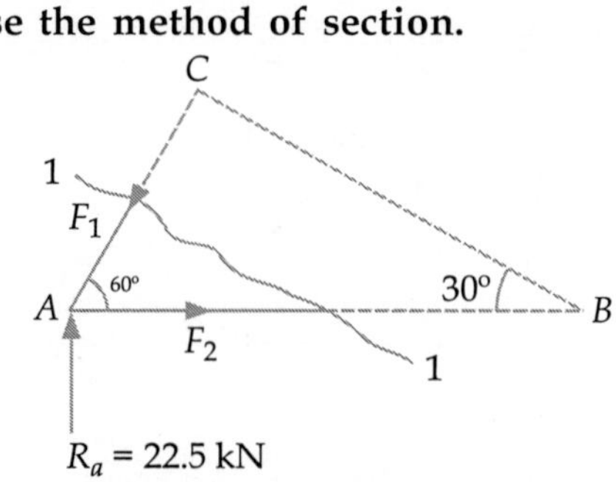

Fig. 5.68

Solution : The reactions at the supports are determined in the usual way as done in Example 5.2.

$$R_a = 22.5 \text{ kN} \quad \text{and} \quad R_b = 7.5 \text{ kN}$$
$$AC = AB \cos 60° = 5 \times 0.5 = 2.5 \text{ m}$$
$$BC = AB \sin 60° = 5 \times 0.866 = 4.33 \text{ m}$$

Let the section 1-1 be passed which divides the truss in two parts. The left part of the truss is in equilibrium under the action of:

(*i*) reaction R_a = 22.5 kN,

(*ii*) unknown force F_1 in member AC (assumed compressive)

(*iii*) unknown force F_2 in member AB (assumed tensile).

Taking moments of these forces about point B,

$$\Sigma M_b = 0: \quad R_a \times AB - F_1 \times BC = 0$$

$$\therefore \quad F_1 = R_a \times \frac{AB}{BC} = 22.5 \times \frac{5}{4.33} = \textbf{22.98 kN} \text{ (compressive)}$$

Taking moments about point C,

$$\Sigma M_c = 0: \quad R_a \times AC \cos 60° - F_2 \times AC \sin 60° = 0$$

$$\therefore \quad F_2 = R_a \times \frac{AC \cos 60°}{AC \sin 60°} = R_a \times \frac{\cos 60°}{\sin 60°} = 22.5 \times \frac{0.5}{0.866} = \textbf{12.99 kN} \text{ (tensile)}$$

To determine force in member BC, another section 2-2 is passed which cuts the members AB and BC (Fig. 5.69)

Consider equilibrium of the right part of the truss under the action of force F_2 = 12.99 kN, force F_3 in member BC (assumed compressive) and reaction R_b = 7.5 kN.

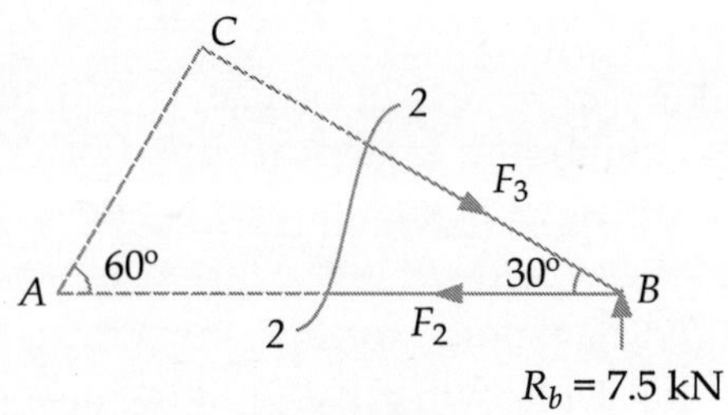

Fig. 5.69

Taking moments about point A,

$$\Sigma M_a = 0: \quad -R_b \times AB + F_3 \times AC = 0$$

$$\therefore \quad F_3 = R_b \times \frac{AB}{AC} = 7.5 \times \frac{5}{2.5} = \textbf{15 kN} \text{ (compressive)}$$

EXAMPLE 5.16

A frame *PQRSTU* is hinged to a rigid support at *P* and is simply supported at *T*. It is loaded as indicated below in Fig. 5.70. Estimate the magnitude and nature of forces in the members *PQ* and *UT*.

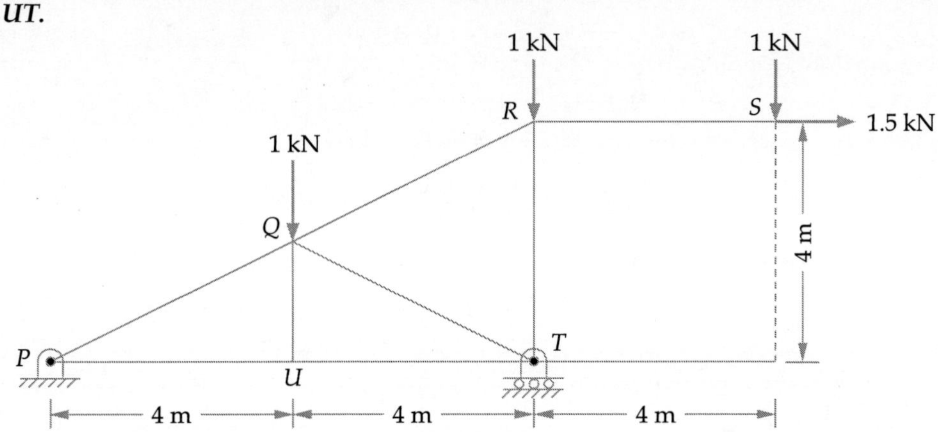

Fig. 5.70

Solution : The reaction at the hinged support P can have two components R_{px} and R_{py} acting in the horizontal and vertical directions respectively. Roller or simple support T is frictionless and provides a reaction R_t at right angles to the roller base. Considering the free body diagram of the entire truss and taking moments about end P,

$$\Sigma M_p = 0: \quad 1 \times 4 + 1 \times 8 + 1 \times 12 + 1.5 \times 4 - R_t \times 8 = 0$$

$$R_t = \frac{4 + 8 + 12 + 6}{8} = 3.75 \text{ kN}$$

From $\Sigma F_y = 0$: $\quad R_{py} = (1 + 1 + 1) - 3.75 = -0.75 \text{ kN} \downarrow$

From $\Sigma F_x = 0$: $\quad R_{px} = 1.5 \text{ kN} \leftarrow$

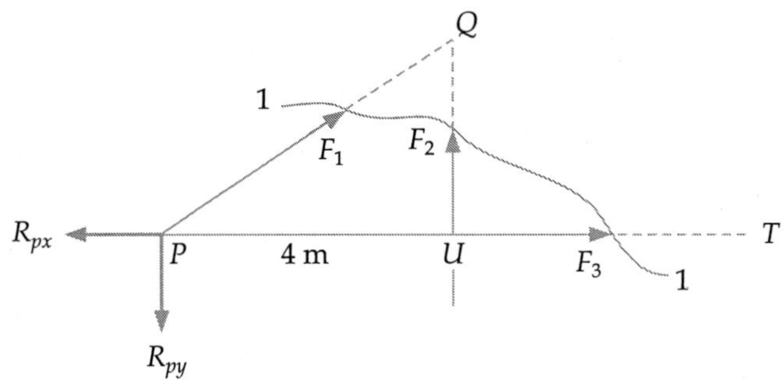

Let a section 1 – 1 be passed which divides the truss in two parts. The left part of the truss is in equilibrium under the action of

(*i*) Reactions R_{px} = 1.5 kN and R_{py} = 0.75 kN acting at point in the directions shown

(*ii*) Unknown force F_1 in member PQ (assumed tensile)

(*iii*) Unknown force F_2 in member U (assumed tensile)

(*iv*) Unknown force F_3 in member UT (assumed tensile)

Taking moment of these forces about point U,

$$R_{py} \times 4 = F_1 \sin\theta \times 4; \quad \theta = \tan^{-1}\left(\frac{4}{8}\right) = 26.56°$$

$$\text{or} \quad F_1 = \frac{R_{py}}{\sin\theta} = \frac{0.75}{\sin 26.56°}$$

$$= 1.667 \text{ kN} \quad \text{(tension)}$$

Taking moments about point Q,

$$R_{px} \times UQ = R_{py} \times PU + F_3 \times UQ$$

$$\text{or} \quad 1.5 \times (4 \tan\theta) = 0.75 \times 4 + F_3 \times 4$$

$$\text{or} \quad F_3 = \frac{1.5 \times 4 \tan 26.56° - 0.75 \times 4}{4} = 0$$

Thus, F_1 (force in member PQ) = 1.677 kN (tension)

F_3 (force in member UT) = 0 (no force)

EXAMPLE 5.17

Determine the forces in the members marked (1), (2) and (3) of the truss loaded and supported as shown in Fig. 5.71.

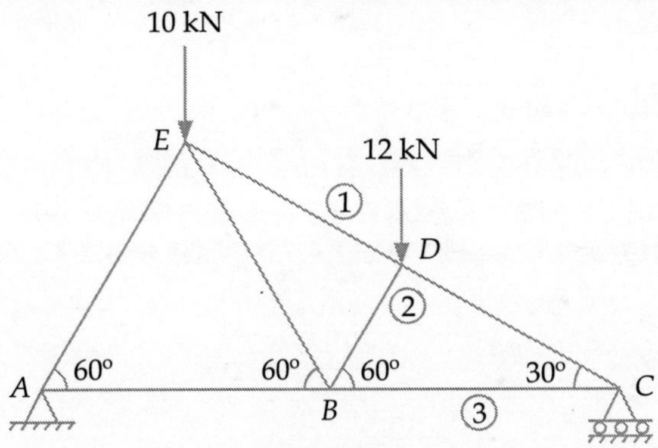

Fig. 5.71

Solution : The reactions at the supports are determined in the usual way as done in Example 5.4

$$R_a = 12 \text{ kN and } R_c = 10 \text{ kN}$$

Let a section x-x be passed in such a way that it cuts the members marked (1), (2), (3) and divides the truss in two parts.

$$CD = BC \cos 30^\circ$$
$$= 2.5 \times 0.866 = 2.165 \text{ m}$$
$$BD = BC \sin 30^\circ = 2.5 \times 0.5 = 1.25 \text{ m.}$$

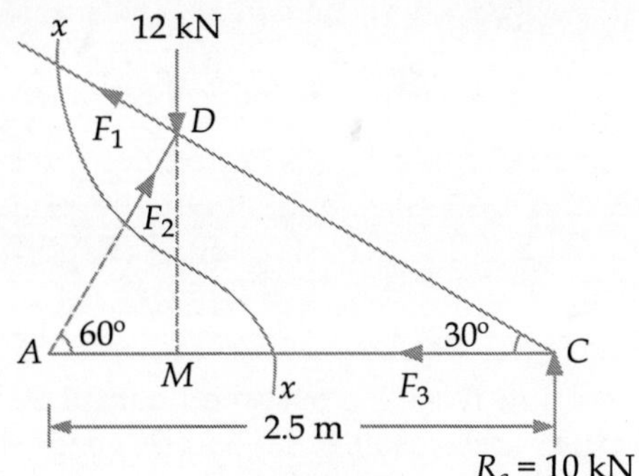

Fig. 5.72

Under equilibrium conditions of right part (Fig. 5.72) of the truss,

(*i*) Taking moments about C,

$$F_2 \times CD - 12 \times MC = 0$$

$$\therefore \quad F_2 = 12 \times \frac{MC}{CD} = 12 \times \frac{CD \cos 30^\circ}{CD}$$
$$= 12 \times \cos 30^\circ = 12 \times 0.866$$
$$= \textbf{10.39 kN} \text{ (compression)}$$

(*ii*) Taking moments B,

$$F_1 \times BD + R_c \times BC - 12 \times BM = 0$$

or $\quad F_1 \times 2.5 \sin 30^\circ + 10 \times 2.5 - 12 \times 1.25 \cos 60^\circ = 0$

or $\quad 1.25\, F_1 + 25 - 7.5 = 0$

$$\therefore \quad F_1 = \frac{7.5 - 25}{1.25} = \textbf{- 14 kN}$$

Negative sign indicates that F_1 is compressive.

(*iii*) Takiong moments about D,

$$F_3 \times DM - R_c \times MC = 0$$

$$\therefore \quad F_3 = R_c \times \frac{MC}{DM} = R_c \times \frac{CD \cos 30^\circ}{CD \sin 30^\circ}$$
$$= 10 \times \frac{0.866}{0.5} = \textbf{17.32 kN} \text{ (tensile)}$$

EXAMPLE 5.18

Make calculations for the load *W* that would produce a force of magnitude 160 kN in the *AB* member of the cantilever truss shown in Fig. 5.73. For this loading, workout forces in members *AE* and *FE* of the truss.

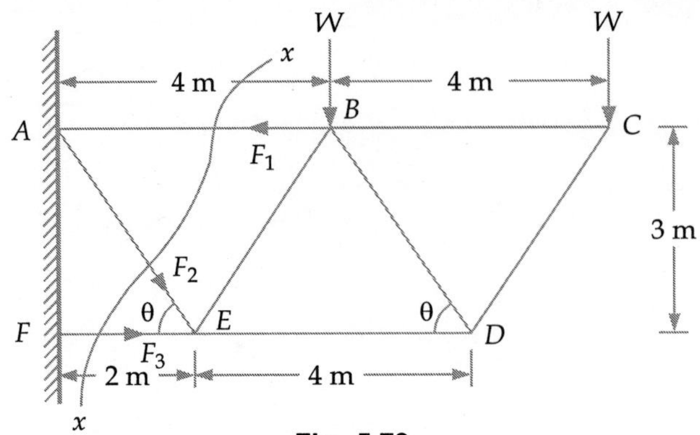

Fig. 5.73

Solution : Let a section x-x be passed in such a way that it cuts the members AB, AE, FE and divides the truss in two parts.

Consider equilibrium of right part of the truss and take moments about the point E,

$$(W \times 2) + (W \times 6) = F_1 \times 3 = 160 \times 3$$

$$\therefore \quad W = \frac{160 \times 3}{8} = \mathbf{60\ kN}$$

To find force F_3 (assumed compressive) in member FE, we consider right portion of the truss and take moments about the point A,

$$(W \times 8) + (W \times 4) = F_3 \times 3$$

$$\therefore \quad F_3 = \frac{12\,W}{3} = 4\,W = 4 \times 60 = \mathbf{240\ kN} \text{ (compressive).}$$

To find force F_2 (assumed compressive) in member AE, we again consider right portion of the truss and resolve the forces vertically.

$$F_2 \sin\theta = W + W\ ; \qquad F_2 = \frac{2\,W}{\sin\theta}$$

From the geometrical configuration,

$$\sin\theta = \frac{AF}{AE} = \frac{3}{\sqrt{3^2 + 2^2}} = 0.832$$

$$\therefore \quad F_2 = \frac{2 \times 60}{0.832} = \mathbf{144.23\ kN} \text{ (compressive)}$$

EXAMPLE 5.19

For the truss loaded and supported as shown in Fig. 5.74, determine the forces in members CD, DG and GH. Use the method of sections.

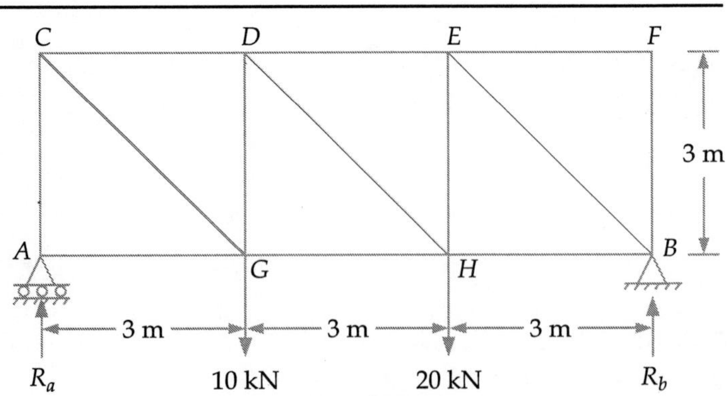

Fig. 5.74

Solution : The reactions at the supports can be determined by considering equilibrium of the truss. Since both the external loads are vertical, only the vertical component of reaction at the hinged end B need to be considered.

Taking moments about end B,

$\Sigma M_b = 0: \quad R_a \times 9 - 10 \times 6 - 20 \times 3 = 0$

$\therefore \quad R_a = \dfrac{10 \times 6 + 20 \times 3}{9} = \mathbf{13.33\ kN}$

Further, $\Sigma F_y = 0: \quad R_a + R_b - 10 - 20 = 0$

$\therefore \quad R_b = 30 - R_a = 30 - 13.33 = \mathbf{16.67\ kN}$

Let section 1-1 be passed in such a way that it cuts the members *CD*, *DG*, *GH* and divides the truss in two parts (Fig. 5.75).

The left part of the truss is in equilibrium under the action of following forces:

(*i*) reaction R_a = 13.33 kN

(*ii*) 10 kN load at joint *G*

(*iii*) force F_1 in member *CD* (assumed compressive), force F_2 in member *GD* (assumed tensile) and force F_3 in member *GH* (assumed tensile)

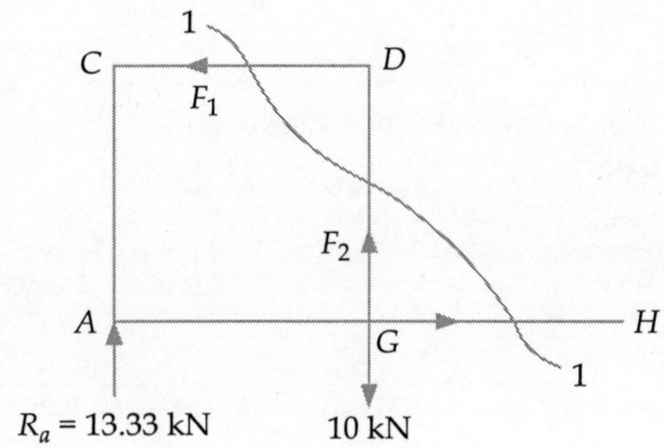

Fig. 5.75

Under equilibrium conditions,

$\Sigma M_a = 0: \quad R_a \times AG - F_3 \times GD = 0$

$\therefore \quad F_3 = R_a \times \dfrac{AG}{GD} = 13.33 \times \dfrac{3}{3} = \mathbf{13.33\ kN}$ (tensile)

$\Sigma M_g = 0: \quad R_a \times AG - F_1 \times AC = 0$

$\therefore \quad F_1 = R_a \times \dfrac{AG}{AC} = 13.33 \times \dfrac{3}{3} = \mathbf{13.33\ kN}$ (compressive)

$\Sigma M_c = 0: \quad 10 \times AG - F_2 \times CD - F_3 \times GD$

$\therefore \quad F_2 = \dfrac{10 \times 3 - 13.33 \times 3}{3} = \mathbf{-\ 3.33\ kN}$

The negative with force F_2 is indicative of the fact that a wrong choice has been made while assuming its direction. Obviously, the force F_2 in member *GD* is 3.33 kN compressive.

EXAMPLE 5.20

Calculate the force in member *DJ* of the Howe roof truss as shown in Fig. 5.76. Neglect any horizontal component of force at the supports.

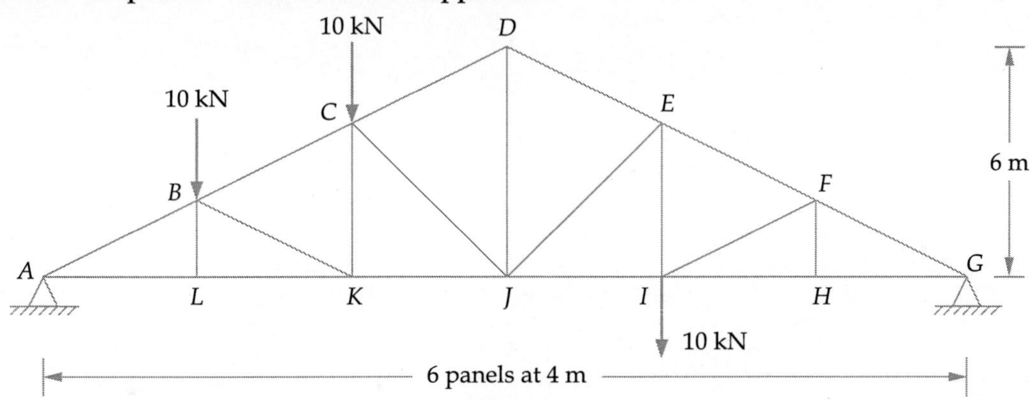

Fig. 5.76

Solution : The reactions at the supports can be determined by considering equilibrium of the entire truss. Since the external loads are vertical and any horizontal component of force at the supports is to be neglected, only the vertical component of the reactions at the support need to be considered.

Taking moments about end *A*,

$$R_g \times 24 = 10 \times 4 + 10 \times 8 + 10 \times 16 = 280$$

$$\therefore \quad R_g = \frac{280}{24} = 11.67 \text{ kN}$$

$$\text{and} \quad R_a = (10 + 10 + 10) - 11.67 = 18.33 \text{ kN}$$

Let section 1–1 be passed which cuts the members *CD*, *CJ* and *KJ* and divides the truss in two parts.

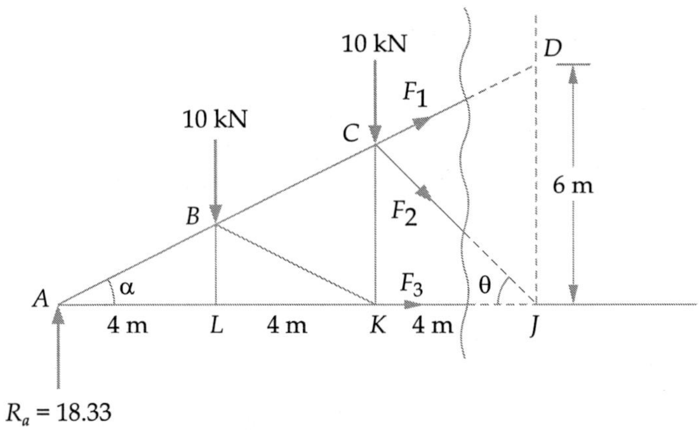

Under equilibrium conditions of left part of the truss

Taking moments about point *A*,

$$10 \times 4 + 10 \times 8 + F_2 \sin\theta \times 12 = 0$$

$$\text{or} \quad F_2 = -\frac{40 + 80}{12 \sin\theta}$$

From the geometrical configuration,

$$\tan\alpha = \frac{CK}{AK} = \frac{DJ}{AJ}; \quad CK = \frac{DJ}{AJ} \times AK = \frac{6}{12} \times 8 = 4 \text{ m}$$

$$\theta = \tan^{-1}\left(\frac{CK}{KJ}\right) = \tan^{-1}\left(\frac{4}{4}\right) = \tan^{-1} = 45°$$

$$\alpha = \tan^{-1}\left(\frac{CK}{AK}\right) = \tan^{-1}\left(\frac{4}{8}\right) = \tan^{-1} 0.5 = 26.56°$$

$$\text{Then} \quad F_2 = -\frac{40 + 80}{12 \sin 45} = -14.14 \text{ kN}$$

The negative sign with force F_2 is indicative of the fact that a wrong choice has been make while assuming its direction. Obviously, the force F_2 in member *CJ* is 14.14 kN compressive.

Taking moments about point *J*,

$$18.33 \times 12 - 10 \times 8 - 10 \times 4 + F_1 \cos 26.56 \times 6 = 0$$

$$\text{That gives:} \quad F_1 = \frac{80 + 40 - 219.96}{6 \cos 26.56} = -18.63 \text{ N}$$

The negative sign implies that force F_1 in member *CD* is compressive.
Taking moments about point *C*,

$$18.33 \times 8 - 10 \times 4 - F_3 \times 4 = 0$$

$\therefore$ Froce in member *KJ*,

$$F_3 = \frac{18.33 \times 8 - 10 \times 4}{4} = 26.66 \text{ kN} \quad \text{(tension)}$$

(*ii*) Now, let a section 2–2 be passed which cuts the member *DE, DJ, JC* and *JK*.

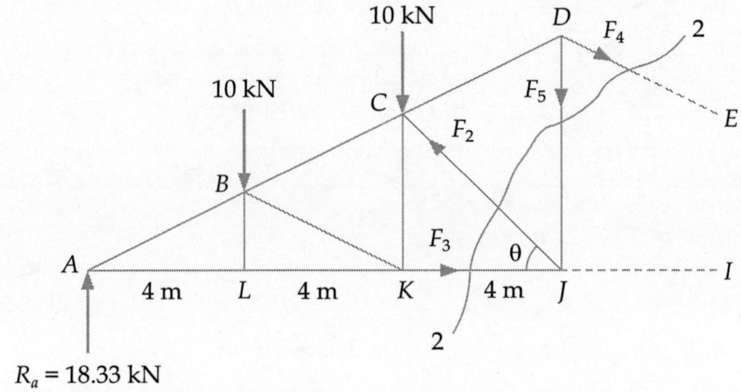

Under equilibrium conditions of the left part of truss, taking moments about point *G*,

$$18.33 \times 24 - 10 \times 20 - 10 \times 16 + F_2 \sin\theta \times 12 - F_5 \times 12 = 0$$

or

$$439.92 - 200 - 160 + 14.14 \times \sin 45° \times 12 - 12\,F_5 = 0$$

$$\therefore \quad F = \frac{439.92 - 200 - 160 + 119.96}{12} = 16.66 \text{ kN}$$

The member *DJ* is thus subjected to a tensile force of 16.66 kN

EXAMPLE 5.21

Using the method of sections, analyse the struss shown in Fig. 5.77 regarding forces in members *ED*, *DF* and *FC*.

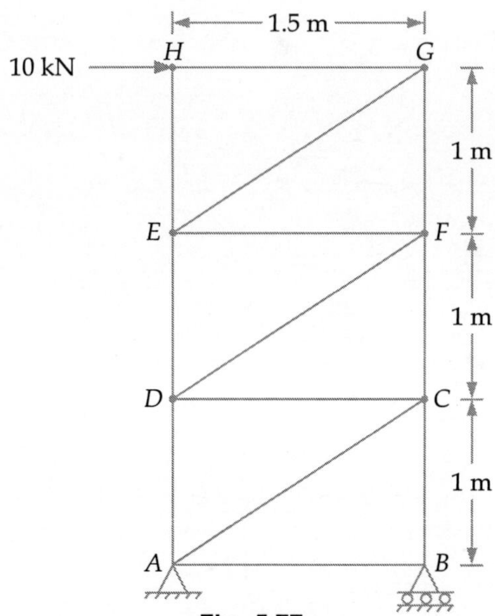

Fig. 5.77

Solution : Let section *x-x* be passed in such a way that it cuts the member *ED*, *DF*, *FC* and divides the truss in two parts. For considering the equilibrium of the upper part of the truss (Fig. 5.78), there is no need to make calculations for the reaction at the supports.

Let F_1, F_2 and F_3 be the forces with their directions assumed as shown in Fig. 5.78.

Under equilibrium conditions,

$\Sigma M_f = 0 : \quad 10 \times GF - F_1 \times EF = 0$

$$\therefore F_1 = 10 \times \frac{GF}{EF} = 10 \times \frac{1}{1.5} = \mathbf{6.67\ kN} \text{ (tensile)}$$

$\Sigma F_d = 0 : \quad 10 \times HD - F_3 \times FE = 0$

$$\therefore F_3 = 10 \times \frac{HD}{FE} = 10 \times \frac{2}{1.5} = \mathbf{13.33\ kN} \text{ (compressive)}$$

Further equating all the forces in horizontal direction we get,

$$10 - F_2 \cos\theta = 0 ; \qquad F_2 = \frac{10}{\cos\theta}$$

From the geometry

$$\cos\theta = \frac{1.5}{\sqrt{1.5^2 + 1^2}} = \frac{1.50}{1.80} = 0.833$$

$$\therefore \qquad F_2 = \frac{10}{0.833} = \mathbf{12\ kN} \text{ (tensile)}$$

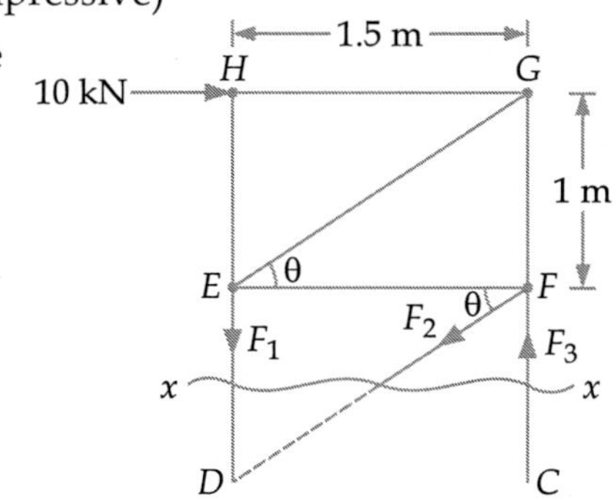

Fig. 5.78

5.3.3. Graphical Methods

The analytical methods, *i.e.*, the method of joints and the method of sections give absolutely correct results regarding forces set up in the members of a perfect truss when it is loaded. However, there arise certain situations where it becomes impossible to get the results from these methods. Recourse is then made to graphical methods which are quite convenient and provide reasonably accurate results.

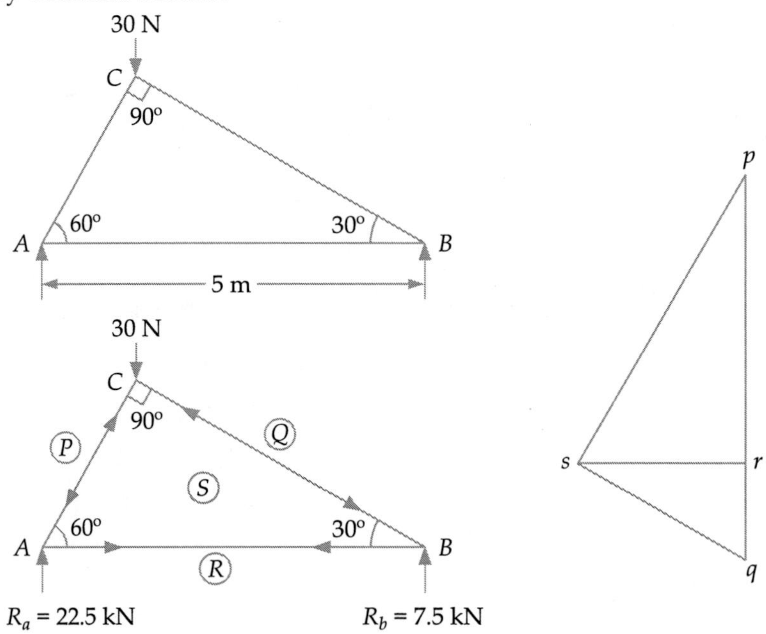

Fig. 5.79

The steps involved for obtaining graphical solution of a frame are explained below with reference to a triangular truss loaded and supported as shown in Fig. 5.79.

The reactions at the supports are determined in the usual way as done in Example 5.2

$$R_a = 22.5 \text{ kN} \quad \text{and} \quad R_b = 7.5 \text{ kN}$$

- **Making a space diagram:** The given truss or frame is drawn accurately according to some linear scale. The loads and support reactions are also indicated on the frame both in magnitude and direction. The naming of the various members of the frame is then done according to Bow's notation. In this technique, a force is designated by two letters which are written on either side of the line of action of the force.

  In the space diagram drawn as shown in Fig. 5.76, the spaces between the forces have been marked as *P*, *Q*, *R* and *S*.

- **Constructing a vector diagram:**

(*i*) Take any point *p* and draw a vertical line *pq* to represent force *PQ* = 30 kN. If scale is taken as 1 cm = 5 kN, then *pq* = 6 cm. From point *q*, take *qr* equal to 1.5 cm to represent the reaction R_b = 7.5 kN acting vertically upwards. Then *rp* represents the reaction R_a.

(*ii*) Choose a joint where not more than two unknown forces are acting (say joint A).

(*iii*) From point *p*, draw a line parallel to *PS* and from point *q* draw a line parallel to *QS*. These two lines meet at point *s*. Measure *ps* and *sr* in the vector diagram.

ps = 5.2 cm ∴ Force *PS* (force in member *AC*) = 5.2 × 5 = **26 kN**

sr = 2.6 cm ∴ Force *SR* (force in member *AB*) = 2.6 × 5 = **13 kN**

For determining the nature of these forces, we move in clockwise direction round the joint *A* in space diagram. First *PS*, then *SR* and then *RP*.

In the vector diagram *p* to *s* means that the arrow must be marked towards the joint *A* in space diagram. Obviously then the member *AC* is subjected to 26 kN compressive force.

Considering *SR* and the corresponding vector *s* to *r* in vector diagram, the joint *A* is being pulled and as such arrow must be marked directed away from the joint. Obviously then member *AC* is subjected to 13 kN tensile force.

(*iv*) Considering joint B: In the vector diagram, the vector *sq* represents force in member *BC* designated by *SQ* in the space diagram

sq = 3 cm ∴ Force *SQ* (force in member *BC*) = 3 × 5 = **15 kN**

Again moving round the joint *B* in clockwise direction, we have first *RS*, then *SQ* and then *QR*. In the vector diagram

rs : arrow is to be marked from *B* towards *A* (away from the joint)

sq : arrow is to be marked from *C* towards *B* (towards the joint)

Obviously then the member *CB* is subjected to 15 kN compressive force.

EXAMPLE 5.22

A truss has been loaded and supported as shown in Fig. 5.80. Determine graphically the forces in the various members of the structure shown in the figure.

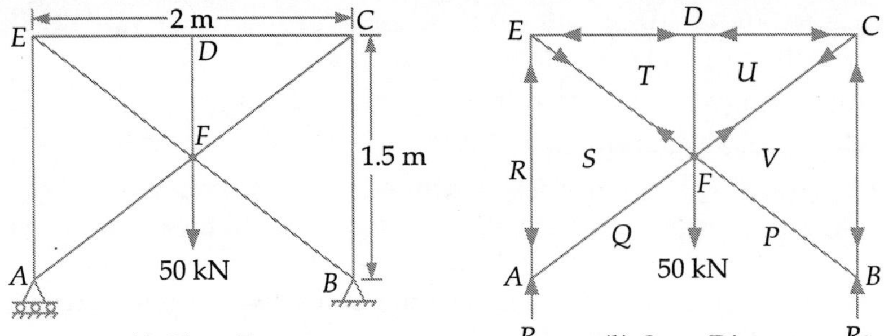

(a) Given Truss (b) Space Diagram

Fig. 5.80

Solution : Draw the space diagram and name the various forces according to Bow's notation (Fig. 5.80 *b*).

To draw the force or vector diagram (Fig. 5.80 *c*), proceed as follows:

(*i*) Draw *pq* to represent force *PQ* = 50 kN. From point *q*, take *qr* to represent the reaction R_a = 25 kN acting vertically upwards. Then *rp* represents the reaction R_b = 25 kN.

(*ii*) Draw *rs* and *qs* parallel to *RS* and *QS* to locate point *s*. The point *s* coincides with point *q*.

(*iii*) Draw *rt* and *st* parallel to *RT* and *ST* respectively and locate the point *t*.

(*iv*) Draw *ru* and *tu* parallel to *RU* and *TU* respectively to locate point *u*. The point *u* coincides with point *t*.

(*v*) Draw *uv* and *rv* parallel to *UV* and *RV* respectively to locate point *v*.

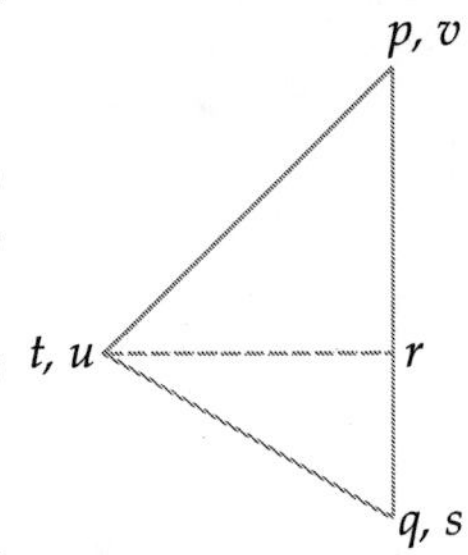

(c) Vector Diagram

Fig. 5.80

Mark arrows on the joints and measure the forces from the force diagram. It may be noted that:

q and *s* coincide ∴ force in member *AF* = 0
t and *u* coincide ∴ force in member *DF* = 0
p and *v* coincide ∴ force in member *BF* = 0

The forces in the various members have been tabulated belows:

Member	*AE*	*ED, DC*	*BC*	*BF*	*AF*	*DF*	*EF, FC*
Bow's notation	*RS*	*RT, RU*	*RV*	*VP*	*QS*	*UT*	*ST, UV*
Force (kN)	25	34	30	0	0	0	42
Nature	*C*	*C*	*C*	-	-	-	*T*

EXAMPLE 5.23

Find graphically or otherwise, the forces in all the members of the truss shown in Fig. 5.81. Present your results in a tabular form, stating whether a member is in tension or compression.

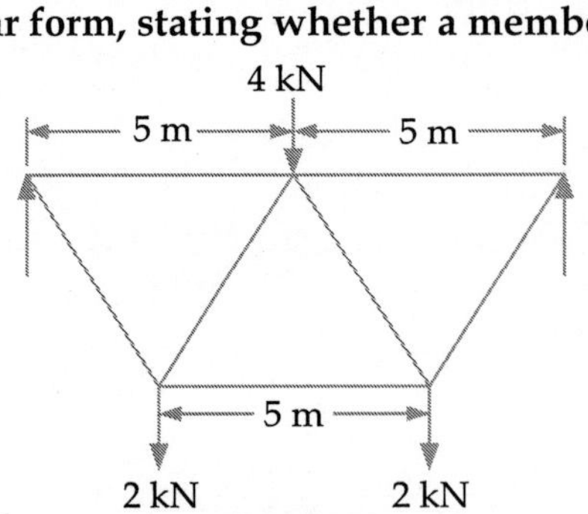

Fig. 5.81

Solution : Since the loading and truss are symmetrical,

$$R_1 = R_2 = \frac{4+2+2}{2} = 4 \text{ kN}$$

Choosing a convenient scale, draw the space diagram and indicate the forces by using Bow's notation. Proceed to draw the force/vector diagram by considering the joints in succession.

Mark arrows on the joints and measure the magnitude of forces from the vector diagram. The forces in the various members are tabulated below:

Member Bow's Notation	*AF*	*BH*	*HC*	*HG*	*GD*	*GF*	*EF*
Force (kN) Nature	2 *C*	2 *C*	4.5 *T*	2.25 *C*	3 *T*	2.25 *C*	4.5 *T*

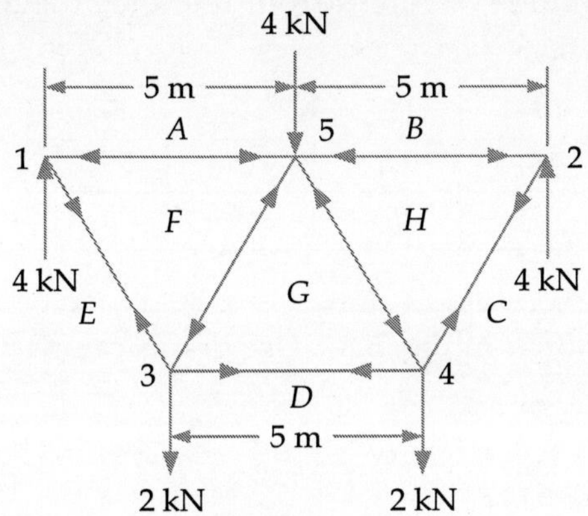

Fig. 5.82

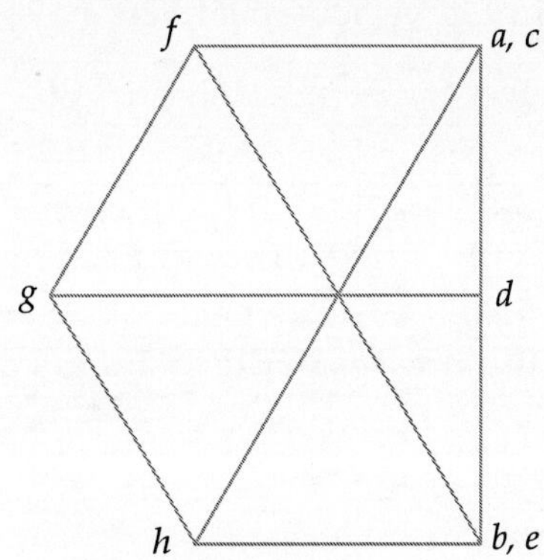

Fig. 5.83

EXAMPLE 5.24

Determine the stresses in all members of the truss shown in Fig. 5.84. Use graphical method.

Solution : Let R_1 = reaction of the left-hand support

R_2 = reaction of the right hand support

To find these reactions, we take moments about joint 1. That gives

$$R_2 \times 16 = (6 \times 12) + (2 \times 8) + (4 \times 4)$$
$$= 72 + 16 + 16 = 104$$

$$\therefore \quad R_2 = \frac{104}{16} = 6.5 \text{ kN}$$

$$R_1 = (6 + 2 + 4) - 6.5 = 5.5 \text{ kN}$$

Choosing a convenient scale, draw the space and vector diagrams as shown in Fig. 5.85 and Fig. 5.86.

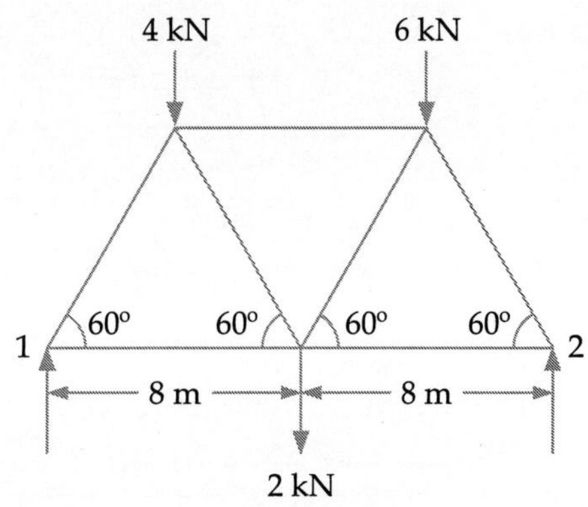

Fig. 5.84

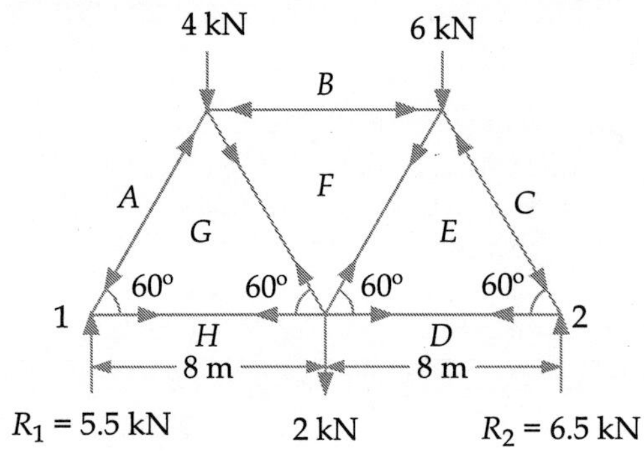

Fig. 5.85

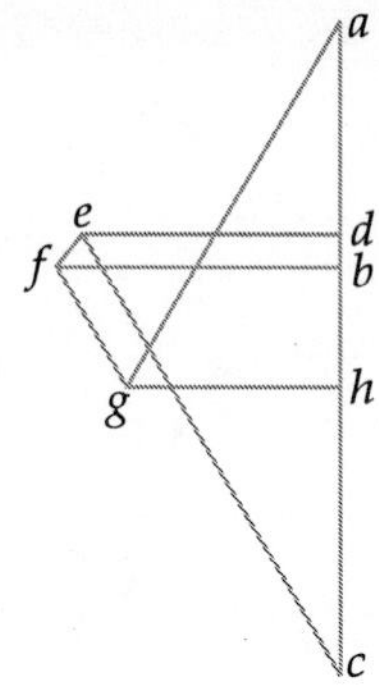

Fig. 5.86

Mark arrows on the joints and measure the magnitude of forces from the force diagram. The forces in the various members have been tabulated below:

Member	*AG*	*GH*	*GF*	*BF*	*FE*	*EC*	*ED*
Forces (kN)	6.3	3.2	1.7	4.0	0.5	7.5	3.75
Nature	C	T	T	C	T	C	T

EXAMPLE 5.25

Determine the stresses in the members of the truss shown in Fig. 5.87. Use graphical method.

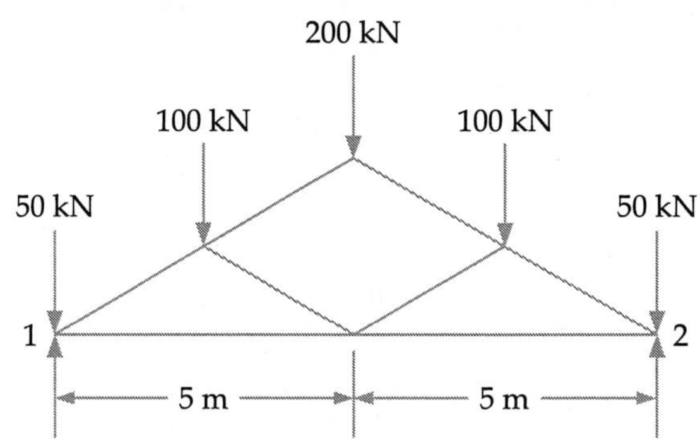

Fig. 5.87

Solution : Let R_1 = reaction of the left-hand support

R_2 = reaction of the right-hand support

Since the truss and the loads are symmetrical,

$$R_1 = R_2 = \frac{50 + 100 + 200 + 100 + 50}{2} = 250 \text{ kN}$$

Choosing a convenient scale, draw the space and vector diagrams as shown in Fig. 5.88 and Fig. 5.89.

Mark arrows on the joints and measure the magnitude of forces from the force diagram. The forces in the various members have been tabulated below:

Member	*BL, HE*	*KC, JD*	*LK, JH*	*GL, GH*	*KJ*
Force (kN)	400	300	100	350	100
Nature	C	C	C	T	T

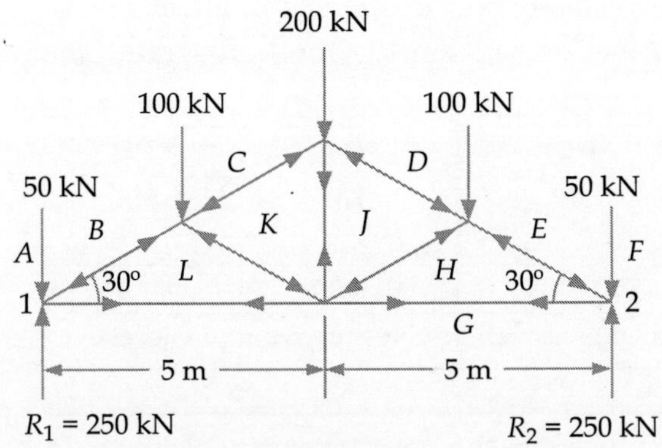

Fig. 5.88

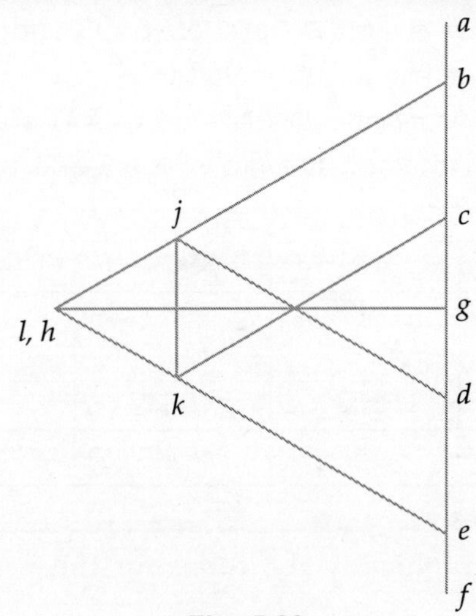

Fig. 5.89

EXAMPLE 5.26

Using graphical method, determine the magnitude and nature of the forces in all members of the cantilever truss loaded as shown in Fig. 5.90.

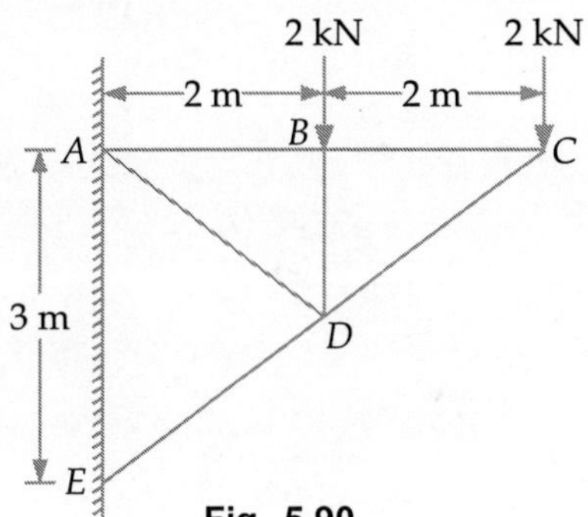

Fig. 5.90

Solution : The steps involved for determining graphically the forces in members of the truss are:

- **Making a space diagram:** Draw the space diagram (Fig. 5.91) for the truss along with loads of 2 kN at joints *B* and *C*. Name the various members according to Bow's notation.

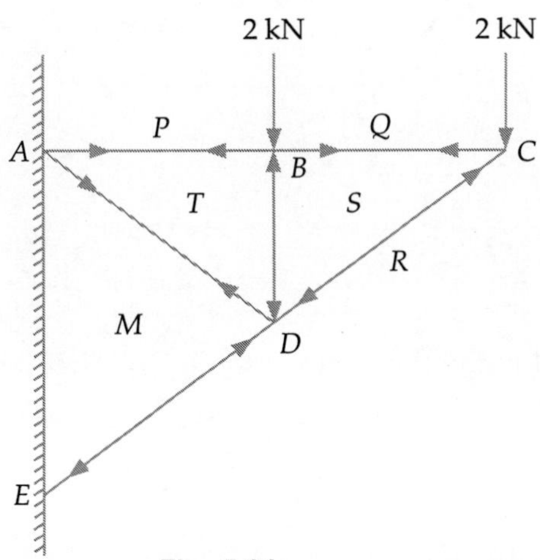

Fig. 5.91

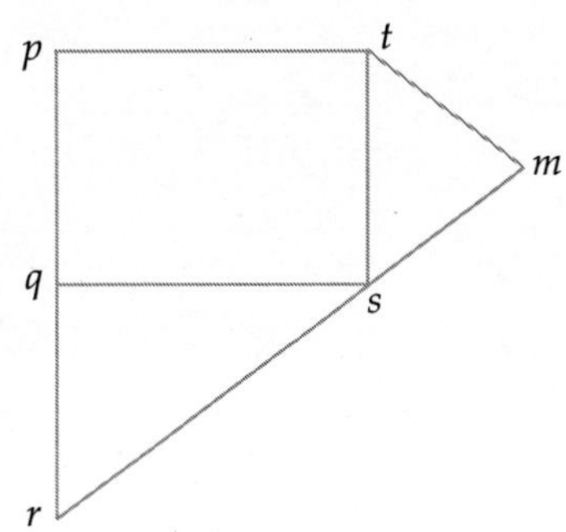

Fig. 5.92

- **Constructing a vector diagram:** The vector/force diagram (Fig. 5.92) is constructed as explained below:
  (*i*) Take any point *p* and adopting a convenient scale, draw *pq* and *qr* vertically to represent the two loads of 2 kN each.

(*ii*) Draw *qs* and *rs* paraller to *QS* and *RS* and obtain the point *s*.

(*iii*) Draw *st* and *pt* parallel to *ST* and *PT* respectively and obtain the point *t*.

(*iv*) Draw *rm* and *tm* parallel to *RM* and *TM* respectively and obtain the point *m*.

Mark arrows on the joints and scale (measure) the forces from the forces diagram. The forces in the various members have been tabulated below:

Member	*AB*	*BC*	*BD*	*DC*	*DA*	*DE*
Bow's notation	*PT*	*QS*	*TS*	*SR*	*TM*	*MR*
Force	2.65 kN	2.65 kN	1.65	3.35 kN	2.0 kN	5.0 kN
Nature	tensile	tensile	tensile	compressive	compressive	compressive

EXAMPLE 5.27

Workout graphically the forces in the various members of the cantilever truss shown in Fig. 5.93.

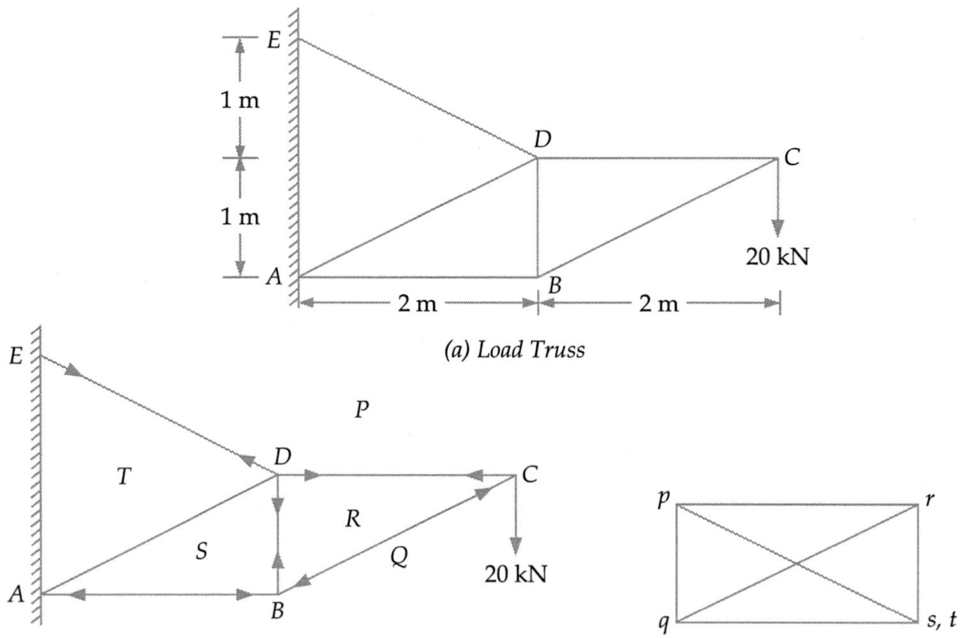

(*a*) *Load Truss*

(*b*) *Space Diagram*

(*c*) *Vector Diagram*

Fig. 5.93

Solution : Draw the space diagram (Fig. 5.93 *b*) and name the various forces following Bow's notation.

To draw the force or vector diagram (Fig. 5.93 *c*), proceed as follows:

(*i*) Draw *pq* to represent the force *PQ* = 20 kN. Draw *pr* and *qr* parallel to *PR* and *QR* respectively and obtain the point *r*.

(*ii*) Draw *rs* and *qs* parallel to *RS* and *QS* respectively and get the point *s*.

(*iii*) Draw *pt* and *st* parallel to *PT* and *ST* respectively and get the point *t*. The point *t* coincides with point *s*.

Mark arrows on the joints and measure the forces from the force diagram. The forces in the various members have been tabulated below:

Member	*AB*	*BC*	*CD*	*DA*	*DE*	*DB*
Bow's notation	QS	QR	PR	ST	PT	RS
Force	40	45	40	0	45	20
Nature	*C*	*C*	*T*	–	*T*	*T*

EXAMPLE 5.28

The loading and the support conditions for a particular cantilever truss are shown in Fig. 5.94. Determine graphically the force in each member of the truss, stating whether it is in tension or compression.

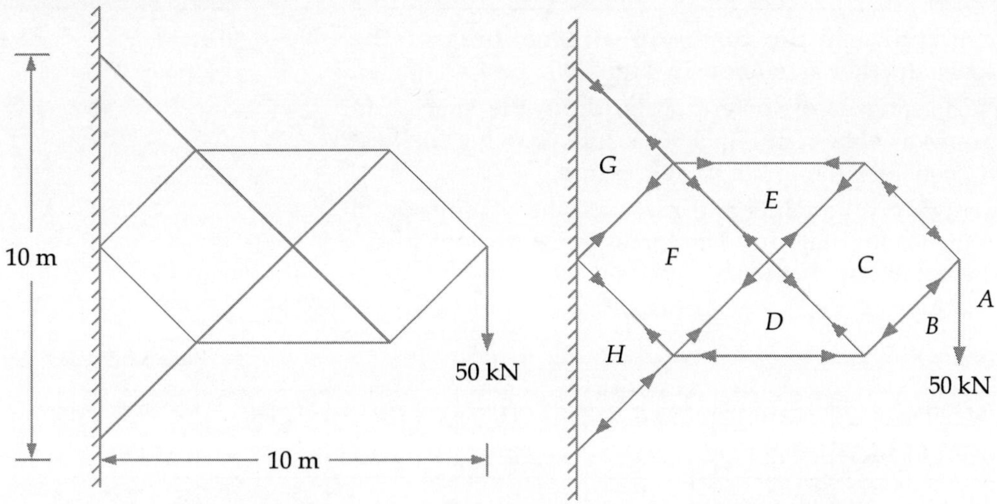

(*a*) *Loaded truss* (*b*) ***Space diagram***

Fig. 5.94

Solution : Draw the space diagram (Fig. 5.94 *b*) and name the various forces following Bow's notation.

To draw the force or vector diagram, (Fig. 5.94 *c*) proceed as follows:

(*i*) Draw *ab* to represent the force *AB* = 50 kN. Draw *ac* and *bc* parallel to *AC* and *BC* respectively and obtain the point *c*.

(*ii*) Draw *cd* and *bd* parallel to *CD* and *BD* respectively and obtain the part *d*.

(*iii*) Draw *ce* and *ae* parallel to *CE* and *AE* respectively and locate the point *e*.

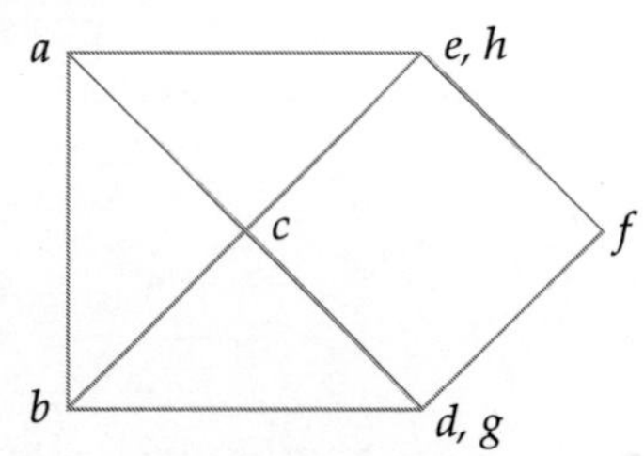

(*c*) *Vector diagram*

Fig. 5.94

(*iv*) Draw *df* and *ef* parallel to *DF* and *EF* respectively and mark the point *f*.

(*v*) Draw *fg* and *ag* parallel to *FG* and *AG* respectively and get the point *g*.

(*vi*) Draw *fh* and *bh* parallel to *FH* and *BH* respectively and obtain the point *h*.

Mark arrows on the joints and measure the forces from the force diagram. The forces in the various members have been tabulated below:

Member	Force (kN)	Nature	Member	Force (kN)	Nature
AG	72	*T*	*HF*	35	*C*
AE	50	*T*	*FG*	35	*T*
AC	35	*T*	*FE*	35	*T*
BC	35	*C*	*FD*	35	*C*
BD	50	*C*	*EC*	35	*C*
BH	72	*C*	*DC*	35	*T*

EXAMPLE 5.29

Determine graphically the forces in all members of the cantilever truss loaded as shown in Fig. 5.95.

Solution : Choosing a convenient scale, draw the space and vector diagrams as shown in Fig. 5.96. While drawing the vector diagram, consider the joints in succession.

Mark arrows on the joints and measure the magnitude of forces from the vector diagram. The forces in the various members have been tabulated below:

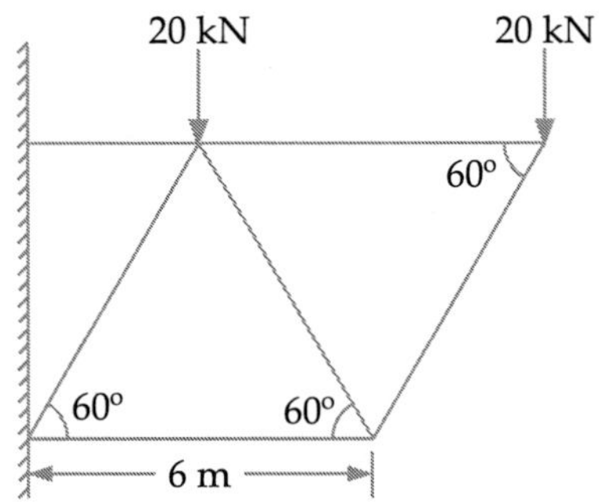

Fig. 5.95

Member	*AC*	*BC*	*CD*	*DB*	*DE*	*EF*
Force (kN)	12	24	24	24	48	48
Nature	*T*	*C*	*T*	*C*	*C*	*T*

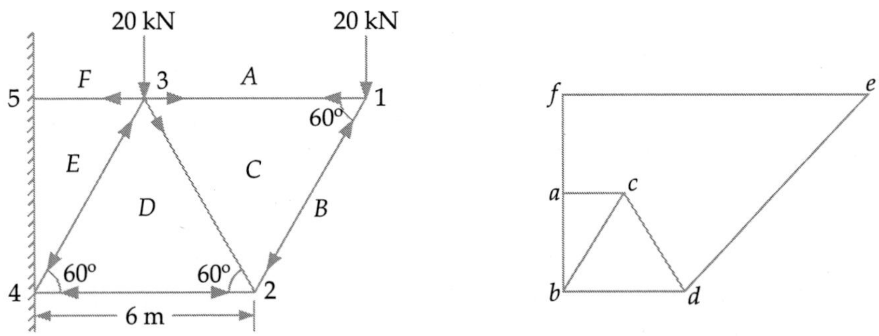

(a) Space diagram *(b) Vector diagram*

Fig. 5.96

EXAMPLE 5.30

Determine graphically the forces in all members of the cantilever frame loaded as shown in Fig. 5.97.

Solution : Choosing a convenient scale, draw the space and vector diagrams as shown in Fig. 5.98. While drawing the vector diagram, the joints are to be considered in succession.

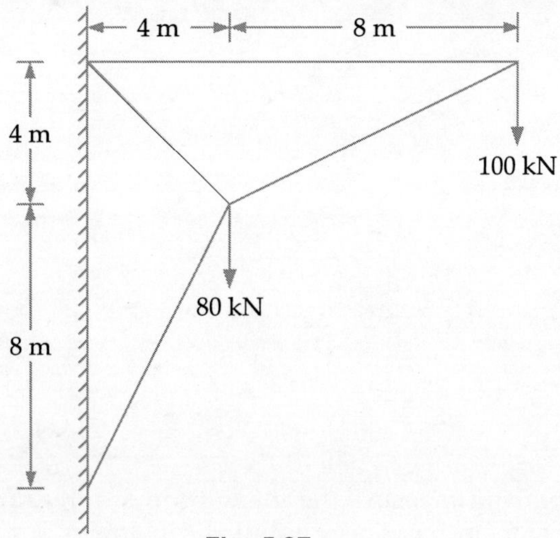

Fig. 5.97

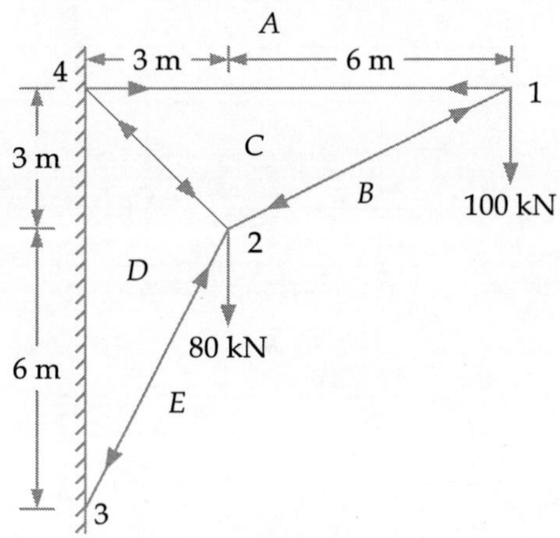

(a) Space diagram

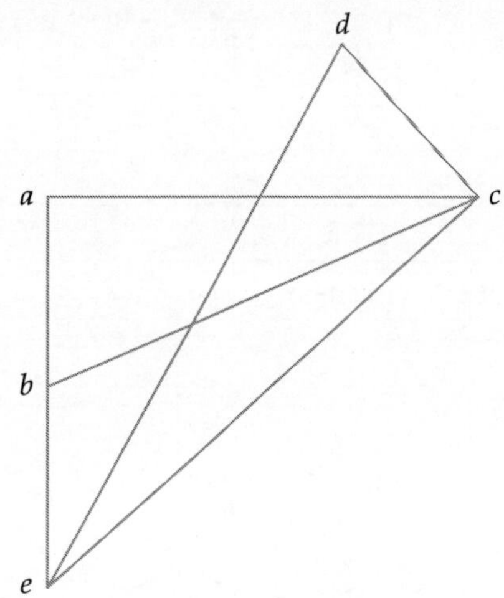

(b) Vector diagram

Fig. 5.98

Mark arrows on the joints and measure the magnitudes of forces from the vector diagram. The forces in the various members have been tabulated below:

Member	*AC*	*BC*	*CD*	*DE*
Force (kN)	200	225	100	280
Nature	*T*	C	C	C

EXAMPLE 5.31

A cantilever frame has been loaded as shown in Fig. 5.99. Determine graphically the forces in all the members of the frame.

Solution : Choosing a convenient scale, draw the space and vector diagrams as shown in Fig. 5.100. While drawing the vector diagram, consider the joints in succession.

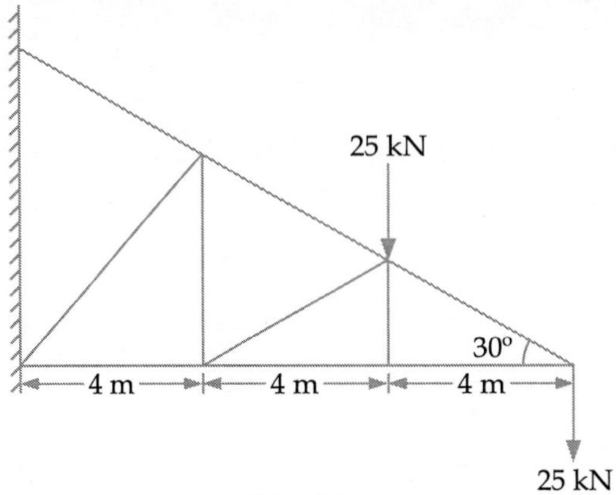

Fig. 5.99

Mark arrows on the joints and measure the magnitude of forces from the vector diagram. The forces in the various members have been tabulated below:

Member	Force (kN)	Nature	Member	Force	Nature
BC	44	*C*	*ED*	25	*C*
BD	44	*C*	*DC*	0	-
BF	65	*C*	*CA*	50	*T*
FG	11.5	*C*	*HE*	75	*T*
FE	12.5	*T*	*HG*	85	*T*

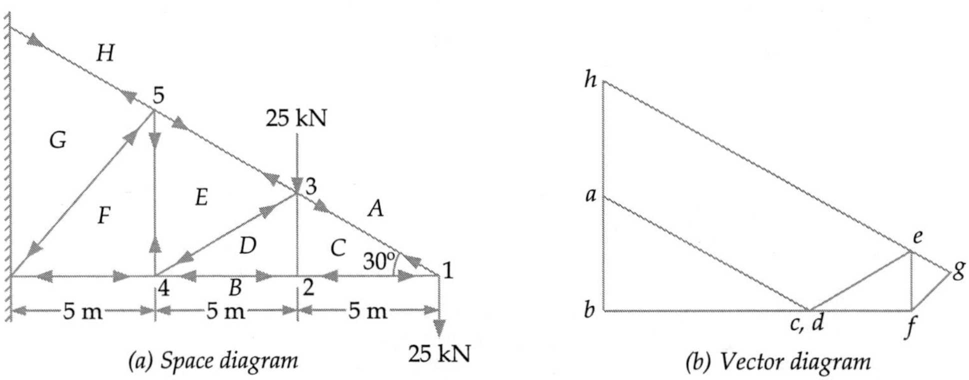

(a) Space diagram　　*(b) Vector diagram*

Fig. 5.100

REVIEW AND SUMMARY

1. A frame is a rigid structure made up of three or more members welded or riveted together. The framed structure finds application in the roofs of the sheds at railway platforms, workshops and industrial buildings, and bridge trusses etc.
2. The bars used in the framed strucutre are slender and are usually of angle section, *T*-section and *I*-section.
3. The bars are called two-force members because they are either in tension or incompression. the member under tension is called a tie and the member under compression is called a strut.
4. The frame is called plane frame if all the members lie in the same plane and the forces act along the plane of the frame.

A frame in which all the members do not lie in the same plane is called a space frame.

5. A structure is said to be perfect if the number of members is just sufficient to prevent its distorsion of shape when subjected to external loads. For a perfect frame, the number of members (m) and the number of joints are correlated by the expression :

$$m = 2j - 3$$

The structure is termed imperfect or deficient if the number of members in it is less than that required for a perfect frame. The structure would be called redundant if it has members more than that required for a perfect frame.

6. The statically determinate frame are the frames whose force analysis can be made by applying the equations of statics only. Such frames normally have roller or hinged supports. Roller frames are frictionless and provide a reaction at right angles to the roller base. The line action of the reaction for a hinged support depends upon the load system on the frame.
7. The analysis of a frame involves the determination of reaction at the supports and internal stresses induced in the members due to external loads.
8. The stresses in the members of a framed structure are worked out by the following methods :
   (*i*) *Analytical method :* method of joints and method of sections
   (*ii*) *Graphical method*

REVIEW QUESTIONS

A. Conceptual and conventional questions :

1. What is a framed structure? List its applications.
2. Define and differentiate between
   (*i*) plane frame and space frame,
   (*ii*) statically determinate and statically indeterminate frame,
   (*iii*) perfect, deficient and redundant frame.
3. What distinguishes a roller support from a hinged support?
4. State the assumptions made while making an analysis of a framed structure.
5. Name the different methods of finding out the forces in the members of a perfect frame. Which one is used where and why?
6. Which one of the followings is a redundant frame and what is its degree of redundancy?

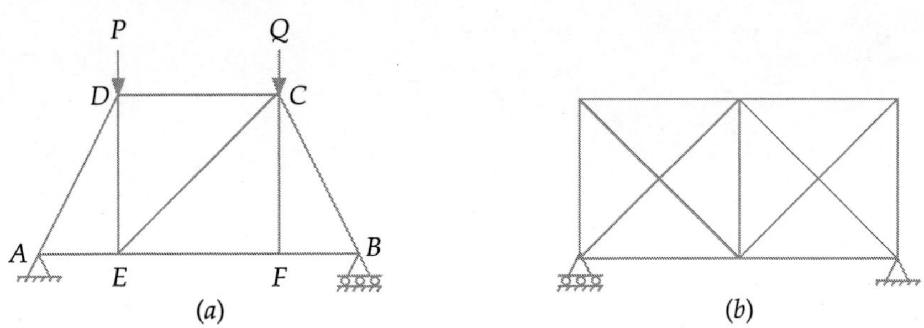

Fig. 5.101

7. Outline the steps involved while making an analysis of a simple truss by the (*i*) method of joints, and (*ii*) method of sections.
8. How will you find the forces in the members of truss joint by graphical method. Illustrate with an example.

9. With reference to the trusses shown in Fig. 5.102, identity the members which will not be subjected to any force.

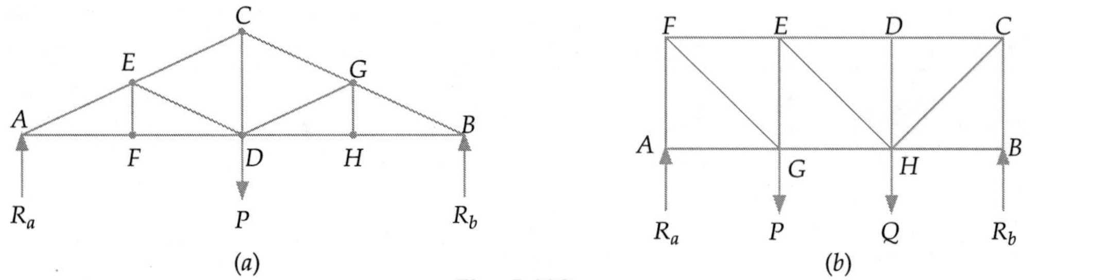

Fig. 5.102

10. Determine the magnitude and nature of forces in members of the truss loaded and supported as shown in Fig. 5.103.
11. Determine the magnitude and nature of forces in the various members of the truss loaded and supported as shown in Fig. 5.104.

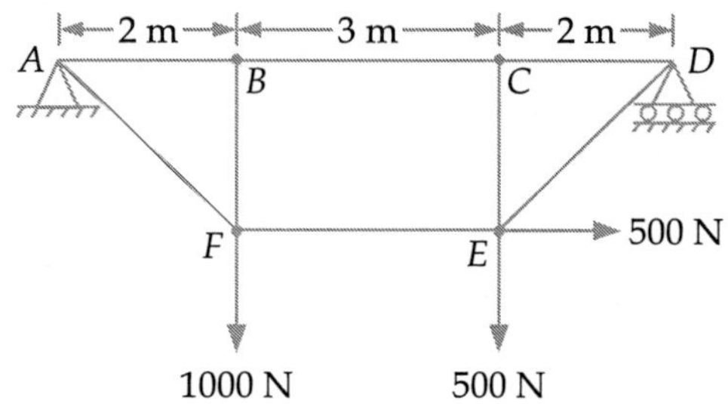

Fig. 5.103

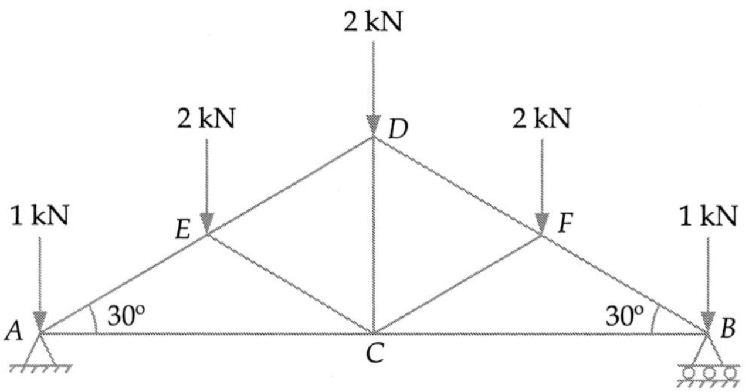

Fig. 5.104

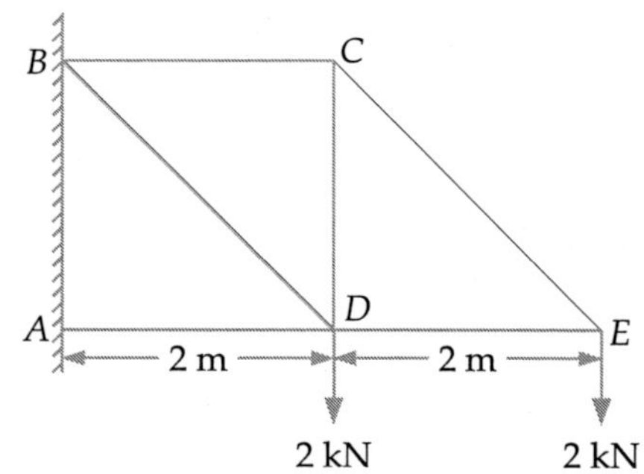

Fig. 5.105

12. Determine the forces in each member of the cantilever truss loaded as shown in Fig. 5.105.
13. Make calculations for the forces induced in various members of a truss which has been loaded and supported as shown in Fig. 5.106.

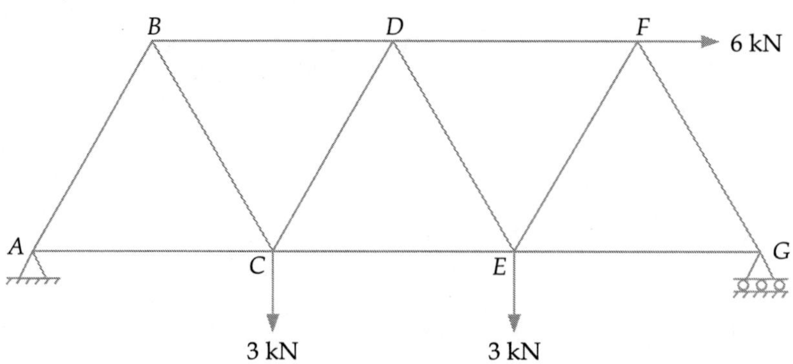

Fig. 5.106

14. Determine the forces in the members of a plane truss which is hinged to the foundation and is loaded as shown in Fig. 5.107.

(F_5 = F_5 = 0 ; F_4 = 2 kN (*T*) ; F_3 = 2.83 kN (*C*) ; F_2 = 2 kN (*C*) and F_1 = 2.83 (*T*))

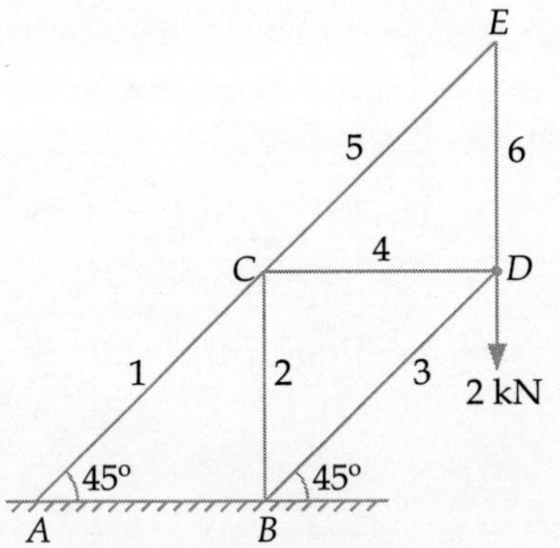

Fig. 5.107

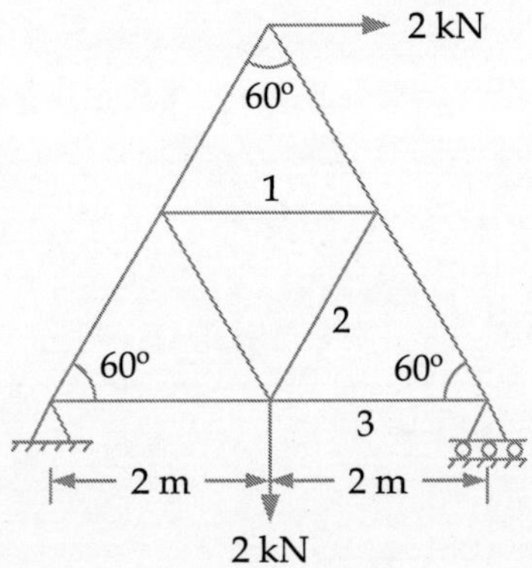

Fig. 5.108

15. A truss is loaded and supported as shown in Fig. 5.108. Make calculations for the axial forces set up in the members 1, 2 and 3.

(F_1 = 1 kN (*C*); F_2 = 1 kN (*T*); and F_3 = 1.58 kN (*T*))

B. Complete the following statements with most appropriate word/words :

1. Frame is a .... structure that is used to .... loads
2. A body that supports another body acted upon by a system of forces is called a ....
3. The frames are generally supported on a .... support or a .... support
4. In a roller support, the line of reaction is .... the plane of rollers.
5. A member of a framed structure under tension is called a ...., and under compression is called a ....
6. The bars of a framed structured are called two-force members because they are either in .... or in ....
7. The analysis of a truss involves the determination of .... at the supports and .... induced in the members due to external loads
8. The .... and .... are the analytical methods for finding out the stresses in the members of a framed structure
9. If only a single force is acting at a joint, then for the equilibrium of this joint, the force must be ....

Answers:

1. rigid, support 2. support 3. roller, hinged 4. normal to 5. tie, strut 6. tension, compression 7. reactions, internal stresses 8. method of joints, method of sections 9. zero

C. Multiple choice questions :

1. A framed structure is said to be perfect if the following correlation exists between the number of joints j and the number of members m

(*a*) $m = 2j - 3$ (*b*) $m = 3j - 3$
(*c*) $m = 2j - 1$ (*d*) $m = j - 2$

2. Which of the structures shown below represents a redundant frame?

(*a*)

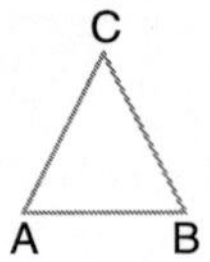

(*b*)

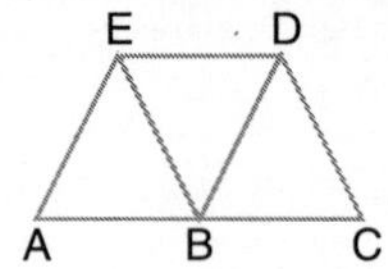

(*c*)

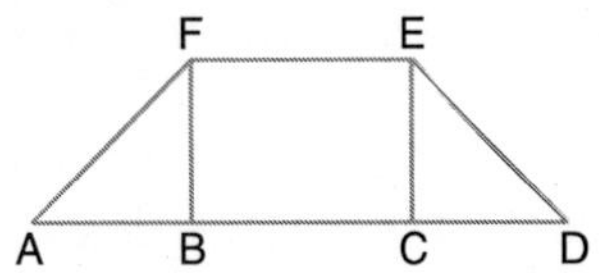

(*d*)

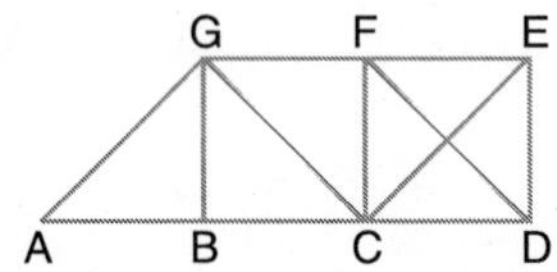

3. Consider the force/forces acting in the member/members meeting at the unloaded joint of a framed structure

1.

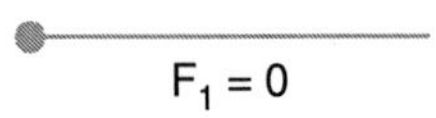

2.

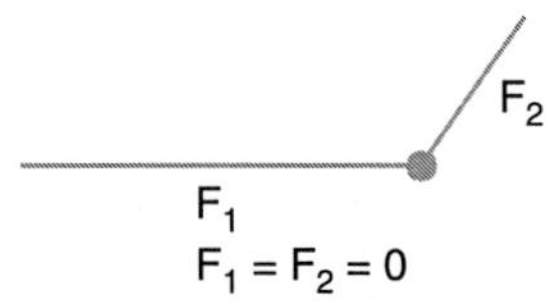

3.

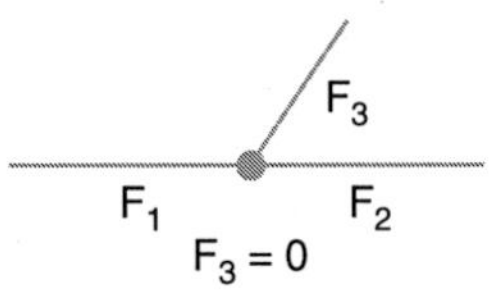

Which of these aspects is correct ?

(*a*) 1, 2 and 3
(*b*) 1 and 2
(*c*) 1 and 3
(*d*) 2 and 3

Answers :

1. (*a*) 2. (*d*) 3. (*a*)

❑❑❑

UNIT - III

Chapter-6

Centre Points: Centroid and Centre of Gravity

Chapter-7

Moment of Inertia: Area and Mass

CHAPTER

6

CENTRE POINTS: CENTROID AND CENTRE OF GRAVITY

Learning objectives : A study of the subject matter presented in this chapter will enable the reader to

- *define centre of gravity, centre of mass and centroidal axis ; differentiate between centre of gravity and centroid*
- *state the equations for locating the centroid/centre of gravity of a body*
- *determine the position of centroid/centre of gravity of different plane sections commonly used in engineering practice*
- *state and appreciate the utility of Pappus and Guldinus theorems*

A body is comprised of a number of particles. Each particle is attracted by force due to gravity of earth towards its centre. Since the body is negligible in comparison with the earth, all forces are considered to be parallel with each other. The resultant of all these forces acts at a point which is known as centre of gravity. There are many engineering situations where it is essential to know the position of centre of gravity of bodies. Attention has been focussed in this chapter to discuss the concept of centre of gravity and different aspects related to it.

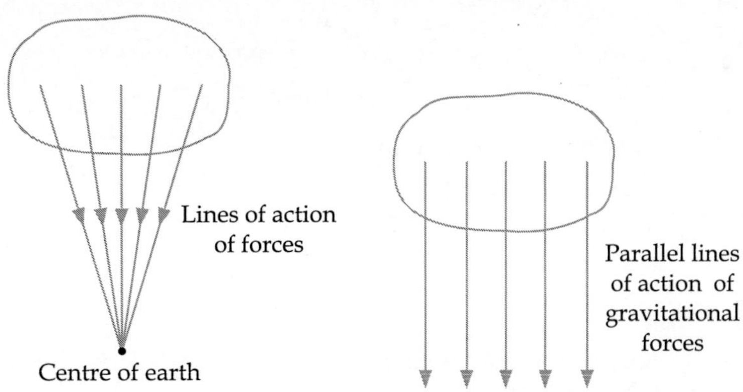

Fig. 6.1

6.1. CENTRE OF GRAVITY AND CENTROID

Centre of gravity of a body is defined as the point through which resultant of the gravitional force (weight) acts for any orientation of the body. The position of *CG* depends upon the shape of the body and this may or may not necessarily be within the boundary of the body. Further,

(*i*) A body has only one centre of gravity.

(*ii*) Its location does not change even with a change in the orientation of the solid body. Whatever may be the orientation of the body, there is no change in the position of mass particles relative to each other. Obviously then the resultant of the gravitational forces on the body acts through the same point.

(*iii*) It lies in a plane of symmetry, if any, of a body. The plane of symmetry of a body is a plane through the body such that the mass distribution is symmetrical on either side of the plane.

(*iv*) It is an imaginary point which may occur inside or outside the body.

The plane figures such as rectangle/parallelogram, triangle/polygon, circle and line etc. have only the length, area and volume and no mass or weight. The point where the entire length, area or volume is assumed to be concentrated is called the *centroid.* This point is called the centroid of line, centroid of area and centroid of volume as the case may be. The moment of area about any axis passing through the centroid is zero.

Further if value of gravity is uniform for each part and the solid is of uniform density throughout, then the centroid, centre of gravity and centre of mass are coincident.

The term centre of gravity applies to bodies with mass and weight, and the centroid applies to plane figures which have area only but no mass. When thickness, *i.e.*, mass of the body is not considered, the *CG* and centroid are synonymous and pass through the same point.

6.2. LOCATION OF CENTROID/CENTRE OF GRAVITY

The co-ordinates $(\overline{x}, \overline{y})$ of the centroid are determined from the fact that moments of all small areas dA comprising the area A about an axis must be equal to the moment of area A itself about the same axis. With reference to Fig. 6.2, moment of areas of all the strips about y-axis

$$= a_1x_1 + a_2x_2 + a_3x_3 + \ldots = \Sigma ax$$

moment of total area A about the y-axis

$$= A\overline{x}$$

Equating these moments,

$$A\overline{x} = \Sigma a x \;; \quad \overline{x} = \frac{\Sigma a x}{A} \qquad \ldots(6.1)$$

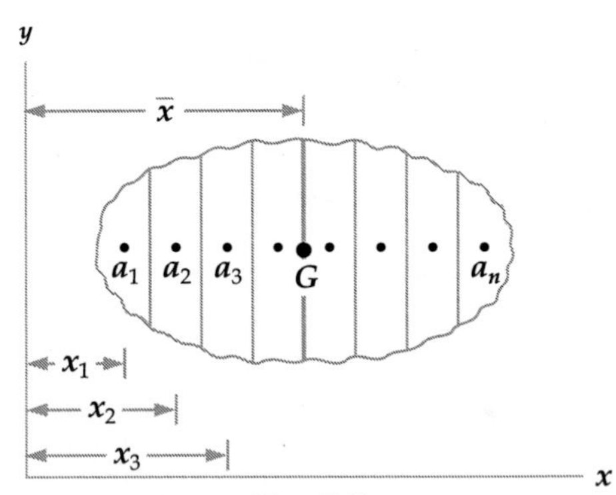

Fig. 6.2

where $\Sigma\, a\, x$ is the algebraic sum of the moments of areas about y-axis (considering areas acting parallel to y-axis) and A is the total area.

Similarly when the moments are taken about x-axis, we get

$$A\overline{y} = \Sigma\, a\, y \;; \quad \overline{y} = \frac{\Sigma\, a y}{A} \qquad \ldots(6.2)$$

where $\Sigma\, a\, y$ is the algebraic sum of the moments of areas about x-axis (considering areas acting parallel to x-axis) and A is the total area.

The location of *CG* of solids is accomplished on the same lines with the exception that moments are taken of masses, rather than of areas, of the individual parts of the solids. The *CG* will then have the co-ordinates

$$\bar{x} = \frac{\Sigma\, mx}{\Sigma\, m} \quad \text{and} \quad \bar{y} = \frac{\Sigma\, my}{\Sigma\, m} \qquad ...(6.3)$$

If all the sections are of the same material having density ρ, then

$$m_1 = \rho\, V_1;\quad m_2 = \rho\, V_2;\quad m_3 = \rho\, V_3 \quad \text{etc}$$

and $$V = V_1 + V_2 + V_3 + \ldots$$

The identity 6.3 may then be rewritten as

$$\bar{x} = \frac{\Sigma \rho V x}{\Sigma \rho V} = \frac{\Sigma V x}{\Sigma V}$$

$$\bar{y} = \frac{\Sigma \rho V y}{\Sigma \rho V} = \frac{\Sigma V y}{\Sigma V} \qquad ...(6.4)$$

These expressions show that if the sections/bodies are made of the same material and are of the same density through out, their centre of gravity is their centre of volume.

Further, if the segments/bodies are of same cross-section but of different lengths, them

$$V_1 = al_1;\quad V_2 = al_2;\quad V_3 = al_3 \text{ etc}$$

and $$V = al_1 + al_2 + al_3 + \ldots$$

By substitution of these results in expression 6.4, we obtain

$$\bar{x} = \frac{\Sigma alx}{\Sigma al} = \frac{\Sigma lx}{\Sigma l}$$

$$\bar{y} = \frac{\Sigma aly}{\Sigma al} = \frac{\Sigma ly}{\Sigma l} \qquad ...(6.5)$$

Apparently the centre of gravity of a wire, pipe or rod of constant cross-section corresponds to their centre of lengths.

It is worthwhile to note that the expressions 6.1 to 6.5 are equally valid when calculating the centroid/centre of gravity of composite laminas and solids. The area of a hole or any cut or punched section existing in the plane figure will be treated as negative. The moment of a negative area will also be negative provided the entire figure lies in the first quadrant.

Information is given below on the position of centroid/centre of gravity of different laminas (plane sections) most commonly used in engineering practice:

(*i*) For a rectangular section, *CG* lies at the intersection of diagonals, *i.e.*, at the mid point of its breadth as well as the depth. (Fig. 6.3 *a*).

(*ii*) For the triangular section, *CG* lies at the point of intersection of the three medians. If height of the triangle is h, then its *CG* is at a distance of $h/3$ from the base along vertical axis (Fig. 6.3 *b*).

(*iii*) For a circular section, *CG* lies at the centre of circle (Fig. 6.3 *c*). The *CG* of a semi-circle of radius R is at a distance of $\frac{4R}{3\pi}$ from its base measured along vertical axis (Fig. 6.3 *d*).

(*iv*) The *CG* of a sphere of radius R lies at a distance of R from every point. For the hemi-sphere, *CG* would be at a distance of $3R/8$ from its base measured along the vertical axis. (Fig. 6.3 *e*).

(*v*) The *CG* of a uniform rod is at its mid point (Fig.6.3 *f*).

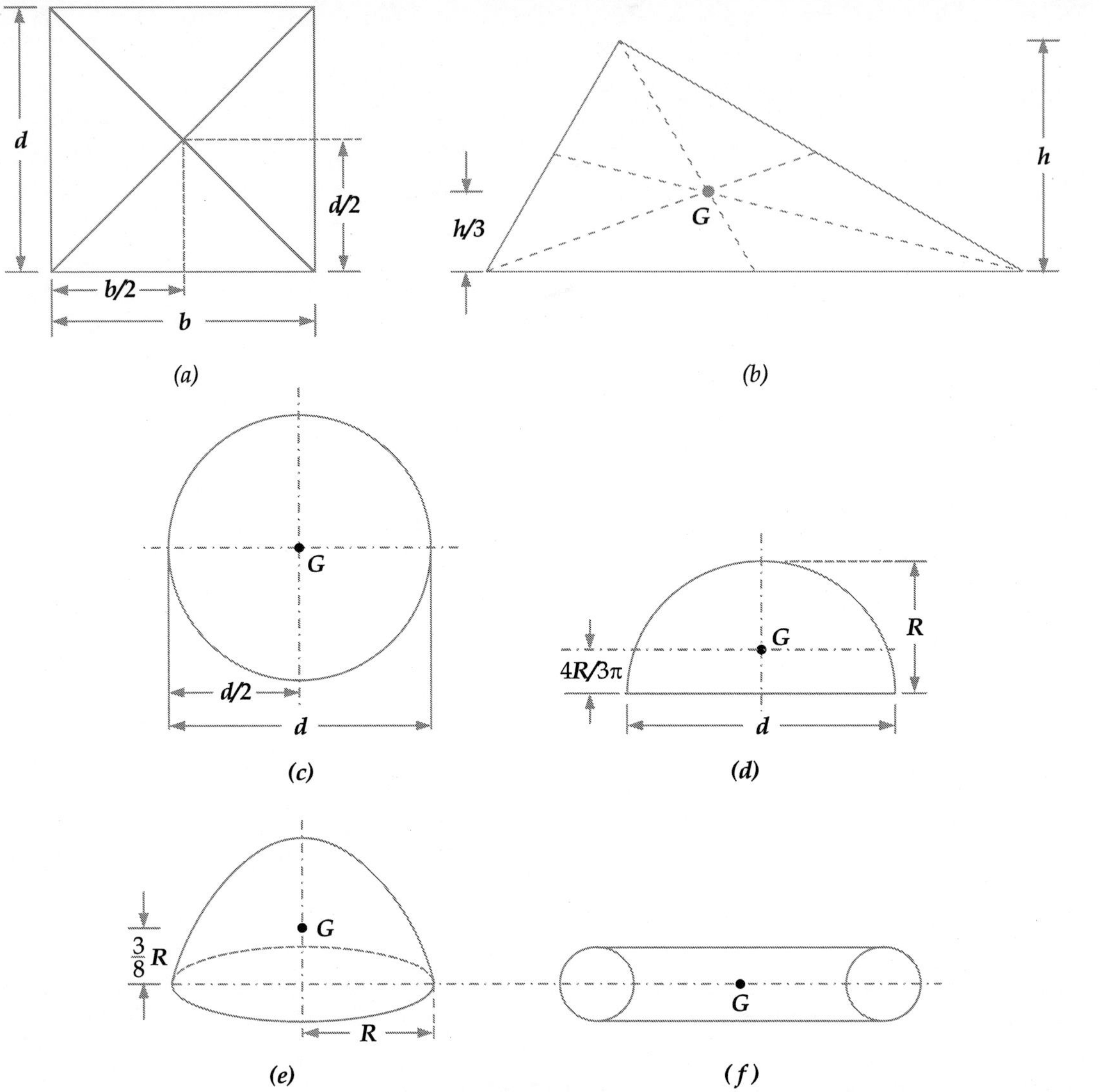

Fig. 6.3

EXAMPLE 6.1

Find the centroid of a uniform wire of length L

Solution : The centroid of a wire, pipe or rod of constant cross-section corresponds to their centre of lengths, and is given by

$$\overline{x} = \frac{\Sigma\, x\, dl}{\Sigma\, dl},\ \overline{y} = \frac{\Sigma\, y\, dl}{\Sigma\, dl}$$

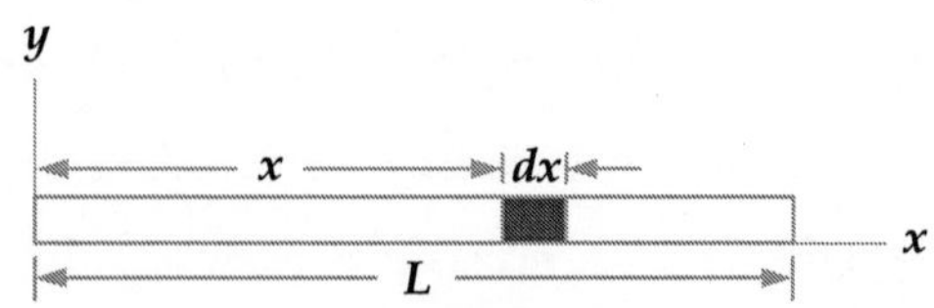

Fig. 6.4

When the x-axis is so chosen that it passes through the centre of the wire and along its length, $\overline{y} = 0$

Further, $$\Sigma\, x\, dl = \int_0^L x\, dx = \left|\frac{x^2}{2}\right|_0^L = \frac{L^2}{2}$$

$$\therefore \quad \Sigma\, dl = \int_0^L dx = L$$

$$\therefore \quad \bar{x} = \frac{L^2/2}{L} = \frac{L}{2}$$

EXAMPLE 6.2

Locate the centroid of the bent wire shown in Fig. 6.5

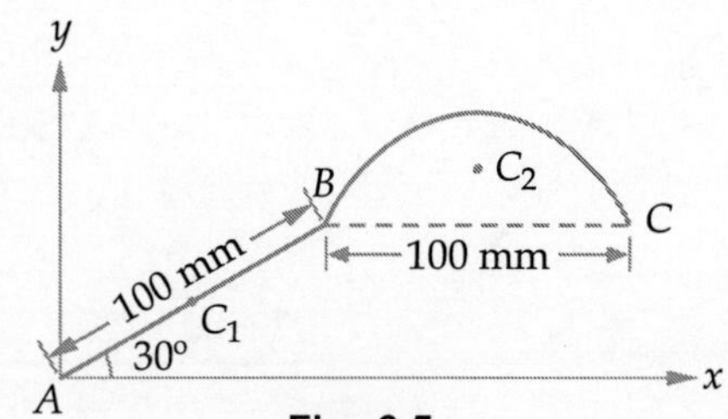

Fig. 6.5

Solution : The composite wire consists of two parts *AB* and *BC* :

(*i*) Length of portion $AB = l_1 = 100$ mm
Its centroid C_1 is at the centre of the line

$$x_1 = 50 \cos 30° = 43.30 \text{ mm}$$
$$y_1 = 50 \sin 30° = 25 \text{ mm}$$

(*ii*) Length of semi-circular portion *BC*,

$$l_2 = \pi r = \pi \times 50 = 157 \text{ mm}$$

The centroid C_2 of a semi-circular line is at $\frac{2r}{\pi} = \frac{2 \times 50}{\pi} = 31.83$ mm from its base line *BC*. Accordingly the co-ordinates of centroid C_2 of the semi-circular portion are :

$$x_2 = 100 \cos 30° + 50 = 136.6 \text{ mm}$$
$$y_2 = 100 \sin 30° + 31.83 = 81.83 \text{ mm}$$

Let the centroid of the bent wire be located at distance $\bar{x}$ from the *x*-axis and at distance $\bar{y}$ from the *y*-axis. Then

$$\bar{x} = \frac{l_1x_1 + l_2x_2}{l_1 + l_2} = \frac{100 \times 43.30 + 157 \times 136.6}{100 + 157} = \mathbf{100.3\ mm}$$

and

$$\bar{y} = \frac{l_1y_1 + l_2y_2}{l_1 + l_2} = \frac{100 \times 25 + 157 \times 81.83}{100 + 157} = \mathbf{59.72\ mm}$$

EXAMPLE 6.3

Locate the centroid of the area shown in Fig. 6.6. All the dimensions are in mm.

Solution : The given area can be considered as a rectangle of size 150 × 150 mm minus a semicircle of radius 50 mm minus a triangle of base 50 mm and height 100 mm.

The areas and coordinates of centroid of those segments with respect to the given axis are :

Rectangle : $a_1 = 150 \times 150 = 22500 \text{ mm}^2$

$$x_1 = \frac{150}{2} = 75 \text{ mm and}$$

$$y_1 = \frac{150}{2} = 75 \text{ mm}$$

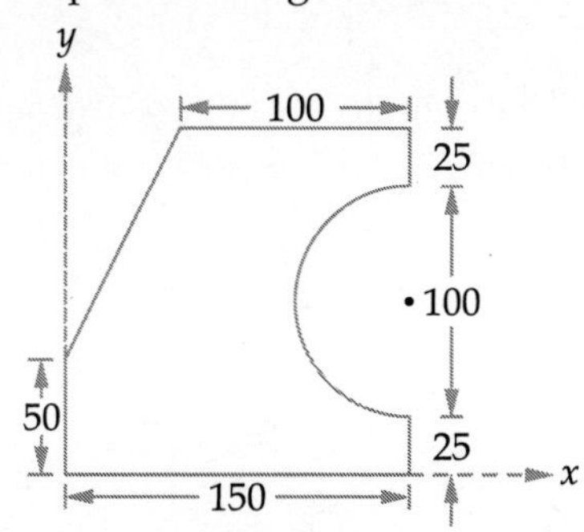

Fig. 6.6

Semicircle : $a_2 = \frac{\pi r^2}{2} = \frac{\pi \times 50^2}{2} = 3925 \text{ mm}^2$

$$x_2 = 150 - \frac{4r}{3\pi} = 150 - \frac{4 \times 50}{3\pi} = 128.77 \text{ m}$$

$$y_2 = 25 + r = 25 + 50 = 75 \text{ mm}$$

Triangle : $a_3 = \frac{1}{2} bh = \frac{1}{2} \times 50 \times 100 = 2500 \text{ mm}^2$

$$x_3 = \frac{b}{3} = \frac{50}{3} = 16.67 \text{ mm}$$

$$y_3 = 150 - \frac{h}{3} = 150 - \frac{100}{3} = 116.67 \text{ mm}$$

Then : $$\bar{x} = \frac{a_1x_1 - a_2x_2 - a_3x_3}{a_1 - a_2 - a_3}$$

$$= \frac{22500 \times 75 - 3925 \times 128.77 - 2500}{22500 - 3925 - 2500} \qquad = \mathbf{70.94 \text{ mm}}$$

and $$\vec{y} = \frac{a_1y_1 - a_2y_2 - a_3y_3}{a_1 - a_2 - a_3}$$

$$= \frac{22500 \times 75 - 3925 \times 75 - 2500 \times 1}{22500 - 3925 - 2500} \qquad = \mathbf{68.52 \text{ mm}}$$

EXAMPLE 6.4

Locate the centroid of the area shown in Fig. 6.7 with respect to the axes indicate in the figure.

Solution : The composite area has been divided into three segments namely a triangle, a rectangle and a semi-circle. The areas and the co-ordinates of centroid of these segments with respect to the given axes are:

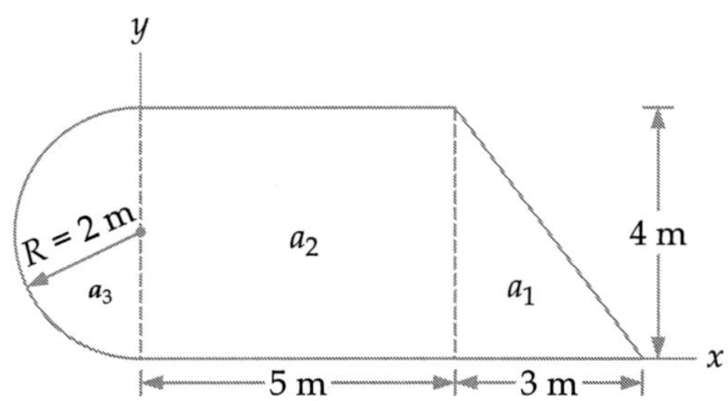

Fig. 6.7

Triangular segment:

$$a_1 = \frac{1}{2} \times 3 \times 4 = 6 \text{ m}^2$$

$$x_1 = 5 + \frac{1}{3} \times 3 = 6 \text{ m} \; ; \quad y_1 = \frac{4}{3} = 1.33 \text{ m}$$

Rectangular segment: $a_2 = 5 \times 4 = 20 \text{ m}^2$

$$x_2 = \frac{5}{2} = 2.5 \text{ m} \; ; \quad y_2 = \frac{4}{2} = 2 \text{ m}$$

Semi-circular segment: $a_3 = \frac{1}{2} \times \pi \times 2^2 = 6.28 \text{ m}^2$

$$x_3 = -\frac{4R}{3\pi} = -\frac{4 \times 2}{3 \times \pi} = -0.849 \text{ m}$$

$$y_3 = \frac{4}{2} = 2 \text{ m}$$

The negative with co-ordinate x_3 stems from the fact that this co-ordinate lies on the left of y-axis

Then: $$\bar{x} = \frac{a_1x_1 + a_2x_2 + a_3x_3}{a_1 + a_2 + a_3} = \frac{6 \times 6 + 20 \times 2.5 + 6.28(-0.849)}{6 + 20 + 6.28} = \mathbf{2.5 \text{ m}}$$

and $$\bar{y} = \frac{a_1y_1 + a_2y_2 + a_3y_3}{a_1 + a_2 + a_3} = \frac{6 \times 1.33 + 20 \times 2 + 6.28 \times 2}{6 + 20 + 6.28} = \mathbf{1.875 \text{ m}}$$

EXAMPLE 6.5

A triangular plate in the form of an isosceles triangle *ABC* has base *BC* = 10 cm and altitude = 12 cm. From this plate, a portion in the shape of an isosceles triangle *OBC* is removed. If *O* is the mid point of the altitude of triangle *ABC*, then determine the distance of *CG* of the remainder section from the base.

Solution : Refer Fig 6.8

For a triangle of height h, the CG lies on the axis at a distance $h/3$ from the base.

For triangle ABC,

$$\text{Area } A_1 = \frac{1}{2}\times 10\times 12 = 60 \text{ cm}^2$$

$$y_1 = 12/3 = 4 \text{ cm from base } BC$$

For triangle OBC,

$$A_2 = \frac{1}{2}\times 10\times 6 = 30 \text{ cm}^2$$

$$y_2 = 6/3 = 2 \text{ cm from base } BC.$$

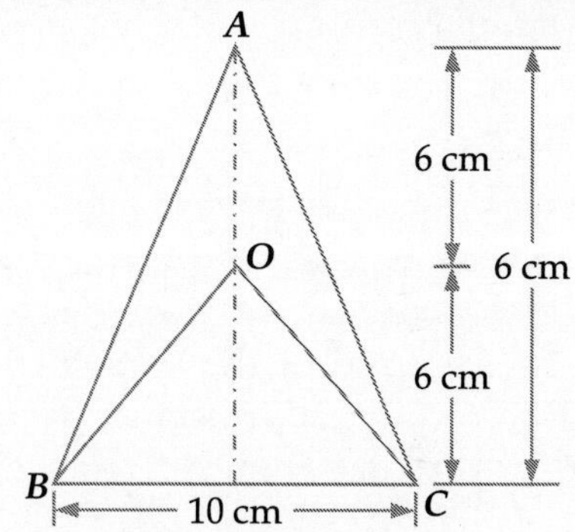

Fig. 6.8

Let y be the distance of CG of the section $ABOCA$ from the base line BC.

$$y = \frac{A_1y_1 + A_2y_2}{A_1 + A_2} = \frac{60\times 4 + (-30)\times 2}{60 + (-30)}$$

The negative sign with A_2 stems from the fact that this area has been removed.

$$y = \frac{240 - 60}{60 - 30} = \mathbf{6 \text{ cm}}$$

Alternatively : Taking the moments of different areas about the base line BC, we get

$$64 \times 4 = 30 \times 2 + (60 - 30)\, y \;; \quad y = 6 \text{ cm}$$

EXAMPLE 6.6

From a rectangular plate whose cross-section is 8 cm × 6 cm, a circular disc of 3 cm² in area is removed. If CG of the remainder is 1 mm from CG of the rectangular plate, work out the distance of the centre of disc from the centre of rectangular plate.

Solution : Refer to Fig 6.9.

Let G_1 and G_2 be the CG's of the rectangular plate and disc respectively . The centre of gravity G of the remainder lies on that side of G_1 which is opposite to G_2.

Taking moment of areas about G_1,

$$A_1 \times 0 = A_2 \times G_1G_2 + (A_1 - A_2)\, G_1G$$

$$(8 \times 6) \times 0 = 3 \times G_1 G_2 + (48 - 3) \times 0{\cdot}1$$

$$\therefore \quad G_1 G_2 = -\frac{45\times 0{\cdot}1}{3} = \mathbf{-\ 1{\cdot}5 \text{ cm}}$$

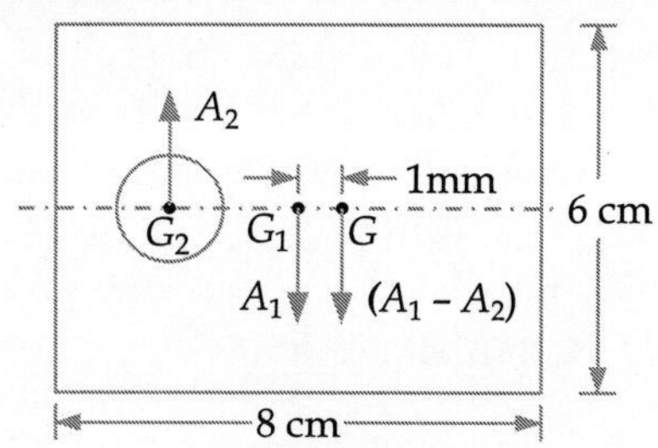

Fig. 6.9

The negative sign stems from the fact that G_2 is to the left of G_1.

EXAMPLE 6.7

A rectangular lamina $ABCD$ 20 cm × 25 cm has a rectangular hole of 5 cm × 6 cm as shown in Fig. 6.10. Locate the centroid of the section.

Solution : For the rectangular lamina $ABCD$,

$$A_1 = 25 \times 20 = 500 \text{ cm}^2$$

$$y_1 = \frac{25}{2} = 12.5 \text{ cm from } AB$$

$$x_1 = \frac{20}{2} = 10 \text{ cm from } AD$$

For the cut rectangular hole,

$$A_2 = 5 \times 6 = 30 \text{ cm}^2$$

$$y_2 = 3 + \frac{6}{2} = 6 \text{ cm from } AB$$

$$x_2 = 10 + 2 + \frac{5}{2} = 14.5 \text{ cm from } AD$$

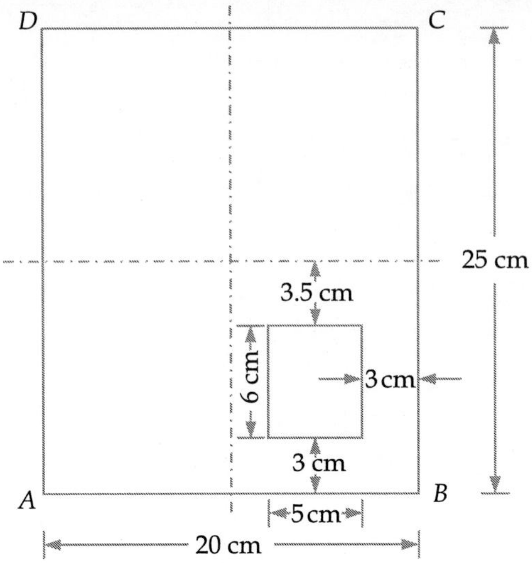

Fig. 6.10

Let centroid of the section be located at a point with co-ordinates $\bar{x}$ (distance from reference axis AD) and $\bar{y}$ (distance from reference axis AB). Then

$$\bar{x} = \frac{A_1 x_1 + A_2 x_2}{A_1 + A_2}$$

and $$\bar{y} = \frac{A_1 y_1 + A_2 y_2}{A_1 + A_2}$$

Considering area A_2 of the cut hole as negative, we get

$$\bar{x} = \frac{500 \times 10 - 30 \times 14{\cdot}5}{500 - 30} = \frac{5000 - 435}{470} = \mathbf{9.71 \text{ cm}}$$

$$\bar{y} = \frac{500 \times 12.5 - 30 \times 6}{500 - 30} = \frac{6250 - 180}{470} = \mathbf{12.91 \text{ cm}}$$

EXAMPLE 6.8

From a circular lamina of diameter *d*, a square hole has been punched out. If one diagonal of the square coincides with the radius of the circle, determine the distance of the centre of the remainder from the centre of the circle.

Solution : The section consists of two parts which have been numbered 1 and 2 as shown in Fig. 6.11. Since the remainder segment is symmetrical about x-axis, $\bar{y} = 0$. For finding out location of centroid, the areas of the components and their centroidal distances from line LMN are as given below:

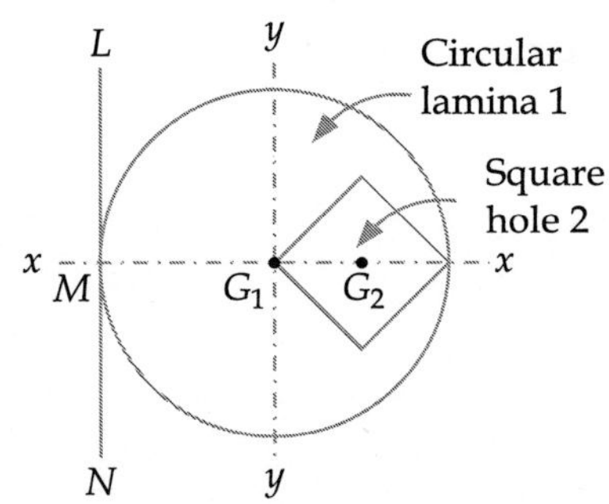

Fig. 6.11

1. Circular lamina

$$\text{area } A_1 = \frac{\pi}{4} d^2$$

centroidal distance $x_1 = MG_1 = d/2$

2. Square hole (removed)

length of diagonal $= d/2$

$$l^2 + l^2 = \frac{d^2}{4} \quad \text{where } l \text{ is the length of each side}$$

$$\text{area } A_2 = l^2 = \frac{d^2}{8} \quad \text{(it will be negative since it has been removed)}$$

$$\text{centroidal distance } x_2 = MG_2 = MG_1 + G_1G_2 = \frac{d}{2} + \frac{d}{4} = \frac{3d}{4}$$

If $\bar{x}$ is the location of centroid of the remainder as measured from LMN, then from moment principle

$$(A_1 - A_2)\bar{x} = A_1 x_1 - A_2 x_2$$

$$\left(\frac{\pi}{4}d^2 - \frac{d^2}{8}\right)\bar{x} = \frac{\pi}{4}d^2 \times \frac{d}{2} - \frac{d^2}{8} \times \frac{3d}{4}$$

or $$d^2\left(\frac{\pi}{4} - \frac{1}{8}\right)\bar{x} = \frac{\pi}{8}d^3 - \frac{3d^3}{32} = \frac{d^3}{8}\left(\pi - \frac{3}{4}\right)$$

$$\bar{x} = \frac{d}{8}\,\frac{4\pi - 3}{4} \times \frac{8}{2\pi - 1} = \frac{d}{4}\left(\frac{4\pi - 3}{2\pi - 1}\right)$$

When the centroidal distance is measured from the centre of circle, then

$$\bar{x} = \frac{d}{2} - \frac{d}{4}\left(\frac{4\pi - 3}{2\pi - 1}\right) = \frac{d}{4}\left[2 - \frac{4\pi - 3}{2\pi - 1}\right]$$

$$= \frac{d}{4}\left[\frac{4\pi - 2 - 4\pi + 3}{2\pi - 1}\right] = \frac{d}{4(2\pi - 1)}$$

EXAMPLE 6.9

Determine the centroid of the lamina shown in Fig. 6.12 given below:

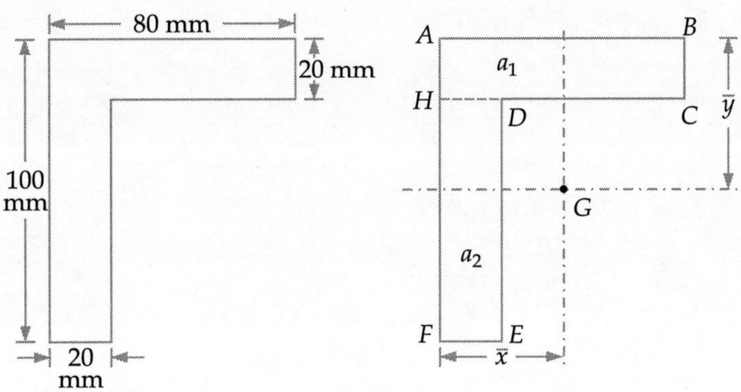

Fig. 6.12

Solution : Let it be presumed that *CG* is located of *G* with the co-ordinates $\bar{x}$ (distance from reference axis *AF*) and $\bar{y}$ (distance from reference axis *AB*). Then areas and co-ordinates (from the chosen reference axis *AF* and *AB*) of the centroids of the two rectangles comprising the given lamina are:

Rectangle ABCH $a_1 = 80 \times 20 = 1600 \text{ mm}^2$

$$x_1 = \frac{80}{2} = 40 \text{ mm}; \quad y_1 = \frac{20}{2} = 10 \text{ mm}$$

Rectangle DEFH $a_2 = 80 \times 20 = 1600 \text{ mm}^2$

$$x_2 = \frac{20}{2} = 10 \text{ mm}; \quad y_2 = 20 + \frac{80}{2} = 60 \text{ mm}$$

Then: $$\bar{x} = \frac{a_1 x_1 + a_1 x_2}{a_1 + a_2} = \frac{1600 \times 40 + 1600 \times 10}{1600 + 1600} = \mathbf{25 \text{ mm}}$$

$$\bar{y} = \frac{a_1 y_1 + a_2 y_2}{a_1 + a_2} = \frac{1600 \times 10 + 1600 \times 60}{1600 + 1600} = \mathbf{35 \text{ mm}}$$

Hence *CG* of the section is at a distance of 25 mm from face *AF* and 35 mm from face *AB*.

EXAMPLE 6.10

Locate the centroid of *T*-section as shown in Fig 6.13 given below:

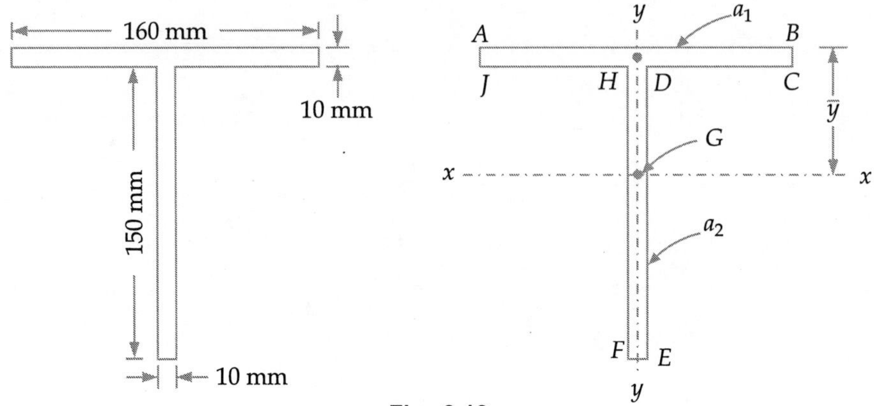

Fig. 6.13

Solution : Let the reference axes be chosen as shown in Fig.6.13. Due to symmetry of the section, its *CG* lies at point *G* on the *y*-axis, *i.e.*, $\bar{x} = 0$.

Further the top fibre *AB* has been chosen as the reference axis to locate the centroid $\bar{y}$.

The areas and co-ordinates of the centroid of the two rectangles comprising the given section are:

Rectangle ABCJ $a_1 = 160 \times 10 = 1600 \text{ mm}^2$

$$y_1 = \frac{10}{2} = 5 \text{ mm}$$

Rectangle DEFH $a_2 = 150 \times 10 = 1500 \text{ mm}^2$

$$y_2 = 10 + \frac{150}{2} = 85 \text{ mm}$$

Then:
$$\bar{y} = \frac{a_1 y_1 + a_2 y_2}{a_1 + a_2} = \frac{1600 \times 5 + 1500 \times 85}{1600 + 1500} = \mathbf{43.71 \text{ mm}}$$

Hence *CG* of the given *T*-section lies on they *y*-axis ($\bar{x} = 0$) and at 43.71 mm from the top face of its flange.

EXAMPLE 6.11

(*a*) Find the centroid of the truss Gusset plate shown in Fig. 6.14 given below:

(*b*) Find the centroid of a semicircular section having outer and inner radii 200 mm and 160 mm respectively.

Solution : The given area can be considered as rectangle *ABFE* minus triangle *CFD*. The areas and co-ordinates of these segments with respect to given axis are:

Rectangle: $a_1 = 1100 \times 1100$

$$= 121 \times 10^4 \text{ mm}^2$$

$$x_1 = \frac{1100}{2} = 550 \text{ mm}$$

$$y_1 = \frac{1100}{2} = 550 \text{ mm}$$

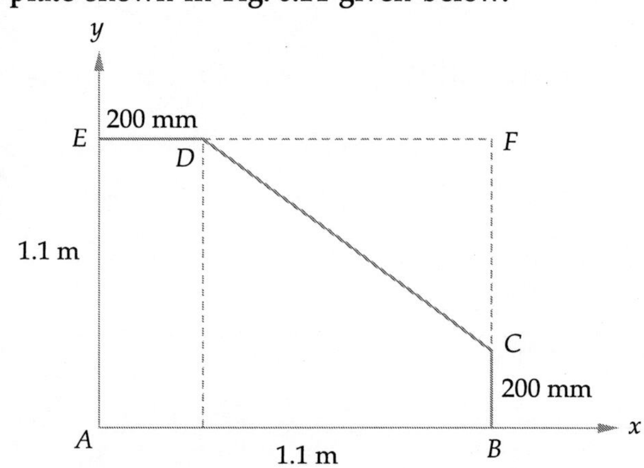

Fig. 6.14

Triangle: $a_2 = \frac{1}{2}\times(1100-200)\times(1100-200) = 40.5 \times 10^4 \text{ mm}^2$

$$x_2 = ED + \frac{2DF}{3} = 200 + \frac{2(900)}{3} = 800 \text{ mm}$$

$$y_2 = BC + \frac{2FC}{3} = 200 + \frac{2(900)}{3} = 800 \text{ mm}$$

Then,
$$\bar{x} = \frac{a_1x_1 - a_2x_2}{a_1 - a_2} = \frac{121\times10^4\times550 - 40.5\times10^4\times800}{121\times10^4 - 40.5\times10^4}$$
$$= \mathbf{424.22\ mm}$$

$$\bar{y} = \frac{a_1y_1 - a_2y_2}{a_1 - a_2} = \frac{121\times10^4\times550 - 40.5\times10^4\times800}{121\times10^4 - 40.5\times10^4}$$
$$= \mathbf{424.22\ mm}$$

(*b*) With reference to Fig. 6.15.

$r_1 = 200$ mm; $r_2 = 160$ mm

The areas and co-ordinates of the segments 1 and 2 with respect to axis x and y are:

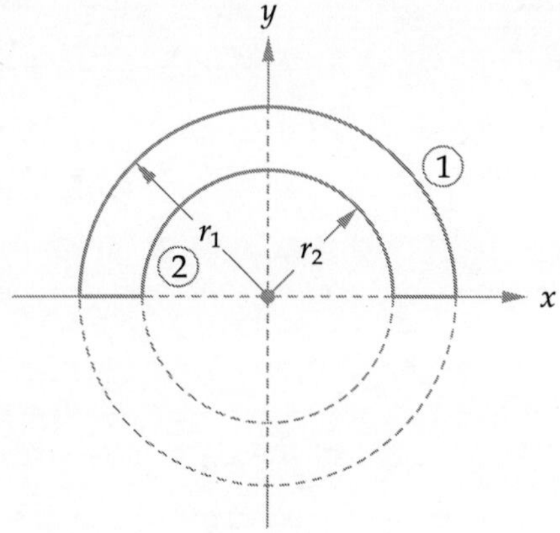

Fig. 6.15

$$a_1 = \frac{1}{2}\left(\pi r_1^2\right) = \frac{\pi}{2}\times200^2 = 20000\pi \text{ mm}^2$$

$$x_1 = 0 \text{ and } y_1 = \frac{4r_1}{3\pi} = \frac{4\times200}{3\pi} = 84.92 \text{ m}$$

$$a_2 = \frac{1}{2}\left(\pi r_2^2\right) = \frac{\pi}{2}160^2 = 12800\pi \text{ mm}^2$$

$$x_2 = 0 \text{ and } y_2 = \frac{4r_2}{3\pi} = \frac{4\times160}{3\pi} = 67.94 \text{ mm}$$

Then:
$$\bar{x} = \frac{a_1x_1 - a_2x_2}{a_1 - a_2} = \frac{20000\pi\times0 - 12800\pi\times0}{20000\pi - 12800\pi} = 0$$

$$\bar{y} = \frac{a_1y_1 - a_2y_2}{a_1 - a_2} = \frac{20000\pi\times84.92 - 12800\pi\times67.94}{20000\pi - 12800\pi}$$
$$= \mathbf{115.11\ mm}$$

EXAMPLE 6.12

(*a*) Find the centroid of the airplane's vertical stabilizer shown in Fig. 6.16.

(*b*) Would you imagine that the moment of inertia of the earth around its own axis is negligible fraction of its moment of inertia about the axis of rotation around the sun? Take mean radius of the earth as 6371 km and the mean radius of rotation around the sun as 149.7 × 10⁶ km.

Solution : The given area can be considered as a rectangle *AECF* minus triangles *BEC* and *ADF*. The areas and coordinates of these segments with respect to given axis are:

Rectangle AECF: $a_1 = 6 \times 5 = 30 \text{ m}^2$

$$x_1 = \frac{AE}{2} = \frac{5}{2} = 2.5 \text{ m}$$

$$y_1 = \frac{EC}{2} = \frac{6}{2} = 3 \text{ m}$$

Triangle BEC:

$$a_2 = \frac{1}{2} \times 1 \times 6 = 3 \text{ m}^2$$

$$x_2 = AB + \frac{2BE}{3}$$

$$= 4 + \frac{2 \times 1}{3} = 4.66 \text{ m}$$

$$y_2 = \frac{1}{3} EC = \frac{1}{3} \times 6 = 2 \text{ m}$$

Triangle ADF:

$$a_3 = \frac{1}{2} \times 3 \times 6 = 9 \text{ m}^2$$

$$x_3 = \frac{FD}{3} = \frac{3}{3} = 1 \text{ m}$$

$$y_3 = \frac{2AF}{3} = \frac{2}{3} \times 6 = 4 \text{ m}$$

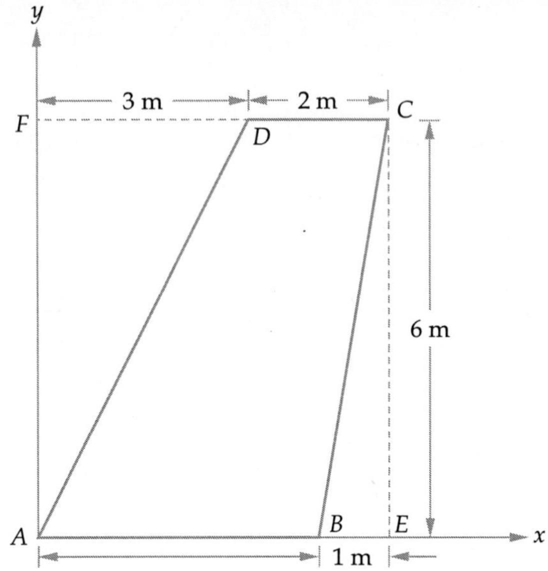

Fig. 6.16

Then:

$$\bar{x} = \frac{a_1 x_1 - a_2 x_2 - a_3 x_3}{a_1 - a_2 - a_3}$$

$$= \frac{30 \times 2.5 - 3 \times 4.66 - 9 \times 1}{30 - 3 - 9} = \mathbf{2.89 \text{ m}}$$

and

$$\bar{y} = \frac{a_1 y_1 - a_2 y_2 - a_3 y_3}{a_1 - a_2 - a_3} = \frac{30 \times 3 - 3 \times 2 - 9 \times 4}{30 - 3 - 9} = \mathbf{2.67 \text{ m}}$$

(*b*) Moment of inertia of the earth about its axis,

$$I_1 = \frac{2}{5} MR^2 = \frac{2}{5} M(6371)^2 = 16.23 \times 10^6 M$$

Moment of inertia of the earth about the axis of rotation around the sun,

$$I_2 = I_1 + Md^2 = 16.23 \times 10^6 M + M \times (149.7 \times 10^6)^2$$

$$= 16.23 \times 10^6 M + 22410.09 \times 10^{12} M$$

$$= 22410.09002 \times 10^{12} M$$

$$\text{Ratio } \frac{I_1}{I_2} = \frac{16.23 \times 10^6}{22410.09002 \times 10^{12}} = 7.24 \times 10^{-10}$$

Since the ratio is negligible, the moment of inertia of the earth around its oun axis can be imagined to be a negligible fraction of its moment of inertia about the axis of rotation around the sun.

EXAMPLE 6.13

Locate the position of the centroid of the plane shaded area depicted in Fig. 6.17 given below:

Solution : Let the reference axes be chosen as shown in Fig. 6.17. Due to symmetry of the section about y-axis, centroid lies on y-axis and hence $\bar{x} = 0$.

The areas and co-ordinates of the centroid of the different sections comprising the given section are:

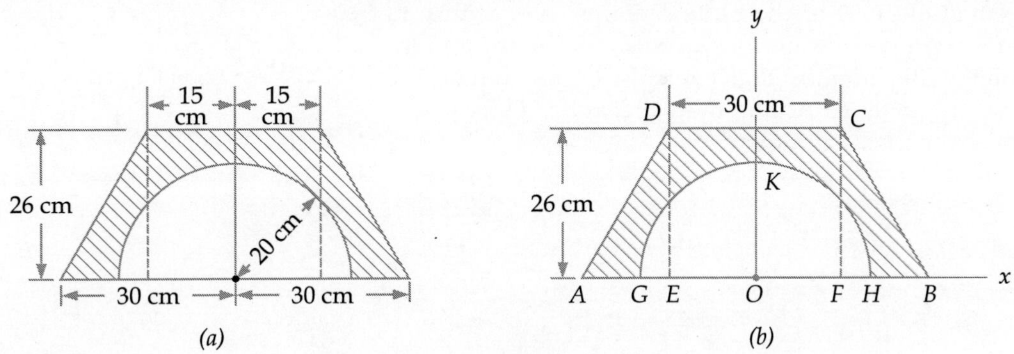

Fig. 6.17

Triangle ADE : $A_1 = \frac{1}{2} AE \times ED = \frac{1}{2} \times (30 - 15) \times 26 = 195 \text{ cm}^2$

$$y_1 = \frac{ED}{3} = \frac{26}{3} = 8.67 \text{ cm}$$

Rectangle EDCF : $A_2 = DC \times ED = 30 \times 26 = 780 \text{ m}^2$

$$y_2 = \frac{ED}{2} = \frac{26}{3} = 13 \text{ cm}$$

Triangle FCB : $A_3 = \frac{1}{2} FB \times FC = \frac{1}{2}(30 - 15) \times 26 = 195 \text{ cm}^2$

$$y_3 = \frac{FC}{3} = \frac{26}{3} = 8.67 \text{ cm}$$

Semi circle GKH : $A_4 = \frac{1}{2}\pi r^2 = \frac{1}{2}\pi \times (20)^2 = 628 \text{ cm}^2$

$$y_4 = \frac{4r}{3\pi} = \frac{4 \times 20}{3\pi} = 8.49 \text{ cm}$$

Let the centroid be located at distance $\overline{y}$ from the x-axis. Then

$$\overline{y} = \frac{A_1y_1 + A_2y_2 + A_3y_3 + A_4y_4}{A_1 + A_2 + A_3 + A_4}$$

$$= \frac{(195 \times 8 \cdot 67) + (780 \times 13) + (195 \times}{195 + 780 + 195}$$

The –ve sign with A_4 stems from the fact that this area has been included though it is non-existent

$$\overline{y} = \frac{1690 \cdot 65 + 10140 \cdot 00 + 169.65 - 53}{542} = \mathbf{15.11 \text{ cm}}$$

EXAMPLE 6.14

An isosceles triangle is to be cut from one edge of a square plate of side 1 m such that the remaining part of the plate remains in equilibrium in any position when suspended from the apex of the triangle. Calculate the area of the triangle to be removed.

Solution : Refer Fig. 6.18. It is to be noted that

(*i*) the section consists of two parts; square plate (1) and the triangular segment (2) which has been removed.

(*ii*) the remaining part of the plate remains in equilibrium when suspended from the apex E of the triangular segment. This implies that centroid of the remaining part of the plate coincides with the apex. Thus if h is height of the triangle, then

$$\bar{y} = h \quad \text{with respect to base } AB.$$

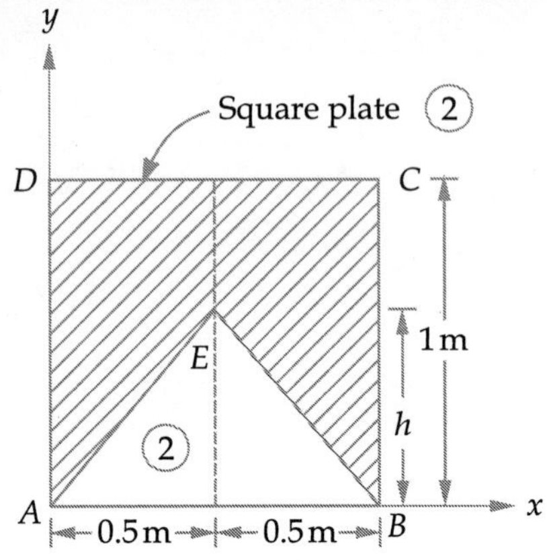

Fig. 6.18

Now, $A_1 = 1 \times 1 = 1 \text{ m}^2; \quad y_1 = \dfrac{1}{2} = 0.5 \text{ m}$

$$A_2 = \frac{1}{2}bh = \frac{h}{2}; \qquad y_1 = \frac{h}{3}$$

Then $$\bar{y} = \frac{A_1 y_1 - A_2 y_2}{A_1 - A_2} = \frac{1 \times 0.5 - h/2 \times h/3}{1 - h/2}$$

$$= \frac{0.5 - 0.167\,h^2}{1 - 0.5h}$$

$\therefore$ $$h = \frac{0.5 - 0.167\,h^2}{1 - 0.5h} \qquad (\because \ \bar{y} = h)$$

Upon simplication: $0.3\,h^2 - h + 0.5 = 0$

Solution of this quadratic equation gives

$$h = \frac{1 \pm \sqrt{1^2 - 4(0.5 \times 0.333)}}{2 \times 0.333} = \frac{1 \pm 0.578}{0.666}$$

$$= 2.37 \text{ m} \quad \text{or} \quad 0.63.4 \text{ m}$$

The value h = 2.37 m is not possible. As such the height of the triangle to be removed is 0.634 m

$\therefore$ Area of triangle to be removed $= \dfrac{1}{2} \times 1 \times 0.634 = \mathbf{0.317 \ m^2}$

EXAMPLE 6.15

Determine the coordinates of the centroid of the lamina as shown in Fig. 6.19. The shaded area is the opening in the lamina.

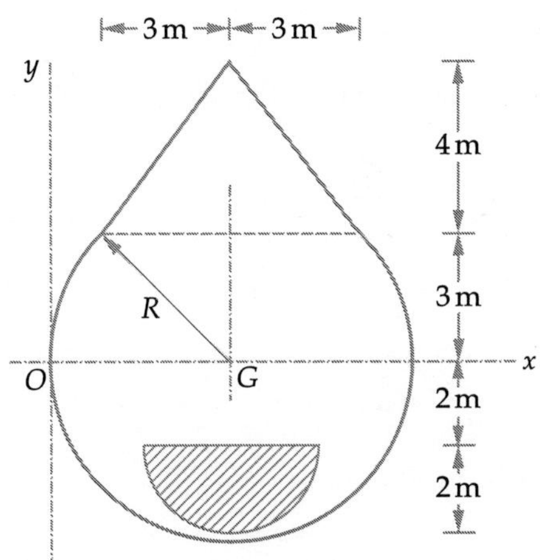

Fig. 6.19

Solution : The given lamina can be split up into the following basic figures as shown in Fig. 6.20

(*i*) Circle of radius R

$$R = BG = \sqrt{(BC)^2 + (CG)^2} = \sqrt{3^2 + 3^2} = 4.243 \text{ m}$$

$$A_1 = \pi R^2 = \pi \times (4.243)^2 = 56.53 \text{ m}^2$$

$$y_1 = 0$$

(*ii*) Triangle BDG

$$A_2 = \frac{1}{2}bh = \frac{1}{2} \times 6 \times 3 = 9 \text{ m}^2$$

$$y_2 = \frac{2}{3}h = \frac{2}{3} \times 3 = 2 \text{ m}$$

(*iii*) Triangle BDE

$$A_3 = \frac{1}{2} \times 6 \times 4 = 12 \text{ m}^2$$

$$y_3 = 3 + \frac{1}{3} \times 4 = 4.33 \text{ m}$$

(*iv*) Semi circular opening (minus)

$$A_4 = \frac{1}{2} \times (\pi \times 2^2) = 6.28 \text{ m}^2$$

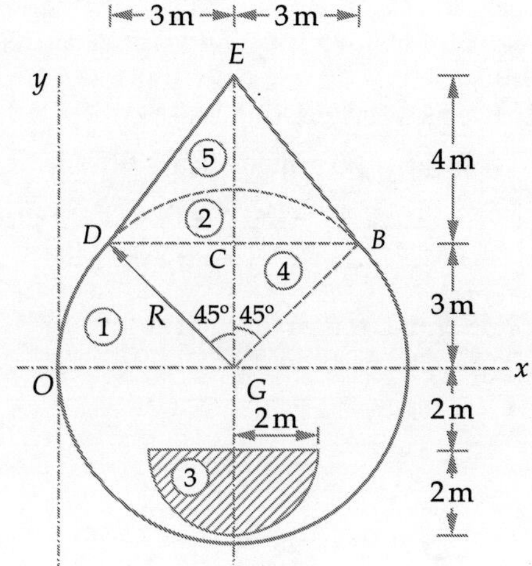

Fig. 6.20

For a semicircular segment, the centroid is at a distance of $4R/3\pi$ from base.

$$\therefore \qquad y_4 = -\left[2 + \frac{4R}{3\pi}\right] = -\left[2 + \frac{4 \times 2}{3\pi}\right] = -2.85 \text{ m}$$

It is to be noted that y_4 is negative because it is below x-axis.

(*v*) Sector BGD (minus)

$$A_5 = \frac{\pi}{4}R^2 = \frac{\pi}{4} \times (4.243)^2 = 14.14 \text{ m}^2$$

For a sector, the centroid is at a distance of $\frac{2}{3}\frac{R\sin\alpha}{\alpha}$ from the apex point.

$$\alpha = \text{semi vestex angle in radians} = 45 \times \frac{\pi}{180} = 0.785$$

$$\therefore \qquad y_5 = \frac{2}{3} \times \frac{4.243 \sin 45°}{0.785} = 2.55 \text{ m}$$

Thus for the given lamina,

$$\bar{y} = \frac{A_1y_1 + A_2y_2 + A_3y_3 - A_4y_4 - A}{A_1 + A_2 + A_3 - A_4 - A_5}$$

$$= \frac{56.53 \times 0 + 9 \times 2 + 12 + 4.33 - 6.}{56.53 + 9 + 12 - 6}$$

$$= \frac{+18 + 51.96 + 17.9 - 36.06}{57.11} = \frac{51.8}{57.11} = \mathbf{0.907 \text{ m}}$$

Due to symmetry of the given lamina $\bar{x} = R =$ **4.243 m.** Hence coordinate of centroid of the lamina are (4.243 m, 0.907 m)

EXAMPLE 6.16

An area in the shape of a semi-circle of 10 cm radius has been punched from a square plate of 50 cm × 50 cm in size as shown in Fig. 6.21. If the centroid of the remainder piece is to be kept the same as for the original piece, make calculations for the width *w* of the area, shown hatched, that needs to be removed.

Solution : The section consists of three parts which have been numbered 1, 2 and 3 as shown in the figure. The areas of these components and their centroidal distances from x-axis and y-axis are as worked out below:

1. Square 50 cm × 50 cm in size

   area $A_1 = 50 \times 50 = 2500 \text{ cm}^2$

   centroid is at (25, 25)

2. Semi-circular area (removed)

   $$\text{area } A_2 = -\frac{\pi}{2} \times (10)^2 = -157 \text{ cm}^2$$

   $$\text{centroid is at } \left(\frac{4 \times 10}{3\pi}, 25\right)$$

3. Rectangular area (removed)

   area $A_3 = -50 \times w = -50\, w \text{ cm}^2$

   $$\text{centroid is at } \left(50 - \frac{w}{2}, 25\right)$$

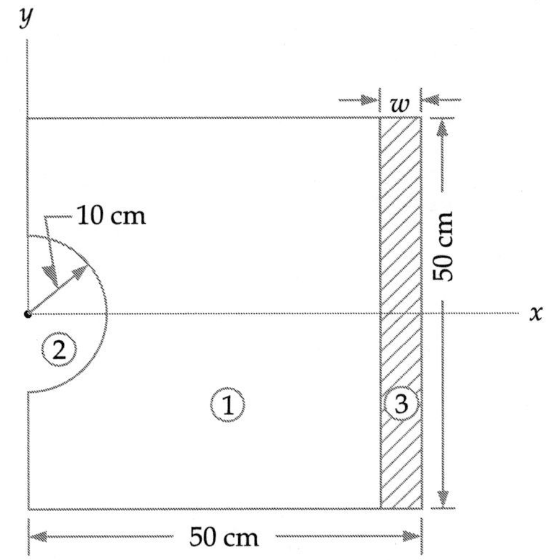

Fig. 6.21

Considering the areas parallel to y-axis, we can write the following identity by applying the principle of moments.

$$A\bar{x} = A_1x_1 + A_2x_2 + A_3x_3$$

Since the centroid of the remainder piece is to be the same as for the original square plate, $\bar{x}$ = 25 cm. Then

$$(2500 - 157 - 50\, w) \times 25 = 2500 \times 25 - 157 \times \frac{4 \times 10}{3\pi} - 50w\left(50 - \frac{w}{2}\right)$$

or $\quad 58575 - 1250\, w = 62500 - 665.67 - 2500\, w + 25\, w^2$

or $\quad 25w^2 - 1250\, w + 3258.33 = 0$

or $\quad w^2 - 50\, w + 130.33 = 0$

Solution of this quadratic equation gives

$$w = \frac{50 \pm \sqrt{50^2 - 4 \times 130.33}}{2} = \frac{50 \pm 44.48}{2} = 47.24 \text{ cm and } 2.76 \text{ cm}$$

w = 47.24 cm is not possible

$\therefore \quad w$ = **2.76 cm**

EXAMPLE 6.17

The frustrum of a right circular cone has a bottom radius 5 cm, top radius 3 cm and height 8 cm. A co-axial cylindrical hole of 4 cm diameter is made throughout the frustrum. Locate the position of centre of gravity of the remaining solid.

Solution : Let the sides *AD* and *BC* be extended to meet at *E* and form a full solid cone. The given solid can then be considered as the solid cone *EAB* minus the solid cone *EDC* above the frustrum minus the axial hole in the frustrum portion.

Due to symmetry of the section, the *CG* lies at a point on the *y*-axis i.e. $\bar{x} = 0$. Further, let the bottom surface *AB* be chosen as the reference axis to locate the centroid $\bar{y}$.

The volumes and the co-ordinates of the centroid of the different segments comprising the given section are :

Cone *EAB* : Since the triangles *EAB* and *EDC* are equiangular

$$\frac{5}{3} = \frac{8+h}{h} \quad \text{where } h \text{ is the height of the cone } EDC$$

$$5h = 24 + 3h \; ; \quad h = 12 \text{ cm}$$

Height of cone *EAB* = 8 + 12 = 20 cm

$$V_1 = \frac{1}{3}\pi \times (5)^2 \times 20 = 523.3 \text{ cm}^3; \quad y_1 = \frac{20}{4} = 5 \text{ cm}$$

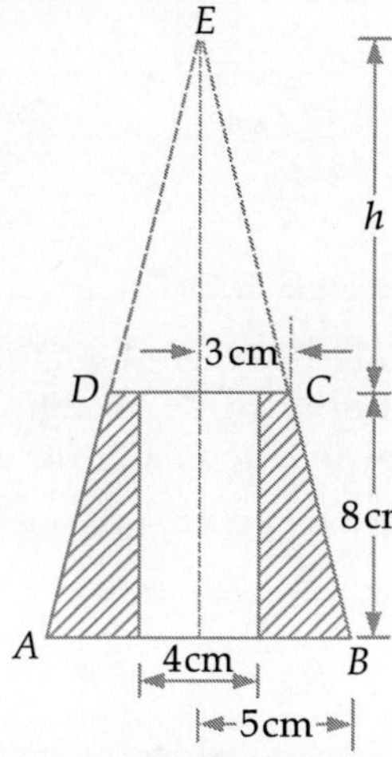

Fig. 6.22

Cone EDC : $h = 12$ cm

$$V_2 = \frac{1}{3}\pi \times (3)^2 \times 12 = 113 \text{ cm}^3; \quad y_2 = 8 + \frac{12}{4} = 11 \text{ cm}$$

Cylindrical hole : $V_3 = \frac{\pi}{4} \times (4)^2 \times 8 = 100.48 \text{ cm}^3; \quad y_3 = \frac{8}{2} = 4 \text{ cm}$

Let the centroid of the given solid be located at distance $\bar{y}$ from the *x*-axis. Then

$$\bar{y} = \frac{V_1 y_1 - V_2 y_2 - V_3 y_3}{V_1 - V_2 - V_3} = \frac{523.3 \times 5 - 113 \times 11 - 100.48 \times 4}{523.3 - 113 - 100.48} = \mathbf{3.139 \text{ cm}}$$

Hence *CG* of the given section lies on the *y*-axis ($\bar{x} = 0$) and at distance 3.139 cm from the bottom face *AB*.

EXAMPLE 6.18

A body consists of a solid hemisphere of radius 4 cm and a right circular solid cone of height 12 cm. The hemisphere and cone have a common base and are made of the same material. Locate the position of *CG* of the composite body.

Solution : Refer Fig. 6.23 for the composite body. Mass of body equals the product of its density and volume. Since both the sections are made of the same material, calculations can be made on the basis of volume.

For the solid cone *ABC*

$$\text{Volume } V_1 = \frac{1}{3}\pi r^2 h = \frac{1}{3}\pi \times 4^2 \times 12 = 200{\cdot}96 \text{ cm}^3$$

Distance of *CG* of cone from its apex *A*,

$$y_1 = 12 - \frac{12}{4} = 9 \text{ cm}$$

The distance y_1 has been so calculated because for a solid cone of height *h*, the *CG* lies on the axis at distance *h*/4 from the base.

For the hemisphere *BCD*

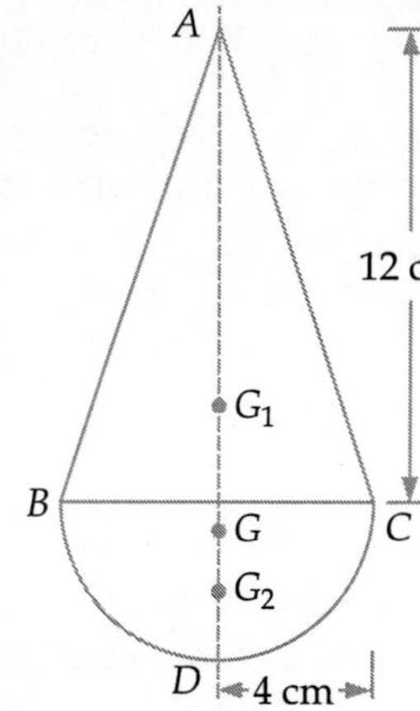

Fig. 6.23

$$\text{Volume } V_2 = \frac{1}{2} \times \frac{4}{3}\pi r^3 = \frac{1}{2} \times \frac{4}{3}\pi \times 4^3 = 133{\cdot}97 \text{ cm}^3$$

Distance of *CG* of hemisphere from point *A*,

$$y_2 = 12 + \frac{3}{8}r = 12 + \frac{3}{8} \times 4 = 13.5 \text{ cm}$$

Let point *G* represent the *CG* of the composite body with distance $AG = \bar{y}$

Then $$\bar{y} = \frac{V_1 y_1 + V_2 y_2}{V_1 + V_2}$$

$$= \frac{(200{\cdot}96 \times 9) + (133{\cdot}97 \times 13{\cdot}5)}{200{\cdot}96 + 133{\cdot}97} = 10{\cdot}8 \text{ cm for vertex } A$$

EXAMPLE 6.19

A hemisphere and a cone have their bases joined together; the two bases being of same size. What should be the greatest height of the cone in terms of radius of the base so that the combined body of the cone and hemisphere may stand upright?

Solution : With reference to Fig. 6.24

Volume of cone

$$V_1 = \frac{1}{3}\pi r^2 h$$

Position of its centroid from point C,

$$CG_1 = \frac{1}{4}h$$

Volume of hemisphere $V_2 = \frac{2}{3}\pi r^3$

Position of its centroid from point *C*,

$$CG_2 = -\frac{3}{8}r \quad (-\text{ ve because it lies below point } C)$$

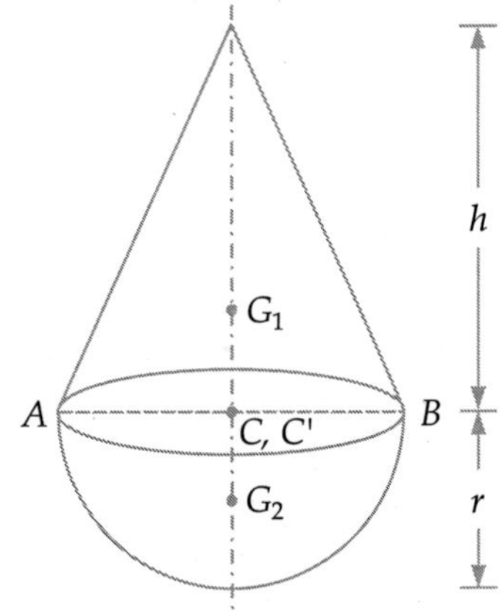

Fig. 6.24

Then from the principle of moments,

$$(V_1 + V_2)\,CC' = V_1 \times CG_1 + V_2 \times CG_2$$

$$= \frac{1}{3}\pi r^2 h \times \frac{1}{4}h + \frac{2}{3}\pi r^3\left(-\frac{3}{8}r\right) = \frac{1}{12}\pi r^2 h^2 - \frac{1}{4}\pi r^4$$

The composite body will stand upright if the position of its centre of gravity coincides with the centre of common base, i.e., $CC' = 0$

That gives

$$0 = \frac{1}{12}\pi r^2 h^2 - \frac{1}{4}\pi r^4$$

or $h^2 = 3\,r^2$ $\quad\therefore\quad$ $h = \mathbf{1.732}\ \boldsymbol{r}$

EXAMPLE 6.20

The distance between *CG* of two masses m_1 and m_2 is *l*. Work out the distance of *CG* of the composite system from mass m_1.

Solution : Refer to Fig 6.25.

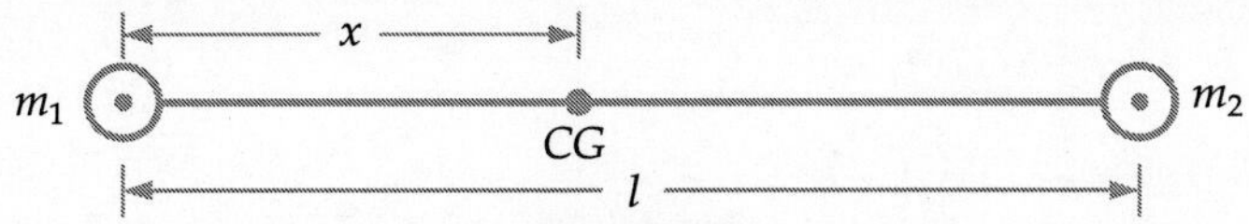

Fig. 6.25

Let x be the distance of CG of the composite system from mass m_1

Taking moments of masses m_1, m_2 and $(m_1 + m_2)$ about the centre of gravity of mass m_1,

$$m_2 l = (m_1 + m_2)\, x$$

$$\therefore \quad x = \frac{m_2}{m_1 + m_2} l$$

EXAMPLE 6.21

A rod of length 5 m is bent so as to form five sides of a regular hexagon. Make calculations for the distance of its centre of gravity from either end of the rod.

Solution : Reference Fig 6.26, $ABCDEF$ is the rod in which

$$AB = BC = CD = DE = EF = 1 \text{ m}$$

If W is the weight of each portion, then total weight of the rod is 5 W and this acts at point G which represents the centre of gravity of the rod. The distance AG is to be determined.

Let a segment FA of 1 m length and weight W be introduced and the regular hexagon be completed. For this hexagon, the CG will be at point O.

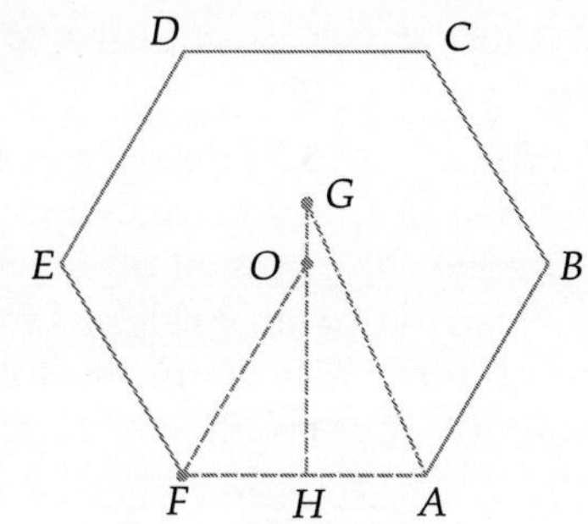

Fig. 6.26

Taking moments of the weights about line FA, we get

$$6W \times OH - W \times 0 = 5W \times GH$$

From the geometrical configuration,

$$OH = 1 \sin 60^\circ = 0.866 \text{ m and } AH = 0.5 \text{ m}$$

That gives

$$6W \times 0.866 = 5W \times GH$$

$$GH = \frac{6 \times 0 \cdot 866}{5} = 1 \cdot 039 \text{ m}$$

$$\therefore \quad AG = \sqrt{(AH)^2 + (GH)^2} = \sqrt{0 \cdot 5^2 + 1 \cdot 039^2} = \mathbf{1.15 \text{ m}}$$

EXAMPLE 6.22

The density at point of a slender rod varies with the first power of the distance of the point from one end of the rod. Locate the mass centre. How does the mass centre compares with the length centroid?

Solution : The mass centroid (centre) of a wire, pipe or rod of constant cross-section is given by

$$\bar{x} = \frac{\Sigma x dm}{\Sigma dm} \text{ and } \bar{y} = \frac{\Sigma y dm}{\Sigma dm}$$

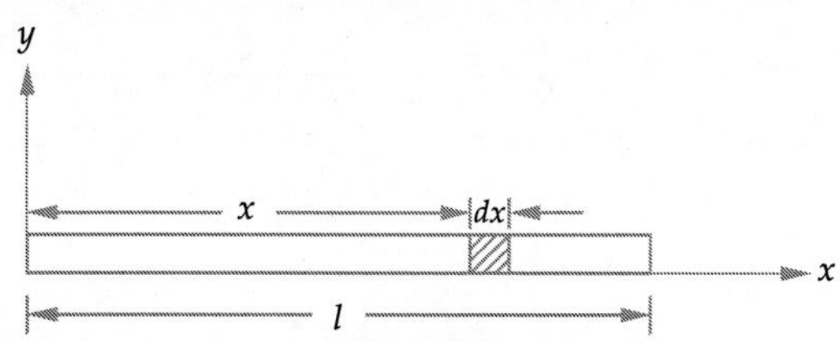

Fig. 6.27

When the x-axis is so chosen that it passes through the centre of the slender rod and along its length, $\bar{y} = 0$

Let the density (taken as mass per unit length) at distance x from the left end of the rod be $\rho = kx$. The mass of an element dx located at distance x from the origin (left end of the rod) is

$$dm = \rho\, dx = kx\, dx$$

Then :
$$\Sigma x\, dm = \int_o^l k x^2 dx = k\frac{l^3}{3}$$

$$\Sigma dm = \int_o^l k x dx = \frac{kl^2}{2}$$

$$\therefore \quad \bar{x} = \frac{kl^3/3}{kl^2/2} = \frac{2}{3}l$$

Thus the mass centre of a slender rod is located at two-third of the length from the end chosen as reference.

The centroid of the length of the rod, however, is at its mid point (Refer to Example 6.1).

EXAMPLE 6.23

Three identical boxes, each havingg length l and weight w, are placed as shown in Fi.g 6.28. Find out the maximum possible distance m through which the top box can extend out from the bottom so that there is no possibility of the toppling of the stack.

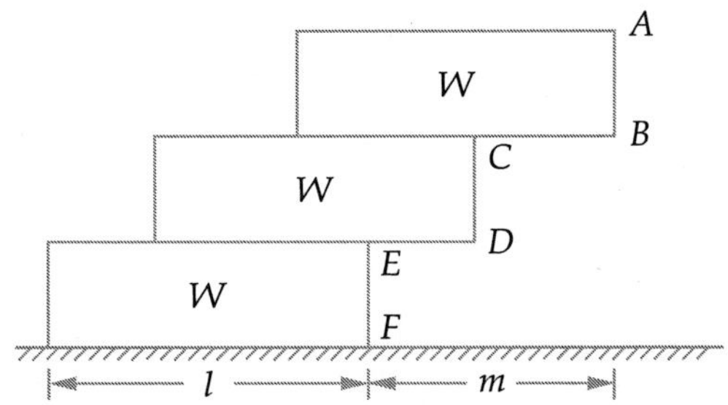

Fig. 6.28

Solution : It is to be noted that to avioid toppling over of the top box, its centre of grovity is in line with the outer edge CD of the middle box. This means that the line of action of the weight W of the top box passes through edge CD of the middle box. Obviously the maximum permissible overhang of the box is $l/2$.

$\therefore$ Distance $CB = l/2$

From free body diagram of middle and bottom box taken together, let x be the overhang of middle slab. Then its weight acts at distance $\left(\frac{l}{2} - x\right)$ from edge EF

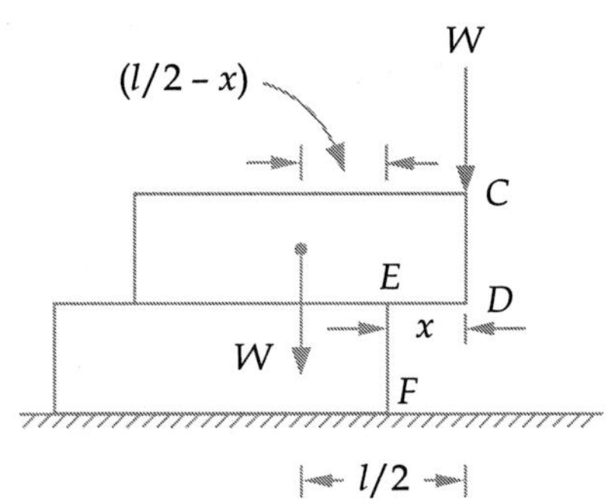

Fig. 6.29

Under equilibrium conditions : $\Sigma M_e = 0$

That is :
$$Wx - W\left(\frac{l}{2} - x\right) = 0$$

$$\text{or} \quad x = \frac{l}{2} - x \; ; \; x = \frac{l}{4}$$

$$\therefore \quad \text{Distance } m = \frac{l}{2} + \frac{l}{4} = \frac{3l}{4}$$

EXAMPLE 6.24

A right circular cone of 20 cm height weighs 1000 N. A cone of 8 cm height and 64 N weight is removed from the top. Determine the distance of CG of the frustrum from the base.

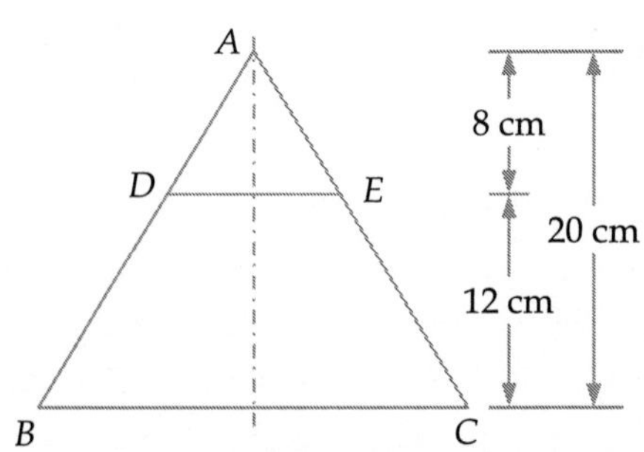

Fig. 6.30

Solution : Refer to Fig 6.30

For a solid cone of height h, the CG lies on the axis at distance $h/4$ above the base.

Distance of CG of full cone from BC

$$y_1 = \frac{20}{4} = 5 \text{ cm}$$

Distance of *CG* of removed conical section from base line *BC*

$$y_2 = 12 + \frac{8}{4} = 14 \text{ cm}$$

Let y be the distance of *CG* of frustrum *DECB* from the base line *BC*.

Now, $$\bar{y} = \frac{m_1 y_1 + m_2 y_2}{m_1 + m_2}$$

Considering mass m_2 as negative since it has been removed,

$$\bar{y} = \frac{1000 \times 5 + (-64) \times 14}{1000 + (-64)} = \frac{5000 - 896}{936} = \mathbf{4.385 \text{ cm}}$$

Alternatively : Taking moments of the weights of different sections about the base line *BC*, we get

$$1000 \times 5 = 64 \times 4 + (1000 - 64)\, y\,; \quad y = 4.385 \text{ cm}$$

EXAMPLE 6.25

A thin wire of homogeneous material is bent to form an isosceles triangle *ABC* as shown in Fig. 6.31. Determine the base angle θ for which the centre of mass of the wire coincides with the centroid of the area enclosed by the wire.

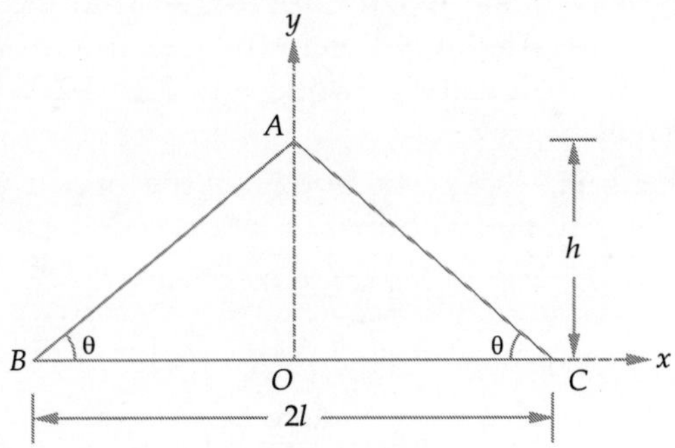

Fig. 6.31

Solution : The triangular area *ABC* enclosed by the wire is symmetrical about the *y*-axis and accordingly the centroid must lie on it. The co-ordinates of the centroid with respect to base are :

$$x_c = 0 \quad \text{and} \quad y_c = \frac{h}{3}$$

The composite wire consists of three segments

(*i*) **Base *BC* :** mass= $2\rho l$ where ρ is the density per unit length

$$y_{c_1} = 0 \text{ (with respect to base } BC)$$

(*ii*) **Segment *AB* :** length $= \sqrt{(l^2 + h^2)}$

$$\text{mass} = \rho\sqrt{(l^2 + h^2)} \quad \text{and} \quad y_{c_2} = \frac{h}{2}$$

(*iii*) **Segment *AC* :** length $= \sqrt{(l^2 + h^2)}$

$$\text{mass} = \rho\sqrt{(l^2 + h^2)} \quad \text{and} \quad y_{c_3} = \frac{h}{2}$$

Then y-coordinate of the centre of mass of the wires forming the triangle is

$$y_c = \frac{\Sigma m \times y_c}{\Sigma m} = \frac{2\rho l \times 0 + \rho\sqrt{(l^2 + h^2)} \times \frac{h}{2} + \rho\sqrt{(l^2}}{2\rho l + \rho\sqrt{(l^2 + h^2)} + \rho\sqrt{(l^2 +}}$$

$$= \frac{\rho h\sqrt{(l^2 + h^2)}}{2\rho l + 2\rho\sqrt{(l^2 + h^2)}} = \frac{h\sqrt{(l^2 + h^2)}}{2[l + \sqrt{(l^2 + h^2)}\,]}$$

According to the given condition,

$$\frac{h}{3} = \frac{h\sqrt{(l^2 + h^2)}}{2[l + \sqrt{(l^2 + h^2)}]}$$

or $$2l + 2\sqrt{(l^2 + h^2)} = 3\sqrt{(l^2 + h^2)}$$

or $$2l = \sqrt{(l^2 + h^2)} \quad ; \quad 4l^2 = l^2 + h^2$$

That gives : $h = \sqrt{3}\,l$

$$\therefore \quad \theta = \tan^{-1}\left(\frac{h}{l}\right) = \tan^{-1}(\sqrt{3}) = \mathbf{60^\circ}$$

EXAMPLE 6.26

A wooden plank of size 10 cm width and 1 cm thick is fixed on top of a semi-circular steel plate of radius 5 cm and thickness 1 cm as shown in Fig. 6.32. Determine the maximum height *h* of the wooden plank so that *CG* of the composite body lies in the junction of the two. Take density of wood equal to 1/10 that of steel.

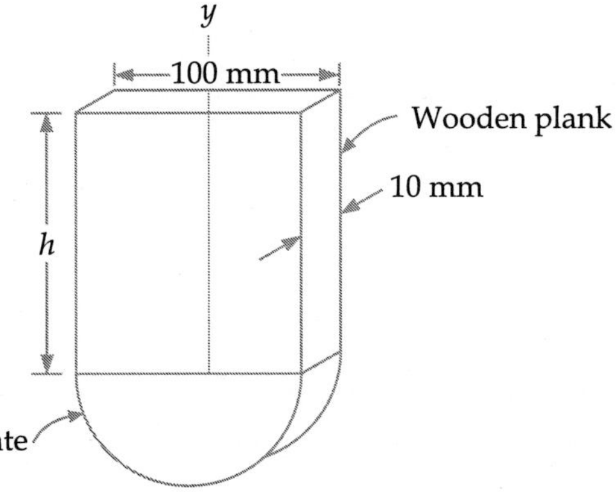

Fig. 6.32

Solution : Mass of body equals the product of its density and volume.

$$m_w = \rho_w \times h\,t\,b$$
$$= \rho_w \times h \times 10 \times 100 = 1000\,h\,\rho_w$$

$$m_s = \rho_w \times \frac{\pi r^2}{2} \times t = \rho_s \times \frac{\pi}{2}(50)^2 \times 10 = 39250\,\rho_s$$

With the co-ordinate axes as shown:

$$y_w = \frac{h}{2} \quad \text{and} \quad y_s = -\frac{4r}{3\pi} = -\frac{4 \times 50}{3\pi} = -21.33 \text{ mm}$$

The negative sign with y_s stems from the fact that centroid for the steel plate lies below the *x*-axis.

Then $$\bar{y} = \frac{m_w y_w + m_s y_s}{m_w + m_s} = \frac{1000\,h\rho_w \times \frac{h}{2} + 39250\,\rho_s(-21.23}{1000\,h\rho_w + 39250\,\rho_s}$$

Since *CG* of the composite body is given to lie on the junction through which passes the *x*-axis, $\bar{y} = 0$.

That gives: $1000\,h\,\rho_w \times \frac{h}{2} + 39250\,\rho_s(-21.23) = 0$

Further $\rho_s = 10\,\rho_w$

$$\therefore \quad 1000\,\frac{h^2}{2}\rho_w + 39250 \times 10\,\rho_w(-21.23) = 0$$

or $$h^2 = \frac{2 \times 39250 \times 10 \times 21.23}{100} = 16665.56$$

$\therefore$ Height of wooden plank, $h = \sqrt{16665.56} = \mathbf{129.09\ mm}$

EXAMPLE 6.27

Determine the centroid of area of a rectangle of breadth *b* and height *h*.

Solution : Let *ABCD* be a rectangle of breadth *b* and height *h*. To determine *y*-coordinate of the centroid of this rectangle, consider a strip of thickness *dy* located at distance *y* from side *AB* of the rectangle. For this elemental strip,

$$\text{area} = b\,dy$$

$$\text{moment about } x\text{-axis} = b\,dy \times y = by\,dy$$

$$\text{area of rectangle } ABCD = bh$$

If $\bar{y}$ is the distance of the centroid from the base, then from the moment principle

$$bh \times \bar{y} = \int_o^h by\,dy = b\left|\frac{y^2}{2}\right|_o^h = \frac{bh^2}{2}$$

$$\therefore \quad \bar{y} = \frac{bh^2}{2} \div bh = \frac{h}{2}$$

Thus the centroid of a rectangle of height *h* is at a distance *h*/2 from the base.

Similarly, we can determine the *x*-coordinate of the centroid of this rectangle by considering a strip of thickness *dx* located at distance *x* from side *AD* of the rectangle. Then

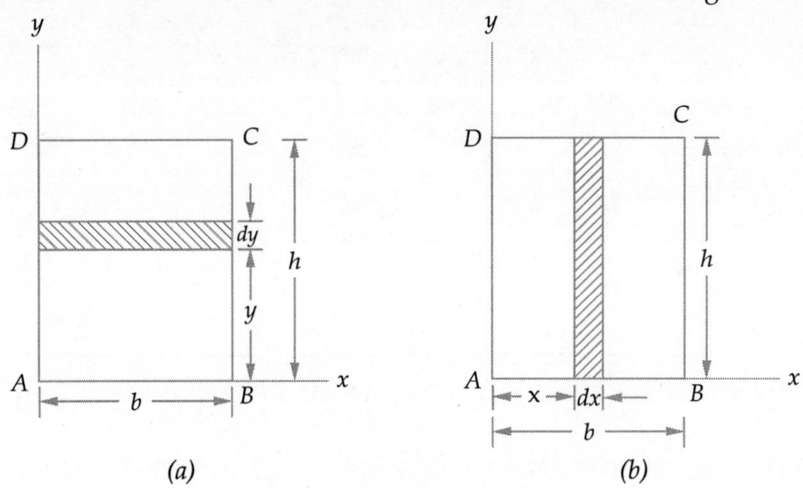

Fi.g. 6.33

$$bh \times \bar{x} = \int_o^b h\,dx \times x = \int_o^b hx\,dx = \frac{hb^2}{2}$$

$$\therefore \quad x = \frac{b}{2}$$

EXAMPLE 6.28

Determine the centroid of the area of a triangle with respect to its base.

Solution : Let *ABC* be the triangle of base width *b* and height *h*. Consider an elementary strip of width l, thickness dy and located at distance y from base BC of the triangle. For this elemental strip,

$$\text{area} = l\,dy$$

$$\text{moment about } x\text{-axis} = l\,dy \times y$$

Since the integration is to be done with respect to y within the limits 0 to h, it is necessary to express l in terms of y. For that we obtain the following correlation from the similarity of triangles ADE and ABC,

$$\frac{l}{b} = \frac{h-y}{h}; \quad l = b\left(1-\frac{y}{h}\right)$$

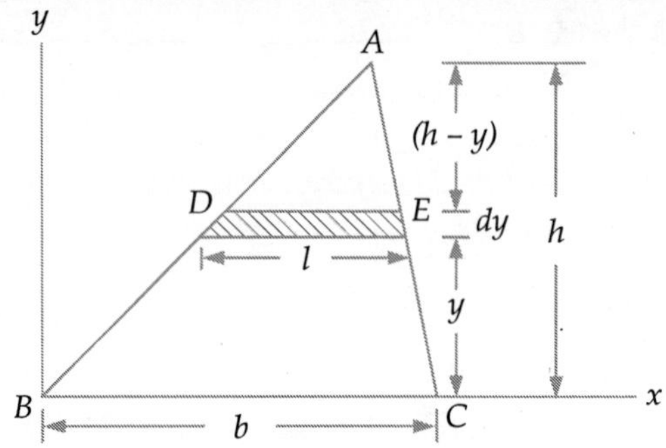

Fig. 6.34

$\therefore$ Moment of elemental strip about x-axis $= b\left(1-\frac{y}{h}\right)y\,dy$

$$\text{area of triangle } ABC = \frac{1}{2}bh$$

If $\bar{y}$ is the distance of the centroid from the base, then from the moment principle

$$\frac{1}{2}bh \times \bar{y} = \int_0^h b\left(1-\frac{y}{h}\right)y\,dy = b\int_0^h \left(y-\frac{y^2}{h}\right)dy$$

$$= b\left|\frac{y^2}{2}-\frac{y^3}{3h}\right|_0^h = b\left(\frac{h^2}{2}-\frac{h^2}{3}\right) = \frac{bh^2}{6}$$

$$\therefore \quad \bar{y} = \frac{bh^2}{6}\times\frac{2}{bh} = \frac{h}{3}$$

Thus the centroid of a triangle of height h is at a distance $h/3$ from the base or $2h/3$ from the apex.

EXAMPLE 6.29

Determine the centre of gravity of a solid right circular cone of height h and base circle radius R.

Solution : Let ABC be the right circular cone of height h and base circle radius R.

Consider an elemental ring of radius r between two planes parallel to the base BC and at distance y and $(y + dy)$ from it. Presuming that the material of the cone is homogeneous (*i.e.*, its density is same through out), the centre of gravity will be its centre of volume.

For the elemental ring,

$$\text{volume} = \pi r^2\,dy$$

$$\text{moment about } x\text{-axis} = \pi r^2 dy \times y$$

Since the integration is to be done with reference to y within the limits 0 to h, it is necessary to express r in terms of y. For that we obtain the following correlation from the similarity of triangles ADE and ABC

$$\frac{r}{R} = \frac{h-y}{h}; \quad r = R\left(1-\frac{y}{h}\right)$$

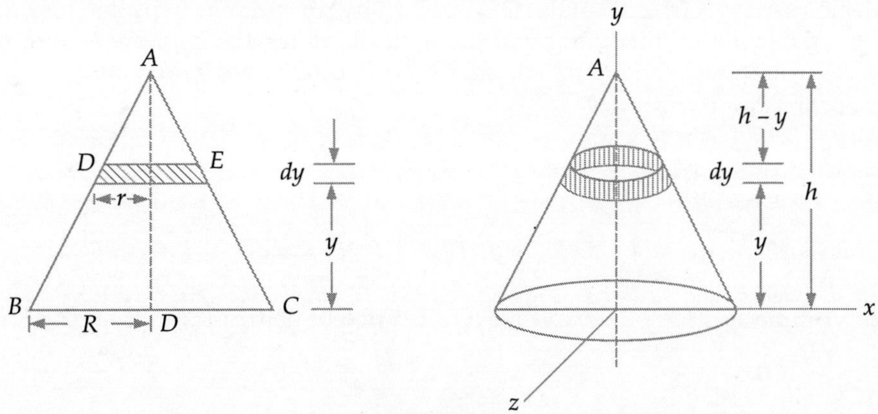

Fig. 6.35

$\therefore$ moment of elemental volume about x-axis $= \pi R^2 \left(1-\frac{y}{h}\right)^2 dy \times y$

$$\text{Volume of cone ABC} = \frac{1}{3}\pi R^2 h$$

If $\overline{y}$ is the distance of the centre of volume from the base, then from the moment principle

$$\frac{1}{3}\pi R^2 h \times \overline{y} = \overline{y} \int_o^h \pi R^2 \left(1-\frac{y}{h}\right)^2 y\, dy = \pi R^2 \int_o^h \left(y + \frac{y^3}{h^2} - 2\frac{y^2}{h}\right) dy$$

$$= \pi R^2 \left| \frac{y^2}{2} + \frac{y^4}{4h^2} - \frac{2}{3}\frac{y^3}{h} \right|_o^h = \pi R^2 \times \frac{h^2}{12}$$

$$\therefore \quad \overline{y} = \left(\pi R^2 \times \frac{h^2}{12}\right) \times \frac{3}{\pi R^2 h} = \frac{h}{4}$$

Thus the centre of gravity of right circular cone of height h is at a distance $h/4$ from the base or $3h/4$ from the apex.

EXAMPLE 6.30

Locate the centre of gravity of a solid hemisphere of radius *R*. How this value would be affected if the hemisphere is hollow with negligible thickness?

Solution : Let *ACB* be a solid hemisphere of radius *R*.

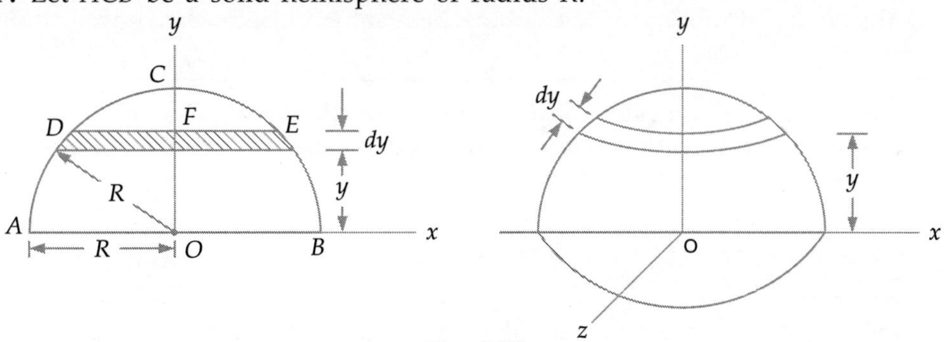

Fig. 6.36

Consider an elementary circular plate *DE* cut off by the planes parallel to *AB* and at distances *y* and $(y + dy)$ from it. Presuming that the material of the hemisphere is homogeneous (*i.e.*, same density throughout) the centre of gravity will be its centre of volume.

For the elementary circular plate *DE*,

$$\text{volume} = \pi\,(DF)^2 \times dy = \pi\,(R^2 - y^2)\,dy$$

$$\text{moment about x-axis} = \pi\,(R^2 - y^2)\,dy \times y$$

$$\text{Volume of the hemisphere} = \int_0^R \pi\,(R^2 - y^2)\,dy = \pi\left| R^2 \times y - \frac{y^3}{3}\right|_0^R = \frac{2}{3}\pi R^3$$

The centre of volume lies on *y*-axis and let its distance be $\bar{y}$ from the base *AB*. Then from the moment principle,

$$\frac{2}{3}\pi R^3 \times \bar{y} = \int_0^R \pi\,(R^2 - y^2)\,y\,dy = \pi\int_0^R (R^2 y - y^3)\,dy$$

$$= \pi\left| R^2\frac{y^2}{2} - \frac{y^4}{4}\right|_0^R = \pi\frac{R^4}{4}$$

$$\therefore \quad \bar{y} = \frac{\pi R^4}{4} \times \frac{3}{2\pi R^3} = \frac{3R}{8}$$

Hence centre of gravity of a solid hemisphere of radius *R* lies on the central radius at distance $3R/8$ from the plane base.

(*b*) If the hemisphere is hollow of negligible thickness, then *DE* is ring with

$$\text{elemental area} = 2\pi R\,dy$$

$$\text{moment about } x\text{-axis} = 2\pi R\,dy \times y$$

$$\text{Area of thin hollow hemisphere} = \int_0^R 2\pi R\,dy = 2\pi\left| R \times y\right|_0^R = 2\pi R^2$$

The centre of area lies on *y*-axis and let its distance from base *AB* be $\bar{y}$. Then from the moment principle

$$2\pi R^2 \times \bar{y} = \int_0^R 2\pi R y\,dy = 2\pi R\left|\frac{y^2}{2}\right|_0^R = \pi R^3$$

$$\therefore \quad \bar{y} = \frac{\pi R^3}{2\pi R^2} = \frac{R}{2}$$

Apparently the centre of gravity of the hollow hemisphere bisects the central radius.

EXAMPLE 6.31

Determine the coordinates of the centroid of a lamina in the shape of a circular sector of radius *r* and central angle 2α. How this result can be used to locate the centroid of a semicircle and quarter of a circle ?

Solution : Refer Fig 6.37 which shows a lamina *OAB* in the shape of a circular sector of radius *r* and central angle 2α. When the axes are chosen as shown, the *x*-axis is an axis of symmetry and accordingly $\bar{y} = 0$

Consider an elementary strip *oab* that subtends an angle $d\theta$ at the centre and has base *ab* $= r\,d\theta$. As angle $d\theta$ is very small, the elementary strip can be considered to be a triangle.

area of elementary strip

$$= \frac{1}{2}(r\,d\theta) \times r = \frac{r^2 d\theta}{2}$$

Centroid of this triangular strip lies on a line that joins O to the mid point of ab and at a distance $\frac{2}{3}r$ from O.

distance x of centroid from y-axis

$$= \frac{2}{3} r \cos\theta$$

moment of area of elementary strip about y-axis

$$= \frac{r^2 d\theta}{2} \times \frac{2}{3} r \cos\theta$$

$$= \frac{1}{3} r^3 \cos\theta \, d\theta$$

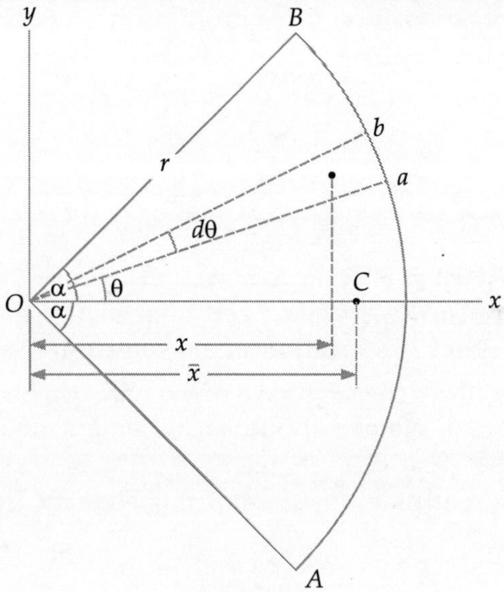

Fig. 6.37

If $\bar{x}$ is the coordinate of the centroid of the lamina from y-axis, then from the moment principle,

$$\bar{x} \times \int_{-\alpha}^{+\alpha} \frac{r^2 d\theta}{2} = \int_{-\alpha}^{+\alpha} \frac{1}{3} r^3 \cos\theta \, d\theta$$

It is to be noted that the variable θ varies from $-\alpha$ to $+\alpha$

$$\bar{x} \times \frac{r^2}{2} \left| \theta \right|_{-\alpha}^{+\alpha} = \frac{r^3}{3} \left| \sin\theta \right|_{-\alpha}^{+\alpha}$$

or $$\bar{x} \times \frac{r^2}{2}(2\alpha) = \frac{r^3}{2} \times 2\sin\alpha$$

$$\therefore \quad \bar{x} = \left[\frac{r^3}{3} \times 2\sin\alpha\right] \times \left[\frac{2}{r^2(2\alpha)}\right] = \frac{2r\sin\alpha}{3\alpha}$$

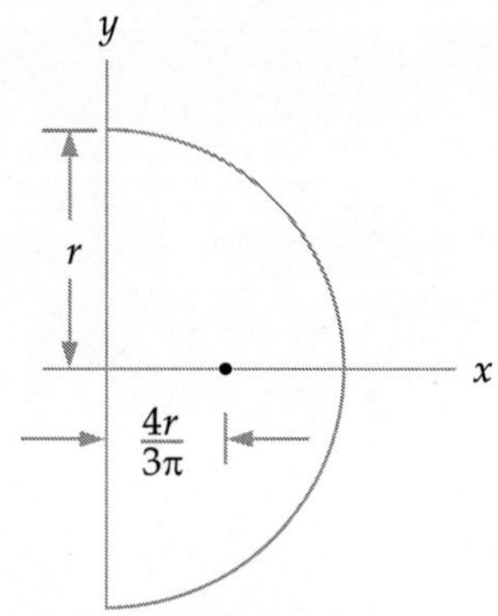

Fig. 6.38

This is the location of the centroid of any sector. The angle α has to be taken in radians.

(*b*) **In case of a semicircle**

$$2\alpha = 180^\circ = \pi \text{ radians}$$

$$\bar{x} = \frac{2r\sin\pi/2}{3\times\pi/2}$$

$$= \frac{4r}{3\pi} = 0.424r$$

In case of quadrant of a circle

$$2\alpha = 90^\circ = \frac{\pi}{2} \text{ radians}$$

$$\bar{x} = \frac{2r\sin\pi/4}{3\times\pi/4}$$

$$= \sqrt{2}\,\frac{4r}{3\pi} \qquad \text{(distance } OC\text{)}$$

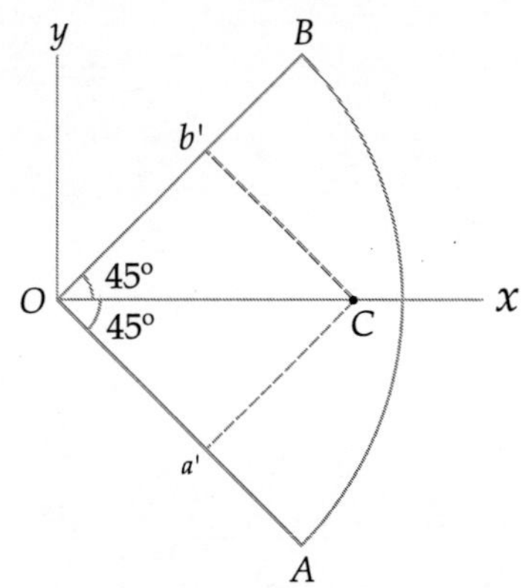

Fig. 6.39

The position of centroid with respect to the radii OA and OB will be

$$oa' = \sqrt{2}\ \frac{4r}{3\pi} \times \cos 45^\circ = \frac{4r}{3\pi}$$

$$ob' = \sqrt{2}\ \frac{4r}{3\pi}\ \cos 45^\circ = \frac{4r}{3\pi}$$

EXAMPLE 6.32

A uniform wire has been bent in the form of a quadrant of the arc of a circle of radius *r*. Workout the location of its centroid.

Solution : The length AB of the wire can be divided into a number of small elements; one such element of length dl subtends an angle $d\theta$ at 0.

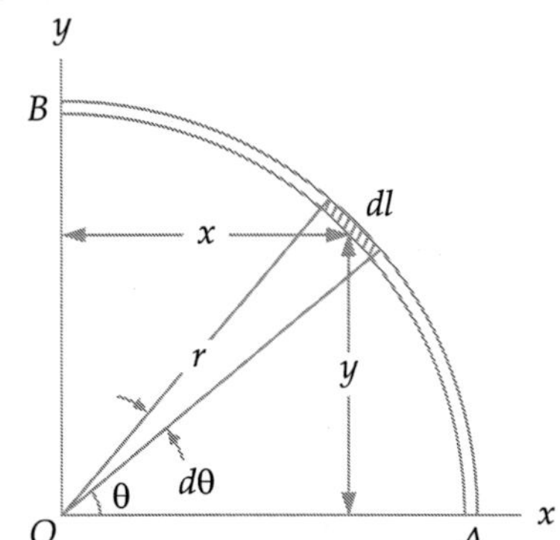

Fig. 6.40

Centroidal distance of this element from x-axis is

$$\bar{y} = \frac{\int y\, dl}{\int dl}$$

In terms of polar coordinates,

$$y = r \sin\theta \text{ and } dl = r\, d\theta$$

Then : $$\int y\, dl = \int_0^{\pi/2} (r \sin\theta)\, r\, d\theta$$

$$= r^2 \int_0^{\pi/2} \sin\theta\, d\theta$$

$$= r^2 \left| \cos\theta \right|_0^{\pi/2} = r^2$$

$$\int dl = \int_0^{\pi/2} r\, d\theta = r\left|\theta\right|_0^{\pi/2} = \frac{r\pi}{2}$$

$$\therefore \quad \bar{y} = \frac{r^2}{\frac{r\pi}{2}} = \frac{2r}{\pi}$$

Since the wire bent in the form of a quadrant of the arc of a circle is symmetrical about the x-axis and y-axis,

$$\bar{x} = \bar{y} = \frac{2r}{\pi}$$

EXAMPLE 6.33

Locate the centroid of a circular arc as shown below in Fig. 6.41.

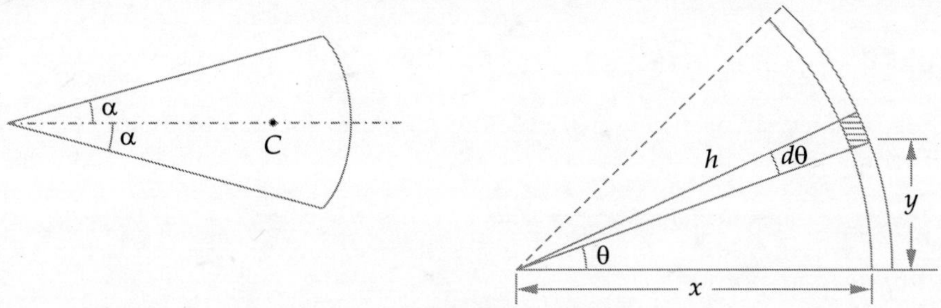

Fig. 6.41

Solution : For a small element with included angle $d\theta$

$$dl = h\,d\theta$$

$$y = h\sin\theta \quad \text{and} \quad x = h\cos\theta$$

$$\bar{y} = \frac{\int_0^{2\alpha} y\,dl}{\int_0^{2\alpha} dl} = \frac{\int_0^{2\alpha} h\sin\theta \times h\,d\theta}{\int_0^{2\alpha} h\,d\theta} = \frac{\int_0^{2\alpha} h^2\sin\theta\,d\theta}{\int_0^{2\alpha} h\,d\theta}$$

$$= \frac{h^2\left|\cos\theta\right|_0^{2\alpha}}{h\left|\theta\right|_0^{2\alpha}} = \frac{h^2\cos 2\alpha}{h\times 2\alpha} = \frac{h}{2}\,\frac{\cos 2\alpha}{\alpha}$$

and

$$\bar{x} = \frac{\int_0^{2\alpha} x\,dl}{\int_0^{2\alpha} dl} = \frac{\int_0^{2\alpha} h\cos\theta \times h\,d\theta}{\int_0^{2\alpha} h\,d\theta} = \frac{\int_0^{2\alpha} h^2\cos\theta\,d\theta}{\int_0^{2\alpha} h\,d\theta}$$

$$= \frac{h^2\left|\sin\theta\right|_0^{2\alpha}}{h\left|\theta\right|_0^{2\alpha}} = \frac{h^2\sin 2\alpha}{h\times 2\alpha} = \frac{h}{2}\,\frac{\sin 2\alpha}{\alpha}$$

EXAMPLE 6.34

Locate the centroid of the area of the parabolic shaded portion shown in Fig. 6.42.

Solution : When $x = a$, $y = h$. Substituting these values in the equation of the curve, we get,

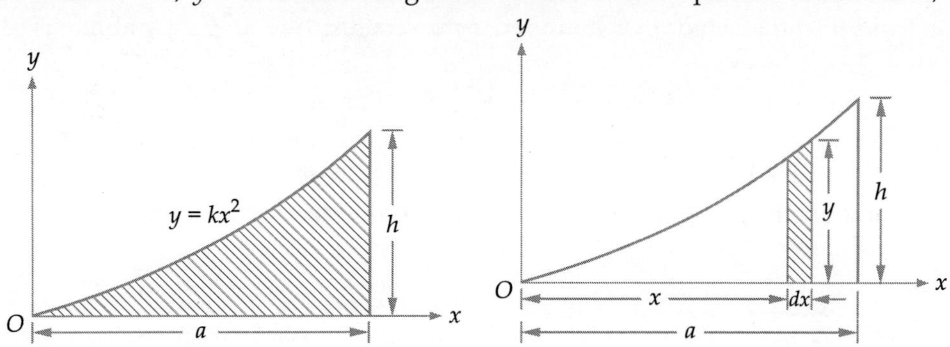

Fig. 6.42

$$h = ka^2 \quad \text{or} \quad k = h/a^2$$

∴ Equation of the curve is $y = \dfrac{h}{a^2}x^2$

Consider an elementary strip of height y, thickness dx and located at distance x from the y-axis. For this elemental strip

$$\text{area} = y\,dx = \frac{h}{a^2}x^2\,dx$$

∴ area of the entire shaded portion

$$= \int_o^a \frac{h}{a^2}x^2dx = \frac{h}{a^2}\left|\frac{x^3}{3}\right|_0^a = \frac{h}{a^2}\times\frac{a^3}{3} = \frac{ah}{3}$$

Position of Centroid from y-axis

Moment of the elemental area about y-axis $= y\,dx \times x$

$$= \frac{h}{a^2}x^2dx \times x = \frac{h}{a^2}x^3dx$$

If $\bar{x}$ is the distance of the centroid from y-axis, then from the moment principle,

$$\frac{ah}{3}\times\bar{x} = \int_o^a \frac{h}{a^2}x^3dx = \frac{h}{a^2}\left|\frac{x^4}{4}\right|_o^a = \frac{ha^2}{4}$$

$$\therefore \quad \bar{x} = \frac{ha^2}{4}\div\frac{ah}{3} = \frac{3}{4}a$$

Position of centroid above x-axis

Moment of the elemental area about x-axis,

$$= y\,dx\times\frac{y}{2} = y^2\frac{dx}{2} = \left(\frac{h}{a^2}x^2\right)^2\frac{dx}{2} = \frac{h^2}{2a^4}x^4dx$$

If $\bar{y}$ is the distance of the centroid above x-axis, then from the moment principle

$$\frac{ah}{3}\bar{y} = \int_o^a \frac{h^2}{2a^4}x^4dx = \frac{h^2}{2a^4}\left|\frac{x^5}{5}\right|_o^a = \frac{h^2a}{10}$$

$$\therefore \quad \bar{y} = \frac{h^2a}{10}\div\frac{ah}{3} = \frac{3h}{10}$$

EXAMPLE 6.35

Locate the centroid of the shaded area bounded by a straight line and a parabola as shown in Fig. 6.43 (*a*).

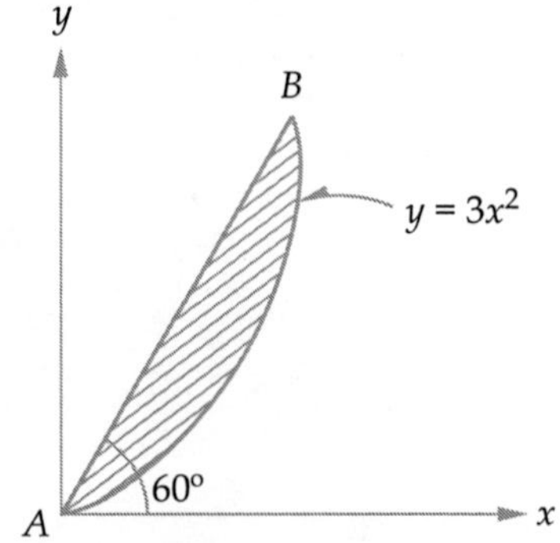

Fig. 6.43 (a)

Solution : The equations of the parabola and the straight line are

$$y = 3x^2 \quad ...(i)$$

and $$y = mx = (\tan 60^\circ)\times x = \sqrt{3}\,x \quad ...(ii)$$

The point B is lying on the straight line as well as on the given parabola. Hence both the expressions (*i*) and (*ii*) hold good for point B. If B has the co-ordinates (a, b), then

$$b = 3a^2 \quad \text{and} \quad b = \sqrt{3}\,a$$

That gives : $$a = \frac{\sqrt{3}}{3} = \frac{1}{\sqrt{3}} \text{ and } b = 1$$

Hence, point B has the co-ordinates $\left(\frac{1}{\sqrt{3}}, 1\right)$.

Consider an elemental vertical strip of thickness dx at a distance x from y-axis (Fig. 6.43(b))

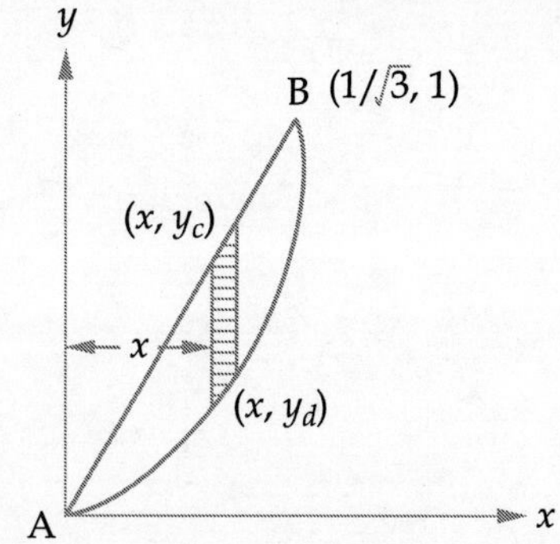

Fig. 6.43 (b)

Area of the elemental strip, dA

$$= (y_c - y_d)dx$$

As $\quad y_c = \sqrt{3}\,x \quad$ and $\quad y_d = 3x^2$

$$dA = (\sqrt{3}x - 3x^2)dx$$

The x and y-coordinates of a centroid are prescribed by the relations :

$$\bar{x} = \frac{\int x\,dA}{\int dA} \quad \text{and} \quad \bar{y} = \int \frac{y\,dA}{dA}$$

Now, $$\int dA = \int_o^a (\sqrt{3}x - 3x^2)\,dx = \left|\sqrt{3}\frac{x^2}{2} - 3\frac{x^3}{3}\right|_o^a = \frac{\sqrt{3}}{2}a^2 - a^3$$

$$\int x\,dA = \int_o^a x(\sqrt{3}x - 3x^2)dx = \left|\sqrt{3}\frac{x^3}{3} - 3\frac{x^4}{4}\right|_o^a = \frac{a^3}{\sqrt{3}} - \frac{3}{4}a^4$$

$$\int y\,dA = \int \frac{y_c + y_d}{2}\,dA = \frac{1}{2}\int_o^a (\sqrt{3}x + 3x^2)(\sqrt{3}x - 3x^2)\,dx$$

$$= \frac{1}{2}\int_o^a (3x^2 - 9x^4)\,dx = \frac{1}{2}\left|3\frac{x^3}{3} - 9\frac{x^5}{5}\right|_o^a = \frac{a^3}{2} - \frac{9}{10}a^5$$

$\therefore$ $$\bar{x} = \frac{\frac{a^3}{\sqrt{3}} - \frac{3}{4}a^4}{\frac{\sqrt{3}}{2}a^2 - a^3} = \frac{\frac{a}{\sqrt{3}} - \frac{3}{4}a^2}{\frac{\sqrt{3}}{2} - a}$$

Substituting the value $a = \frac{1}{\sqrt{3}}$

$$\bar{x} = \frac{1/3 - 3/4 \times 1/3}{\sqrt{3}/2 - 1/\sqrt{3}} = \frac{\sqrt{3}}{6} = \mathbf{0.289}$$

and $$\bar{y} = \frac{\frac{a^3}{2} - \frac{9}{10}a^5}{\frac{\sqrt{3}}{2}a^2 - a^3} = \frac{\frac{a}{2} - \frac{9}{10}a^3}{\frac{\sqrt{3}}{2} - a} = \frac{1/2 \times 1/\sqrt{3} - 9/10 \times (1/\sqrt{3})^3}{\sqrt{3}/2 - 1/\sqrt{3}} = \frac{2}{5} = \mathbf{0.4}$$

6.3. PAPPUS-GULDINUS THEOREMS

When a plane curve rotates about a fixed axis, it generates a surface called the *surface of revolution*.

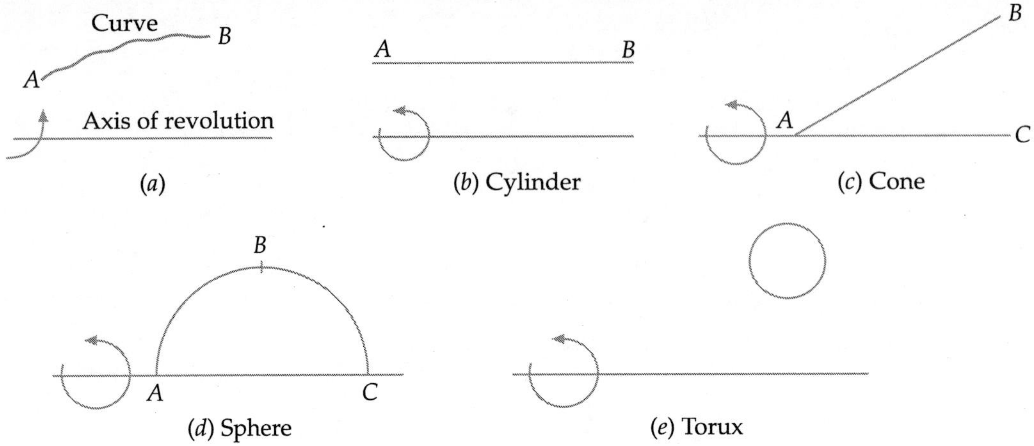

Fig. 6.44 Generation of surface of revolution

Examples (*i*) the surface of a cylinder is generated when straight line *AB* is rotated about an axis.

(*ii*) the surface of a cone gets generated when a straight line *AB* is rotated about an axis *AC*

(*iii*) the surface of a sphere is obtained when a semi-circular arc *ABC* is rotated about the diameter *AC*

(*iv*) the surface of a ring or torus gets formed when the circumference of a circle is rotated about a non-intersecting axis.

The generation/formation of these surfaces of revolution has been indicated in Fig. 6.44.

When a plane area is rotated about a fixes axis, it generates a body called the *body of revolution*.

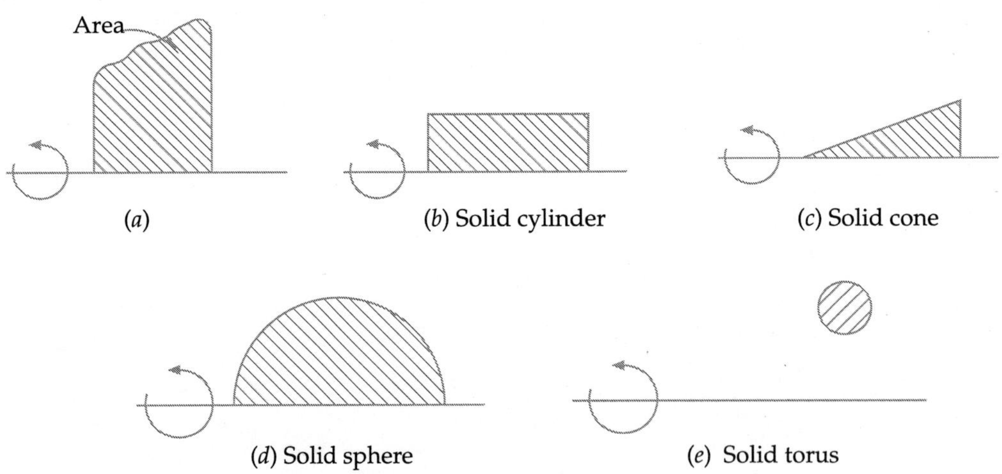

Fig. 6.45 Generation of solid volume of revolution

Examples: A solid cylinder may be generated by rotating a rectangular area, a cone by rotating a triangular area, a solid sphere by rotating a semi-circular area, and a solid torus by rotating a full circular area.

The generation/formation of these bodies of revolution has been indicated in Fig. 6.45.

Theorem 1. The area of the surface generated by rotating any plane curve about a non-intersecting axis in its plane is equal to the product of the length of the curve and the distance 4travelled by its centroid.

Let AB be any plane curve of length l that lies in the x-y plane and does not intersect the x-axis. Consider an elemental length dl_i so short that it may be thought to be a minute portion of a straight line. Let y-coordinate of its mid point be y_i and that of the centroid of the entire curve be $\bar{y}$.

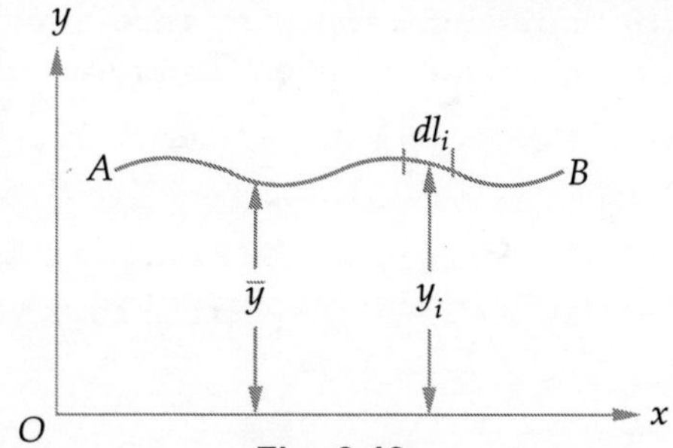

Fig. 6.46

When this curve is rotated about the x-axis, then area generated by the elementary length $= 2\pi y_i\, dl_i$

$$\text{area generated by the entire line} = \int 2\pi y_i\, dl_i$$

The integral $\int y_i dl_i$ is equal to $\bar{y}\,l$ and therefore

$$\text{area generated} = l \times 2\pi\bar{y}$$

where $2\pi\bar{y}$ is the distance travelled by the centroid of the curve

$\therefore$ area generated = (length of the generating curve) × (distance travelled by centroid of the curve)

It is to be noted that the generating curve does not cross the axis about which it is rotated. If it does so, then the two sections on either side of the axis would generate areas of opposite signs and the theorem would not apply.

Theorem II. The volume of the solid generated by rotating any plane figure about a non-intersecting axis in its plane is equal to the product of area of the figure and the distance travelled by its centroid.

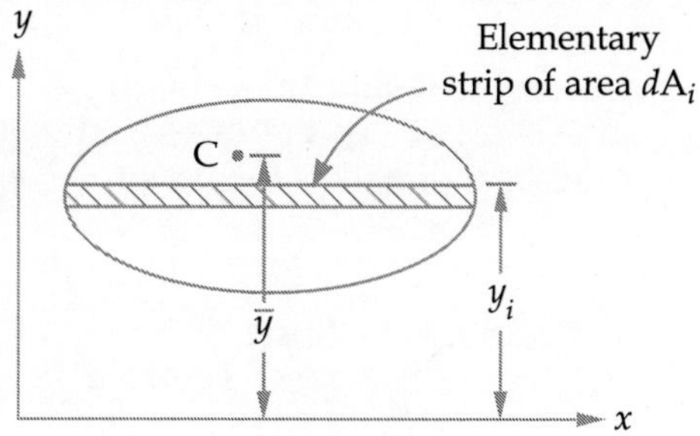

Fig. 6.47

Consider area A of any plane figure divided into a large number of very thin strips parallel to the x-axis. Let the y-coordinate of the middle of any strip of area dA_i be y_i and the centroid of the entire curve be $\bar{y}$. When the plane of the figure is rotated about the x-axis, then

$$\text{volume generated by the elementary area} = 2\pi y_i\, dA_i$$

$$\text{volume generated by the entire area} = \int 2\pi y_i\, dA_i$$

The integral $\int y_i dA_i$ is equal to $\bar{y}A$ and therefore

$$\text{volume generated } V = A \times 2\pi\bar{y}$$

The parameter $(2\pi\bar{y})$ is the distance travelled by the centroid of the area being rotated.

$\therefore$ volume generated = (area of the figure) × (distance travelled by the centroid)

Utility : The theorems of Pappus - Guldinus offer a simple way to compute the areas of surfaces of revolution and the volumes of bodies of revolution.

Conversely their application lies in determining the centroid of a plane curve when the area of the surface generated by the curve is known or to determine the centroid of a plane area when the volume of the body generated by the area is known.

EXAMPLE 6.36

Using Pappus - Guldinus theorems, calculate the

(*i*) area of a surface of revolution obtained by revolving a quarter circular arc of radius *r* about *o* - *y* (Fig. 6.48 a)

(*ii*) volume generated by revolving a quarter ellipse about *o* - *x* (Fig. 6.48 *b*)
(*iii*) volume of a sphere of radius *r*.

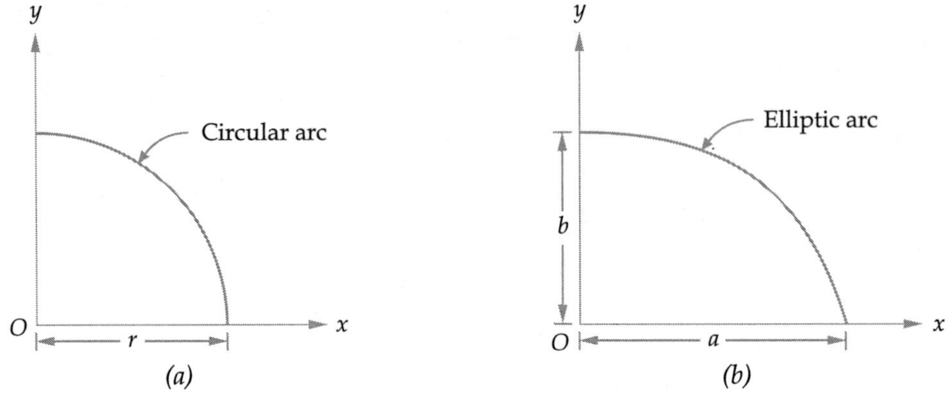

Fig. 6.48

Solution : (*i*) For a quarter circle of radius *r*,

distance of centroid from *y*-axis $= 2r/\pi$

distance travelled by centroid $= 2\pi \times \dfrac{2r}{\pi} = 4r$

length of revolving arc $= \pi r/2$

From Theorem I of Pappus-Guldinus

area generated = length of arc × distance travelled by centroid

$$= \frac{\pi r}{2} \times 4r = 2\pi r^2$$

(*ii*) For a quarter ellipse,

distance of centroid from *x*-axis $= 4b/3\pi$

distance travelled by centroid $= 2\pi \times \dfrac{4b}{3\pi} = \dfrac{8b}{3}$

area of the surface being revolved $= \dfrac{\pi ab}{4}$

From Theorem II of Pappus - Guldinus,

volume generated = area revolved × distance travelled by centroid

$$= \frac{\pi ab}{4} \times \frac{8b}{3} = \frac{2\pi ab^2}{3}$$

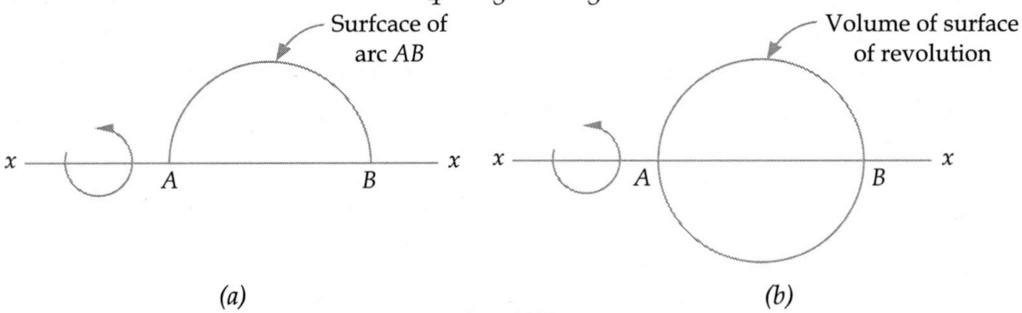

Fig. 6.49

(*iii*) A solid sphere is obtained by revolving a semi-circular area about its diameter *AB* (Fig. 6.49).

distance of centroid from *x*-axis $= 4r/3\pi$

distance travelled by centroid = $2\pi \times 4r/3\pi = 8r/3$

area of semi-circle = $\pi r^2/2$

From Theorem II of Pappus-Guldinus

volume of solid sphere = area of semi-circle × distance travelled by entroid

$$= \frac{\pi r^2}{2} \times \frac{8r}{3} = \frac{4}{3}\pi r^3$$

EXAMPLE 6.37

Apply Pappus-Guldinus theorem to obtain:

(*i*) the centroid of a semi-circular arc of radius *r*

(*ii*) the centroid of a quarter circular area of radius *r*.

Solution: (*i*) A sphere results when a semi-circular arc is rotated about it axis

area of sphere = $4\pi r^2$

length of semi-circular arc = πr

According to Pappus-Guldinus theorem, the area of the sphere equals the product of the length of the generating semicircle and the distance travelled by its centroid in one revolution. That is

$$4\pi r^2 = \pi r \times 2\pi \bar{y}$$

$$\therefore \quad \bar{y} = \frac{4\pi r^2}{\pi r \times 2\pi} = \frac{2r}{\pi}$$

Thus the centroid of a line in the form of semi-circular arc is $2r/\pi$

(*ii*) A hemisphere results when a quarter circular area is rotated about its axis.

$$\text{volume of hemisphere} = \frac{2}{3}\pi r^3$$

$$\text{area of quarter circle} = \frac{\pi r^2}{4}$$

According to Pappus-Guldiness theorem, the volume generated equals the product of area of the revolving body and the distance travelled by its centroid while it is being revolved. That is

$$\frac{2}{3}\pi r^3 = \frac{\pi r^2}{4} \times 2\pi \bar{y}$$

$$\therefore \quad \bar{y} = \frac{2}{3}\pi r^3 \div \left(\frac{2\pi r^2}{4} \times 2\pi\right) = \frac{4r}{3\pi}$$

Thus the distance of the centroid of a quarter circular area from its vertical edge is $4r/3\pi$

EXAMPLE 6.38

A solid circular metal ring has a rectangular cross-section of size 4 cm × 6 cm. If its external and internal radii are 8 cm and 6 cm respectively, calculate the total surface area and volume of the ring.

Solution: According to 1st theorem of Pappus-Guldinus,

Surface area generated = length of revoluing shape × distance travelled by its centroid

= (circumference of rectangular cross-section) × $2\pi\bar{y}$

$$= 2(4+6) \times 2\pi\left(\frac{8+6}{2}\right) = \mathbf{879.2\ cm^2}$$

(*b*) According to 2nd theorem of Pappus-Guldinus,

Volume of ring = area of revolving shape × distance travelled by its centroid

$$= (4 \times 6) \times 2\pi \left(\frac{8+6}{2} \right) = \mathbf{1055.04\ cm^3}$$

EXAMPLE 6.39

(*a*) A line of length 2*l* unit is rotated about *y*-axis as shown in Fig. 6.46. Apply Pappus-Guldinus theorem to determine the surface area of the line.

(*b*) The frustrum of a right circular cone has bottom diameter 50 cm, top diameter 20 cm and height 30 cm. Determine the area of its lateral surface.

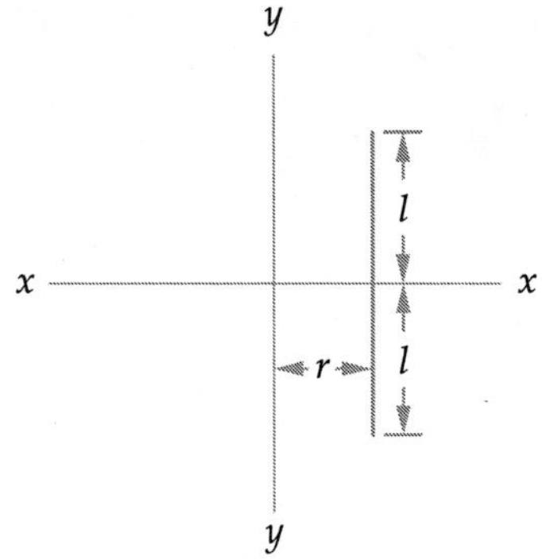

Fig. 6.50

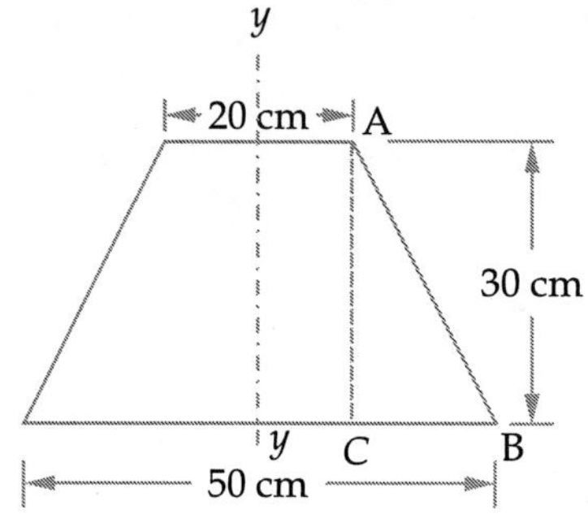

Fig. 6.51

Solution : From 1st theorem of Pappus-Guldinus

Surface area generated = length of revolving line × distance travelled by its centroid

$$= 2l \times (2\pi \times r) = 4\pi r l$$

(*b*) The lateral surface of the frustrum of the cone is generated when the generator line *AB* is revolved about the axis *yy*.

$$\text{length of line } AB = \sqrt{(AC)^2 + (BC)^2} = \sqrt{30^2 + 15^2} = 33.54 \text{ cm}$$

Distance of centroid of the line from the axis of rotation

$$= \frac{10 + 25}{2} = 17.5 \text{ cm}$$

From 1st theorem of Pappus-Guldinus,

area of lateral surface generated = length of the revolving line × distance travelled by its centroid

$$= 33.54 \times (2\pi \times 17.5) = \mathbf{3686\ cm^2}$$

REVIEW AND SUMMARY

1. Centre of gravity of a body is that point through which the entire weight of the body acts irrespective of its orientation. A body has only one centre of gravity.
2. Centre of mass is the central point for the distribution of mass, *i.e.*, the point at which the entire mass of the body is assumed to be concentrated.
3. The term centre of gravity is applied to solids with mass and weight, while centroid applies to length, area and volume of plane figures which have no mass.

   When thickness, *i.e.*, mass of the body is not considered, the centre of gravity and centroid are synonymous and pass through the same point. Further, centroid and centre of mass coincide if the density of the material is uniform throughout the body.

4. Centroidal axis is that axis which passes through the centre of gravity of the body or through the centroid of an area.
5. The co-ordinates of the centroid of an area are prescribed by the relation.

$$\bar{x} = \frac{\Sigma ax}{A} \quad \text{and} \quad \bar{y} = \frac{\Sigma ay}{A}$$

The centre of gravity of a solid has the co-ordinates

$$\bar{x} = \frac{\Sigma mx}{M} \quad \text{and} \quad \bar{y} = \frac{\Sigma my}{M}$$

6. Theorems of Pappus and Guldiness
   (*i*) The area of a surface of revolution is equal to the product of length of the generating curve and the distance travelled by the centroid of the curved line while the surface is being generated,
   (*ii*) The volume of a body of revolution is equal to the product of the generating area and the distance travelled by the centroid of the area while the body is being generated.

REVIEW QUESTIONS

A. Conceptual and conventional questions :

1. Define the terms centroid and centre of gravity.
2. Differentiate between centre of gravity and centroid. Under what conditions these will coincide?
3. The centroid of an area always lies on the **axis of symmetry**, if any, of an area. What does the term axis of symmetry imply?
4. The centre of gravity of a body is the point at which the resultant gravitational force on the body acts, irrespective of the orientation of the body. Point out the significance of the phase 'irrespective of the orientation of the body'.
5. Give the position of centroid of the following standard sections:
   (*i*) rectangle (*ii*) triangle (*iii*) uniform rod and (*iv*) semicircle.
6. Write down the positions of the centre of gravity for the following regular solids: cylinder, sphere, hemisphere and right circular cone.
7. What is meant by an axis/plane of symmetry of a body in the context of centroid/centre of gravity of bodies?
8. Derive from first principles the centroids of
   (*a*) cone and (*b*) hemisphere
9. When will be centroid and centre of mass coincide?
10. Define first moment of an area about an axis
11. Determine the centroid of the area remaining after a circle of diameter r is removed from a circle of radius r as shown in Fig. 6.52. **(r/6 to the left of y-axis)**

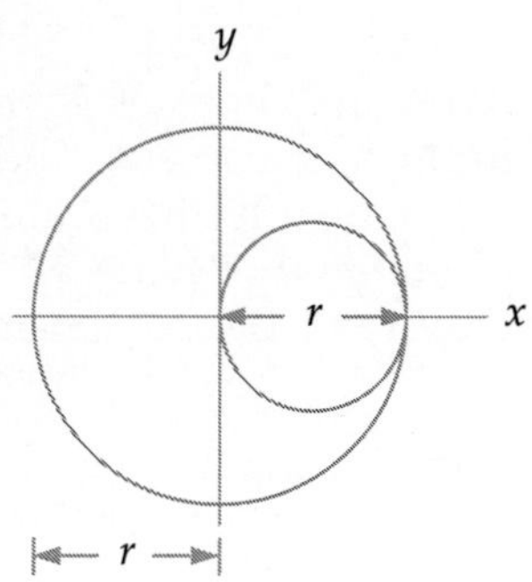

Fig. 6.52

12. A sphere of radius r is cut from a larger sphere of radius R. If the distance between their centres is a, locate the centroid of the remaining volume.

$$\left(\frac{ar^3}{R^3 - r^3}\right)$$

13. Locate the centroid of area *ABCDE* shown in Fig. 6.53. **(7.51 cm, 5.83 cm)**

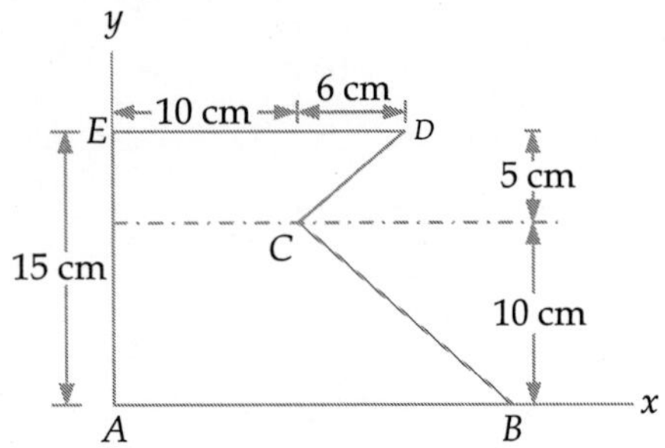

Fig. 6.53

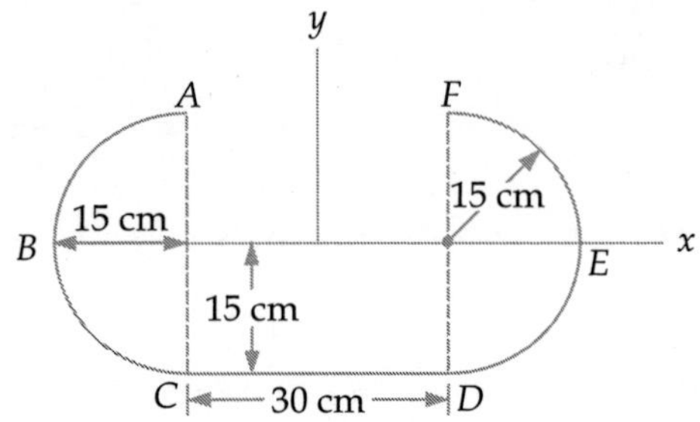

Fig. 6.54

14. A rod has been bent into the shape *ABCDEF* as shown in Fig. 6.54. Determine the position of its centroid. **(0, – 3.62 cm)**
15. Find the centre of gravity of the channel section whose flanges are 40 mm × 10 mm, and web is 60 mm × 10 mm. The total height of the channel is 80 mm. **(15 mm, 30 mm)**
16. Determine the centroid of z-shaped lamina shown in Fig. 6.55.

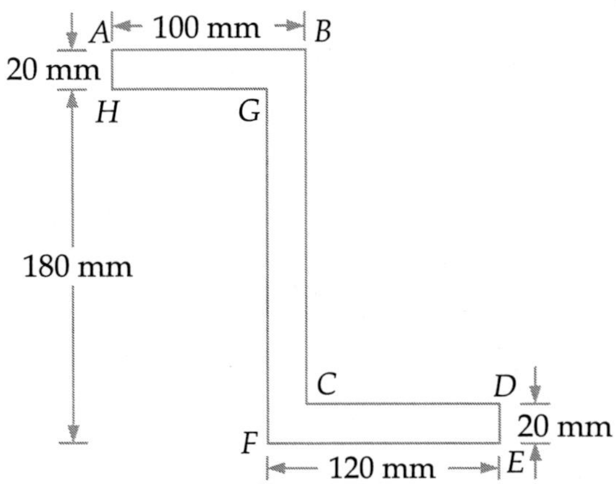

$$\left(\begin{array}{l}\bar{x} = 104.74 \text{ mm from DE };\\ \bar{y} = 95.26 \text{ mm from FE}\end{array}\right)$$

Fig. 6.55

17. A circular hole of 20 mm diameter has been drilled in a 100 diameter circular plate of uniform thickness. The centre of hole lies at 26 mm distance from the centre of plate. Subsequently the hole is filled with a material having density three times that of density of material of plate. Make calculations for the centre of gravity of the composite plate. **(1.93 mm from the centre of plate)**
18. A right circular cylinder of 12 cm diameter is attached to a hemisphere of the same diameter face to face. The density of the material of the hemisphere is twice that of the cylinder material. Make calculations for the greatest height of the cylinder so that the centre of gravity of the composite body lies on the plane joining the hemisphere with the cylinder. **(6 cm)**
19. An isosceles triangle AEB is to be cut from a square ABCD of side *a* as shown in Fig. 6.56. Compute the height *h* of this triangle so that its vertex E will be the centroid of the remaining shaded area.

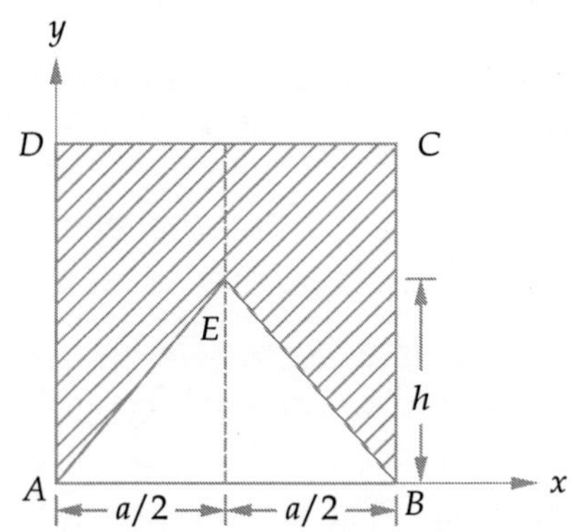

Fig. 6.56

20. A solid consists of a cylinder and a hemisphere of equal radius fixed base to base. Find the ratio of radius to height of the cylinder so that the combined solid may stand in equilibrium with the hemisphere touching the table.

B. Complete the following statements with most appropriate word/words :

1. Centroid relates to .... of line or curve, ... of a surface and .... of a body
2. .... is the point where the entire weight of the body is assumed to the concentrated
3. The first moment of area about any axis passing through the .... is zero
4. The distance of the *CG* of a solid hemisphere from the base is .... where r is the radius of the sphere
5. The *CG* of a solid circular cone divides the axis in the ratio ....
6. The centroid of a triangle is the point of .... of its medians
7. First moment of an area about x-axis is ....
8. If an area is symmetric about an axis, the .... lies on that axis
9. Surface of revolution is the surface generated by rotating a .... about a fixed axis
10. When a plane area is rotated about a fixed axis, ... gets generated

Answers :

1. length, area, volume 2. centre of gravity 3. centroid 4. $3r/8$ 5. 3 : 1 6. concurrence or intersection 7. $\int y\,dA$ 8. centroid 9. plane curve 10. body of revolution

C. Multiple choice questions :

1. Centre of mass of a body
   (*a*) must lie somewhere inside the body
   (*b*) is located at the geometric centre of the body
   (*c*) lies at the geometric centre of the body provided it is of uniform density
   (*d*) is synonymous with centre of gravity
2. The *CG* of a plane lamina will not be at its geometrical centre if it is a
   (*a*) circle (*b*) square
   (*c*) rectangle (*d*) right angled triangle
   (*e*) equilateral triangle
3. The *CG* of a triangle lies at the point of intersection of
   (*a*) diagonals (*b*) altitudes
   (*c*) bisector of angles (*d*) medians
4. For a solid cone of height h, the *CG* lies on the axis at a distance above the base equal to
   (*a*) $h/4$ (*b*) $h/3$ (*c*) $2h/3$ (*d*) $3h/8$
5. The position of *CG* of a solid hemisphere of radius r lies on the central radius at a distance
   (*a*) $\frac{2}{3}r$ (*b*) $\frac{3}{4}r$ (*c*) $\frac{3}{8}r$ (*d*) $\frac{1}{2}r$
   from the plane base
6. Identify the wrong statement/statements
   (*a*) For a cylinder and prism, the *CG* lies on the axis at half the height from the base
   (*b*) For cones and pyramids, the CG lies at one-fourth of the height of the axis from the base

(*c*) The centroid of a semicircle lies on the radius perpendicular to the base at $4r/3\pi$ where r is the radius of the semicircle

(*d*) The *CG* of a triangular plane figure lies at the point of intersection of altitudes and its distance is $2h/3$ from the base

7. Refer Fig. 6.57, for the geometry of a trapezium. The distance of its *CG* from base line is

(*a*) $\frac{h}{4}\left(\frac{2a+b}{a+b}\right)$ (*b*) $\frac{h}{3}\left(\frac{2a+b}{a+b}\right)$

(*c*) $\frac{h}{2}\left(\frac{a+2b}{a+b}\right)$ (*d*) $\frac{h}{3}\left(\frac{a+2b}{a+b}\right)$

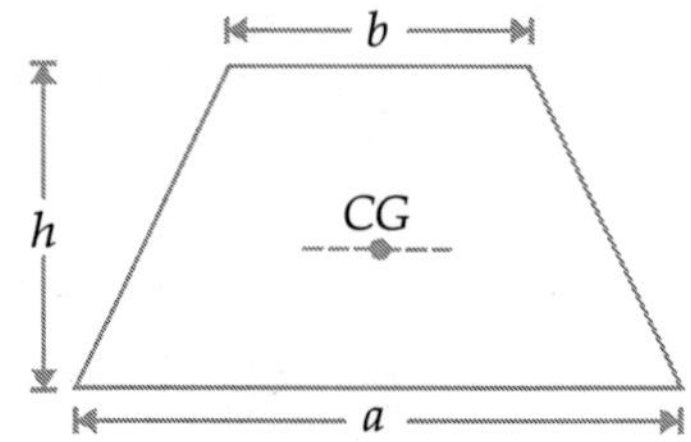

Fig. 6.57

8. If the distance between *CG* of two masses m_1 and m_2 is l, then the distance of *CG* of the composite system from mass m_1 will be

(*a*) $\frac{m_2\, l}{m_1 + m_2}$ (*b*) $\frac{m_1\, l}{m_1 + m_2}$

(*c*) $\frac{(m_1 - m_2)\, l}{m_1 + m_2}$ (*d*) $\frac{m_1}{l\,(m_1 + m_2)}$

9. A right circular cone of 20 cm height weighs 1000 N. A cone of 8 cm height and 64 N weight is removed from the top. The distance of *CG* of the frustrum from the base is approximately

(*a*) 2.5 cm (*b*) 3.25 cm (*c*) 4.25 cm (*d*) 6 cm

10. A triangular plate in the form of an isosceles triangle *ABC* has base *BC* = 10 cm and altitude = 12 cm. From this plate, a portion in the shape of an isosceles triangle *OBC* is removed. If *O* is the mid point of the altitude of triangle *ABC*, then distance of *CG* of the remainder section from the base is

(*a*) 3.5 cm (*b*) 4.8 cm (*c*) 6 cm (*d*) 8 cm

11. For what of the following configurations of plane lamina, the centre of gravity does not lie at its geometrical centre

(*a*) circle (*b*) square

(*c*) rectangle (*d*) right angled triangle

Answers :

1. (*c*) 2. (*d*) 3. (*d*) 4. (*a*) 5. (*c*) 6. (*d*)

7. (*b*) 8. (*a*) 9. (*c*) 10. (*c*) 11. (*d*)

CHAPTER

7

MOMENT OF INERITA: AREA AND MASS

Learning objectives : After having made a study of the subject matter presented in this chapter, the reader will be able to

- *define inertia, area/mass moment of inertia and radius of gyration of a lamina*
- *state and prove the parallel axis theorem and perpendicular axis theorem*
- *determine the area/mass moment of inertia of lamina of different shapes and composite sections*
- *define polar moment of inertia and state its application*
- *explain the terms principal moments of inertia and product of inertia*

It may be recalled that the moment of force about a point is the product of force (F) and the perpendicular distance (x) between the point and the line of action of force.

Moment of force = Fx

If this moment Fx is further multiplied by the distance x, then a quantity Fx^2 is obtained which is referred to as the moment of moment or the second moment of force

Moment of moment = $Fx \times x = Fx^2$

If the term force F in the above identity is replaced by area or mass of the body, the resulting parameter is called the moment of inertia (*MOI*). Thus

Moment of inertia of a plane area = Ax^2

Mass moment of inertia of a body = mx^2

where A and m respectively denote the area and mass of the body.

Inertia refers to the property of a body by virtue of which the body resists any change in its state of rest or of uniform motion. Area moment of inertia is essentially a measure of resistance to bending, and is applied while dealing with the deflection or deformation of members in bending.

The mass moment of inertia gives a measure of the resistance that body offers to change in angular velocity and accordingly is used in conjunction with rotation of rigid bodies.

7.1. MOMENT OF INERTIA AND RADIUS OF GYRATION

Moment of inertia (*MOI*) of any lamina is the second moment of all elemental areas dA comprising the lamina. With refence to Fig. 7.1.

$$I_{xx} = \text{moment of inertia about } x\text{-axis} = \Sigma\,(y\,dA)\,y$$

where $y\,dA$ is the first moment of area dA about x-axis and $(y\,dA)\,y$ is the moment of first moment (called second moment) of area dA about x-axis.

$$I_{xx} = \Sigma\, y^2\, dA$$

Likewise:

$$I_{yy} = \text{moment of inertia about } y\text{-axis}$$
$$= \Sigma\, x^2\, dA$$

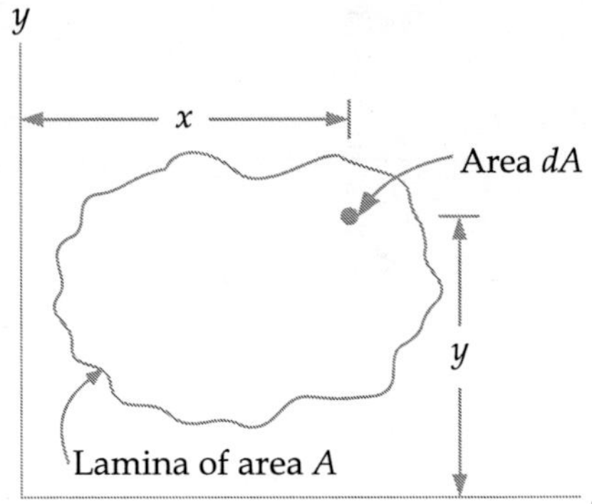

Fig. 7.1

Obviously moment of inertia of a section about an axis is prescribed by the cummulative product of area and square of the distance from that axis.

The units of moment of MOI are the fourth power of length. When the measurements are in mm, MOI has units of mm^4.

7.1.1. Parallel axis Theorem

The moment of inertia of a plane lamina about any axis is equal to the sum of its MOI about a parallel axis through its centre of gravity G and the product of its area (mass) and the square of the distance between the two axes. With reference to Fig. 7.2.

$$I_{AA} = I_{gxx} + Ah^2 \qquad \text{...(7.3)}$$

where I_{gxx} is MOI of the lamina about an axis x-x passing through its CG and I_{AA} is the MOI about any axis AA which is parallel to x-x and at a distance h from it.

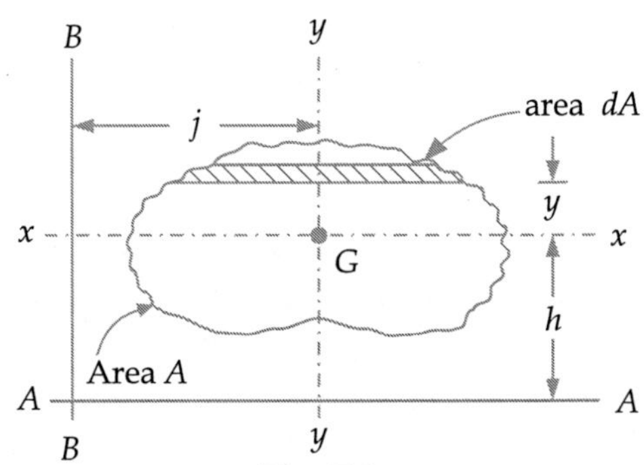

Fig. 7.2

Proof : The lamina consists of an infinite number of small elemental components parallel to the x-axis. Let one such elemental component of area dA be located at distance y from the x-axis. Obviously then its distance from the axis AA will be $(h + y)$.

Moment of inertia of the elemental component about axis AA will be

$$= dA\,(h + y)^2$$

Then moment of inertia of the entire lamina about axis AA

$$= \Sigma\, dA(h+y)^2 = \Sigma\, dAh^2 + \Sigma\, dAy^2 + \Sigma\, dA(2hy)$$

$$= h^2 \sum dA + \sum dA\,y^2 + 2h \sum dA\,y$$

Now, $\qquad h^2\, \Sigma\, dA = Ah^2 \qquad (\because\ \Sigma\, dA = A)$

$\Sigma\, dA\,y^2$ = moment of inertia of the lamina about the axis x-x.

$\Sigma\, dA\,y$ = 0 because $x - x$ is centroidal axis.

That gives: $\qquad I_{AA} = I_{xx} = Ah^2 \qquad$ ...(7.4)

Also $\qquad I_{BB} = I_{yy} = Aj^2 \qquad$ ...(7.5)

7.1.2. Perpendicular Axis Theorem

The moment of inertia of a plane lamina about an axis perpendicular to the plane of the lamina is equal to the sum of the moments of inertia of the lamina about the two axes at right angles to each other and intersecting each other at the point where the perpendicular axis passes through it.

Proof: With reference to Fig. 7.3, ox and oy are the two mutually perpendicular axes lying in the plane of lamina, and oz is the axis normal to the lamina and passing through o (the point of intersection of the axes ox and oy). The distance of an elemental component of area dA from oz, *i.e.*, from point o is r.

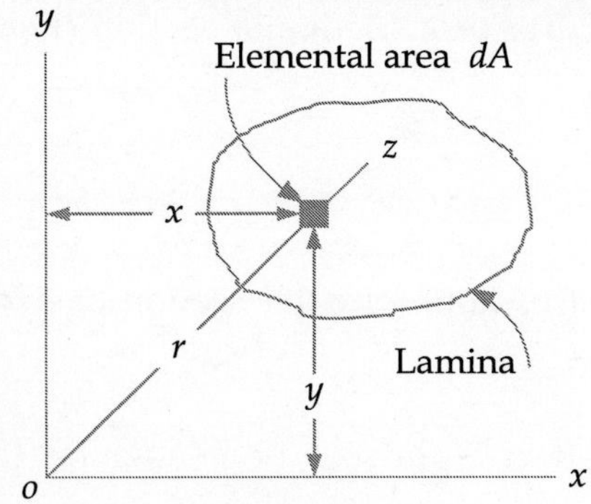

Fig. 7.3

Moment of inertia of the elemental component about axis oz

$$= dA\, r^2 = dA\,(x^2 + y^2)$$

Moment of inertia of the elemental component about axis oz

$$I_{zz} = \Sigma\, dA\,(x^2 + y^2) = \Sigma\, dA\, x^2 + \Sigma\, dA\, y^2$$

But, $\Sigma\, dA\, x^2$ = moment of inertia of the lamina about the axis $oy = I_{yy}$

$\Sigma\, dA\, y^2$ = moment of inertia of the lamina about the axis $ox = I_{xx}$

$$\therefore \quad I_{zz} = I_{xx} + I_{yy} \qquad \text{...(7.6)}$$

7.1.3. Radius of Gyration

If the entire area (or mass) of a lamina is considered to be concentrated at a point such that there is no change in the moment of inertia about a given axis, then distance of that point from the given axis is called the ***radius of gyration***.

The relation between radius of gyration k and moment of inertia I can be put in the form

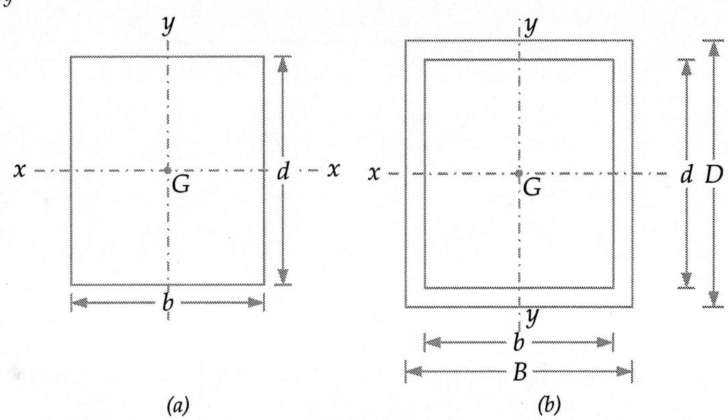

Fig. 7.4

$$I = Ak^2\ ;\quad k = \sqrt{\frac{I}{A}} \qquad \text{...(7.7)}$$

Apparently the radius of gyration of a lamina is the square root of the ratio of its moment of inertia to its area.

The moment of inertia of common standard sections are presented below:

(*i*) For a rectangular section with breadth b and depth d (Fig. 7.4 *a*).

$$I_{xx} = \frac{bd^3}{12};\quad I_{yy} = \frac{db^3}{12}$$

For a hollow rectangular section (Fig. 7.4 *b*)

$$I_{xx} = \frac{BD^3 - bd^3}{12}\ ;\quad I_{yy} = \frac{DB^3 - db^3}{12}$$

(*ii*) For a triangular section (Fig. 7.5)

$$I_{xx} = \frac{bh^3}{36}$$

$$I_{AB} = \frac{bh^3}{12} \qquad ...(7.7)$$

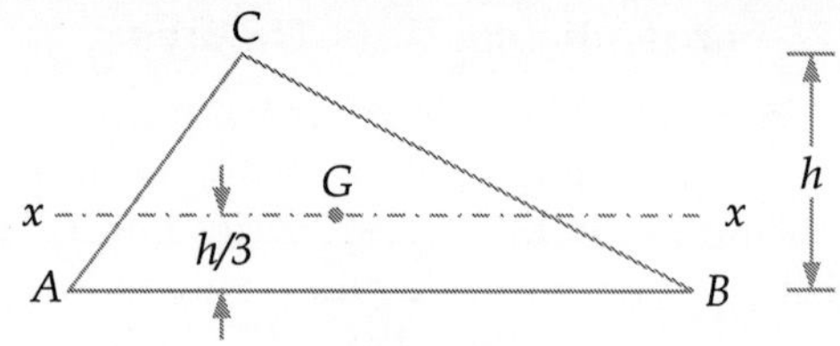

Fig. 7.5

(*iii*) For a circular section of diameter *d* (Fig. 7.6*a*)

$$I_{xx} = I_{yy} = \frac{\pi}{64} d^4$$

If *zz* is the axis perpendicular to lamina and passing through *CG*, then

$$I_{zz} = I_{xx} + I_{yy} = \frac{\pi}{32} d^4$$

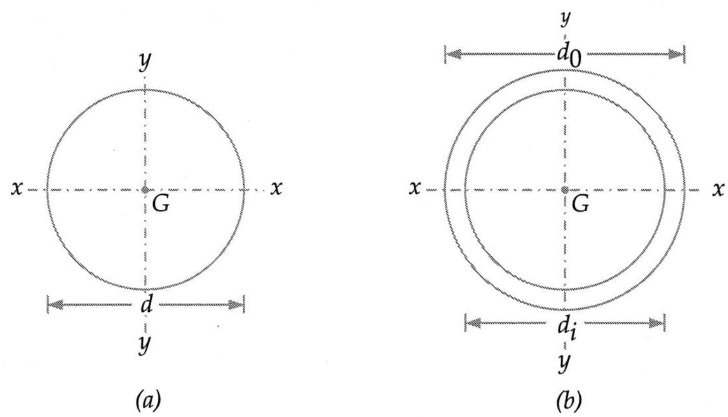

Fig. 7.6

For a hollow circular section with outer diameter d_0 and inner diameter d_i

$$I_{xx} = I_{yy} = \frac{\pi}{64}\left(d_0^4 - d_i^4\right)$$

If *zz* is the axis perpendicular to the plane of lamina and passing through *CG* then

$$I_{zz} = I_{xx} + I_{yy} = \frac{\pi}{32}\left(d_0^4 - d_i^4\right)$$

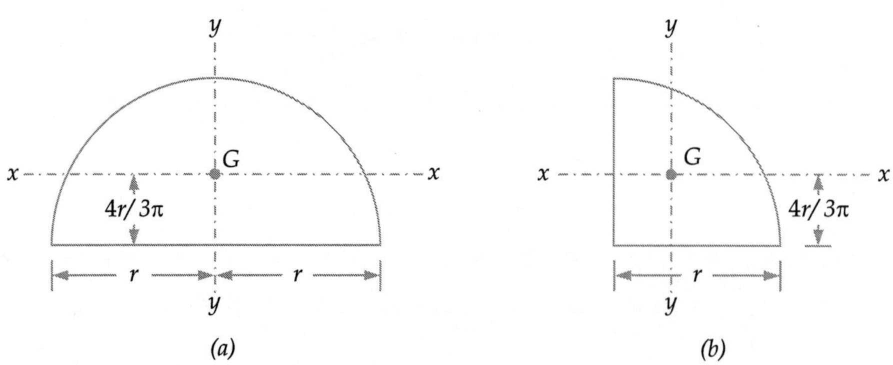

Fig. 7.7

(*iv*) For a semi-circle (Fig. 7.7 *a*)

$$I_{xx} = 0.11\ r^4 \text{ and}$$

$$I_{yy} = \pi r^4/8$$

For a quarter circle (Fig. 7.7*b*)

$$I_{xx} = I_{yy} = 0.055\ r^4$$

(*v*) For an ellipse (Fig 7.8),

$$I_{xx} = \frac{\pi ab^3}{4},\ I_{yy} = \frac{\pi ba^3}{4}$$

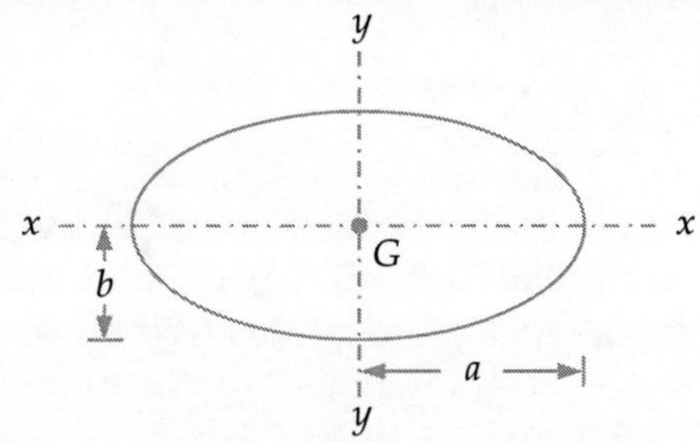

Fig. 7.8

7.2. MOMENT OF INERTIA OF LAMINAE OF DIFFERENT SHAPES

7.2.1. Rectangular Lamina

Consider a rectangular lamina *ABCD* of width *b* and depth *d*. Let *xx* and *yy* be the axis which pass through the centroid of the area and are parallel to the sides of the lamina. The centroid lies at the mid point of the width as well as the depth.

Consider a small strip of thickness *dy* located at a distance *y* from the axis *xx*.

Area of the elemental strip = *b dy*

Moment of inertia of the elemental component about the axis *xx*,

$$= dA \times y^2 = b\,dy \times y^2 = by^2\,dy$$

Moment of inertia of the entire lamina about the axis *xx*,

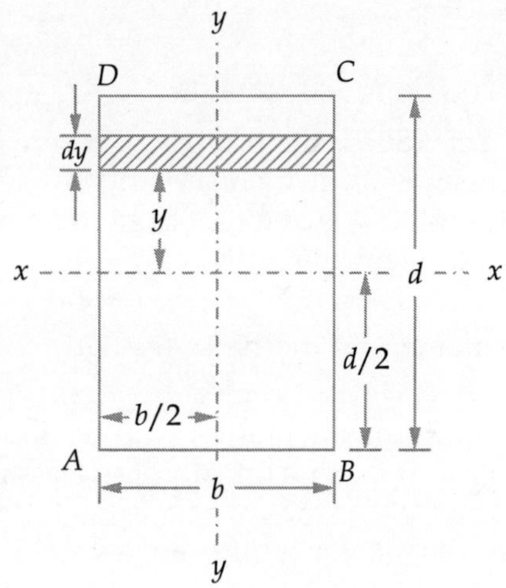

Fig. 7.9

$$I_{xx} = \int_{-d/2}^{d/2} by^2 dy = b\left|\frac{y^3}{3}\right|_{-d/2}^{d/2}$$

$$= b\left[\frac{d^3}{24} + \frac{d^3}{24}\right] = \frac{bd^3}{12} \qquad ...(7.8\ a)$$

Similarly the moment of inertia of the lamina about the axis *yy* is

$$I_{yy} = \frac{db^3}{12} \qquad ...(7.8b)$$

Let I_{AB} be the moment of inertia of the lamina about its bottom face *AB*. Then from the theorem of parallel axis

$$I_{AB} = I_{xx} + Ah^2 = \frac{bd^3}{12} + bd\left(\frac{d}{2}\right)^2 = \frac{bd^3}{12} + \frac{bd^3}{4} = \frac{bd^3}{3} \qquad ...(7.9a)$$

Similarly the moment of inertia of the lamina about the face *AD* would be

$$I_{AD} = \frac{db^3}{3} \qquad ...(7.9b)$$

From the theorem of perpendicular axis, the polar moment of inertia I_p of the lamina is

$$I_p = I_{xx} + I_{yy} = \frac{bd^3}{12} + \frac{db^3}{12} \quad ...(7.10)$$

The polar moment of inertia is the inertia about the polar axis; an axis which passes through the centroid of the lamina and is normal to it.

For a rectangular lamina ($B \times D$) with a rectangular hole ($b \times d$) made centrally (Fig. 7.10) the moment of inertia about any centroidal axis is

= MOI of bigger rectangle
- MOI of smaller rectangle

Thus:

$$I_{xx} = \frac{BD^3}{12} - \frac{bd^3}{12} \quad ...(7.11)$$

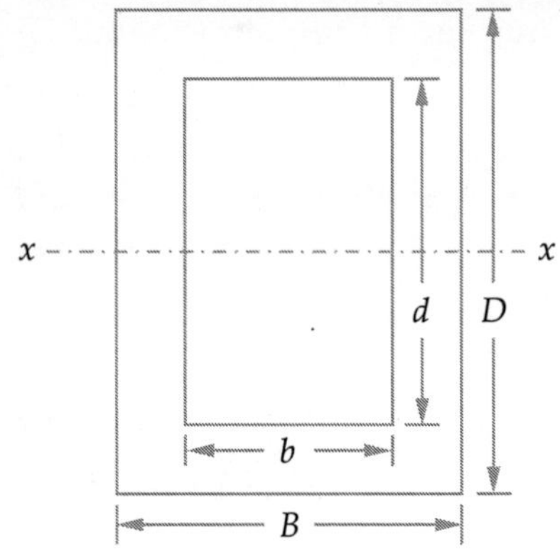

Fig. 7.10

7.2.2. Triangular Lamina

Let ABC be the triangle of base width b and height h. Consider an elementary strip of width l, thickness dy and located at distance y from the base of the triangle. For this elemental strip,

$$\text{area} = l\, dy$$

moment of inertia of this strip about base BC

$$= y^2\, dA = y^2 l\, dy$$

Since the integration is to be done with respect to y within the limits 0 to h, it is necessary to express l in terms of y. For that, we have the following correlation from the similarity of triangles ADE and ABC,

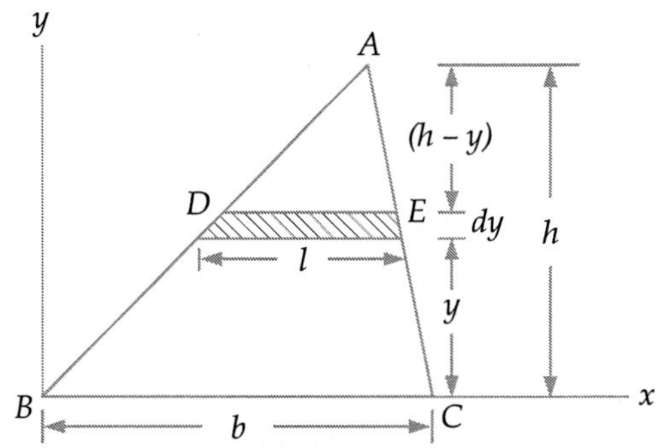

Fig. 7.11

$$\frac{l}{b} = \frac{h-y}{h}; \quad l = b\,(1 - y/h)$$

$\therefore$ Moment of inertia of the triangle about the base

$$I_{\text{base}} = \int_o^h y^2 b\left(1 - \frac{y}{h}\right) dy$$

$$= b\int_o^b\left(y^2 - \frac{y^3}{h}\right) dy = b\left|\frac{y^3}{3} - \frac{y^4}{4h}\right|_o^h = b\left(\frac{h^3}{3} - \frac{h^3}{4}\right) = \frac{bh^3}{12} \quad ...(7.12)$$

For a triangle, the centroidal axis I_{xx} is at a distance of $y_c = h/3$ from the base. Then from the theorem of parallel axis : $I_{\text{base}} = I_{xx} + Ay_c^2$, we have

$$\therefore \quad I_{xx} = I_{\text{base}} - Ay_c^2 = \frac{bh^3}{12} - \left(\frac{1}{2}bh\right) \times \left(\frac{h}{3}\right)^2 = \frac{bh^3}{12} - \frac{bh^3}{18} = \frac{bh^3}{36} \quad ...(7.13)$$

7.2.3. Circular Lamina

Consider an element of sides $r\,d\theta$ and dr within a circular lamina of radius R. Moment of this elemental area about the diametical axis x-x,

$$= y^2 dA = (r \sin\theta)^2 \times r d\theta\, dr = r^3 \sin^2\theta\, d\theta\, dr$$

∴ Moment of inertia of the entire circular lamina

$$I_{xx} = \int_o^R \int_o^{2\pi} r^3 \sin^2\theta \, d\theta \, dr$$

$$= \int_o^R \int_o^{2\pi} r^3 \frac{1-\cos 2\theta}{2} \, d\theta \, dr$$

$$= \int_o^R \frac{r^3}{2}\left[\theta - \frac{\sin 2\theta}{2}\right]_o^{2\pi} dr$$

$$= \int_o^R \frac{r^3}{2}(2\pi)\, dr = 2\pi\left|\frac{r^4}{8}\right|_o^R = \frac{\pi}{4}R^4$$

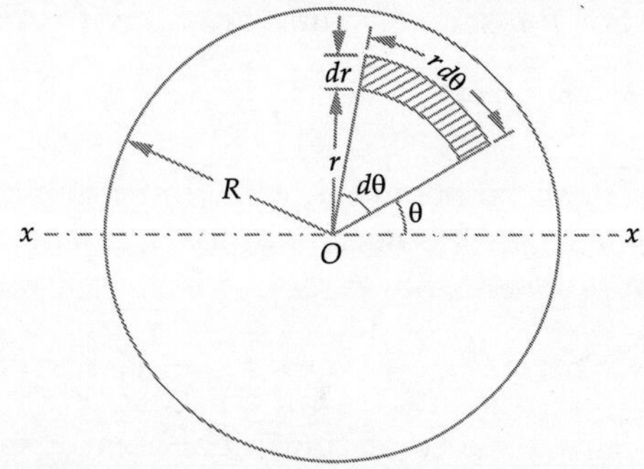

Fig. 7.12

If d is the diameter of the circular lamina, then

$$I_{xx} = \frac{\pi}{4}\left(\frac{d}{2}\right)^4 = \frac{\pi}{64}d^4 \qquad ...(7.14)$$

Likewise $\qquad I_{yy} = \frac{\pi}{64}d^4$

If zz is the axis through the centroid and normal to the plane of the lamina, then

$$I_{zz} = I_{xx} + I_{yy} = \frac{\pi}{64}d^4 + \frac{\pi}{64}d^4$$

$$= \frac{\pi}{32}d^4 \qquad ...(7.15)$$

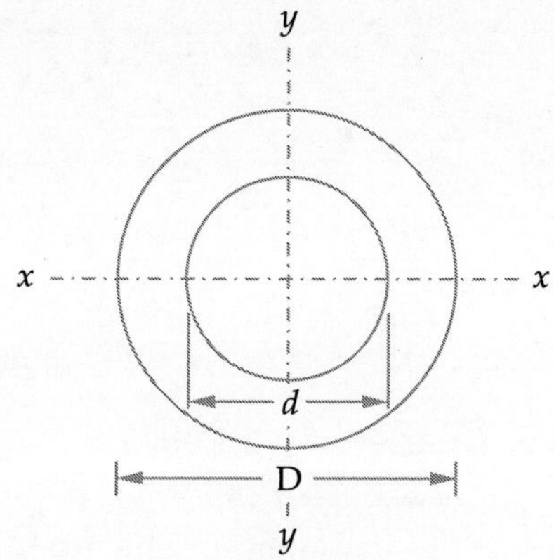

Fig. 7.13

The axis zz is called the polar axis and I_{zz} is referred to as the *polar moment of inertia*.

Polar moment of inertia has application in problems relating to torsion of cylindrical shafts and rotation of slabs.

For a circular lamina of diameter D with a central circular hole of diameter d (Fig. 7.13) the moment of inertia about any centroidal axis is

$$I_{xx} = I_{yy} = \frac{\pi}{64}(D^4 - d^4) \qquad ...(7.16)$$

The corresponding polar moment of inertia is

$$I_{zz} = I_p = \frac{\pi}{32}(D^4 - d^4) \qquad ...(7.17)$$

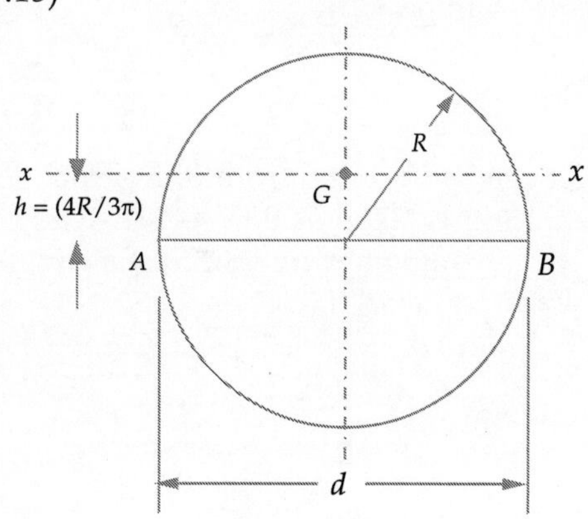

Fig. 7.14

7.2.4. Semi-circular Lamina

The moment of inertia of a circular lamina having diameter d about its diameteral axis AB is

$$= \frac{\pi}{64}d^4$$

For the semi-circular lamina with *AB* as its base, the moment of inertia about *AB* would be

$$I_{AB} = \frac{1}{2} \times \left(\frac{\pi}{64} d^4 \right) = \frac{\pi}{128} d^4$$

It can be obtained from first principles if the limit of integration is taken as 0 to π instead of 0 to 2π in the derivation of moment of inertia of a circular lamina about its diameteral axis. That is

$$I_{AB} = \int_o^R \int_o^\pi r^3 \sin^2\theta \, d\theta \, dr = \frac{\pi}{8} R^4 = \frac{\pi}{128} d^4 \qquad \text{...(7.18)}$$

The distance of centroidal axis *xx* of the semicircle from its base *AB is*

$$h = \frac{4R}{3\pi} = \frac{2d}{3\pi}$$

$$\text{Area of semicircle} = \frac{1}{2}\left(\frac{\pi}{4} d^2 \right) = \frac{\pi d^2}{8}$$

$$\text{or} \quad \frac{\pi d^4}{128} = I_{xx} + \frac{\pi d^2}{8} \times \left(\frac{2d}{3\pi} \right)^2 = I_{xx} + \frac{d^4}{18\pi}$$

$$\therefore \quad I_{xx} = \frac{\pi d^4}{128} - \frac{d^4}{18\pi} = 0.00686\, d^4 = 0.11 R^4 \qquad \text{...(7.19)}$$

7.2.5. Quarter of a Circle

Reference Fig. 7.15, *LAB* is the quadrant of a circular lamina of diameter *d*. The moment of inertia of a quadrant equals 1/4th of the moment of inertia of the circular lamina.

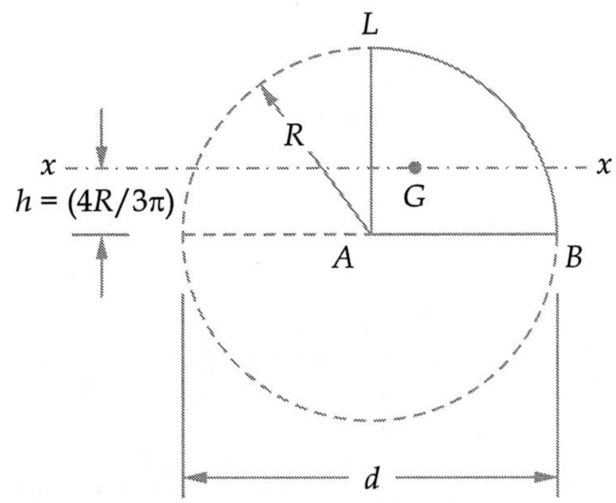

Fig. 7.15

$$\therefore \quad I_{AB} = \frac{1}{4} \times \left(\frac{\pi}{64} d^4 \right) = \frac{\pi}{256} d^4$$

It can be obtained from first principles if the limit of integration is taken as 0 to π/2 instead of 0 to 2π in the derivation of moment of inertia of a circular lamina about its diametral axis. That is

$$I_{AB} = \int_o^R \int_o^{\pi/2} r^3 \sin^2\theta \, d\theta \, dr$$

$$= \frac{\pi}{16} R^4 = \frac{\pi}{256} d^4 \qquad \text{...(7.20)}$$

The distance of the centroid of the quadrant *LAB* from *AB* is

$$h = \frac{4R}{3\pi} = \frac{2d}{3\pi}$$

$$\text{Area of quadrant} = \frac{1}{4} \times \left(\frac{\pi}{4} d^2 \right) = \frac{\pi}{16} d^2$$

From parallel axis theorem,

$$I_{AB} = I_{xx} + Ah^2$$

or $$\frac{\pi d^4}{256} = I_{xx} + \frac{\pi d^2}{16} \times \left(\frac{2d}{3\pi}\right)^2 = I_{xx} + \frac{d^4}{3\pi}$$

∴ $$I_{xx} = \frac{\pi d^4}{256} - \frac{d^4}{36\pi} = 0.00343\, d^4 = 0.055\, R^4 \qquad ...(7.21)$$

EXAMPLE 7.1

The moment of inertia of rectangular section beam about x-x and y-y axes passing though the centroid are 250×10^6 mm^4 and 40×10^6 mm^4 respectively. Calculate the size of the section.

Solution : Let b and d denote the breadth and depth respectively of the rectangular section beam. Then

$$I_{xx} = \frac{bd^3}{12}; \quad 250 \times 10^6 = \frac{bd^3}{12} \qquad ...(i)$$

and $$I_{yy} = \frac{db^3}{12}; \quad 40 \times 10^6 = \frac{db^3}{12} \qquad ...(ii)$$

Dividing expression (*i*) by expression (*ii*)

$$5.25 = \left(\frac{d}{b}\right)^2 \quad \text{or} \quad \frac{d}{b} = 2.5$$

Substituting $d = 2.5\, b$ in expression (*i*), we get

$$\frac{b}{12}(2.5\, b)^3 = 250 \times 10^6$$

or $$b^4 = \frac{250 \times 10^6 \times 12}{(2.5)^3} = 1.92 \times 10^8$$

That gives: $b = 117.7$ mm and $d = 2.5 \times 117.7 = 294.25$ mm

Therefore required size of the section is:

= 117.3 mm (breadth) × 294.25 mm (depth)

EXAMPLE 7.2

Find the moment of inertia of a rolled steel joist girder of symmetrical I section shown in Fig. 7.16.

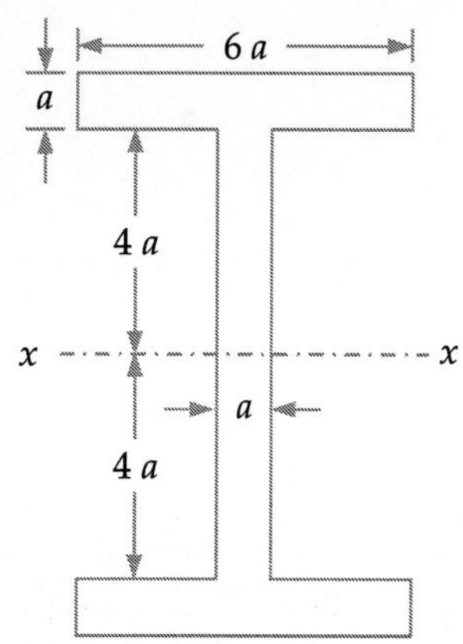

Fig. 7.16

Solution : The areas of the three rectangles comprising the I-section are:

upper flange $A_1 = 6a \times a = 6a^2$

web $A_2 = 8a \times a = 8a^2$

lower flange $A_3 = 6a \times a = 6a^2$

MOI of upper flange about x-axis (using parallel axis theorem)

$$= \frac{6a \times a^3}{12} + 6a^2 \times \left(4a + \frac{a}{2}\right)^2$$

$$= \frac{a^4}{2} + \frac{243\, a^4}{2} = 122a^4$$

MOI of web about *x*-axis $= \dfrac{a \times (8a)^3}{12} = \dfrac{128a^4}{3}$

MOI of lower flange about *x*-axis (using parallel axis theorem)

$$= \frac{6a \times a^3}{12} + 4a^2\left(4a + \frac{a}{2}\right)^2 = \frac{a^4}{2} + \frac{243\,a^4}{2} = 122\,a^4$$

∴ Total *MOI* of the given I-section about x-axis

$$= 122\,a^4 + \frac{128a^4}{3} + 122\,a^4$$

$$= \frac{860}{3}a^4$$

The *MOI* of the given I-section could also be worked out with reference to Fig. 7.17.

$$I_{xx} = I_{x1} - I_{x2}$$

$$= \frac{6a \times (10a)^3}{12} - \frac{5a \times (8a)^3}{12}$$

$$= 500\,a^4 - \frac{640}{3}a^4 = \frac{860}{3}a^4$$

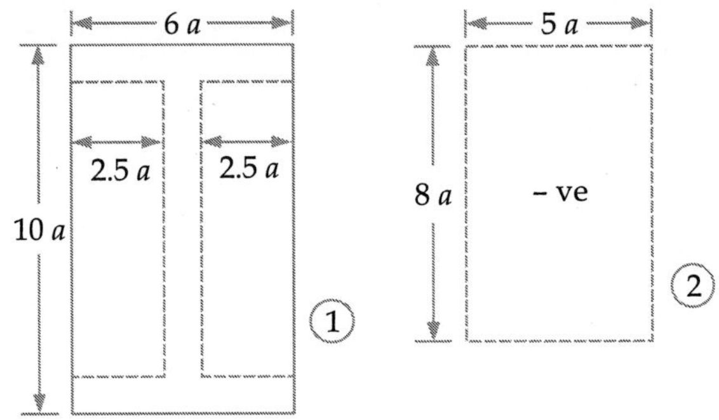

Fig. 7.17

EXAMPLE 7.3

Determine the moment of inertia of the T-section shown in Fig. 7.13 about an axis passing through the centroid and parallel to top most fibre of the section. Proceed to determine the moment of inertia about axis of symmetry and hence find out the radii of gyration.

Solution : From the calculations made in Example 7.10 the *CG* of the given T-section lies on the *y*-axis and at distance 43.71 mm from the top face of its flange

$$\bar{x} = 0 \quad \text{and} \quad \bar{y} = 43.71 \text{ mm}$$

Refering to this centroidal axis, the centroid of a_1 is (0.0, 38.71 mm) and that of a_2 is (0.0, 41.29 mm).

Moment of inertia of the section about centroid axis is

I_{xx} = *MOI* of area a_1 about centroidal axis + *MOI* of area a_2 about centroidal axis

$$= \left[\frac{160 \times 10^3}{12} + 1600 \times (38.71)^2\right] + \left[\frac{10 \times 150^3}{12} + 1500 \times (41.29)^2\right]$$

$$= 7780672 \text{ mm}^4$$

Similarly $\quad I_{yy} = \dfrac{10 \times 160^3}{12} + \dfrac{150 \times 10^3}{12} = 3425833 \text{ mm}^4$

The radius of gyration is given by $k = \sqrt{\dfrac{I}{A}}$

$$\therefore \quad k_{xx} = \sqrt{\frac{7780672}{3100}} = \mathbf{50.1\ mm}$$

$$k_{yy} = \sqrt{\frac{3425833}{3100}} = \mathbf{34.24\ mm}$$

EXAMPLE 7.4

Determine the moment of inertia of the area shown shaded in Fig. 7.18 about axis *xx* which coincides with the base edge *AB*.

Solution : The given section comprises the full rectangle *ABCD* minus the semi-circle *DEC*.

Moment of cnertia of rectangle *ABCD* about *AB*

$$I_1 = I_{G1} + A_1 h_1^2$$

$$= \frac{2 \times 2.5^3}{12} + (2 \times 2.5) \times 1.25^2$$

$$= 2.604 + 7.812 = 10.416 \text{ cm}^4$$

Moment of inertia of semi-circle about *AB*

$$I_2 = I_{G2} + A_2 h_2^2$$

$$= 0.11\, r^2 + \frac{1}{2}\pi r^2 \times \left(2.5 - \frac{4r}{3\pi}\right)^2$$

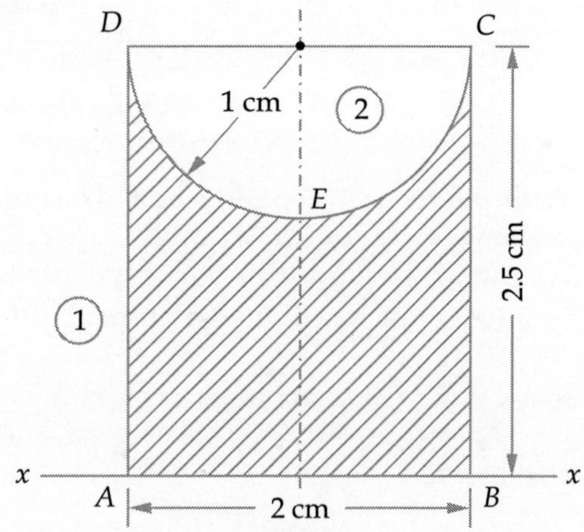

Fig. 7.18

The parameter $\frac{4r}{3\pi}$ is the distance of centroid of semi-circle from *DC*.

$$\therefore \quad I_2 = 0.11 \times 1^2 + \frac{1}{2}\pi \times (1)^2 \times \left(2.5 - \frac{4 \times}{3\pi}\right. = 0.11 + 6.76 = 6.87 \text{ cm}^4$$

∴ Moment of inertia of shaded area bout *AB* = 10.416 – 6.87 = 3.546 cm^4

EXAMPLE 7.5

Determine the polar moment of inertia of the I-section shown in Fig. 7.19 (*a*). Also make calculations for the radius of gyration with respect to *x*-axis and *y*-axis.

Solution : The I-section is symmetrical about *y*-axis and accordingly its *CG* lies at point *G* on the *y*-axis, *i.e.*, *x* = 0. Further, the bottom fibre of lower flange has been chosen as reference axis to locate the centroid $\bar{y}$.

The areas and co-ordinates of centroids of the three rectangles comprising the given section are:

Lower flange: $a_1 = 10 \times 1 = 10 \text{ cm}^2$

$$y_1 = \frac{1}{2} = 0.5 \text{ cm}$$

Web: $a_2 = 12 \times 1 = 12 \text{ cm}^2$

$$y_2 = 1 + \frac{12}{2} = 7 \text{ cm}$$

Upper flange: $a_3 = 8 \times 1\ 8 \text{ cm}^2$

$$y_3 = 1 + 12 + \frac{1}{2}$$

$$= 13.5 \text{ cm}$$

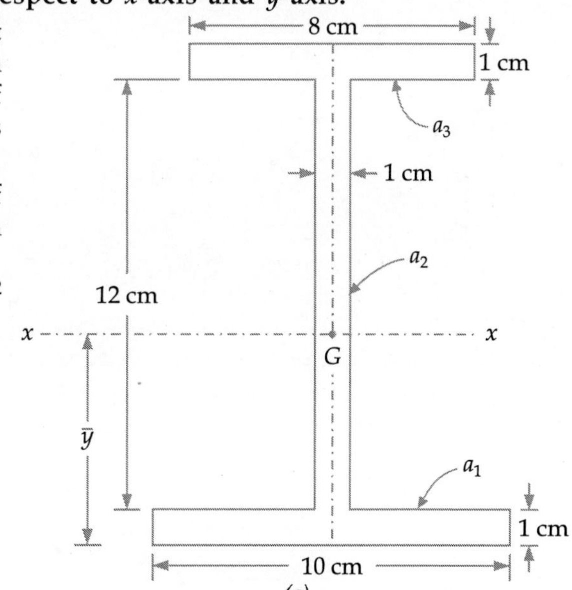

(*a*)

Fig. 7.19 (a)

Then: $\bar{y} = \dfrac{a_1y_1 + a_2y_2 + a_3y_3}{a_1 + a_2 + a_3}$

$= \dfrac{10 \times 0.5 + 12 \times 7 + 8 \times 13.5}{10 + 12 + 8}$

$= \dfrac{5 + 84 + 108}{30} = 5.57$ cm

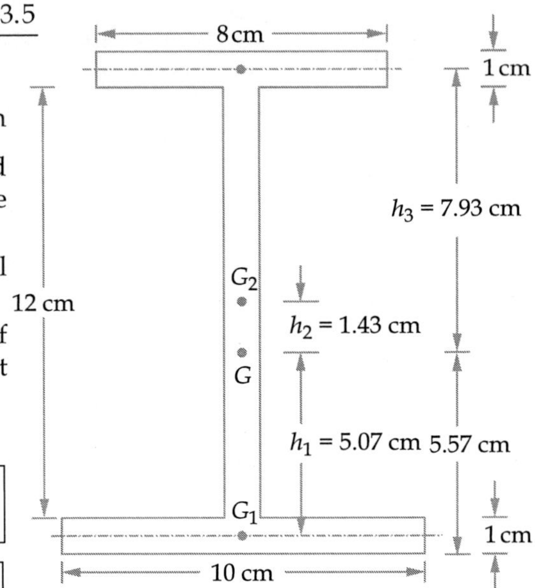

Fig. 7.19 (b)

With reference to the centroidal axes, the centroid of the lower flange, web and upper flange are (0, 5.07), (0, 1.43) and (0, 7.93) respectively.

Moment of inertia of the I-section about centroidal axis is

= *MOI* of area a_1 about centroidal axis + *MOI* of area a_2 about centroidal axis + *MOI* of area a_3 about centroidal axis.

$= (I_{G1} + A_1h_1^2) + (I_{G2} + A_2h_2^2) + (I_{G3} + A_3h_3^2)$

$= \left[\dfrac{10 \times 1^3}{12} + 10 \times (5.07)^2\right] + \left[\dfrac{1 \times 12^3}{12} + 12 \times (1.43)^2\right] + \left[\dfrac{8 \times 1^3}{12} + 8 \times (7.93)^2\right]$

$= (0.833 + 257.05) + (144 + 24.54) + (0.67 + 503.08) = 930.17 \text{ cm}^4$

and $I_{yy} = \dfrac{1 \times 10^3}{12} + \dfrac{12 \times 1^3}{12} + \dfrac{1 \times 8^3}{12} = 83.33 + 1 + 42.64 = 127 \text{ cm}^4$

Polar moment of inertia $= I_{xx} + I_{yy} = 930.17 + 127 =$ **1057.17 cm⁴**

(*b*) The radius of gyration is given by: $k = \sqrt{\dfrac{I}{A}}$

$\therefore \quad k_{xx} = \sqrt{\dfrac{930.17}{30}} =$ **5.567 cm**

$k_{yy} = \sqrt{\dfrac{127}{30}} =$ **2.057 cm**

EXAMPLE 7.6

Determine the moment of inertia about centroidal axes *x-x* and *y-y* of the channel section shown in Fig. 7.20.

Solution : The section is divided into three rectangles with areas

$A_1 = 10 \times 1.5 = 15 \text{ cm}^2$

$A_2 = (40 - 1.5 - 1.5) \times 1$

$= 37 \text{ cm}^2$

$A_3 = 10 \times 1.5 = 15 \text{ cm}^2$.

$\Sigma A = A_1 + A_2 + A_3$

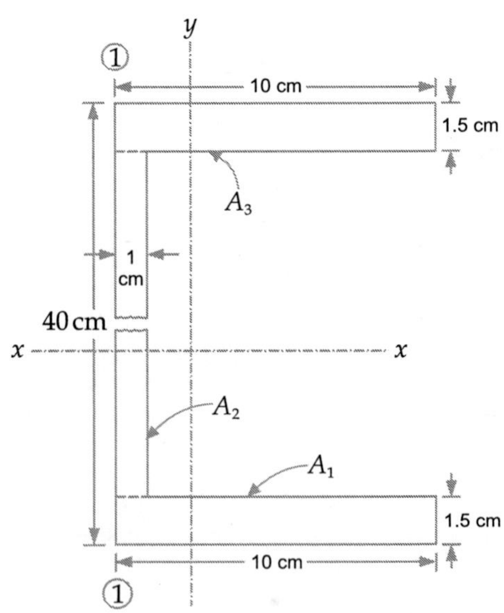

Fig. 7.20

$= 15 + 37 + 15$

$= 67 \text{ cm}^2$

The given section is symmetrical about the horizontal axis passing through the centroid of rectangle A_2.

The distance of the centroid of the section with reference to section 1 - 1 is

$$\frac{\Sigma Ax}{\Sigma A} = \frac{(15 \times 5) + (37 \times 1/2) + (15 \times 5)}{67} = 2.51 \text{ cm}$$

With reference to the centroidal axes x-x and y-y, the centroids of the rectangles are:

$[(5 - 2.51), (40/2 - 1.5/2)]$ or (2.49, 19.25) for rectangle A_1

[(2.51 - 1/2), 0.0] or (2.01, 0.0) for rectangle A_2

$[(5 - 2.51), (40/2 - 1.5/2)]$ or (2.49, 19.25) for rectangle A_3

Then invoking parallel axis theorem, the moment of inertia of areas A_1, A_2, and A_3 about x-x,

$$I_{xx} = \left[\frac{10 \times 1.5^3}{12} + 15 \times 19.25^2\right] + \left[\frac{1 \times 37^3}{12}\right] + \left[\frac{10 \times 1.5^3}{12} + 15 \times 19.25^2\right]$$

$$= (2.812 + 5558.437) + (4221.083) + (2.812 + 558.437) = \mathbf{155343.58 \text{ cm}^4}$$

Similarly,

$$I_{yy} = \left[\frac{1.5 \times 10^3}{12} + 15 \times 2.49^2\right] + \left[\frac{37 \times 1^3}{12}\right] + \left[\frac{1.5 \times 10^3}{12} + 15 \times 2.49^2\right]$$

$$= (125 + 93.00) + 3.08 + (125 + 93.00) = \mathbf{439.08 \text{ cm}^4}$$

EXAMPLE 7.7

Determine I_{xx} and I_{yy} of the cross-section of a cast iron beam shown in Fig. 7.21.

Solution : The *MOI* of the given sections can be worked out by looking it as a rectangle minus two semi-circles.

$\therefore$ $I_{xx} = I_{xx}$ of rectangle $- I_{xx}$ of circular part

$$= \frac{bd^3}{12} - \frac{\pi r^4}{4}$$

$$= \frac{12 \times 15^3}{12} - \frac{\pi \times 50^3}{4}$$

$= 33.75 - 490.87$

$= \mathbf{2884.13 \text{ cm}^4}$

Likewise : $I_{yy} = I_{yy}$ of rectangle $- I_{yy}$ of semi-circular parts

$$I_{yy} \text{ of rectangle} = \frac{15 \times 12^3}{12} = 2160 \text{ cm}^4$$

For the semi-circular part ACB;

MOI about its diameter, $I_{AB} = \frac{1}{2} \times \frac{\pi \times 5^4}{4} = 245.43 \text{ cm}^4$

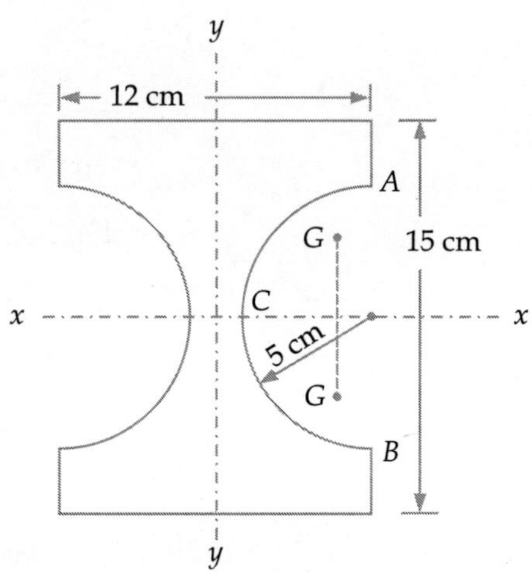

Fig. 7.21

Distance of its *CG* from the diameter,

$$h = \frac{4r}{3\pi} = \frac{4 \times 5}{3\pi} = 2.12 \text{ cm}$$

$$\text{Area } A = \frac{1}{2}\pi r^2 = \frac{1}{2}\pi \times 5^2 = 39.27 \text{ cm}^2$$

From the correlation, $I_{AB} = I_{GG} + Ah^2$, the moment of inertia of semi-circular part about its centroidal axis

$$I_{GG} = 245.43 - 39.27 \times (2.12)^2 = 68.94 \text{ cm}^4$$

Again from the parallel axis theorem,

$$I_{yy} = I_{GG} + Ah_1^2$$

where h_1 = distance between axis and *G*-axis, = 6 – 2.12 = 3.88 cm

$$\therefore \quad I_{yy} = 68.94 + 39.27 \times 3.88^2 = 660.13 \text{ cm}^4$$

Since there are two semi-circular parts,

I_{yy} for two semi-circular parts = 2 × 660.13 = 1320.26 cm^4

∴ I_{yy} for the section = 2160 – 1320.26 = **839.74 cm^4**

EXAMPLE 7.8

Determine the moments of inertia about the *x* and *y* centroidal axis of a beam whose cross-sectional area is as shown in Fig. 7.22. All dimensions are in cm.

Solution : The given section has been divided into three segments marked 1, 2 and 3

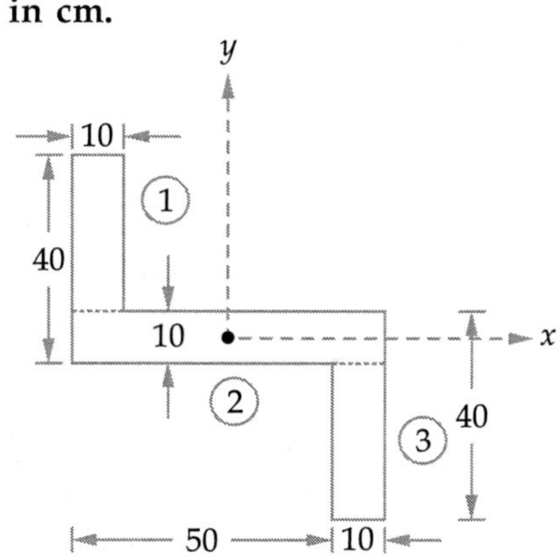

Fig. 7.22

$$(I_{xx})_1 = I_{G_1} + A_1h_1^2 = I_{G_1} + A_1(\bar{y} - y_1)^2$$

$$= \frac{1}{12} \times 10 \times 30^3 + (30 \times 10)(35 - 15)^2$$

$$= 1.425 \times 10^5 \text{ cm}^4$$

$$(I_{xx})_2 = I_{G_2} + A_2h_2^2 = I_{G_2} + A_2(\bar{y} - y_2)^2$$

$$= \frac{1}{12} \times 10^3 \times 60 + (60 \times 10) \times 0$$

$$= 0.05 \times 10^5 \text{ cm}^4$$

$$(I_{xx})^3 = I_{G_3} + A_3h_3^2 = I_{G_3} + A_3(\bar{y} - y_3)^2$$

$$= \frac{1}{12} \times 10 \times 30^3 + (30 \times 10)(35 - 15) = 1.425 \times 10^5 \text{ cm}^4$$

$$\therefore \quad I_{xx} = 1.425 \times 10^5 + 0.05 \times 10^5 + 1.425 \times 10^5 = \mathbf{2.90 \times 10^5 \text{ cm}^4}$$

$$(I_{yy})_1 = I_{G_1} + A_1h_1^2 = I_{G_1} + A_1(\bar{x} - x_1)^2$$

$$= \frac{1}{12} \times 30 \times 10^3 + (30 \times 10) \times (30 - 5 = 1.9 \times 10^5 \text{ cm}^4$$

$$(I_{yy})_2 = I_{G_2} + A_2h_2^2 = I_{G_2} + A_2(\bar{x} - x_2)^2$$

$$= \frac{1}{12} \times 60^3 \times 10 + (60 \times 10) \times 0 = 1.8 \times 10^5 \text{ cm}^4$$

$$(I_{yy})_3 = I_{G_3} + A_3h_3^2 = I_{G_3} + A_3(\bar{x} - x_3)^2$$

$$= \frac{1}{12} \times 30 \times 10^3 + (30 \times 10) \times (30 - 5$$
$$= 1.9 \times 10^5 \text{ cm}^4$$
$$\therefore \quad I_{yy} = 1.9 \times 10^5 + 1.8 \times 10^5 + 1.9 \times 10^5 = \mathbf{5.6 \times 10^5 \text{ cm}^4}$$

EXAMPLE 7.9

Find the moment of inertia about the centroid horizontal axis of the area shown shaded in Fig. 7.23. The section consists of triangle *ABC*, semi-circle on *BC* as diameter, and a circular hole of diameter 4 cm with its centre on *BC*.

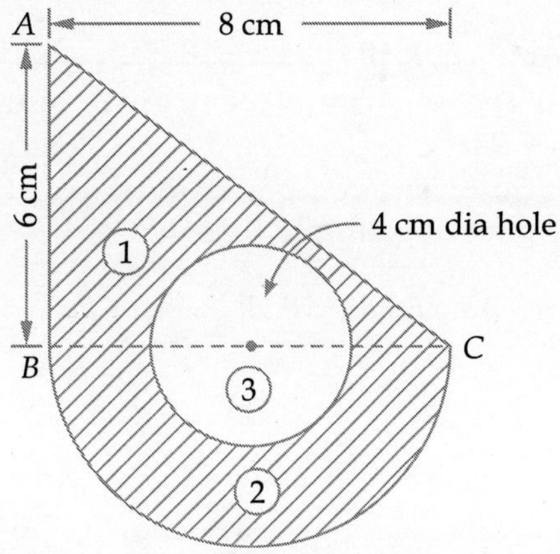

Fig. 7.23

Solution : The shaded area can be considered as a triangle (1), semicircle (2) and a circular hole (3)

Location of Centroid : For the triangular element,

$$a_1 = \frac{1}{2} \times 6 \times 8 = 24 \text{ cm}^2$$

y_1 (distance of centroid from *BC*)

$$= \frac{6}{3} = 2 \text{ cm}$$

For the semi-circular element,

$$a_2 = \frac{1}{2} \pi r^2$$
$$= \frac{1}{2} \pi \times 4^2 = 25.12 \text{ cm}^2$$

$$y_2 \text{(distance of centroid from } BC) = \frac{-4r}{3\pi} = \frac{-4 \times 4}{3\pi} = -1.7 \text{ cm}$$

The negative sign stems from the fact that it lies below *BC*.

For the circular hole

$$a_3 = \pi r^2 = \pi \times 2^2 = 12.56 \text{ cm}^2 \text{ (this area is removed)}$$
$$y_3 = 0 \text{ (centroid lies on } BC)$$

$\therefore$ Distance of the centroid of the shaded area from *BC*

$$= \frac{\Sigma ay}{\Sigma a} = \frac{a_1 y_1 + a_2 y_2 - a_3 y_3}{a_1 + a_2 - a_3} = \frac{24 \times 2 + 25.12 \times (-1.7) - 12.56 \times 0}{24 + 25.12 - 12.56} = 0.145 \text{ cm}$$

Moment of Inertia

$$I_1 = \text{moment of inertia of triangle } ABC \text{ about base } BC = \frac{1}{12} bh^3 = \frac{1}{12} \times 8 \times 6^3$$
$$= 144 \text{ cm}^4$$

$$I_2 = \text{moment of inertia of semi-circle about } BC = \frac{1}{128} \pi d^4 = \frac{1}{128} \times \pi \times 8^4$$
$$= 100.48 \text{ cm}^4$$

$$I_3 = \text{moment of inertia of circular hole about } BC = \frac{\pi}{64} d^4 = \frac{\pi}{64} \times 4^4 = 12.56 \text{ cm}^2$$

∴ Moment of inertia of the shaded area about BC

$$= 144 + 100.48 - 12.56 = 231.92 \text{ cm}^4$$

Area of the shaded portion = 24 + 25.12 – 12.56 = 36.56 cm

Invoking parallel axis theorem,

Moment of inertia of shaded area about centroidal axis

$$I_G = I_{BC} - Ah^2$$
$$= 231.92 - 36.56 \times 0.145^2$$
$$= \mathbf{231.15 \text{ cm}^4}$$

EXAMPLE 7.10

Find the horizontal centroidal moment of inertia of the lamina ABCDEFG shown in Fig. 7.24.

Solution : The composite figure is divided into the following simple figures:

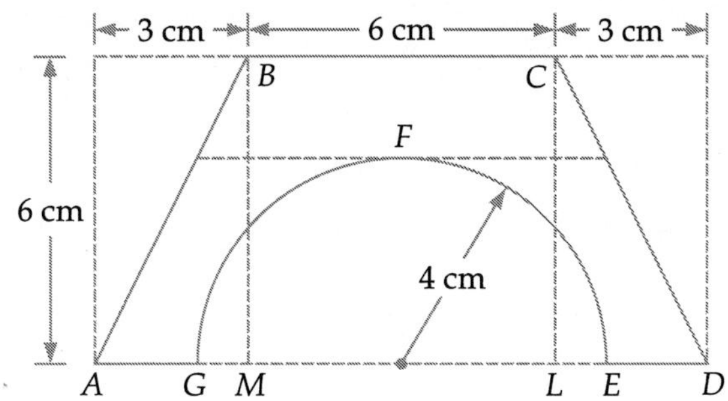

Fig. 7.24

1. A triangle ABM: $A_1 = \frac{1}{2} bh$

$$= \frac{1}{2} \times 3 \times 6 = 9 \text{ cm}^2$$

$$I_{G_1} = \frac{bh^3}{36} = \frac{3 \times 6^3}{36} = 18 \text{ cm}^4$$

y_1 (centroidal distance from line AD) $= \frac{6}{3} = 2$ cm

2. A square $BCLM$: $A_2 = 6 \times 6 = 36 \text{ cm}^2$

$$I_{G_2} = \frac{6 \times 6^3}{12} = 108 \text{ cm}^4$$

y_2 (centroidal distance from line AD) $= \frac{6}{2} = 3$ cm

3. A triangle CDL: $A_3 = \frac{1}{2} bh = \frac{1}{2} \times 3 \times 6 = 9 \text{ cm}^2$

$$I_{G_3} = \frac{bh^3}{36} = \frac{3 \times 6^3}{36} = 18 \text{ cm}^4$$

y_3 (centroidal distance from line AD) $= \frac{6}{3} = 2$ cm

4. A semi-circle GFE to be subtracted: $A_4 = \frac{\pi r^2}{2} = \frac{\pi \times 4^2}{2} = 25.12 \text{ cm}^2$ (– ve)

$$I_{G_4} = 0.11 \text{ r}^4 = 0.11 \times 4^4 = 28.16 \text{ cm}^4$$

y_4 (centroidal distance from line AD) $= \frac{4r}{3\pi} = \frac{4 \times 4}{3\pi} = 1.698$ cm

For the composite section

$$\bar{y} = \frac{\Sigma Ay}{\Sigma A} = \frac{(9 \times 2) + (36 \times 3) + (9 \times 2) - (25.1}{9 + 36 + 9 - 25.12}$$

$$= \frac{18 + 108 + 18 - 42.65}{28.88} = 3.51 \text{ cm}$$

Then $I_{xx} = I_{xx1} + I_{xx2} + I_{xx3} - I_{xx4}$

$$= [18 + 9 \times (3.51 - 2)^2] + [108 + 36 \times (3.51 - 3)^2] + [18 + 9 \times (3.51 - 2)^2] - [28.16 + 25.12\,(3.51 - 1.698)^2]$$

The above relation has been written by applying the parallel axis theorem:

$$I_{xx} = I_{GG} + Ah^2$$

$$\therefore \quad I_{xx} = 38.52 + 117.36 + 38.52 - 110.64 = \mathbf{83.76\ cm^4}$$

EXAMPLE 7.11

For the shaded area shown in Fig. 7.25, determine:

(*a*) co-ordinates of the centroid, and

(*b*) moment of inertia about the centroidal axes

The removed area is semicircular and all the given dimensions are in cm.

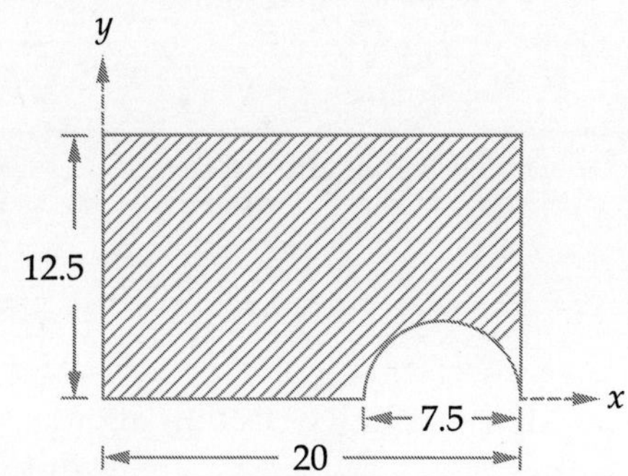

Fig. 7.25

Solution: The shaded area is the area of the rectangle 20 × 12.5 cm minus the area of semicircle of radius 3.75 cm.

For the rectangle:

$$\text{area } a_1 = 20 \times 12.5 = 250 \text{ cm}^2$$

$$y_1 = \frac{12.5}{2} = 6.25 \text{ cm}; \quad x_1 = \frac{20}{2} = 10 \text{ cm}$$

For the semicircle:

$$\text{area } a_2 = \frac{1}{2}\pi r^2 = \frac{1}{2} \times \pi \times 3.75^2 = 22.08 \text{ cm}^2$$

$$y_2 = \frac{4r}{3\pi} = \frac{4 \times 3.75}{3\pi} = 1.59 \text{ cm}$$

$$x_2 = (20 - 3.75) = 16.25 \text{ cm}$$

∴ For the shaded section,

$\bar{x}$ (distance of centroid from axis *oy*)

$$= \frac{a_1x_1 - a_2x_2}{a_1 - a_2} = \frac{250 \times 10 - 22.08 \times 16.25}{250 - 22.08} = \mathbf{9.39\ cm}$$

$\bar{y}$ (distance of centroid from axis *ox*)

$$= \frac{a_1y_1 - a_2y_2}{a_1 - a_2} = \frac{250 \times 6.25 - 22.08 \times 1.59}{250 - 22.08} = \mathbf{6.70\ cm}$$

Thus the co-ordinates of the centroid are: $\bar{x}$ = 9.39 cm and $\bar{y}$ = 6.70 cm

Moment of inertia of the shaded section = MOI of rectangle – MOI of semicircle

(*i*) With reference to centroidal *x*-axis

$$\text{MOI of rectangle} = \frac{20 \times 12.5^3}{12} + (20 \times 12.5) \times (6.7 - 6.$$

$$= 3255.21 + 50.62 = 3305.83 \text{ cm}^4$$

$$\text{MOI of semicircle} = 0.11r^4 + \frac{1}{2}\pi r^2(\bar{y} - y_2)^2$$

$$= 0.11 \times 3.75^4 + \frac{1}{2}\pi(3.75)^2 \times (6.70 -$$

$$= 21.75 + 576.50 = 598.25 \text{ cm}^4$$

∴ MOI of shaded section about centroidal x-axis

$$= 3305.83 - 598.25 = \mathbf{2707.58\ cm^4}$$

(*ii*) With reference to centroidal y-axis

$$\text{MOI of rectangle} = \frac{12.5 \times 20^3}{12} + (20 \times 12.5)(10 - 9.3$$

$$= 8333.33 + 93.02 = 8426.35 \text{ cm}^4$$

$$\text{MOI of semicircle} = \frac{\pi r^4}{8} + \frac{1}{2}\pi r^2(20 - 3.75 - 9.39)^2$$

$$= \frac{\pi \times 3.75^4}{8} + \frac{1}{2}\pi(3.75)^2 \times (6.86)^2$$

$$= 77.62 + 1038.98 = 1116.6 \text{ cm}^4$$

∴ MOI of shaded section about centroidal y-axis

$$= 8426.35 - 1116.6 = \mathbf{7309.75\ cm^4}$$

EXAMPLE 7.12

Determine the moment of inertia of the plane area shown in Fig. 7.26, about its centroidal axis.

Solution : The given section has been divided into three segments marked as 1, 2 and 3.

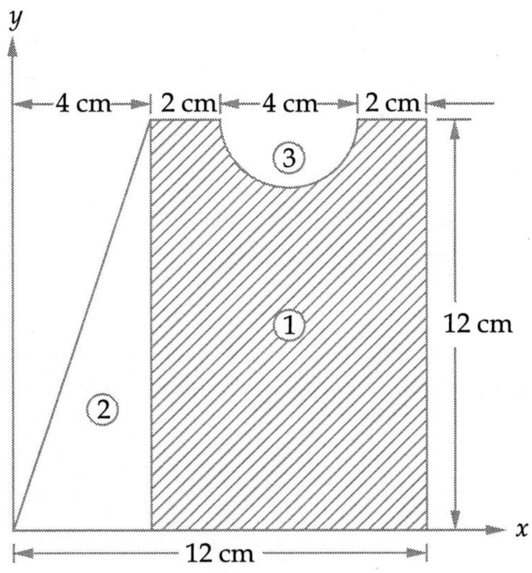

Fig. 7.26

1. Rectangle

$$a_1 = 8 \times 12 = 96 \text{ cm}^2$$

$$x_1 = 4 + \frac{8}{2} = 8 \text{ cm}; \quad y_1 = \frac{12}{2} = 6 \text{ cm}$$

2. Triangle

$$a_2 = \frac{1}{2} \times 4 \times 12 = 24 \text{ cm}^2$$

$$x_2 = \frac{2}{3} \times 4 = 2.67 \text{ cm};$$

$$y_2 = \frac{12}{3} = 4 \text{ cm}$$

3. Semicircle

$$a_3 = \frac{1}{2} \times \left[\frac{\pi}{4} \times (4)^2\right] = 6.28 \text{ cm}^2$$

$$x_3 = 4 + 2 + \frac{4}{2} = 8 \text{ cm}; \qquad y_3 = 12 - \left(\frac{4 \times 2}{3\pi}\right) = 11.15 \text{ cm}$$

Then the coordinates of the centroid of the given plane area are :

$$\bar{x} = \frac{a_1x_1 + a_2x_2 - a_3x_3}{a_1 + a_2 - a_3} = \frac{96 \times 8 + 24 \times 2.67 - (6.28 \times 8)}{96 + 24 - 6.28} = \frac{768 + 64.08 - 50.24}{113.72}$$

$$= 6.87 \text{ cm}$$

$$\bar{y} = \frac{a_1y_1 + a_2y_2 - a_3y_3}{a_1 + a_2 - a_3} = \frac{96 \times 6 + 24 \times 4 - 6.28 \times 11.15}{96 + 24 - 6.28} = \frac{576 + 96 - 70.02}{113.72}$$

$$= 5.29 \text{ cm}$$

The centroid of the given lamina have been shown in Fig. 7.27.

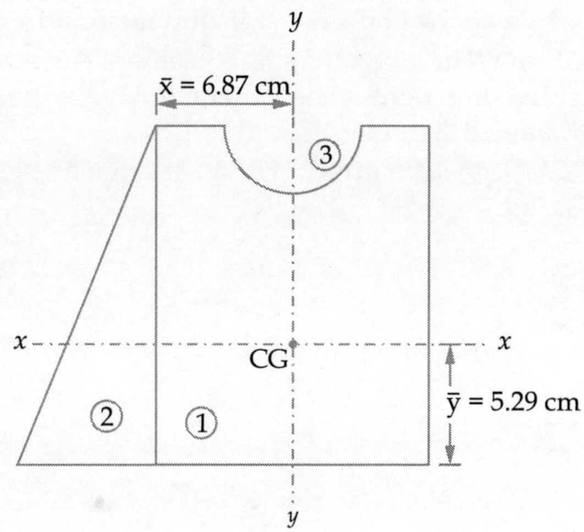

Fig. 7.27

Moment of inertia about horizontal centroidal axis:

$$I_{xx} = I_1 + I_2 - I_3$$

where

$$I_1 = I_{G_1} + A_1h_1^2$$

$$= I_{G_1} + A_1(\bar{y} - y_1)^2$$

$$= \frac{8 \times 12^3}{12} + (8 \times 12)(5.29 - 6)^2$$

$$= 1200.39 \text{ cm}^4$$

$$I_2 = I_{G_2} + A_2h_2^2$$

$$= I_{G_2} + A_2(\bar{y} - y_2)^2$$

$$= \frac{4 \times 12^3}{36} + \left(\frac{1}{2} \times 4 \times 12\right)(5.29 - 4)^2 = 231.94 \text{ cm}^4$$

$$I_3 = I_{G_3} + A_3h_3^2 = I_{G_3} + A_3(\bar{y} - y_3)^2$$

$$= 0.11r^4 + \frac{1}{2}\pi r^2\left(\bar{y} - \frac{4r}{3\pi}\right)^2$$

$$= 0.11 \times 2^4 + \frac{1}{2}\pi \times 2^2\left(5.29 - \frac{4 \times 2}{3\pi}\right) = 125.62 \text{ cm}^4$$

$$\therefore \quad I_{xx} = 1200.39 + 231.94 - 125.62 = \mathbf{1306.71\ cm^4}$$

Moment of inertia about vertical centroidal axis

$$I_{yy} = I_1 + I_2 - I_3$$

where

$$I_1 = I_{G_1} + A_1h_1^2 = I_{G_1} + A_1(\bar{x} - x_1)^2$$

$$= \frac{12 \times 8^3}{12} + (12 \times 8) \times (6.87 - 8)^2 = 634.58 \text{ cm}^4$$

$$I_2 = I_{G_2} + A_2h_2^2 = I_{G_2} + A_2(\bar{x} - x_2)^2$$

$$= \frac{12 \times 4^3}{36} + \left(\frac{1}{2} \times 12 \times 4\right) \times \left(6.87 - \frac{8}{3}\right) = 444.69 \text{ cm}^4$$

$$I_3 = I_{G_3} + A_3 h_3^2 = \frac{\pi r^4}{8} + \frac{1}{2}\pi r^2(\bar{x} - x_3)^2$$

$$= \frac{\pi \times 2^4}{8} + \frac{1}{2}\pi(2)^2(6.87 - 8)^2 = 14.3 \text{ cm}^4$$

$$\therefore \quad I_{yy} = 634.58 + 444.69 - 14.3 = \mathbf{1064.97 \text{ cm}^4}$$

EXAMPLE 7.13

Make calculations for the moment of inertia of the triangular element *ABC* shown in Fig. 7.28, about its centroidal axes. All dimensions in cm.

Solution : The given triangular element *ABC* can be considered as right angled triangle *ABD* minus the right angled triangle *ACD*.

Location of centroid : For the triangular element *ABD*,

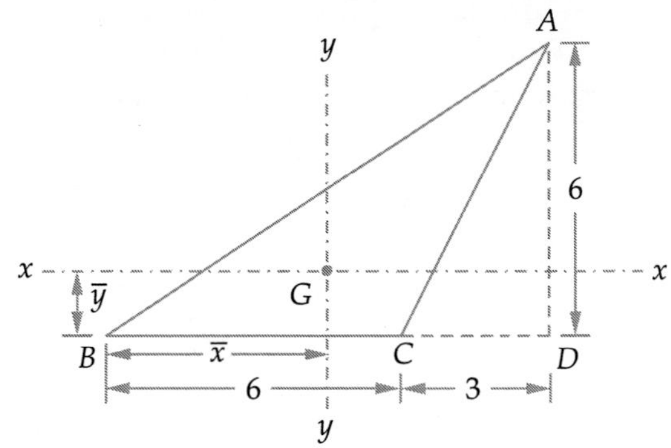

Fig. 7.28

$$a_1 = \frac{1}{2} \times 6 \times 9 = 27 \text{ cm}^2$$

$$x_1 \text{ (measured from } B) = \frac{2}{3} \times 9 = 6 \text{ cm}$$

$$y_1 \text{ (measured from base } BC) = \frac{1}{3} \times 6 = 2 \text{ cm}$$

For the triangular element *ACD*,

$$a_2 = \frac{1}{2} \times 3 \times 6 = 9 \text{ cm}^2$$

$$x_2 \text{ (measured from } B) = 6 + \frac{2}{3} \times 3 = 8 \text{ cm}$$

$$y_2 \text{ (measure from base } BC) = \frac{1}{3} \times 6 = 2 \text{ cm}$$

Then for the given triangular section,

$$\bar{x} = \frac{a_1 x_1 - a_2 x_2}{a_1 - a_2} = \frac{27 \times 6 - 9 \times 8}{27 - 9} = 5 \text{ cm}$$

$$\bar{y} = \frac{a_1 y_1 - a_2 y_2}{a_1 - a_2} = \frac{27 \times 2 - 9 \times 2}{27 - 9} = 2 \text{ cm}$$

Moment of inertia

$(I_{xx})_1$ = moment of inertia of triangle *ABD* about centroidal axis *xx*

$$= I_{G_1} + A_1 h_1^2 = \frac{9 \times 6^3}{36} + \frac{1}{2} \times 6 \times 9 \times (2 - 2)^2 = 54 \text{ cm}^4$$

$(I_{xx})_2$ = moment of inertia of triangle *ACD* about centroidal axis *xx*

$$= I_{G_2} + A_2 h_2^2 = \frac{3 \times 6^3}{36} + \frac{1}{2} \times 3 \times 6 \times (2 - 2)^2 = 18 \text{ cm}^4$$

∴ Moment of inertia of the given section *ABC* about centroidal axis *xx*

$$I_{xx} = 54 - 18 = \mathbf{36 \text{ cm}^4}$$

$(I_{yy})_1$ = moment of inertia of triangle ABD about centroidal axis yy

$$= I_{G_1} + A_1 h_1^2 = \frac{6 \times 9^3}{36} + \frac{1}{2} \times 6 \times 9 \times (6-5)^2 = 148.5 \text{ cm}^4$$

$(I_{yy})_2$ = moment of inertia of triangle ACD about centroidal axis yy

$$= I_{G_2} + A_2 h_2^2 = \frac{6 \times 3^3}{36} + \frac{1}{2} \times 3 \times 6 \times (8.5)^2 = 85.5 \text{ cm}^4$$

∴ Moment of inertia of the given section ABC about centroidal axis yy

$$I_{yy} = 148.5 - 86.5 = \mathbf{63\ cm^4}$$

EXAMPLE 7.14

Two equal channels are kept as shown in Fig. 7.29 at a distance *d* between them to from the cross-section of a column. Calculate the value of distance *d* if the centroidal moments of inertia I_{xx} and I_{yy} of the cross-section are equal.

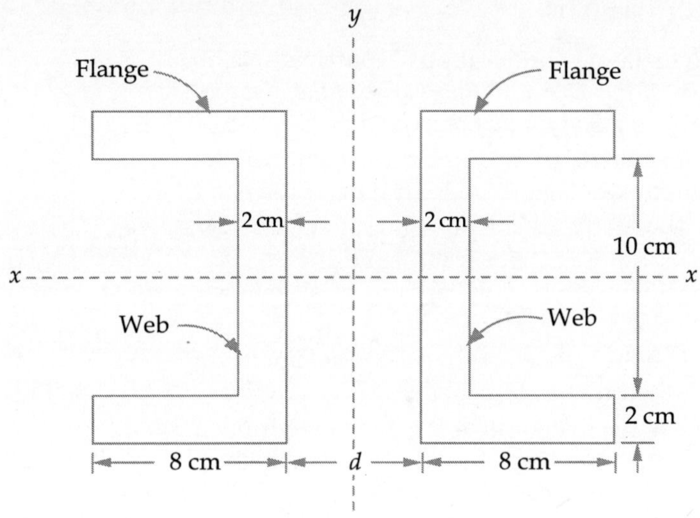

Fig. 7.29

Solution : I_{xx} = moment of inertia of 4 flanges + moment of inertia of 2 webs

$$= 4\left[\frac{8 \times 2^3}{12} + 8 \times 2 \times \left(5 + \frac{2}{2}\right)^2\right] + 2\left[\right.$$

$$= 4\,(5.33 + 576) + 333.33 = 2658.65 \text{ cm}^4$$

Here (5 + 2/2) is the distance of the centroid of the flange with area (8 × 2) from the x-axis.

I_{yy} = moment of interlia of 4 flanges + moment of inertia of 2 webs

$$= 4\left[\frac{2 \times 8^3}{12} + 8 \times 2(4 + d/2)^2\right] + 2\left[\frac{1(}{}\right.$$

$$= 4\left[85.33 + 16\left(16 + \frac{d^2}{4} + 4d\right)\right] + 2\left[6\right.$$

$$= 341.32 + 1024 + 16\,d^2 + 256d + 13.34 + 40 + 5d^2 + 40d$$
$$= 21\,d^2 + 296\,d + 1418.66$$

But it is given that $I_{xx} = I_{yy}$

$\therefore\ 21d^2 + 296d + 1418.66 = 2658.65$

or $\quad 21d^2 + 296d - 1240 = 0$

or $\quad d^2 + 14.09d - 59.05 = 0$

Solution of the quadratic equation gives

$$d = \frac{-14.09 \pm \sqrt{(14.09)^2 - 4 \times 1(-59.0}}{2 \times 1} = \frac{-14.09 \pm 20.85}{2}$$

$$= \mathbf{3.28\ cm}$$

(Taking + *ve* velue)

Hence the distance between the two shannel sections is 3.38 cm.

7.3. MASS MOMENT OF INERTIA

The mass moment of inertia of a body about a particular axis is defined as *"the product of the mass and the square of the distance between the mass centre of the body and the axis"*.

The mass moment of inertia is an important term for the study of the rotational motion of a rigid body. It gives a measure of the resistance that the body offers to change in angular velocity.

The body can be considered to be split up into small masses. Let

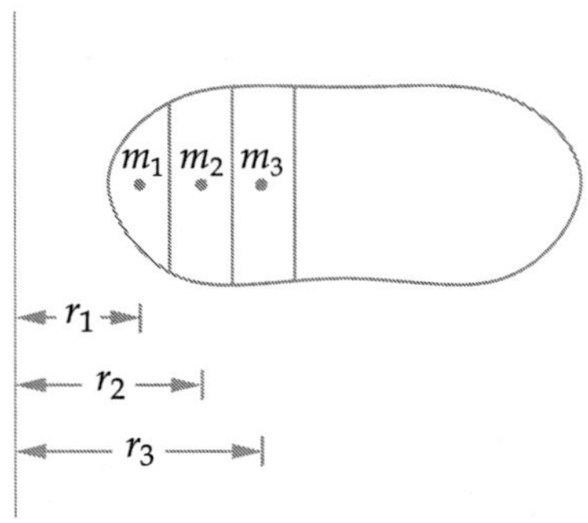

Fig. 7.30

$m_1, m_2 \ldots m_n$ be the masses of the various elements of the body

and $\quad r_1, r_2 \ldots r_n$ be the distances of the above mentioned elements from the axis about which mass moment of inertia is to be determined.

The mass moment of inertia of the body can be written as

$$I = m_1 r_1^2 + m_2 r_2^2 + \ldots m_n r_n^2$$
$$= \Sigma\, mr^2$$

The summation of a large number of terms in the above expression can be replaced by integration. Consider a small mass *dm* rotating about and at distance *r* from the axis of rotation, then

$$I = \int r^2 dm$$

The radius of gyration *k* of the body with respect to the prescribed axis is defined by the relation

$$I = k^2 M\,; \quad k = \sqrt{\frac{I}{M}}$$

where M is the mass of the body.

The radius of gyration gives a measure of the distance at which the entire mass of the body is assumed to be concentrated.

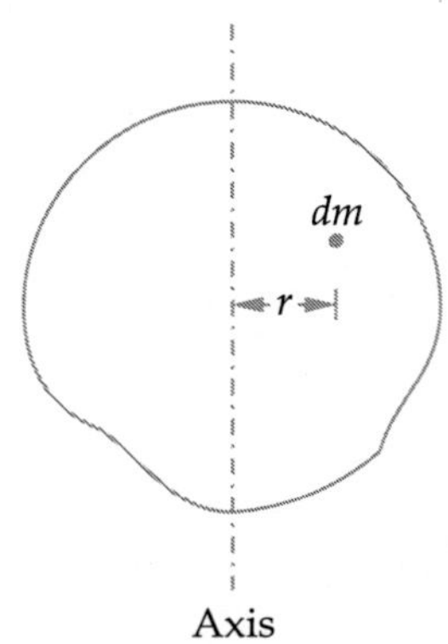

Fig. 7.31

Unit of mass moment of inertia is kg-m^2 and that of radius of gyration is m.

Parallel axis theorem : The mass moment of inertia of a body about an axis parallel to the centroidal axis is equal to the moment of inertia about the centroidal axis plus the product of mass and the square of the shortest distance between centre of gravity and the axis concerned. With reference to Fig. 7.32.

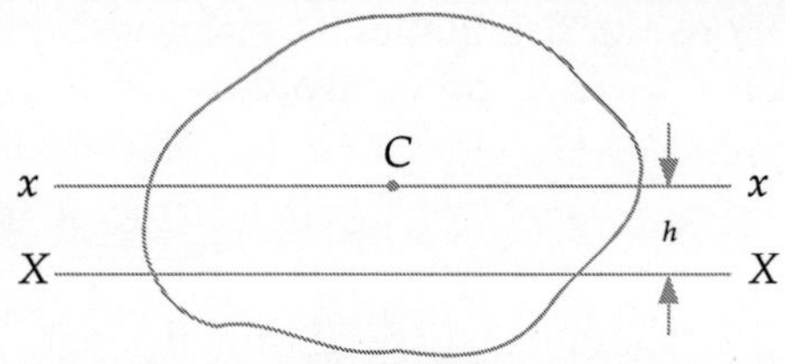

Fig. 7.32

where m is the mass of the body and h is the distance between the two parallel axis.

$$I_{XX} = I_{xx} + mh^2$$

7.4. MASS MOMENT OF INERTIA FOR SPECIFIED CASES

7.4.1. Thin Uniform Rod

Fig. 7.33 shows a thin uniform rod of length l. Consider a small elemental length dx located at distance x from the centroidal axis y-y normal to the rod.

Mass of the segmental element

$$dm = m\,dx$$

where m is the mass per unit of length.

Mass moment of inertia of the element about axis y-y

$$= dm\,x^2 = mx^2dx$$

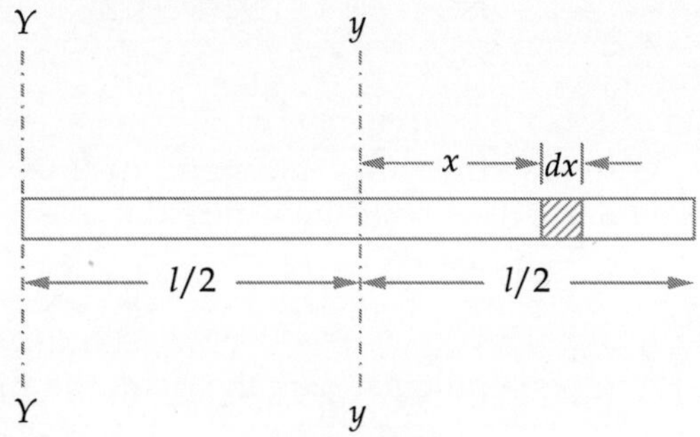

Fig. 7.33

The mass moment of inertia of the entire rod about axis y-y can be worked out by integrating the above expression between the limits $-l/2$ to $l/2$. That is

$$I_{yy} = m\int_{-l/2}^{l/2} x^2dx = m\left|\frac{x^3}{3}\right|_{-l/2}^{l/2} = m\left(\frac{l^3}{24}+\frac{l^3}{24}\right)$$

$$= \frac{ml^3}{12} = \frac{Ml^2}{12}$$

where $M = ml$ is the mass of the whole rod.

If it is required to determine the mass moment of inertia of the rod about axis YY at the left end of the rod, we can use the parallel axis theorem

$$I_{YY} = I_{yy} + Mh^2$$

where h is the distance between the axis YY and the centroidal axis yy.

$$I_{YY} = \frac{Ml^2}{12} + M\left(\frac{l}{2}\right)^2 = \frac{Ml^2}{3}$$

7.4.2. Rectangular Plate

Figure 7.34, shows a rectangular plate of width b, depth d and uniform thickness t. If ρ is the density of the plate material, then mass of the plate

$$M = \text{density} \times \text{volume}$$
$$= \rho\, b\, t\, d$$

Consider an elemental strip of depth dy located at distance y from the centroidal axis xx

mass of the elementa strip $dm = \rho\, b\, t\, dy$

mass moment of inertial of the strip about axis xx

$$= dm\, y^2 = \rho\, b\, t\, y^2 dy$$

The moment of inertia for the entire mass of plate about axis xx can be worked out by integrating the above expression between the limits $-d/2$ to $d/2$. That is

$$I_{xx} = \rho bt \int_{-d/2}^{d/2} y^2 dy = \rho bt \left| \frac{y^3}{3} \right|_{-d/2}^{d/2}$$

$$= \rho bt \left(\frac{d^3}{24} + \frac{d^3}{24} \right) = (\rho btd) \times \frac{d^2}{12}$$

$$= \frac{1}{12} Md^2 \qquad ...(7.23\ a)$$

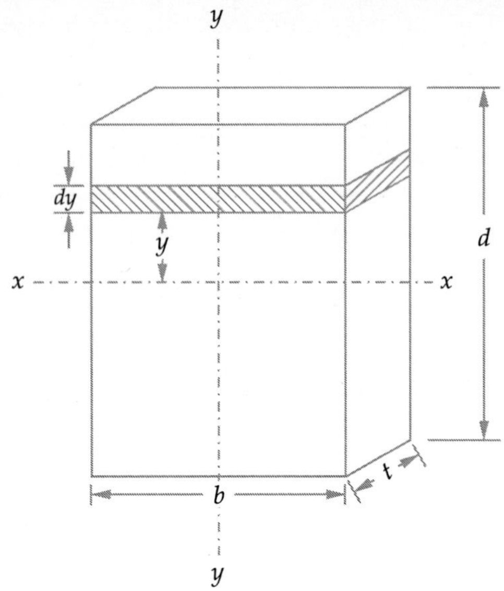

Fig. 7.34

Likewise the mass moment of inertia of the rectangular plate about the centroidal axis yy is

$$I_{yy} = \frac{1}{12} Mb^2 \qquad ...(7.23\ b)$$

From perpendicular axis theorem, the moment of inertia about axis zz is

$$I_{zz} = I_{xx} + I_{yy}$$

$$= \frac{1}{12} Md^2 + \frac{1}{12} Mb^2 = \frac{1}{12} M(d^2 + b^2) \qquad ...(7.24)$$

7.4.3. Triangular Plate

Figure 7.35 shows a triangular plate of base width b, height h and uniform thickness t. If ρ is the density of the plate material, then

mass of the plate,

$$M = \text{density} \times \text{volume}$$

$$= \text{density} \times (\text{area} \times \text{thickness})$$

$$= \rho \times \frac{1}{2} b\, h\, t$$

Consider an elemental strip (assumed rectangle) of width l and depth dy located at distance y from the base line.

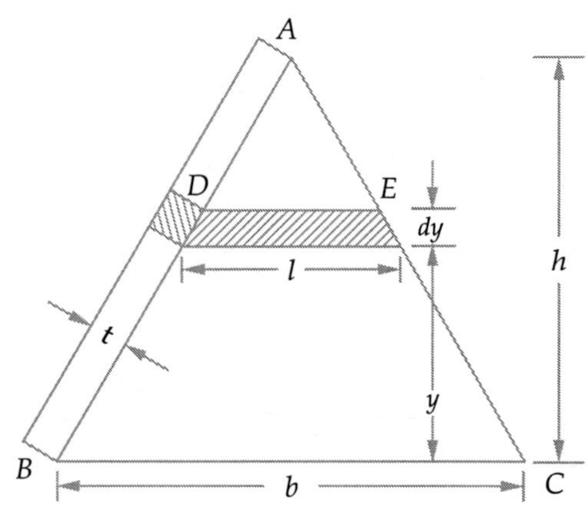

Fig. 7.35

mass of the elemental strip $dm = \rho\, l t\, dy$

mass moment of inertia of the strip about base

$$= dm\, y^2 = \rho\, l\, t\, y^2 dy$$

Since the integration is to be done with respect to y within the limits 0 to h, it is necessary to express l in terms of y. For that we have the following correlation from the similarity of triangles ADE and ABC,

$$\frac{l}{b} = \frac{h-y}{h}, \quad l = b\,\frac{h-y}{h}$$

∴ mass moment of inertia of the triangular plate about base line

$$I_{\text{base}} = \int_o^h \rho b\left(\frac{h-y}{h}\right) t\, y^2 dy$$

$$= \frac{\rho bt}{h}\int_o^h (h-y)\,y^2 dy$$

$$= \frac{\rho bt}{h}\left|\frac{hy^3}{3} - \frac{y^4}{4}\right|_o^h = \frac{\rho bt}{h}\left(\frac{h^4}{3} - \frac{h^4}{4}\right) = \frac{\rho bt\,h^3}{12}$$

$$= \frac{\rho bht}{2} \times \frac{h^2}{6} = \frac{1}{6} Mh^2 \qquad \text{...(7.25)}$$

7.4.4. Circular Lamina

Figure 7.36 shows a thin circular plate of radius R and uniform thickness t. If ρ is the density of the plate material, then

mass of the plate M = density × volume
= density × (area × thickness) = $\rho\pi R^2 t$

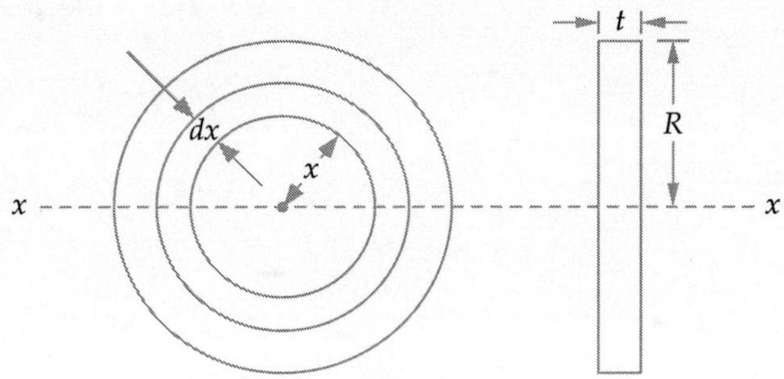

Fig. 7.36

Consider an elementary ring of radius r and width dr.

mass of elemental ring $dm = \rho\,[\pi(r + dr)^2 - \pi r^2]t$
$= \rho\,(2\pi r dr)\,t = 2\pi\, t\, \rho\, r\, dr$

Mass moment of inertia of this elementary ring about the polar axis zz
$= dm\,r^2 = 2\pi\, t\, \rho r^3 dr$

Mass moment of inertia of the circular plate about polar axis zz

$$= 2\pi t\rho \int_o^R r^3 dr = 2\pi t\rho \left|\frac{r^4}{4}\right|_o^R$$

$$= \rho\pi R^2 t \times \frac{R^2}{2} = \frac{MR^2}{2} \qquad \text{...(7.26)}$$

where $M = \rho\,\pi\,R^2 t$ is the mass of the circular lamina

Invoking the theorem of perpendicular axis, the mass moment of inertia of a circular lamina about xx or yy axis is

$$I_{xx} = I_{yy} = \frac{I_{zz}}{2} = \frac{1}{4} MR^2 \quad ...(7.27)$$

7.4.5. Solid Sphere

Figure 7.37 shows a solid sphere of radius R with O as centre. If ρ is the density of the material of the sphere, then

mass of the sphere = density × volume

$$= \rho \times \frac{4}{3} \pi R^3$$

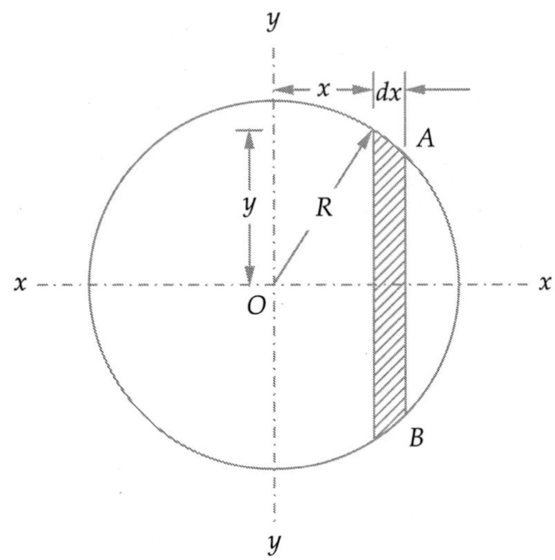

Fig. 7.37

Let attention be focussed on a thin disc AB of thickness dx and at radius x from the centre.

radius of the disc $y = \sqrt{R^2 - x^2}$

mass of the disc $dm = \rho \times \pi y^2 dx$

$= \rho\pi (R^2 - x^2)\, dx$

Mass moment of inertia of this elementary disc about the polar axis zz

$= dm\, y^2 = \rho\pi (R^2 - x^2)dx \times (R^2 - x^2)$

$= \rho\pi (R^2 - x^2)^2\, dx$

$= \rho\pi (R^4 + x^4 - 2R^2x^2)dx$

The mass moment of inertia of the whole sphere can be worked out by integrating the above expression between the limits $- R$ to R.

∴ Mass moment of inertia of the sphere about polar axis zz

$$I_{zz} = \rho\pi \int_{-R}^{R} (R^4 + x^4 - 2R^2x^2)\, dx$$

$$I_{zz} = \rho\pi \left| R^4 x + \frac{x^5}{5} - 2R^2 \frac{x^3}{3} \right|_{-R}^{R} = \frac{16\rho\pi R^5}{15} = \frac{4}{5} MR^2 \quad ...(7.28)$$

where $M = \frac{4}{3}\rho\pi R^3$ is the mass of the solid sphere

Invoking the theorem of perpendicular axis, the mass moment of inertia of a solid sphere about xx or yy axis is,

$$I_{xx} = I_{yy} = \frac{I_{zz}}{2} = \frac{2}{5} MR^2$$

7.4.6. Solid Cylinder

Figure 7.38 shows a solid cylinder of radius R and height h. If ρ is the density of the material of the cylinder, then

mass of the cylinder = density × volume

$$M = \rho \times \pi R^2 h$$

Consider a thin disc of thickness dy located at distance y from the centroidal axis xx.

mass of the elemental disc, $dm = \rho \times \pi R^2 dy$

It may be recalled that mass moment of inertia of a circular lamina about its diametral axis is given by

$$= \frac{1}{4} MR^2$$

∴ mass moment of inertia of the elemental disc about its diameteral axis is

$$I_{dia} = \frac{1}{4} dm\, R^2$$

Invoking parallel axis theorem, the mass moment of inertia of elemental disc about axis xx is

$$I_{xx} = I_{dia} + (dm)\, y^2$$

$$= \frac{1}{4} dm\ R^2 + dm\, y^2$$

$$= \frac{1}{4} (\rho\pi R^2 dy) R^2 + (\rho\pi R^2 dy) y^2$$

$$= \frac{1}{4} \rho\pi R^4 dy + \rho\pi R^2 y^2 dy$$

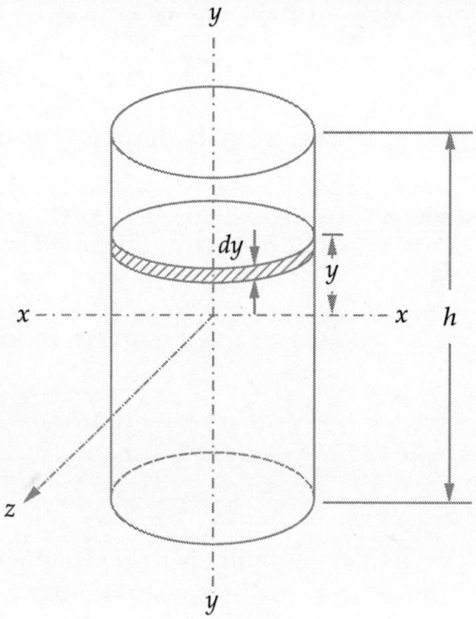

Fig. 7.38

The mass moment of inertia of the entire solid cylinder can be worked out by integrating the above expression between the limits $-h/2$ to $h/2$. Thus,

$$I_{xx} = \frac{1}{4} \rho\pi R^4 \int_{-h/2}^{h/2} dy + \rho\pi R^2 \int_{-h/2}^{h/2} y^2 dy$$

$$= \frac{1}{4} \rho\pi R^2 h + \frac{1}{12} \rho\pi R^2 h^3$$

$$= \rho\pi R^2 h \left(\frac{R^2}{4} + \frac{h^2}{12} \right) = M \left(\frac{R^2}{4} + \frac{h^2}{12} \right) = \frac{1}{12} M(3R^2 + h^2) \quad ...(7.29\ a)$$

where $M = \rho\pi R^2 h$ is the mass of the cylinder.

Similarly

$$I_{yy} = \frac{1}{12} M(3R^2 + h^2) \quad ...(7.29\ b)$$

and

$$I_{zz} = I_{xx} + I_{yy} = \frac{1}{6} M(3R^2 + h^2) \quad ...(7.30)$$

Note : For a thin cylinder, $R = 0$. That gives :

$$I_{xx} = I_{yy} = \frac{1}{12} Mh^2 = \frac{1}{12} Ml^2$$

For a thin disc, $h = 0$. That gives :

$$I_{xx} = I_{yy} = \frac{1}{4} MR^2 \quad \text{and} \quad I_{zz} = \frac{1}{2} MR^2$$

7.4.7. Right Circular Cone

Consider a solid cone of height h and radius R. If ρ is the density of the material of the cone, then mass of the cone M = density × volume

$$= \rho \times \frac{1}{3} \pi R^2 h$$

Consider an element of thickness dy and radius r at distance y from the apex A

mass of the elemental strip, $dm = \rho \pi r^2 dy$

mass moment of inertia of the elemental strip about axis yy

$$= \frac{1}{2} \times \text{mass moment of inertia about polar axis}$$

$$= \frac{1}{2} dmr^2 = \frac{1}{2} (\rho \pi r^2 dy) r^2$$

$$= \frac{1}{2} \rho \pi r^4 dy$$

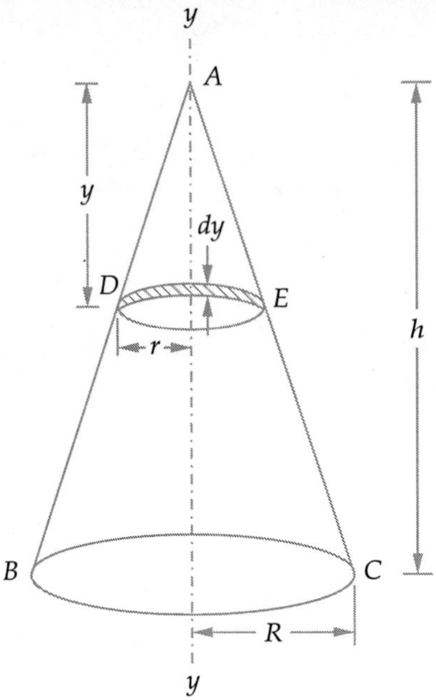

Fig. 7.39

Since the integration is to be done with respect to y within the limits 0 to h, it is necessary to express r in terms of y. For that we have the following correlation from the similarity of triangles ADE and ABC

$$\frac{r}{R} = \frac{y}{h}, \quad r = R\frac{y}{h}$$

∴ mass moment of inertia of the cone about axis yy

$$I_{yy} = \int_o^h \frac{1}{2} \rho \pi (Ry/h)^4 dy$$

$$= \frac{\rho \pi R^4}{2h^4} \left| \frac{y^5}{5} \right|_o^h = \frac{\rho \pi R^4 h}{10} = \frac{\rho \pi R^2 h}{3} \times \frac{3}{10} R^2 = \frac{3}{10} MR^2 \quad \text{...(7.31)}$$

where $M = \frac{1}{3} \pi \rho R^2 h$ is the mass of the right circular cone.

EXAMPLE 7.15

Would you imagine that the moment of inertia of the earth around its own axis is negligible fraction of its moment of inertia about the axis of rotation around the sun? Take mean radius of the earth as 6371 km and the mean radius of rotation around the sun as 149.7 × 10⁶ km.

Solution : (*b*) Moment of inertia of the earth about its axis,

$$I_1 = \frac{2}{5} MR^2 = \frac{2}{5} M(6371)^2 = 16.23 \times 10^6 M$$

Moment of inertia of the earth about the axis of rotation around the sun,

$$I_2 = I_1 + Md^2 = 16.23 \times 10^6 M + M \times (149.7 \times 10^6)^2$$

$$= 16.23 \times 10^6 M + 22410.09 \times 10^{12} M$$

$$= 22410.09002 \times 10^{12} M$$

$$\text{Ratio } \frac{I_1}{I_2} = \frac{16.23 \times 10^6}{22410.09002 \times 10^{12}} = 7.24 \times 10^{-10}$$

Since the ratio is negligible, the moment of inertia of the earth around its oun axis can be imagined to be a negligible fraction of its moment of inertia about the axis of rotation around the sun.

EXAMPLE 7.16

(*a*) Set up a relation between the mass moment of inertia and the moment of inertia of a plate about an axis lying in the plane of the plate.

(*b*) Determine the mass moment of inertia of a steel rectangular plate 8 cm × 16 cm and thickness 1 cm about the centroidal axis parallel to the 8 cm side. Take mass density of steel as 8000 kg/m^3.

(*c*) Determine the mass moment of inertia of a triangular plate of base 50 cm and height 100 cm about its base. The plate has a uniform thickness of 1 cm and density of the plate material is 7500 kg/m^3.

Solution : Figure 7.40 shows a thin plate of uniform thickness t and homogeneous mass density ρ.

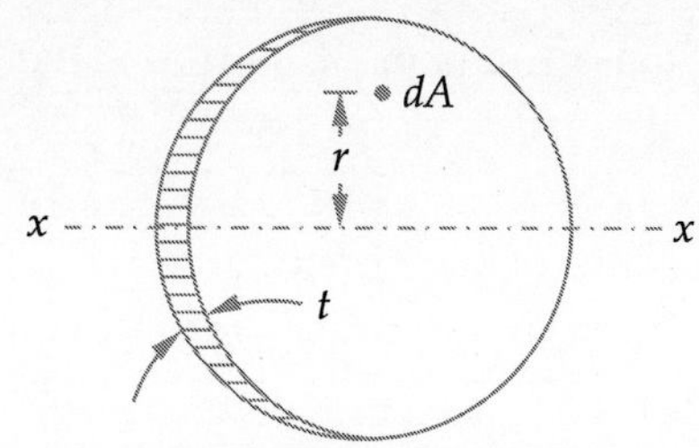

Fig. 7.40

Consider an elemental area dA located at distance r from the axis xx

mass of the elemental area, dm = density × volume

$$= \rho \times (dA \times t)$$

$$= \rho t dA$$

mass moment of inertia of this elemental area

$$= dm \times r^2 = \rho t r^2 dA$$

For the entire plate,

$$(I_{xx})_{mass} = \int \rho t r^2 \, dA = \rho t \int r^2 \, dA$$

The integral $\int r^2 dA$ is the moment of inertia of the area of the plate about the axis xx

$$\therefore \quad (I_{xx})_{mass} = \rho t \, (I_{xx})_{area}$$

(*b*) Refer Fig. 7.41 for the configuration of the rectangular plate

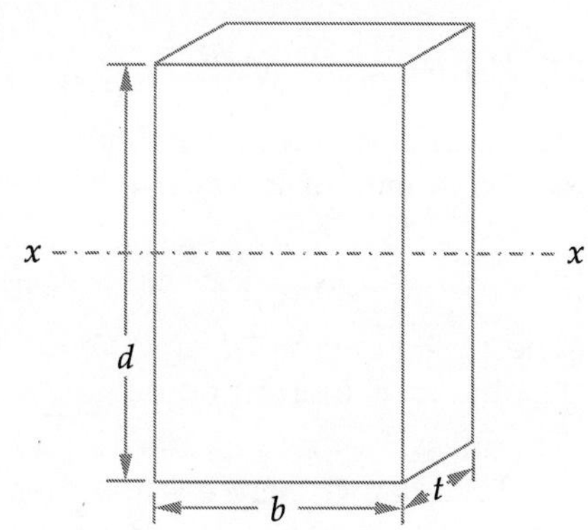

Fig. 7.41

$$b = 8 \text{ cm}; \quad d = 16 \text{ cm} \quad \text{and} \quad t = 1 \text{ cm}$$

mass of the plate M= density × volume

$$= 8000 \times \left(\frac{8 \times 16 \times 1}{10^6}\right)$$

$$= 1.024 \text{ kg}$$

Mass moment of inertia about axis xx is

$$I_{xx} = \frac{1}{12} Md^2 = \frac{1}{12} \times 1.024 \times \left(\frac{16}{100}\right)^2 = \mathbf{2.184 \times 10^{-3}} \text{ kg m}^2$$

(*c*) For the triangular plate,

base width b = 50 cm; height h = 100 cm and thickness t = 1 cm

mass of the triangular plate, M = density × volume

= density × (area × thickness)

$$= 7500 \times \left(\frac{1}{2} \times 0.5 \times 1\right) \times 0.01 = 18.75 \text{ kg}$$

Mass moment of inertia of a triangular plate about its base is

$$= \frac{1}{6} Mh^2 = \frac{1}{6} \times 18.75 \times 1^2 = \mathbf{3.125} \text{ kg m}^2$$

EXAMPLE 7.17

A rectangular column of section 30 cm × 50 cm and height 4 m is centrally cast over a concrete bed which measures 3 m × 5 m and thickness 40 cm. If the mass density of concrete is 2500 kg/m^3, make calculations for the mass moment of inertia of the column and bed combination about a pole.

Solution : Mass of column M_c = density × volume

$$= 2500 \times (0.3 \times 0.5 \times 4) = 1500 \text{ kg}$$

mass moment of inertia of the column about the pole

$$= \frac{1}{2} M(d^2 + b^2) = \frac{1}{12} \times 1500 \times (0.3^2 + 0.5^2) = 42.5 \text{ kg m}^2$$

Mass of the concrete bed = 2500 × (3 × 5 × 0.4) = 15000 kg

mass moment of inertia of concrete bed about the pole

$$= \frac{1}{12} \times 15000 \times (3^2 + 5^2) = 42500 \text{ kg m}^2$$

The mass moment of inertia of the composite body is given by the sum of mass moment of inertia of the column and the mass moment of inertia of the bed.

Mass moment of inertia of composite section

$$= 42.5 + 42500 = \mathbf{42542.5} \text{ kg m}^2$$

REVIEW AND SUMMARY

1. The moment of an area about any axis in its plane is called the first moment of area and the second moment is known as moment of inertia of the area. Mathematically

$$I_{xx} = \int_A x^2 dA \quad \text{and} \quad I_{yy} = \int_A y^2 dA$$

2. The first moment of area about centroidal axis is zero, while the second moment of area (moment of inertia) about centroidal axis has a minimum non-zero value.
3. The mass moment of inertia about an axis is the product of elemental mass and the square of the distance between the mass centre of the elemental mass and the axis.

$$\text{Mass moment of inertia} = \int x^2 dm$$

4. Radius of gyration (k) of a body or a given body (or lamina) is defined as the distance from the axis of reference to where the whole mass (or area) of a body is assumed to be concentrated, and that is not to make any change in the moment of inertia about the given axis.

$$k_x = \sqrt{I_{xx}/A} \quad ; \quad k_y = \sqrt{I_{yy}/A}$$

5. The modulus of section or section modulus of a figure represents the quantity that results when the moment of inertia of the figure about its centre of gravity is divided by the distance of the extremes fibre from the centroidal axis.

6. Parallel axis theorem: The moment of inertia of a plane lamina about any axis is equal to the sum of its MOI about a parallel axis through its centre of gravity and product of area (or mass) and the square of the distance between the two axis.

$$I_{AA} = I_{gxx} + Ah^2$$

7. Perpendicular axis theorem: The moment of inertia of a plane lamina about the axis perpendicular to the plane of the lamina is equal to the sum of moments of inertia of the lamina about the two axes at right angles and intersecting each other at the point where the perpendicular axis passes through it

$$I_{zz} = I_{xx} + I_{yy}$$

8. Polar moment of inertia is defined as the moment of inertia of the lamina or plane about the axis perpendicular to the plane of section. Invoking perpendicular axis theorem,

$$\text{Polar moment of inertia } I_{zz} \text{ (or } J) = I_{xx} + I_{yy}$$

Polar moment of inertia has application in problems relating to torsion of cylindrical shafts and rotation of slabs.

REVIEW QUESTIONS

A. Conceptual and conventional questions :

1. Define moment of inertia of an area. Why is it called the second moment of area?
2. Define radius of gyration. How is it related to moment of inertia?
3. State and explain
   (*a*) parallel axes theorem
   (*b*) perpendicular axes theorem concerning moment of inertia for a plane area.
4. What is meant by polar moment of inertia? State its application
5. Define moment of inertia of a mass. How radius of gyration of a body is expressed in terms of its mass moment of inertia
6. State parallel axis theorem as applied to mass moment of inertia.
7. (*a*) Define product of inertia. Mention its salient properties.
   (*b*) State and explain the parallel axis theorem for product of inertia.
   (*c*) State and explain the concepts of principal axis and principal moment of inertia.
8. Set up an expression for the mass moment of inertian of a uniform rod of length l about its
   (*i*) centroidal axis normal to the rod
   (*ii*) axis at the end of the rod and normal to it
9. Consider a semi-circular area of diameter 4 cm. Determine its moment of inertia about a tangent parallel to the diameter (Fig. 7.42).

(10.15 cm^4)

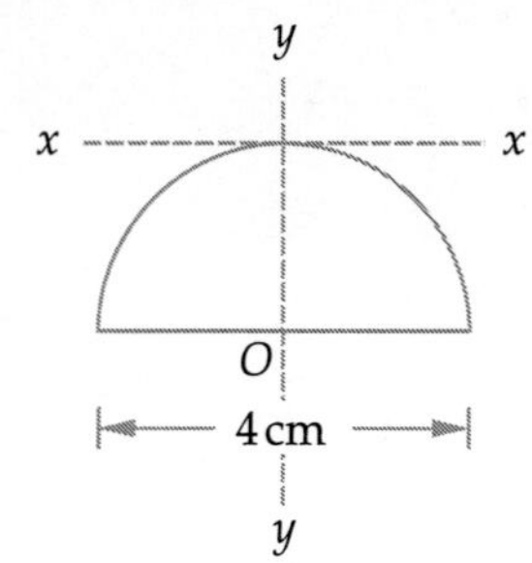

Fig. 7.42

10. Consider a square section *ABCD* having each side equal to 10 cm. What will be the ratio of the moment of inertia of this section about centroidal axis parallel to the side to that about the diagonal *AC*.

(1)

11. Determine the moment of inertia of the section shown in Fig. 7.43, about the centroidal axis parallel to *yy*-axis. All dimensions are in cm.

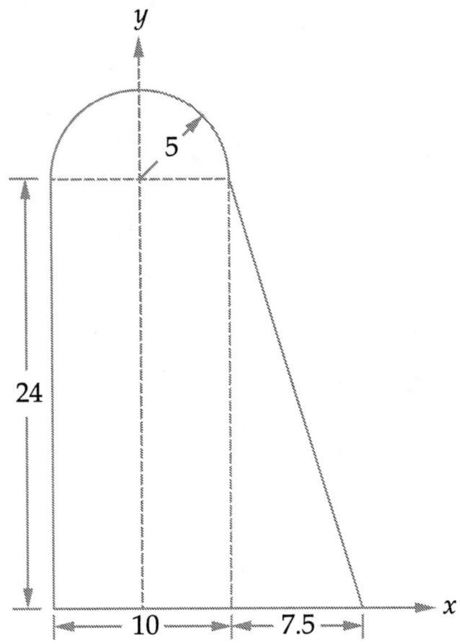

Fig. 7.43

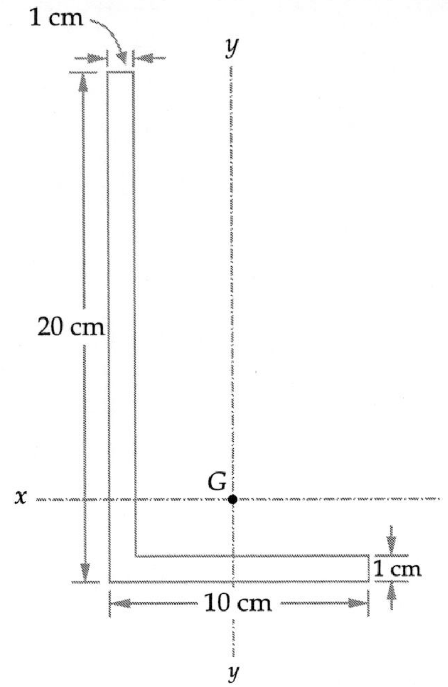

Fig. 7.44

12. Determine the moment of inertia about the centroidal axis for the unequal angle section shown in Fig. 7.44. **(1227.6 cm^4, 218.6 cm^4)**
13. Determine the moment of inertia of I-section about the centroidal axis parallel to the flange.

Top flange = 100 mm × 10 mm

Bottom flange = 200 mm × 10 mm

Web = 100 mm × 10 mm **($\bar{y}$ = 45.25 mm, I = 843 × 10^4 mm^4)**

14. A rectangular hole is made in a triangular section as shown in Fig. 7.45. Determine the moment of inertia of the section :
    (*a*) about base *AB*
    (*b*) about the centroidal axis parallel to base *AB*.

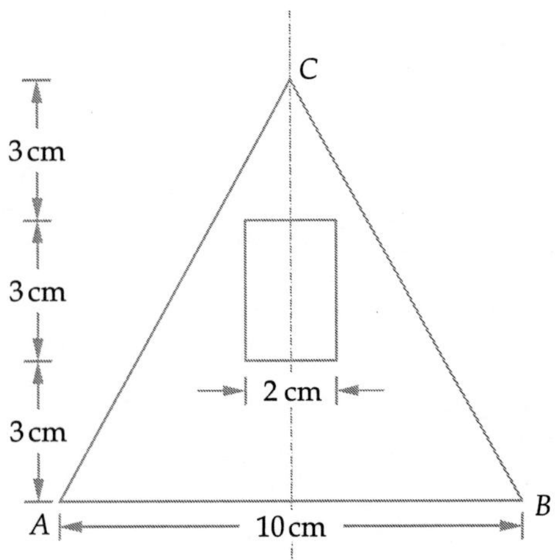

Fig. 7.45

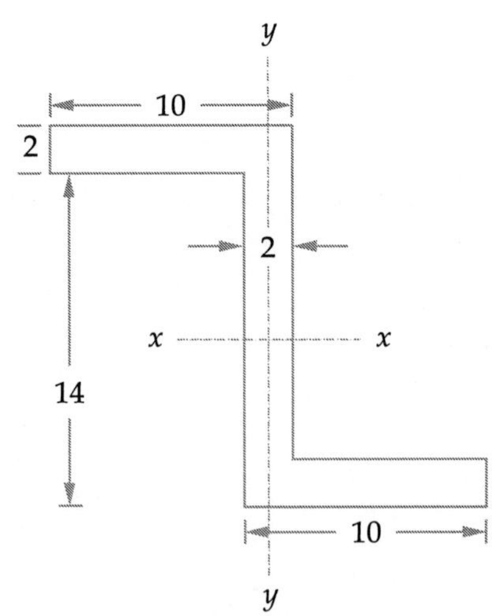

Fig. 7.46

15. Determine the moment of inertia of the z-section (Fig. 7.46) about the centroidal axis x - x and y - y. All dimensions are in cm.
16. For the area shown in Fig. 7.47 calculate the moment of inertia about x-axis. What will the corresponding radius of gyration?
17. Find the moment of inertia of the lamina with a circular hole of 15 mm diameter about the axes AB and CD as shown in Fig. 7.48. All the dimensions are in mm.

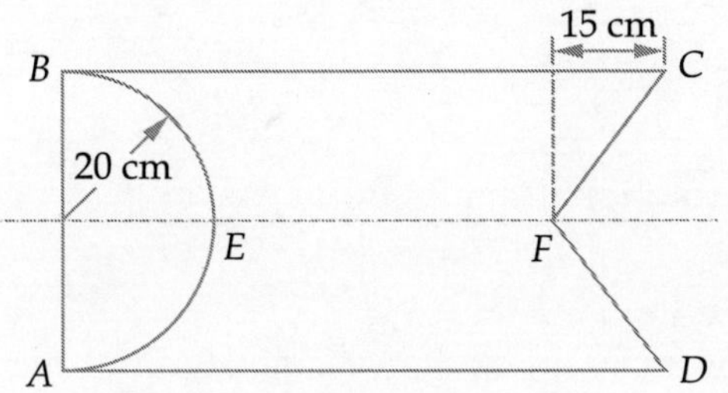

Fig. 7.47

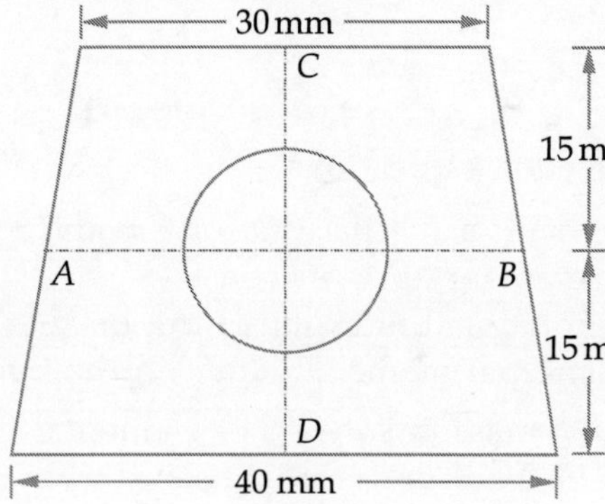

Fig. 7.48

18. Determine the distance d at which the two 2 cm × 6 cm rectangles shown in Fig. 7.49 be spaced so that $I_{xx} = I_{yy}$. **(5.3 cm)**

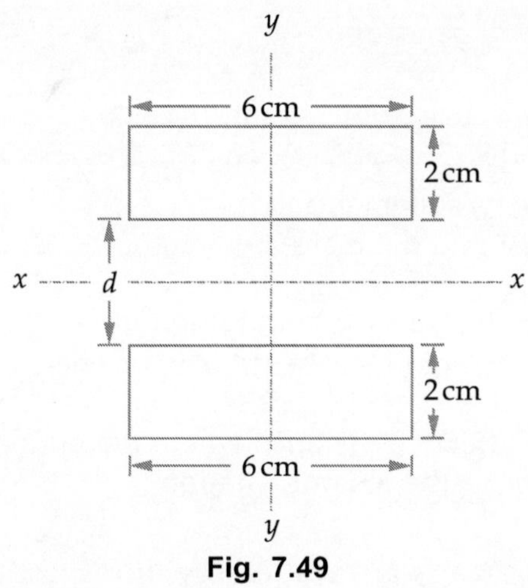

Fig. 7.49

B. Complete the following statements with most appropriate word/words :

1. Moment of inertia of an area dA at a distance r from an axis is ....
2. Moment of inertia of a triangle of base (b) and height (h) about an axis through the apex and parallel to the base is ....
3. Moment of inertia of a semi-circle of radius (r) with respect to centroidal axis parallel to the diameter is ....
4. Moment of inertia of an area is minimum about its .... axis
5. Moment of inertia (I) and radius of gyration (k) are related by the expression ....
6. A rectangular section has width b and depth d. Its polar moment of inertia would be ....

7. Mass moment of inertia of a mass *dm* at a distance *r* from the axis is ....
8. The mass moment of inertia for a thin spherical shell is .... and that for a solid sphere is ....

Answers:

1. $\int r^2 dA$ 2. $\frac{bh^3}{4}$ 3. $0.11\, r^4$ 4. centroidal 5. $I = Ak^2$ 6. $\frac{bd^3}{12} + \frac{db^3}{12}$ 7. $\int r^2 dm$

8. $\frac{2}{3}mr^2$ and $\frac{2}{5}mr^2$

C. Multiple choice questions :

1. The moment of inertia MOI of a body is
   (*a*) the moment of its inertia (*b*) the rotational analogue of mass
   (*c*) the rotational moment acting on the body
   (*d*) the inertial moment acting on the body
2. A body of mass *M* consists of elemental portions of masses m_1, m_2 .... and these masses lie at distances x_1, x_2 .... from a fixed straight line called axis. The MOI of the body about that axis is
   (*a*) $\sum m\,x$ (*b*) $\sum m\,x^2$ (*c*) $\frac{\sum m\,x^2}{\sum m\,x}$ (*d*) $\frac{\sum m^2\,x}{\sum m}$
3. The MOI of a body does not depend upon
   (*a*) shape of the body
   (*b*) mass of the body and its distribution within the body
   (*c*) axis of rotation of the body (*d*) angular velocity of the body
4. Identify the incorrect statement/statements:
   (*a*) The area moment of inertia and the mass moment of inertia are two entirely unrelated concepts
   (*b*) An axis passings through the centroid and lying in the plane is called centroidal axis
   (*c*) Moment of inertia of a body equals twice the kinetic energy of rotation when angular velocity of the body is unity
   (*d*) Radius of gyration is the effective distance where the entire mass may be considered to be located with respect to the axis of rotation
   (*e*) None of the above
5. The moment of inertia of an area is always least with respect to
   (*a*) centroidal axis (*b*) vertical axis
   (*c*) radius of gyration (*d*) depends upon configuration of the area
6. The moment of inertia of rectangular element ($b \times h$) about an axis coincident with side *b* would be
   (*a*) $b\,h^3/12$ (*b*) $b\,h^3/6$ (*c*) $b\,h^3/3$ (*d*) $2\,b\,h^3/3$
7. For a rectangular element with sides *b* and *h*, the moment of inertia about its centroidal axis parallel to side *b* would be
   (*a*) $b\,h^3/3$ (*b*) $b\,h^3/12$ (*c*) $b\,h^3/24$ (*d*) $b\,h^3/36$
8. The moment of inertia of a triangle of base width *b* and height *h* with respect to its base would be
   (*a*) $b\,h^3/8$ (*b*) $b\,h^3/12$ (*c*) $b\,h^3/24$ (*d*) $b\,h^3/36$

9. The ratio of moment of inertia of a rectangle and that of a triangle, having same base and height, with respect to their bases would be
   (*a*) 2 : 1 (*b*) 3 : 1 (*c*) 4 : 1 (*d*) 6 : 1
10. The ratio of the moment of inertia of a triangle of base width b and height h with respect to an axis coinciding with its base and a centroidal axis parallel to the base would be
   (*a*) 2 : 1 (*b*) 3 : 1 (*c*) 1 : 4 (*d*) 6 : 1
11. The radius of gyration of a circular area of radius r with respect to centroidal axis is
   (*a*) 0.1 r (*b*) 0.2 r (*c*) 0.5 r (*d*) 0.7 r
12. The ratio of moment of inertia of a circular body about its x-axis to that about y-axis is
   (*a*) 0.25 (*b*) 0.5 (*c*) 1.0 (*d*) 2.0
13. The second moment of a plane area about any axis compared to its second moment about the neutral axis is always
   (*a*) more (*b*) less (*c*) equal
   (*d*) relationship depends upon the dimensions of the plane area
14. The axis X-X and Y-Y for a plane area A are separated by distance r. Then according to parallel axis theorem:
   (*a*) $I_x I_y = A r^2$ (*b*) $I_y / I_x = A r^2$ (*c*) $I_y = I_x + A r^2$ (*d*) $I_x + I_y = A r^2$
15. If I_x and I_y denote the moment of inertia of a lamina about x-axis and y-axis respectively, then the moment of inertia of the lamina about z-axis would be
   (*a*) $I_x + I_y$ (*b*) $I_x \times I_y$ (*c*) $\sqrt{I_x^2 + I_y^2}$ (*d*) $I_x - I_y$
16. Polar moment of inertia
   (*a*) applies to masses whereas moment of inertia is applicable to areas only
   (*b*) is the moment of inertia for an area with respect to a line or axis parallel to centroidal axis
   (*c*) is the moment of inertia for an area with respect to a line or axis perpendicular to the plane of the area
   (*d*) is the moment of inertia for an area relative to a line or axis lying outside the plane of area.

Answers :

1. (*c*)	2. (*b*)	3. (*d*)	4. (*e*)	5. (*a*)	6. (*c*)
7. (*b*)	8. (*b*)	9. (*c*)	10. (*b*)	11. (*c*)	12. (*c*)
13. (*a*)	14. (*c*)	15. (*a*)	16. (*c*)		

❏❏❏

UNIT - IV

Chapter-8

Rectilinear and Curvilinear Motion

Chapter-9

Rotation and General Plane Motion

CHAPTER

8

RECTILINEAR AND CURVILINEAR MOTION

Learning objectives : The main objective of the subject matter presented in this chapter is to make the reader familiar with following aspects of kinematics
- *rectilinear and curvilinear motion*
- *displacement, velocity, acceleration and graphical representation of their interrelationship*
- *equation of rectilinear motion with uniform acceleration and motion under gravity*
- *motion along a curved path, normal and tangential components of acceleration*
- *relative velocity*

When a body changes its position with respect to other bodies, then the body is said to be in motion. Thus, a relative change in the position of a body is its motion. Motion involves both space and time.

Trajectory is a path followed by a body during its motion. It may be a straight line or a curve. Straight line path is observed in *rectilinear motion,* whereas curved path is observed in *curvilinear motion*. If the curved path lies in a plane, its is called plane curvilinear motion. *Rotary* or *circular* motion is a special case of curvilinear motion where all particles of a body move in a concentric circle.

8.1. DISPLACEMENT, VELOCITY AND ACCELERATION

Displacement: "*The change of position of a particle or a body with respect to a certain fixed reference point is termed as displacement.*"

Consider a body that moves along a curved path and takes time t to move from position A to B. The total distance covered by the body along the path followed is called *distance* and the linear (shortest) distance between the two positions is called *displacement*.

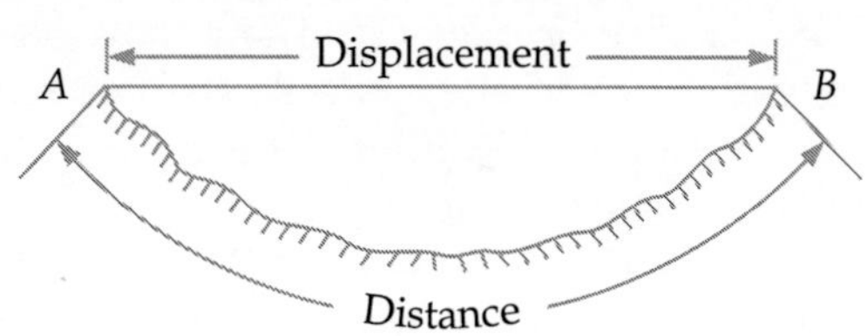

Fig. 8.1

Again, with reference to Fig 8.2, let the body move from reference point O to position A and then from A to position B.

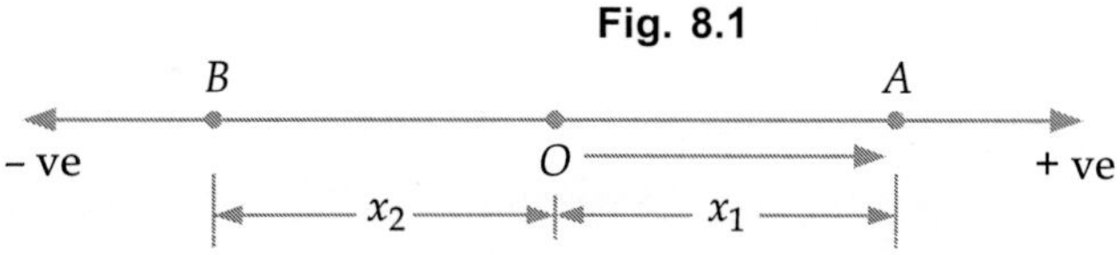

Fig. 8.2

Then:

Displacement of body = $-\ x_2$

Distance travelled by the particle = $x_1 + x_1 - x_2 = 2x_1 - x_2$

It is to be noted that:

(*i*) Displacement is the shortest distance covered from the particle from the reference point.

(*ii*) Displacement is taken positive to the right of reference point and negative to its left.

(*iii*) Displacement has both magnitude and direction and so is a vector quantity. Distance is scalar quantity since it has only magnitude.

Velocity: *The rate of change of position of a body with respect to time is called velocity.*

With reference to Fig. 8.3, the position A of a body at time t is at distance x from the reference point (origin O). At time $(t + \Delta t)$, the body shifts to position B which is at a distance $(x + \Delta x)$ from O. Then average velocity of the body over time interval Δt is

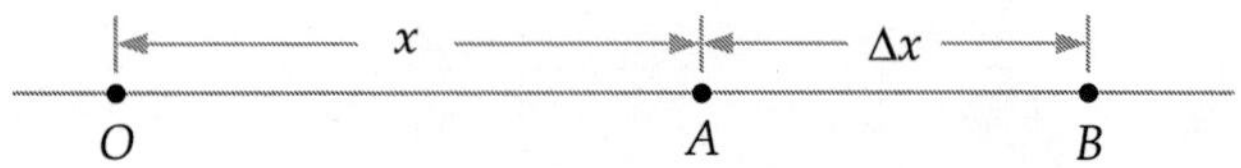

Fig. 8.3

$$v_{av} = \frac{\Delta x}{\Delta t} \quad \text{...(8.1)}$$

The velocity at a particular instant of time is called instantaneous velocity. Choosing the time interval Δt and the displacement Δx to be very very small,

$$\text{Instantaneous velocity } v = \lim_{\Delta x \to 0} \frac{\Delta x}{\Delta t} = \frac{dx}{dt} \quad \text{...(8.2)}$$

It is to be noted that velocity of a body is a measure of the rate of change of its position in a particular direction, and hence is a vector quantity. Speed is defined as rate of change of position of the body (with respect to its surroundings) irrespective of direction; speed is a scalar quantity.

The units of velocity or speed for defining their magnitude are taken as m/s or km/hr.

Acceleration: *The rate of change of velocity of a body with respect to time is called acceleration.*

Let v be the velocity of the body at any time t, and let it become $(v + \Delta t)$ at a later time $(t + \Delta t)$. Then average acceleration of the body over time interval Δt is

$$\text{Average acceleration } a = \frac{\Delta v}{\Delta t} \quad \text{...(8.3)}$$

If the time interval Δt and the change in velocity Δv are chosen to be very very small, then

$$\text{acceleration } a = \lim_{\Delta t \to 0} \frac{\Delta v}{\Delta t} = \frac{dv}{dt} \quad \text{...(8.4)}$$

The units for defining the magnitude of acceleration are m/s^2 or km/hr^2. The acceleration may be positive or negative. The positive acceleration is simply referred to as acceleration and the negative acceleration is called as retardation or deceleration.

Acceleration is positive if the velocity is increasing and the body is moving further in the positive direction. A negative acceleration implies a slow down in the motion of the body.

When the body moves with constant velocity (*i.e.* when acceleration is zero) the body is said to have *uniform motion.* The motion of a body moving with constant acceleration is referred to as *uniformly accelerated motion.*

8.2. GRAPHICAL REPRESENTATION

The motion of the bodies is often graphically represented by the following plots:

8.2.1. Distance-time graph (x-t curve)

The displacement of the body is plotted with respect to time; displacement x lies along ordinate and time t is taken along abscissa;

Since $v = dx/dt$, the slope of x - t curve at any instant gives the velocity of the body at that instant. Then, with reference to Fig. 8.4.

(*i*) At time t_1, the curve has a positive slope and accordingly the velocity is positive. A positive velocity implies that the displacement is increasing and that the body is moving to the right of origin. If the slope is positive and constant, the displacement would increase by equal amount in equal intervals of time, *i.e.*, the body would be moving with a uniform velocity.

(*ii*) At time 2, the slope is zero and so is the velocity. The displacement is then not changing with time and the body is at rest.

(*iii*) At time 3, the slope is negative and so is the velocity.

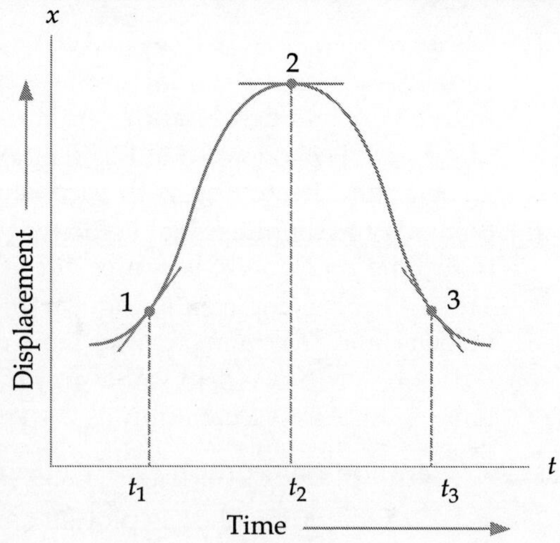

Fig. 8.4

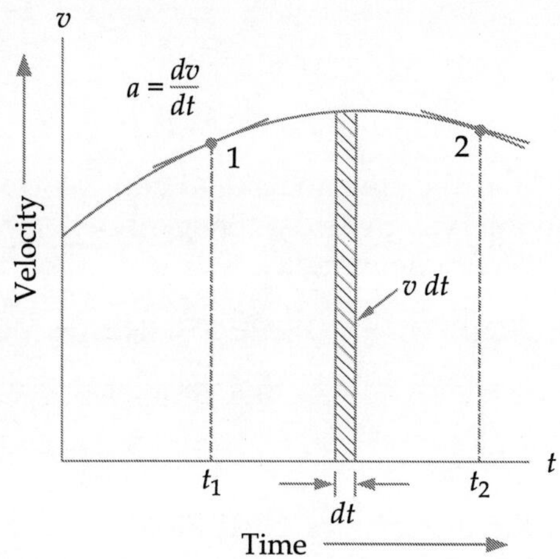

Fig. 8.5

8.2.2. Velocity-time graph (v – t curve)

The velocity of the body is plotted as a function of time. Since $a = dv/dt$, the slope of the $v - t$ curve at any instant gives the acceleration of the body at that instant.

Further,
$$v = \frac{dx}{dt}$$
$$\int dx = \int v\,dt$$
$$(x_2 - x_1) = \int v\,dt \qquad ...(8.5)$$

The above identity implies that area under the $v - t$ curve corresponding to a given time interval gives the change in displacement or the distance travelled by the body during the same time interval.

With reference to Fig. 8.6:

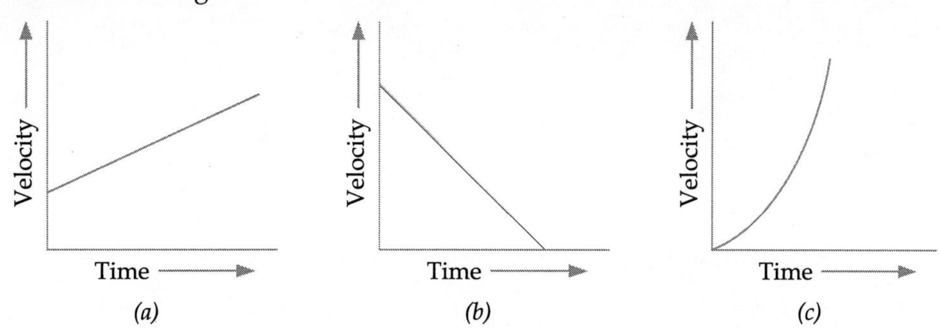

Fig. 8.6

(*a*) the velocity increases linearly with time. Since the $v - t$ curve has a constant slope, the velocity increases by equal amounts at equal intervals of time. The body is then moving with a constant acceleration. At the start (when $t = 0$), the body has a finite velocity. The $v - t$ curve depicted in Fig. 8.6 (*a*) thus corresponds to a body which has finite initial velocity and is moving with a constant acceleration.

(*b*) the body has a finite initial velocity. With passage of time, the velocity decreases linearly with time and finally becomes zero, *i.e.*, the body comes to rest. The $v - t$ curve depicted in Fig. 8.6 (*b*) thus corresponds to a body which has finite initial velocity and is moving with a constant retardation (negative slope of graph).

(*c*) the slope of the velocity-time graph is different at different times. The body does not have a uniform acceleration as its velocity is not changing at constant time.

8.2.3. Acceleration-time graph (a – t curve)

The acceleration of the body is plotted as function of time.

Since $a = \dfrac{dv}{dt}$, we have

$$\int dv = \int a\,dt$$

$$\text{or} \quad v_2 - v_1 = \int a\,dt \qquad \text{...(8.6)}$$

Thus the area under $a - t$ curve corres-ponding to a given time interval gives the change in velocity of the body during the some time interval.

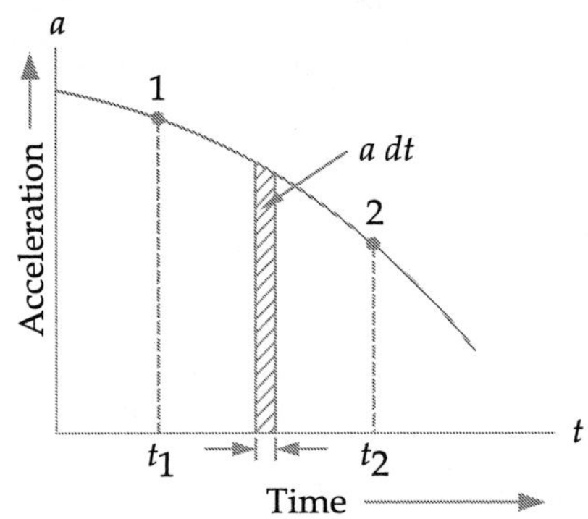

Fig. 8.7

8.3. EQUATIONS OF RECTILINEAR MOTION

When a body moves in a straight line with uniform acceleration, the equations of motion are:

$$v = u + at\,; \quad v^2 - u^2 = 2\,as\,; \quad s = ut + \frac{1}{2}\,at^2$$

and the distance travelled in *n*th second is

$$s_n = u + \frac{a}{2}(2n - 1)$$

where u is the initial velocity, v is the final velocity, a is the acceleration of the body and s represents the distance travelled in time t.

(*i*) acceleration a = rate of change of velocity

$$= \frac{\text{change of velocity}}{\text{time}} = \frac{v - u}{t}$$

$$v - u = at \qquad \therefore\; v = u + at \qquad \text{...(8.7)}$$

Alternatively:

$$a = \frac{dv}{dt}$$

$$dv = a\,dt$$

Integrating both sides within the limit $v = u$ at $t = 0$ and $v = v$ at $t = t$

$$\int_u^v dv = a\int_0^t dt \qquad (\because\ a \text{ is a constant})$$

$v - u = at \qquad \therefore\ v = u + at$

(*ii*) Distance travelled = average velocity × time

$$s = \frac{u+v}{2} \times t = \frac{u+u+at}{2} \times t \qquad (\because v = u + at)$$

$$= \left(u + \frac{at}{2}\right) \times t = ut + \frac{1}{2}at^2 \qquad ...(8.8)$$

Alternatively:

$$v = \frac{dx}{dt}$$

$$dx = v\,dt = (u + at)\,dt$$

Upon integration

$$\int_0^s dx = \int_0^t (u+at)\,dt \quad \text{assuming } x = 0 \text{ at } t = 0 \text{ and } x = s \text{ at } t = t$$

or $\quad s = ut + \frac{1}{2}at^2$

(*iii*) Distance = average velocity × time

$$s = \frac{u+v}{2} \times t = \frac{u+v}{2} \times \frac{v-u}{a} \qquad \left(\because \begin{array}{r} v = u + at \\ t = \dfrac{v-u}{a} \end{array}\right)$$

$$= \frac{v^2 - u^2}{2a} \qquad \therefore \quad v^2 - u^2 = 2as \qquad ...(8.9)$$

Alternatively:

$$a = \frac{dv}{dt} = \frac{dv}{dx} \times \frac{dx}{dt} = \frac{dv}{dx} \times v$$

or $\quad a\,dx = v\,dv$

Upon integration

$$a\int_0^s dx = \int_u^v v\,dv \qquad \text{assuming that } v = u \text{ at } x = 0 \text{ and } v = v \text{ at } x = s$$

$$as = \left|\frac{v^2}{2}\right|_u^v; \quad as = \frac{v^2 - u^2}{2}$$

$\therefore \quad v^2 - u^2 = 2as$

(*iv*) Distance covered in n seconds,

$$s_n = un + \frac{1}{2}an^2$$

Distance covered in $(n - 1)$ seconds

$$s_{n-1} = u(n-1) + \frac{1}{2}a(n-1)^2 = u(n-1) + \frac{1}{2}a(n^2 + 1 - 2n)$$

$\therefore$ Distance covered in nth second,

$$s_{nth} = s_n - s_{n-1}$$

$$= un + \frac{1}{2}an^2 - u(n-1) - \frac{1}{2}a(n^2 + 1$$

$$= un + \frac{1}{2}an^2 - un + u - \frac{1}{2}an^2 - \frac{1}{2}a$$

$$= u - \frac{a}{2} + an = u + \frac{a}{2}(2n-1) \qquad ...(8.10)$$

EXAMPLE. 8.1

A car travels from one station to another along a straight road. First half of the distance is covered with velocity 60km/hr and the second half is covered with velocity 90km/hr. Determine the average speed of the motor.

Solution: Let t be the total time taken to cover the entire distance s. Then

$$t = t_1 + t_2 = \frac{s/2}{v_1} + \frac{s/2}{v_2} = \frac{s}{2}\left(\frac{v_1 + v_2}{v_1 v_2}\right)$$

$$v_{av} = \frac{s}{t} = s \div \frac{s}{2}\left(\frac{v_1 + v_2}{v_1 v_2}\right) = \frac{2v_1 v_2}{v_1 + v_2} = \frac{2 \times 60 \times 90}{60 + 90} = \mathbf{72\ km/hr}$$

EXAMPLE 8.2

A train crosses a tunnel in 30 seconds time. The speeds of the train at entry to and at exit from the tunnel are 36 km/hr and 54 km/hr respectively. If the acceleration remains constant, then make calculations for the length of tunnel.

Solution: Average speed $= \dfrac{36 + 54}{2} = 45\ \text{km/hr} = \dfrac{45 \times 1000}{3600} = 12.5\ \text{m/s}$

Length of tunnel = average speed x time = $12.5 \times 30 =$ **375 m**

EXAMPLE 8.3

A toy car accelerates from rest at a constant rate of 2 m/s² for some time. Then it retards at a constant rate of 4 m/s² and comes to rest. If the car remains in motion for 3 seconds, determine the maximum speed attained and the total distance travelled by the car.

Solution: Let V be the maximum speed attained by the car. Then

$$V = a_1 t_1 = a_2 t_2$$

where t_1 and t_2 are time periods for acceleration and retardation respectively.

$$\therefore \quad t_1 = \frac{a_2}{a_1} t_2 = \frac{4}{2} t_2 = 2\, t_2$$

Also $\quad t = t_1 + t_2$

$3 = 2\, t_2 + t_2; \quad t_2 = 1\ \text{s} \quad \text{and} \quad t_1 = 2\ \text{s}$

$\therefore \quad V = a_1 t_1 = 2 \times 2 = 4\ \text{m/s}$

Now applying the correlation $s = ut + \frac{1}{2}at^2$, we get

$$s_1 = 0 \times 2 + \frac{1}{2} \times 2 \times 2^2 = 4\ \text{m}$$

$$s_2 = 4 \times 1 + \frac{1}{2} \times (-4) \times 1^2 = 2\ \text{m}$$

$\therefore$ Total distance covered = 4 + 2 = 6 m

EXAMPLE. 8.4

Two cars *A* and *B* move at 54 km/hr in the same direction and the car *B* is 300 m ahead of car *A*. If the car *A* is accelerated at 6 m/s^2 while the car *B* continues to move with the same velocity, calculate the time taken by car *A* to overtake car *B*.

Solution: Speed of cars *A* and *B* = 54 km/hr

$$= \frac{54 \times 1000}{3600} = 15 \text{ m/s}$$

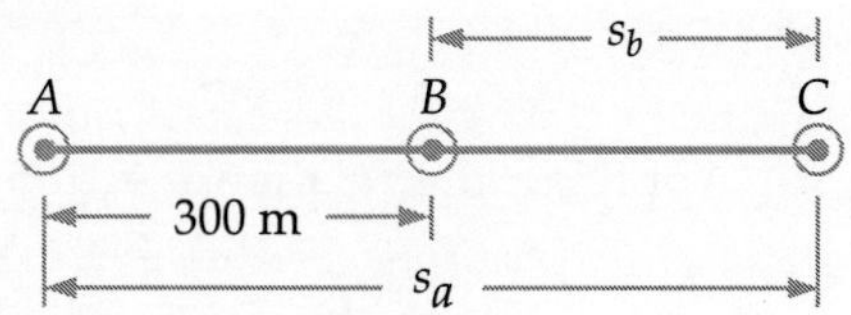

Fig. 8.8

Let the cars *A* and *B* meet at location *C* after time *t* covering distances s_a and s_b respectively.

Then

$$s_a = s_b + 300$$

$$15t + \frac{1}{2}(6)t^2 = 15t + 300$$

$$3t^2 = 300 \; ; \quad t = \sqrt{100} = \mathbf{10\ s}$$

EXAMPLE 8.5

A particle moves with uniform acceleration along a straight line *ABC*. The speeds of the particle at positions *A* and *C* are 5 cm/s and 25 cm/s respectively. If the point *B* lies midway between *A* and *C*, what will be the ratio of times taken by the particle to travel distances *AB* and *BC* ?

Solution: Refer Fig. 8.9

Invoking the relation $v^2 - u^2 = 2as$ where a is the uniform acceleration

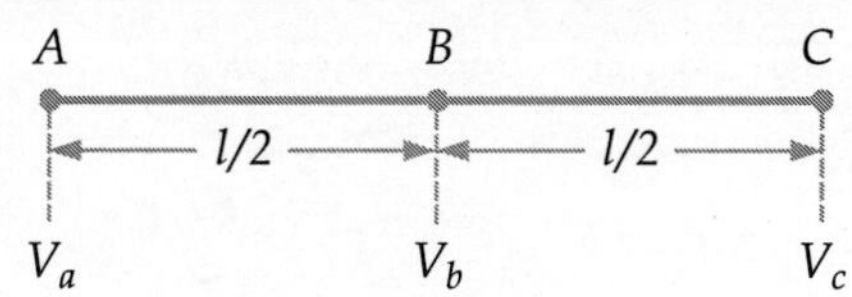

Fig. 8.9

$$V_c^2 - V_a^2 = 2al \; ;$$

$$25^2 - 5^2 = 2al \text{ or } a = \frac{300}{l}$$

$$V_b^2 - V_a^2 = 2al/2 \; ; \quad V_b^2 - 5^2 = 2 \times \frac{300}{l} \times l/2 = 300$$

$$V_b = \sqrt{325} \simeq 18 \text{ m/s}$$

The times t_1 and t_2 to cover distances *AB* and *BC* are given by

$$t_1 = \frac{V_b - V_a}{a} = \frac{18 - 5}{a} = \frac{13}{a}$$

$$t_2 = \frac{V_c - V_b}{a} = \frac{25 - 13}{a} = \frac{12}{a}$$

$\therefore$ Required ratio $t_1/t_2 = 13/12 \simeq 1:1$

EXAMPLE 8.6

On turning a corner, a motorist driving at 54 km/hr finds a child on the road 50 m ahead. He instantaneously stops the engine, applies the brakes and the vehicle is brought to halt within 5 m of the child. Make calculations for the uniform rate of deceleration and the time for which the brakes were applied.

Solution: Speed of vehicle $u = 54 \text{ km/hr} = \dfrac{54 \times 1000}{3600} = 15 \text{ m/s}$

$$s = 50 - 5 = 45 \text{ m}; \quad v = 0$$

(*i*) Invoking the relation $v^2 - u^2 = 2as$

$$a = \frac{v^2 - u^2}{2s} = \frac{0^2 - 15^2}{2 \times 45} = -\mathbf{2.5\ m/s^2} \text{ (retardation)}$$

(*ii*) Applying the relation $v = u + at$

$$t = \frac{v - u}{a} = \frac{0 - 15}{(-2.5)} = \mathbf{6\ s}$$

EXAMPLE 8.7

A motorist is driving his car at 60 km/hr when he observes that a traffic light 250 m ahead turns red. The traffic light is timed to remain red for 20 seconds before it turns green. The motorist wishes to pass the traffic lights without stopping to wait for it to turn green. Make calculations for (*a*) the required uniform acceleration of the car, and (b) the speed of the car as it passes the traffic light.

Solution: Speed of car $u = 60 \text{ km/hr} = \dfrac{60 \times 1000}{3600} = 16.67 \text{ m/s}$

$$s = 250 \text{ m and } t = 20 \text{ s}$$

(*a*) Applying the relation $s = ut + \frac{1}{2}at^2$

$$250 = 16.67 \times 20 + \frac{1}{2}a(20)^2$$

$$a = \frac{250 - 333.34}{\frac{1}{2}(20)^2} = -0.417 \text{ m/s}^2 \text{ (retardation)}$$

(*b*) Using the relation $v = u + at$

$$v = 16.67 + (-0.417) \times 20 = \mathbf{8.33\ m/s}$$

$$= \frac{8.33 \times 3600}{1000} \simeq \mathbf{30\ km/hr}$$

EXAMPLE 8.8

Three marks *A*, *B* and *C* spaced at a distance of 100 m are made along a straight road. A car starting from rest and accelerating uniformly passes the mark *A* and takes 10 seconds to reach the mark *B*, and further 8 seconds to reach the mark *C*. Make calculations for

(*a*) the magnitude of the acceleration of the car,

(*b*) the velocity of the car at *A*,

(*c*) the velocity of the car at *B*, and

(*d*) the distance of the mark *A* from the starting point.

Solution: Refer Fig. 8.10 for the position of the starting point *O* and the marks *A*, *B* and *C*.

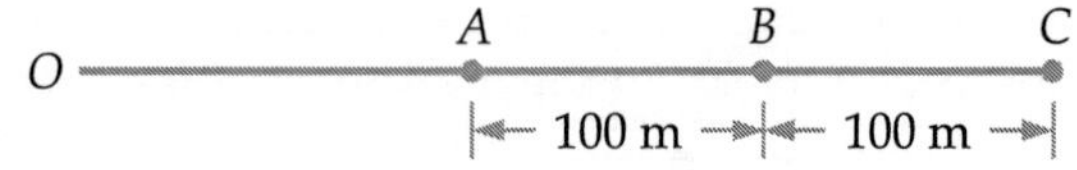

Fig. 8.10

Let u_a and u_b be the velocities at A and B respectively.
Considering motion from A to B,

$$s = u_a t + \frac{1}{2}at^2$$

$$100 = u_a \times 10 + \frac{1}{2}\, a \times (10)^2$$

$$100 = 10\,u_a + 50a \qquad ...(i)$$

Considering motion from A to C

$$200 = u_a \times 18 + \frac{1}{2}\, a \times (18)^2$$

$$200 = 18u_a + 162a \qquad ...(ii)$$

Solving identities (*i*) and (*ii*), we get

$$u_a = 8.61 \text{ m/s} \quad \text{and} \quad a = 0.278 \text{ m/s}^2$$

(*c*) Considering motion from B to C and using the relation $s = ut + \frac{1}{2}at^2$

$$100 = (u_b \times 8) + \left(\frac{1}{2} \times 0.278 \times 8^2\right) = 8u_b + 8.896$$

∴ Velocity of car at mark B, $u_b = \dfrac{100 - 8.896}{8}$ = **11.388 m/s**

(*d*) Considering motion from O to A and using the relation $v^2 - u^2 = 2s$

$$u_a^2 - u_0^2 = 2as$$

$$8.61^2 - 0^2 = 2 \times 0.278 \times s$$

∴ Distance of mark A from the starting point 0,

$$s = \frac{8.61^2}{2 \times 0.278} = \mathbf{133.3\ m}$$

EXAMPLE 8.9

A vehicle having rectilinear motion is moving with a velocity of 36 km/hr and accelerates uniformly to 72 km/hr over a distance of 200 m. Calculate the acceleration and the time taken. How much distance will be covered by the vehicle in the 5th second?

Solution: $u = 36 \text{ km/hr} = \dfrac{36 \times 1000}{3600} = 10 \text{ m/s}$

$$v = 72 \text{ km/hr} = \frac{72 \times 1000}{3600} = 20 \text{ m/s}$$

If a is the acceleration and t is the time taken, then using the relations $v^2 - u^2 = 2as$ and $v = u + at$, we get

$$20^2 - 10^2 = 2a \times 200\ ; \quad a = \frac{400 - 100}{2 \times 200} = \mathbf{0.75\ m/s^2}$$

$$20 = 10 + 0.75t\ ; \quad t = \frac{20 - 10}{0.75} = \mathbf{13.33\ s}$$

(*b*) Distance moved in the 5th second,

$$s_n = u + \frac{a}{2}(2n - 1) = 10 + \frac{0.75}{2}(2 \times 5 - 1) = \mathbf{13.375\ m}$$

EXAMPLE 8.10

A car moving in a straight line starts from rest, accelerates uniformly at 10 m/s² for the first 3*s*, then travels at uniform speed for the next 2*s* and finally decelerates uniformly at 5 m/s² for the next 4*s*. Draw the acceleration-time, velocity-time graphs and determine the distance travelled by the car during the given time period.

Solution: Refer to Fig. 8.11 for the acceleration time plot

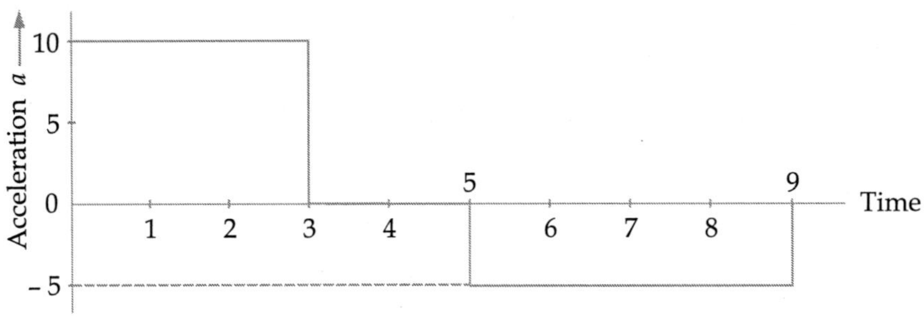

Fig. 8.11

Using the relation $v = u + at$

velocity after 3*s* = 0 + 10 × 3 = 30 m/s

velocity after 9*s* = 30 + (– 5) × 4 = 10 m/s

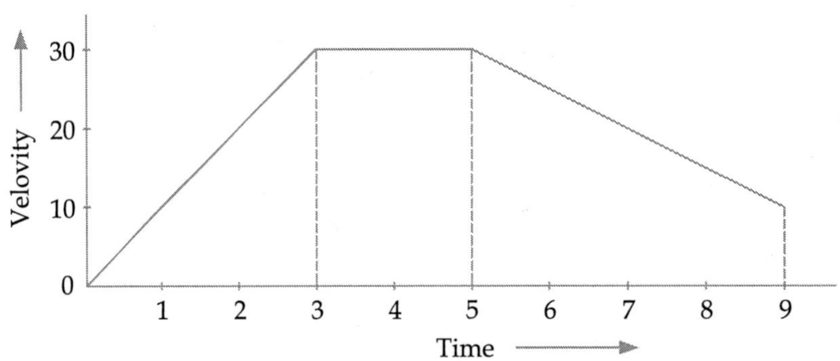

Fig. 8.12

Refer to Fig. 8.12 for the velocity-time graph.

Area under $v - t$ graph gives the distance travelled by the moving car.

$$\therefore \quad \text{Distance travelled} = \left(\frac{1}{2} \times 3 \times 30\right) + (30 \times 2) + \frac{30 + 10}{2}$$

$$= 45 + 60 + 80 = \mathbf{185\ m}$$

EXAMPLE 8.11

Two stations *A* and *B* lie a distance 3.6 km apart. An automobile starts from rest from station *P* and speeds upto 12 m/s with a constant acceleration of 1.2 m/s². The speed is maintained constant for a certain time and then the brakes are applied. The automobile finally comes to

rest at station B with a uniform deceleration of 1.5 m/s^2. Determine the total time required to perform this journey between the given stations.

Solution: Maximum velocity attained by the automobile during the course of motion is 12 m/s.

Refer Fig. 8.13 which represents velocity time graph for the automobile. The area under the velocity-time graph represents the distance travelled.

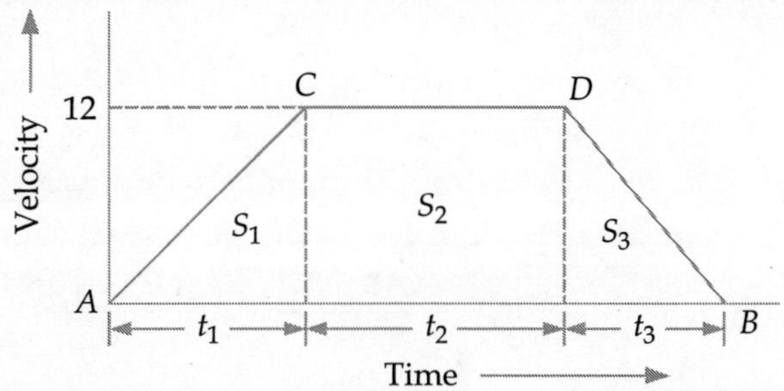

Fig. 8.13

$$S = S_1 + S_2 + S_3$$

$$3600 = \left(\frac{1}{2}\times 12\times t_1\right)+(12\times t_2)+\left(\frac{1}{2}\times 1\right. \qquad = 6t_1 + 12t_2 + 6t_3 \quad ...(i)$$

Considering the acceleration period (A to C)

$$v = u + at\,; \quad 12 = 0 + 1.2\,t, \quad \therefore\ t_1 = 10s$$

Considering the deceleration period (D to B)

$$v = u + at\,; \quad 0 = 12 + (-1.5) \times t_3 \quad \therefore\ t_3 = 8s$$

Substituting $t_1 = 10s$ and $t_3 = 8s$ in expression (i), we get

$$3600 = 6 \times 10 + 12t_2 + 6 \times 8$$

$$t_2 = \frac{3600 - 60 - 48}{12} = 291\text{ s}$$

$\therefore$ Total time taken to cover the distance = 10 + 291 + 8 = **309 s**

EXAMPLE 8.12

Establish the equations of linear motion by refering to velocity-time diagram. Presume that the motion is with uniform acceleration.

Solution: Refer Fig. 8.14 for the velocity-time diagram. At t = 0, the velocity has a finite value u and it becomes v at time t. Since the acceleration is uniform, the velocity has a linear variation *i.e.*, the slope of the curve is constant.

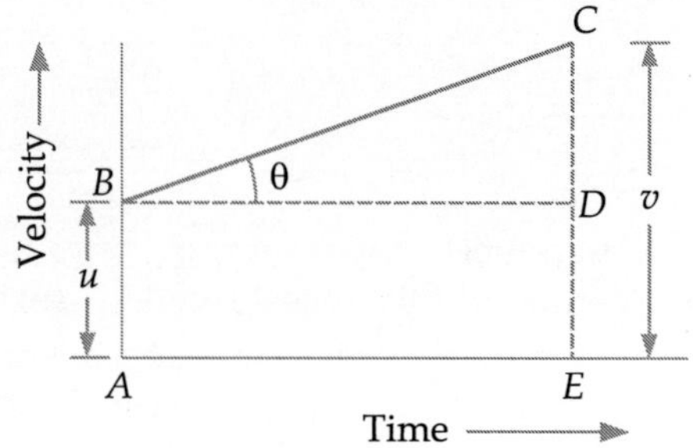

Fig. 8.14

acceleration a = slope of the diagram

$$= \tan\theta = \frac{CD}{BD}$$

$$= \frac{CE - DE}{AE}$$

$$= \frac{CE - BA}{t - 0} = \frac{v - u}{t}$$

$$\therefore \quad v = u + at$$

(ii) The area under the velocity time graph represents the distance travelled

$$s = \text{area ABCE}$$

$$= \text{area of rectangle ABDE} + \text{area of triangle BDC}$$

$$= AE \times AB + \frac{1}{2} BD\times CD$$

$$= AE \times AB + \frac{1}{2} BD \times BD \tan\theta$$

$$= (t \times u) + \frac{1}{2} t \times t \times a \;=\; ut + \frac{1}{2}at^2$$

(iii) s = area $ABCE$ $= \dfrac{AB + EC}{2} \times AE = \dfrac{u+v}{2} \times t = \dfrac{u+v}{2} \times \dfrac{v-u}{a}$ $\left(\because \; v = u + at, \; t = \dfrac{v-u}{a}\right)$

$$= \frac{v^2 - u^2}{2a}$$

$\therefore \quad v^2 - u^2 = 2as$

EXAMPLE 8.13

Train *A* starts with uniform acceleration of 0.25 m/s^2, attains a speed of 45 kmph which subsequently remains constant. One minute after the start of train *A*, another train *B* steams off on a parallel track with uniform acceleration of 0.45 m/s^2. If the maximum speed gained by train is 60 km/hr, how much time does it take for the train *B* to overtake train *A*?

Solution : Let the train *B* overtake train *A* after *t* seconds from the start of train *A*. Since the train *B* starts 60 seconds after the start of train *A*, it also implies that the train *B* will catch up with train *A* at the end of $(t - 60)$ seconds after its own start.

Motion of train A : Initial velocity $u = 0$; acceleration $a = 0.25 \text{ m/s}^2$

Final velocity $v = 45 \text{ km/hr} = \dfrac{45 \times 1000}{3600} = 12.5 \text{ m/s}$

Time taken to attain speed of 12.5 m/s,

$$t_1 = \frac{v-u}{a} = \frac{12.5 - 0}{0.25} = 50 \text{ sec}$$

For 50 seconds of its travel, the train is having accelerated motion. Distance travelled during this period is

$$s_1 = ut_1 + \frac{1}{2}at_1^2 = \frac{1}{2}at_1^2 \qquad (\because u = 0)$$

$$= \frac{1}{2} \times 0.25 \times 50^2 = 312.5 \text{ m}$$

Subsequently the train moves with uniform velocity of 12.5 m/s for $(t - 50)$ seconds when it is overtaken. Distance travelled during this period is

$$s_2 = \text{uniform velocity} \times \text{time} = 12.5\,(t - 50)$$

$\therefore$ Distance covered by train *A* before it is overtaken by train *B*

$$= 312.5 + 12.5(t - 50) \qquad ...(a)$$

Motion of train *B* :

$$u = 0, \quad a = 0.45 \text{ m/s}^2$$

$$v = \frac{60 \times 1000}{3600} = 16.67 \text{ m/s}$$

Time taken to attain speed of 16.67 m/s,

$$t_2 = \frac{16.67 - 0}{0.45} = 37.04 \text{ sec}$$

Distance travelled during this period of accelerated motion

$$= \frac{1}{2} \times 0.45 \times (37.04)^2 = 308.69 \text{ m}$$

Subsequently the train moves with uniform velocity of 16.67 m/s for [t - 60 - 37.04] sec when it overtakes train A. Distance travelled during this period is

$$= 16.67 \times (t - 97.04)$$

Distance covered by train B when it overtakes train A

$$= 308.69 + 16.67\,(t - 97.04) \quad \text{...(b)}$$

When the train B overtakes train A, the distance covered by both the trains will be equal. That is

$$312.5 + (t - 50) \times 12.5 = 308.69 + (t - 97.04) \times 16.67$$

or
$$312.5 + 12.5\,t - 625 = 308.69 + 16.67\,t - 1617.6$$

That gives
$$t = \frac{312.5 - 625 - 308.69 + 1617.6}{16.67 - 12.5} = \mathbf{238.95 \text{ sec}}$$

The solution can be obtained easily from velocity-time graphs for trains A and B shown in Fig. 8.15

V_a = maximum speed attained by train A = 12.5 m/s

t_1 = time taken by train A to attain speed of 12.5 m/s = 50 sec

V_b = maximum speed of train B = 60 km/hr = 16.67 m/s

t_2 = time taken by train B to attain speed of 16.67 m/s = 37.04 sec

At the time of overtake, the distance travelled by both train A and B is same. The distance covered is given by the area under velocity time graph. Accordingly

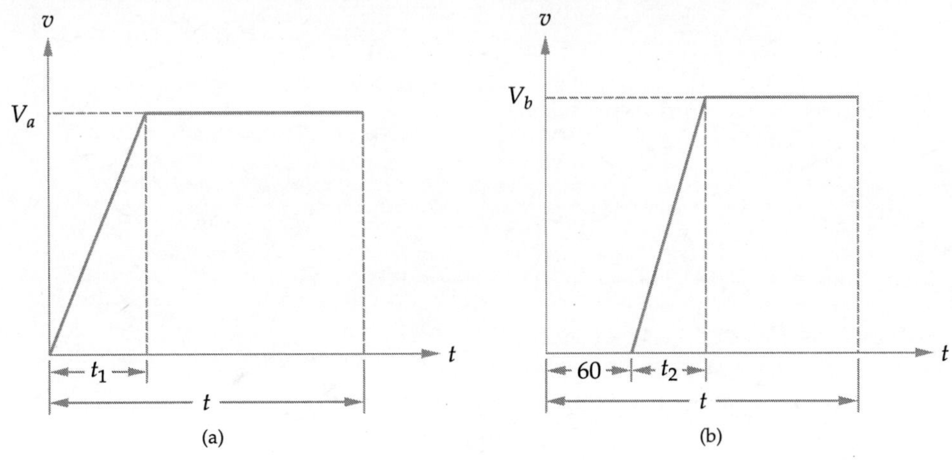

Fig. 8.15

area for car A (Fig. 8.15 a) = area for car B (Fig. 8.15 b)

$$\frac{1}{2} \times t_1 \times V_a + (t - t_1) V_a = \frac{1}{2} t_2 \times V_b + (t - t_2 - 60)\, V_b$$

or
$$\frac{1}{2} \times 50 \times 12.5 + (t - 50) \times 12.5 = \frac{1}{2} \times 37.04 \times 16.67 + (t - 37.04 - 6($$

or $\quad 312.5 + 12.5t - 625 = 308.73 + 16.67t - 1617.66$

$$\therefore \quad t = \frac{312.5 - 625 - 308.73 + 1617.66}{16.67 - 12.5} = \mathbf{238.95\ sec}$$

Hence car B will overtake car A at t = 238.95 sec

EXAMPLE 8.14

A train takes 2/3 hour while going from station *A* to station *B* and also from station *B* to *C*. The distance between stations *A* and *B* is 25 km and, the stations *B* and *C* are 40 km apart. Calculate :

(*a*) the velocity of train at stations *A* and *C*, and

(*b*) the magnitude of acceleration with which the train is running.

Assume that the acceleration is uniform.

Solution : Refer to Fig 8.16 for the velocity-time graph

The area of the velocity-time graph gives the distance travelled during any time interval. Thus,

$$s_1 = \text{area of trapezium } OABE$$

$$= \frac{1}{2} \times \text{sum of parallel sides} \times \text{altitude}$$

$$= \frac{1}{2}(V_a + V_b) \times t_1$$

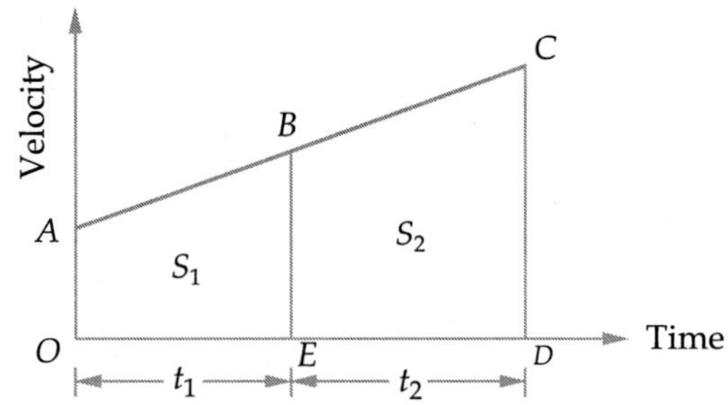

Fig. 8.16

where V_a and V_b are the velocities of the running train as it passes stations A and B respectively.

$$\therefore \quad 25 = \frac{1}{2}(V_a + V_b) \times \frac{2}{3} \; ; \quad V_a + V_b = 75 \qquad ...(i)$$

Likewise :

$$40 = \frac{1}{2}(V_b + V_c) \times \frac{2}{3} \; ; \quad V_b + V_c = 120 \qquad ...(ii)$$

Subtracting expression (*i*) from expression (*ii*),

$$V_c - V_a = 45 \qquad ...(iii)$$

Further, $\quad s_1 + s_2 = \frac{1}{2}(V_a + V_c) \times (t_1 + t_2)$

$$25 + 40 = \frac{1}{2}(V_a + V_c) \times \frac{4}{3} \; ; \quad V_a + V_c = 97.5 \qquad ...(iv)$$

Adding expressions (*iii*) and (*iv*), we get

$2V_c = 142.5 \; ; \quad V_c =$ **71.25 km/hr**

$V_a = 71.25 - 45 =$ **26.25 km/hr**

(*b*) Acceleration $\; a$ = rate of change of velocity

$$= \frac{V_c - V_a}{t_1 + t_2} = \frac{71.25 - 26.25}{2/3 + 2/3} = \mathbf{33.75\ km/hr^2}$$

EXAMPLE 8.15

An electric train runs between two stations which are 2.5 km apart and takes 6 minutes from start to stop. The train has a constant running speed of 36 km/hr between the end of

acceleration and beginning of retardation. If the acceleration and retardation are both uniform and numerically equal to each other, make calculations for their values.

Solution : Refer to Fig 8.17 for the velocity-time graph.

The slope of velocity-time graph represents acceleration.

Since acceleration and retardation are numerically equal to each other, we have :

$$\tan\theta_1 = \tan\theta_2 ; \quad \theta_1 = \theta_2$$

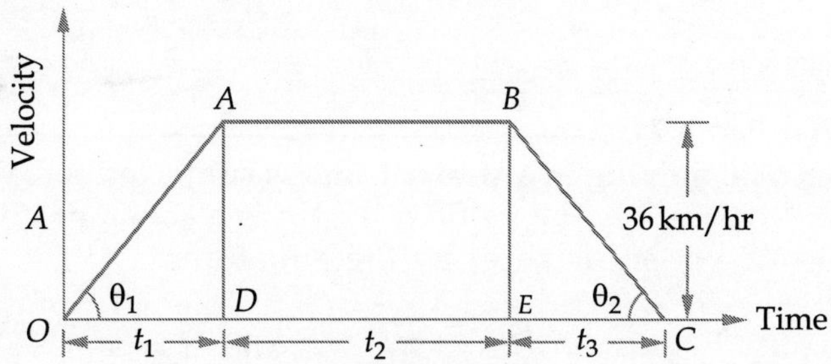

Fig. 8.17

Obviously the triangles *OAD* and *BCE* are similar in all respects. That gives

$$OD = EC, \quad \text{i.e.,} \quad t_1 = t_3$$

The total travel time from start to stop is given to be 6 minutes, *i.e.*, 0.1 hour

$$\therefore \quad t_1 + t_2 + t_3 = 0.1 \quad \text{or} \quad 2t_1 + t_2 = 0.1 \qquad \text{...}(i)$$

The area of the velocity-time graph gives the distance travelled during any time interval. Thus

$$s = s_1 + s_2 + s_3 = \frac{1}{2} \times 36 \times t_1 + 36 \times t_2 + \frac{1}{2} \times 36 \times t_3$$

$$\therefore \quad 2.5 = 18t_1 + 36t_2 + 18t_3 = 36t_1 + 36t_2 \qquad (\because t_1 = t_3)$$

$$\text{or} \quad t_1 + t_2 = 0.0694 \qquad \text{...}(ii)$$

From expressions (*i*) and (*ii*),

$$2t_1 + (0.0694 - t_1) = 0.1; \quad t_1 = 0.1 - 0.0694 = 0.0306 \text{ hr}$$

From the triangle *OAD* representing accelerating motion,

acceleration a = rate of change of velocity

$$= \frac{V_a - V_o}{t_1} = \frac{36 - 0}{0.0306} = \mathbf{1176.47\ km/hr^2}$$

Since retardation is stated to be numerically equal to acceleration, the magnitude of retardation is also 1176.47 km/hr^2.

EXMAPLE 8.16

A particle starts with velocity v_o and the acceleration-velocity relationship is prescribed as $a = -kv$ where k is a constant. Set up an expression that prescribes the displacement-time relation for the particle.

Solution : $a = -kv$; $\dfrac{dv}{dt} = -kv$ or $\displaystyle\int_{v_o}^{v} \frac{dv}{v} = -\int_o^t k\,dt$

$$\text{or} \quad \log_e \frac{v}{v_o} = -kt \quad \therefore \quad v = v_o e^{-kt}$$

Further $v = \dfrac{dx}{dt}$ and therefore

$$\frac{dx}{dt} = v_o e^{-kt} \quad \text{or} \quad \int_o^x dx = v_o \int_o^t e^{-kt} dt$$

$$\text{or} \quad x = \frac{v_o}{-k} \left| e^{-kt} \right|_o^t = \frac{v_o}{-k}(e^{-kt} - 1)$$

That gives the displacement-time relationship as

$$x = \frac{v_0}{k}(1 - e^{-kt})$$

EXAMPLE 8.17

A particle moving in a straight line is subjected to a resistance which produces a retardation of kv^3 where v is the velocity and k is a constant. Show that v and the time t are given in terms of x (the distance) by the equations

$$v = \frac{u}{1+kux} \quad \text{and} \quad t = \frac{kx^2}{2} + \frac{x}{u}$$

Solution : From the relation $a = \frac{dv}{dt}$ and $v = \frac{dx}{dt}$, we have

$$a = \frac{dv}{dt} = \frac{dv}{dx} \times \frac{dx}{dt} = \frac{dv}{dx} \times v$$

$\therefore$ $v\frac{dv}{dx} = -k\,v^3$; – ve sign because of retardation

or $\frac{dv}{v^2} = -k\,dx$

upon integration : $\left|-\frac{1}{v}\right|_u^v = -k\,x$

or $-\frac{1}{v} + \frac{1}{u} = -k\,x \;;\; \frac{1}{v} = \frac{1}{u} + kx = \frac{1+kux}{u}$

$\therefore$ $$v = \frac{u}{1+kux}$$

Further $v = \frac{dx}{dt}$ and therefore

$$(1 + k\,u\,x)\,dx = u\,dt$$

Upon integration : $x + \frac{kux^2}{2} = ut$

Dividing throughout by u, we get

$$t = \frac{kx^2}{2} + \frac{x}{u}$$

EXAMPLE 8.18

A projectile enters a resisting medium at $x = 0$ with an initial velocity $V_0 = 0.4$ m/s and travels 0.1 m before coming to rest. The velocity of the projectile varies with distance according to the relation $V = V_0 - k\,x$ where k is a constant. Calculate the time taken by the projectile to penetrate 0.08 m into the resisting medium.

Solution : $V_0 = 0.4$ m/s and $V = 0$ at $x = 0.1$. Accordingly from the given relation $V = V_0 - k\,x$, we have

$$0 = 0.4 - k \times 0.1 \;;\; k = 4$$

$\therefore \quad V = 0.4 - 4x$

Now, $\quad a = \dfrac{dV}{dt} = -4\dfrac{dx}{dt} = -4\,V$. That gives

$$\frac{dV}{V} = -4\,dt$$

Integrating both sides within the prescribed limits

$$\int_{0.4}^{V} \frac{dV}{V} = -4\int_0^t dt$$

or $\quad \left|\log_e V\right|_{0.4}^{V} = -4\,t \; ; \quad \log_e V - \log_e 0.4 = -4t$

$\therefore \quad \log_e V = -4t + \log_e 0.4 = -4t - 0.916$

At $x = 0.08$ m, $V = 0.4 - 4 \times 0.08 = 0.02$ m/s

$\therefore \quad \log_e(0.02) = -4t - 0.916$

or $\quad -3.912 = -4t - 0.916 \; ; \quad t =$ **0.749 sec**

EXAMPLE 8.19

The rectilinear motion of a motor car starting from rest is prescribed by the relation $a = \dfrac{6}{1.5\,V + 2}$ **where** a **is the acceleration in m/s^2 and** V **is the velocity in m/s. Calculate the time taken and distance covered by the motor car to attain the velocity of 6 m/s.**

Solution : Using the relation $a = dV/dt$, the given equation can be written as

$$\frac{dV}{dt} = \frac{6}{1.5V + 2}; \quad (1.5\,V + 2)\,dV = 6\,dt$$

Integrating both sides within the limits

At $t = 0$, $V = 0$ and at $t = t$, $V = 6$ m/s, we get

$$\int_o^6 (1.5\,V + 2)\,dV = 6\int_o^t dt \quad \text{or} \quad \left|1.5\frac{V^2}{2} + 2V\right|_0^6 = 6t$$

or $$1.5 \times \frac{6^2}{2} + 2 \times 6 = 6t; \quad t = \mathbf{6.5\ sec}$$

(*b*) Using the relation $a = V\,dV/dx$, the given equation can be written as

$$V\frac{dV}{dx} = \frac{6}{1.5\,V + 2}; \quad (1.5V^2 + 2V)\,dV = 6\,dx$$

Integrating both sides within the limits

At $x = 0$, $V = 0$ and at $x = x$, $V = 6$ m/s

$$\int_0^6 (1.5\,V^2 + 2V)\,dV = 6\int_o^x dx \quad \text{or} \quad \left|1.5\frac{V^3}{3} + 2\frac{V^2}{2}\right|_o^6 = 6x$$

or $$1.5 \times \frac{6^3}{3} + 2 \times \frac{6^2}{2} = 6x \; ; \quad x = \mathbf{24\ m}$$

Thus the distance travelled by the motor car is 24 m

EXAMPLE 8.20

A train travels along a straight track such that the distance travelled is directly proportional to the cube of the time interval. If it travels 480 m during the first minute, determine the displacement, velocity and acceleration 15 seconds after the start.

Solution : Since the distance travelled is directly proportional to cube of time interval, we may write

$$x \propto t^3 \; ; \quad x = c\,t^3 \quad \text{where } c \text{ is constant}$$

Substituting the given data : $x = 480$ m at $t = 1$ min $= 60$ sec

$$480 = c \times 60^3 \; ; \quad c = 2.22 \times 10^{-3}$$

Thus the relationship between displacement and time is

$$x = 2.22 \times 10^{-3}\, t^3$$

Differentiating with respect to time t,

$$\frac{dx}{dt} = v = 2.22 \times 10^{-3} \times (3t^2) = 6.66 \times 10^{-3}\, t^2$$

and differentiating once again,

$$\frac{d^2x}{dt^2} = \frac{dv}{dt} = a = 6.66 \times 10^{-3} \times (2t) = 13.32 \times 10^{-3}\, t$$

The expressions for the displacement, velocity and acceleration are, therefore,

Displacement $x = 2.22 \times 10^{-3}\, t^3$; velocity $v = 6.66 \times 10^{-3}\, t^2$ and acceleration $a = 13.32 \times 10^{-3}\, t$

Substituting $t = 15$ seconds after the start

$$x = 2.22 \times 10^{-3} \times (15)^3 = \mathbf{7.492\ m}$$

$$v = 6.66 \times 10^{-3} \times (15)^2 = \mathbf{1.4985\ m/s}$$

$$a = 13.32 \times 10^{-3} \times (15) = \mathbf{0.1998\ m/s^2}$$

EXAMPLE 8.21

A particle moves along a straight line and its motion is represented by the equation

$$s = 16t + 4t^2 - 3t^3$$

where s is in metres and t is in seconds. Determine :

(a) displacement, velocity and acceleration 2 seconds after start

(b) displacement and acceleration when velocity is zero

(c) displacement and velocity when acceleration is zero

Solution : $s = 16t + 4t^2 - 3t^3$

Differentiating with respect to t,

$$\frac{ds}{dt} = 16 + 8t - 9t^2 \; ; \quad v = 16 + 8t - 9t^2$$

Again differentiating with respect to t,

$$\frac{d^2s}{dt^2} = 8 - 18t \; ; \quad a = 8 - 18t$$

(*a*) when $t = 2$, i.e., 2 seconds after start

$$s = 16 \times 2 + 4 \times (2)^2 - 3 \times (2)^3 = 32 + 16 - 24 = \mathbf{24\ m}$$

$$v = 16 + 8 \times 2 - 9 \times (2)^2 = 16 + 16 - 36 = \mathbf{-\,4\ m/s}$$

$$a = 8 - 18 \times 2 = 8 - 36 = \mathbf{-28\ m/s^2}$$

(*b*) At condition of zero velocity,

$$0 = 16 + 8t - 9t^2; \quad 9t^2 - 8t - 16 = 0$$

$$t = \frac{8 \pm \sqrt{64 - 4 \times 9 \times (-16)}}{2 \times 9} = \frac{8 \pm 25.3}{18}$$

The particle will have zero velocity at $t = \dfrac{8 + 25.3}{18} = 1.85$ s

Displacement at 1.85 s after start

$$= 16 \times 1.85 + 4 \times (1.85)^2 - 3 \times (1.85)^3$$

$$= 29.6 + 13.69 - 18.99 = \mathbf{24.3\ m}$$

Acceleration at 1.85 s after start,

$$= 8 - 18 \times 1.85 = 8 - 33.3 = \mathbf{-25.3\ m/s^2}$$

(*c*) Corresponding to zero acceleration

$$8 - 18t = 0\ ; \quad t = 0.444 \text{ s}$$

Displacement at 0.444 s after start

$$= 16 \times 0.444 + 4 \times (0.444)^2 - 3 \times (0.444)^3$$

$$= 7.104 + 0.788 - 0.262 = \mathbf{7.63\ m}$$

Velocity at 0.444 s after start

$$= 16 + 8 \times 0.444 - 9 \times (0.444)^2$$

$$= 16 + 3.552 - 1.774 = \mathbf{17.775\ m/s}$$

It is worthwhile to remark that the particle will have maximum velocity when acceleration is zero. Thus

$$V_{max} = 17.775 \text{ m/s}$$

EXAMPLE 8.22

A particle moves along a straight line with an acceleration prescribed by the relation $a = (4t^2 - 3t + 2)$ where a is in m/s² and t is in seconds. The particle has a velocity of 10 m/s at $t = 3$ seconds, and it is located 12 m to the right of origin at $t = 2$ seconds. Determine the position and velocity of the particle after 5 seconds.

Solution : From the relation $a = dv/dt$, we have

$$dv = a\,dt = (4t^2 - 3t + 2)\,dt$$

$$v = \int (4t^2 - 3t + 2)\,dt = 4\frac{t^3}{3} - 3\frac{t^2}{2} + 2t + C_1$$

From the given condition : $v = 10$ m/s at $t = 3$ seconds, we have

$$10 = 4 \times \frac{3^3}{3} - 3 \times \frac{3^2}{2} + 2 \times 3 + C_1 = 36 - 13.5 + 6 + C_1$$

or $$C_1 = 10 - 36 + 13.5 - 6 = -18.5$$

∴ $$v = \frac{4}{3}t^3 - \frac{3}{2}t^2 + 2t - 18.5 \qquad ...(i)$$

(*a*) When $t = 5$ seconds, the velocity of the particle becomes

$$v = \frac{4}{3} \times (5)^3 - \frac{3}{2} \times (5)^2 + 2 \times 5 - 18.5$$

$$= 166.67 - 37.5 + 10 - 18.5 = \mathbf{120.67\ m/s}$$

(*b*) Using the relation $v = dx/dt$, equation (*i*) can be written as

$$dx = \left(\frac{4}{3}t^3 - \frac{3}{2}t^2 + 2t - 18.5\right)dt$$

Upon integration: $x = \int\left(\frac{4}{3}t^3 - \frac{3}{2}t^2 + 2t - 18.5\right)dt = \frac{t^4}{3} - \frac{t^3}{2} + t^2 - 18.5t + C_2$

From the given condition $x = 12$ m at $t = 2$ seconds, we have

$$12 = \frac{2^4}{3} - \frac{2^3}{2} + 2^2 - 18.5 \times 2 + C_2 = 5.33 - 4 + 4 - 37 + C_2$$

or $$C_2 = 12 - 5.33 + 4 - 4 + 37 = 43.67$$

$\therefore$ $$x = \frac{t^4}{3} - \frac{t^3}{2} + t^2 - 18.5t + 43.67$$

Hence when $t = 5$ seconds, the displacement of the particle from origin becomes

$$x = \frac{5^4}{3} - \frac{5^3}{2} + 5^2 - 18.5 \times 5 + 43.67$$

$$= 208.33 - 62.5 + 25 - 92.5 + 43.67 = \mathbf{122\ m}$$

EXAMPLE 8.23

The acceleration of a particle is expressed as $a = 10 - x$. The particle starts with no initial velocity at the position $x = 0$. Determine (*i*) the velocity of the particle when $x = 8$ m, (*ii*) the position of the particle where the velocity is again zero, and (*iii*) the velocity of the particle when acceleration becomes zero.

Solution : From the relation $a = \frac{dV}{dt}$ and $V = \frac{dx}{dt}$, we have

$$a = \frac{dV}{dt} = \frac{dV}{dx} \times \frac{dx}{dt} = \frac{dV}{dx} \times V$$

$\therefore$ $$V\,dV = a\,dx = (10 - x)dx$$

upon integration : $$\frac{V^2}{2} = \left[10x - \frac{x^2}{2}\right] + C$$

The constant of integration C can be worked out from the condition that the particle starts from rest. That is

At t = 0, $V = 0$ and $x = 0$

That gives C = 0

$\therefore$ $$V^2 = 20x - x^2 \quad \text{and} \quad a = 10 - x$$

(*i*) Velocity of the particle when $x = 8$ m

$$V^2 = 20 \times 8 - 8^2 = 96\,; \quad V = \mathbf{9.798\ m/s}$$

(*ii*) Position of the particle when the velocity becomes zero

$$V^2 = 20x - x^2 = x(20 - x) = 0$$

$\therefore$ $$x = 0 \quad \text{and} \quad x = \mathbf{20\ m}$$

(*iii*) Velocity of the particle when acceleration is zero

$$a = 10 - x = 0\,; \quad x = 10\text{ m}$$

$$V^2 = 20 \times 10 - 10^2 = 100\,; \quad V = \mathbf{10\ m/s}$$

EXAMPLE 8.24

A particle moves along horizontal direction and its position at any instant is prescribed by the relation

$$x = 3t^3 - 5t^2$$

where x is in metres and t is in seconds.

Make calculations for the following parameters :

(*a*) displacement during t = 2 sec to 5 sec

(*b*) average velocity during t = 2 sec to 5 sec and instantaneous velocity at t = 2 sec and 5 sec.

(*c*) average acceleration during t = 2 sec to 5 sec and instantaneous acceleration at t = 2 sec and 5 sec.

(*d*) distance travelled in the first four seconds (t = 0 to t = 4 sec).

Solution : $x = 3t^3 - 5t^2$

(*a*) Displacement during t = 2 sec to 5 sec

$$x_5 - x_2 = (3 \times 5^3 - 5 \times 5^2) - (3 \times 2^3 - 5 \times 2^2)$$
$$= (375 - 125) - (24 - 20) = \textbf{246 m}$$

(*b*) Average velocity and instantaneous velocity

Let subscripts f and i denote the final and initial values

$$V_{av} = \frac{x_f - x_i}{t_f - t_i} = \frac{246}{5-2} = \textbf{82 m/s}$$

Upon differentiation of displacement x with respect to time t, one obtains the expression for instantaneous velocity.

$$V = \frac{dx}{dt} = \frac{d}{dt}(3t^3 - 5t^2) = 9t^2 - 10t$$

Instantaneous velocity at t = 2 sec and t = 5 sec then works out as

$$V_2 = 9 \times 2^2 - 10 \times 2 = \textbf{16 m/s}$$
$$V_5 = 9 \times 5^2 - 10 \times 5 = \textbf{175 m/s}$$

(*c*) Average acceleration and instantaneous acceleration

$$a_{av} = \frac{V_f - V_i}{t_f - t_i} = \frac{175 - 16}{5-2} = \textbf{53 m/s}^2$$

Upon differentiation of velocity V with respect to time t, one obtains the expression for instantaneous acceleration :

$$a = \frac{dV}{dt} = \frac{d}{dt}(9t^2 - 10t) = 18t - 10$$

Instantaneous acceleration at t = 2 sec and t = 5 sec then works out as

$$a_2 = 18 \times 2 - 10 = \textbf{26 m/s}^2$$
$$a_5 = 18 \times 5 - 10 = \textbf{80 m/s}^2$$

(*d*) Distance travelled in the first 4 seconds :

From the given expression $x = 3t^3 - 5t^2$, it is evident that upto certain value of t, the displacement is negative. At the instant when change in sign occurs, dx/dt is zero. That gives

$$9t^2 - 10t = 0; \quad t = 0, 1.11$$

$$x \text{ at } 1.11 = 3 \times (1.11)^3 - 5 \times (1.11)^2 = -2.057 \text{ m}$$

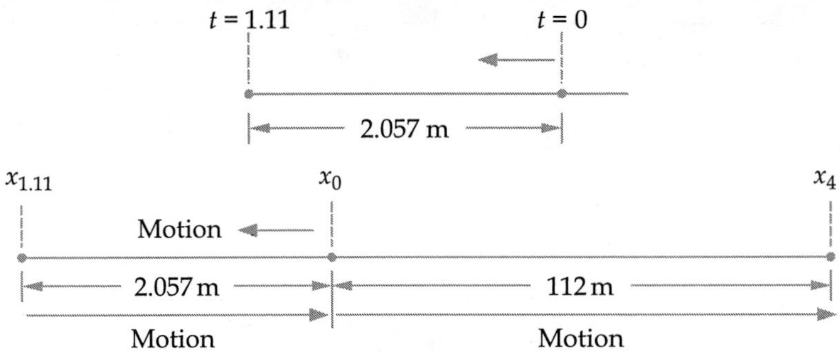

Fig. 8.18

From $t = 0$ to $t = 1.11$, (dx/dt) is negative, hence the motion is towards the left.

When $t > 1.11$, (dx/dt) is positive; hence the motion is towards right.

At $t = 4 : x = 3 \times 4^3 - 5 \times 4^2 = 112$ m

$\therefore$ Total distance travelled = 2.057 + 2.057 + 112 = **116.114 m**

EXAMPLE 8.25

The velocity-time graph for a motor car on a straight road is as shown in Fig. 8.19. What will be the acceleration of the car at $s = 25$ m and $s = 75$ m?

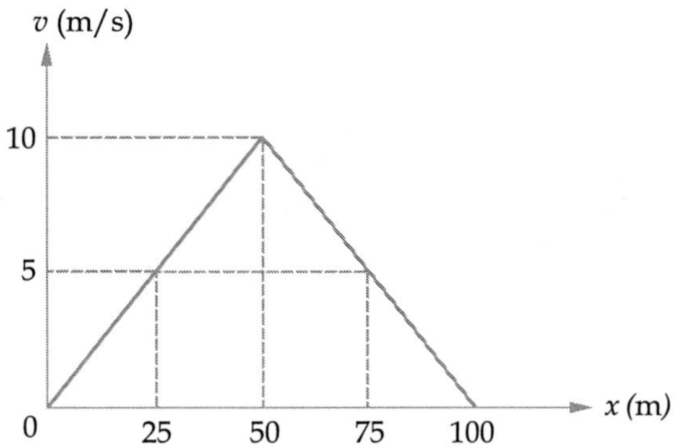

Fig. 8.19

Solution : Acceleration $a = \dfrac{dv}{dt} = \dfrac{dv}{ds} \times \dfrac{ds}{dt}$

Since $ds/dt = v$, we have

$$a = v\frac{dv}{ds}$$

Further dv/ds represents slope of v-s curve

(*i*) At $s = 25$ m : $v = 5$ m/s and

$$\frac{dv}{ds} = \frac{5-0}{25-0} = 0.2$$

$\therefore$ acceleration $a = 5 \times 0.2 =$ **1 m/s²**

(*ii*) At $s = 75$ m : $v = 5$ m/s and $\dfrac{dv}{ds} = \dfrac{0-5}{100-75} = -0.2$

$\therefore$ acceleration $a = 5 \times (-0.2) = -1 \text{ m/s}^2$

The negative sign implies that at $s = 75$ m, the motor car will be undergoing deceleration.

EXMAPLE 8.26

The variation of acceleration with time for a particle undergoing motion is as depicted in Fig. 8.20. The particle starts from rest and moves along a straight horizontal path.

Plot the velocity-time and the displacement-time graphs. Proceed to calculate the maximum speed attained and the maximum distance traversed by the particle.

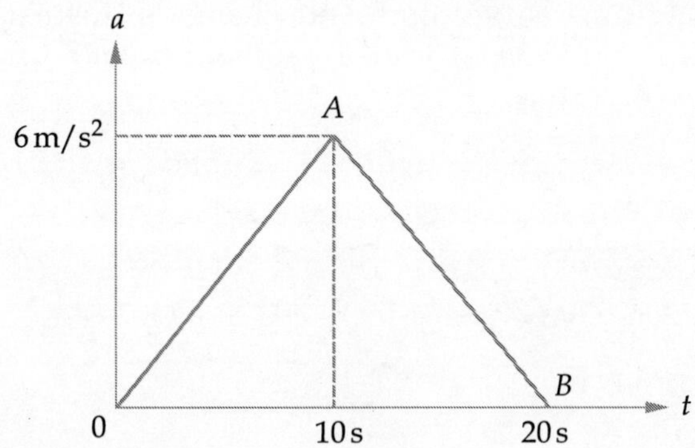

Fig. 8.20

Solution : The area under acceleration-time curve represents change in velocity.

$$V_0 = 0 \quad \text{(given)}$$

$$V_{10} - V_0 = \frac{1}{2} \times 10 \times 6 = 30$$

$$\therefore \quad V_{10} = 30 \text{ m/s}$$

$$V_{20} - V_{10} = \frac{1}{2} \times (20 - 10) \times 5 = 30$$

$$\therefore \quad V_{20} = 60 \text{ m/s}$$

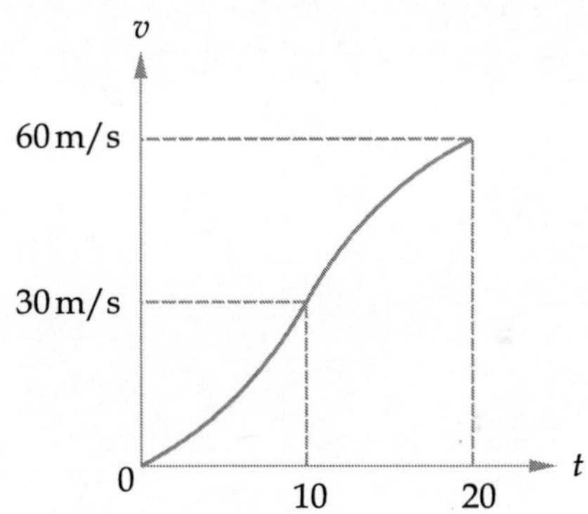

Fig. 8.21

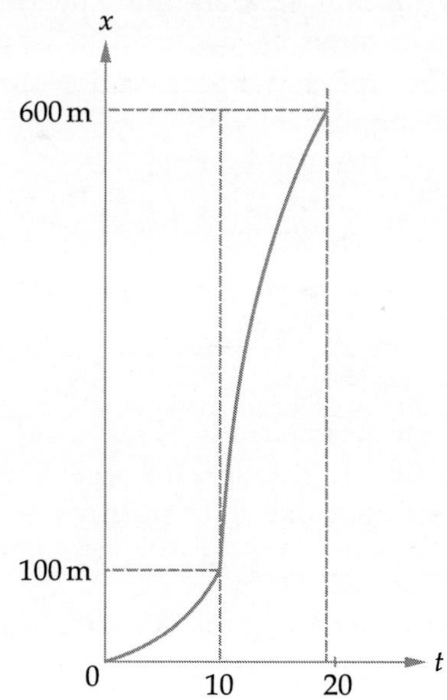

Fig. 8.22

It is to be noted that :

(*i*) for $0 < t < 10$ seconds ; the acceleration-time curve is a straight line with positive slope. Obviously the velocity-time graph is a parabola convex downwards.

(*ii*) for $10 < t < 20$ seconds, the acceleration-time curve is a straight line with negative slope. The corresponding velocity-time variation is a parabola convex upwards.

The velocity-time variation will then plot as shown in Fig. 8.21.

(*b*) The area under velocity-time curve represents change in displacement. Further for a parabola convex downwards $A = 1/3\,bh$ and for a parabola convex upwards $A = 2/3\,bh$

$$x_0 = 0 \quad \text{(given)}$$

$$x_{10} - x_0 = \frac{1}{3} \times 10 \times 30 = 100$$

$$\therefore \quad x_{10} = 100 \text{ m}$$

$$x_{20} - x_{10} = \frac{2}{3} \times 10 \times 30 + 10 \times 30 = 500$$

$$\therefore \quad x_{20} = 600 \text{ m}$$

It is to be noted that :

(*i*) for $0 < t < 10$ seconds and for $10 < t < 20$, the slope of v-t curve is positive

(*ii*) for $0 < t < 10$, cubical parabola is convex downwards

(*iii*) for $10 < t < 20$, cubical parabola is convex upwards

The displacement-time will then plot as shown in Fig. 8.22.

Further : Maximum velocity = **60 m/s**

Maximum distance traversed = **600 m**

EXAMPLE 8.27

A train starts from rest and moves on a straight road along the horizontal direction. The variation of its acceleration with time depicts as shown in Fig. 8.23. Calculate the speed attained and distance covered by the train in 25 seconds.

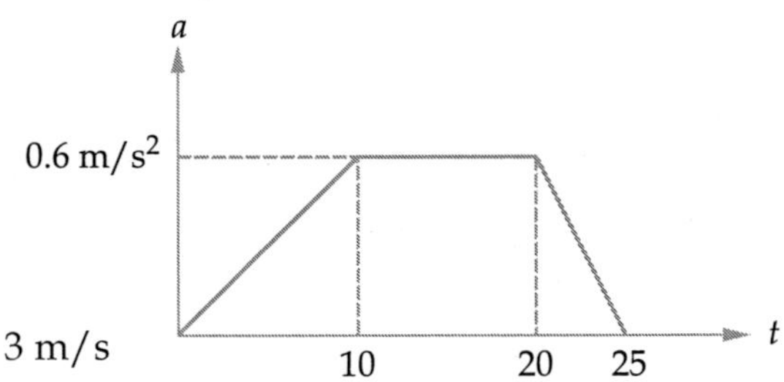

Fig. 8.23

Solution : The area under acceleration-time curve represents change in velocity,

$$V_0 = 0 \text{ (given)}$$

$$V_{10} - V_0 = \frac{1}{2} \times 10 \times 0.6 = 3; \quad \therefore \quad V_{10} = 3 \text{ m/s}$$

$$V_{20} - V_{10} = (20 - 10) \times 0.6 = 6; \quad \therefore \quad V_{20} = 9 \text{ m/s}$$

$$V_{25} - V_{20} = \frac{1}{2} \times (25 - 20) \times 0.6 = 1.5; \quad \therefore \quad V_{25} = 10.5 \text{ m/s}$$

It is to be noted that :

(*i*) for $0 < t < 10$ seconds, the acceleration-time curve is a straight line with positive slope. Obviously the velocity-time graph is a parabola convex downwards.

(*ii*) for $10 < t < 20$ seconds, the a-t curve is a horizontal straight line. The corresponding velocity time variation is a straight line with positive slope.

(*iii*) for $20 < t < 25$ seconds, the a-t curve is a straight line with negative slope. The velocity-time variation would then be a parabola convex upwards.

The velocity time variation would then plot as shown in Fig. 8.24.

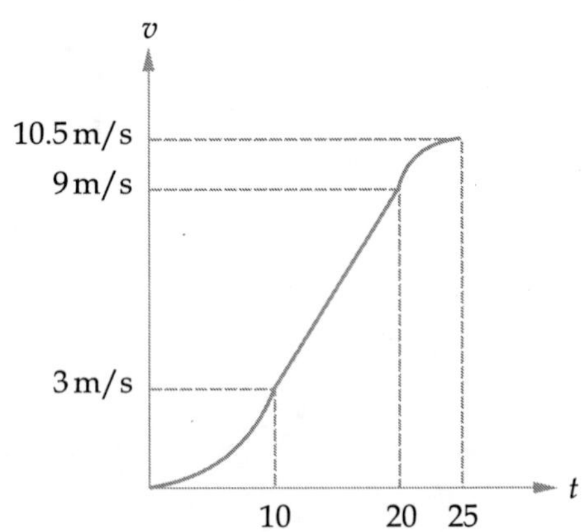

Fig. 8.24

Distance travelled = area of velocity-time graph

$$S = s_1 + s_2 + s_3$$

Recalling that for a parabola convex downwards $A = \frac{1}{3}bh$ and for a parabola convex upwards $A = \frac{2}{3}bh$, we have

$$s_1 = \frac{1}{3} \times 10 \times 3 = 10 \text{ m}$$

$$s_2 = \frac{1}{2} \times (20 - 10) \times (9 - 3) + (20 - 10) \quad = 60 \text{ m}$$

$$s_3 = \frac{2}{3} \times (25 - 20) \times (10.5 - 9) + (25 - \quad = 12.5$$

∴ Total distance travelled by the train 25 seconds = 10 + 60 + 12.5 = **82.5 m**

8.4. MOTION UNDER GRAVITY

When a body falls freely, then its velocity increases as it approaches the earth. The increase in the velocity of falling body is due to gravitational acceleration g whose value is normally taken to be 9.81m/s^2 . When motion is considered under gravity, the acceleration, a in the equations of motion for rectilinear motion is replaced by g and we obtain

$$v = u + gt \; ; \; h = ut + \frac{1}{2}gt^2 \; ; \quad v^2 - u^2 = 2gh$$

and distance travelled in the n th second is

$$s_n = u + \frac{g}{2}(2n - 1)$$

(*i*) When the body falls from height h, then $u = 0$ and hence

$$v = gt \; ; \quad h = \frac{1}{2}gt^2 \; ; \quad v^2 = 2gh$$

and $$s_n = \frac{g}{2}(2n - 1)$$

(*ii*) When the bodies are thrown vertically upwards, then g becomes a retarding parameter. The velocity of the body then decreases gradually till it becomes zero. For a body thrown in vertically upward direction, the equations of motion are

$$v = u - gt \; ; \quad h = ut - \frac{1}{2}gt^2 \; ; \quad v^2 - u^2 = -2gh$$

and $$s_n = u - \frac{g}{2}(2n - 1)$$

(*iii*) At the point of maximum height attained by a body thrown vertically upwards, the velocity v of body becomes zero. Then from expression $v = u - gt$ and $v^2 - u^2 = -2gh$, we have

Time taken in attaining maximum height $= \frac{u}{g}$

Maximum height attained $= \frac{u^2}{2g}$

(*iv*) After attaining the maximum height, the body starts falling down. Then from the relation $v^2 - u^2 = 2gh$, the velocity of the body on reaching the ground will be $v = \sqrt{2gh}$. Neglecting air friction, the velocity with which the body reaches the ground is same with which it is thrown upwards.

(*v*) From the relation $v = u - gt$, the time taken by the body to fall from position of maximum height to ground is $t = \frac{v}{g} = \frac{u}{g}$

∴ Total time taken by the body to fall back to ground is $\frac{2u}{g}$.

EXAMPLE 8.28

A body is allowed to fall from the top of a tower. It falls through half the height in 2 seconds. What will be the total time to reach the ground?

Take $g = 10$ m/s².

Solution:

$$s = ut + \frac{1}{2}gt^2 \quad \text{and} \quad u = 0$$

$$\frac{s}{2} = 0 \times t + \frac{1}{2} \times 10 \times (2)^2 = 20$$

∴ Height of tower = 40 m

Again,

$$40 = \frac{1}{2} \times 10 \times t^2 \;; \quad t \simeq \mathbf{2.8\ s}$$

EXAMPLE 8.29

A stone takes 6 seconds to reach the ground after it is dropped from the tower. If the stone is stopped after 2 seconds of its fall and then released again, how much more time it will take to reach the ground. Take $g = 10$ m/s².

Solution: $s = ut + \frac{1}{2}gt^2$ and $u = 0$

∴ Height of tower $s = 0 \times 6 + \frac{1}{2} \times 10 \times 6^2 = 180$ m

Distance covered in 2 seconds $= \frac{1}{2} \times 10 \times 2^2 = 20$ m

Let t be the time taken by the stone to cover the remaining 160 m, starting from rest.

$$160 = \frac{1}{2} \times 10 \times t^2 \;; \quad t = \mathbf{5.6\ s}$$

EXAMPLE 8.30

A body is dropped from rest at height *h*. It covers a distance of 9*h*/25 in the last second. Determine the height *h*. Take $g = 10$ m/s².

Solution: Let t be the time taken to reach the ground level.

$$h = \frac{1}{2}gt^2 \qquad (\because u = 0)$$

Also $$\left(1-\frac{9}{25}\right)h = \frac{1}{2}g(t-1)^2$$

Dividing the above identities and solving for t, we get

$$\frac{16}{25} = \frac{(t-1)^2}{t^2} \; ; \quad \frac{4}{5} = \frac{t-1}{t}$$

$$t = 5 \text{ s}$$

$$\therefore \quad \text{Height } h = \frac{1}{2}\times 10\times 5^2 = \mathbf{125\ m}$$

EXAMPLE 8.31

Water drips from a tap at uniform rate onto the floor which is 1.5 m below. The first drop strikes the floor at the instant the fourth drop is on the verge of falling from the tap. What would be the location of the different drops when the first drop hits the floor?

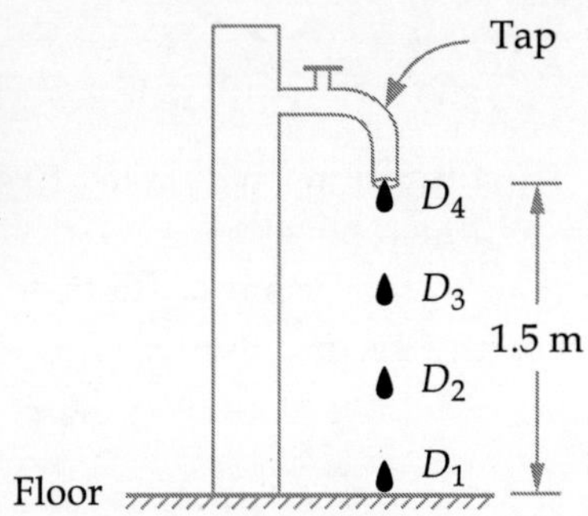

Fig. 8.25

Solution : The drops fall at regular intervals of time say t. When the first drop D_1 strikes the floor, the fourth drop D_4 begins to fall from the tap. This implies that the time interval between the first and 4th drop is $3t$, between the first and third drop is $2t$ and between the first and second drop is t.

Applying the relation $s = ut + \frac{1}{2}gt^2$ and with $u = 0$, we have

$$\text{Vertical distance } D_1D_4 = 1.5 = \frac{1}{2} \times 9.81 \times (3t)^2$$

$$t = \sqrt{\frac{2\times 1.5}{9\times 9.81}} = 0.1843 \text{ s}$$

Then : $$D_3D_4 = \frac{1}{2}\times 9.81\times(0.1843)^2 = \mathbf{0.167\ m}$$

$$D_2D_4 = \frac{1}{2}\times 9.81\times(2\times 0.1843)^2 = \mathbf{0.667\ m}$$

Accordingly : height of drop D_3 above the floor = (1.5 – 0.167) = **1.333 m**
height of drop D_2 above the floor = (1.5 – 0.667) = **0.833 m**

EXAMPLE 8.32

A ball of mass 2 kg was thrown vertically upwards and at a height of 5 m above the ground, its kinetic energy was equal to its potential energy. How much total time does the ball take to get back to the point from where it was thrown? Take g = 10 m/s².

Solution: At a height of 5 m above the ground;

$$mgh = \frac{1}{2}mv^2 \; ; \quad v = \sqrt{2gh} = \sqrt{2\times 10\times 5} = 10 \text{ m/s}$$

Also $$v^2 - u^2 = 2gh \; ; \quad 10^2 - u^2 = 2 \times (-10) \times 5$$

That gives initial velocity of ball $u = 10\sqrt{2}$ m/s

At maximum height, the ball will have zero velocity. Accordingly

$$0 = u - gt\ ; \quad t = \frac{u}{g} = \frac{10\sqrt{2}}{10} = \sqrt{2}\ s$$

Hence time taken in ascending the maximum height will be $\sqrt{2}$ s. The time taken for return path will also be same. Hence total time taken will be $2\sqrt{2}$ seconds.

EXAMPLE 8.33

A stone was thrown vertically upwards from the ground with a velocity of 100 m/s. After 2 seconds, another stone was thrown vertically upwards from the same place. If both the stones strike the ground at the same time, calculate the velocity with which the second stone was thrown. Take g = 10 m/s^2.

Solution: At maximum height, the stone will have zero velocity. Accordingly considering upward motion of first stone.

$$0 = u - gt\ ; \quad t = \frac{u}{g} = \frac{100}{10} = 10\ \text{s}$$

Thus the stone will take 10 s in ascending to maximum height. The time taken for return path will also be same.

∴ total time of flight = 10 + 10 = 20 s

For the second stone:

time of flight = 20 – 2 = 18 s

$$\text{time taken to reach maximum height} = \frac{18}{2} = 9\ \text{s}$$

Then $0 = u - 10 \times 9\ ; \quad u =$ **90 m/s**

EXAMPLE 8.34

A stone is dropped into a well and the sound of splash is heard after 4 seconds. Assuming velocity of sound to be 350 m/s, make calculations for the depth of well.

Solution: Let h = the depth of well

t_1 = time taken by stone to strike water

t_2 = time taken by sound to reach from surface of water to top of well

Then $t_1 + t_2 = 4$ ...(*i*)

Considering downward motion of stone and using the relation $s = ut + \frac{1}{2}gt^2$, we have

$$h = 0 \times t_1 + \frac{1}{2} \times 9.81 \times t_1^2\ ; \quad h = 4.905\ t_1^2$$

Considering the motion of sound, the time taken by the sound to reach from surface of water to top of well is given by

$$t_2 = \frac{\text{depth of well}}{\text{speed of sound}} = \frac{h}{350} = \frac{4.905 t_1^2}{350} \qquad (\because h = 4.905 t_1^2)$$

Then from identity (*i*),

$$t_1 + \frac{4.905 t_1^2}{350} = 4\ ; \quad 350\ t_1 + 4.905 t_1^2 = 1400$$

$\therefore \quad 4.905\, t_1^2 + 350\, t_1 - 1400 = 0$

Solution of this quadratic equation gives

$$t_1 = \frac{-350 \pm \sqrt{350^2 + 4 \times 4.905 \times 1400}}{2 \times 4.905} = \frac{-350 \pm 387.26}{9.81}$$

Taking the + ve root; $t_1 = 3.798$ s

$\therefore$ Depth of well $h = 4.905 t_1^2 = 4.905 \times 3.798^2 =$ **70.75 m**

EXAMPLE 8.35

A stone is dropped from the top of a tower 40 m height. At the same instant, another stone is thrown upward from the foot of tower with an initial velocity of 20 m/s. At what distance from the top and after how much time the two stones cross each other ? Further proceed to calculate the relative velocity with which the stones cross.

Solution: Let the two stones cross each other at distances h from the top of tower after t seconds.

Applying the relation $s = ut + \frac{1}{2}at^2$

For the downward motion of first stone,

$$u = 0\,;\quad s = h \text{ and } a = g = 9.81 \text{ m/s}^2$$

$$\therefore \quad h = 0 \times t + \frac{1}{2} \times 9.81 \times t^2$$

$$= 4.905t^2 \qquad ...(i)$$

For the upward motion of second stone

$u = 20$ m/s;

$s = (40 - h)$

and $a = -g = -9.81$ m/s^2

$$\therefore \quad 40 - h = 20t + \frac{1}{2} \times (-9.81)t^2$$

$$= 20t - 4.905t^2 \qquad ...(ii)$$

Adding expressions (*i*) and (*ii*), we get

$$40 = 20t\,; \quad \therefore \ t = \mathbf{2\ s}$$

$$h = 4.905t^2 = 4.905 \times 2^2 = \mathbf{19.62\ m}$$

The stones thus cross each other at a distance of 19.62 m from the top of the tower.

(*b*) velocity of first stone at the crossing point

$$v = u + gt = 0 + 9.81 \times 2 = 19.62 \text{ m/s} \downarrow$$

velocity of second stone at the crossing point

$$v = u - gt = 20 - 9.81 \times 2 = 0.38 \text{ m/s} \uparrow$$

Relative velocity = 19.62 – (– 0.38) = **20 m/s**

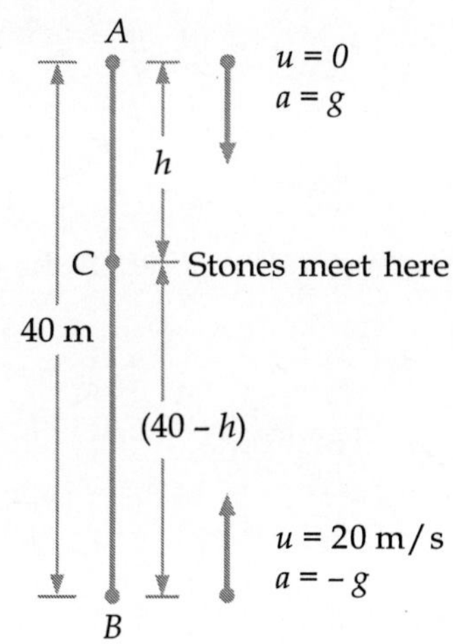

Fig. 8.26

EXAMPLE 8.36

A stone is thrown vertically upwards with a velocity 20 m/s from the top of a tower 25 m high. Make calculations for the following parameters:

(*i*) the maximum height to which the stone will rise in its flight
(*ii*) velocity of the stone during its downward travel at a point in the same level as the point of projection,
(*iii*) time required for the stone to reach the ground.
Take $g = 10$ m/s^2

Solution:

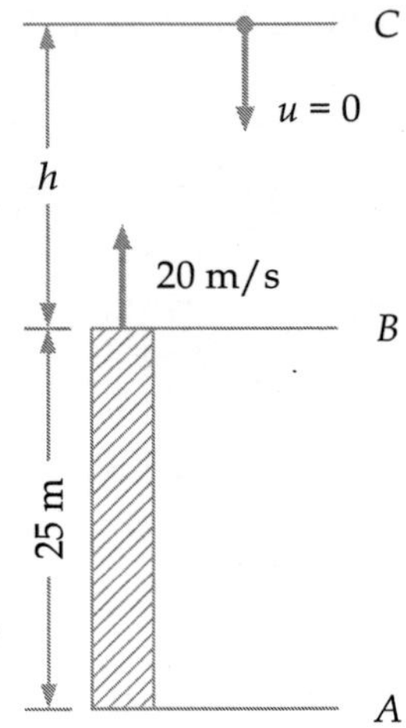

Fig. 8.27

(*i*) **Motion *B* to *C*:** Let t be the time taken by the stone to reach the maximum height (level C).

Using the relation: $v = u + at$

$$u = 20 \text{ m/s}, \quad v = 0$$

and $$a = -g = -10 \text{ m/s}^2$$

$$\therefore \quad 0 = 20 - 10t\ ; \quad t = 2 \text{ s}$$

Also $$v^2 - u^2 = 2ah$$

$$0^2 - 20^2 = 2 \times (-10) \times h\ , \quad h = 20 \text{ m}$$

Total height from ground = 20 + 25 = **45 m**

(*ii*) **Motion *C* to *B***

The ball took 2 s while ascending from level B to maximum height (level C). It will take same time during return journey from level C to B.

Velocity at B, $v_b = u + gt$

$= 0 + 10 \times 2 =$ **20 m/s**

Apparently the velocity with which the stone reaches level B (top of tower) is same with which it was thrown upwards.

(*iii*) **Motion *B* to *A***

$$h = ut + \frac{1}{2}gt^2$$

$$25 = 20t + \frac{1}{2} \times 10 \times t^2\ ; \quad 25 = 20t + 5t^2$$

$$\therefore \quad t^2 + 4t - 5 = 0$$

Solution of this quadratic equation gives

$$t = \frac{-4 \pm \sqrt{4^2 - 4 \times 1 \times (-5)}}{2} = \frac{-4 \pm 6}{2} = 1 \text{ s}$$

$\therefore$ Total time during which the stone is in flight

= time from B to C + time from C to B + time from B to A

= 2 + 2 + 1 = **5 s**

EXAMPLE 8.37

A particle projected vertically upwards is seen at *P* after t_1 seconds. If it takes t_2 second more from this point *P* to the point of projection, prove that

(*i*) the height of *P* above the point of projection $= \dfrac{1}{2} g\, t_1\, t_2$

(*ii*) the maximum height attained $= \dfrac{1}{2} g \left[\dfrac{t_1 + t_2}{2}\right]^2$

Solution: Time of travel from A to $P = t_1$

Time of travel from P to B to $A = t_2$

Since time of ascent equals the time of descent,

Time of travel from P to $B = \dfrac{t_2 - t_1}{2}$

$\therefore$ Time taken by the particle in ascending to maximum height

$$= t_1 + \frac{t_2 - t_1}{2}$$

$$= \frac{t_1 + t_2}{2}$$

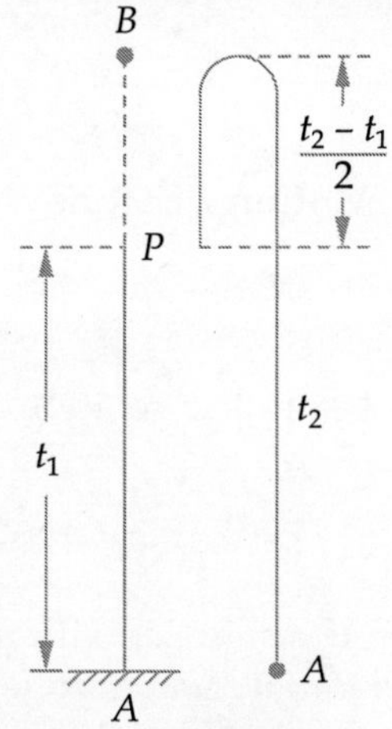

Fig. 8.28

At position B corresponding to maximum height, the particle will have zero velocity. Accordingly considering upward motion of particle from A to B,

$$0 = u - g\left(\frac{t_1 + t_2}{2}\right); \quad u = g\left(\frac{t_1 + t_2}{2}\right)$$

(*i*) Using the relation $s = ut - \frac{1}{2}gt^2$,

Height of P above the point of projection A

$$s = g\left(\frac{t_2 + t_2}{2}\right)t_1 - \frac{1}{2}g\,t_1^2 = \frac{gt_1^2}{2} + \frac{g\,t_1t_2}{2} - \frac{g\,t_1^2}{2} = \frac{1}{2}g\,t_1\,t_2$$

(*ii*) Using the relation $v_2 - u^2 = -2gs$,

Maximum height attained by the particle,

$$s = \frac{u^2}{2g} \qquad (\because \ v = 0 \text{ at the position of maximum height})$$

$$= g^2\left(\frac{t_1 + t_2}{2}\right)^2 \times \frac{1}{2g} = \frac{1}{2}g\left(\frac{t_1 + t_2}{2}\right)^2$$

EXAMPLE 8.38

Two stones are dropped simultaneously from rest from two points both 200 m above the ground. One of the stone during its downward motion hits a fixed inclined plane located at 80 m distance from the ground. The plane is inclined at an angle to the horizontal as shown in Fig. 8.29. As a result of this impact, the direction of velocity of the stone becomes horizontal. Compare the times of fall of the two stones to reach the ground.

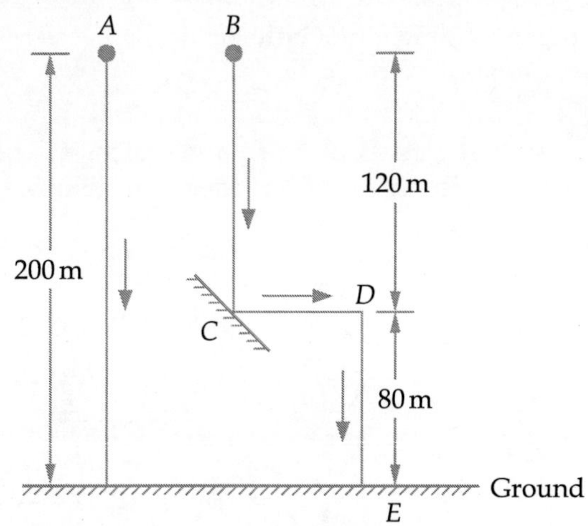

Fig. 8.29

Solution :

Motion of stone A :

$u = 0; \quad s = 200$ m

t = time taken to reach the ground

Using the relation : $\quad s = ut + \frac{1}{2}gt^2$, we have

$$200 = 0 \times t + \frac{1}{2} \times 9.81 \times t^2; \quad t = \sqrt{\frac{200 \times 2}{9.81}} = 6.385 \text{ sec}$$

Motion of stone B: From the first part of the journey from B to C,

$u = 0; \quad s = 120$ m

t_1 = time taken to reach the inclined plane at C

Using the relation : $s = ut + \frac{1}{2} gt^2$, we have

$$120 = 0 \times t_1 + \frac{1}{2} \times 9.81 \times t_1^2; \quad t_1 = \sqrt{\frac{120 \times 2}{9.81}} = 4.946 \text{ sec}$$

After impact with the inclined plane at point C, the direction of motion of stone B becomes horizontal. This implies that

(*i*) the velocity of stone B becomes zero in the vertical direction.

(*ii*) the journey for the remaining distance of (200 – 120) = 80 m is performed with zero initial velocity.

$$80 = 0 \times t_2 + \frac{1}{2} \times 9.81 \times t_2^2; \quad t_2 = \sqrt{\frac{80 \times 2}{9.81}} = 4.038 \text{ sec}$$

∴ Time taken by the stone B to reach the ground

= 4.946 + 4.038 = 8.984 sec

Hence ratio of times taken by stones B and A to reach the ground

$$= \frac{8.984}{6.385} = \mathbf{1.407}$$

EXAMPLE 8.39

An elevator whose floor to ceiling height is 2.75 m starts ascending with constant upward acceleration of 1.2 m/s^2. Two seconds after the start, a loose bolt drops from the ceiling towards the floor of the elevator. Calculate :

(*a*) time taken by the bolt to reach the elevator floor.

(*b*) displacement of the bolt during that time, and

(*c*) the distance travelled by the bolt during the same time.

Solution : Using kinematic relation $V = u + at$, the velocity of the elevator at two seconds after the start is

$$V = 0 + 2 \times 1.2 = 2.4 \text{ m/s}$$

The bolt travels up with a velocity of 2.4 m/s and then drops under gravity. If t denotes the time to reach the floor, then the distance moved by the bolt is

$$s_1 = ut - \frac{1}{2} gt^2 = 2.4t - \frac{1}{2} \times 9.81 t^2$$

Distance moved by the bolt during the same time t is

$$s_2 = ut + \frac{1}{2} gt^2 = 2.4t + \frac{1}{2} \times 1.2 t^2$$

But $s_2 = (s_1 + 2.75)$ and therefore

$$2.4t + \frac{1}{2} \times 1.2t^2 = 2.4t - \frac{1}{2} \times 9.81 t^2 + 2.75$$

or $(0.6 + 4.905)t^2 = 2.75$; $t =$ **0.707 sec**

The bolt takes 0.707 sec to reach the floor.

(*b*) Displacement of bolt during t = 0.707 sec

$$= ut - \frac{1}{2}gt^2 = 2.4 \times 0.707 - \frac{1}{2} \times 9.81 \times 0.707^2$$
$$= 1.697 - 2.452 = -0.755 \text{ m}$$

The negative sign implies that this displacement is downwards with respect to initial position.

(*c*) The velocity of the bolt is zero when it reaches the floor. If t is the time taken to reach the floor, then

$$v = u - gt \;;\quad 0 = 2.4 - 9.81\,t \;;\quad t = 0.2496$$

Distance moved by bolt during this time is

$$s = ut - \frac{1}{2}gt^2 = 2.4 \times 0.2446 - \frac{1}{2} \times 9.81 \times (0.2446)^2$$
$$= 0.2935 \text{ m}$$

Time during which the bolt moves up = (0.707 – 0.2446) = 0.4624 sec

Distance moved by the bolt during this period is

$$s = ut - \frac{1}{2}gt^2 = 0 \times 0.4624 + \frac{1}{2}(-9.81) \times (0.4624)^2$$
$$= -1.0487 \text{ m} = 1.049 \text{ m in magnitude}$$

∴ Total distance moved by the bolt

$$= 0.2935 + 1.0487 = \mathbf{1.3422 \text{ m}}$$

8.5. CURVILINEAR MOTION

The motion of a particle along a curved path other than a straight line is known as curvilinear motion.

Examples of curvilinear motion are :

(*i*) an automobile vehicle negotiating a turn on the road

(*ii*) projectile motion of a bullet fired from a gun

(*iii*) motion of bob of pendulum oscillating in vertical plane

(*iv*) motion of satellite around the earth.

The curvilinear motion may be two-dimensional (plane curvilinear) or three-dimensional (space curvilinear). The study of the plane curvilinear motion is made with reference to the following coordinate systems :

8.5.1. Rectangular-Cartesian x, y Coordinates

The position of the particle on the curved path at any instant is defined by the position vector $\vec{r}$.

$\vec{r} = x\boldsymbol{i} + y\boldsymbol{j}$ where $\boldsymbol{i}$ and $\boldsymbol{j}$ are the unit vectors

Magnitude $r = |\vec{r}| = \sqrt{x^2 + y^2}$

Velocity vector $\vec{V} = \frac{d\vec{r}}{dt} = \frac{d}{dt}(x\boldsymbol{i} + y\boldsymbol{j})$

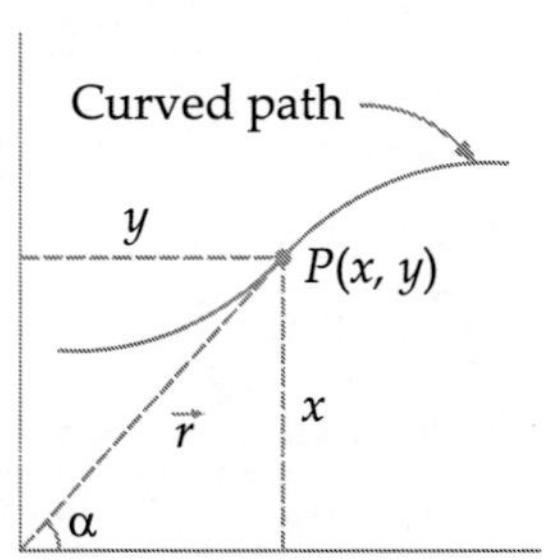

Fig. 8.30

Since i and j are constant terms, we get

$$\vec{V} = i\frac{dx}{dt} + j\frac{dy}{dt} = V_x i + V_y j$$

Resultant velocity $V = \sqrt{V_x^2 + V_y^2}$

The direction of velocity is tangential to the path of motion of the particle. If α is the angle made by the resultant with x-axis, then

$$\tan\alpha = \frac{V_y}{V_x} \quad ; \quad \alpha = \tan^{-1}\left(\frac{V_y}{V_x}\right)$$

Further, acceleration $\vec{a} = \dfrac{d\vec{v}}{dt} = \dfrac{d}{dt}(V_x i + V_y j) = \dfrac{d}{dt}\left(\dfrac{dx}{dt}i + \dfrac{dy}{dt}j\right)$

$$= \frac{d^2x}{dt^2}i + \frac{d^2y}{dt^2}j = a_x i + a_y j$$

Resultant acceleration $a = \sqrt{a_x^2 + a_y^2}$

If β is the angle made by resultant acceleration with x-axis, then

$$\tan\beta = \frac{a_y}{a_x} \quad ; \quad \beta = \tan^{-1}\left(\frac{a_y}{a_x}\right)$$

For a curvilinear motion in space, we may write

$$\vec{r} = x i + y j + z k$$

$$\vec{V} = V_x i + V_y j + V_z k$$

$$\vec{a} = a_x i + a_y j + a_z k$$

EXAMPLE 8.40

The motion of a particle is defined by the relations : $x = t^2 + 3t$ and $y = t^3 - 8t^2 + 3$ where x and y are in meters and y is in seconds. (*a*) Write the equations defining the motion of the particle in vectorial form (*b*) Calculate the velocity and acceleration of the particle at time t = 2 seconds.

Solution :

$$x = t^2 + 3t \qquad\qquad y = t^3 - 8t^2 + 3$$

$$V_x = \frac{dx}{dt} = 2t + 3 \qquad\qquad V_y = \frac{dy}{dt} = 3t^2 - 16t$$

$$a_x = \frac{dV_x}{dt} = 2 \qquad\qquad a_y = \frac{dV_y}{dt} = 6t - 16$$

At any time t, the motion of the particle in vectorial form can be written as

Position vector $\vec{r} = x i + y j = (t^2 + 3t) i + (t^3 - 8t^2 + 3) j$

Velocity vector $\vec{V} = V_x i + V_y j = (2t + 3) i + (3t^2 - 16t) j$

Acceleration $\vec{a} = a_x i + a_y j = 2i + (6t - 16) j$

(*b*) At time t = 2 seconds

$$\vec{V} = (2 \times 2 + 3) i + (3 \times 2^2 - 16) j = 7i - 20j = V_x i + V_y j$$

∴ $V_x = 7$ m/s and $V_y = -20$ m/s.

Since V_x is positive and V_y is negative, the resultant lies in the fourth quadrant.

Resultant velocity $V = \sqrt{V_x^2 + V_y^2} = \sqrt{7^2 + (-20)^2} = $ **21.19 m/s**

$$\tan\alpha = \frac{V_y}{V_x} = \frac{20}{7} = 2.857 \;\; ; \;\; \alpha = \mathbf{70.71^\circ}$$

Further, $\vec{a} = 2i + (6 \times 2 - 16)j = 2i - 4j = a_x i + a_y j$

∴ $a_x = 2$ m/s² and $a_y = -4$ m/s²

Since a_x is positive and a_y is negative, the resultant acceleration lies in the fourth quadrant.

Resultant acceleration $a = \sqrt{a_x^2 + a_y^2} = \sqrt{2^2 + (-4)^2} = $ **4.47 m/s²**

$$\tan\beta = \frac{a_y}{a_x} = \frac{4}{2} = 2 \;\; ; \;\; \beta = \mathbf{63.43^\circ}$$

The velocity and acceleration vectors have been shown below in Fig. 8.31

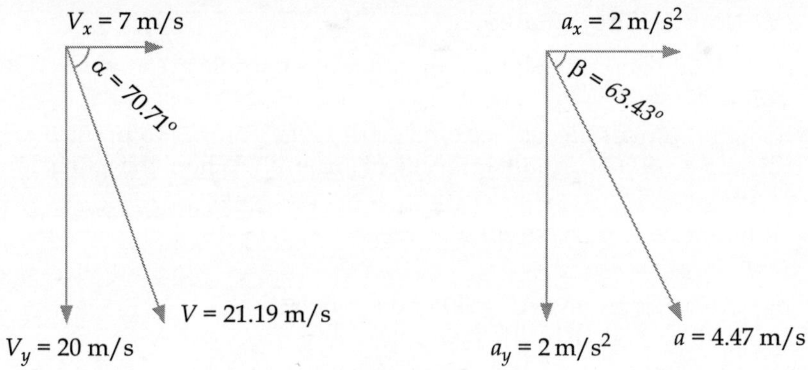

Fig. 8.31

EXAMPLE 8.41

A particle moves in x-y plane and its coordinates are defined by the relations

$$x = -3\sin t \quad \text{and} \quad y = 2 - 3\cos t$$

where x and y are in metres and time t is in seconds:

(a) State the equation of path along which the particle moves, and

(b) Write expressions for position vector, velocity vector and acceleration vector for the particle at the instant when it starts motion.

Solution : $x = -3\sin t$ and $y = 2 - 3\cos t$

or $$\sin t = -\frac{x}{3} \quad \text{and} \quad \cos t = \frac{2-y}{3}$$

$$\sin^2 t + \cos^2 t = \left(\frac{-x}{3}\right)^2 + \left(\frac{2-y}{3}\right)^2$$

∴ $$1 = \frac{x^2}{3^2} + \frac{(2-y)^2}{3^2} \quad \text{or} \quad x^2 + (2-y)^2 = 3^2$$

Apparently the curved path is a circle of radius 3 m and origin (0, 2) (*b*) Position vector $\vec{r} = x\boldsymbol{i} + y\boldsymbol{j} = (-3 \sin t)\boldsymbol{i} + (2 - 3 \cos t)\boldsymbol{j}$

$$\text{Velocity vector } \vec{V} = \frac{d\vec{V}}{dt} = \frac{d}{dt}[(-3 \sin t)\,\boldsymbol{i} + (2 - 3 \cos t)\boldsymbol{j}]$$
$$= (-3 \cos t)\,\boldsymbol{i} + (3 \sin t)\,\boldsymbol{j}$$

$$\text{Acceleration vector } \vec{a} = \frac{d\vec{V}}{dt} = \frac{d}{dt}[(-3 \cos t)\boldsymbol{i} + (3 \sin t)\,\boldsymbol{j}]$$
$$= (3 \sin t)\boldsymbol{i} + (3 \cos t)\boldsymbol{j}$$

At the instant of start, *i.e.* at $t = 0$

$$\vec{r} = (-3 \sin 0)\boldsymbol{i} + (2 - 3 \cos 0)\boldsymbol{j} = -1\boldsymbol{j}$$
$$\vec{V} = (-3 \cos 0)\boldsymbol{i} + (3 \sin 0)\boldsymbol{j} = -3\boldsymbol{i}$$
$$\vec{a} = (3 \sin 0)\boldsymbol{i} + (3 \cos 0)\boldsymbol{j} = 3\boldsymbol{j}$$

Obviously at start, the particle has a velocity of -3 m/s^2 along $-$ *ve* x-axis and an acceleration of 3 m/s^2 along the + *ve* y-axis.

8.5.2. Tangential and Normal Coordinates

In this system, the velocity and acceleration of a moving particle are expressed in tangential (t) and normal (n) components.

Consider a particle that moves along a curved path from point A to point B and traverses an infinitely small distance δS in a small interval of time δt. When the particle moves, the distance S changes with time.

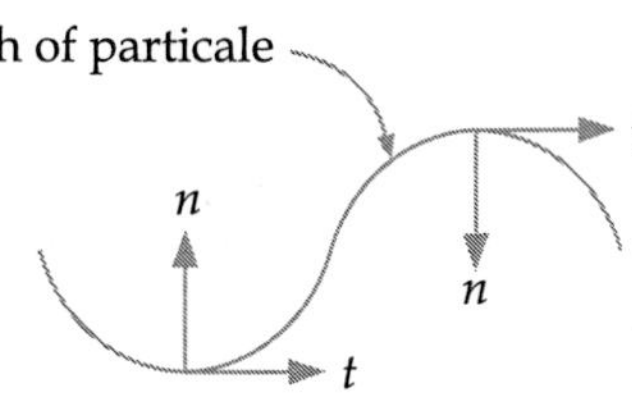

When the time interval δt is very small,

chord AB = arc AB = δS

Further, it is to be noted that angle $BAC = \delta\theta$ and as such the components of distance δS travelled by the particle along the tangential and normal directions are $\delta S \cos d\theta$ and $\delta S \sin \delta\theta$.

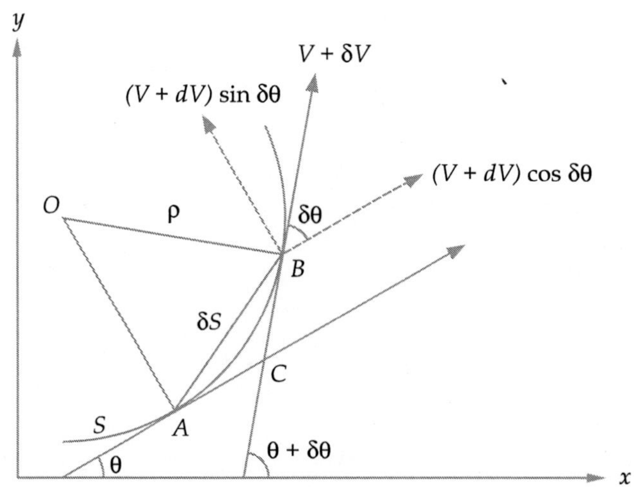

Fig. 8.32

The velocity of the particle at point A in the tangential direction is,

$$V_t = \frac{\text{distance moved along the tangent}}{\text{time interval}}$$

$$= \underset{\delta t \to 0}{\delta t} \frac{\delta S \cos\delta\theta}{\delta t} = \underset{\delta t \to 0}{\delta t} \frac{\delta S}{\delta t} \cos \delta\theta$$

As $\delta\theta$ is small, $\cos \delta\theta \to 1$ and therefore

$$V_t = \underset{\delta t \to 0}{\delta t} \frac{\delta S}{\delta t} = \frac{dS}{dt}$$

The velocity of the particle in the normal direction is,

$$V_n = \frac{\text{distance moved along the tangent}}{\text{time interval}}$$

$$= \underset{\delta t \to 0}{\delta t} \frac{\delta S \sin \delta\theta}{\delta t}$$

As $\delta\theta$ is small, $\sin \delta\theta \to \delta\theta$. Neglecting the product $(\delta S \times \delta\theta)$, we get

$$V_n = 0$$

Thus the normal component of velocity of the moving particle is zero. It implies that the velocity of a particle moving along a curved path is equal to the tangential velocity and that the direction of velocity is tangential to the path at any point.

Further, it may be noted that

(*i*) At point *A*, the velocity *V* is directed along the tangent that is inclined at an angle θ with the horizontal. The velocity in a direction normal to the tangent is zero.

(*ii*) At point *B*, the velocity $(V + \delta V)$ is again directed along the tangent at that point. This tangent makes an angle $\delta\theta$ with the tangent at point *P*.

The components of velocity vector $(V + \delta V)$ are;

$(V + \delta V) \cos d\theta$ in the direction of tangent at *P*, and

$(V + \delta V) \sin d\theta$ in the direction normal to tangent at *P*

$$\text{Tangential acceleration } a_t = \frac{\text{change in tangential velocity}}{\text{time interval}}$$

$$= \underset{\delta t \to 0}{\delta t} \frac{(V + \delta V) \cos \delta\theta - V}{\delta t}$$

When the time interval δt is very small, the arc *AB* will be very small and so will be $\delta\theta$ and then $\cos \delta\theta = 1$

$$\therefore \quad a_t = \underset{\delta t \to 0}{\delta t} \frac{(V + \delta V) - V}{\delta t} = \frac{dV}{dt}$$

Thus the tangential acceleration is due to change in the magnitude of velocity.

$$\text{Normal acceleration } a_n = \frac{\text{change in normal velocity}}{\text{time interval}} = \underset{\delta t \to 0}{\delta t} \frac{(V + \delta V) \sin \delta\theta - 0}{\delta t}$$

For small $d\theta$, $\sin \delta\theta = \delta\theta$. Neglecting the product $(\delta V \times \delta\theta)$ we may write

$$a_n = \frac{Vd\theta}{dt} = V \frac{d\theta}{dS} \times \frac{dS}{dt} = \frac{V\,dS/dt}{dS/d\theta} = \frac{V \times V}{\rho} = \frac{V^2}{\rho}$$

where $\rho = \dfrac{dS}{d\theta}$ is the radius of curvature of the path. When the body moves in a circular orbit of radius *r*, then

$$a_n = \frac{V^2}{r}$$

The normal acceleration is due to change in the direction of velocity. It is directed towards the centre of curvature and is known as centripetal acceleration.

Resultant acceleration $a = \sqrt{a_t^2 + a_n^2}$

If the resultant acceleration makes an angle ϕ with the normal, then

$$\tan \phi = a_t / a_n$$

It is worthwhile to note that

(*i*) If the magnitude of velocity is constant between two points, then the tangential acceleration $a_t = 0$.

(*ii*) If the direction of motion of particle does not change, *i.e.,* if the motion of the particle is along a straight line, then $\rho \to \infty$. In that case the normal acceleration $a_n = 0$.

(*iii*) When the particle is moving along a circle, the radius of curvature ρ is equal to radius of the circle r. Hence the normal acceleration is

$$a_n = \frac{v^2}{r}$$

The normal acceleration is directed towards the centre of curvature and hence is known as **centripetal acceleration**.

EXAMPLE 8.42

A motorist is driving at 80 km/hr on the curved position of a highway of 400 m radius. He suddenly applies the brakes and that causes the speed to decrease to 45 km/hr at a constant rate in 8 seconds. Determine the tangential and normal components of acceleration immediately after the application of brakes and 4 seconds later.

Solution : Initial velocity of vehicle $u = 80 \text{ km/hr} = \frac{80 \times 1000}{3600} = 22.22 \text{ m/s}$

Final velocity of vehicle $v = \frac{45 \times 1000}{3600} = 12.5 \text{ m/s}$

Tangential acceleration is the rate of change of velocity

$\therefore$ Tangential acceleration $a_t = \frac{12.5 - 22.22}{8} = -1.215 \text{ m/s}^2$

The – ve sign implies retardation

At the instant of application of brakes, $V = 22.22$ m/s

Normal acceleration $a_n = \frac{V^2}{r} = \frac{(22.22)^2}{r} = 1.234 \text{ m/s}^2$

$\therefore$ Total acceleration $a = \sqrt{a_n^2 + a_t^2} = \sqrt{(1.234)^2 + (-1.215)^2} = \mathbf{1.732} \text{ m/s}^2$

(*b*) Velocity of vehicle 4 seconds after the application of brakes,

$$V = u + at = 22.22 + -(1.215 \times 4) = 17.36 \text{ m/s}$$

$$a_n = \frac{(17.36)^2}{400} = 0.753 \text{ m/s}^2 \quad ; \quad a_t = -1.215 \text{ m/s}^2$$

$$\therefore \qquad a = \sqrt{(0.753)^2 + (-1.215)^2} = \mathbf{1.429 \text{ m/s}^2}$$

EXAMPLE 8.43

An automobile enters a curved road at 30 km/hr and then leaves at 48 km/hr. The curved road is in the form of quarter of a circle and has a length of 400 m. If the car travels at constant acceleration along the curve, calculate the resultant acceleration at both ends of the curve.

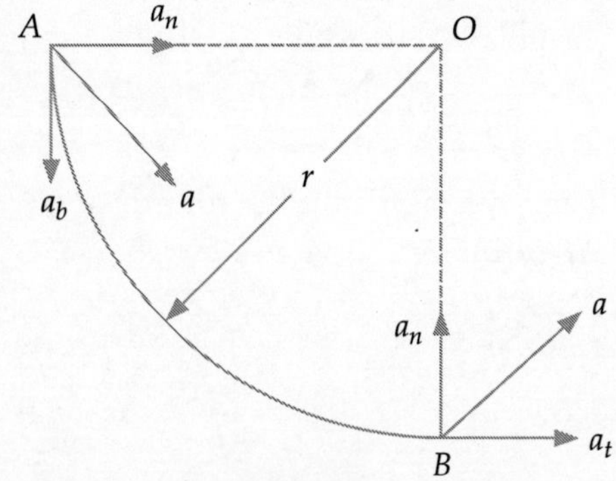

Fig. 8.33

Solution : With reference to Fig. 8.33, let the automobile travel from A to B.

$$V_a = 30 \text{ km/hr} = \frac{30 \times 1000}{3600} = 8.33 \text{ m/s}$$

$$V_b = 48 \text{ km/hr} = \frac{48 \times 1000}{3600} = 13.33 \text{ m/s}$$

Using the kinematic equation $v^2 - u^2 = 2as$,

$$\text{Tangential acceleration } a_t = \frac{13.33^2 - 8.33^2}{2 \times 400} = 0.135 \text{ m/s}^2$$

If r is the radius of the circular track, then

$$\frac{1}{4} \times (2\pi r) = 400 \quad ; \quad r = \frac{400 \times 2}{\pi} = 254.77 \text{ m}$$

At end A :

$$a_n = \frac{V_a^2}{r} = \frac{8.33^2}{254.77} = 0.272 \text{ m/s}^2$$

$$a_t = 0.135 \text{ m/s}^2$$

$$\therefore \text{ Resultant acceleration } a = \sqrt{(0.272)^2 + (0.135)^2} = \mathbf{0.304} \text{ m/s}^2$$

If α is the direction of resultant with tangential acceleration, then

$$\tan\alpha = \frac{a_n}{a_t} = \frac{0.272}{0.135} = 2.0148 \quad ; \quad \alpha = \mathbf{63.60}$$

At end B :

$$a_n = \frac{V_b^2}{r} = \frac{13.33^2}{254.77} = 0.697 \text{ m/s}^2$$

$$a_t = 0.135 \text{ m/s}^2$$

$$\text{Resultant acceleration } a = \sqrt{(0.697)^2 + (0.135)^2} = \mathbf{0.71} \text{ m/s}^2$$

If α is the direction of resultant with tangential acceleration, then

$$\tan\alpha = \frac{a_n}{a_t} = \frac{0.697}{0.135} = 5.163 \quad ; \quad \alpha = \mathbf{79.04^\circ}$$

EXAMPLE 8.44

On a bumpy road, the motorist applies brakes and the speed of the vehicle reduces uniformly from 60 km/hr at point A to 40 km/hr at point B; the points A and B are 120 m apart. If the vehicle has a resultant acceleration of 2.5 m/s² at point A, estimate the radius of curvature of the road at this point. Proceed to calculate the acceleration of the vehicle at point B where the road has a radius of curvature equal to 90 m.

Solution :

$$V_a = 60 \text{ km/hr} = \frac{60 \times 1000}{3600} = 16.67 \text{ m/s}$$

$$V_b = \frac{40 \times 1000}{3600} = 11.11 \text{ m/s}$$

Using the kinematic relation $v^2 - u^2 = 2as$,

$$\text{Tangential acceleration } a_t = \frac{11.11^2 - 16.67^2}{2 \times 120} = -0.643 \text{ m/s}^2$$

The - ve sign implies retardation

At point A : $a = 2.5 \text{ m/s}^2 \;;\; a_t = -0.643 \text{ m/s}^2$

$$\therefore \quad a_n = \sqrt{a^2 - a_t^2} = \sqrt{(2.5)^2 - (-0.643)^2} = 2.416 \text{ m/s}^2$$

Now $a_n = \dfrac{V^2}{r}$ or $r = \dfrac{V^2}{a_n} = \dfrac{16.67^2}{2.416} = \mathbf{115 \text{ m}}$

Thus the radius of curvature of the road at point A is 115 m

At point B : $a_n = \dfrac{V^2}{r} = \dfrac{11.11^2}{90} = 1.371 \text{ m/s}^2$

$a_t = -0.643 \text{ m/s}^2$

$$\text{Resultant acceleration } a = \sqrt{a_n^2 + a_t^2} = \sqrt{(1.371)^2 + (-0.643)^2} = \mathbf{1.514} \text{ m/s}^2$$

Hence the acceleration of the vehicle at point B is 1.514 m/s^2.

EXAMPLE 8.45

A motorist is driving at a constant speed of 36 km/hr on the curved portion of a road of 200 m radius. Determine the velocity and acceleration component of the vehicle along (*a*) *x* and *y* directions (*b*) normal and tangential directions 10 seconds after it has entered the curved portion of the road.

Solution: Refer to Fig. 8.34.

$$\text{Speed } V = 36 \text{ km/hr} = \frac{36 \times 1000}{3600} = 10 \text{ m/s}$$

$$\text{Angular velocity } \omega = \frac{V}{r} = \frac{10}{200} = 0.05 \text{ rad/s}$$

At any time instant t, the vehicle has the coordinates

$$x = r\cos\theta = r\cos\omega t$$

$$y = r\sin\theta = r\sin\omega t$$

The velocity and acceleration components of the vehicle in x and y-directions at $t = 10$ second then work out as

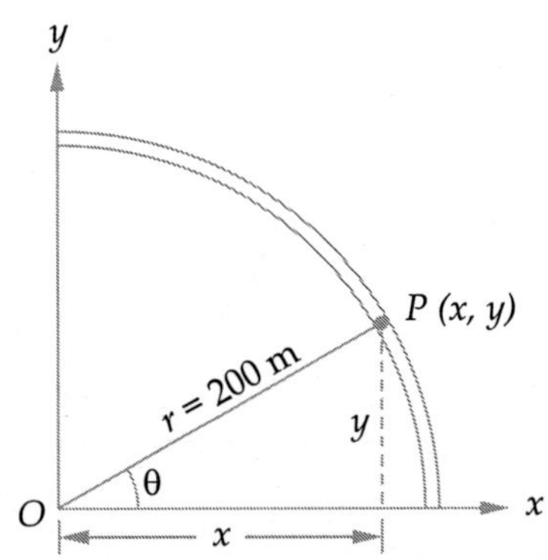

Fig. 8.34

$$V_x = \frac{dx}{dt} = \frac{d}{dt}(r\cos\omega t) = -\omega r\sin\omega t$$

$$V_y = \frac{dy}{dt} = \frac{d}{dt}(r\sin\omega t) = \omega r\cos\omega t$$

$$a_x = \frac{dV_x}{dt} = \frac{d}{dt}(-\omega r\sin\omega t) = -\omega^2 r\cos\omega t$$

$$a_y = \frac{dV_y}{dt} = \frac{d}{dt}(\omega r\cos\omega t) = -\omega^2 r\sin\omega t$$

Substituting t = 10 s, r = 200 m and ω= 0.05 rad/s

$$= 0.05 \times \frac{180}{\pi} = 2.866 \text{ degrees}$$

We get:

$$V_x = -0.05 \times 200\,[\sin(2.866 \times 10)] = -4.796 \text{ m/s } (\leftarrow)$$
$$V_y = 0.05 \times 200\,[\cos(2.866 \times 10)] = 8.775 \text{ m/s } (\uparrow)$$
$$a_x = -(0.05)^2 \times 200\,[\cos(2.866 \times 10)] = -0.439 \text{ m/s}^2 (\leftarrow)$$
$$a_y = -(0.05)^2 \times 200\,[\sin(2.866 \times 10)] = -0.24 \text{ m/s}^2 (\downarrow)$$

(*ii*) The velocity and acceleration components in tangential and normal directions are:

$$V_n = 0 \quad \text{and} \quad V_t = V = 10 \text{ m/s}$$

$$a_n = \frac{V^2}{r} = \frac{10 \times 10}{200} = 0.5 \text{ m/s}^2$$

$$a_t = \frac{dV}{dt} = 0 \qquad (\because \; V \text{ is constant})$$

Angular position θ of the vehicle after 10 seconds

$$= \omega t = 2.866 \times 10 = 28.66°$$

8.5.3. Radial and transverse coordinates

In this system:

(*i*) the position of the particle is defined by the polar coordinates r and θ.

(*ii*) the velocity and acceleration of the particle are expressed in the radial and transverse directions.

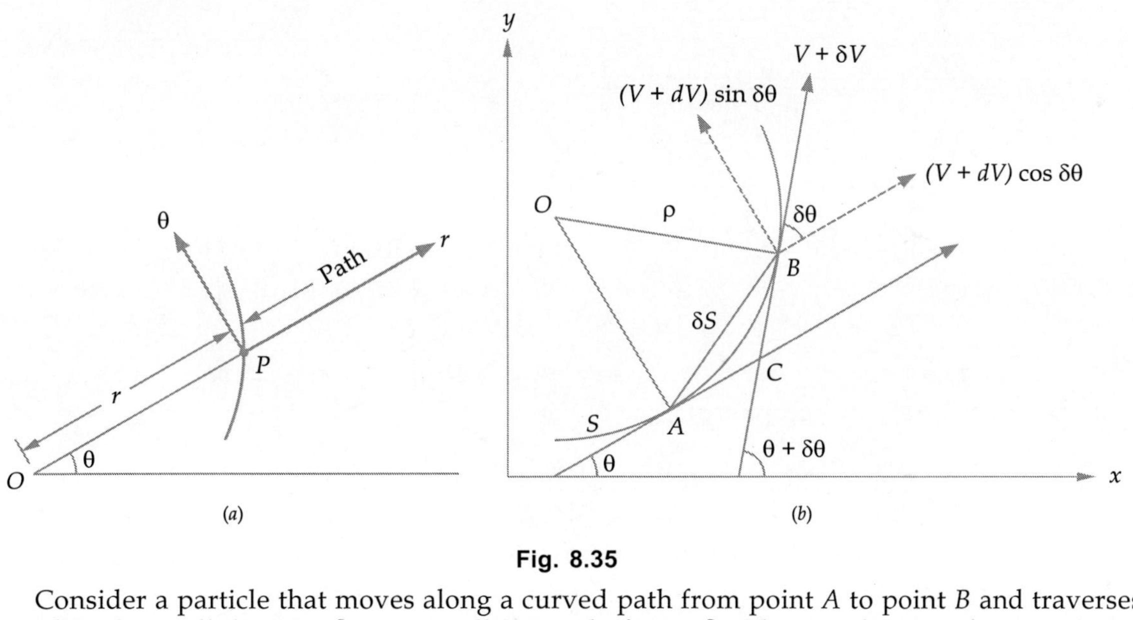

Fig. 8.35

Consider a particle that moves along a curved path from point A to point B and traverses an infinitely small distance δs in a small interval of time δt. The coordinates of points A and B are (r, θ) and $(r + \delta r, \theta + \delta\theta)$ respectively.

Velocity components : The component of velocity along OA is known as radial velocity and it is given by

$$V_r = \frac{\text{radial displacement, } i.e., \text{ displa}}{\text{time interval}}$$

$$= \underset{\delta t \to 0}{\text{Lt}} \frac{OC - OA}{\delta t} = \underset{\delta t \to 0}{\text{Lt}} \frac{(r + \delta r) \cos \delta\theta - r}{\delta t}$$

When the time interval δt is very small, the arc AB will be very small and so will be $\delta\theta$ and then $\cos \delta\theta = 1$

$$V_r = \underset{\delta t \to 0}{\text{Lt}} \frac{(r + \delta r) - r}{\delta t} = \frac{dr}{dt}$$

The component of velocity in the transverse direction (perpendicular to the radial direction OA) is known as transverse velocity and is given by

$$V_\theta = \frac{\text{transverse displacement, } i.e., \text{ dis}}{\text{time ir}}$$

$$= \underset{\delta t \to 0}{\text{Lt}} \frac{BC}{\delta t} = \underset{\delta t \to 0}{\delta t} \frac{(r + \delta r) \sin \delta\theta}{\delta t}$$

As $\delta\theta$ is very small, $\sin \delta\theta \to \delta\theta$. Neglecting the product $(\delta r \times \delta\theta)$, we get

$$V_\theta = \underset{\delta t \to 0}{\text{Lt}} \frac{r\,\delta\theta}{\delta t} = \frac{r\,d\theta}{dt}$$

Acceleration components : Let V_r and V_θ be the velocity components in the radial and transverse directions at point A, and the corresponding components at point B be $(V_r + \delta V_r)$ and $(V_\theta + \delta V_\theta)$.

The acceleration of the particle along the radial direction is called **radial acceleration**, and it is given by

$$a_r = \frac{\text{change in velocity along radial d}}{\text{time interval}}$$

$$= \underset{\delta t \to 0}{\text{Lt}} \frac{[(V_r + \delta V_r) \cos \delta\theta - (V_\theta + \delta}{\delta t}$$

As $\delta\theta$ is very small, $\cos \delta\theta \to 1$ and $\sin \delta\theta = \delta\theta$. Further neglecting the product $(\delta V_\theta \times \delta\theta)$, we get

$$a_r = \underset{\delta t \to 0}{\text{Lt}} \frac{V_r + \delta V_r - V_\theta\, \delta\theta - V_r}{\delta t} = \underset{\delta t \to 0}{\text{Lt}} \frac{\delta V_r - V_\theta\, \delta\theta}{\delta t} = \frac{dV_r}{dt} - V_\theta \frac{d\theta}{dt}$$

Substituting for velocity components, we obtain

$$a_r = \frac{d}{dt}\left(\frac{dr}{dt}\right) - \left(r\frac{d\theta}{dt}\right) \times \frac{d\theta}{dt} = \frac{d^2 r}{dt^2} - r\left(\frac{d\theta}{dt}\right)^2$$

The acceleration of the particle along the transverse direction is called **transverse acceleration**, and it is given by

$$a_\theta = \frac{\text{change in velocity along transve}}{\text{time interval}}$$

$$= \underset{\delta t \to 0}{\text{Lt}} \frac{[(V_r + \delta V_r) \sin \delta\theta + (V_\theta + \delta}{\delta t}$$

$$= \underset{\delta t \to 0}{\text{Lt}} \frac{(V_r + \delta V_r)\,\delta\theta + (V_\theta + \delta V_\theta) \times}{\delta t}$$

Neglecting the product ($\delta V_r \times \delta\theta$), we obtain

$$a_\theta = \underset{\delta t \to 0}{\text{Lt}} \frac{V_r\,\delta\theta + \delta V_\theta}{\delta t} = V_r \frac{d\theta}{dt} + \frac{dV_\theta}{dt}$$

Upon making substitution for velocity components, we get

$$a_\theta = \frac{dr}{dt}\frac{d\theta}{dt} + \frac{d}{dt}\left(r\frac{d\theta}{dt}\right) = \frac{dr}{dt}\frac{d\theta}{dt} + \frac{d\theta}{dt}\frac{dr}{dt} + r\frac{d^2\theta}{dt^2} = 2\frac{dr}{dt}\frac{d\theta}{dT} + r\frac{d^2\theta}{dt^2} = \frac{1}{r}\frac{d}{dt}\left(r\frac{d^2\theta}{dt^2}\right)$$

Resultant acceleration $a = \sqrt{(a_r)^2 + (a_t)^2}$

EXAMPLE 8.46

A particle moves along a curved path defined by

$$r = 5\theta \quad \text{and} \quad \theta = \frac{t^2}{3}$$

where r is in metres and t is in seconds.

Make calculation for the velocity and acceleration of the particle when $\theta = \pi/2$.

Solution :

$$\theta = \frac{t^2}{3} \qquad r = 5\theta = 5 \times \frac{t^2}{3}$$

$$\frac{d\theta}{dt} = \frac{2}{3}t \qquad \frac{dr}{dt} = \frac{5}{3} \times 2t = \frac{10}{3}t$$

$$\frac{d^2\theta}{dt^2} = \frac{2}{3} \qquad \frac{d^2r}{dt^2} = \frac{10}{3}$$

When $\theta = \pi/2$; $\frac{\pi}{2} = \frac{t^2}{3}$; $t = \sqrt{\frac{3\pi}{2}} = 2.17$ sec

Velocity of particle at t = 2.17 sec

$$\text{radial velocity } V_r = \frac{dr}{dt} = \frac{10}{3}t = \frac{10}{3} \times 2.17 = 7.23 \text{ m/s}$$

$$\text{transverse velocity } V_\theta = r\frac{d\theta}{dt} = \frac{5t^2}{3} \times \frac{2}{3}t = \frac{10}{9}t^3 = \frac{10}{9} \times (2.17)^3 = 11.35 \text{ m/s}$$

Resultant velocity $V = \sqrt{(V_r)^2 + (V_\theta)^2} = \sqrt{(7.23)^2 + (11.35)^2} = \mathbf{13.46}$ m/s

The angle made by the resultant velocity with the radial direction is

$$\tan\alpha = \frac{11.35}{7.23} = 1.57\,; \quad \alpha = 57.5^\circ$$

Acceleration of particle at t = 2.17 sec

$$\text{radial acceleration } a_r = \frac{d^2r}{dt^2} - r\left(\frac{d\theta}{dt}\right)^2 = \frac{10}{3} - \frac{5}{3}t^2 \times \left(\frac{2}{3}t\right)^2 = \frac{10}{3} - \frac{20}{27}t^4$$

$$= \frac{10}{3} - \frac{20}{27}(2.17)^4 = -\,13.09 \text{ m/s}^2$$

transverse acceleration $a_\theta = 2\frac{dr}{dt}\frac{d\theta}{dt} + r\frac{d^2\theta}{dt^2} = 2 \times \frac{10}{3}t \times \frac{2}{3}t + \frac{5t^2}{3} \times \frac{2}{3} = \frac{50}{9}t^2$

$$= \frac{50}{9} \times (2.17)^2 = 26.19 \text{ m/s}^2$$

Resultant acceleration $a = \sqrt{(a_r)^2 + (a_\theta)^2} = \sqrt{(-13.09)^2 + (26.16)^2} = 29.25 \text{ m/s}^2$

The angle made by the resultant acceleration with the radial direction is

$$\tan \beta = \frac{29.25}{13.09} = 2.234 \quad ; \quad \beta = 65.88°$$

8.6. RELATIVE VELOCITY

Consider two trains A and B moving in parallel tracks and in the same direction with velocities 60 km/hr and 45 km/hr respectively. These are the absolute velocities with respect to a fixed frame of reference. To a passenger sitting in train B, the train A will appear to be moving with a speed of (60 – 45) = 15 km/hr.

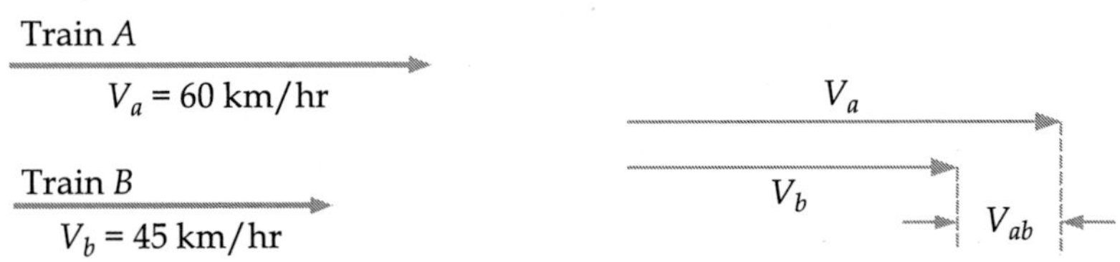

Fig. 8.36

This implies that relative velocity of A with respect to B is

$$V_{ab} = V_a - V_b = 60 - 45 = 15 \text{ km/hr}$$

When the motions of the trains are along parallel tracks but in opposite directions, then relative velocity of A with respect to B is

$$V_{ab} = V_a - (-V_b) = 60 - (-45) = 105 \text{ km/hr}$$

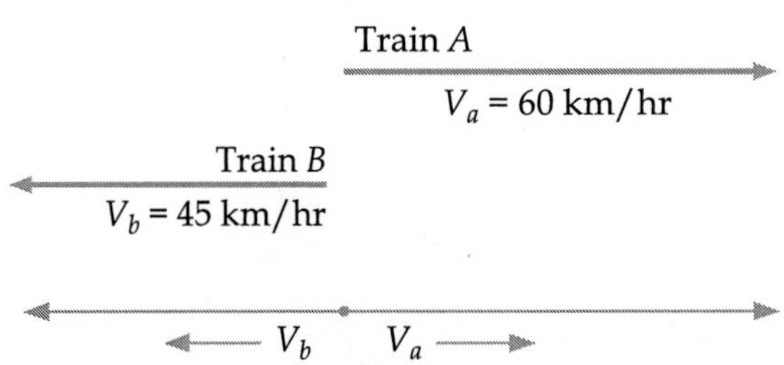

Fig. 8.37

Quite often the motion of two bodies is along different directions rather than being along parallel lines (Fig. 8.38 *a*). However, the relative velocity of A with respect to B will always be equal to the difference between the velocities of A of B. Since velocity is a vector quantity, the relative velocity V_{ab} will be the vector difference between V_a and V_b. This vectorial difference can be determined graphically as show in Fig. 8.38 *b* and 8.38 *c*. It is to be noted that V_{ab} (relative velocity of A with respect to B) and V_{ba} (relative velocity of B with respect to A) have the same magnitude but opposite directions.

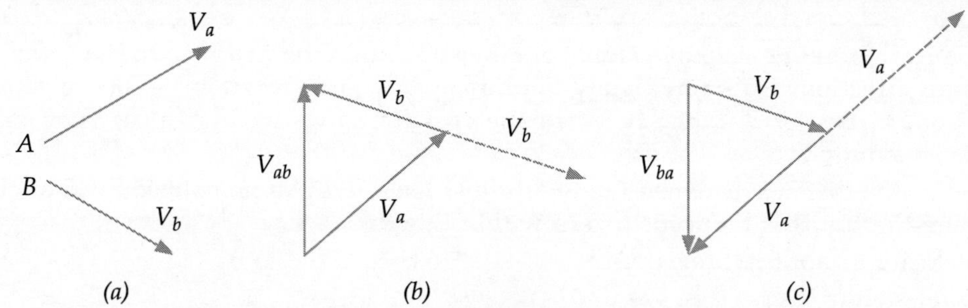

Fig. 8.38

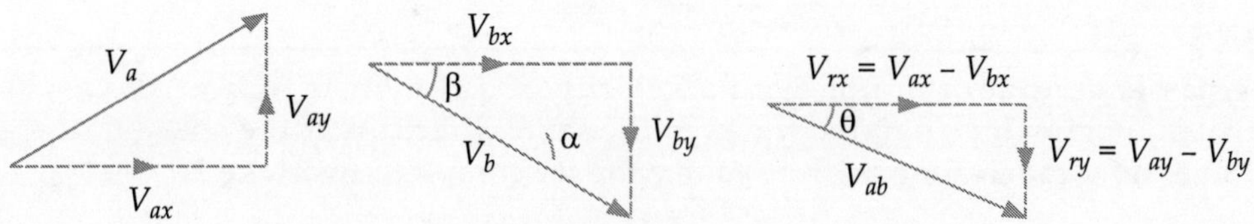

Fig. 8.39

The method of resolution of vectors can also be used to determine the magnitude and direction of relative velocity. The steps involved in this analytical method are:

(*i*) Workout the horizontal and vertical components of the given velocity vectors.

$$V_{ax} = V_a \cos\alpha \; ; \quad V_{ay} = V_a \sin\alpha$$
$$V_{bx} = V_b \cos\beta \; ; \quad V_{by} = V_b \sin\beta$$

(*ii*) Obtain the components of relative velocity by taking the difference in the component velocities in each direction

$$V_{rx} = V_{ax} - V_{bx} \quad \text{and} \quad V_{ry} = V_{ay} - V_{by}$$

The relative velocity V_{ab} is then determined by combining the components of relative velocity. That is

$$V_{ab} = \sqrt{V_{rx}^2 + V_{ry}^2}$$

and its inclination with horizontal axis is

$$\theta = \tan^{-}\left(\frac{V_{ry}}{V_{rx}}\right)$$

EXAMPLE 8.47

Two cars are moving in the same direction with a speed of 45 km/hr and a distance of 10 km separates them. A car coming from the opposite direction meets these two cars at an interval of 6 minutes. Determine the speed of this car.

Solution: Let V be the speed of the car coming from the opposite direction. Then velocity of this car relative to either of the two cars is

$$V - (-45) = (V + 45) \text{ km/hr}$$

$$\therefore \quad 10 = (V + 45) \times \frac{6}{60} \qquad (\because s = vt)$$

$$V = \frac{10 \times 60}{6} - 45 = \mathbf{55\ km/hr}$$

EXAMPLE 8.48

The driver of a train moving at 20 m/s sights another train moving at 4 m/s on the same track and in the same direction. He immediately applies brakes and the train begins to retard at 1 m/s². What should be the distance between the two trains at the instant of first sight if collision is to be avoided?

Solution: Let s be the distance between the two trains. There will be no collision if the relative velocity between the trains is reduced to zero within this distance.

Relative velocity of approaching train = 20 – 4 = 16 m/s

Using the relation $v^2 - u^2 = 2as$, we find

$$0 - 16^2 = 2 \times 1 \times s \qquad \therefore \quad s = \mathbf{128\ m}$$

EXAMPLE 8.49

A 200 m long passenger train running with a velocity of 72 km/hr is to overtake a 150 long goods train that is moving on a parallel track in the same direction. If the speed of the goods train is 36 km/hr, how much time will be taken for its complete overtake ?

Solution: Velocity of passenger train $V_a = 72 \text{ km/hr} = \dfrac{72 \times 1000}{3600} = 20 \text{ m/s}$

Velocity of goods train $V_b = 36 \text{ km/hr} = \dfrac{36 \times 1000}{3600} = 10 \text{ m/s}$

Relative velocity of passenger train with respect to goods train,

$$V_{ab} = V_a - V_b = 20 - 10 = 10 \text{ m/s}$$

Total distance to be covered = 200 + 150 = 350 m

Then $\quad 350 = 10 \times t \qquad (\because s = vt)$

$$\therefore \quad t = \mathbf{35\ s}$$

Hence the passenger train will take 35 seconds to pass the goods train.

EXAMPLE 8.50

Two train *A* (length 125 m) and *B* (length 100 m) move in opposite directions along parallel tracks. At the instant of complete pass over, the train A is moving at 10 m/s with a constant acceleration of 0.1 m/s², and train *B* has a uniform speed. If the trains take 12 seconds to pass one another, determine the uniform speed of train *B*.

Solution: Total distance to be passed = 125 + 100 = 225 m

From the relation, $\quad s = ut + \dfrac{1}{2}at^2$

$$225 = V \times 12 + \frac{1}{2} \times 0.1 \times 12^2 = 12V + 7.2$$

where V is the relative velocity of train A with respect to train B

$$V = \frac{225 - 7.2}{12} = 18.15 \text{ m/s}$$

If V_b is the velocity of train B, them

$$V = V_a - (-V_b)\ ; \quad 18.5 = 10 + V_b$$

$$\therefore \quad V_b = \mathbf{8.5\ m/s}$$

EXAMPLE 8.51

Two roads cross at 90° angle at the intersection point *O* as shown in Fig. 8.40. At an instant of time, a car *B* passes the intersection travelling east at a constant speed of 36 km/hr. At the same instant, a car *A* at a distance of 35 m north of the intersection starts from the rest and moves south with a constant acceleration of 1.2 m/s². Determine the position, velocity and acceleration of car *B* relative to car *A* six seconds after car *A* crosses the intersection of loads.

Solution:

Car A: $u = 0,\ a = 1.2\ \text{m/s}$ and $t = 6\ \text{s}$

$$v = u + at = 0 + 1.2 \times 6 = 7.2\ \text{m/s}$$

$$s = ut + \frac{1}{2}at^2 = 0 + \frac{1}{2} \times 1.2 \times 6^2 = 21.6\ \text{m}$$

The car *A* will thus be at a distance of (35 – 21.6) = 13.4 m from the origin along *y*-axis.

Car B: $\text{speed} = 36\ \text{km/hr} = \dfrac{36 \times 1000}{3600}$

$$= 10\ \text{m/s} \quad \text{and} \quad a = 0$$

$$s = ut + \frac{1}{2}at^2 = 10 \times 6 = 60\ \text{m}$$

The car *B* will thus be at a distance of 60 m from the origin along *x*-axis.

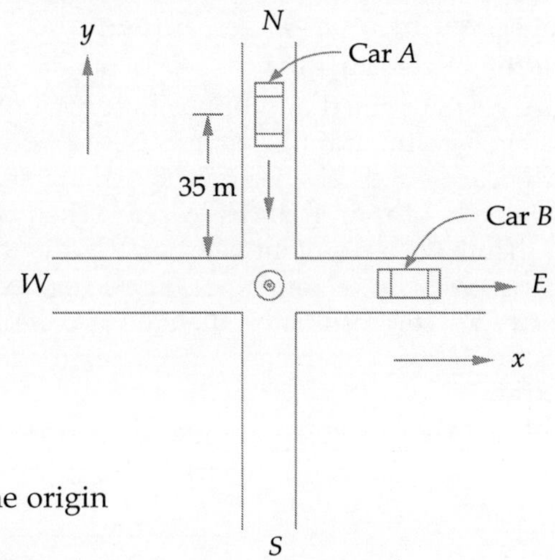

Fig. 8.40

(*i*) Relative position of car *A* with respect to car *B*

$$\bar{r}_{ab} = \bar{y}_a - \bar{x}_b$$

and this is prescribed by vector $\overline{BA}$

$$\overline{BA} = \sqrt{OA^2 + OB^2}$$

$$= \sqrt{13.4^2 + 60^2} = 61.478$$

$\therefore \quad \bar{r}_{AB} = \mathbf{61.478\ m}$ and

$$\alpha = \tan^{-1}\frac{13.4}{60} = \mathbf{12.587°}$$

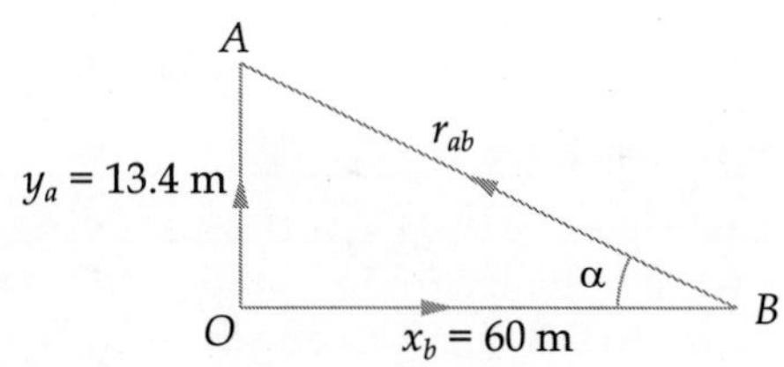

(*ii*) Relative velocity of car *A* with respect to car *B*

$$\bar{V}_{ab} = \bar{v}_a - \bar{v}_b$$

and this is represented by vector $\overline{BA}$

$$\overline{BA} = \sqrt{OA^2 + OB^2}$$

$$= \sqrt{7.2^2 + 10^2} = 12.32$$

$\therefore \quad \bar{V}_{ab} = 12.32\ \text{m/s}$ and

$$\beta = \tan^{-1}\frac{7.2}{10} = 35.75°$$

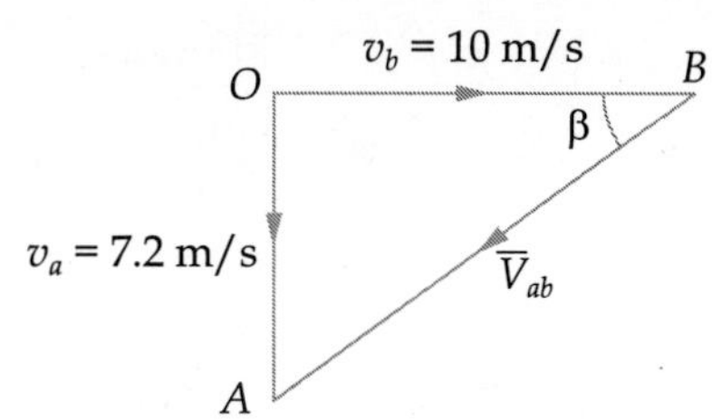

(*iii*) Relative acceleration of car *A* with respect to car *B*

$$\bar{a}_{ab} = \bar{a}_a - \bar{a}_b = 1.2 - 0 = 1.2\ \text{m/s}^2$$

EXAMPLE 8.52

A car moving along a South-East road at 70 km/hr crosses a railway crossing. At the same time a train also passes the same crossing at 50 km/hr along the railway track which is due East. Determine the magnitude and direction of the apparent velocity of car as observed by a passenger in the train.

Solution: Refer Fig. 8.41

A velocity equal and opposite to that of passenger in the train is superimposed both on the car as well as the train. By doing so, the velocity of train reduces to zero. The car will then have a resultant velocity V_{ct} due to its own velocity V_c and the superimposed velocity V_t. This resultant is then the apparent velocity of car with respect to passenger in the train.

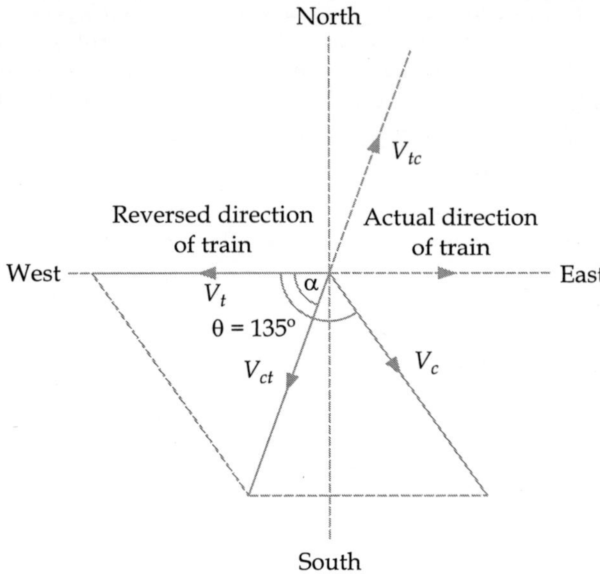

Fig. 8.41

From the parallelogram law of vectors,

$$V_{ct} = \sqrt{V_c^2 + V_t^2 + 2V_cV_t \cos\theta}$$

$$= \sqrt{70^2 + 50^2 + 2 \times 70 \times 50 \times \cos 13} = \mathbf{49.50\ m/s}$$

$$\tan\alpha = \frac{V_c \sin\theta}{V_t + V_c \cos\theta} = \frac{70 \sin 135°}{50 + 70 \cos 135} = \frac{70 \times 0.707}{50.0 - 49.50} = 98.98$$

or α = 89.42° with West, *i.e.*, 0.58° West of South

Note. The apparent velocity of train as observed by the driver of the car will have the same magnitude but opposite in direction (shown dotted)

Magnitude = 49.50 m/s

Direction = 0.58° East of North.

EXAMPLE 8.53

A passenger sitting in a train moving at 54 km/hr is hit by a stone thrown at right angles to it with a velocity of 18 km/hr. Calculate the velocity and the direction with which the stone appears to hit the passenger.

Solution: Refer Fig 8.42

Velocity of train $V_t = 54 \text{ km/hr}$

$$= \frac{54 \times 1000}{3600} = 15 \text{ m/s}$$

Velocity of stone $V_s = 18 \text{ km/hr}$

$$= \frac{18 \times 1000}{3600} = 5 \text{ m/s}$$

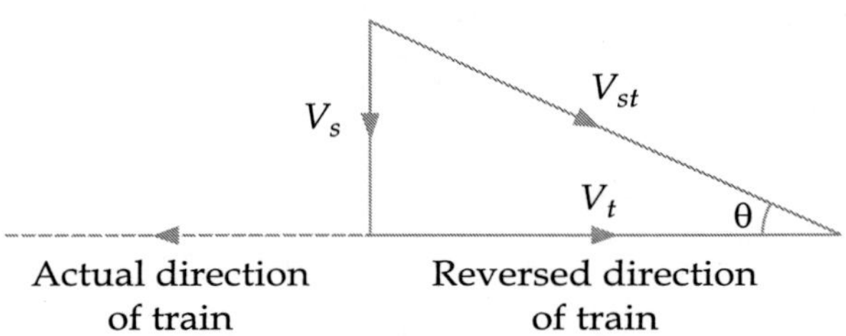

Fig. 8.42

A velocity equal and opposite to that of train is superimposed both on the stone and the train. By doing so, the velocity of train reduces to zero. The stone will then have a resultant velocity V_{st} due to its own velocity V_s and the superimposed velocity V_t. This resultant is then the apparent velocity with which the stone hits the passenger.

$$V_{st} = \sqrt{V_s^2 + V_t^2} = \sqrt{5^2 + 15^2} = \sqrt{250}$$

$$\theta = \tan^{-1}\frac{V_s}{V_t} = \tan^{-1}\frac{5}{15} = \mathbf{18.43°}$$

EXAMPLE 8.54

A motor cyclist travelling West with speed of 10 m/s feels the rain coming down at 45° to the vertical. On reducing speed to 6 m/s, he feels the rain to be at 30° with the vertical. Make calculations for the magnitude and direction of the actual velocity of rain.

Solution: Refer Fig. 8.43

Let CO represent the magnitude and direction of the actual velocity of rain.

From ΔADC:

$$\tan 45° = \frac{DC}{AD}$$

$$1 = \frac{y}{10 - x} \;; \quad y = 10 - x \qquad ...(i)$$

From ΔBDC:

$$\tan 60° = \frac{DC}{BD}$$

$$1.732 = \frac{y}{6 - x} \;; \quad y = 10.392 - 1.732x \qquad ...(ii)$$

From expressions (*i*) and (*ii*)

$$10 - x = 10.392 - 1.732x$$

$$x = \frac{10.392 - 10}{0.732} = 0.535 \text{ m/s}$$

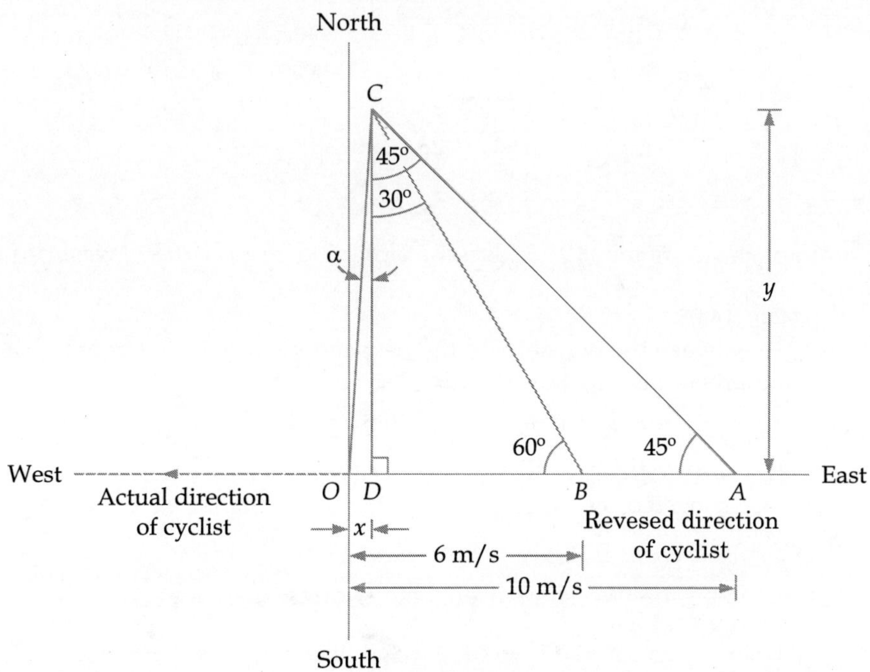

Fig. 8.43

Substituting $x = 0.535$ m/s in expression (*i*), one obtains

$$y = 10 - 0.535 = 9.465 \text{ m/s}$$

$$\tan \alpha = \frac{x}{y} = \frac{0.535}{9.465} = 0.0565 \; ; \quad \alpha = \mathbf{3.23^\circ}$$

Magnitude of velocity of rain,

$$OC = \sqrt{OD^2 + CD^2} = \sqrt{0.535^2 + 9.465^2}$$

$$= \sqrt{0.286 + 89.586} = \mathbf{9.48 \text{ m/s}}$$

EXAMPLE 8.55

Two ships, S_1 and S_2, at a given instant are 3 km apart; S_2 being south-east of S_1. The ship S_1 is travelling 12 km/hr due east and ship S_2 is travelling 20 km/hr due north. Make calculations for (*a*) the velocity of ship S_2 relative to ship S_1, (*b*) the shortest distance between the two ships, and (*c*) the time when the two ships are nearest.

Solution: Refer Fig. 8.44

Velocity of ship S_2 relative to ship S_1,

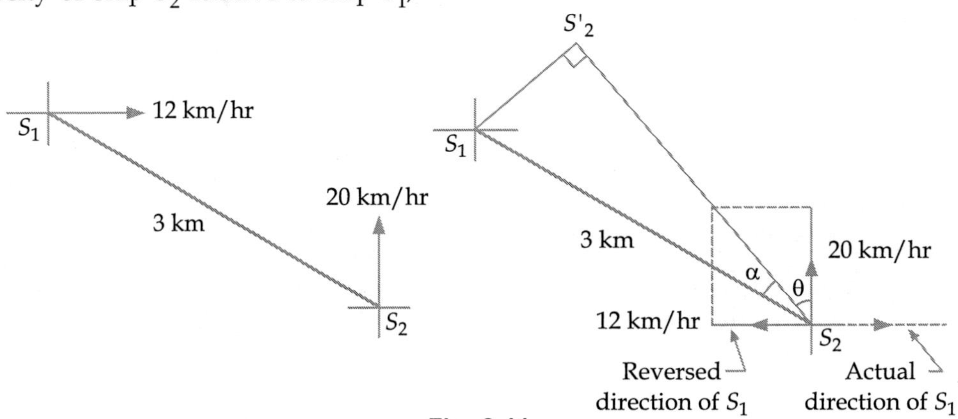

Fig. 8.44

$$V_r = \sqrt{20^2 + 12^2} = \sqrt{544} = 23.32 \text{ km/hr}$$

$$\tan \theta = \frac{12}{20} = 0.6 \; ; \quad \theta = 30.97^\circ$$

The ship S_2 will appear to travel at 23.32 km/hr along a direction 30.97° west of north to a passenger on ship S_1.

(*b*) $\alpha = 45 - 30.97 = 14.03^\circ$

The shortest distance between the two ships is the perpendicular distance from the apparently stationary ship S_1 to the apparent direction of ship S_2

$$\text{Shortest distance} = S_1 S_2' = S_1 S_2 \sin \alpha = 3 \sin 14.03' = \mathbf{0.7275 \text{ km}}$$

(*c*) Apparent (relative) velocity of ship S_2 = 23.32 km/hr

The apparent distance travelled by ship S_2 when it reaches the shortest distance point S_2' is

$$S_2 S_2' = S_1 S_2 \cos \alpha = 3 \cos 14.03' = 2.911 \text{ km}$$

∴ Interval of time when the two ships are at the shortest distance,

$$t = \frac{s}{v} = \frac{2.911}{23.32} = 0.1248 \text{ hr} = \mathbf{7.488 \text{ min}}$$

REVIEW AND SUMMARY

1. When a body changes its position with respect to its surroundings, then the body is said to be in motion. Thus relative change in position is motion.
2. Trajectory is a path followed by the body during its motion. It may be a straight line or a curve. Straight line path is observed in rectilinear motion, whereas curved path is observed in curvilinear motion.
3. When a body moves in a straight line with uniform velocity, the equations of motion are:
$$v = u + at \; ; \; v^2 - u^2 = 2as \; ; \; s = ut + \frac{1}{2}at^2$$
and distance travelled in nth second is
$$s_n = u + \frac{a}{2}(2n - 1)$$
where u is initial velocity, v is final velocity, a is acceleration of body and s represents distance travelled in time t.
4. When the body moves with constant velocity (*i.e.* when acceleration is zero) the body is said to have *uniform motion*. The motion of a body moving with constant acceleration is referred to as *uniformly accelerated motion*.
5. When motion is considered under gravity, the acceleration a in the equations of motion for rectilinear motion is replaced by gravitational acceleration g, and we obtain
$$v = u + gt; \quad h = ut + \frac{1}{2}gt^2;$$
$$v^2 - u^2 = 2gh$$
and the distance travelled in nth second is
$$s_n = u + \frac{g}{2}(2n - 1)$$
6. When a body moves in a circular orbit of radius r, then

Normal acceleration $a_n = \dfrac{V^2}{r}$

The normal acceleration is due to change in the direction of velocity. It is directed towards the centre of curvature and is known as centripetal acceleration.

The tangential acceleration is due to change in magnitude of velocity and it equals
$$a_t = \frac{\text{change in tangential velocity}}{\text{time interval}} = \frac{\iota}{}$$
Resultant acceleration $a = \sqrt{a_t^2 + a_n^2}$

It the resultant acceleration makes an angle ϕ with the normal, then
$$\tan\phi = \frac{a_t}{a_n}$$
It is worthwhile to note that:

(*i*) If the magnitude of velocity is contant between two points, then tangential acceleration $a_t = 0$

(*ii*) If the direction of motion of the particle does not change, *i.e.*, the motion of the particle is along a straight line, then $r \to \infty$. In that case normal acceleration $a_n = 0$

7. Relative velocity is the velocity of an object as observed from another moving object, and it equals as vector difference of the velocities of the two objects.
8. Resultant velocity is the combined effect obtained as the vector addition of the velocities caused by different forces acting on an object.

REVIEW QUESTIONS

A. Conceptual and conventional questions :

1. Differentiate between kinematics and kinetics.
2. When is the motion of a particle called rectilinear, and when curvilinear?
3. Differentiate between (*a*) displacement and distance, (b) average velocity and instantaneous velocity, (*c*) average acceleration and instantaneous acceleration.
4. How a uniform motion differs from a uniformly accelerated motion?
5. What parameters define rectilinear motion? State the relation between those parameters?
6. Draw and interpret the following graphs pertaining to rectilinear motion
   (*a*) Displacement-time curve, (*b*) Velocity-time curve and
   (*c*) Acceleration-time curve
7. What is meant by relative motion?
8. Define normal acceleration and tangential acceleration in connection with curvilinear motion.
9. A car travels along a straight track between two points at a mean speed of 80 m/s. It returns at a mean speed of 60 km/hr. What is the mean speed for the round trip? **(66.55 km/hr)**
10. A train accelerates from rest with constant acceleration α m/s^2 and after some time decelerates at a uniform rate of β m/s^2 till it comes to rest. If the total time taken between the two stations is *t*, work out an expression for the maximum velocity acquired by the train during the course of its motion. $\left(V_{max} = \dfrac{\alpha\beta}{\alpha+\beta}t\right)$
11. The velocity-time graph of a particle moving along a straight line is as shown in Fig. 8.45. What would be the ratio of the distances covered during the time intervals *BD* and *DC*? **(3 : 1)**

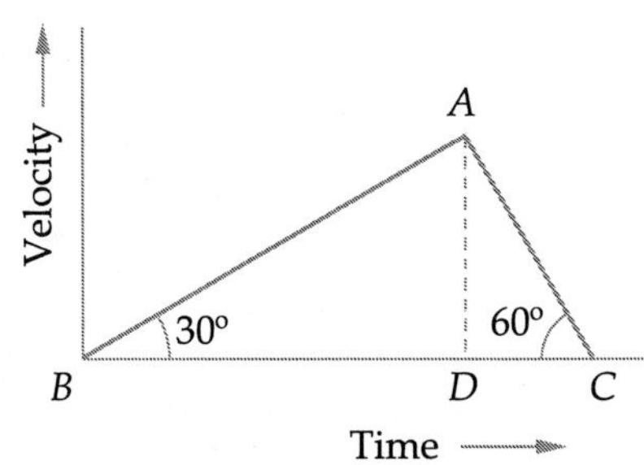

Fig. 8.45

12. The displacement time graph for two particles A and B are straight lines in clined at angles of 30° and 45° with the time axis. What will be the ratio of their velocities? **(V_a/V_b = 0.577)**
13. A vehicle having rectilinear motion is moving with a velocity of 54 km/hr. The vehicle is brought to halt in a distance of 100 m by the application of brakes. Determine the uniform rate of deceleration and the time for which the brakes were applied? **(1.125 m/s^2, 13.33s)**
14. A body in reclinear motion is found to travel 15 m in 5 th second and 25 m in the 10 th second. What distance will it travel in 15 seconds from the staring point if its acceleration is constant throughout? **(315 m)**
15. Two trains *A* and *B* leave the same station on parallel tracks. The train *A* starts with a uniform acceleration of 2.0 m/s^2 and attains a speed of 45 km/hr which remains constant. The train *B* leaves 1 minute later, with a uniform acceleration of 0.4 m/s^2 to attain a maximum speed of 72 km/hr. When will the train *B* overtatke train *A*? **(114.6 *s*)**
16. A ball is projected upwards with a velocity of 24 m/s. Two seconds later, a second ball is projected vertically upwards with a velocity of 18 m/s. At what point above the surface of earth will they meet? **(15.95 m)**
17. To a man walking with a speed of 2 m/s, the rain appears to fall vertically. If the velocity of rain is 4 m/s, calculate its direction with respect to the direction of man's velocity.

18. A car is travelling east at a constant speed 7 m/s. At the same time, another car starts from rest towards north with a constant acceleration of 2 m/s^2. Determine the position, velocity and acceleration of second car relative to the first car after 5 seconds.

(s_{21} = 43 m at 35.5°; V_{21} = 12.2 m/s at 55°; a_{21} = 2 m/s^2)

19. A particle moves in the x-y plane and its x-coordinate is defined as

$$x = 5t^3 - 105t$$

where x is in cm and t is in seconds. When t = 2 sec, the total acceleration equals 75 cm/s^2. Presuming that y-component of acceleration is constant and the particle starts from rest at the origin when t = 0, make calculation for the velocity when t = 4 second.

B: Complete the following statements with most appropriate word/words :

1. Motion of particle on a straight line is termed as .... motion and on curved path as .... motion.
2. Plane motion means motion with .... acceleration.
3. The change in the position of a particle with respect to a certain fixed reference is termed as ....
4. The motion of a body moving with constant acceleration is called .... motion.
5. The slope of distance-time graph at any instant gives the .... of the body at that instant.
6. The area of velocity-time graph with time axis represents the .... by the body.
7. The area under the acceleration-time graph corresponding to a given time interval gives the change in .... of the body during the same time interval.
8. For a body thrown vertically upwards with velocity u, the maximum height attained is ....
9. For a body thrown vertically upwards with velocity u, the total time taken by the body to fall back to ground is ....
10. A stone is made to fall from rest from the top of a tower. The distance travelled in the first, second and third second would be in the ratio ....
11. The normal acceleration of a body moving in a circular orbit is directed .... the centre of curvature and is called .... acceleration.
12. For a body moving in a circular orbit, the magnitude of velocity remains constant. Then its .... will be zero.
13. Relative velocity is the velocity of an object as observed from another moving object and it equals as ....of the velocities of the two objects.
14. Starting from rest when a body moves with uniform acceleration, then the distances travelled by the body in successive seconds are in the ratio ....

Answers:

1. linear, curvilinear 2. uniform 3. displacement 4. uniformly accelerated 5. velocity 6. distance covered 7. velocity 8. $\frac{u^2}{g}$ 9. $\frac{2u}{g}$ 10. 1 : 3 : 5 11. inwards towards, centripetal 12. tangential acceleration 13. vector difference 14. 1 : 4 : 9

C. Multiple choice questions :

1. For first half of time a car travels with velocity v_1 and for the second half of time it travels with velocity v_2. The average velocity of the car is

(a) $\frac{v_1 + v_2}{2}$ (b) $\frac{2 v_1 v_2}{v_1 + v_2}$ (c) $\sqrt{v_1 v_2}$ (d) $\frac{v_1 v_2}{2(v_1 + v_2)}$

2. Sandeep goes to office at a speed of 6 km/hr and returns to his home at a speed of 4 km/hr. If the total time taken is 2 hr, the distance between his office and home is closest to

(*a*) 5 km (*b*) 8 km (*c*) 10 km (*d*) 12 km

3. Mandeep is late for his office by 16 minutes when he walks at 3/4 of his usual speed. What is then the usual time taken by him to cover the stipulated distance ?
(*a*) 30 min (*b*) 36 min
(*c*) 48 min (*d*) 60 min

4. A passenger train takes 2 hours less for a journey of 300 km if its speed is increased by 5 km/hr over its usual speed. What is then the usual speed?
(*a*) 15 km/hr (*b*) 29 km/hr (*c*) 25 km/hr (*d*) 30 km/hr

5. A cyclist starts from rest with uniform acceleration. If he covers 2.6 m in the 7th second, the uniform acceleration is
(*a*) 0.3 m/s^2 (*b*) 0.4 m/s^2 (*c*) 0.5 m/s^2 (*d*) 0.6 m/s^2

6. A particle starts from rest and moving with a constant acceleration covers a distance x_1 in the third second and distance x_2 in the 5th second. The ratio x_1/x_2 is
(*a*) 3/5 (*b*) 5/9 (*c*) 9/25 (*d*) 25/81

7. A car moving with uniform acceleration covers 450 m in a 5 second interval, and covers 700 m in the next 5 second interval. The acceleration of the car is
(*a*) 7 m/s^2 (*b*) 10 m/s^2 (*c*) 25 m/s^2 (*d*) 50 m/s^2

8. A car moving with speed u can be stopped in minimum distance x when brakes are applied. If the speed becomes n times, the minimum distance over which the car can be stopped would take the value
(*a*) x/n (*b*) nx (*c*) x/n^2 (*d*) n^2x

9. When two bodies move uniformly towards each other, the distance between them decreases by 6 m/s. If both bodies move in the same direction with the same speeds, the distance between them increases by 4 m/s. Then the speeds of the two bodies are:
(*a*) 3 m/s and 3 m/s (*b*) 4 m/s and 2 m/s
(*c*) 5 m/s and 1 m/s (*d*) 8 m/s and 5 m/s

10. A particle starts from rest with a constant acceleration α m/s^2 and after some time it decelerates at a uniform rate of β m/s^2 till it comes to rest. If the total time taken between two rest positions is *t*, the maximum velocity acquired by the particle would be
(*a*) $\frac{\alpha+\beta}{2}t$ (*b*) $\frac{\alpha-\beta}{2}t$ (*c*) $\left(\frac{\alpha\beta}{\alpha+\beta}\right)t$ (*d*) $\left(\frac{\alpha+\beta}{\alpha-\beta}\right)t$

11. A train accelerates from rest at a constant rate α for some time and then it retards to rest at constant rate β. If total distance covered by the train is *x*, then maximum velocity attained by the train during the course of its motion would be
(*a*) $\left[\frac{\alpha+\beta}{2\,\alpha\beta}x\right]^{1/2}$ (*b*) $\left[\frac{2\alpha\beta}{\alpha+\beta}x\right]^{1/2}$ (*c*) $\left[\frac{\alpha+\beta}{\alpha-\beta}x\right]^{1/2}$ (*d*) $\left[\frac{\alpha\beta}{\alpha-\beta}x\right]^{1/2}$

12. A body is allowed to fall from the top of a tower. It falls through half the height in 2 seconds. The total time taken to reach the ground is nearly
(*a*) 4.5 sec (*b*) 4 sec (*c*) 3.2 sec (*d*) 2.8 sec

13. A body is dropped from rest at height *h*. It covers a distance 9/25 *h* in the last second. The height *h* is about
(*a*) 100 m (*b*) 115 m (*c*) 125 m (*d*) 150 m

14. A stone takes 6 seconds to reach the ground after it is dropped from a tower. If the stone

is stopped after 2 seconds of its fall and then released again, it will take .... more seconds to reach the ground.
(*a*) 3.4 (*b*) 4.0 (*c*) 4.8 (*d*) 5.6
Take $g = 10$ m/s^2

15. Two metallic blocks having masses in the ratio 2 : 3 are made to slide down a frictionless inclined plane starting initially from rest position. When these blocks reach the bottom of the inclined plane, they will have their kinetic energies in the ratio
(*a*) 2 : 3 (*b*) 4 : 9 (*c*) 3 : 2 (*d*) 9 : 4

16. The time-acceleration graph of a body moving with uniform acceleration is a straight line
(*a*) passing through the origin (*b*) inclined to the time axis
(*c*) parallel to time axis (*d*) parallel to acceleration axis

17. The displacement-time graph for two particles *A* and *B* are straight lines inclined at angles of 30° and 45° with the time axis. The ratio of velocities $V_a : V_b$ will be about
(*a*) 0.33 (*b*) 0.5 (*c*) 0.58 (*d*) 0.87

18. The time variation of the position of a particle in rectilinear motion is given by
$$x = 2t^3 + t^2 + 2t$$
If v is the velocity and a the acceleration of particle in consistent units, the motion started with
(*a*) $v = 0$; $a = 0$ (*b*) $v = 0$; $a = 2$ (*c*) $v = 2$; $a = 0$ (*d*) $v = 2$, $a = 2$

19. A boatman can row at 6 km/hr in still water. If the river flows steadily at 3 km/hr, along which direction with the bank of river the boatman should row in order to reach a point on the other bank directly opposite to the point from where he started ?
(*a*) 30° (*b*) 60° (*c*) 120° (*d*) 150°
The angle θ is measured from the river bank in counter clockwise direction.

20. A body moves with a speed of 10 m/s in the curved path of 25 m radius of curvature. If the tangential acceleration is 3 m/s^2, then the total acceleration for the body will be
(*a*) 3.3 m/s^2 (*b*) 4 m/s^2 (*c*) 5 m/s^2 (*d*) 5.8 m/s^2

21. The distance covered by a particle travelling on a circular path and starting from rest is given by the equation: $s = kt^2$ where k is a constant. The tangental acceleration of the particle will be
(*a*) $k/2$ (*b*) k (*c*) $2k$ (*d*) $4k$

Answers :

1. (*a*)	2. (*a*)	3. (*c*)	4. (*c*)	5. (*b*)	6. (*b*)
7. (*b*)	8. (*d*)	9. (*b*)	10. (*c*)	11. (*b*)	12. (*d*)
13. (*c*)	14. (*d*)	15. (*a*)	16. (*c*)	17. (*c*)	18. (*d*)
19. (*c*)	20. (*c*)	21. (*c*)			

CHAPTER

9

ROTATION AND GENERAL PLANE MOTION

Learning objectives : The key concepts included in this chapter relate to the following aspects of rotation and general plane motion

- *angular displacement, velocity and acceleration; relationship between circular motion and linear motion*
- *equations of motion is a circular path with uniform angular acceleration*
- *general plane motion and Chasle's theorem*
- *instantaneous centre and its applications*

The motion of a particle in curvilinear motion is on a curved path. The particle does not repeat the journey once it has passed over a point on the curved path. Its centre of rotation also goes on changing.

In the rotary (circular) motion, the movement of the particle is along a circular path. The particle repeats its journey along the same circular path about the same centre of rotation which remains fixed. The shafts, pulleys and flywheels etc. undergo circular motion when they rotate about their geometric axis.

9.1 TERMS RELATED TO CIRCULAR MOTION

The terms associated with circular motion are:

(*i*) *Angular displacement:* The displacement of a body in rotation is called angular displacement, and it is measured in terms of the angle through which the body moves from the initial state. With reference to Fig. 9.1, a body starts from initial position *A* and while moving in circular path reaches the position *B* after a certain time interval. Then $\angle AOB = \theta$ is a measure of the angular displacement of the body. This displacement has magnitude as well as direction, and apparently displacement is a vector quantity. The direction is a rotation-either clockwise or anti-clockwise.

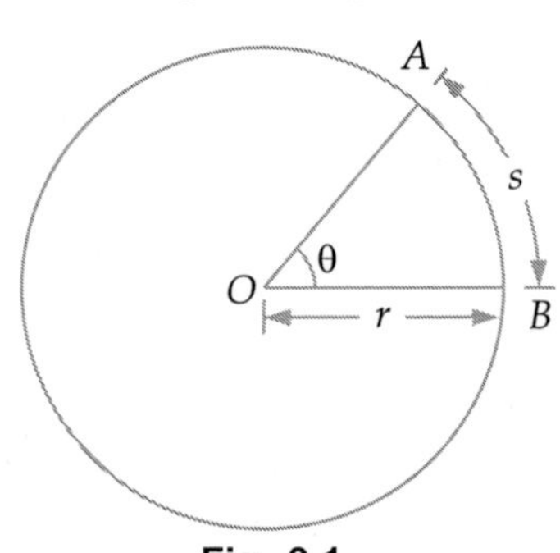

Fig. 9.1

Angular displacement θ can be measured in radians, degrees or revolutions. Various units of angular displacement are connected by the relation:

$$1 \text{ revolution} = 2\pi \text{ radians} = 360 \text{ degrees.}$$

(ii) Angular velocity: The rate of change of angular displacement of a body with respect to time is called angular velocity.

If the body traverses angular distance $d\theta$ over a time interval dt, then the average angular velocity ω is

$$\omega = \frac{d\theta}{dt}$$

The unit of angular velocity for defining its magnitude is usually taken as radians per second (rad/s). Many a times angular velocity is expressed in terms of revolutions per minute (rpm). If the body turns N rev/min, then

$$\text{number of revolutions per sec} = \frac{N}{60}$$

In one revolution, the angular displacement of the body is 360 degrees or 2π radians

Angle traversed by the body is one second

= (angle covered in one revolution) × (number of revolutions per second)

$$= 2\pi \times \frac{N}{60} = \frac{2\pi N}{60}$$

But the angle traversed by the body in one second is a measure of the angular velocity w. Then

$$\omega = \frac{2\pi N}{60} \text{ rad/sec}$$

(iii) Angular acceleration: The rate of change of angular velocity of a body with respect to time is called angular acceleration.

angular acceleration = rate of change of angular velocity

$$\alpha = \frac{d\omega}{dt} = \frac{d}{dt}\left(\frac{\theta}{t}\right) = \frac{d^2\theta}{dt^2}$$

The angular acceleration is sometimes expressed as

$$\alpha = \frac{d\omega}{dt} = \frac{d\omega}{d\theta} \times \frac{d\theta}{dt} = \omega\frac{d\omega}{d\theta} \qquad \left(\because \frac{d\theta}{dt} = \omega\right)$$

The unit of angular acceleration for defining its magnitude is usually taken as radian per second per second (rad/sec^2)

Relationship between circular motion and linear motion: When a body moves from A to B (Fig. 9.1), the distance travelled by it is s. If r is the distance of the body from the centre of rotation, then

$$s = r\theta$$

and
$$\frac{ds}{dt} = \frac{d}{dt}(r\theta) = \frac{r\,d\theta}{dt} \qquad (\because r \text{ is constant})$$

Noting that ds/dt = linear velocity v and $d\theta/dt$ = angular velocity ω, we may write

$$v = \omega r$$

That is, the linear velocity is r-times the angular velocity

When the expression $v = \omega r$ is differentiated with respect to time, we get

$$\frac{dv}{dt} = \frac{d}{dt}(\omega r) = \frac{r\,d\omega}{dt}$$

Since dv/dt = linear acceleration a and $d\omega/dt$ = angular acceleration α, we may write

$$a = \alpha r$$

That is, the linear acceleration is r-times the angular acceleration.

It is worthwhile to point out that when a body moves along a curved path (curvilinear motion), the linear velocity is referred to as tangential velocity. Further, the body undergoes tangential and normal components of acceleration given by

$$\text{tangential acceleration } a_t = \alpha r$$

$$\text{normal acceleration } a_n = \frac{v^2}{r} = \omega^2 r$$

The tangential acceleration is due to change in magnitude of the velocity and the normal acceleration is due to change in direction of the body.

Fig. 9.2 illustrates these components for any point P. The normal component is directed towards the centre of the circle. The total acceleration a is the vector sum of the two components.

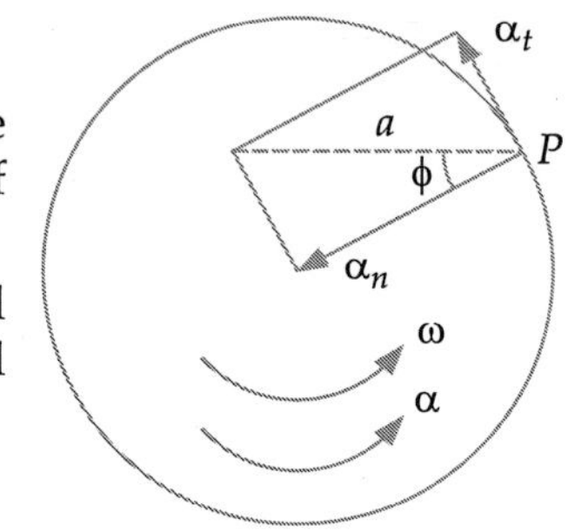

Fig. 9.2

$$a = \sqrt{\alpha_n^2 + \alpha_t^2}$$

and the angle between the total acceleration and the radius is

$$\phi = \tan^{-1}(\alpha_t/\alpha_n)$$

9.2. EQUATIONS OF CIRCULAR MOTION

When a body moves in a circular path with uniform angular acceleration, the equations of motion are:

$$\omega = \omega_0 + \alpha t; \quad \omega^2 - \omega_0^2 = 2\alpha\theta; \quad \theta = \omega_0 t + \frac{1}{2}\alpha t^2$$

where ω_0 is the initial angular velocity, ω is the final angular velocity, α is angular acceleration of the body and θ represents the angle moved in time t.

(*i*) angular acceleration α = rate of change of angular velocity

$$= \frac{\text{change of angular velocity}}{\text{time}} = \frac{\omega - \omega_0}{t}$$

$$\omega - \omega_0 = \alpha t$$

$$\therefore \quad \omega = \omega_0 + \alpha t$$

Alternatively:

$$\alpha = \frac{d\omega}{dt} \; ; \quad d\omega = \alpha\, dt$$

Integrating both sides within the limit $\omega = \omega_0$ at $t = 0$ and $\omega = \omega$ at $t = t$

$$\int_{\omega_0}^{\omega} d\omega = \alpha \int_0^t dt \qquad (\because \alpha \text{ is constant})$$

$$\omega - \omega_0 = \alpha t$$

$$\therefore \quad \omega = \omega_0 + \alpha t \qquad ...(9.1)$$

(*ii*) Angle travelled = average angular velocity × time

$$\theta = \frac{\omega_0 + \omega}{2} \times t = \frac{\omega_0 + \omega_0 + \alpha t}{2} \times t \qquad (\because \omega = \omega_0 + \alpha t)$$

$$= \left(\omega_0 + \frac{\alpha t}{2}\right) t = \omega_0 t + \frac{1}{2}\alpha t^2$$

Alternatively:

$$\omega = \frac{d\theta}{dt} \quad ; \quad d\theta = \omega\, dt = (\omega_0 + \alpha\, t)\, dt$$

Upon integration

$$\int_0^{\theta} d\theta = \int_0^{t} (\omega_0 + \alpha t)\, dt$$

assuming $\theta = 0$ at $t = 0$ and $\theta = \theta$ at $t = t$

$$\theta = \left|\omega_0 t + \frac{\alpha t^2}{2}\right|_0^t = \omega_0 t + \frac{1}{2}\alpha t^2 \qquad \text{...(9.2)}$$

(*iii*) Angle moved = average angular velocity × time

$$\theta = \frac{\omega_0 + \omega}{2} \times \frac{\omega - \omega_0}{\alpha} = \frac{\omega^2 - \omega_0^2}{2\alpha} \qquad \left(\because \omega = \omega_0 + \alpha t \quad t = \frac{\omega - \omega_0}{\alpha}\right)$$

$$\therefore \quad \omega^2 - \omega_0^2 = 2\alpha\theta$$

Alternatively:

$$\alpha = \frac{d\omega}{dt} = \frac{d\omega}{d\theta} \times \frac{d\theta}{dt} = \frac{d\omega}{d\theta} \times \omega$$

or $\quad \alpha\, d\theta = \omega\, d\omega$

upon integration and assuming $\omega = \omega_0$ at $\theta = 0$ and $\omega = \omega$ at $\theta = \theta$

$$\alpha \int_0^{\theta} d\theta = \int_{\omega_0}^{\omega} \omega\, d\omega$$

$$\alpha\,\theta = \left|\frac{\omega^2}{2}\right|_{\omega_0}^{\omega} \quad ; \quad \alpha\theta = \frac{\omega^2 - \omega_0^2}{2}$$

$$\therefore \quad \omega^2 - \omega_0^2 = 2\alpha\theta \qquad \text{...(9.3)}$$

EXAMPLE 9.1.

A flywheel rotates freely on friction less bearings at 240 rev/min. How many revolutions will it make in 10 seconds after the start? Proceed to determine the angular speed if this wheel turns 500 revolutions in 15 seconds.

Solution: There will be no resistance to motion in frictionless bearings. Obviously the wheel will then rotate with uniform motion for which $\theta = \omega t$

(*i*) $\omega = 240$ rev/min; $t = 10$ seconds

$$\therefore \quad \theta = \omega t = 240 \times \frac{10}{60} = 40 \text{ revolutions} = 40 \times 2\pi = 80\pi \text{ radians}$$

(*ii*) $\theta = 500$ revolutions ; $t = 15$ seconds

$$\therefore \quad \omega = \frac{\theta}{t} = \frac{500}{(15/60)} = 2000 \text{ rev/min} = 2000 \times \frac{2\pi}{60} = \mathbf{209.33} \text{ rad/s}$$

EXAMPLE 9.2.

The speed of a flywheel changes from 10 rad/s to 30 rad/s in 5 seconds time. Determine the angular acceleration of the wheel.

How many revolutions the wheel would turn to attain a speed of 600 rev/min?

Solution: Using the kinematic relation $\omega = \omega_0 + \alpha t$, we have

$$\text{angular acceleration } \alpha = \frac{\omega - \omega_0}{t} = \frac{30 - 10}{5} = \mathbf{4} \text{ rad/s}^2$$

(*b*)
$$\omega = \frac{2\pi N}{60} = \frac{2\pi \times 600}{60} = 62.8 \text{ rad/s}$$

Then from the relation, $\omega^2 - \omega_0^2 = 2\alpha\theta$,

$$\text{total angular displacement } \theta = \frac{\omega^2 - \omega_0^2}{2\alpha} = \frac{62.8^2 - 10^2}{2 \times 4} = 480.48 \text{ radians.}$$

Now, 1 revolution $\equiv 2\pi$ radians

$$\therefore \quad \text{Total revolutions made by the wheel} = \frac{480.48}{2\pi} = \mathbf{76}$$

EXAMPLE 9.3.

A roller of a motorised mortar mixer rotates for 5 seconds with a uniform angular acceleration and describes 120 radians during this time. It then rotates with constant angular velocity and covers 100 radians during the next 5 seconds. Make calculations for the initial angular velocity and the angular acceleration.

Solution: Let ω be the constant angular velocity during the later 5 seconds period of motion. Then

$$\omega = \frac{\text{angular displacement}}{\text{time}} = \frac{100}{5} = 20 \text{ rad/s}$$

If ω_0 is the initial angular velocity and α is the constant angular acceleration during the first 5 seconds of motion, then from the kinematic relations

$$\omega = \omega_0 + \alpha t \text{ and } \theta = \omega_0 t + \frac{1}{2} \alpha t^2$$

we have
$$20 = \omega_0 + 5\alpha \quad \text{and} \quad 120 = 5\,\omega_0 + \frac{1}{2}\alpha \times 25$$

From these two expressions, we get

$$120 = 5\,(20 - 5\alpha) + \frac{1}{2}\alpha \times 25 = 100 - 25\alpha + \frac{25\alpha}{2}$$

or
$$\frac{-25\alpha}{2} = 20$$

$$\therefore \quad \alpha = \frac{2 \times 20}{25} = -\mathbf{1.6} \text{ rad/s}^2$$

and
$$\omega_0 = 20 - 5\alpha = 20 - 5\,(-1.6) = \mathbf{28} \text{ rad/s}$$

EXAMPLE 9.4.

A grinding wheel is attached to the shaft of an electric motor of rated speed 1800 rev/min. When the power is switched on, the unit attains the rated speed in 5 seconds. Assuming uniformly accelerated motion, determine the number of revolutions the wheel turns to attain the rated speed.

Solution: $\omega_0 = 0$; $\omega = \dfrac{2\pi N}{60} = \dfrac{2\pi \times 1800}{60} = 188.4$ rad/s

From the kinematic relation: $\omega = \omega_0 + \alpha t$, we have

$$\text{angular acceleration } \alpha = \frac{\omega - \omega_0}{t} = \frac{188.4 - 0}{5} = 37.68 \text{ rad/s}^2$$

Now from the relation: $\omega_2 - \omega_0^2 = 2\alpha\theta$

$$\text{total angular displacement } \theta = \frac{\omega^2 - \omega_0^2}{2\alpha} = \frac{188.4^2 - 0}{2 \times 37.68} = 471 \text{ radians}$$

Since, 1 revolution ≡ 2π radians

$$\text{total revolutions turned by the wheel} = \frac{471}{2\pi} = \mathbf{75}$$

EXAMPLE 9.5.

A 5 m long bar of small cross-section rotates in the horizontal plane about a vertical axis through one end. It accelerates uniformly from 1000 rev/min to 1500 rev/min in 5 seconds interval. Make calculations for the linear velocity of its mid point at the beginning and end of the time interval.

Solution: Let the subscripts *b* and *e* denote the beginning and end of the time interval.

The linear velocity and the angular velocity are connected by the relation:

$$v = \omega r$$

$$\therefore \quad v_b = \omega_b r = \frac{2\pi \times 100}{60} \times \left(\frac{5}{2}\right) = \mathbf{26.17} \text{ m/s} \qquad \left(\because \omega = \frac{2\pi N}{60}\right)$$

$$v_e = \omega_e r = \frac{2\pi \times 1500}{60} \times \left(\frac{5}{2}\right) = \mathbf{39.25} \text{ m/s}$$

It is to be noted that the angular velocity is same for any point of the bar whereas the linear velocity depends on the radius of the point (whose velocity is needed) from the centre of rotation.

EXAMPLE 9.6.

The piston of an engine is stated to make 360 strokes per minute. Determine the angular and linear speed of a point on the periphery of 1.8 m diameter flywheel fitted to the crank shaft of the engine.

Solution: Corresponding to two strokes of the piston, there is one revolution of the crank shaft.

$$\therefore \quad \text{Number of revolutions made by the crank shaft} = \frac{360}{2} = 180 \text{ rev/min}$$

Since the flywheel is rigidly fixed to the crankshaft,

Number of revolutions made by the flywheel = 180 rev/min

$\therefore$ Angular speed of the flywheel $\omega = \dfrac{2\pi N}{60} = \dfrac{2\pi \times 180}{60} = \mathbf{18.84}$ rad/s

The linear velocity and the angular velocity are connected by the relation:

$$v = \omega r$$

$\therefore$ Linear speed $v = 18.84 \times \left(\dfrac{1.8}{2}\right) = \mathbf{16.96}$ m/s

EXAMPLE 9.7.

A flywheel 0.5 m in diameter accelerates uniformly from rest to 360 rpm in 12 seconds. Determine the velocity and acceleration of a point on the rim of the flywheel 0.1 second after it has started from rest.

Solution: $\omega_0 = 0;\quad \omega = \dfrac{2\pi N}{60} = \dfrac{2\pi N}{60} = 37.68$ rad/s

From the kinematic relation: $\omega = \omega_0 + \alpha t$, we have

$$\text{angular acceleration } \alpha = \frac{\omega - \omega_0}{t} = \frac{37.68 - 0}{12} = 3.14 \text{ rad/s}$$

Angular velocity of the wheel 1 second after starting

$$\omega = \omega_0 + \alpha\, t = 0 + 3.14 \times 1 = 3.14 \text{ rad/s}$$

The linear velocity and the angular velocity are connected by the relation:

$$v = wr$$

$\therefore$ linear speed $v = 3.14 \times \dfrac{0.5}{2} = 0.785$ m/s

The linear velocity will be tangential to the rim.

The components of acceleration are:

$$\text{tangential acceleration } \alpha_t = \alpha r = 3.14 \times \frac{0.5}{2} = 0.785 \text{ m/s}^2$$

$$\text{normal acceleration } \alpha_n = \omega^2 r = (3.14)^2 \times \frac{0.5}{2} = 1.232 \text{ m/s}^2$$

Magnitude of total acceleration,

$$\alpha = \sqrt{\alpha_t^2 + \alpha_n^2} = \sqrt{0.785^2 + 1.232^2} = \mathbf{1.461} \text{ m/s}^2$$

Let ϕ be the angle between the total acceleration vector and the radius

$$\phi = \tan^{-1}\frac{\alpha_t}{\alpha_n} = \tan^{-1}\left(\frac{0.785}{1.232}\right) = \mathbf{32.5^\circ}$$

EXAMPLE 9.8

A shaft supported between two bearings carries a wheel of 1.5 m diameter. A constant moment is applied at the rim of the wheel and it attains a speed of 120 revolutions per minute in 10 minutes from rest. Determine:

(*a*) the peripheral velocity of the wheel,

(*b*) the angular acceleration of the wheel, and

(*c*) the number of revolutions made by the wheel

Solution: Angular velocity, $\omega = \dfrac{2\pi N}{60} = \dfrac{2\pi \times 120}{60} = 12.56$ rad/s

Radius of the wheel, $r = \dfrac{1.5}{2} = 0.75$ m

The angular and peripheral velocities are connected by the identity: $v = \omega r$

$\therefore$ peripheral speed of the wheel $v = 12.56 \times 0.75 = \mathbf{9.42}$ m/s

(*b*) From the kinematic relation $\omega = \omega_0 + \alpha t$, we have

$$\text{angular acceleration } \alpha = \frac{\omega - \omega_0}{t} = \frac{12.56 - 0}{10 \times 60} = \mathbf{0.0209} \text{ rad/s}^2$$

(*c*) From the kinematic relation $\omega^2 - \omega_0^2 = 2\alpha\theta$

$$\text{angular displacement } \theta = \frac{\omega^2 - \omega_0^2}{2\alpha} = \frac{12.56^2 - 0^2}{2 \times 0.0209} = 3774 \text{ radians}$$

Since 2π radians $\equiv$ 1 revolution

$$\text{Total number of revolutions make by the wheel} = \frac{3774}{2\pi} = \mathbf{601}$$

EXAMPLE 9.9.

A wheel accelerates uniformly from rest to a speed of 180 rpm in 0.5 seconds. It then rotates at that speed for 2 seconds before decelerating to rest in 0.3 seconds. Workout the revolutions made by the wheel during the entire time interval.

Solution: (*i*) During uniform acceleration (from $t = 0$ to $t = 0.5$ sec)

$$\omega_1 = \frac{2\pi \times 180}{60} = 18.84 \text{ rad/s}$$

$$\alpha_1 = \frac{\omega_1 - \omega}{t} = \frac{18.84 - 0}{0.5} = 37.68 \text{ rad/s}^2$$

From the kinematic relation $\omega^2 - \omega_0^2 = 2\alpha\theta$, we have

$$\theta_1 = \frac{\omega_1^2 - \omega_0^2}{2\alpha_1} = \frac{18.84^2 - 0^2}{2 \times 37.68} = 4.7 \text{ radians}$$

(*ii*) During uniform speed (from $t = 0.5$ sec to 2.5 sec)

$\omega_2 = 18.84$ rad/s

$\theta_2 = \omega_2 t = 18.84 \times 2 = 37.68$ radians.

(*iii*) During uniform deceleration (from $t = 2.5$ sec to 2.8 sec)

$\omega_2 = 18.84$ rad/s ; $\omega_3 = 0$

$$\alpha_3 = \frac{\omega_3 - \omega_2}{t} = \frac{0 - 18.84}{0.3} = -62.8 \text{ rad/s}$$

From the kinematic relation $\omega^2 - \omega_0^2 = 2\alpha\theta$, we have

$$\theta_3 = \frac{\omega_3^2 - \omega_2^2}{2\alpha} = \frac{0^2 - 18.84^2}{2(-62.8)} = 2.83 \text{ radians}$$

$\therefore$ Total angular displacement $= \theta_1 + \theta_2 + \theta_3$

$= 4.7 + 37.68 + 2.83 = 45.22$ radians

Since 1 revolution $\equiv 2\pi$ radians

$$\text{Total revolutions turned by the wheel} = \frac{45.22}{2\pi} = \mathbf{7.2} \text{ revolutions}$$

Alternatively: Refer Fig. 9.3 which represents the velocity-time graph for the wheel. The area under the velocity-time graph gives the angle turned during the entire time-interval.

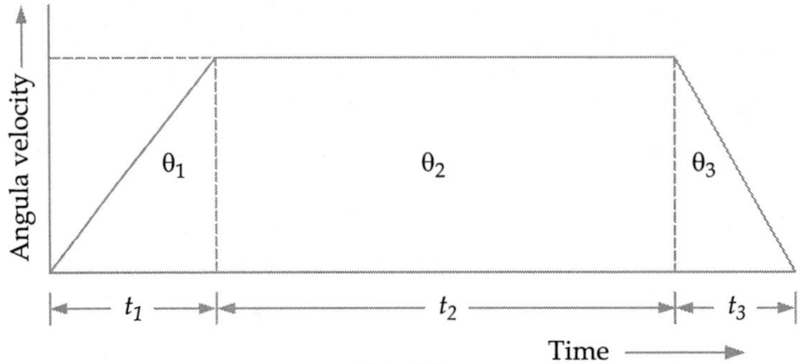

Fig. 9.3

$$\theta = \theta_1 + \theta_2 + \theta_3 = \left(\frac{1}{2} \times 0.5 \times 18.84\right) + (2 \times 18.84) + \left(\frac{1}{2} \times 0.3 \times 18.84\right)$$

$$= 4.71 + 37.68 + 2.83 = 45.22 \text{ radians}$$

$\therefore$ Total revolutions turned by the wheel $= \dfrac{45.22}{2\pi} =$ **7.2** revolutions

EXAMPLE 9.10.

A flywheel had an initial angular speed of 3000 rev/min in clockwise direction. When a constant turning moment was applied to the wheel, it got subjected to a uniform anticlockwise angular acceleration of 3 rev/sec². Determine the angular velocity of the wheel after 20 seconds, and the total number of revolutions made during this period.

Solution: Initial angular velocity, $\omega_0 = \dfrac{2\pi N}{60} = \dfrac{2\pi \times 3000}{60} = 314$ rad/s (clockwise)

Angular acceleration $\alpha = 3 \text{ rev/s}^2 = 3 \times 2\pi = 18.84 \text{ rad/s}^2$ (anticlockwise)

From the kinematic relation, $\omega = \omega_0 + \alpha\, t$, the angular velocity of the wheel after 20 seconds is

$$\omega = 314 + (-18.84) \times 20 = -62.8 \text{ rad/s}^2$$

The negative sign indicates that the wheel is turning anticlockwise.

Let the wheel change its direction of rotation after t_1 seconds. At the instant of change of direction, the velocity becomes zero. Accordingly

$$0 = 314 - 18.84\, t_1 \quad ; \quad t_1 = \frac{314}{18.84} = 16.67 \text{ s}$$

For $t_1 = 16.67$ s, the wheel would turn clockwise and for $t_2 = (20 - t_1) = (20 - 16.67) = 3.33$ seconds, the wheel would turn anticlockwise.

$$\text{Clockwise rotation } \theta_1 = \omega_0 t_1 + \frac{1}{2}\alpha t_1^2 = 314 \times 16.67 + \frac{1}{2}(-18.84) \times 16.67^2$$

$$= 2617.67 \text{ radians}$$

$$= \frac{2617.67}{2\pi} = 416.83 \text{ revolutions}$$

Anticlockwise rotation $\theta_2 = 0 - \frac{1}{2}(-18.84) \times 3.33^2$

$$= 104.46 \text{ radians} = \frac{104.46}{2\pi} = 16.63 \text{ revolutions}$$

$\therefore$ Gross number of revolutions = 416.83 + 16.63 = **433.46** rev

Net revolutions = 416.83 – 16.63 = **400.20** rev (clockwise)

EXAMPLE 9.11.

A flywheel rotating freely at 1800 rev/min in the clockwise direction is subjected to a variable counter clockwise torque which is intially applied at *t* = 0. The torque produces a counter-clockwise acceleration prescribed by the relation $\alpha = 3t$ rad/s^2 where *t* is the time in second. Make calculations for:

(*a*) the time required for the flywheel to reduce its clockwise angular speed to 100 rev/min;

(*b*) the time required for the flywheel to reverse its direction of motion.

Solution: Let it be presumed that the counter clockwise direction is positive.

$$\omega_1 = -\frac{2\pi \times 1800}{60} = -188.4 \text{ rad/s}$$

$$\omega_2 = -\frac{2\pi \times 1000}{60} = -104.67 \text{ rad/s}$$

Angular acceleration $\alpha = \frac{d\omega}{dt} = 3t$ or $d\omega = 3t\,dt$. Integrating both sides within the prescribed limits.

$$\int_{-188.4}^{-104.67} d\omega = 3\int_0^t t\,dt$$

or $$-104.67 + 188.4 = \frac{3t^2}{2};\ t = \sqrt{\frac{2}{3} \times 83.73} = \mathbf{7.47 \text{ sec}}$$

(*b*) At reversal of direction, the angular speed will be zero. Therefore,

$$\int_{-188.4}^{0} d\omega = 3\int t\,dt$$

or $$188.4 = \frac{3t^2}{2};\ t = \sqrt{\frac{2}{3} \times 188.4} = \mathbf{11.21 \text{ sec}}$$

EXAMPLE 9.12.

A shaft is accelerated uniformly from 600 rev/min to 900 rev/min in 2 seconds. It continues accelerating at this rate for a further period of 4 seconds and then continues to rotate at the maximum speed attained. Make calculations for the time taken to complete the first 180 revolutions.

Solution: (*i*) During uniform acceleration from 600 rpm to 900 rpm (*t* = 2 sec)]

$$\omega_0 = \frac{2\pi \times 600}{60} = 62.8 \text{ rad/s} \qquad \left(\because \omega = \frac{2\pi N}{60}\right)$$

$$\omega_1 = \frac{2\pi \times 900}{60} = 94.2 \text{ rad/s}$$

$$\alpha_1 = \frac{\omega_1 - \omega_0}{t} = \frac{94.2 - 62.8}{2} = 15.7 \text{ rad/s}^2$$

The angular displacement θ_1 for the entire accelerating period of 6 seconds can be worked out by using the kinematic relation $\theta = \omega t + \frac{1}{2} \alpha t^2$. That is

$$\theta_1 = 62.8 \times 6 + \frac{1}{2} \times 15.7 \times 6^2 = 659.4 \text{ radians}$$

Since 1 revolution ≡ 2π radians,

Revolutions turned by the wheel during the accelerating period

$$= \frac{659.4}{2\pi} = 105 \text{ revolutions.}$$

(*ii*) Revolutions to be made by the shaft at uniform speed attained at the end of accelerating period = 180 – 105 = 75 revolutions

Corresponding angular displacement = 75 × 2π = 471 radians

Angular velocity at the end of acceleration period,

$$= \omega_0 + \alpha t = 62.8 + 15.7 \times 6 = 157 \text{ rad/s}$$

Time taken to turn 471 radians at uniform speed of 157 rad/s

$$t = \frac{\theta}{w} = \frac{471}{157} = 3 \text{ seconds}$$

∴ Required time = 2 + 4 + 3 = 9 seconds

EXAMPLE 9.13.

The motion of a gear is defined by the relation $\theta = 2t^3 - 5t^2 + 8t + 20$ where θ is in radians and t is in seconds.

Determine the angular displacement, the angular velocity and the angular acceleration at time t = 3 seconds.

Solution: Given that $\theta = 2t^3 - 5t^2 + 8t + 20$ ...(*i*)

Differentiating the above expression with respect to t, we get

$$\frac{d\theta}{dt} = 6t^2 - 10t + 8 \quad \text{or} \quad \omega = 6t^2 - 10t + 8 \qquad ...(ii)$$

where $w = \frac{d\theta}{dt}$ is the angular velocity of the rotating gear.

Differentiation of identity (*ii*) with respect to t gives:

$$\frac{d\omega}{dt} = 12t - 10 \quad \text{or} \quad \alpha = 12t - 10 \qquad ...(iii)$$

where α = angular acceleration of the rotating gear.

Then at t = 5 seconds we have:

Angular displacement $\theta = 2 \times (5)^3 - 5 \times (5)^2 + 8 \times 5 + 20$

$= 250 - 125 + 40 + 20 =$ **185 radians**

Angular velocity $\omega = 6 + (5)^2 - 10 \times 5 + 8 = 150 - 50 + 8 =$ **108 rad/s**

Angular acceleration $\alpha = 12 \times 5 - 10 =$ **50 rad/s²**

9.3 GENERAL PLANE MOTION

Generally a body undergoes the following three types of plane motions :

• **Translation :** The particles constituting the rigid body move in parallel planes and travel the same distance. During translation, the particles have the same velocity and acceleration, and a straight line drawn on the moving body remains parallel to its original position at any time.

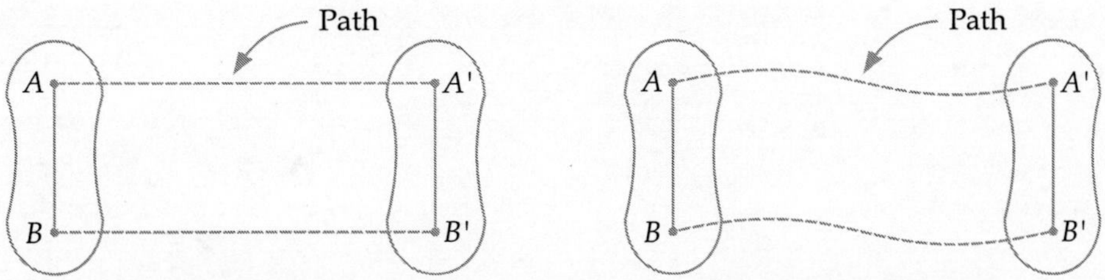

(a) Rectilinear translation *(b) Curvilinear translation*

Fig. 9.4

If the path traced by the particle during motion is a straight line, then the motion is said to be rectilinear translation. If the particle traces a curved path, the motion is called curvilinear translation.

• **Rotation :** The body rotates about a fixed point and all the particles constituting the body move in a circular path. The fixed point about which the body rotates is called the point of rotation and the axis passing through the point of rotation is called the axis of rotation. A point lying on the axis of rotation has a zero velocity and zero acceleration.

The point of rotation may lie inside the body (Fig. 9.5 *a*) or outside the body (Fig. 9.5 *b*)

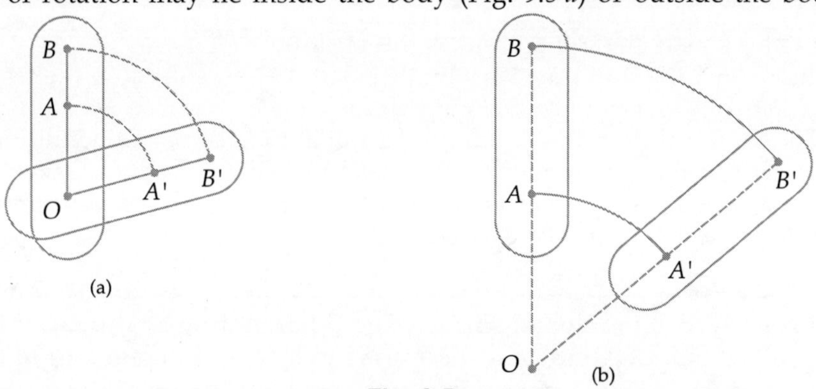

Fig. 9.5

• **General plane motion** (combined motion of translation and rotation) : There exist certain situations where a body possesses both motions of translation and rotation simultaneously at a particular instant. Examples of such a combined motion are :

– motion of roller without slipping

– motion of the wheel of a locomotive train, truck and car etc.

– a rod sliding against a wall at one end and floor at the other end.

The combination of rotation and translation motion is usually referred to as general plane motion.

With reference to Fig. 9.6, a link *AB* shifts to a new position at any instant. A link is a rigid piece of material joining two parts of a machine. This shift of the link can be considered in two parts

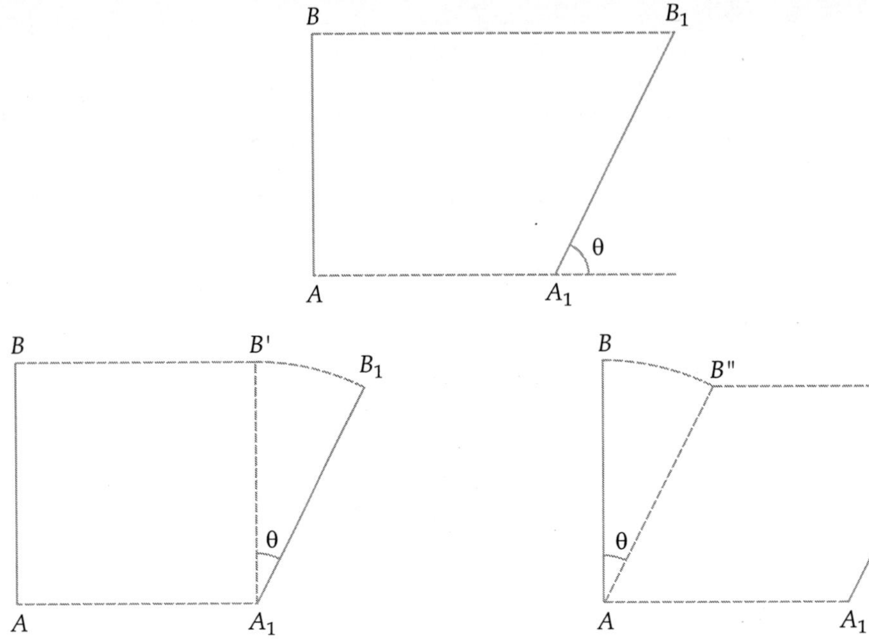

Fig. 9.6.

(*i*) The link translates by a distance AA_1 and takes the position A_1B'

(*ii*) From position A_1B', the link rotates by an angle θ and occupies the final position A_1B_1

OR

(*a*) The link rotates by an angle θ and takes the position AB''

(*b*) From position AB'', the link translates by distance AA_1 and shifts to the final position A_1B_1.

It is to be noted that the shift of the link from the initial to final position will remain same irrespective of the fact whether the motion of translation or that of rotation takes place first. **Chasle's theorem** states *that any general displacement of a rigid body can be represented by a combination of translatory motion and rotational motion.*

EXAMPLE 9.14.

Consider a wheel rolling on a horizontal surface, and a bar sliding against wall at one end and the floor at the other end. Illustrate how these acts can be split into motions of translation and rotation.

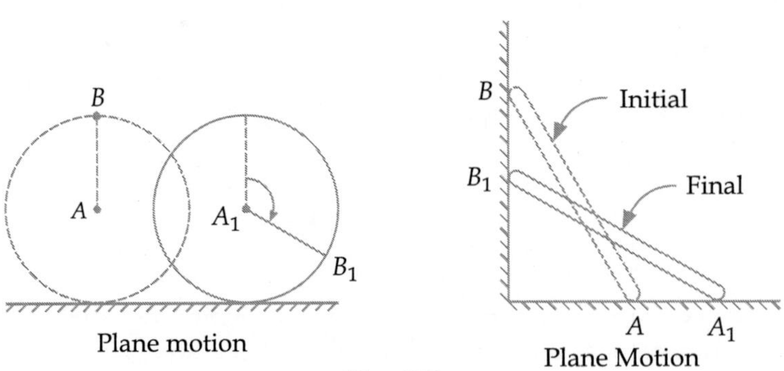

Fig. 9.7

Solution : Case (*a*) : Wheel rolling on a horizontal surface

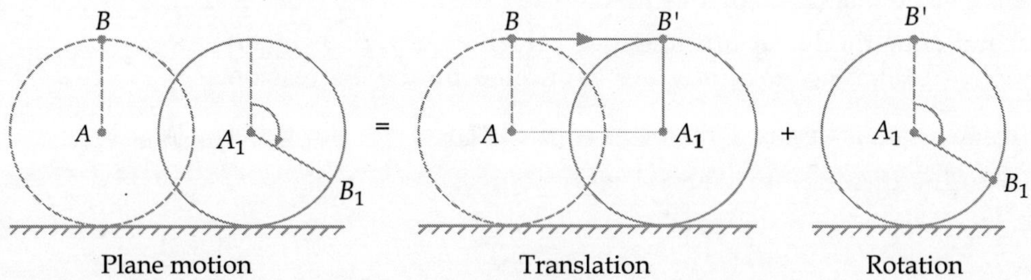

Fig. 9.8

With reference to Fig. 9.8, the general plane motion of a wheel rolling on a horizontal surface from initial position *AB* to final position A_1B_1 can be replaced by the following motions taken in order

(*i*) motion of translation from *AB* to A_1B'

(*ii*) rotation about A_1 from position A_1B' to A_1B_1

Case (*b*) : Bar sliding against wall at one end and floor at the other end

With reference to Fig. 9.9, the general plane motion of a bar sliding against wall at one end and floor at the other end from initial position *AB* to A_1B_1 can be replaced by the following motions taken in order

(*i*) motion of translation of the bar from position *AB* to A_1B'

(*ii*) rotation about A_1 from position A_1B' to A_1B_1

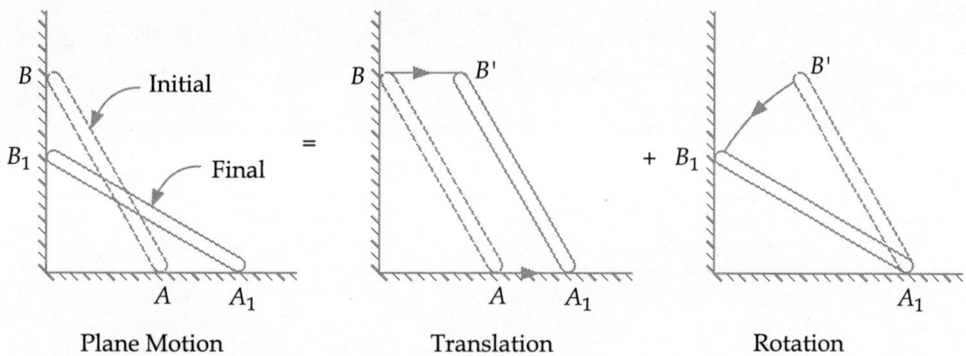

Fig. 9.9

9.4 INSTANTANEOUS CENTRE

While analysing plane motion of a body, a point can be located in the plane which has zero velocity. The plane motion of all the particles constituting the body may then be considered as pure rotation about that point. Such a point is called the *instantaneous centre* or *virtual centre of body*. The axis passing through this point and at right angles to the plane of motion is called *instantaneous axis of rotation*. The instantaneous centre changes every moment, and its locus is called centrods. The surface generated by the instantaneous axis is called the *axode*.

Depending on the given situation, following are some of the methods used to determine the instantaneous centre:

(*i*) Let v_a and ω be the linear and angular velocities of a point A on a rigid body (Fig. 9.10). Since ω and v_a are connected by the expression $v = \omega r$, the instantaneous centre I then lies at a distance $\frac{v_a}{\omega}$ along the perpendicular to the direction of velocity v_a at point A.

$$IA = \frac{v_a}{\omega}$$

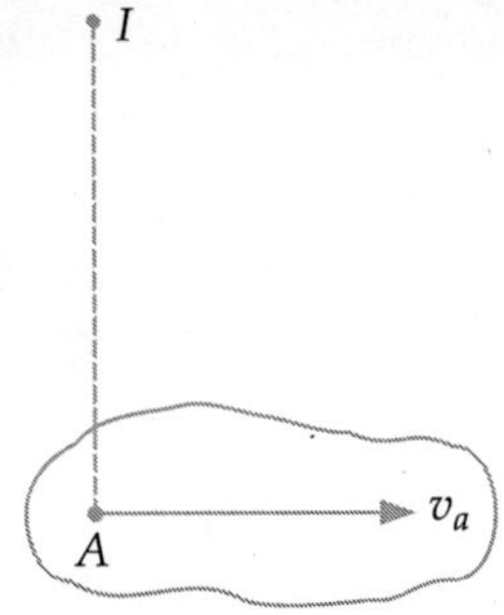

Fig. 9.10

(*ii*) Let v_a and v_b be the linear velocities at points A and B on a rigid body (Fig. 9.11). These velocities are directed along the direction on AA' and BB' respectively. The instantaneous centre I is then the point of intersection of the lines erected perpendicular to the direction of velocities at the given points. Then

$$v_a = \omega \times IA \quad \text{and} \quad v_b = \omega \times I_B$$

where ω is the angular velocity with which the body shall appear to rotate about the instantaneous centre I.

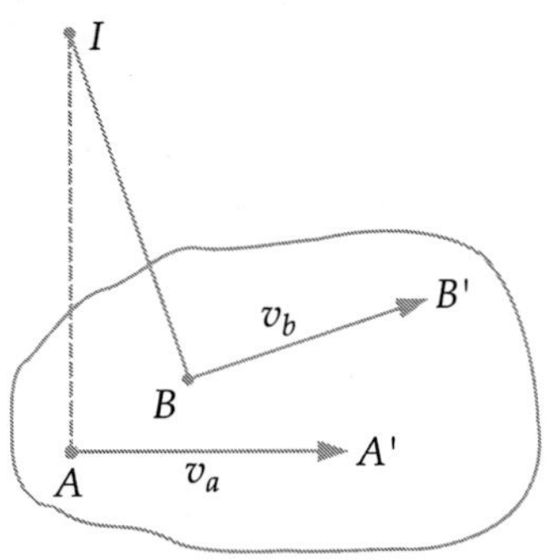

Fig. 9.11

(*iii*) Consider a solid circular cylinder in contact with two horizontal conveyer belts running with different velocities in the same or opposite directions as shown in Fig. 9.12.

The instantaneous centre is found by determining the point of intersection of line AB with the line joining the extremities of the velocities vectors (Fig. 9.13). It is to be noted that when the velocities v_a and v_b are in the same direction, I lies outside AB and divides it externally in the ratio v_a to v_b.

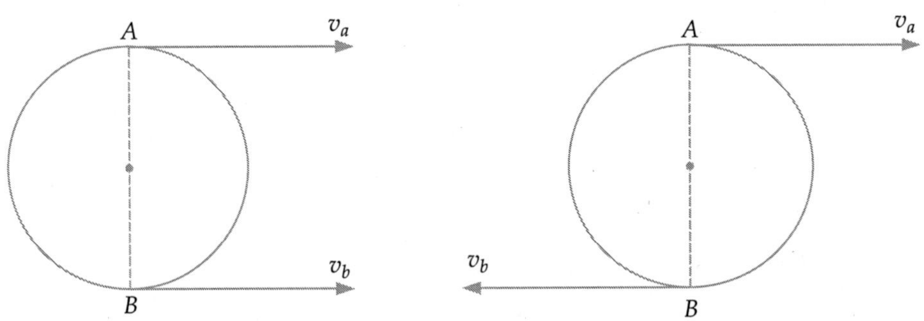

Fig. 9.12

When velocities v_a and v_b are in opposite directions, I lies between A and B, and divides AB internally in the ratio v_a to v_b.

Every point on the cylinder shall appear to rotate about the instantaneous centre I with an angular velocity ω, and therefore

$$\omega = \frac{v_a}{IA} = \frac{v_b}{IB}$$

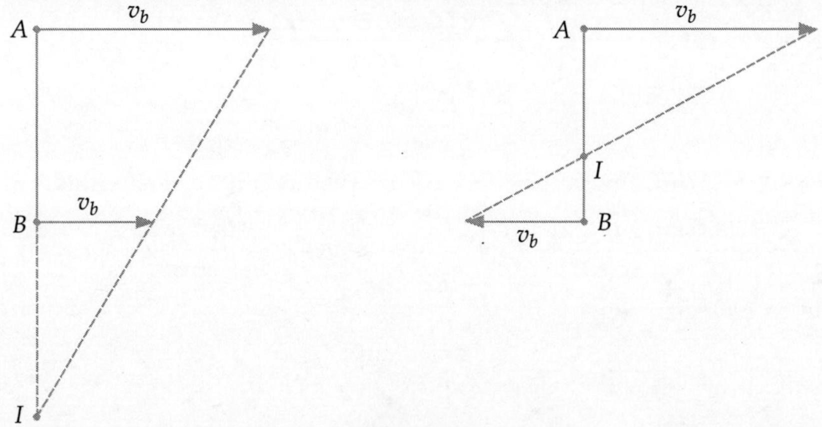

Fig. 9.13

(*iv*) For the piston-connecting rod assembly (Fig. 9.14)

The point B moves in a circle whose centre is O, and the direction of V_b is perpendicular to OB. The velocity V_a of the piston, at any instant, must be horizontal. The instantaneous centre I of the connecting rod AB is the point of intersection of perpendiculars to V_a and V_b through A and B respectively. Then angular velocity ω_{ab} of the connecting rod about I is

$$\omega_{ab} = \frac{V_a}{IA} = \frac{V_b}{IB} = \frac{\omega r}{IB} \qquad ...(i)$$

where ω is angular velocity of the crank and r is its length. The distances IA and IB can be located as under :

$$AO = AC + CO = AB \cos \phi + BO \cos \theta = l \cos \phi + r \cos \theta$$

where l is the length of connecting rod.

$$IA = AO \tan \theta = (l \cos \phi + r \cos \theta) \tan \theta$$
$$= (l \cos \phi \tan \theta + r \sin \theta) \qquad ...(ii)$$

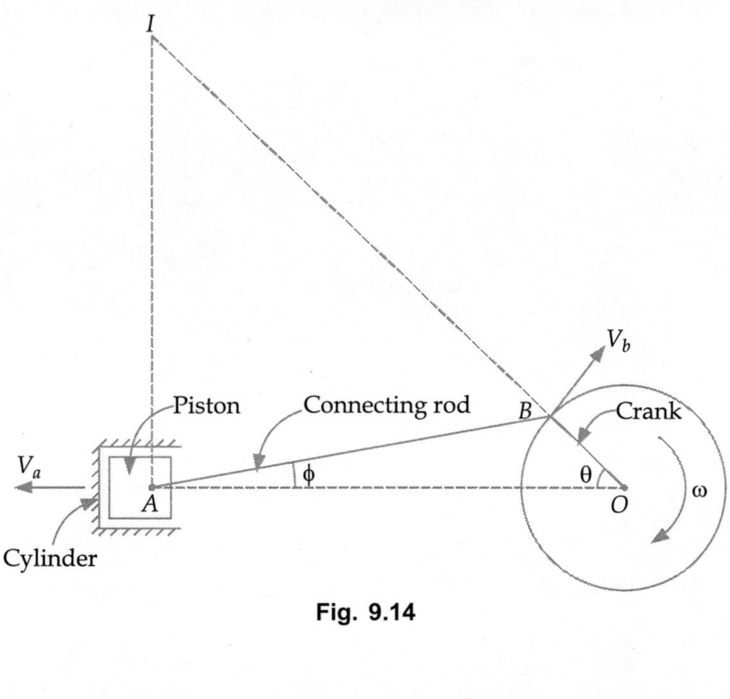

Fig. 9.14

$$IB = IO - BO = \frac{AO}{\cos\theta} - BO = \frac{l\cos\phi + r\cos\theta}{\cos\theta} - r$$

$$= \frac{l\cos\phi}{\cos\theta} + r - r = \frac{l\cos\phi}{\cos\theta} \qquad ...(iii)$$

From the identities (*i*), (*ii*), (*iii*), the velocity V_a of the piston can be written as

$$V_a = \frac{IA}{IB} V_b = \frac{l\cos\phi\tan\theta + r\sin\theta}{l\cos\phi/\cos\theta} \times \omega r = \left(\cos\theta\tan\theta + \frac{r}{l}\frac{\sin\theta\cos\theta}{\cos\phi}\right)\omega r$$

Applying sine rule to $\Delta\, OAB$,

$$\frac{OB}{\sin\phi} = \frac{AB}{\sin\theta}; \quad \frac{r}{\sin\phi} = \frac{l}{\sin\theta}; \quad \frac{r}{l} = \frac{\sin\phi}{\sin\theta}$$

Then $\quad V_a = \omega\,(r\sin\theta + r\cos\theta\tan\phi) = \omega\,(l\sin\phi + r\cos\theta\tan\phi) \qquad ...(9.4)$

EXAMPLE 9.15

At the instant shown in Fig. 9.15, the rod *AB* is rotating clockwise at 2.5 rad/sec. If end *C* of the rod *BC* is free to move on a horizontal surface, make calculations for the angular velocity of rod *BC* and the velocity of its end point *C*.

Solution : Velocity of end *B*, $V_b = \omega r = 2.5 \times 1.5 = 3.75$ m/s

The direction of V_b is normal to *AB* and as *BC* is perpendicular to *AB*, it lies along *BC*. The velocity V_c of end *C*, at any instant, must be horizontal.

The instantaneous centre *I* of rod *BC* is the point of intersection of perpendiculars to V_b and V_c through *B* and *C* respectively (Fig. 9.16)

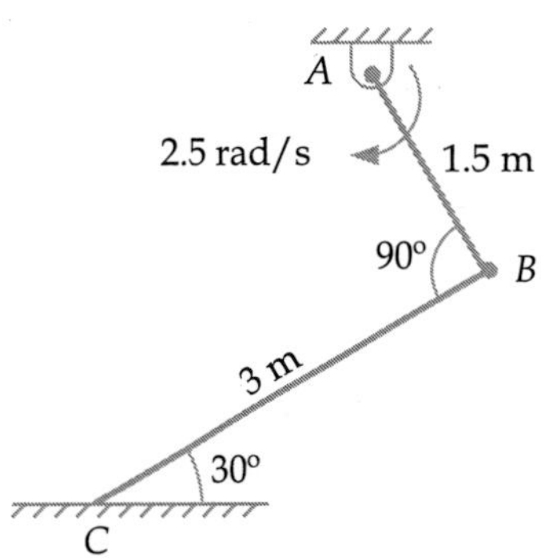

Fig. 9.15

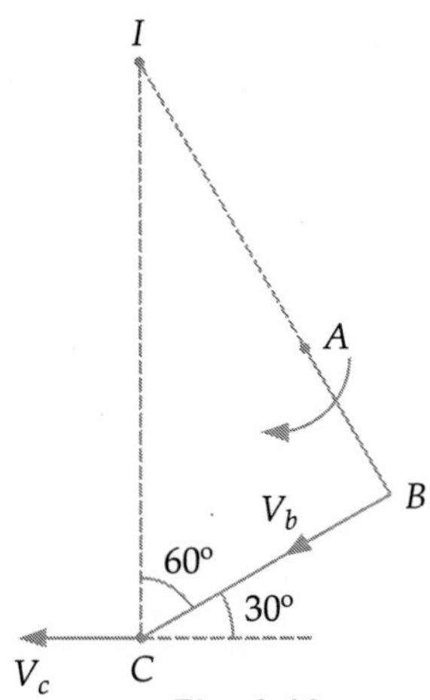

Fig. 9.16

$$IC = \frac{BC}{\cos 60^\circ} = \frac{3}{0.5} = 6 \text{ m}$$

$$IB = IC\sin 60^\circ = 6 \times 0.866 = 5.196 \text{ m}$$

Angular velocity of rod *BC*,

$$\omega = \frac{V_b}{IB} = \frac{V_c}{IC}$$

$$\omega = \frac{3.75}{5.196} = \mathbf{0.722\ rad/s}\ \text{(clockwise)}$$

and $$V_c = \omega \times IC = 0.722 \times 6 = \mathbf{4.33\ m/s}$$

EXAMPLE 9.16.

In a crank and connecting rod mechanism, the crank is 300 mm long and the connecting rod 1500 mm long. If the crank rotates uniformly at 300 rpm, find the velocity of the cross-head when the crank is inclined at 30° with the inner dead centre.

Solution: Refer Fig. 9.17 for the crank and connecting rod mechanism.

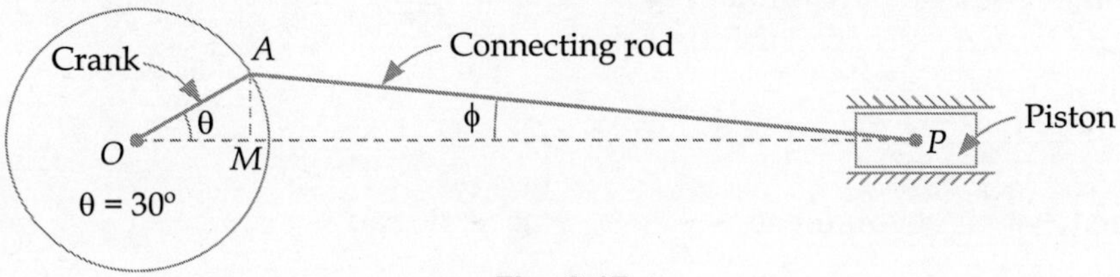

Fig. 9.17

Crank radius r = 300 mm

Length of connecting rod l = 1500 mm

From the law of triangles: $\frac{l}{\sin\theta} = \frac{r}{\sin\phi}$

$$\sin\phi = \frac{r\sin\theta}{l} = \frac{300\times\sin 30°}{1500} = 0.1$$

That gives: ϕ = 5.74°

Let x denote the distance between the crank centre O and the piston point P. Then

$$x = OM + OP = r\cos\theta + l\cos\phi$$

$$\frac{dx}{dt} = r\sin\theta\frac{d\theta}{dt} + l\sin\phi \times \frac{d\phi}{dt}$$

$$= r\sin\theta\frac{d\theta}{dt} + l\sin\phi\ \frac{d\phi}{d\theta}\times\frac{d\theta}{dt}$$

$$= r\sin\theta \times \omega + l\sin\phi \times \left(\frac{r}{l}\frac{\cos\theta}{\cos\phi}\right) \times \omega \qquad ...(a)$$

$$\text{Angular velocity } \omega = \frac{2\pi N}{60} = \frac{2\pi\times 300}{60} = 31.4\ \text{rad/s}$$

Substituting the appropriate values in expression (a), we get velocity of cross-head,

$$\frac{dx}{dt} = 300\sin 30° \times 31.4 + 1500 \times \sin 5.74 \times \left(\frac{300}{1500}\times\frac{\cos 30°}{\cos 5.74}\right) \times 31.4$$

$$= 4710 + 820 = 5530\ \text{mm/s} = \mathbf{5.53\ m/s}$$

EXAMPLE 9.17.

A cylinder of diameter 3 m rolls without slipping along a horizontal surface *PQ* as shown in Fig. 9.18. If its centre has a uniform velocity of 30 m/s, determine the velocities of the points *B* and *D* lying on the rim of the cylinder.

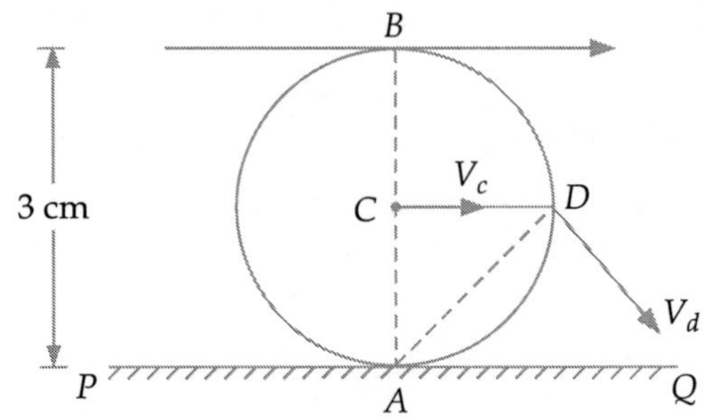

Fig. 9.18

Solution : When the cylinder rolls without slipping, its point of contact with the horizontal surface at any instant has zero velocity. Thus the point *A* is instantaneously at rest and every other point on the cylinder has angular velocity about *A*. Hence *A* is instantaneous centre of the cylinder. Then

$$V_c = \omega \times AC\,; \quad V_B = \omega \times AB\,; \quad V_d = \omega \times AD$$

$$\text{Angular velocity of the cylinder } \omega = \frac{V_c}{AC} = \frac{30}{3/2} = 20 \text{ rad/s}$$

$$\text{Distance } AD = \sqrt{(DC)^2 + (AC)^2} = \sqrt{1.5^2 + 1.5^2} = 2.121 \text{ m}$$

$$\therefore \quad \text{Velocity of point } B, \quad V_b = \omega \times AB = 20 \times 3 = \mathbf{60\ m/s}$$

$$\text{Velocity of point } D, \quad V_d = 20 \times 2.121 = \mathbf{42.42\ m/s}$$

EXAMPLE 9.18.

In a reciprocating engine mechanism (Fig. 9.19), the lengths of the crank *OB* and connecting rod *AB* are 30 cm and 1 metre respectively. If the crank is rotating at a constant angular velocity of 200 rev/min, determine (a) the angular velocity of connecting rod and (b) the velocity of the piston when the crank makes an angle of θ = 45° with the horizontal.

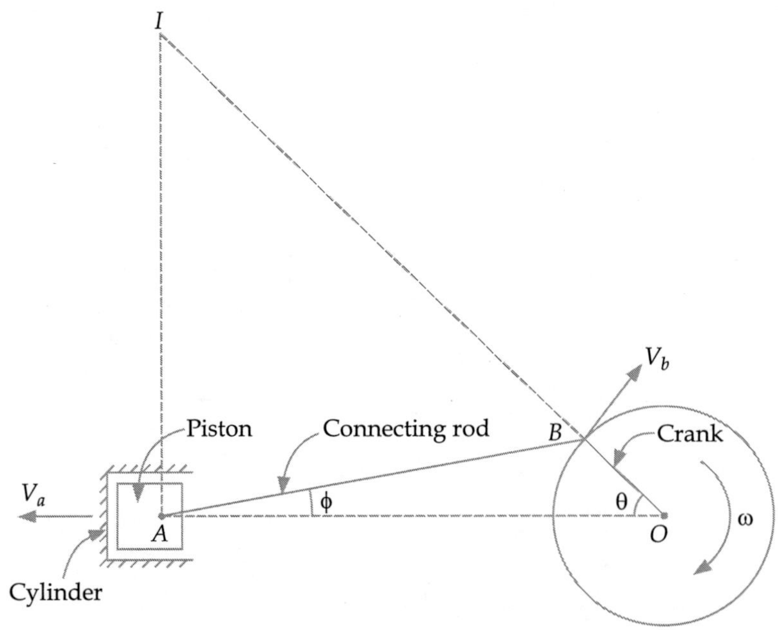

Fig. 9.19

Solution : Angular velocity of crank $\omega = \dfrac{2\pi N}{60} = \dfrac{2\pi \times 200}{60} = 20.93$ rad/s

$$V_b = \omega r = 20.93 \times 0.3 = 6.279 \text{ m/s}$$

The point B moves in a circle whose centre is O, and the direction of V_a is perpendicular to OB. The velocity V_a of the piston, at any instant must be horizontal. The instantaneous centre I of the connecting rod AB is the point of intersection of perpendiculars to V_a and V_b through A and B respectively. Then angular velocity ω_{ab} of the connecting rod about I is

$$\omega_{ab} = \frac{V_a}{IA} = \frac{V_b}{IB} = \frac{\omega r}{IB} \quad \text{...}(i)$$

Let ϕ be the angle between the connecting AB and the line AO. Then applying sine rule to $\Delta\, OAB$

$$\frac{OB}{\sin\phi} = \frac{AB}{\sin\theta}; \quad \sin\phi = \frac{OB}{AB}\sin\theta = \frac{0.3}{1} \times \sin 45^\circ = 0.2121$$

$$\therefore \quad \phi = 12.25^\circ$$

The distance IA and IB are worked out from the relations

$$IA = l\cos\phi\,\tan\theta + r\sin\theta = (1 \times \cos 12.25^\circ \times \tan 45^\circ) + 0.3 \sin 45^\circ$$

$$= 0.9772 + 0.2121 = 1.1843 \text{ m}$$

$$\text{and} \quad IB = \frac{l\cos\phi}{\cos\theta} = 1 \times \frac{\cos 12.25^\circ}{\cos 45^\circ} = 1.382 \text{ m}$$

Then from identity (i)

angular velocity of connecting rod, $\omega_{ab} = \dfrac{6.279}{1.382} = $ **4.543 rad/s**

velocity of piston $V_a = \omega_{ab} \times IA = 4.543 \times 1.1843 =$ **5.38 m/s**

EXAMPLE 9.19.

The ends of a slender beam AB of length 2.5 m are constrained to remain in contact with a horizontal floor and a vertical wall respectively as shown in Fig. 9.20. If the end A has a velocity of 1.5 m/s, determine the angular velocity of the beam and the velocity of its end B at the position shown in the figure.

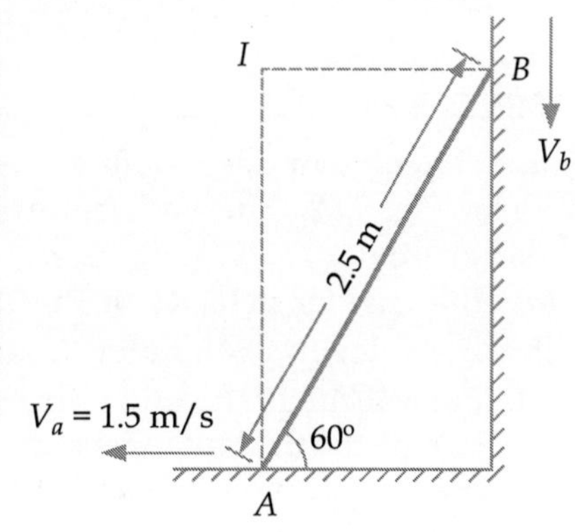

Fig. 9.20

Solution : The velocity V_a at end A is horizontal and the velocity V_b at end B is vertical. The instantaneous centre I of the beam AB is the point of intersection of perpendicular to V_a and V_b through A and B respectively.

$$IA = AB \sin 60^\circ = 2.5 \times 0.866 = 2.165 \text{ m}$$

$$IB = AB \cos 60^\circ = 2.5 \times 0.5 = 1.25 \text{ m}$$

Then angular velocity ω_{ab} of the beam is given by

$$\omega_{ab} = \frac{V_a}{IA} = \frac{1.5}{2.165} = 0.693 \text{ rad/s}$$

$\therefore$ Velocity of end B is

$$V_b = \omega_{ab} \times IB = 0.693 \times 1.25 = \mathbf{0.866\ m/s}$$

EXAMPLE 9.20.

What is kinetics? Define the terms impulse and angular momentum.

A bar *AB* rests at the edge of a wall at some point *C* with its end *A* resting on a horizontal floor as shown in Fig. 9.21. If the end *A* moves with a constant velocity V_a, set up an expression for the angular velocity of the bar in terms of *h* and θ. Proceed to calculate the angular velocity if

$$V_a = 2.5 \text{ m/s}; \quad h = 4 \text{ m} \quad \text{and} \quad \theta = 30°$$

Solution: The velocity of the end *A* is along the horizontal line *OA* and the velocity of *C* on the bar at any instant is along the bar. The position of instantaneous centre *I* is determinod by drawing lines perpendicular to the direction of these velocities.

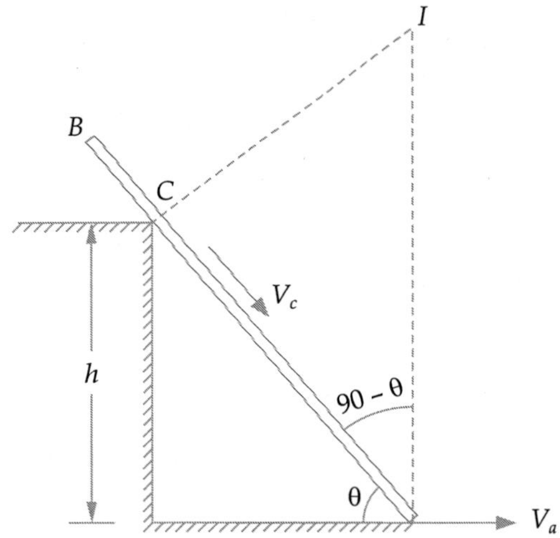

Fig. 9.21

The angular velocity of the bar $\omega = \dfrac{V_a}{IA}$ ...(*i*)

From triangles *OAC* and *AIC*, we have

$$AC = \frac{OC}{\sin\theta};$$

$$AC = IA\cos(90 - \theta) = IA\sin\theta$$

$$\therefore \quad \frac{OC}{\sin\theta} = IA\sin\theta$$

$$\text{or} \quad IA = \frac{OC}{\sin^2\theta} = \frac{h}{\sin^2\theta}$$

Substituting $IA = \dfrac{h}{\sin^2\theta}$ in expression (*i*), we get

$$\omega = \frac{V_a\sin^2\theta}{h} = \frac{2.5\sin^2 30}{4} = \mathbf{0.1562\ rad/s}$$

EXAMPLE 9.21.

A bar of length 2 m has its ends *A* and *B* constrained to move horizontally and vertically as shown in Fig. 9.22. The end *A* moves with a constant velocity of 6 m/s horizontally. Make calculations for:

(*a*) the angular velocity of the bar *AB*,

(*b*) the velocity of the end *B*, and

(*c*) the velocity of the midpoint of the bar at the instant when the bar makes an angle of 30° with the horizontal.

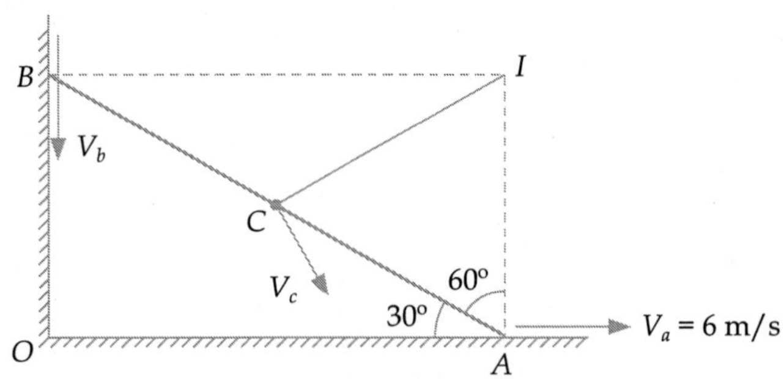

Fig. 9.22

Solution: Velocity of end *A*, $V_a = 6$ m/s

Velocity of end *B*, $V_b = V_a + V_{ba}$

where V_{ba} is the relative velocity of end *B* with respect to end *A* and it is due to rotation of bar *AB* about *A*. This relative is normal to bar *AB* at point *B* and it equals ω*l*.

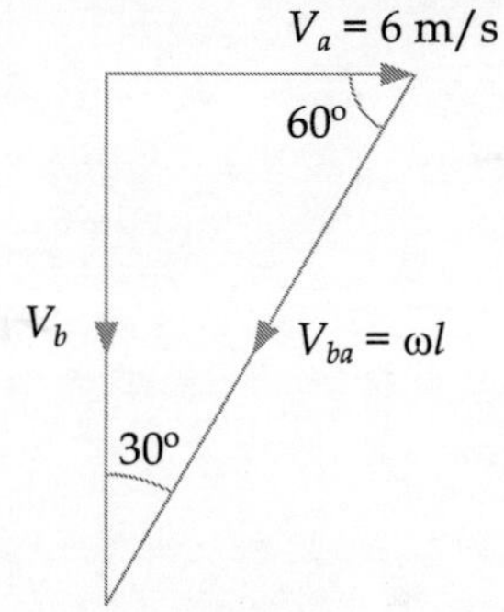

From the vector diagram,

$$\frac{V_a}{\sin 30°} = \frac{V_{ba}}{\sin 90°} = \frac{V_b}{\sin 60°}$$

(*i*) $V_b = \frac{V_a}{\sin 30°} \times \sin 60° = \frac{6}{0.5} \times 0.866$

$= \mathbf{10.39\ m/s}$

(*ii*) $V_{ba} = \frac{V_a}{\sin 30°} \times \sin 90° = \frac{6}{0.5} \times 1 = 12 \text{ m/s}$

$\therefore$ Angular velocity of bar, $\omega = \frac{V_{ba}}{l} = \frac{12}{2} = \mathbf{6\ rad/s}$

(*iii*) Velocity of midpoint *C* of the bar

$$\vec{V}_c = \vec{V}_a + \vec{V}_{ca}$$

$$V_a = 6 \text{ m/s and } V_{ca} = \omega \frac{l}{2} = 6 \times \frac{2}{2} = 6 \text{ m/s}$$

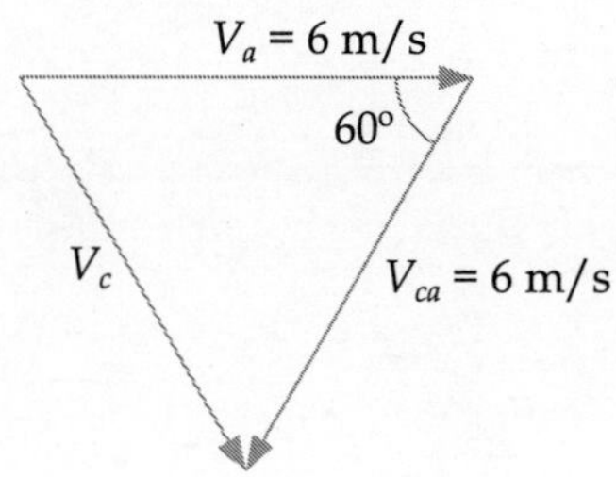

With reference to vector diagram, the magnitude of velocity *V* can be determined by using the cosine relation

$$V_c = \sqrt{V_a^2 + V_{ca}^2 - 2V_aV_{ca}\cos 60°}$$

$$= \sqrt{6^2 + 6^2 - 2 \times 6 \times 6 \times 0.5} = \sqrt{36} = \mathbf{6\ m/s}$$

This is also apparent from the vector diagram which is an equilateral triangle.

Alternatively the problem can be worked out by the instantaneous centre method.

The velocity V_a at end *A* is horizontal and the velocity V_b at end *B* is vertical. The instantaneous centre *I* of the bar *AB* is the point of intersection of perpendicular to V_a and V_b through *A* and *B* respectively

$$IA = BO = AB \sin 30° = 2 \sin 30° = 1 \text{ m}$$

$$IB = AO = AB \cos 30° = 2 \cos 30° = 1.732 \text{ m/s}$$

Then the angular velocity of the bar is given by

$$\omega_{ab} = \frac{V_a}{IA} = \frac{6}{1} = 6 \text{ rad/s}$$

(*i*) Velocity of end $B = \omega \times IB = 6 \times 1.732$

$= \mathbf{10.392\ m/s}$

(*ii*) In ΔIAC; $AC = 1$ m; $IA = 1$ m (calculated above) and $\angle IAC = 60°$

Obviously the triangle *IAC* is an equilibrium triangle. That gives

$$IC = IA = 1 \text{ m}$$

$\therefore$ Velocity of mid point $V_c = \omega \times IC = 6 \times 1 = \mathbf{6\ m/s}$

EXAMPLE 9.22.

The ends *A* and *B* of line 5 m long are restrained to move in vertical and horizontal guides as shown in Fig. 9.23. The end *A* moves 2.5 m/s upward when it is 3 m above *C*. Determine:

(*i*) angular velocity of link *AB*,

(*ii*) velocity of end *B*, and

(*iii*) relative velocity of end *B* with respect to end *A*.

Solution: Velocity of end A, $V_a = 2.5$ m/s

Velocity of end B, $V_b = V_a + V_{ba}$

where V_{ba} is the relative velocity of end B with respect to end A and it is due to rotation of link about A and it equals ωl.

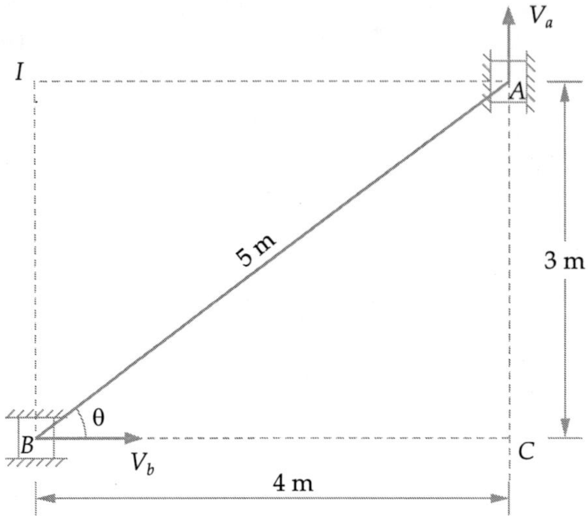

Fig. 9.23

From the vector diagram,

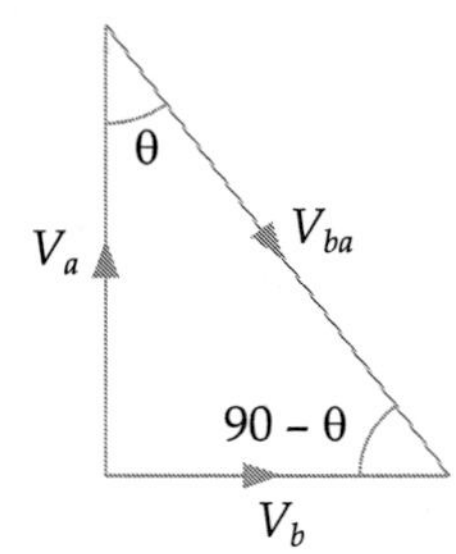

$$\frac{V_a}{\sin(90-\theta)} = \frac{V_{ba}}{\sin 90} = \frac{V_b}{\sin\theta}$$

or
$$\frac{V_a}{\cos\theta} = \frac{V_{ba}}{\sin 90°} = \frac{V_b}{\sin\theta}$$

From the given geometry: $\sin\theta = \frac{3}{5} = 0.6$ and $\cos\theta = \frac{4}{5} = 0.8$

Then: (*i*) $V_b = \frac{V_a}{\cos\theta}\times\sin\theta = \frac{2.5}{0.8}\times 0.6 = $ **1.875 m/s**

(*ii*) $V_{ba} = \frac{V_a}{\cos\theta}\times\sin 90° = \frac{2.5}{0.8}\times 1 = $ **3.125 m/s**

(*iii*) Angular velocity of bar,

$$\omega_{ab} = \frac{V_{ba}}{l} = \frac{3.125}{5} = 0.625 \text{ rad/s}$$

Alternatively, the problem can be worked out by instantaneous centre method

The velocity at end A is horizontal and that at end B is vertical. The point of intersection of the normals to these velocities V_a and V_b locates the instantaneous centre I.

$$IA = 4\text{ m} \quad\text{and}\quad IB = 3\text{ m}$$

Then the angular velocity of the bar is given by

$$\omega_{ab} = \frac{V_a}{IA} = \frac{2.5}{4} = 0.625 \text{ rad/s}$$

Velocity of end $B = \omega \times IB = 0.625 \times 3 = 1.875$ m/s

Velocity of end B relative to end A,

$$V_{ba} = \omega \times l = 0.625 \times 5 = 3.125 \text{ m/s}$$

EXAMPLE 9.23.

A compound wheel of configuration illustrated in Fig. 9.24 rolls without slipping on a horizontal surface *PQ*. If the velocity at the centre of wheel is 1 m/s, determine the velocities at the points *A*, *B* and *D* indicated in the figure.

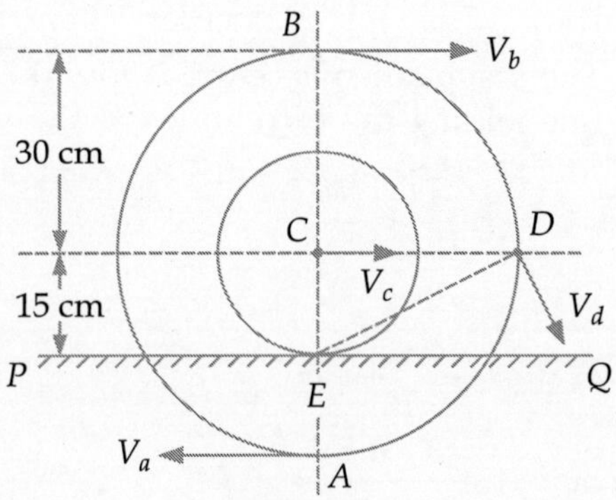

Fig. 9.24

Solution : When the wheel rolls without slipping, its point of contact with the horizontal surface at any instant has zero velocity. Thus the point E is instantaneously at rest and every other point on the wheel has angular velocity about E. Hence E is the instantaneous centre of the wheel. Then

$$V_c = \omega \times CE\,; \quad V_a = \omega \times AE$$

$$V_b = \omega \times BE \quad \text{and} \quad V_d = \omega \times DE$$

Angular velocity of the wheel $\omega = \dfrac{V_c}{CE} = \dfrac{1}{0.15} = 6.67$ rad/s

Velocity of point A : $V_a = \omega \times AE = 6.67 \times 0.15 = 1$ m/s

Velocity of point B : $V_b = \omega \times BE = 6.67 \times 0.30 = 0.45 = 3$ m/s

$$\text{Distance } DE = \sqrt{(CE)^2 + (CD)^2} = \sqrt{(0.15)^2 + (0.3)^2} = 0.364 \text{ m}$$

$\therefore$ Velocity of point $D : V_d = \omega \times DE = 6.67 \times 0.364 =$ **2.43 m/s**

The directions of these velocities at the given points would be as indicated in the figure.

EXAMPLE 9.24.

A cylindrical roller, 50 cm in diameter, is in contact with two horizontal conveyor belts running at uniform speeds of 5 m/s and 3 m/s as shown in Fig 9.25. Assuming that there is no slip at the points of contact, determine

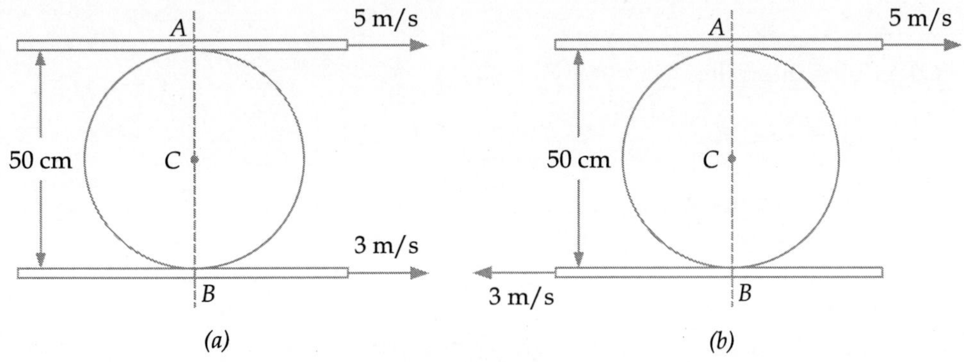

Fig. 9.25

(i) the position of the instantaneous centre of the roller,
(ii) the linear velocity of the centre C, and
(iii) the angular velocity of the roller.

(b) How these parameters would be affected if the velocities of the belts are in opposite direction (Fig. 9.25 b)

Solution : Case (a) : Velocities V_a and V_b are in the same direction.

The instantaneous centre is found by determining the point of intersection of line *AB* with the line joining the extremities of the velocity vectors (Fig. 9.26), then

$$V_a = \omega_{ab} \times IA\,; \quad V_b = \omega_{ab} \times IB$$

or $$\frac{IA}{IB} = \frac{V_a}{V_b} = \frac{5}{3}$$

or $$\frac{IB + AB}{IB} = \frac{5}{3}$$

or $$\frac{IB + 0.5}{IB} = \frac{5}{3}\,;$$

$$IB = 0.75 \text{ m} = 75 \text{ cm}$$

$$IA = IB + AB = 75 + 50 = 125 \text{ cm}$$

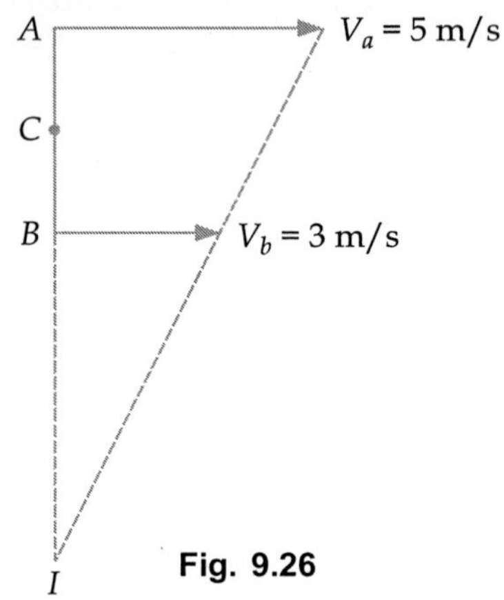

Fig. 9.26

Angular velocity of the roller = angular velocity of *A* about *I*

$$\omega = \frac{V_a}{IA} = \frac{5}{1.25} = \mathbf{4\ rad/s}$$

Linear velocity of the centre of roller,

$$V_c = \omega \times IC = 4 \times (1.25 - 0.25) = \mathbf{4\ m/s}$$

(b) When velocities V_a and V_b are in opposite directions, *I* lies between *A* and *B* (Fig. 9.27), and divides *AB* internally in the ratio V_a to V_b. That is,

$$\frac{IA}{IB} = \frac{V_a}{V_b} = \frac{5}{3}$$

or $$\frac{0.5 - IB}{IB} = \frac{5}{3}\,;$$

$$IB = 0.1875 \text{ m} = 18.75 \text{ cm}$$

$$IA = 50 - 18.75 = 31.25 \text{ cm}$$

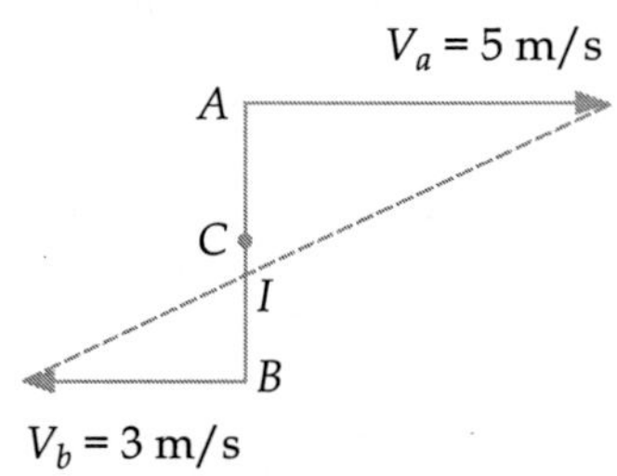

Fig. 9.27

Angular velocity of the roller

$$\omega = \frac{V_a}{IA} = \frac{5}{0.3125} = \mathbf{16\ rad/s}$$

Linear velocity of the centre of roller,

$$V_c = \omega \times IC = \omega \times (IA - AC) = 16 \times (0.3125 - 0.25) = 1 \text{ m/s}$$

REVIEW AND SUMMARY

1. In a circular (rotary) motion, the movement of the particle is along a circular path. The particle repeats its journey along the same circular path about the centre of rotation which remains fixed.

   The shafts, pulleys and flywheels etc. undergo circular motion when they rotate about their geometric axis.

2. The rate of change of angular displacement of a body is called angular velocity.

$$\omega = \frac{d\theta}{dt}$$

If the body turns N revolutions per minute, then

$$\omega = \frac{2\,\pi N}{60}\ \text{rad/s}$$

The linear velocity V and the angular velocity ω are related by the expression: $V = \omega\, r$

3. The rate of change of angular velocity of a body is called angular acceleration.

$$\alpha = \frac{d\omega}{dt} = \frac{d}{dt}\left(\frac{d\theta}{dt}\right) = \frac{d^2\theta}{dt^2}$$

The linear acceleration a and the angular acceleration α are related by the expression: $a = \alpha\, r$.

4. When a body moves along a curved path, the linear velocity is referred to as tangential velocity. Further, the body undergoes tangential and normal components of acceleration given by

$$a_t = \alpha\, r \quad \text{and} \quad a_n = \frac{v^2}{r} = \omega^2 r$$

The tangential acceleration is due to change in magnitude of velocity and the normal acceleration is due to change in direction of the body.

5. When a body moves in a circular path with uniform acceleration, the equations of motion are :

$$\omega = \omega_0 + \alpha t\,;\ \omega^2 - \omega_0^{\,2} = 2\alpha\theta$$

and $$\theta = \omega_0 t + \alpha t^2$$

where ω_0 is the initial angular velocity, ω is the final angular velocity, α is angular acceleration of the body and θ represents the angle moved in time t.

6. A body is said to have motion of translation when the particles constituting the body move in parallel planes and travel the same distance.

7. A body is said to have motion of rotation when the body moves about a fixed point and all the particles constituting the body move in a circular path. The fixed point about which the body rotates is called the point of rotation and the axis passing through the point of rotation is called the axis of rotation. A point lying on the axis of rotation has a zero velocity and zero acceleration.

8. The combination of rotation and translation motion is usually referred to as general plane motion. The body can then be assumed to be rotating about a certain point which is known as instantaneous centre of rotation.

9. *Chasle's theorem* states that any general displacement of a rigid body can be represented by a combination of translatory motion and rotational motion.

REVIEW QUESTIONS

A. Conceptual and conventional questions :

1. State the difference between curvilinear motion and circular (rotary motion)
2. Define the following terms associated with circular motion :
angular displacement, angular velocity and angular acceleration

3. Set up the following relation between linear acceleration (a) and angular acceleration (α)

$$a = \alpha\, r$$

where r is the distance of the body from the centre of rotation.

4. Explain the concept of general plane motion. List some examples of such a motion.
5. State and explain Chasle's theorem.
6. Define the term instantaneous centre or virtual centre in plane motion of rigid bodies.
7. Set up a relation for the velocity of position of a reciprocating heat engine in terms of angular speed of crank, crank angle and ratio of lengths of connecting rod and that of crank
8. A flywheel starting from rest rotates at uniform angular acceleration, and after having rotated 400 revolutions its angular velocity is 600 rpm. Find the angular acceleration and the interval of time to complete 400 revolutions. **(0.785 rad/s^2, 80 sec)**
9. The speed of a flywheel increase from 300 to 600 rev/min in 10 seconds. If the diameter of the wheel is 2 m, determine the angular acceleration and number of revolutions made during this period of 10 seconds. Also workout the normal and tangential acceleration at the rim of the wheel at the end of 10 second. **(0.5 rev/s^2, 75 rev, 394 m/s^2, 3.14 m/s^2)**
10. In a crank and connecting rod mechanism, the lengths of crank and connecting rod are 15 cm and 65 cm respectively. The crank turns 310 rev/min in the clockwise direction. What will be the velocity of the piston when the crank angle is 60° with respect to the inner dead centre position **(4.7 m/s)**
11. A roller of radius 50 mm rides between two horizontal bars moving in opposite directions as shown in Fig. 9.28 Assuming no slip at point of contacts A and B, workout the distance a that defines the position of the path of the instantaneous centre of rotation of the roller. **(42.86 mm)**

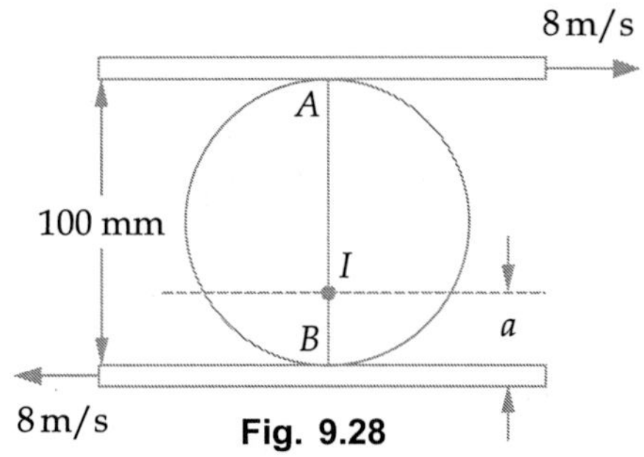

Fig. 9.28

12. A horizontal rod AB is rotating about the vertical axis through the end A at a uniform angular velocity. The linear velocity at the end B is 10 m/s and the linear speed of a point C, 2 metres from A, is 5 m/s. Find the length of the rod and its angular velocity.

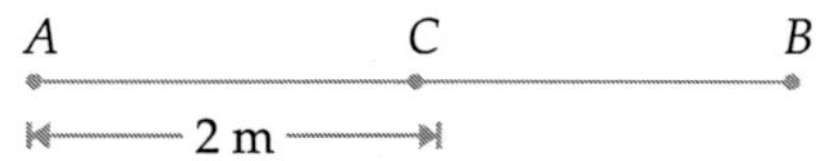

Hint: $V_b = \omega \times AB = 10$

$V_c = \omega \times AC = 5$

$$\therefore \quad \omega = \frac{5}{AC} = \frac{5}{2} = \textbf{2.5 rad/s} \quad \text{and } AB = \frac{10}{\omega} = \frac{10}{2.5} = \textbf{4 m}$$

B. Complete the following statements with most appropriate word/words :

1. The shafts, pulleys and flywheels etc. undergo .... motion when they rotate about their .... axis.
2. The linear acceleration a, the angular acceleration α and the distance r of the body from the centre of rotation are correlated by the identity ....
3. The tangential accelertion is due to change in .... of velocity and the normal acceleration is due to change in .... of the body.
4. A body is said to undergo .... when the particles constituting the rigid body move in parallel planes and travel the same distance.

5. The combination of rotation and translation motion of a body is referred to as .... motion.
6. A rod sliding against a wall at one end and floor at the other end represents a combination of .... motion.
7. .... of the body represents the point about which the plane motion of all the particles constituting the body can be considered as pure rotation.
8. In the absense of external torque, the angular momentum is ....
9. The direction of velocity vector at any point on the circular orbit is .... to the circle at that point.
10. When a bar rotates in horizontal plane, the .... velocity at any point on the bar depends upon its distance from the centre of rotation.

Answers :

1. circular, geometric 2. $a = \alpha r$ 3. magnitude, direction 4. motion of translation 5. general plane motion 6. rotation and translation motion 7. instantaneous centre 8. conserved 9. tangential 10. linear

C. Multiple choice questions :

1. A body moving around a fixed axis constitutes :
   (*a*) curvilinear motion (*b*) circular motion
   (*c*) plane motion (*d*) simultaneous translation and rotation
2. Which of the following units do not conform to the concept of rigid body rotation ?
   (*a*) blades of a fan (*b*) pulley of a belt
   (*c*) phonograph records (*d*) door revolving about a hinge
   (*e*) wheels of an automobile
3. With reference to the adjoining figure, the person A is standing at the centre of a rotating platform facing person B who is riding a bicycle; heading east. The relevant speeds and distances are as given in the figure

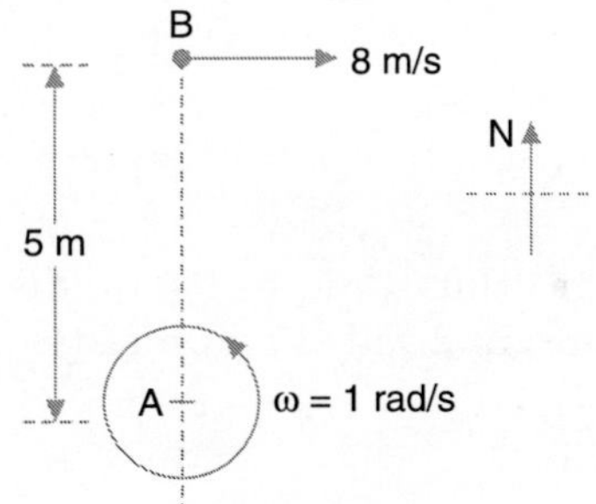

Fig. 9.29.

   At the instant under consideration, the apparent velocity of B as seen by A would be
   (*a*) 3 m/s heading east (*b*) 3 m/s heading west
   (*c*) 8 m/s heading east (*d*) 13 m/s heading east
4. The angular speed of seconds hand of a clock is
   (*a*) π rad/s (*b*) π/6 rad/s
   (*c*) π/15 rad/s (*d*) π/30 rad/s
5. Upon rotation from initially rest position, the radius vector of a body executes angular rotation θ prescribed by the relation $\theta = 4t^3 - 3t^2 + 2t + 6$

The angular acceleration of the body at time t = 2 sec is
(*a*) 38 rad/s^2 (*b*) 42 rad/s^2
(*c*) 52 rad/s^2 (*d*) 30 rad/s^2

6. A stone tied to the end of a string 100 cm long is whirled in a horizontal circle with a constant speed. If the stone makes 15 revolutions in 30 seconds, the acceleration of the stone is closest to
(*a*) 315 cm/s^2 (*b*) 630 cm/s^2
(*c*) 985 cm/s^2 (*d*) 3950 cm/s^2

7. The shaft of a motor starts from rest and attains full speed of 1800 rpm in 10 seconds. The shaft has an angular acceleration of
(*a*) 3 π rad/s^2 (*b*) 6 π rad/s^2
(*c*) 12 π rad/s^2 (*d*) 24 π rad/s^2

8. Two cars are going with constant speeds, round concentric circles of radii r_1 and r_2, and take the same time to complete their circular paths. Their angular speeds will correspond to the ratio
(*a*) $\sqrt{r_1 / r_2}$ (*b*) r_1 / r_2
(*c*) $(r_1 / r_2)^2$ (*d*) 1

9. The speed of a particle moving in a circle of radius 10 cm changes from 1.5π rad/s to 5π rad/s in 2.2 second time. The corresponding linear acceleration of the particle is
(*a*) 25 cm/s^2 (*b*) 32.5 cm/s^2
(*c*) 50 cm/s^2 (*d*) 65 cm/s^2

10. Starting from rest, a particle travels on a circular path and the distance covered is prescribed by the relation $s = kt^2$; where k is constant and t is the time. The particle then has a tangential acceleration of
(*a*) $k/2$ (*b*) k
(*c*) $2k$ (*d*) $4k$

11. An aircraft executes a horizontal loop of radius 1 km with a steady speed of 900 km/hr. In terms of acceleration due to gravity (g = 10 m/s^2), the centripetal acceleration of aircraft would be
(*a*) 3.12 g (*b*) 6.25 g
(*c*) 12.5 g (*d*) 18.75 g

12. For a particle moving in a circular orbit of radius 0.4 m, the angular velocity and angular acceleration at a particular instant are 2 rad/s and 5 rad/s^2. The particle then has a total linear acceleration of
(*a*) 1.9 m/s^2 (*b*) 2.69 m/s^2
(*c*) 3.8 m/s^2 (*d*) 7.24 m/s^2

Answers :

1. (*b*)	2. (*d*) and (*e*)	3. (*d*)	4. (*d*)	5. (*b*)	6. (*c*)
7. (*b*)	8. (*d*)	9. (*c*)	10. (*c*)	11. (*b*)	12. (*b*)

UNIT - V

Chapter-10

Kinetics: Impulse, Momentum, Work and Energy

Chapter-11

Kinetics of Rotary Motion

Chapter-12

Virtual Work

CHAPTER

10

KINETICS: IMPULSE, MOMENTUM, WORK AND ENERGY

Learning objectives : After a study of the subject matter included in this chapter, the reader will become familiar with the following aspects of kinetics

- *impulse, momentum, work and energy*
- *theory of motion of a lift and motion of connected bodies*
- *dynamic equilibrium and D'Alembert's principle*
- *conservative force and principle of conservation of energy ; work-energy principle*
- *work of force by gravity and by a spring*

10.1. FORCE, IMPULSE AND MOMENTUM

Force is an external agent which tends to change the state of rest or of uniform motion of a system. A force is applied whenever the system needs to be accelerated or decelerated. Force is a vector quantity specified completely by its

- magnitude
- point of application
- line of action, and
- direction.

The relationship between motion and force is provided by laws of dynamics; the most prominent being the Newton's second law. Newton's second law of motion states that force is proportional to rate of change of momentum. That is

$$\text{force} \propto \text{rate of change of momentum}$$

Momentum is the product of mass and velocity of a body and represents the energy of motion stored in a moving body.

$$\begin{aligned} \therefore \text{ force} &\propto \text{rate of change of (mass} \times \text{velocity)} \\ &\propto \text{mass} \times \text{rate of change of velocity} \\ &\propto \text{mass} \times \text{acceleration} \end{aligned}$$

Thus $$F \propto ma ; \quad F = \frac{ma}{g_c}$$

where m is the mass, a is acceleration and g_c is proportionality constant.

The SI unit of force is newton (N) which represents the force required to accelerate 1 kg mass with an acceleration of 1 m/s^2.

Thus $\quad 1\text{ N} = \dfrac{1\text{ kg} \times 1\text{ m/s}^2}{g_c} \quad$ and $\quad g_c = \dfrac{1\text{ kg m}}{\text{Ns}^2}$

$$F \xrightarrow{\quad m \quad} a \equiv 1\text{N} \xrightarrow{\quad 1\text{ kg} \quad} 1\text{ m/s}^2$$

Fig. 10.1

The constant of proportionality g_c becomes unity for defining unit of force as Newton and accordingly is dropped. Such a practice corresponds to writing Newton's second law as

Force = mass × acceleration ; $F = ma$

$$\therefore \quad 1\text{ N} = 1\ \frac{1\text{ kg m}}{s^2}$$

Other units of force are:

1 kilo newton (kN) = 10^3 N

1 mega newton (mN) = 10^6 N

1 giga newton (gN) = 10^9 N

Force is also measured in gravitational units expressed as kilogram force (kg_f). A kg_f is defined as that force which when applied to a body of 1 kg mass produces an acceleration equal to 9.81 m/s^2 (gravitational acceleration near the earth's surface).

$$1\text{ kg}_\text{f} = 1\text{ kg} \times 9.81\text{ m/s}^2 = 9.81\text{ kgm/s}^2$$

The newton and kilogram force are connected by the relation:

$$1\text{ kg}_\text{f} = 9.81\text{ N}$$

The identity $F = ma$ may be written as

$$F = m\frac{dV}{dt} = \frac{d}{dt}(mV) = \frac{mV_f - mV_i}{t} \qquad \text{...(10.1)}$$

where V_f and V_i are the final and initial velocities respectively.

The product of mass and velocity (mv) is called momentum of the body. Therefore

mV_i = momentum of the body at the beginning

mV_f = momentum of the body at the end of time t.

Thus "*the change in linear momentum per unit time is proportional to the impressed force and takes place in the direction of the force.*"

Further, $\quad F \times t = mV_f - mV_i = m\,(V_f - V_i) \qquad$ ...(10.2)

The product of force and the time during which it acts ($F \times t$) is called the **impulse** of force. Thus the impulse equals the change in momentum.

Equation of 10.1 is referred to as the impulse-momentum relation.

Momentum and impulse are vector quantities with unit kg m/s or Ns.

The concept of impulse is useful in solving problems when a large force acts for a very short time, e.g., sudden blow of hammer striking a nail.

The *law of conservation of momentum* may be stated as:

"Total momentum of any group of objects always remains the same if no external force acts on it."

Consider that a body A of mass m_1 moving with velocity u_1 collides with another body B of mass m_2 and moving with velocity u_2. Let v_1 and v_2 be their velocities after the collision. Then

Momentum of masses before collision = $m_1u_1 + m_2u_2$

Momentum of masses after collision = $m_1v_1 + m_2v_2$

In accordance with law of conservation of momentum,

momentum before collision = momentum after collision

$$m_1u_1 + m_2u_2 = m_1v_1 + m_2v_2$$

EXAMPLE 10.1.

Determine the force necessary to produce an acceleration of 4 m/s² in a mass of 250 kg.

Solution: From Newton's second law, $F = ma$

$$\therefore \quad F = 250 \times 4 = \mathbf{1000\ N} \text{ (SI units)}$$

$$\text{or} \quad F = \frac{250 \times 4}{9.81} = \mathbf{101.94\ kg_f} \text{ (Gravitational units)}$$

EXAMPLE 10.2.

A body of mass 160 kg is made to move with a velocity of 20 m/s when a force of 80 N acts on it for 60 seconds. Make calculations for the velocity attained by the body when the

(*a*) force acts in the direction of motion

(*b*) force acts in the direction opposite to that of motion.

Solution: Using the relation $F = ma$,

$$\text{Acceleration/retardation of body } a = \frac{F}{m} = \frac{80}{160} = 0.5 \text{ m/s}^2$$

(*a*) When the force acts in the direction of motion, the body will have acceleration. The final velocity v is then given by

$$v = u + at = 20 + 0.5 \times 60 = \mathbf{50\ m/s}$$

(*b*) When the force acts in a direction opposite to that of motion, the body will have retardation. The final velocity v is then given by

$$v = u - at \qquad \text{(– ve sign shows retardation)}$$
$$= 20 - 0.5 \times 60 = \mathbf{-10\ m/s}$$

The – ve sign implies that the body will be moving in the opposite (reversed) direction.

EXAMPLE 10.3.

A body of mass 50 kg rests on a rough horizontal surface (μ = 0.4) and is acted upon by a push applied at an angle of 45° to the horizontal. Determine the magnitude of push if it causes the body to move with an acceleration of 2 m/s².

Solution: Refer Fig. 10.2 for the various force acting on the body

Normal reaction $R = W + P \sin 45°$

$= (50 \times 9.81) + P \sin 45°$

$= 0.707\,P + 490.5$

Force of friction $F = \mu R = 0.4\,(0.707\,P + 490.5)$

$= 0.2828\,P + 196.2$

Unbalanced force causing motion

$= P \cos 45° - (0.2828\,P + 196.2)$

$= 0.707\,P - 0.2828\,P - 196.2 = 0.4242\,P - 196.2$

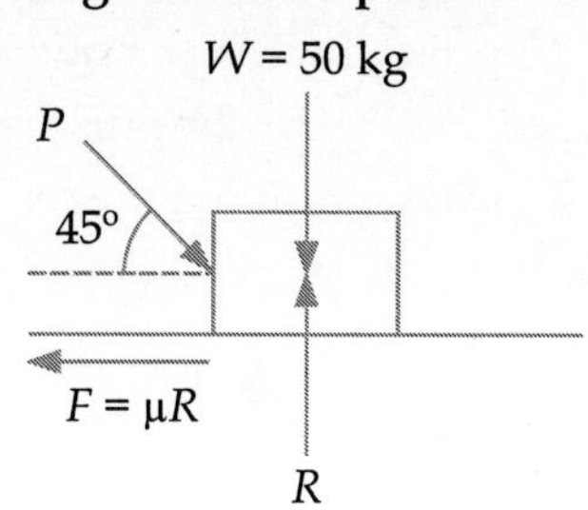

Fig. 10.2

Applying Newton's second law of motion $F = ma$, we have

$$0.4242\,P - 196.2 = 50 \times 2 \quad \therefore \quad \text{Push } P = \frac{100 + 1962.2}{0.4242} = \mathbf{698.25\ N}$$

EXAMPLE 10.4.

A cricket ball of mass 175 gm is moving with a speed of 36 km/hr. What average force will be required to stop the ball in 0.2 second.

Solution: Initial speed of the ball $u = 36 \text{ km/hr} = \dfrac{36 \times 1000}{3600} = 10 \text{ m/s}$

From the impulse-momentum relation,

$$F \times 5 = m\,(v - u)$$

$$F \times 0.2 = \frac{175}{1000}(0 - 10)\ ; \quad F = \mathbf{-\,8.75\ N}$$

The negative sign indicates that the force F is to be applied in a direction opposite to that in which the ball was moving.

EXAMPLE 10.5.

A machine gun fires 180 bullets per minute. Each bullet has a mass of 50 gm and the muzzle velocity of the bullet is 600 m/s. Presuming the firing to be continuous, make calculations for the average reaction of the gun against the support.

Solution: Time taken in firing one bullet $t = \dfrac{60}{180} = 0.333 \text{ s}$

From the impulse–momentum relation,

$$F \times t = m\,(v - u)$$

$$F \times 0.333 = \frac{50}{1000}\,(0 - 600)\ ; \quad F = \mathbf{-\,90.1\ N}$$

The negative sign indicates that this is the reaction of the gun on the bullet. The average reaction of the gun against the support will be numerically equal to it.

EXAMPLE 10.6.

A pile hammer of 250 kg mass is made to fall freely on a pile from a height of 6 m. If the hammer comes to rest in 0.012 second, determine the change in momentum, impulse and average force.

Solution: Since the initial velocity $u = 0$, the velocity with which the hammer falls on the pile is

$$v = \sqrt{2gh} = \sqrt{2 \times 9.81 \times 6} = 10.85 \text{ m/s}$$

Change in momentum = mass × change in velocity = 250 × (10.85 – 0) = **2712.5 Ns**

Impulse equals the change in momentum

∴ Impulse = **2712.5 Ns**

Further, impulse represents the product of force and time during which it acts. Then

$$F \times t = 2712.5$$

$$\therefore \quad \text{Average force } F = \frac{2712.5}{0.012} = 226042 \text{ N} \simeq \mathbf{226.04\ kN}$$

EXAMPLE 10.7.

A glass marble whose weight is 0.2 N falls from a height of 12 m and rebounds to a height of 9 m. The marble and the floor remain in contact for 0.1 second. Determine the impulse and the average force between the marble and the floor.

Solution: Velocity with which the marble strikes the floor

$$= \sqrt{2gh} = \sqrt{2 \times 9.81 \times 12} = 15.34 \text{ m/s (downward)}$$

Velocity of rebound $= \sqrt{2 \times 9.81 \times 9} = 13.29$ m/s (upward)

Impulse = change of momentum = mass × change of velocity

$$= \frac{0.2}{9.81} \times [13.29 - (-15.34)] = \mathbf{0.5837\ Ns}$$

Impulse represents the product of force and the time during which it acts. Then

$$F \times t = 0.5837$$

$$\therefore \text{ Average force } F = \frac{0.5837}{0.1} = \mathbf{5.837\ N}$$

EXAMPLE 10.8.

A man throws a 10 kg suitcase with a horizontal velocity of 4 m/s into a 25 kg platform trolly. Determine the velocity of trolly after the suitcase has slid to stop on trolly.

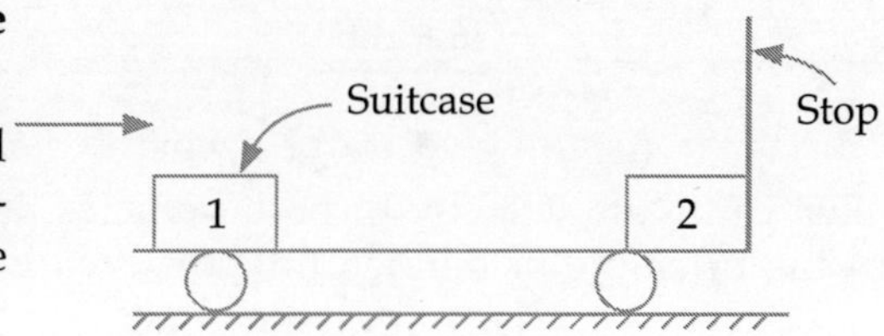

Fig. 10.3

Solution: Let the suffix s and t refer to suitcase and trolly respectively. Applying the principle of conservation of momentum to the trolly-suitcase system, we have

$$m_s (V_s)_1 + m_t (V_t)_1 = m_s (V_s)_2 + m_t (V_t)_2$$
$$= (m_s + m_t) V$$

where V is the common velocity with which the trolly and suitcase move together after the impact.

Substituting the given values, we obtain

$$10 \times 4 + 25 \times 0 = (10 + 25) V$$

$$\therefore \quad V = \frac{40}{35} = \mathbf{1.143\ m/s}$$

EXAMPLE 10.9.

A 150 kg scootorist is travelling with a speed of 36 km/hr on a road that makes an angle of 30° with another road upon which a 60 kg cyclist is travelling at 8 km/hr. When they approach the crossing, they collide and move as one mass. Determine the final velocity; both in magnitude and direction.

Solution: Let suffix 1 and 2 refer to scootorist and cyclist respectively.

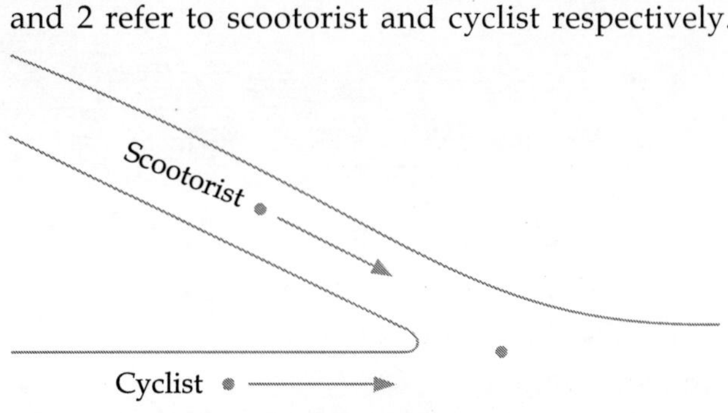

Fig. 10.4

Impact in x-direction:

$(u_1)_x$ = initial velocity of scootorist in x-direction before impact

$$= \frac{36 \times 1000}{3600} \cos 30^\circ = 8.66 \text{ m/s}$$

$(u_2)_x$ = velocity of cyclist in x-direction before impact

$$= \frac{8 \times 1000}{3600} = 2.22 \text{ m/s}.$$

After the impact both will move with the same velocity. Let V_x be the final velocity in x-direction after the impact. Then from the principle of conservation of momentum,

$$m_1 u_1 + m_2 u_2 = (m_1 + m_2) V_x$$

$$150 \times 8.66 + 60 \times 2.22 = (150 + 60) V_x; \quad V_x = 6.82 \text{ m/s}$$

Impact in y-direction:

$(u_1)_y$ = velocity of scootorist in y-direction before impact

$$= \frac{36 \times 1000}{3600} \times \sin 30^\circ = 5 \text{ m/s} (\downarrow)$$

$(u_2)_y$ = velocity of cyclist in y-direction before impact = 0

Let V_y be the velocity of both vehicles after the impact. Then from the principle of momentum conservation,

$$150 \times 5 + 60 \times 0 = (150 + 60) V_y$$

$$V_y = 3.57 \text{ m/s} (\downarrow)$$

Resultant velocity after impact V

$$= \sqrt{V_x^2 + V_y^2} = \sqrt{6.82^2 + 3.57^2} = \mathbf{7.69 \text{ m/s}}$$

and its inclination with the horizontal is

$$\theta = \tan^{-1} \frac{3.57}{6.82} = \mathbf{27.7^\circ}$$

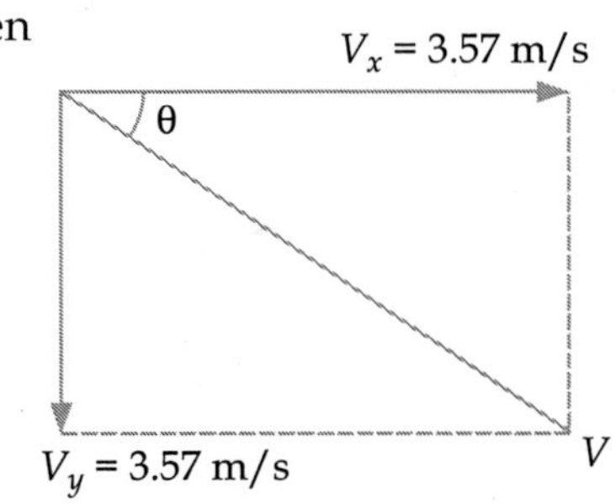

Fig. 10.5

EXAMPLE 10.10.

A block weighing 1.5 kN is initially at rest on a rough horizontal surface (μ = 0.2). It is acted upon by a force *P* which varies with time as shown in Fig. 10.6. Determine the velocity at the end of (*i*) 3 seconds (*ii*) 6 seconds.

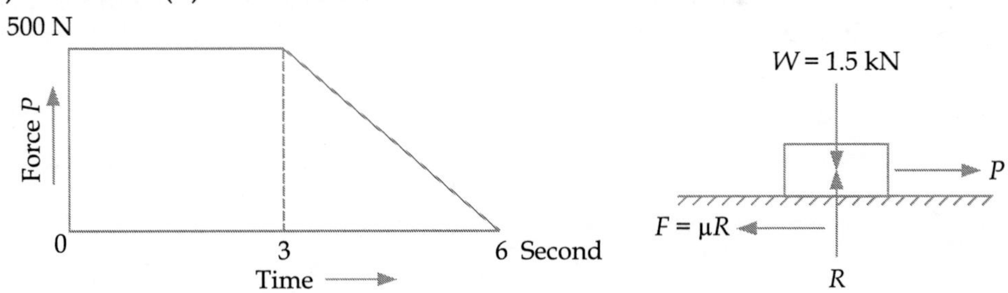

Fig. 10.6

Solution: Frictional force $F = \mu R = \mu W = 0.2 \times 1.5 = 0.3 \text{ kN} = 300 \text{ N}$

Case (*i*) From the impulse-momentum relation,

impulse = change in momentum

or $\quad F \times t = m \times dV$

or $\quad (500 - 300) \times 3 = \dfrac{1.5 \times 10^3}{9.81} (V_3 - 0); \quad V_3 = \mathbf{3.924 \text{ m/s}}$

Thus the block will have a velocity of 3.924 m/s at the end of 3 seconds.

Case (*ii*) Impulse during 3rd and 6th second

$$= \text{area of triangular portion of force-time diagram}$$

$$= \frac{1}{2} \times 500 \times (6 - 3) = 750 \text{ Ns}$$

$$\text{Impulse due to friction} = 300 \times (6 - 3) = 900 \text{ Ns}$$

$$\text{Net impulse} = 750 - 900 = -150 \text{ Ns}$$

Again using the relation,

$$\text{impulse} = \text{change in momentum}$$

$$-150 = \frac{1.5 \times 10^3}{9.81}(V_6 - V_3); \quad V_6 - V_3 = -0.981$$

∴ Velocity of block at the end of 6 seconds,

$$V_6 = V_3 - 0.981 = 3.924 - 0.981 = \mathbf{2.943 \text{ m/s}}$$

EXAMPLE 10.11.

An electric train weighing 10 MN is driven by electric motors which provide a tractive effort equal to 250 kN. Calculate the time taken to accelerate the train to a speed of 80 km/hr. It may be assumed that resistance to motion is 1 percent of the weight of train.

Solution: Speed to be attained $v = 80 \text{ km/hr} = \frac{80 \times 1000}{3600} = 22.22 \text{ m/s}$

Resistance to motion $R = 0.01 \times (10 \times 10^6) = 100 \times 10^3$ N

Tractive effort $P = 250 \text{ kN} = 250 \times 10^3$ N

Force available for acceleration $F = P - R = 150 \times 10^3$ N

Then from Newton's second law $F = ma$, we have

$$150 \times 10^3 = \frac{10 \times 10^6}{9.81} a; \quad \text{acceleration } a = 0.14 \text{ m/s}^2$$

Using kinematic relation $v = u + at$,

Time taken to accelerate the train from rest to 22.22 m/s velocity,

$$t = \frac{v - u}{a} = \frac{22.22 - 0}{0.14} = \mathbf{158.71 \text{ second}}$$

EXAMPLE 10.12.

An automobile is travelling along a straight level highway. When the brakes are applied, the vehicle slides for 2 seconds and covers a distance of 10 m before coming to rest. Assuming that the automobile moves with constant deceleration during this period, determine the coefficient of friction between the tyres and the road.

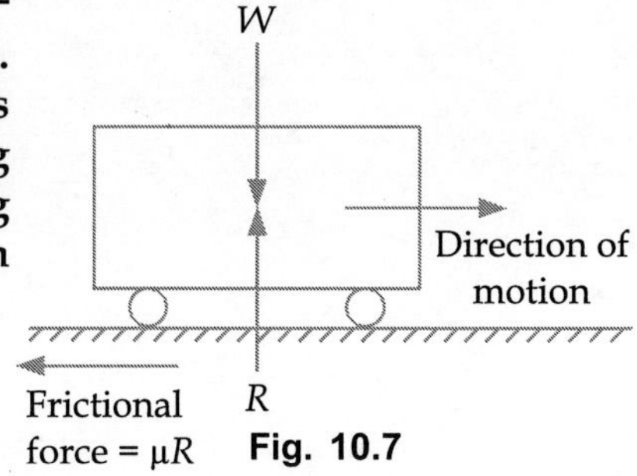

Fig. 10.7

Solution: From the kinematic relations

$v = u + at$ and $s = ut + \frac{1}{2}at^2$, we have

$0 = u + 2a; \quad u = -2a$ ...(*i*)

and $10 = u \times 2 + \frac{1}{2} a \times 2^2; \quad 24 + 2a = 10$ ...(*ii*)

From expressions (*i*) and (*ii*),

$a = -5 \text{ m/s}^2$ (retardation)

Let μ be the coefficient of friction between the tyres and the road. Then

Retarding force $= \mu W$

where W is the weight of automobile.

Applying Newton's second law of motion,

Retarding force = mass × deceleration

$$\mu W = \frac{W}{g} \times 5 \; ; \quad \mu = \frac{5}{9.81} = \mathbf{0.51}$$

EXAMPLE 10.13.

A bullet of mass 0.1 kg and travelling at a speed of 180 m/s penetrated 10 cm when fired into a wooden log. Determine the velocity with which this bullet would emerge when fired with the same velocity into a similar 5 cm thick wooden plank. Also determine the force of resistance assuming it to be uniform.

Solution: From the kinematic relation $v^2 - u^2 = 2as$, we have

$$0 - 180^2 = 2a \times 0.1 \; ; \quad a = -162000 \text{ m/s}^2 \text{ (retardation)}$$

Let the bullet emerge with velocity V from the 5 cm thick plank of wood,

$$V^2 - 180^2 = 2 \times (-162000) \times 0.05$$

The retardation has been taken same as the wooden planks are similar and the resistance offered by wood is uniform

$$V^2 = 32400 - 16200 = 16200 \; ; \quad V = \mathbf{127.28 \text{ m/s}}$$

Then from Newton's second law $F = ma$, we have

Force resistance $F_r = 0.1 \times 162000 = \mathbf{16200 \text{ N}}$

EXAMPLE 10.14.

A shell of mass 60 kg is fired horizontally with a velocity of 250 m/s by a gun of 3000 kg mass. Make calculations for (*a*) the velocity with which the gun recoils, (*b*) the uniform force required to stop the gun in 0.5 m distance, and (*c*) the time required to stop the gun.

It may be presumed that momentum of the system comprising the gun and the shell is conserved.

Solution: Refer Fig. 10.8 for the given system comprising the gun and the shell.

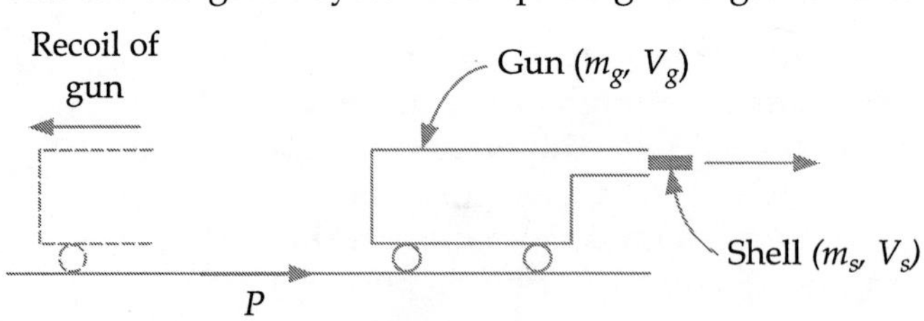

Fig. 10.8

Let suffix s and g denote the shell and gun respectively.

From the principle of conservation of momentum, we have

$$(m_s V_s + m_g V_g)_{\text{Before firing}} = (m_s V_s + m_g V_g)_{\text{After firing}}$$

Before firing, both the shell and the gun have zero velocity. As such initially the momentum is zero and therefore

$$m_s V_s + m_g V_g = 0 \quad \text{or} \quad m_s V_s = -m_g V_g$$

Obviously the velocities of the shell and the gun are in the opposite directions

$$60 \times 250 = -3000 \times V_g \,; \quad V_g = \mathbf{-5\ m/s^2}$$

(*b*) The gun moving with a velocity of 5 m/s² is brought to rest in a distance of 0.5 m. From the kinematic relation $v^2 - u^2 = 2as$, we have

$$a = \frac{v^2 - u^2}{2s} = \frac{0^2 - 5^2}{2 \times 0.5} = -25\ \text{m/s}^2 \quad \text{(retardation)}$$

From Newton's second law; $F = ma$, we have

Force required to stop the gun $= 3000 \times 25$

$$P = 75000\ \text{N} = \mathbf{75\ kN}$$

(*c*) From the kinematic relation $v = u + at$, we have

Time required to stop the gun $t = \dfrac{v - u}{a} = \dfrac{0-(-5)}{25} = \mathbf{0.2\ second}$

EXAMPLE 10.15.

A man weighing 800 N dives into a swimming pool from a height of 20 m. It is estimated that he goes down in water by 2 m and then starts rising. Neglecting air resistance, determine the average resistance offered by water to the motion of the man.

Solution: Let v be the velocity of the man when he reaches the water surface. Using the relation $v^2 - u^2 = 2\,gh$ and noting that $u = 0$, we have

$$v = \sqrt{2\,gh} = \sqrt{2 \times 9.81 \times 20} = 19.81\ \text{m/s}$$

When the man rises upward through a distance of 2 m inside the water, his velocity changes from 19.81 m/s to zero. This occurs because of retardation due to water resistance. From the kinematic relation $v^2 - u^2 = 2\,gh$, we get

$$0^2 - 19.81^2 = 2a \times 2$$

$$a = -\frac{19.81^2}{2 \times 2} = -98.11\ \text{m/s}^2 \quad \text{(retardation)}$$

Let F_r be the average resistance of water acting on the man in the upward direction. Then

Net force acting on the man in the upward direction

$$F_n = F_r - W \quad \text{where } W \text{ is the weight of the man}$$

Applying Newton's second law $F = ma$, we have

$$(F_r - 800) = \frac{800}{9.81} \times 98.11\,; \quad \mathbf{F_r = 8800\ N}$$

EXAMPLE 10.16.

A body of weight 100 N falls from a height of 12 m on a sand bed. It is estimated that the body penetrates 1.2 m into sand before coming to rest. Make calculations for the average thrust exerted by the sand on the body.

Solution: Velocity of body after falling from 12 m height,

$$v = \sqrt{2gh} = \sqrt{2 \times 9.81 \times 12} = 15.34\ \text{m/s}$$

After penetration in the sand, the velocity of the body becomes zero. Using the kinematic relation $v^2 - u^2 = 2as$, we have

$$0^2 - 15.34^2 = 2a \times 1.2$$

$$a = -\frac{15.34^2}{2 \times 1.2} = -98.05 \text{ m/s}^2 \quad \text{(retardation)}$$

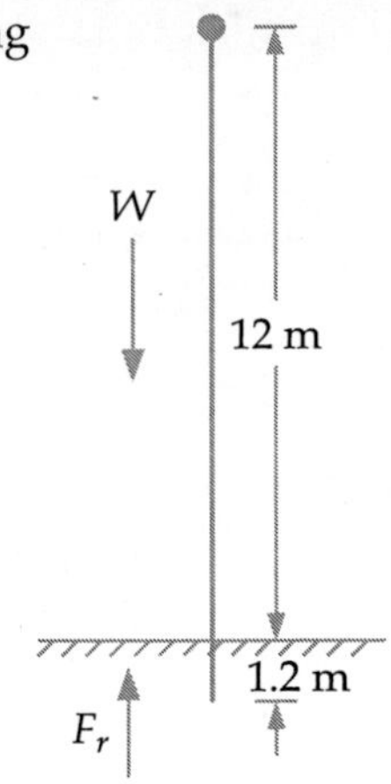

Fig. 10.9

Let F_r be the average resistance exerted by the sand on the body.

Then net force acting on the body in the upward direction,

$F_n = F_r - W$ where W is the weight of the body

Applying Newton's second law $F = ma$, we have

$$(F_r - 100) = \frac{100}{9.81} \times 98.05; \quad F_r = \mathbf{1100\ N}$$

EXAMPLE 10.17.

A train weighing 4000 kN has a frictional resistance of 5 N/kN of weight. Determine the steady pull which the locomotive must exert if the speed of the train is to be increased from 30 km/hr to 60 km/hr within a period of 1.5 minutes.

Solution: Initial velocity of train $u = 30 \text{ km/hr} = \frac{30 \times 1000}{3600} = 8.33 \text{ m/s}$

Final velocity $v = 60 \text{ km/hr} = \frac{60 \times 1000}{3600} = 16.66 \text{ m/s}$

Using the relation $v = u + at$, the acceleration of train is

$$\text{acceleration } a = \frac{v-u}{t} = \frac{16.66 - 8.33}{(2.5 \times 60)} = 0.092 \text{ m/s.}$$

Frictional resistance F_f = 5 N per kN of train weight = 5 × 4000 = 20000 N

The steady pull, *i.e.*, the tractive force F_t acts in the direction of motion, while the frictional resistance acts opposite to the direction of motion.

∴ Net force in the direction of motion

$$F_n = F_t - F_f = (F_t - 20000) \text{ N}$$

Then from Newton's second law $F = ma$, we have

$$(F_t - 20000) = \frac{4000 \times 10^3}{9.81} \times 0.092 = 37512$$

∴ Steady (average) pull to be exerted by the locomotive,

$$F_t = 37512 + 20000 = 57512 \text{ N} = \mathbf{57.512\ kN}$$

EXAMPLE 10.18

An automobile weighing 12 kN is moving with a uniform speed of 72 km/hr over a vertical curve *ABC* of parabolic shape as indicated in the figure given below:

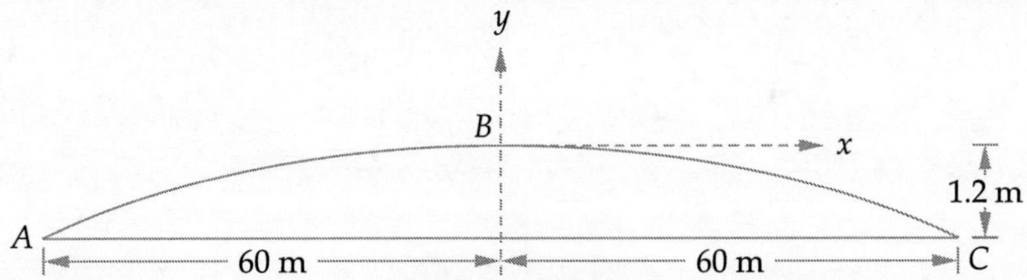

Fig. 10.10

Determine the total pressure exerted by the wheels of the automobile as it passes the topmost point *B*.

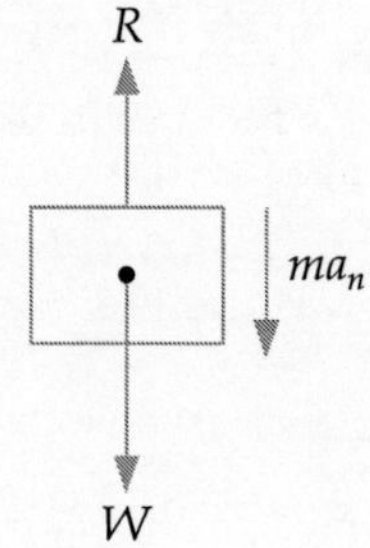

Solution: Let R be the reaction exerted by the wheels. Then

$$\text{Net force} = \text{mass} \times \text{acceleration}$$

$$W - R = m\,a_n$$

where a_n is the normal acceleration.

$$\therefore \quad R = W - m\,a_n = mg - m\frac{V^2}{\rho}$$

$$= mg\left[1 - \frac{V^2}{\rho g}\right] \qquad \text{...}(i)$$

$$\text{Velocity of automobile } V = \frac{72 \times 1000}{3600} = 20 \text{ m/s}$$

The radius of curvature ρ of the road can be worked out by assuming that the equation of parabola is

$$x^2 = ky; \quad 60^2 = k \times 1.2, \quad k = 3000$$

$$\therefore \quad x^2 = 3000y \quad \text{and} \quad \frac{dy}{dx} = \frac{2x}{3000}$$

$$\frac{d^2y}{dx^2} = \frac{2}{3000}$$

$\frac{dy}{dx} = 0$ at point B

$$\text{Now,} \quad \rho = \frac{\left[1 + \left(\frac{dy}{dx}\right)^2\right]^{3/2}}{\frac{d^2y}{dx^2}} = \frac{(1+0)^{3/2}}{2/3000} = 1500 \text{ m}$$

Substituting the relevant values in identity (*i*), we get

$$\text{Reaction exerted on wheels, } R = 12\left(1 - \frac{20^2}{1500 \times 9.81}\right)$$

$$= \mathbf{11.67\ kN}$$

10.2. MOTION OF LIFT

Consider a person of mass m inside a lift. The weight mg of the person acts vertically downwards while the reaction R acts vertically upwards

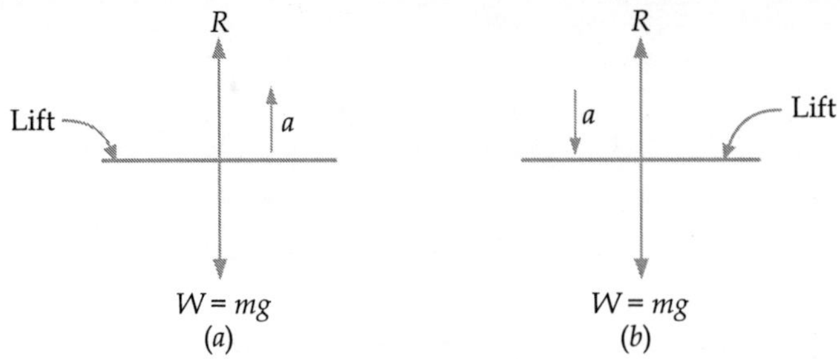

Fig. 10.11

Case 1: When the lift moves **upwards** with an acceleration a, the force causing motion is $(R - mg)$

$$\therefore \quad (R - mg) = ma$$

$$R = m(g + a) = mg\left(1 + \frac{a}{g}\right) \qquad ...(i)$$

Case 2: When the lift moves **downwards** with an acceleration a, the force causing motion is $(mg - R)$

$$\therefore \quad mg - R = ma$$

$$R = m(g - a) = mg\left(1 - \frac{a}{g}\right) \qquad ...(ii)$$

It is evident from identities (*i*) and (*ii*) that weight of the person apparently increases by the factor (a/g) when the platform moves up and decreases by the same factor when the platform moves down. This also implies that the person exerts more force on the floor of lift when moving upwards, and exerts less force when moving in downward direction.

Further, it may be noted that tension T in the cable supporting the load equals the reaction R. That is

$$T = W\left(1 + \frac{a}{g}\right) \quad \text{when lift moves upward}$$

and $$T = W\left(1 - \frac{a}{g}\right) \quad \text{when lift moves downwared} \qquad ...(10.3)$$

EXAMPLE 10.19

A lift carrying a load of 1000 N is moving with uniform acceleration of 2.5 m/s^2. Calculate the tension in the cable supporting the load when (a) lift is moving upward, (b) lift is moving downward.

Solution: Let T be the tension in the cables supporting the load

(*a*) *When lift moves upwards*

$$T - W = ma = \frac{W}{g}a \qquad \left(\because m = \frac{W}{g}\right)$$

$$T = W\left(1 + \frac{a}{g}\right) = 1000\left(1 + \frac{2.5}{9.81}\right) = \mathbf{1254.84\ N}$$

(*b*) *When lift moves downward*

$$W - T = ma = \frac{W}{g}\ a$$

$$T = W\left(1 - \frac{a}{g}\right) = 1000\left(1 - \frac{2.5}{9.81}\right) = \mathbf{745.16\ N}$$

EXAMPLE 10.20

An elevator of weight 25 kN is moving vertically downwards with a constant acceleration. Determine the limits of cable tension.

Starting from rest, the elevator is stated to travel a distance of 25 m during an interval of 8 seconds. What will be the cable tension during this time ?

Solution: Let T be the tension in the cables supporting the elevator. When the elevator moves downward,

$$W - T = ma = \frac{W}{g}\ a$$

$$T = W\left(1 - \frac{a}{g}\right)$$

The cable tension has two limits: when acceleration a is zero and when it has a maximum value equal to $g = 9.81\ \text{m/s}^2$.

Cable tension when lift moves with zero acceleration,

$$T = 25 \times 10^3\left(1 - \frac{0}{g}\right) = \mathbf{25 \times 10^3\ N}$$

Cable tension when lift moves with acceleration equal to g,

$$T = 25 \times 10^3\left(1 - \frac{g}{g}\right) = 0$$

(*b*) Applying kinematic equation, $s = ut + \frac{1}{2}at^2$, to the motion of elevator

$$25 = 0 \times t + \frac{1}{2}\ a \times 8^2 = 32\ a$$

$\therefore$ Constant acceleration of elevator, $a = \frac{25}{32} = 0.781\ \text{m/s}^2$

Again, using the relation

$$T = W\left(1 - \frac{a}{g}\right) = 25 \times 10^3\left(1 - \frac{0.781}{9.81}\right) = \mathbf{23 \times 10^3\ N}$$

EXAMPLE 10.21

An elevator cage of a mine shaft, weighing 6 kN, is lifted or lowered by means of a wire rope. Starting from rest, it moves upwards with constant acceleration and acquires a velocity of 2.5 m/s with in a distance of 3 m. Calculate the tensile force in the cable during the accelerated motion.

When a man weighing 800 N enters this elevator during its upward motion with a velocity of 2.5 m/s, the elevator is uniformly declerated to halt in 2 seconds. Determine the force exerted by the man on the floor of elevator.

Solution: Let T be the tensile force in the cable. Then considering upward motion of elevator,

$$T = mg\left(1+\frac{a}{g}\right) = W\left(1+\frac{a}{g}\right)$$

The acceleration acquired by the elevator after travelling 2.5 m is obtained by using the kinematic equation $v^2 - u^2 = 2as$. Then

$$2.5^2 - 0 = 2\,a \times 3 \qquad (u = 0 \text{ as elevator starts from rest})$$

$$a = \frac{2.5 \times 2.5}{6} = 1.042 \text{ m/s}^2$$

$$\therefore \quad T = 6 \times 1000\left(1+\frac{1.042}{9.81}\right) = \mathbf{6637.3\ N}$$

Considering motion of man,

$$v = u + at\,; \quad u = 2.5 \text{ m/s} \quad \text{and } v = 0$$

$$0 = 2.5 + a \times 2\,; \quad a = -1.25 \text{ m/s}^2$$

The negative sign with acceleration suggests that it is opposite to the direction of velocity or is downward.

$$\text{Reaction } R = W\left(1+\frac{a}{g}\right) = 800\left[1+\frac{(-1.25)}{9.81}\right] = \mathbf{523.55\ N}$$

Thus the pressure at the floor of elevator under the feet of the man is 523.55 N.

EXAMPLE 10.22

An elevator of mass 600 kg is ascending with an acceleration of 2.5 m/s^2. During this ascent, its operator who weighs 700 N is standing on the scales placed on the floor of the elevator. Calculate the scale reading and total tension in the cables of elevator during this motion.

Solution: Weight of elevator = 600 × 9.81 = 5886 N. When the operator stands on the scale placed on the floor of the elevator, the reaction offered by the floor equals the reading on scale.

Considering motion of man,

$$R - W = ma = \frac{W}{g}\,a$$

$$R = W\left(1+\frac{a}{g}\right) = 700\left(1+\frac{2.5}{9.81}\right) = \mathbf{878.4\ N}$$

Considering motion of elevator and man taken together,

$$T - W = ma = \frac{W}{g}\,a$$

$$T = W\left(1+\frac{a}{g}\right) = (5886 + 700)\left(1+\frac{2.5}{9.81}\right) = \mathbf{8264.4\ N}$$

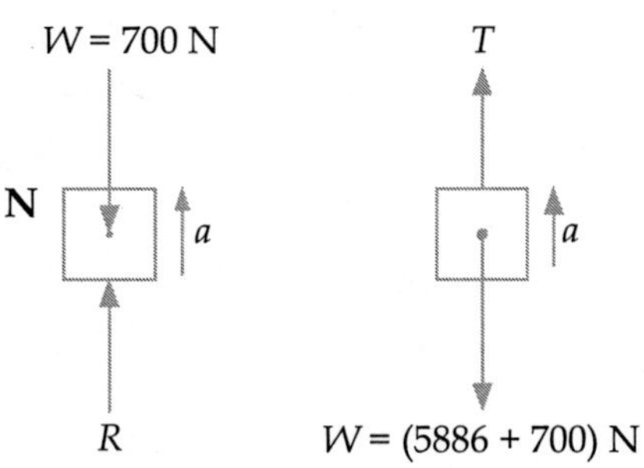

Fig. 10.12

10.3 MOTION OF CONNECTED BODIES

Attention has been focussed in the following section to examine the motion of a body when it is connected by a string to another body which is subjected to some external force. Both the bodies would be moving with the same velocity and acceleration.

10.3.1 Motion of two bodies connected by a string passing over a smooth pulley

Consider two bodies of masses m_1 and m_2 connected at the ends of a light inextensible string that passes over a smooth frictionless pulley (Fig. 10.13). If $m_1 > m_2$, then the mass m_1 moves downwards and the mass m_2 moves upwards.

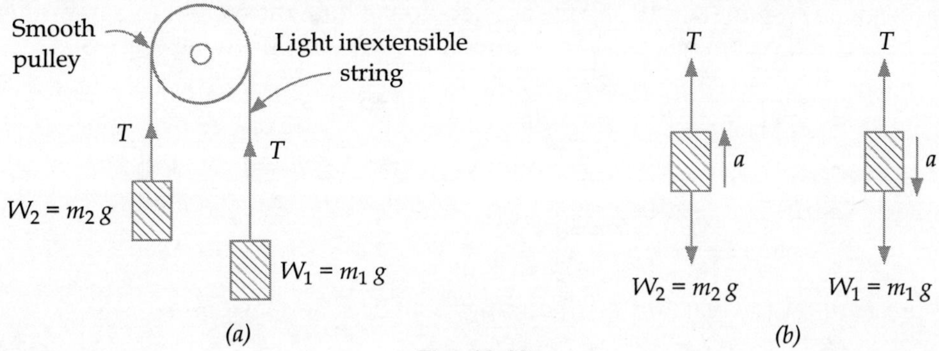

Fig. 10.13

It is to be noted that.

(*i*) the acceleration of both the weights is same otherwise the cord would break or become slack.

(*ii*) for a smooth frictionless pulley and continuous cord, the tension induced in the string will be same throughout.

With reference to the free body diagram [Fig 10.13 (*b*)], the equations of motion for the two masses are as follows:

Downwand motion of mass m_1 : $m_1 g - T = m_1 a$ ...(*i*)

Upward motion of mass m_2 : $T - m_2 g = m_2 a$ ...(*ii*)

Adding expressions (*i*) and (*ii*)

$$m_1 g - m_2 g = m_1 a + m_2 a$$

∴ Acceleration of the system, $a = \dfrac{m_1 - m_2}{m_1 + m_2} g = \dfrac{W_1/g - W_2/g}{W_1/g + W_2/g} g = \dfrac{W_1 - W_2}{W_1 + W_2} g$ ...(10.4)

Upon substitution of this value of acceleration *a* in expression (*i*), we get

$$m_1 g - T = m_1 \times \frac{m_1 - m_2}{m_1 + m_2} g$$

$$\therefore \quad T = m_1 g \left[1 - \frac{m_1 - m_2}{m_1 + m_2}\right] = m_1 g \times \frac{2m_2}{m_1 + m_2}$$

$$= \frac{2m_1 m_2}{m_1 + m_2} g = 2\frac{W_1/g \times W_2/g}{W_1/g + W_2/g} g = \frac{2W_1 W_2}{W_1 + W_2} \quad ...(10.5)$$

Pressure on (or reaction of) the pulley equals the vector sum of the tensions T_1 and T_2 induced in the string lying on two sides of the pulley. Here $T_1 = T_2 = T$ and they act along the same direction, *i.e.*, angle θ between then is zero. Then from the parallelogram law of forces

$$R = \sqrt{P^2 + Q^2 + 2PQ \cos\theta} = \sqrt{T^2 + T^2 + 2TT \cos 0^\circ} = \sqrt{4T^2} = 2T$$

Thus the pressure on the pulley is twice the tension in the string.

$$\therefore \quad R = 2T = \frac{4W_1 W_2}{W_1 + W_2} \quad ...(10.6)$$

EXAMPLE 10.23

Two objects of weights 50 N and 30 N respectively are connected by a light cord passing over a smooth frictionless pulley. Determine (*a*) the acceleration with which the weights move, (*b*) the tension in the cord, and (*c*) the pressure on the pulley.

Solution: For the given system (Fi.g 10.13), the acceleration of both the weights is same otherwise the cord would break or become slack. The common acceleration of the system is

$$a = \frac{W_1 - W_2}{W_1 + W_2} g = \frac{50 - 30}{50 + 30} \times 9.81 = 2.45 \text{ m/s}^2$$

Further, tension in the cord is same throughout (because friction of the pulley is negligible) and is given by

$$T = \frac{2W_1 W_2}{W_1 + W_2} = \frac{2 \times 50 \times 30}{50 + 30} = 37.5 \text{ N}$$

Also, the pressure on (or reaction of) the pulley is twice the tension.

$$R = 2T = 2 \times 37.5 = \mathbf{75\ N}$$

EXAMPLE 10.24

Two blocks *A* and *B* of weights 600 N and 400 N respectively are connected to the two ends of cord which passes over pulley. Neglecting inertia of the pulley and assuming that the cord is inextensible, find the acceleration of block *A*. How much weight should be added to or taken away from block *A*, if the acceleration of the block is required to be *g*/3 downwards?

Solution: Let W_1 = 600 N and W_2 = 400 N.

Then for the given arrangement, the acceleration of block *A* or that of the system is

$$a = \frac{W_1 - W_2}{W_1 + W_2} g = \frac{600 - 400}{600 + 400} g = \frac{200}{1000} g = \frac{g}{5}$$

Let *W* be the weight of block *A* if its acceleration is to be *g*/3 downwards. Then

$$\frac{g}{3} = \frac{W - W_2}{W + W_2} g = \frac{W - 400}{W + 400} g$$

or $W + 400 = 3(W - 400) = 3W - 1200; \quad W = 800$ N

Hence a weight (800 – 600), *i.e.*, **200 N** must be added to block *A*.

EXAMPLE 10.25

A system of frictionless pulleys carries two weights hung by inextensible cords as shown is Fig. 10.14. Make calculations for the tension in the cords and acceleration of the weights

Proceed to workout the displacement and velocity of weight W_1 after 4 seconds from start. Presume that the system is released from rest.

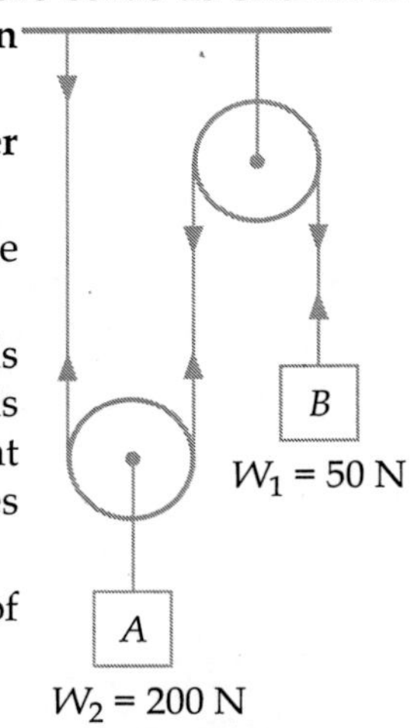

Fig. 10.14

Solution: Since the pulleys are frictionless and the cord is continuous, the tension *T* induced in the cord will be same throughout.

It is evident from the given arrangement that when weight W_1 travels a unit distance, the weight W_2 travels half the distance. Acceleration is proportional to distance travelled, and therefore if *a* is acceleration of weight W_1, the weight W_2 will have acceleration equal to *a*/2. The weight W_1 moves upward and the weight W_2 moves downward.

With reference to the free body diagrams (Fig. 10.15), the equations of motion for the two weights are as follows:

$$T - 50 = \frac{50}{9.81} \times a \quad \text{(upward motion of block } B\text{)}$$

$$\text{or} \quad T = 50 + \frac{50}{9.81} a \qquad ...(i)$$

Also $200 - T = \frac{200}{9.81} \times \frac{a}{2}$

(downward motion of block A)

or $2T = 200 - \frac{100}{9.81}a$

or $T = 100 - \frac{50}{9.81}a$ ...(*ii*)

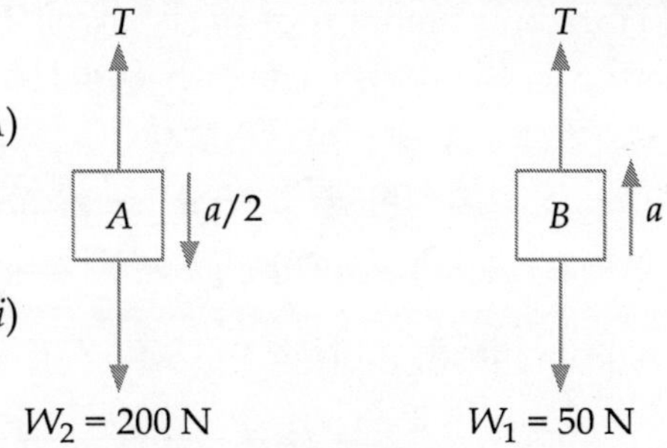

Fig. 10.15

From expression (*i*) and (*ii*), we obtain

$$50 + \frac{50}{9.81}a = 100 - \frac{50}{9.81}a$$

or $\frac{100}{9.81}a = 50 \quad \therefore \quad a = \frac{50 \times 9.81}{100} = 4.905 \text{ m/s}^2$

Thus the acceleration of blocks A and B are 2.452 m/s^2 and 4.905 m/s^2 respectively.

From expression (*i*)

$$T = 50 + \frac{50}{9.81} \times 4.905 = 75 \text{ N}$$

(*b*) Applying kinematic equations to the motion of weight B ($W_1 = 50$ N)

$$v = u + at = 0 + 4.905 \times 4 = 19.62 \text{ m/s}$$

$$s = ut + \frac{1}{2}at^2 = 0 \times 4 + \frac{1}{2} \times 4.905 \times 4^2 = 39.24 \text{ m}$$

10.3.2 Motion of two bodies connected at the edge of a horizontal surface

Consider a block of mass m_2 resting on a rough horizontal surface. The block is attached to a light inextensible string which passes over a smooth frictionless pulley at the end of the surface. The portion of the string, beyond the pulley, hangs vertically down and carries a mass m_1 at the end.

Let the mass m_1 move downwards and the mass m_2 move along the surface of the plane. It is to be noted that:

(*i*) the acceleration of both the masses is same otherwise the string will break or become loose

(*ii*) for a smooth friction-less pulley and continuous string, the tension induced in the string is same throughout

With reference to the free body diagrams [10.16 (*b*)], the equations of motion for the two masses are as follows:

Downward motion of mass m_1: $m_1 g - T = m_1 a$ ...(*i*)

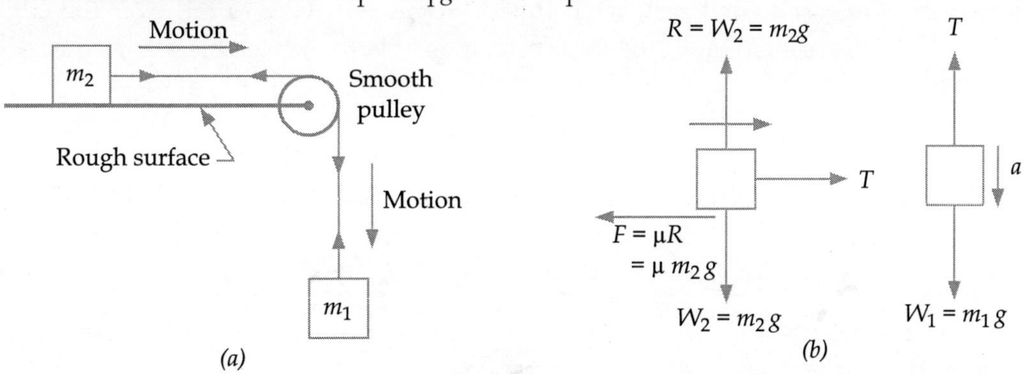

Fig. 10.16

Horizontal motion of mass m_2: $T - \mu m_2 g = m_2 a$ ...(ii)

where μ is the coefficient of friction between the block and the rough surface.

Adding expressions (i) and (ii),

$$m_1 g - \mu m_2 g = m_1 a + m_2 a$$

∴ acceleration of the system, $a = \dfrac{m_1 g - \mu m_2 g}{m_1 + m_2} = \dfrac{W_1 - \mu W_2}{W_1 + W_2} g$ ...(10.7)

Upon substitution of the value of acceleration a in expression (i), we get

$$m_1 g - T = m_1 \times \frac{m_1 g - \mu m_2 g}{m_1 + m_2}$$

$$T = m_1 g - m_1 \frac{m_1 g - \mu m_2 g}{m_1 + m_2} = m_1 g \left[1 - \frac{m_1 - \mu m_2}{m_1 + m_2}\right] = m_1 g \left[\frac{m_2 (1 + \mu)}{m_1 + m_2}\right]$$

$$= W_1 \left[\frac{W_2 (1 + \mu)}{W_1 + W_2}\right] = \frac{W_1 W_2 (1 + \mu)}{W_1 + W_2} \quad ...(10.8)$$

For a smooth horizontal surface, μ = 0. In that case the equations 10.7 and 8.38 get modified to

$$a = \frac{W_1}{W_1 + W_2} g \quad \text{and} \quad T = \frac{W_1 W_2}{W_1 + W_2} \quad ...(10.9)$$

Pressure on (or reaction of) the pulley equals the vector sum of tensions T_1 and T_2 induced in the string lying on two sides of the pulley. Here $T_1 = T_2 = T$ and they act perpendicular to each other, *i.e.*, angle θ between them is 90°. Then from the parallelogram law of forces,

$$R = \sqrt{P^2 + Q^2 + 2PQ \cos \theta}$$

$$= \sqrt{T^2 + T^2 + 2TT \cos 90^\circ} = \sqrt{2T^2} = \sqrt{2} T \quad ...(10.10)$$

Thus the pressure on the pulley is $\sqrt{2}$ times the tension induced in the string.

EXAMPLE 10.26

A block of weight 10 N is resting on a rough horizontal table 1.5 m from its edge. This block A is attached to a string which passes over a pulley at the end of the table. The portion of the string beyond the pulley hangs vertically down and carries a block B of weight 8 N at the end. If the coefficient of friction between the block A and table is 0.2, make calculations for the acceleration of the system, tension induced in the string and pressure on the pulley. Neglect friction in the pulley and extension of the string.

If the system is released from rest, what time block *A* will take to fall over the edge ?

Solution: For the given system (Fig. 10.16), the acceleration of both the weights is same otherwise the string would break or become loose. The common acceleration of the system is

$$a = \frac{W_1 - \mu W_2}{W_1 + W_2} g = \frac{8 - 0.2 \times 10}{8 + 10} \times 9.81 = \mathbf{3.27\ m/s^2}$$

Further, tension in the cord is same throughout (because friction of pulley is negligible) and is given by

$$T = \frac{W_1 W_2 (1 + \mu)}{W_1 + W_2} = \frac{8 \times 10 (1 + 0.2)}{8 + 10} = \mathbf{5.33\ N}$$

Pressure on the pulley $= \sqrt{2} T = \sqrt{2} \times 5.33 = \mathbf{7.538\ N}$

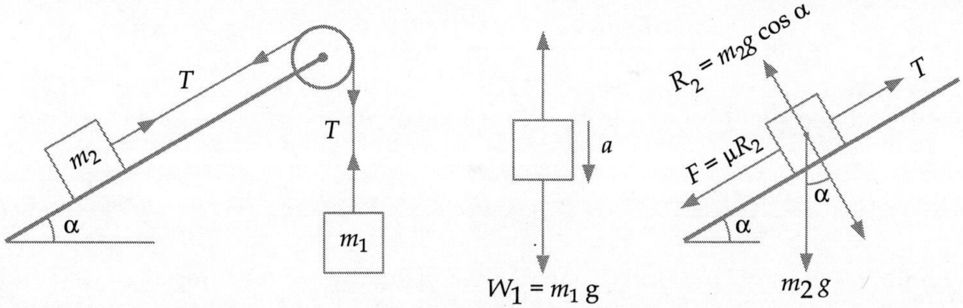

(a) (b)

Fig. 10.17

(*b*) Applying kinematic equations to the motion of weight *A*,

$$v^2 - u^2 = 2as \; ; \quad u = 0 \text{ as the system is released from rest.}$$

$$\therefore \quad v = \sqrt{2as} = \sqrt{2 \times 3.27 \times 1.5} = 3.13 \text{ m/s}$$

Also, $$v = u + at \; ; \quad t = \frac{v}{a} = \frac{3.13}{3.27} = \mathbf{0.957\ s}$$

10.3.3 Motion of two bodies connected by a string; one hangs free and the other on a rough inclined plane.

Consider two bodies of masses m_1 and m_2 connected at the ends of a light inextensible string that passes over a smooth frictionless pulley. Further, mass m_1 hangs free and the mass m_2 lies on a rough surface inclined at an angle α with the horizontal.

It is to be noted that:

(*i*) the acceleration of both the masses is same, otherwise the cord could break or become slack

(*ii*) for a smooth frictionless pulley and continuous cord, the tension induced in the string will be same throughout

Let mass m_1 move downwards. Then with reference to free body diagrams [(Fig. 10.16 (*b*)], the equations of motion for the two masses are as follows:

$$m_1g - T = m_1a \qquad \text{...(i)}$$

$$T - m_2g \sin \alpha - \mu\, m_2 g \cos \alpha = m_2a \qquad \text{...(ii)}$$

Adding expressions (*i*) and (*ii*),

$$m_1g - m_2g \sin \alpha - \mu\, m_2g \cos \alpha = (m_1 + m_2)\, a$$

∴ Acceleration of the system,

$$a = \frac{m_1g - m_2g\sin\alpha - \mu\, m_2g\cos\alpha}{m_1 + m_2} = \frac{W_1 - W_2 \sin\alpha - \mu\, W_2 \cos\alpha}{W_1 + W_2} \times g \qquad \text{...(10.11)}$$

From expression (*i*), the tension *T* in the string is

$$T = m_1 (g - a) = m_1 \left[g - \frac{m_1g - m_2g\sin\alpha - \mu\, m_2\, g}{m_1 + m_2} \right.$$

$$= m_1 \left[\frac{m_1g + m_2g - m_1g + m_2g\sin(}{m_1 + m_2} \right.$$

$$= m_1 \left[\frac{m_2 g(1 + \sin\alpha + \mu\cos\alpha)}{m_1 + m_2} \right] = \frac{W_1 W_2 (1 + \sin\alpha + \mu\cos\alpha)}{W_1 + W_2} \quad ...(10.12)$$

For a smooth inclined plane, $\mu = 0$. In that case the equations 10.11 and 10.12 get modified to

$$a = \frac{W_1 - W_2 \sin\alpha}{W_1 + W_2} g \quad \text{and} \quad T = \frac{W_1 W_2 (1 + \sin\alpha)}{W_1 + W_2} \quad ...(10.13)$$

Pressure on (or reaction of) the pulley equals the vector sum of tensions T_1 and T_2 induced in the string lying on two sides of the pulley. Here ($T_1 = T_2$) and the angle θ between them is $(90 - \alpha)$. Then from the parallelogram law of forces

$$R = \sqrt{P^2 + Q^2 + 2PQ\cos\theta} = \sqrt{T^2 + T^2 + 2TT\cos(90 - \alpha)}$$

$$= \sqrt{2T^2(1 + \sin\alpha)} = T\sqrt{2(1 + \sin\alpha)} \quad ...(10.14)$$

Thus the pressure on the pulley is $\sqrt{2(1 + \sin\alpha)}$ times the tension induced in the string.

EXAMPLE 10.27

For the mass-pulley system depicted in Fig. 10.18, the mass m_2 = 5 kg is placed on a smooth inclined plane of inclination α where as the mass m_1 = 5 kg is hanging free. If acceleration of the system is 1.5 m/s², determine inclination of the plane, tension in the string and reaction at the pulley.

How the acceleration of the system would be affected if each mass is doubled?

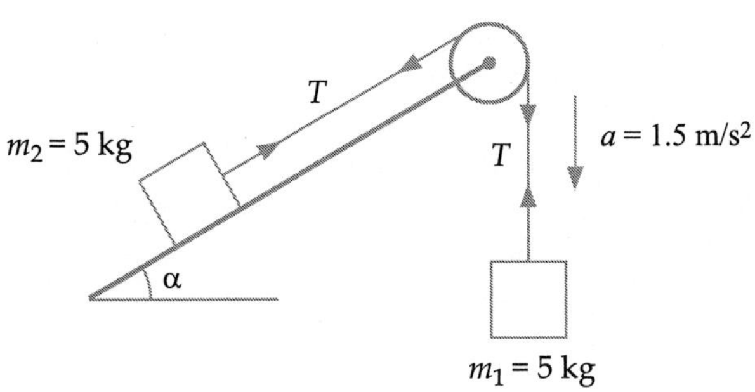

Fig. 10.18

Solution: Component of mass m_2 along the plane $= m_2 \sin\alpha$. If the mass m_1 moves downward with acceleration a and T is the tension in the string, then equations of motion for the two masses are:

$$m_1 g - T = m_1 a \quad ...(i)$$

and

$$T - m_2 g \sin\alpha = m_2 a \quad ...(ii)$$

Upon adding these identities and solving, we get

$$\text{Acceleration } a = \frac{m_1 - m_2 \sin\alpha}{m_1 + m_2} \times g$$

$$1.5 = \frac{5 - 5\sin\alpha}{5 + 5} \times 9.81$$

$$\sin\alpha = \left(5 - \frac{1.5 \times 10}{9.81}\right) \times \frac{1}{5} = 0.692; \quad \alpha = \mathbf{43.96^\circ}$$

From expression (*i*), the tension T in the string is

$$T = m_1 (g - a) = 5(9.81 - 1.5) = \mathbf{41.55\ N}$$

$$\text{Reaction at the pulley } R = T\sqrt{2(1 + \sin\alpha)} = 41.55\sqrt{2(1 + \sin 43.96^\circ)}$$

$$= \mathbf{76.48\ N}$$

(*b*) Acceleration $a = \dfrac{m_1 - m_2 \sin \alpha}{m_1 + m_2} \times g$

An examination of the above expression shows that acceleration of the system would remain unaffected when each mass is doubled.

$\therefore \qquad a = \mathbf{1.5\ m/s^2}$

EXAMPLE 10.28

Two bodies of weights 50 N and 10 N are connected to the two ends of a light inextensible string which passes over a smooth pulley. The weight 10 N is placed on a rough inclined plane of angle of inclination 20° while the weight of 50 N hangs vertically downward. If the coefficient of friction between the body and the plane is 0.2, make calculations for the acceleration of the system, tension induced in the string, reaction at the pulley and the distance moved by the body in 3 seconds starting from rest.

Solution: Refer Fig. 10.19 $W_1 = 50$ N, $W_2 = 10$ N and $\mu = 0.2$

The acceleration of both the weights is equal and tension induced in the string is same throughtout. The weight $W_2 = 10$ N is stated to move up the plane. Then with reference to the free body diagrams, the equations of motion for the two weights are as follows:

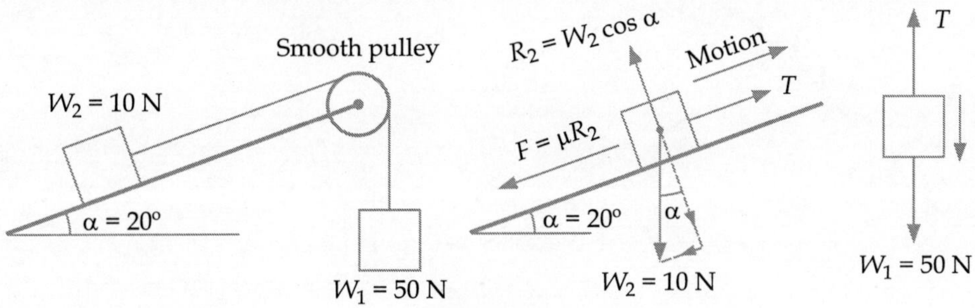

Fig. 10.19

$$W_1 - T = \frac{W_1}{g} a \qquad ...(i)$$

$$T - W_2 \sin \alpha - \mu W_2 \cos \alpha = \frac{W_2}{g} a \qquad ...(ii)$$

Adding expressions (*i*) and (*ii*)

$$W_1 - W_2 \sin \alpha - \mu W_2 \cos \alpha = \left(\frac{W_1}{g} + \frac{W_2}{g}\right) a$$

$$a = \frac{W_1 - W_2 \sin\alpha - \mu W_2 \cos\alpha}{W_1 + W_2} \times g$$

$$= \frac{50 - 10 \sin 20° - 0.2 \times 10 \cos 20°}{50 + 10} \times 9.81$$

$$= \frac{50 - 10 \times 0.342 - 0.2 \times 10 \times 0.939}{60} \times 9.81$$

$$= \frac{50 - 3.42 - 1.878}{60} \times 9.81 = \mathbf{7.31\ m/s^2}$$

(*b*) From expression (*i*), tension T in the string is

$$T = W_1 - \frac{W_1}{g}\,a = 50 - \frac{50}{9.81} \times 7.31 = \mathbf{12.74\ N}$$

(*c*) Reaction at the pulley,

$$R = T\sqrt{2(1+\sin\alpha)} = 12.74\sqrt{2(1+\sin 20°)} = \mathbf{20.87\ N}$$

(*d*) Applying kinematic equation to the motion of weight W_2 = 10 N,

$$s = ut + \frac{1}{2}at^2\ ; \quad u = 0 \text{ as the system starts from rest}$$

$$s = 0 + \frac{1}{2} \times 7.31 \times 3^2 = \mathbf{32.89\ m}$$

10.3.4 Motion of two bodies connected over inclined planes

Consider two blocks A and B of masses m_1 and m_2 respectively connected by a weightless inextensible rope passing over a frictionless pulley and placed on inclined planes making α_1 and α_2 with the horizontal as shown in Fig. 10.20.

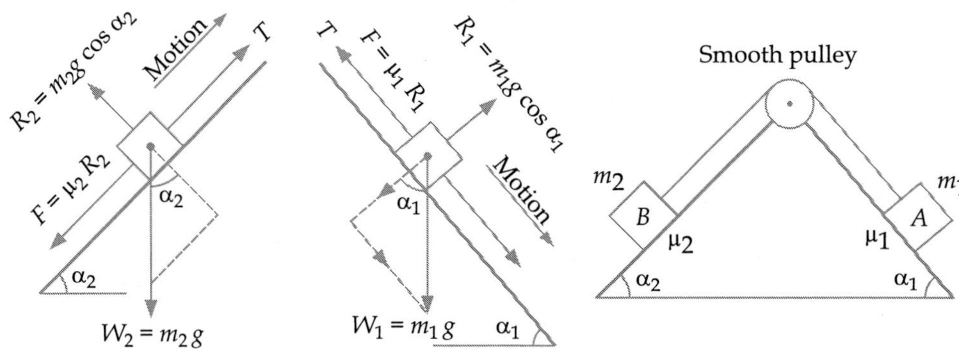

Fig. 10.20

It is to be noted that

(*i*) the acceleration of both the weights is same otherwise the rope would break or become slack

(*ii*) for a smooth frictionless pulley and continuous rope, the tension induced in the rope is same throughout

Let mass m_1 move downwards. Then with reference to free body diagrams, the equations of motion for the two masses are as follows:

$$m_1 g \sin\alpha_1 - T - \mu_1 m_1 g \cos\alpha_1 = m_1 a \qquad ...(i)$$

$$T - m_2 g \sin\alpha_2 - \mu_2 m_2 g \cos\alpha_2 = m_2 a \qquad ...(ii)$$

Adding expressions (*i*) and (*ii*),

$$m_1 g \sin\alpha_1 - \mu_1 m_1 g \cos\alpha_1 - m_2 g \sin\alpha_2 - \mu_2 m_2 g \cos\alpha_2 = (m_1 + m_2)\,a$$

∴ Acceleration of the system,

$$a = \frac{m_1 g \sin\alpha_1 - m_2 g \sin\alpha_2 - \mu_1 m_1}{m_1 + m_2}$$

$$= \frac{W_1 \sin\alpha_1 - W_2 \sin\alpha_2 - \mu_1 W_1}{W_1 + W_2} \times g \qquad ...(10.15)$$

From expression (*i*), the tension T in the string is

$$T = m_1 g \sin\alpha_1 - \mu_1 m_1 g \cos\alpha_1 - m_1 a$$

$$= m_1 g \sin\alpha_1 - \mu_1 m_1 g \cos\alpha_1 - m_1 \left[\frac{m_1 g \sin\alpha_1 - m_2 g \sin\alpha_2 - \mu_1 m}{m_1 + m}\right.$$

$$= \frac{1}{m_1 + m_2}\left[\begin{array}{l}(m_1 + m_2)(m_1 g \sin\alpha_1 \\ - m_1 (m_1 g \sin\alpha_1 - m_2\end{array}\right.$$

$$= \frac{1}{m_1 + m_2}[m_1^2 g \sin\alpha_1 - \mu_1 m_1^2 g \cos$$

$$- m_1^2 g \sin\alpha_1 + m_1 m_2 g \sin\alpha_2 + \mu_1$$

$$= \frac{1}{m_1 + m_2}[m_1 m_2 g \sin\alpha_1 + m_1 m_2$$

$$= \frac{1}{m_1 + m_2}[g\, m_1 m_2 (\sin\alpha_1 + \sin\alpha$$

$$= \frac{g\, m_1 m_2}{m_1 + m_2}[(\sin\alpha_1 + \sin\alpha_2) - (\mu_1$$

In terms of weights : $m_1 = \frac{W_1}{g}$ and $m_2 = \frac{W_2}{g}$

$$T = \frac{W_1 W_2}{W_1 + W_2}[(\sin\alpha_1 + \sin\alpha_2) - (\mu \qquad ...(10.16)$$

For smooth horizontal surfaces, $\mu_1 = \mu_2 = 0$. In that case, the equations 10.15 and 10.16 get modified to

$$a = \frac{W_1 \sin\alpha_1 - W_2 \sin\alpha_2}{W_1 + W_2}\, g$$

$$T = \frac{W_1 W_2}{W_1 + W_2}(\sin\alpha_1 + \sin\alpha_2) \qquad ...(10.17)$$

Pressure on (or reaction of) the pulley equals the vector sum of tensions T_1 and T_2 induced in the string lying on two sides of the pulley. Here $T_1 = T_2 = T$ and the angle θ betwen them is $180 - (\alpha_1 + \alpha_2)$. Then from the parallelogram law of forces

$$R = \sqrt{P^2 + Q^2 + 2PQ\cos\theta} = \sqrt{T^2 + T^2 + 2TT\cos\{180 - (\alpha_1 +}$$

$$= \sqrt{2T^2 - 2T^2\cos(\alpha_1 + \alpha_2)} = \sqrt{2T^2[1 - \cos(\alpha_1 + \alpha_2)]}$$

$$= \sqrt{2T^2 \times 2\sin^2\frac{\alpha_1 + \alpha_2}{2}} = 2T\sin\frac{\alpha_1 + \alpha_2}{2} \qquad ...(10.18)$$

Thus the pressure on the pulley is $2\sin\frac{\alpha_1 + \alpha_2}{2}$ times the tension induced in the string.

EXAMPLE 10.29

Two smooth inclined planes whose inclinations with the horizontal are 20° and 30°, are placed back to back. Two bodies of weight 100 N and 60 N are placed on them and are connected by a light inextensible string passing over a smooth pulley as shown in Fig. 10.21. Calculate the common acceleration of the system, tension in the string and pressure on the pulley.

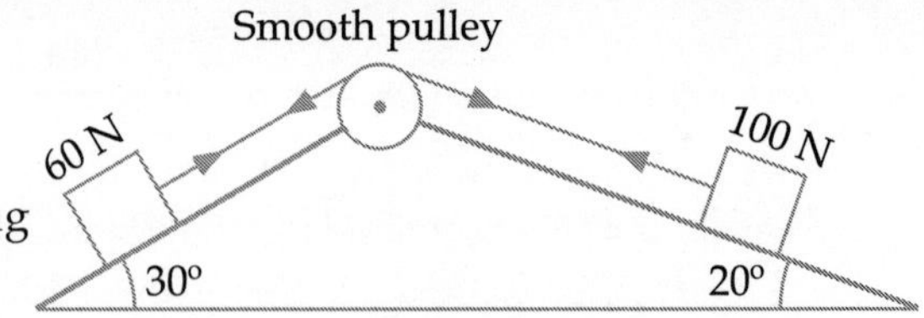

Fig. 10.21

Solution: With reference to Fig. 10.21

$$W_1 = 100 \text{ N} \quad \text{and} \quad \alpha_1 = 20°$$
$$W_2 = 60 \text{ N} \quad \text{and} \quad \alpha_2 = 30°$$

Components of weights W_1 and W_2 along respective planes are

$$W_1 \sin \alpha_1 \quad \text{and} \quad W_2 \sin \alpha_2$$

If weight W_1 moves downward with acceleration a and T is the tension in the string, then equations of motion for the two weights are

$$W_1 \sin \alpha_1 - T = m_1 a = \frac{W_1}{g} a \qquad ...(i)$$

and
$$T - W_2 \sin \alpha_2 = m_2 a = \frac{W_2}{g} a \qquad ...(ii)$$

Adding expression (*i*) and (*ii*)

$$W_1 \sin \alpha_1 - W_2 \sin \alpha_2 = \frac{a}{g}(W_1 + W_2)$$

∴ Acceleration of the system,

$$a = \frac{g(W_1 \sin\alpha_1 - W_2 \sin\alpha_2)}{W_1 + W_2} = \frac{9.81(100 \sin 20° - 60 \sin 30°)}{100 + 60}$$
$$= \frac{9.81(100 \times 0.342 - 60 \times 0.5)}{160} = 0.2575 \text{ m/s}^2$$

Then from expression (*i*),

$$\text{Tension in the string, } T = W_1 \sin \alpha_1 - \frac{W_1}{g} a = 100 \sin 20° - \frac{100}{9.81} \times 0.2575$$
$$= 34.20 - 2.62 = \mathbf{31.58\ N}$$

$$\text{Pressure on the pulley, } P = 2T \sin \frac{\alpha_1 + \alpha_2}{2} = 2 \times 31.58 \sin \frac{20 + 30}{2} = \mathbf{26.69\ N}$$

10.4. D'ALEMBERT'S PRINCIPLE

The following section deals with the motion of a body when it is connected by a string to another body which is subjected to some external force. Solution to the system is worked out by applying D' Alembert's principle which states that

> **"the system of forces acting on a body in motion is in dynamic equilibrium with the inertia force of the body"**

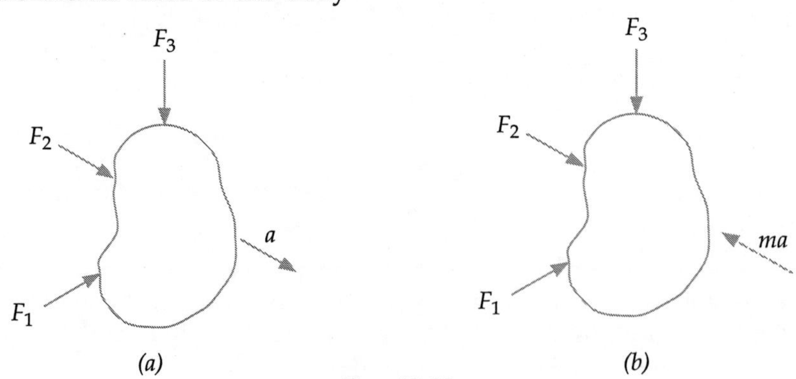

Fig. 10.22

Consider a body subjected to a system of forces F_1, F_2 and F_3 as shown in Fig. 10.21 (*a*). The resultant of these forces will cause the body to move along its direction with an acceleration *a*. The body can be made to remain static if a force equal to *ma* is made to act in the reversed direction. Then for dynamic equilibrium of the body, the sum of the resultant force and the reversed force should be zero. That is

$$\Sigma F = 0; \quad R - ma = 0 \quad \text{or} \quad R = ma$$

The force (– *ma*) is termed as the **inertia force** or **reverse effective force.**

EXAMPLE 10.30

Determine the acceleration produced in a body of 50 kg mass when it is acted upon by a force of 125 N. Use D'Alembert's principle.

Solution:

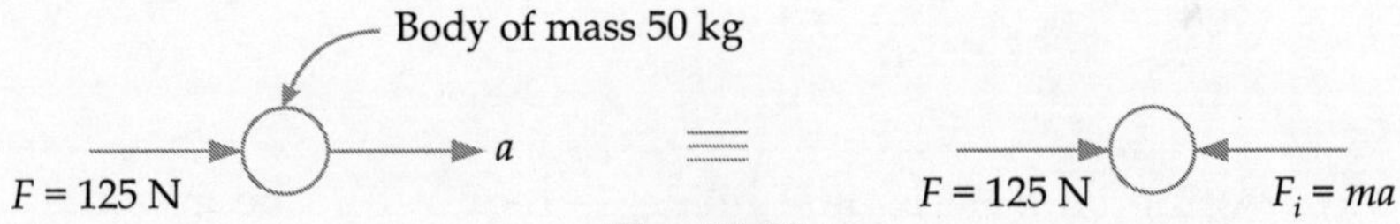

Fig. 10.23

The problem is solved using D'Alembert's principle by applying the inertial force $F_i = ma$ in a direction opposite to that of motion. The body is then in dynamic equilibrium and accordingly

$$\Sigma F_x = 0; \qquad 125 - ma = 0$$

$$125 - 50 \times a = 0 \qquad \therefore \quad a = \frac{125}{50} = \mathbf{2.5\ m/s^2}$$

EXAMPLE 10.31

A man weighing 750 N stands on the floor of a lift. Determine the pressure exerted on the floor when (*a*) the lift moves upward with on acceleration of 2.5 m/s^2 (*b*) the lift moves downward with an acceleration of 2.5 m/s^2.

If 900 N pressure is to be exerted on the floor, then with what acceleration the lift should move upward? Work-out your solution by using D'Alembert's principle

Solution: (*i*) *When the lift moves upwards.* The lift can be brought into dynamic equilibrium by applying an inertia force $F_i = ma$ in the downward direction [Fig. 10.23 (*a*)]. Accordingly,

$$\Sigma F_y = 0; \qquad T - mg - ma = 0$$

where *T* is the tension in the cable supporting the lift or the reaction of the lift or the pressure exerted by the man on the floor of the lift.

$$T = mg + ma = \frac{W}{g}g + \frac{W}{g}a$$

$$= \frac{W}{g}(g + a) = \frac{750}{9.81}(9.81 + 2.5) = \mathbf{941.13\ N}$$

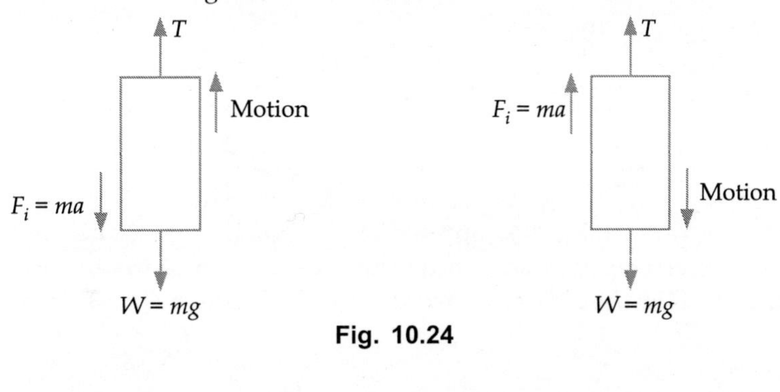

Fig. 10.24

(*ii*) *When the lift moves downwards* The lift is brought into dynamic equilibrium by applying the inertia force $F_i = ma$ in the upward direction [Fig. 10.23 (*b*)] Accordingly

$$\Sigma F_y = 0; \qquad T + ma - mg = 0$$

$$T = m(g - a) = \frac{W}{g}(g - a) = \frac{750}{9.81}(9.81 - 2.5) = \mathbf{558.87\ N}$$

(*iii*) Let a_u be the upward acceleration required to cause a pressure of 900 N on the floor. Then with reference to Fig. 10.23 (*a*)

$$T - mg - ma_u = 0$$

$$900 - \frac{750}{9.81} \times 9.81 - \frac{750}{9.81} a_u = 0$$

$$\therefore \qquad a_u = (900 - 750) \times \frac{9.81}{750} = \mathbf{1.962\ m/s^2}$$

EXAMPLE 10.32

A block of mass 50 kg, resting on a horizontal plane, is required to be given an acceleration of 2 m/s towards right by applying a push *P* at an angle of 45 degree with the horizontal. Assuming that coefficient of friction between the block and plane is 0.4, work-out the magnitude of push *P*. Obtain your solution by applying D'Alembert's principle.

Solution: Refer Fig. 10.25

Let an inertia force $F_i = ma$ be applied in a direction opposite to that of motion. The block is then in dynamic equilibrium under the action of following set of forces :

(*i*) Weight $W = mg$ of the block acting downwards

(*ii*) Inertia force $F_i = ma$ acting towards left

(*iii*) Frictional force $F = \mu R = \mu W$ acting opposite to direction of motion

(*iv*) Push *P* applied at angle of 45 degree with the horizontal.

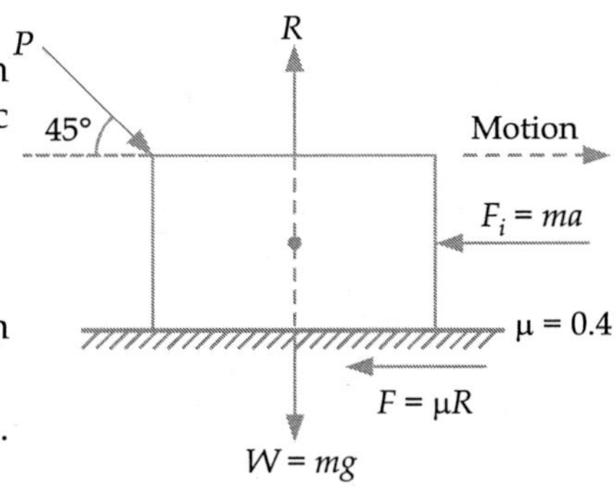

Fig. 10.25

Writing the equations for equilibrium, we get

$$\Sigma F_x = 0; \qquad P \cos 45° - \mu R - ma = 0$$

$$P \cos 45° - 0.4R - 50 \times 2 = 0 \qquad \text{...}(i)$$

$$\Sigma F_y = 0; \qquad -P \sin 45° - mg + R = 0$$

$$R = P \sin 45° + 50 \times 9.81 \qquad \text{...}(ii)$$

From expressions (*i*) and (*ii*),

$$P \cos 45° - 0.4\,(P \sin 45° + 50 \times 9.81) - 50 \times 2 = 0$$

or $$P\,(\cos 45° - 0.4 \sin 45°) - 0.4 \times 50 \times 9.81 - 100 = 0$$

or $$0.424\,P = 196.2 + 100 = 296.2$$

$$\therefore \qquad \text{Necessary push } P = \frac{296.2}{0.424} = \mathbf{698.58\ N}$$

EXAMPLE 10.33

Two blocks of mass 60 kg and 15 kg are connected by a string and move along a rough horizontal surface when a force of 300 N is applied to the block of 60 kg mass as shown in Fig. 10.26. Apply D'Alembert's principle to determine the acceleration of the blocks and tension in the string. Presume that coefficient of friction between the sliding surface of the blocks and the plane is 0.25.

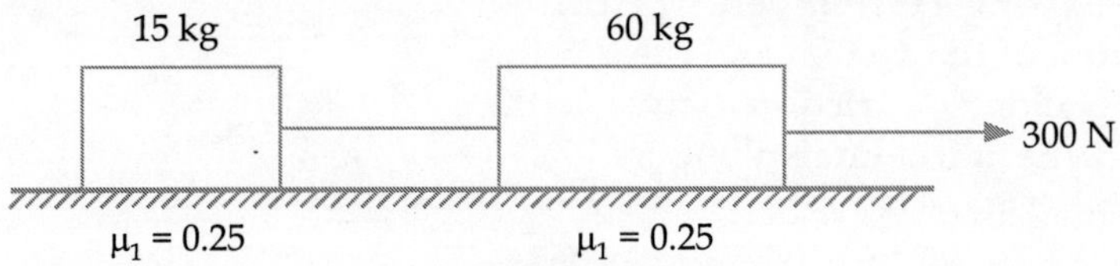

Fig. 10.26

Solution: Since the system moves towards right

(*i*) the friction force would act towards left, and

(*ii*) the inertia force needs to be applied towards left to bring the system in a state of dynamic equilibrium

$$F = \mu_1 R_1 + \mu_2 R_2 = \mu(W_1 + W_2) = \mu g(m_1 + m_2)$$
$$= 0.25 \times 9.81 \times (15 + 60) = 183.94 \text{ N } (\leftarrow)$$
$$F_i = ma = (m_1 + m_2)\, a = (15 + 60)a = 75a\ (\leftarrow)$$

Writing the equation for dynamic equilibrium (D'Alembert's principle), we get

$$300 - 183.94 - 75\,a = 0$$

$\therefore$ Acceleration of the system $a = \dfrac{300 - 183.94}{75}$ = **1.547 m/s²**

(*iii*) Considering dynamic equilibrium of the block of mass 60 kg.

$\Sigma F_x = 0;$ $\quad 300 - T - F_i - F = 0$

or $\quad 300 - T - 60 \times 1.547 - 0.25 \times (60 \times 9.81) = 0$

or $\quad 300 - T - 92.82 - 147.15 = 0$

$\therefore$ Tension in the string $T = 300 - 92.82 - 147.15 =$ **60.30 N**

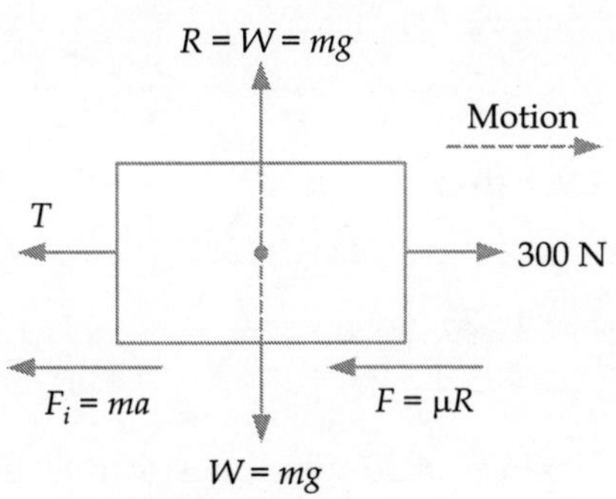

Fig. 10.27

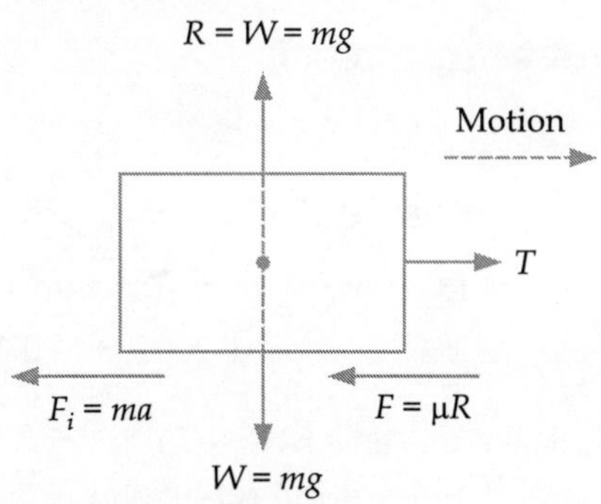

Fig. 10.28

The tension T in the string could also be worked out by considering dynamic equilibrium of the block of 15 kg mass.

$\Sigma F_x = 0;$

$$T - F_i - F = 0$$

or $\quad T - 15 \times 1.547 - 0.25 \times (15 \times 9.81)$

$\therefore \quad T = 15 \times 1.547 + 0.25 \times (15 \times 9.81) =$ **59.98 N**

EXAMPLE 10.34

A block weighing 40 N is accelerated along the horizontal plane by means of a 20 N weight attached to it through flexible, inextensible and weightless rope passing over a smooth pulley

as shown in Fig. 10.29. Using D'Alembert's principle, determine the acceleration of the system and the tension in the rope. Take coefficient of friction between the 40 N weight block and the horizontal plane as 0.3.

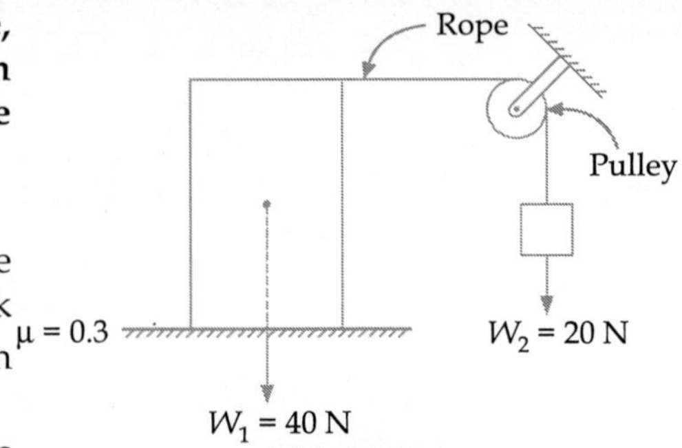

Fig. 10.29

Solution : It is to be noted that

(*i*) the acceleration of both the weights is same otherwise the rope would break or become slack

(*ii*) for a smooth frictionless pulley, the tension induced in the rope will be same throughout.

Considering motion of 20 N *Weight:* The weight is brought into dynamic equilibrium by applying inertia force F_i in a direction opposite to that of its motion which is downwards. Then

$$20 - T - F_i = 0$$

where $$F_i = ma = \frac{20}{9.81}a = 2.038\,a$$

or $$T = 20 - F_i = 20 - 2.038\,a \qquad ...(i)$$

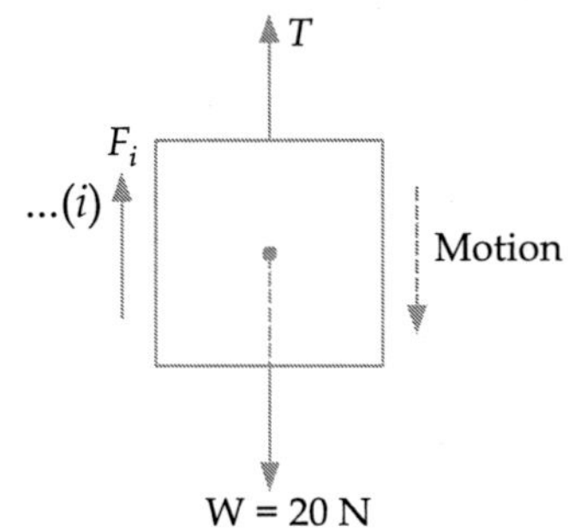

Fig. 10.30

Considering motion of block of 40 N *weight*

$$\text{Frictional force } F = \mu R = \mu W = 0.3 \times 40 = 12 \text{ N} \quad (\leftarrow)$$

$$\text{Inertia force } F_i = ma = \frac{W}{g}a = \frac{40}{9.81}a = 4.077a \quad (\leftarrow)$$

Then by D' Alembert's principle,

$$T - F - F_i = 0$$

or $$T = F + F_i = 12 + 4.077a \qquad ...(ii)$$

From expressions (*i*) and (*ii*),

$$20 - 2.038a = 12 + 4.077a$$

$\therefore$ Acceleration of the system $$a = \frac{20-12}{2.038+4.077} = 1.038 \text{ m/s}^2$$

Substituting $a = 1.038 \text{ m/s}^2$ in expression (*ii*), we obtain

Tension in the rope $T = 12 + 4.077 \times 1.038 =$ **16.23 N**

10.5. WORK, POWER AND ENERGY

Work: Work is said to be done when the point of application of force moves in the direction of force. The amount of work done equals the product of force and the resulting displacement in the direction of force.

Fig. 10.31

With reference to Fig. 10.31, F is the force acting an a body and the body moves a distance x in the direction of force. Then the work done on the body is

$$\text{Work done} = \text{force} \times \text{distance}\,; \quad W = Fx$$

Most of the time, the force and the displacement are not collinear. Work done by the force is then expressed as the product of the component of force in the direction of motion and the distance moved. With reference to Fig. 10.31, consider the force to be inclined at an angle θ with the direction of motion. Then

$$\text{Component of force in the direction of motion} = F \cos\theta$$

$$\text{Work done} = F \cos\theta \times x$$

When θ = 90°, cos θ = 0 and accordingly no work will be done. Hence if the force and displacement are at right angles, then the work done will be zero.

In case the force changes with displacement, then the work done can be obtained by the expression

$$W = \int_1^2 F(x)\, dx$$

If the force versus displacement curve is drawn (Fig. 10.32), then area under the curve gives the work done.

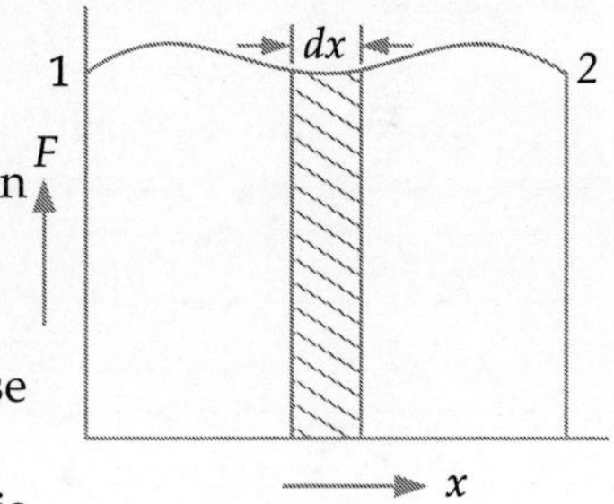

Fig. 10.32

Note:

(*i*) work done is a scalar quantity; it has a magnitude and sense but no direction.

(*ii*) work done by a force is zero if either the displacement is zero or the force acts normal to displacement.

(*iii*) work done is taken to be positive if the point of application of force moves in the direction of the force. Work done is considered negative when the direction of force is opposite to that of displacement.

The unit of work is joule (J) which is defined as the work done by a constant force of one newton acting on a body and moving it through a distance of one metre in its direction.

$$1\text{ J} = 1\text{ Nm}$$

• **Power:** Power is the rate of doing work and is obtained by dividing the work done by time. The unit of power is watt (W), kilowatt (kW) or megawatt (MW). Watt represents a work equivalent of one joule done per second

$$\text{Power} = \text{work done per unit time} = \text{J/s}$$

Thus $1\text{ W} = 1\text{ J/s} = 1\text{ Nm/s}$

$$1\text{ kW} = 10^3\text{ W} \quad \text{and} \quad 1\text{ MW} = 10^6\text{ W}$$

• **Potential energy and kinetic energy:** Energy is defined as the capacity to produce a change from existing conditions, *i.e.*, capacity to exert force through a distance and do work.

Potential energy is the energy possessed by a body due to its position or elevation; it represents the ability to do work against the body weight in lifting it from the surface of earth.

When a body of mass m kilogram is elevated through a distance h meters above the surface of earth. Then

$$\text{Weight of the body} = mg$$

$$\text{Potential energy} = \text{work done against gravity}$$

$$= \text{weight} \times \text{height raised} = mgh$$

Potential energy of a body is taken zero when its mass rests upon the earth. When the height of the mass changes from elevation h_1 to h_2, the change in potential (ΔPE) equals the work done on the body in raising it from lower elevation to higher elevation, and is given by

$$\Delta PE = mg\,(h_2 - h_1)$$

Consider a body of mass m shifting from position 1 (elevation h_1) to position 2 (elevation h_2) along path A.

Gravitational force on the body at position 1 = $-mg$

The negative sign is due to the fact that the weight of the body is acting downwards whilst the y-axis is chosen upwards.

Work done as the body moves from position 1 to position 2

$$W_{1\text{-}2} = \int F\,dy = \int_1^2 (-mg)\,dy = -mg\,(h_2 - h_1)$$

The work done against the force of gravity is stored as potential energy of the body, that is

$$W_{1\text{-}2} = mg\,h_1 - mg\,h_2 = (PE)_1 - (PE)_2 = \text{change in potential energy}$$

The work done is positive when $(h_2 - h_1)$ is negative, *i.e.*, the body moves downward. In that case, the weight of the body and its displacement are in the same direction. Further, the work done against gravitational force is independent of the path followed and depends on the initial and final values of the parameter mgh, *i.e.*, potential energy. This implies that the work done will be same and equal to $mg\,(h_1 - h_2)$ even when the body follows a path B which is different from the previous path.

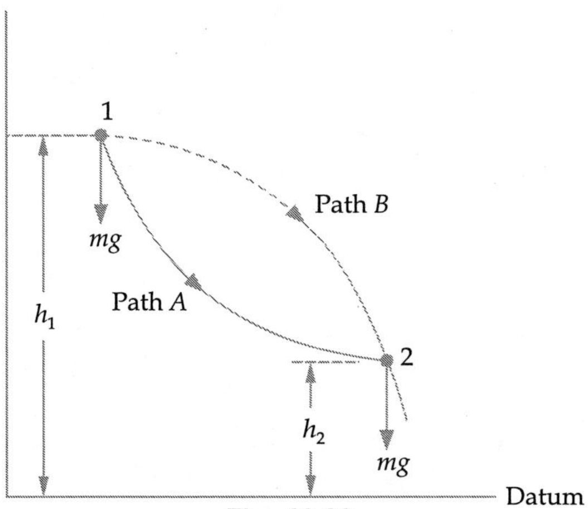

Fig. 10.33

Conservative force : If the work of force in moving an object between two positions is independent of the path followed by the object and can be expressed as a change in its potential energy, then such a force is called a *conservative force.* The notable examples of conservative force are the gravity force, elastic force, spring force etc.

Kinetic energy is the energy possessed by a body by virtue of its motion.

Consider a body of constant mass m which moves through a distance dx in the direction of force acting upon it.

Elementary work done = force × elemental distance moved

$$= \left(m\frac{dV}{dt}\right) \times dx = mV\,dV \qquad \left(\because \frac{dx}{dt} = V\right)$$

where V is the instantaneous velocity in the direction of distance moved. Then

Kinetic energy gained by the body = total work done = $m\int V\,dV$

If the body experiences a change in velocity from V_1 to V_2, then

$$\text{change in kinetic energy, } \Delta KE = m\int_1^2 V\,dV = \frac{1}{2}m\left(V_2^2 - V_1^2\right)$$

If the kinetic energy of the body at rest with respect to the earth is arbitrarily assigned a zero value, then

$$\text{Kinetic energy, } KE = \frac{1}{2}\,mV^2$$

where V is the velocity of the body relative to earth.

Since energy represents the capacity to do work, its units are same as those of work: Joule (J) or kilo joule (kJ).

EXAMPLE 10.35

A box of mass 80 kg is moving at a speed of 10 m/s on a ropeway. If the box is 40 m above the ground level, estimate the potential energy and kinetic energy of the box.

Solution:

$$\text{Potential energy} = mgh = 80 \times 9.81 \times 40 = 41392 \text{ J} = \mathbf{41.39 \text{ kJ}}$$

$$\text{Kinetic energy} = \frac{1}{2} mV^2 = \frac{1}{2} \times 80 \times 10^2 = 4000 \text{ J} = \mathbf{4 \text{ kJ}}.$$

EXAMPLE 10.36

A vehicle accelerates a glider of 125 kg mass from rest to a speed of 50 km/hr. Make calculations for the work done on the glider by the vehicle. What change would occur in the kinetic energy of the glider if subsequently its velocity reduces to 20 km/hr on the application of brakes?

Solution: Initial velocity of glider $V_1 = 0$ m/s

$$\text{Velocity of glider after acceleration } V_2 = 50 \text{ km/hr} = \frac{50 \times 10^3}{3600} = 13.89 \text{ m/s}$$

The work done on the glider in accelerating it from velocity V_1 to velocity V_2 equals the change in kinetic energy of the glider.

$$\therefore \text{ Work done on the glider } W = \frac{1}{2} m\left(V_2^2 - V_1^2\right) = \frac{1}{2} \times 125 \times 13.89^2 = \mathbf{12058 \text{ J}}$$

Upon application of brakes,

$$\text{Velocity of glider } V_3 = 20 \text{ km/hr} = \frac{20 \times 10^3}{3600} = 5.56 \text{ m/s}$$

$$\text{Change in kinetic energy } \Delta KE = \frac{1}{2} m\left(V_3^2 - V_2^2\right) = \frac{1}{2} \times 125(5.56^2 - 13.89^2) = \mathbf{10126 \text{ J}}$$

EXAMPLE 10.37

What will be the kinetic energy in kWh of an aeroplane which has a mass of 30 tons and is travelling at 1000 km/hr speed? If this plane is made to nose vertically upwards at this speed with power off, calculate the vertical distance through which the plane will move. Take g = 9.80 m/s².

Solution: Speed of the plane $V = 1000 \text{ km/hr} = \dfrac{1000 \times 10^3}{3600} = 277.78 \text{ m/s}$

$$\text{Kinetic energy of the plane} = \frac{1}{2} mV^2 = \frac{1}{2} \times (30 \times 10^3) \times (277.78)^2 = 77161743 \text{ J} = 77162 \text{ kJ}$$

1 kWh equals 3600 kJ

$$\therefore \text{ Kinetic energy of the plane} = \frac{77162}{3600} = 21.43 \text{ kWh}$$

When the power is off and the plane is made to move vertically upwards, the kinetic energy gets converted to potential energy. Thus

$$77161743 = mgh$$

∴ Vertical distance through which the plane moves,

$$h = \frac{77161743}{(1000 \times 10^3) \times 9.80} = \mathbf{7.87 \text{ m}}$$

EXAMPLE 10.38

An artificial satellite of mass 500 kg is moving towards the moon. Make calculations for the kinetic and potential energies

(*i*) relative to the earth when 50 km from launching and travelling at 2500 km/hr. Take earth's gravitational field equal to 7.9 m/s².

(*ii*) relative to the moon when travelling at the same velocity and 50 km from its destination where 1 kg mass has a weight of 3N.

Solution: Speed of the satellite, $V = 2500 \text{ km/hr} = \dfrac{2500 \times 10^3}{3600} = 694.45 \text{ m/s}$

(*i*) Kinetic energy $= \dfrac{1}{2} \times 500 \times (694.45)^2 = 120.6 \times 10^6 \text{ J} = 120.6 \text{ MJ}$

Potential energy $= mgh = 500 \times 7.9 \times (50 \times 10^3) = 197.5 \times 10^6 \text{J} = \mathbf{197.5 \text{ MJ}}$

(*ii*) 1 kg mass weighs 3 N, and hence 500 kg will weigh (relative to the moon) as 1500 N

Potential energy $= mgh = 1500 \times (50 \times 10^3) = 75 \times 10^6 \text{ J} = \mathbf{75 \text{ MJ}}$

Kinetic energy $= \dfrac{1}{2} mV^2 = \dfrac{1}{2} \times 500 \times (694.45)^2 = 120.6 \times 10^6 = \mathbf{120.6 \text{ MJ}}$

It is to be noted that kinetic energy is not dependent on gravitional field of the moon.

EXAMPLE 10.39

A block weighing 400 N is resting on a smooth horizontal surface. What work will be done if the block is to be moved through 2 m distance by applying

(*a*) a horizontal force of 150 N

(*b*) a force of 150 N whose line of action makes an angle of 30° with the horizontal.

Solution:

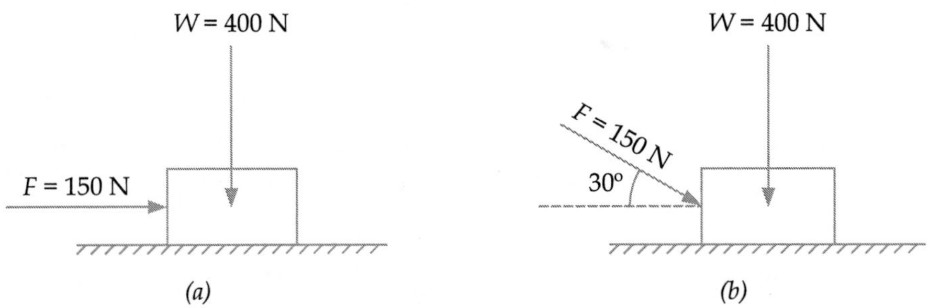

Fig. 10.34

Case (*i*): Work done = force × distance moved = 150 × 2 = **300** Nm

Case (*ii*): Work done = component of force in the direction of motion × distance moved

$= F \cos\theta \times x = 150 \cos 30° \times 2 = \mathbf{259.81} \text{ Nm}$

Note: The weight of the block does not contribute towards work done as its line of action is at right angles to that of displacement.

EXAMPLE 10.40

A fire engine lifts 30 × 10³ litres of water to a height of 5 m and discharges it with a velocity of 8 m/s. What is the amount of energy spent during this process?

If the task is accomplished in quarter of an hour, determine input power of the fire engine which has an overall efficiency of 75 percent.

Solution: Quantity of water lifted = 30×10^3 litres = $30\ m^3$

Since $1\ m^3$ of water has a mass of 1000 kg,

$$\text{mass of water lifted} = 30 \times 10^3 \text{ kg}$$

$$\text{Potential energy} = mgh = 30 \times 10^3 \times 9.81 \times 5 = 1471.5 \times 30^3 \text{ Nm}$$

$$\text{Kinetic energy at delivery} = \frac{1}{2} mV^2 = \frac{1}{2} \times (30 \times 10^3) \times 8^2 = 960 \times 10^3 \text{ Nm}$$

∴ Total energy required to be spent by the fire engine

$$= 1471.5 \times 10^3 + 960 \times 10^3 = 2431.5 \times 10^3 \text{ Nm}$$

This energy is spent by the fire engine in quarter an hour, *i.e.*, 900 seconds,

output power of fire engine = energy spent per second

$$= \frac{2431.5 \times 10^3}{900} = 2.70 \times 10^3 \text{ W} = 2.70 \text{ kW}$$

$$\text{Input power} = \frac{\text{output power}}{\text{overall efficiency}} = \frac{2.70}{0.75} = \mathbf{3.6\ kW}$$

EXAMPLE 10.41

A train weighing 5000 kN starts from rest and accelerates uniformly to 75 km/hr in 40 seconds. If the frictional resistance is estimated to 3 kN per 1000 kN of weight of trains, work out the maximum power required and the power required to maintain the speed of 75 km/hr.

Solution: $u = 0; \quad v = \dfrac{75 \times 1000}{3600} = 20.83 \text{ m/s}; \quad$ and $\quad t = 40$ seconds

Then from the kinematic relation $v = u + at$,

$$\text{acceleration } a = \frac{v - u}{t} = \frac{20.83 - 0}{40} = 0.521 \text{ m/s}^2$$

frictional resistance to motion R = 3 kN per 1000 kN of weight of train

$$= 3 \times \frac{5000}{1000} = 15 \text{ kN}$$

Let P be tractive effort in newtons. Then the force available for traction,

$$= P - R = P - 15000$$

Applying Newton's second law of motion $F = ma$, we have

$$(P - 15000) = \frac{5000 \times 10^3}{9.81} \times 0.521 = 265545$$

$$P = 280545 \text{ N}$$

$$\text{Maximum power required} = P \times v = 280545 \times 20.83 \text{ Nm/s}$$

$$= 5843752 \text{ W} = \mathbf{5.844\ MW}$$

(*b*) At any time after 40 seconds, the force is required only to overcome frictional resistance of 15 kN

$$\text{Power required} = (15 \times 10^3) \times 20.83 \text{ Nm/s} = 312450 \text{ W} = \mathbf{312.45\ kW}$$

EXAMPLE 10.42

A wooden block of 750 N weight is placed on an inclined plane which makes an angle of 30° with the horizontal. What work will be done if the block is to be moved upwards the plane for 2.5 m distance and the (*a*) plane is smooth, (*b*) plane is rough with coefficient of friction 0.2.

Solution:

Case (*a*): When the plane is smooth, the force of friction will be zero. Then the force P on the block when it is on the verge of moving upwards is given by

$$P = W \sin\theta = 750 \times \sin 30^\circ = 375 \text{ N}$$

Work done in pulling the block up

$= P \times s$ where s is the distance moved

$= 375 \times 2.5 =$ **937.5** Nm

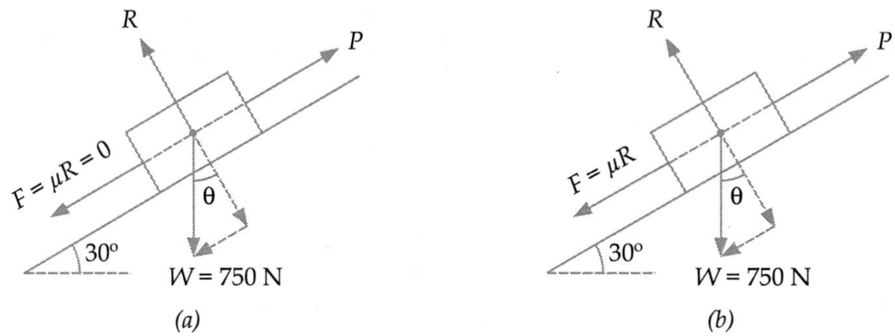

Fig. 10.35

Case (*b*): Refer Fig. 10.35 for the various forces acting on the block. For equilibrium of block,

Normal reaction $R = W\cos\theta = 750 \times \cos 30^\circ = 649.5$ N

Force of friction $F = \mu R = 0.2 \times 649.5 = 129.9$ N

Then the force P on the block when it is on the verge of moving upwards is given by

$$P = W\sin\theta + F = 750 \sin 30^\circ + 129.9 = 504.9 \text{ N}$$

Work done in pulling the block up

$= P \times s$ where s is the distance moved

$= 504.9 \times 2.5 =$ **1262.25** Nm

EXAMPLE 10.43

A 750 kW power engine, working at full power propells a 2500 kN train up an incline of 1 in 100 at a speed of 60 km/hr. If the track resistance is 5 N per kN weight of train, determine the acceleration with which the train is moving.

Solution: Speed of train $= 60 \text{ km/hr} = \dfrac{60 \times 1000}{3600} = 16.67 \text{ m/s}$

Frictional resistance of track $= 5 \times 2500 = 12.5 \times 10^3$ N

Slope of the track = 1 in 100

$$\therefore \quad \sin\theta = \tan\theta = \frac{1}{100}$$

Refer Fig. 10.36 for the various forces acting on the train. The component of weight $W\sin\theta$ acts down the plane and so does the frictional resistance F of the track (it has to be in a direction opposite to that of the motion of train).

∴ Net force on the train in the direction of motion

$$F_n = P - F - W\sin\theta$$

where P is the force exerted by the engine up the plane.

Power of the engine $= 750 \text{ kW} = 750 \times 10^3$ Nm/s

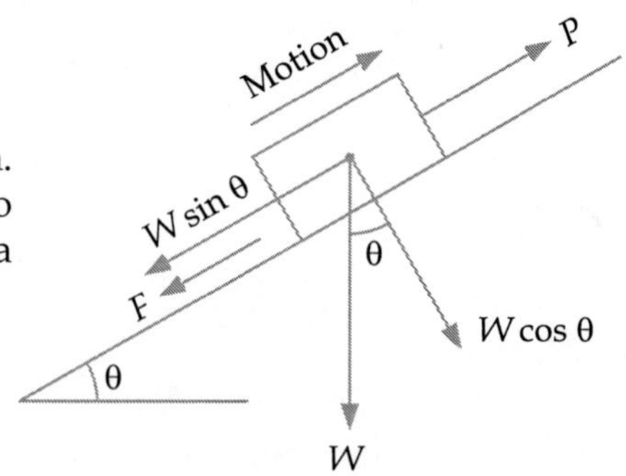

Fig. 10.36

$\therefore \quad P \times 16.67 = 750 \times 10^3; \quad P = 44.99 \times 10^3$ N

Thus $\quad F_n = 44.99 \times 10^3 - 12.5 \times 10^3 - 2500 \times 10^3 \times \frac{1}{100} = 7.49 \times 10^3$ N

Applying Newton's second law of motion $F = ma$, we have

$$7.49 \times 10^3 = \frac{2500 \times 10^3}{9.81} a$$

$$\therefore \quad \text{Acceleration of train } a = \frac{7.49 \times 10^3 \times 9.81}{2500 \times 10^3} = \mathbf{0.0294\ m/s}$$

EXAMPLE 10.44

A truck weighing 5 kN just moves freely (engine is not running) at 30 km/hr down a slope of 1 in 50, and the road resistance at this speed is just sufficient to prevent any acceleration. Estimate the road resistance per kN weight of truck.

What power will the engine have to exert to run up the same slope at double the speed when the road resistance remains the same? Take efficiency of engine as 40 percent.

Solution:

Case 1: *Truck moves freely down the plane.*

Refer Fig. 10.37 for the various forces acting on the truck.

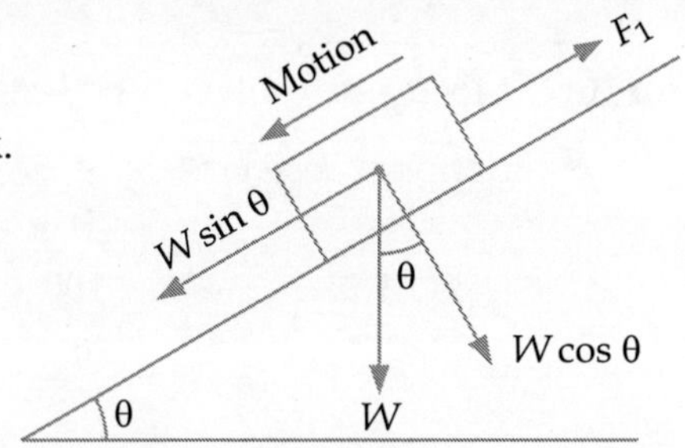

Fig. 10.37

Speed of truck = 30 km/hr

$$u = \frac{30 \times 1000}{3600} = 8.33 \text{ m/s}$$

slope of the truck = 1 in 50

$$\therefore \quad \sin\theta = \tan\theta = \frac{1}{50}$$

The component of weight $W \sin\theta$ acts down the plane and the road resistance F_1 acts up the plane (opposite to the direction of motion of truck).

Net force on the truck in the direction of motion,

$$= W\sin\theta - F_1$$

As the truck moves freely with uniform velocity, its acceleration is zero and so will be the net force ($F = ma$).

$$\therefore \quad W\sin\theta - F_1 = 0; \quad F_1 = W\sin\theta = (5 \times 10^3) \times \frac{1}{50} = 100 \text{ N}$$

$$\therefore \quad \text{Road resistance per kN weight of truck} = \frac{100}{5} = \mathbf{20\ N}$$

Case 2: *Truck moves upward with double the speed.*

Speed of truck $v = 2u = 2 \times 8.33 = 16.66$ m/s

Road resistance $F = 100$ N (same as before)

$$\sin\theta = \frac{1}{50}$$

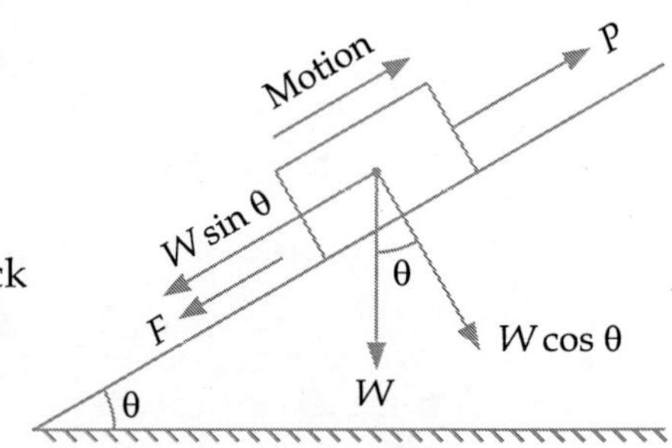

Fig. 10.38

Fig. 10.38 shows the system of forces acting on the truck during its upward motion.

Net force on the truck in the direction of motion

$$= P - F - W\sin\theta$$

where P is the force exerted by engine up the plane. Again the truck moves upward with uniform speed, *i.e.*, no acceleration and accordingly

$$P - F - W \sin\theta = 0$$

$$\therefore \quad P = F + W\sin\theta = 100 + 5 \times 10^3 \times \frac{1}{50} = 200 \text{ N}$$

Output power required = $P \times v$

$= 200 \times 16.66 = 3332 \text{ W} = 3.332 \text{ kW}$

$$\therefore \quad \text{Input power of engine} = \frac{\text{output power}}{\text{efficiency of engine}} = \frac{3.332}{0.4} = \mathbf{8.33\ kW}$$

EXAMPLE 10.45

Calculate the power required to pull a 400 kN train up an incline of 1 in 100 at a steady speed of 60 km/hr. Take frictional resistance equal to 5 N per kN weight of train.

At a particular instant while ascending the incline, the supply of steam to the locomotive pulling the train is shut off. How far the train will move before comming to rest? Assume that track resistance remains unchanged.

Solution: Velocity of train $v = 60 \text{ km/hr} = \dfrac{60 \times 1000}{3600} = 16.67 \text{ m/s}$

Frictional resistance F = 5 N per kN weight of train

$= 5 \times 4000 = 2000 \text{ N}$

Slope of plane = 1 in 100

$$\therefore \quad \sin\theta = \tan\theta = \frac{1}{100}$$

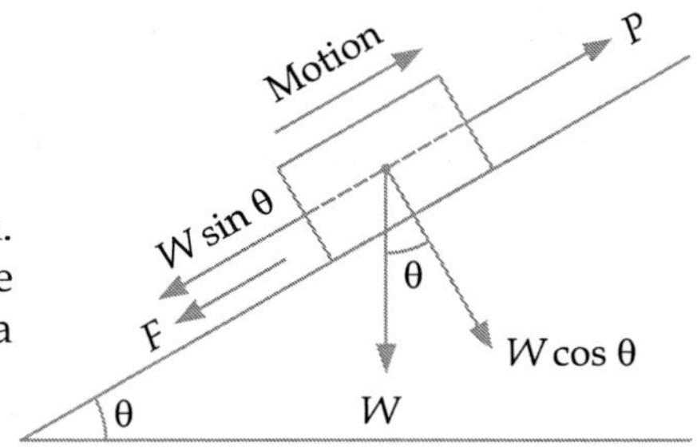

Fig. 10.39

Refer Fig. 10.39 for the various forces acting on the train. The component $W \sin\theta$ of the weight of train acts down the plane and so does the frictional resistance (it has to be in a direction opposite to that of the motion of the train).

Net force on the train in the direction of motion,

$$F_n = P - F - W\sin\theta$$

where P is the force exerted by the engine up the plane. As the train moves with uniform speed, its acceleration is zero and so will be the net force ($F = ma$)

$$\therefore \quad P - F - W\sin\theta = 0$$

$$P = F + W\sin\theta = 2000 + (400 \times 10^3) \times \frac{1}{100} = 6000 \text{ N}$$

Power required to pull the engine = $P \times v = 6000 \times 16.67$

$= 100.02 \times 10^3 \text{ Nm/s} = \mathbf{100.02\ kW}$

(*b*) When the steam supply is shut off, there will be no pull exerted by the engine. The force acting down the plane is provided by frictional resistance and the component $W \sin\theta$ of the weight of train.

Total retarding force = $F + W\sin\theta$

$$F_r = 2000 + (400 \times 10^3) \times \frac{1}{100} = 6000 \text{ N}$$

Let the train travel a distance s before coming to rest.

initial velocity = 16.67 m/s and final velocity = 0

Then from the work – energy equation,

$$-F_r \times s = \frac{1}{2} m (v^2 - u^2)$$

The negative sign with F_r stems from the fact that retarding force is acting opposite to the direction of motion.

$$-6000 \times s = \frac{1}{2} \times \frac{400 \times 10^3}{9.81} \times (0^2 - 16.67^2)$$

$$\therefore \text{ Distance moved } s = \frac{400 \times 10^3}{2 \times 9.81} \times \frac{16.67^2}{6000} = \mathbf{944.24\ m}$$

EXAMPLE 10.46

A force of 500 N applied initially and horizontal is just sufficient to move a body of 1250 N weight placed on a smooth horizontal surface. The force applied does not remain constant but varies linearly with distance moved by the body, the linear variation is prescribed by the relation

$$P = P_0 + 8x$$

where P_0 is the initial force and x is the distance moved by the body in metres. Depict the force-distance diagram and determine the work done when the body has moved through 20 m distance.

Solution: Initial force P_0 = 500 N

Force after the body has moved a distance of 20 m

$$P = P_0 + 8 \times 20$$
$$= 500 + (8 \times 20) = 660 \text{ N}$$

The force-distance variation has been depicted in Fig. 10.40.

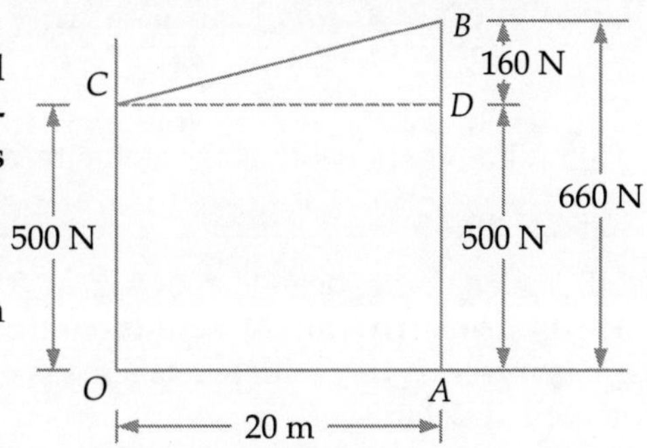

Fig. 10.40

The area of this diagram gives the work done.

Work done = area of traperium $OABC$

= area of rectangle $OADC$ + area of triangle BCD

$$= (500 \times 20) + \frac{1}{2} \times 20 \times 160 = \mathbf{11600\ Nm}$$

Alternatively:

$$\text{average force} = \frac{500 + 660}{2} = 580 \text{ N}$$

work done = average force × distance moved

= 580 × 20 = **11600 Nm**

EXAMPLE 10.47

When a body of mass m moves through a liquid at low velocity, the force of resistance due to viscosity is given by $F = kv$ where k is the resisting force at unit velocity. Show that velocity would decrease exponentially with time and linearly with displacement.

Solution: Retarding force $F = -kv$

$$\therefore \quad \text{Retardation } \frac{dv}{dt} = -\frac{kv}{m}$$

The negative sign stems from the fact that it is negative acceleration.
Separating the variables and upon integration, one obtains

$$\int \frac{dv}{v} = -\frac{k}{m}\int dt\,; \quad \log_e v = -\frac{kt}{m}$$

That gives: $\quad v = e^{-kt/m}$

The above relation shows that the velocity decreases exponentially with time.

(*ii*) Again $\dfrac{dv}{dt} = -\dfrac{kv}{m}$

Since $v = \dfrac{ds}{dt}$, we can write

$$\frac{d}{dt}\left(\frac{ds}{dt}\right) = -\frac{k}{m}\left(\frac{ds}{dt}\right)$$

or $$\frac{d^2s}{dt^2} = -\frac{k}{m}\frac{ds}{dt}$$

Upon integration:

$$\frac{ds}{dt} = -\frac{k}{m}s\,; \quad v = -\frac{k}{m}s$$

This relation shows that the velocity decreases linearly with displacement.

EXAMPLE 10.48

A high speed land racer is moving at a speed of 100 m/s. The resistance to motion of the vehicle is primarily due to aerodynamic drag which can be approximated as 0.2 V^2 newton with V is m/sec. If the vehicle has a mass of 4000 kg, what distance will it coast before its speed is reduced to 70 m/sec?

Solution: From Newton's law,

$$F = m\frac{dV}{dt}; \quad \frac{dV}{dt} = \frac{F}{m} = \frac{-0.2V^2}{4000} = -5 \times 10^{-5}V^2$$

The negative sign is because of drag.

Separating the variables and integrating, we have

$$\int \frac{dV}{V^2} = -5 \times 10^{-5}dt; \quad -\frac{1}{V} = -5 \times 10^{-5}t + C_1$$

The constant of integration C_1 can be evaluating by noting that at $t = 0$, $V = 100$ m/s. That gives $C_1 = 0.01$ and accordingly

$$-\frac{1}{V} = -5 \times 10^{-5}t + 0.01$$

Replacing $V = dx/dt$ and then separating the variables

$$\frac{dt}{5\times 10^{-5} + 0.01} = dx$$

For integration of above expression, let

$$\eta = 5 \times 10^{-5}t + 0.01; \quad d\eta = 5 \times 10^{-5}dt$$

Making these substitutions

$$\frac{d\eta}{\eta} = 5 \times 10^{-5}dx$$

or $\log_e \eta = 5 \times 10^{-5}x + C_2$

or $\log_e (5 \times 10^{-5}t + 0.01) = 5 \times 10^{-5}x + C_2$

For evaluating the constant for integration, we use the condition $x = 0$ at $t = 0$. That gives $C_2 = \log_e 0.01$.

Accordingly $\log_e (5 \times 10^{-5}t + 0.01) = 5 \times 10^{-5} + \log_e 0.01$

or $\log_e 0.01 + \log_e (5 \times 10^{-3}t + 1) = 5 \times 10^{-5}x + \log_e 0.01$

$\therefore \quad \log_e (5 \times 10^{-3}t + 1) = 5 \times 10^{-5}x$ ...(b)

Substituting $V = 70$ m/s in expression (a),

$$\frac{1}{70} = 5 \times 10^{-5}t + 0.01 \; ; \quad t = \left(\frac{1}{70} - 0.01\right) \div 5 \times 10^{-5}$$

$$= 85.71 \text{ sec}$$

The distance travelled by the racer is then obtained by substituting $t = 85.71$ sec in expression (b)

$$x = \frac{\log_e (5 \times 10^{-3} \times 85.71)}{5 \times 10^{-5}} = 7132.5 \text{ m}$$

$$= \mathbf{7.132 \text{ km}}$$

EXAMPLE 10.49

A particle of mass 1 kg moves in a straight line under the influence of a force which increases linearly with time at the rate of 60 N per second. At time $t = 0$, initial force may be taken as 50 N. Determine the acceleration and velocity of a particle 4 sec after it started from rest at the origin.

Solution: Let the variation of force with time be prescribed by the relation

$$F = at + b, \quad \text{where } a \text{ and } b \text{ are constants.}$$

$$\frac{dF}{dt} = a\,; \qquad 60 = a$$

Since $F = 50$ N at $t = 0$, we get

$$50 = 60 \times 0 + b\,; \qquad b = 50$$

Hence the force-time relation may be written as

$$F = 60\,t + 50$$

Putting $t = 4$ in the above relation, we get the value of force acting on the particle 4 sec after it started from rest. That is

$$F = 60 \times 4 + 50 = 290 \text{ N}$$

From the relation $F = ma$,

Acceleration after 4 sec, $a = \frac{290}{1} = \mathbf{290 \text{ m/s}^2}$

At any time t, the acceleration of the particle will be

$$a = \frac{F}{m} = \frac{60t + 50}{1} = 60\,t + 50$$

$$\frac{dv}{dt} = 60\,t + 50 \qquad \left(\because a = \frac{dv}{dt}\right)$$

Upon integration : $\int_0^v dv = (60\,t + 50)\,dt$

$$v = \frac{60t^2}{2} + 50\,t = 30\,t^2 + 50\,t$$

∴ velocity of particle 4 sec after start

$$v = 30 \times (4)^2 + 50 \times 4 = \mathbf{680\ m/s}$$

EXAMPLE 10.50

A hammer of mass 200 kg is made to fall freely from 2 m height on the head of a pile of 1500 kg mass. The pile is driven 5 cm into the ground in one blow. Determine (*a*) the common velocity of the pile and hammer after the impact, (*b*) the energy lost in the impact, and (*c*) the average resistance of the ground to penetration.

Solution: Let v be the velocity of the hammer when it reaches the pile. Then using the kinematic relation

$$v^2 - u^2 = 2\,gh\,;$$
$$v^2 - 0^2 = 2 \times 9.81 \times 2$$
$$v = \sqrt{2 \times 9.81 \times 2} = 6.26\ \text{m/s}$$

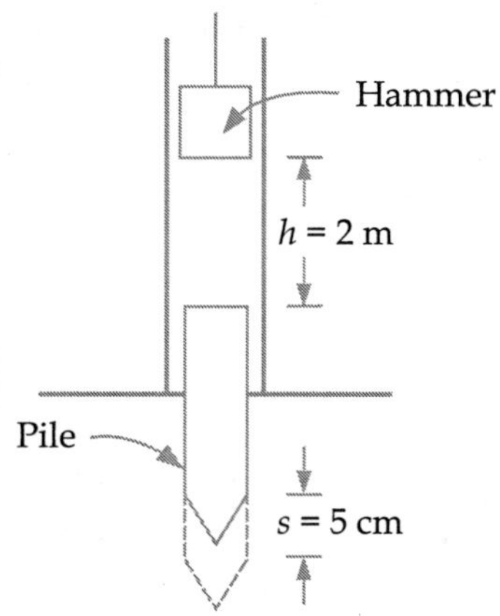

Fig. 10.41

Let V be the common velocity of the total mass $(M + m)$ in motion after the impact. Then from the principle of conservation of momentum,

momentum before impact = momentum after impact

$$mv = (M + m)\,V$$
$$200 \times 6.26 = (1500 + 200)\,V\,;$$
$$V = 0.736\ \text{m/s}$$

(*b*) Kinetic energy before impact $= \frac{1}{2}mV^2 = \frac{1}{2} \times 200 \times (6.26)^2 = 3918.76$ Nm

Kinetic energy after impact $= \frac{1}{2}(M + m)\,V^2$

$$= \frac{1}{2}(1500 + 200) \times (0.736)^2 = 460.44\ \text{Nm}$$

∴ Loss of kinetic energy at impact = 3918.76 – 460.44 = **3458.32 Nm**

(*c*) After penetration in the ground, the velocity of the system (pile + hammer) becomes zero. Using the kinematic relatio $v^2 = u^2 = 2as$, we have

$$0^2 - 0.736^2 = 2a \times (0.05)$$
$$a = -\frac{0.736^2}{2 \times 0.01} = -5.407\ \text{m/s}^2 \quad \text{(retardation)}$$

Let F_r be the average resistance exerted by the ground. Then net force acting on the system in the upward direction,

$$F_n = F_r - W$$

where W is the weight of hammer and pile.

Applying Newton's second law of motion $F = ma$, we have

$$F_r - (1500 + 200) \times 9.81 = (1500 + 200) \times 5.417$$
$$F_r - 16677 = 9208.8\,;\quad F_r = \mathbf{25885.8\ N}$$

EXAMPLE 10.51

A hammer of mass 0.6 kg and moving with a velocity of 6 m/s drives a nail of 20 gm mass horizontally into a fixed wooden block. The nail is stated to penetrate 25 mm per blow of hammer. Workout the resistance offered by the block to the penetration of nail and the energy lost in driving the nail 75 mm inside the block.

Solution: Refer Fig. 10.42 which shows the system for the various parameters inserted.

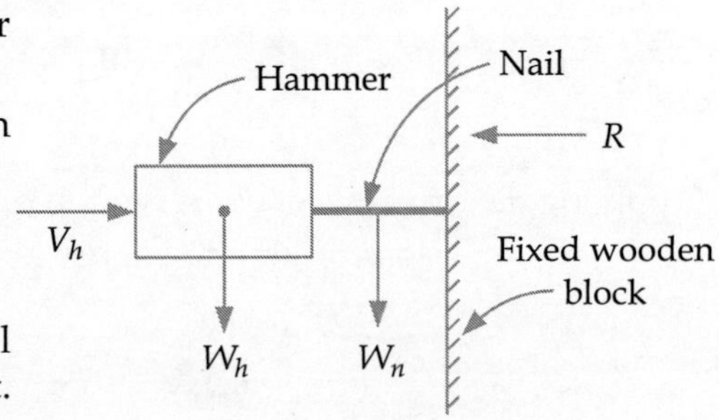

Fig. 10.42

Applying the principle of conservation of momentum

momentum before impact

= momentum after impact

$$m_h V_h + m_n V_n = (m_h + m_n)\, V$$

where suffix h and n refer to hammer and nail respectively and V is the common velocity after impact.

$0.6 \times 5 + 0.02 \times 0 = (0.6 + 0.02)\, V \quad ; \quad V = 4.838$ m/s.

The resistance offered by the block will retard the motion and bring the nail and hammer to rest after blow has been given.

Then from the work energy relation,

$$-R \times s = \frac{1}{2}\, m\, (v^2 - u^2)$$

The negative sign with R stems from the fact that resistance force acts opposite to the direction of motion of nail

$$-R \times 0.025 = \frac{1}{2} \times (0.6 + 0.02)\,(0 - 4.838^2)$$

Solution gives: R = **290.24 N**

The weight of the hammer and the nail do not contribute towards work done because their line of action is normal to that of displacement.

(*b*) There is 25 mm of penetration in 1 blow

∴ 75 mm of peneration would take place in 3 blows.

$$\text{Kinetic energy before impact} = \frac{1}{2} \times 0.6 \times 6^2 = 10.8 \text{ Nm}$$

$$\text{Kinetic energy after impact} = \frac{1}{2}\,(0.6 + 0.02) \times 4.838^2 = 7.26 \text{ Nm}$$

Loss in kinetic energy at each blow,

$$= 10.8 - 7.26 = 3.54 \text{ Nm}$$

∴ Total loss in energy = 3.54 × 3 = **10.62 Nm**

EXAMPLE 10.52

A body of mass 10 kg is suspended by a string of length 1 m. It is struck by a bullet travelling horizontally with a velocity of 450 m/s. The bullet weighs 30 gm and gets embedded into the body after striking it. Determine the maximum angle through which the body swings.

Solution: Refer Fig. 10.43 which shows the given system with the various parameters inserted.

Applying the principle of conservation of momentum,

momentum before impact

= momentum after impact

$m_a V_a + m_b V_b = (m_a + m_b)\, V$

where V is the common velocity with which the body and the bullet together after impact.

$$\frac{30}{1000} \times 450 + 10 \times 0 = \left(\frac{30}{1000} + 10\right) V$$

or $13.5 = 10.03\, V$; $\therefore\ V = 1.346$ m/s

Applying the principle of conservation of energy, we have

loss of kinetic energy = gain of potential energy

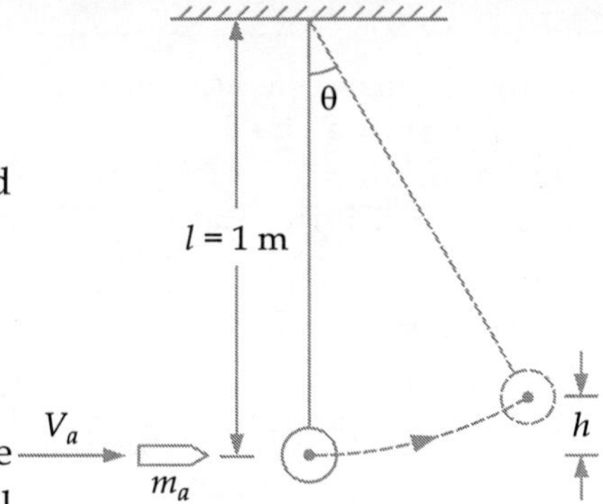

Fig. 10.43

$$\frac{1}{2}\,(m_a + m_b)\, V^2 = (m_a + m_b)\, gh$$

where h is the height to whcih the body rises

$$h = \frac{V^2}{2g} = \frac{1.346^2}{2 \times 9.81} = 0.0923 \text{ m}$$

From the geometry of Fig. 8.95,

$$\cos\theta = \frac{l - h}{l} = \frac{1 - 0.0923}{1} = 0.9077$$

$$\therefore \quad \theta = \cos^{-1}(0.9077) = \mathbf{24.81^\circ}$$

10.6 WORK-ENERGY PRINCIPLE

Consider a force F acting on an object which may be displaced from position 1 to position 2 over the course of action covering distance s as shown in Fig. 10.44. When the object is subjected to more than one force, then F represents the net force or the resultant force acting on the object.

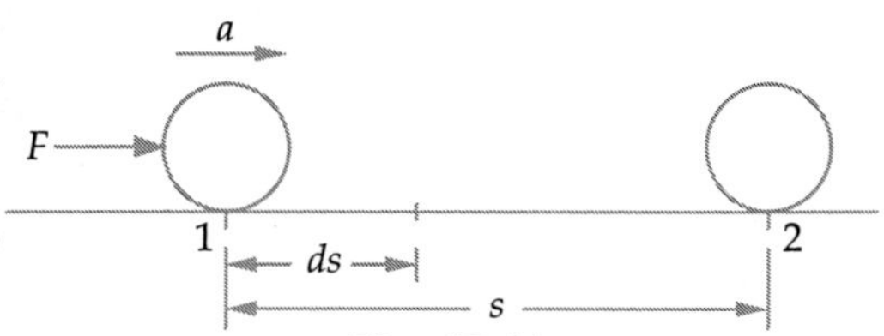

Fig. 10.44

For an elementary distance ds travelled by the objected in time dt, the work done by the force would be

$$dW = F\, ds \qquad \text{...(i)}$$

Newtons second law of motion provides that

$$F = \text{mass} \times \text{acceleration} = ma$$

Further, $$a = \frac{dv}{dt} = \frac{dv}{ds} \times \frac{ds}{dt} = v\frac{dv}{ds}$$

Accordingly expression (i) may be rewritten as

$$dW = \left(m \times v\frac{dv}{ds}\right) \times ds = mv\, dv \qquad \text{...(ii)}$$

Integrating the above identity with in appropriate limits

$$\int_1^2 dW = m\int_1^2 v\, dv$$

or $$W_{1-2} = m\left|\frac{v^2}{2}\right|_1^2 = \frac{1}{2}mv_2^2 - \frac{1}{2}mv_1^2 = (KE)_2 - (KE)_1 \qquad \text{...(10.19)}$$

Thus the work done on the object equals the change in kinetic energy of the object Equation 10.19 represents the well-known **work-energy principle** in mechanics.

10.7 CONSERVATION OF MECHANICAL ENERGY

In the conservative force field, the work done in moving an object from position 1 to position 2 is independent of the path followed and depends on a change in the potential energy of the end states only. That is

$$W_{1-2} = (PE)_1 - (PE)_2$$

Equating this expression for work done with that provided by the work-energy equation, we get

$$(PE)_1 - (PE)_2 = (KE)_2 - (KE)_1$$

or $$(PE + KE)_1 = (PE + KE)_2 = \text{constant} \qquad \text{...(10.20)}$$

The above relation implies that the sum of *PE* and *KE* remains constant for all positions of the object during its motion in the conservative force field.

The function (*PE* + *KE*), *i.e.*, the sum of potential energy and kinetic energy of an object is called mechanical energy.

Equation 10.20 is referred to as the principle of conservation of mechanical energy.

In general, the law of conservation of energy may be stated as

"the total energy possessed by an object remains constant provided no energy is added to or subtracted from it"

Alternatively:

"the energy can neither be created nor destroyed though it can be transformed form one form to another".

Consider a body of mass *m* resting on the top of a tower of height *h*. At this position

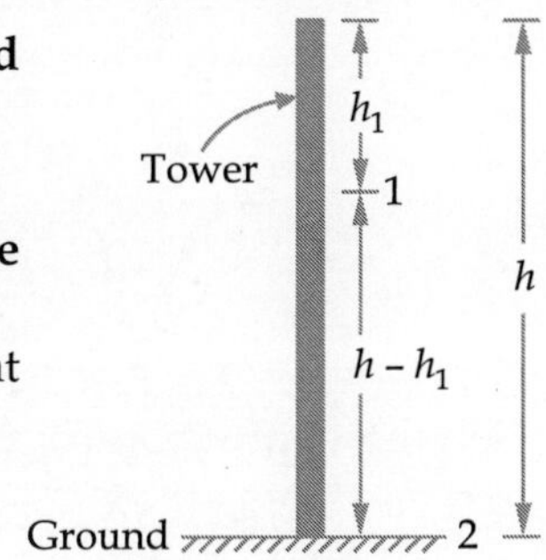

Fig. 10.45

$$KE \text{ of body} = \frac{1}{2}mv^2 = 0 \qquad (\because \ v = 0)$$

PE of body with respect to ground = mgh

$\therefore$ Total energy of the body at the top of tower is

$$= KE + PE = 0 + mgh = mgh \qquad \text{...}(i)$$

Energy at position 1:

Let the body fall to position 1 which is at a distance h_1 from the top of the tower. The velocity of the body at this location can be worked out from the kinematic equation $v^2 - u^2 = 2as$. That gives

$$v_1^2 - 0^2 = 2gh_1; \quad v_1^2 = 2gh_1$$

$$KE \text{ of the body at position 1} = \frac{1}{2}mv_1^2 = \frac{1}{2}m \times 2gh_1 = m\,gh_1$$

PE of the body at position 1 = $mg\,(h - h_1)$

$\therefore$ Total energy of the body at position 1

$$= mgh_1 + mg(h - h_1) = mgh \qquad \text{...}(ii)$$

Energy at position 2:

Finally when the body has the complete fall from the top of tower to ground level (position 2), we have

$$v_2^2 - 0^2 = 2gh\,; \quad v_2^2 = 2gh$$

$$KE \text{ at ground level} = \frac{1}{2}mv_2^2 = \frac{1}{2}m \times 2gh = mgh$$

PE at ground level = $mg \times 0 = 0$

∴ Total energy of the body at ground level

$$= mgh + 0 = mgh \qquad ...(iii)$$

The above analysis does illustrate that the sum of *KE* and *PE* of a body remains constant under the action of gravity force (a conservative force). Thus, though the *KE* and *PE* individually change throughout the motion, their sum always remains constant.

EXAMPLE 10.53

A shell of 80 kg mass in travelling with a velocity of 500 m/s. During its motion, it splits into two equal portions which continue to travel in the same direction. If an energy equivalent to 1.5×10^6 Nm is released at the instant of explosion, determine the subsequent velocities of the two portions.

Solution: Let V_1 and V_2 be the velocities of the two portions. By conservation of momentum,

momentum before explosion = momentum after explosion

$$80 \times 500 = (m_1 V_1 + m_2 V_2) = 40\ V_1 + 40\ V_2$$

$$V_1 + V_2 = 1000 \qquad ...(i)$$

Through conservation of energy,

initial kinetic energy + kinetic energy of explosion = final kinetic energy

or $$\frac{1}{2} \times 80 \times 500^2 + 1.5 \times 10^6 = \frac{1}{2} \times 40 \times V_1^2 + \frac{1}{2} \times 40 \times V_2^2$$

or $$10 \times 10^6 + 1.5 \times 10^6 = 20\ V_1^2 + 20\ V_2^2$$

or $$V_1^2 + V_2^2 = 0.575 \times 10^6 \qquad ...(ii)$$

Solving expression (*i*) and (*ii*), we get

$$V_1^2 + (1000 - V_1)^2 = 0.575 \times 10^6$$

or $$V_1^2 + 1 \times 10^6 + V_1^2 - 2000\ V_1^2 = 0.575 \times 10^6$$

or $$2\ V_1^2 - 2000\ V_1^2 + 0.425 \times 10^6 = 0$$

∴ $$V_1 = \frac{-(-2000) \pm \sqrt{(-2000)^2 - 4 \times 2 \times (}}{2 \times 2}$$

$$= \frac{2000 \pm 774.6}{4} = 693.65 \quad \text{or} \quad 306.35$$

Thus V_1 = **693.65 m/s** and V_2 = **306.35 m/s**

EXAMPLE 10.54

A body of 5 kg mass is initially at rest on a rough horizontal surface (μ = 0.2) and is acted upon by a 20 N pull applied horizontally. Calculate:

(*a*) the work done by the net force on the body in 5 seconds

(*b*) change in kinetic energy of the body in 5 seconds. Comment on the results.

Solution: Refer Fig. 10.46 for the various forces acting on the body.

Normal reaction R = weight of the body

$$= 5 \times 9.81 = 49.05 \text{ N}$$

Force of friction $F_f = \mu R = 0.2 \times 49.05 = 9.81$

Net force causing motion = $20 - 9.81 = 10.19$ N

Applying Newton's second law of motion $F = ma$, we have

$$10.19 = 5 \times a$$

∴ Acceleration of the body $a = \frac{10.19}{5} = 2.038 \text{ m/s}^2$

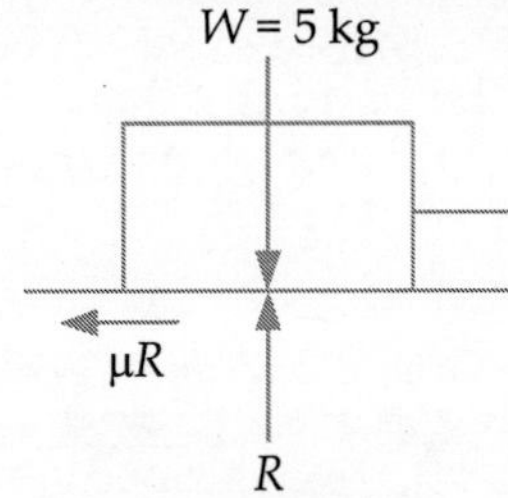

Fig. 10.46

From kinetmatic equation $s = ut + \frac{1}{2}at^2$, the distance moved by the body in 5 seconds is

$$s = 0 \times 5 + \frac{1}{2} \times 2.038 \times 5^2 = 25.475 \text{ m}$$

∴ Work done by the net force = 10.19 × 25.475 = **259.59 Nm**

(*b*) The body starts from rest ($u = 0$) and accordingly initial kinetic energy of the body is zero ($KE_1 = 0$)

Final kinetic energy $KE_2 = \frac{1}{2}mV^2$

The velocity v can be worked out from the relation $v = u + at$,

$$v = 0 + 2.038 \times 5 = 10.19 \text{ m/s}$$

$$KE_2 = \frac{1}{2} \times 5 \times (10.19)^2 = 259.59$$

∴ Change in kinetic energy = $KE_2 - KE_1$ = 259.59 − 0 = **259.59 Nm**

Comment: Work done by net force on the body is equal to the change in kinetic energy of the body. This is in accordance with the work-energy principle.

EXAMPLE 10.55

A block of weight 2000 N rests on a rough horizontal surface (μ = 0.2) and is pulled by a force of 800 N applied at an angle of 30° to the horizontal. Determine the velocity attained by the block after it has moved 20 m starting from rest. Proceed to calculate the further distance moved by the body if the pull is removed. Use work–energy relation.

Solution: Refer Fig. 10.47 for the various forces acting on the block.

Normal reaction

$$R = W - P \sin\theta$$
$$= 2000 - 800 \sin 30^\circ = 1600 \text{ N}$$

Force of friction

$$F = \mu R = 0.2 \times 1600 = 320 \text{ N}$$

Unbalanced force causing motion

$$= P \cos 30^\circ - F$$
$$= 800 \times 0.866 - 320 = 372.82 \text{ N}$$

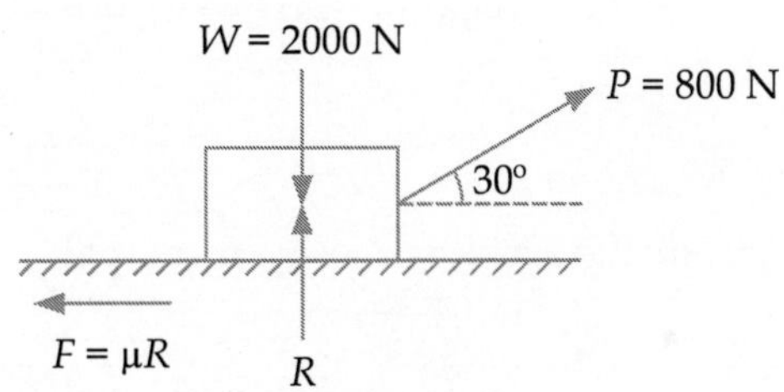

Fig. 10.47

Applying the work-energy relation,

work done = change in kinetic energy

force × distance moved = change in kinetic energy

$$372.82 \times 20 = \frac{1}{2} m (v^2 - u^2) = \frac{1}{2} \times \frac{2000}{9.81} (v^2 - 0^2)$$

$$v^2 = \frac{372.82 \times 20 \times 2 \times 9.81}{2000} = 73.15 \; ; \; v = \mathbf{8.55 \text{ m/s}}$$

(*b*) When the pull is removed, there will be only force of friction which retards the motion and brings the body to rest.

$$F = \mu R = \mu W = 0.2 \times 2000 = 400 \text{ N}$$

initial velocity = 8.55 m/s

final velocity = 0

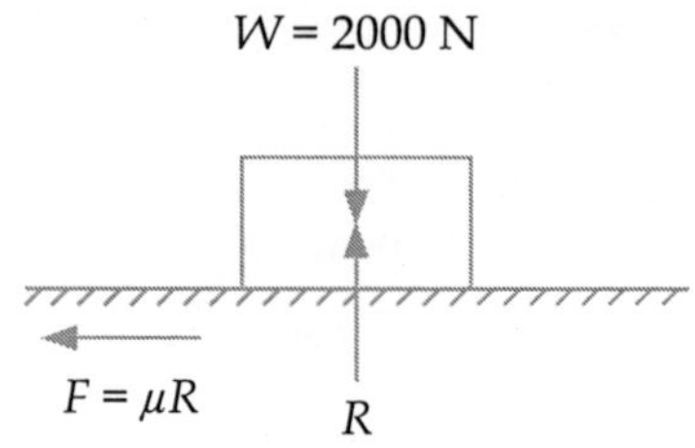

Fig. 10.48

Then from the work - energy equation,

$$-F \times s = \frac{1}{2} m (v^2 - u^2)$$

The negative sign with F stems from the fact that frictional force acts opposite to the direction of motion.

$$-400 \times s = \frac{1}{2} \times \frac{2000}{9.81} (0^2 - 8.55^2)$$

$$\therefore \quad \text{Distance moved } s = \frac{2000}{2 \times 9.81} \times \frac{8.55^2}{400} = \mathbf{18.63 \ m}$$

EXAMPLE 10.56

A train weighing 2 × 10⁶ N starts from rest with an acceleration of 0.8 m/s² and acquires a speed of 90 km/hr. Determine the kinetic energy corresponding to final speed and the average power required.

Subsequently the power is shut off and the train is subjected to a retarding force equal to 8% of the weight of train. Calculate the distance the train will travel before coming to rest.

Solution: Final speed of train $V = 90 \text{ km/hr} = \dfrac{90 \times 1000}{3600} = 25 \text{ m/s}$

$$\text{Kinetic energy of the train} = \frac{1}{2} mV^2 = \frac{1}{2} \times \frac{2 \times 10^6}{9.81} \times 25^2 = \mathbf{63.71 \times 10^6 \ Nm}$$

From the kinematic relation $v = u + at$, the time taken to acquire the speed of 25 m/s is given by

$$t = \frac{v - u}{a} = \frac{25 - 0}{0.8} = 31.25 \text{ sec}$$

From the work - energy principal,

work done = change in kinetic energy

$= 63.71 \times 10^6$ Nm ($\because$ initial $KE = 0$)

This change in kinetic energy is in a period of 31.25 second.

$$\text{Work done per second} = \frac{63.71 \times 10^6}{31.25} = 2.039 \times 10^6 \text{ Nm/s}$$

$$\therefore \text{ Average power} = 2.039 \times 10^6 \text{ J/s} = 2.039 \times 10^6 \text{ W} = \mathbf{2.039 \ MW}$$

(*b*) Retarding force = 8% of weight of train $= \dfrac{8}{100} \times (2 \times 10^6) = 0.16 \times 10^6$ N

Again from the work - energy principle,

force × distance = change of kinetic energy

$(0.16 \times 10^6) \times s = 63.71 \times 10^6$

∴ Distance travelled by the train before coming to rest,

$$s = \frac{63.71 \times 10^6}{0.16 \times 10^6} = \mathbf{398.19\ m}$$

EXAMPLE 10.57

A wooden block of weight 40 N rests on a rough horizontal plane having friction coefficient μ = 0.35. The block is struck by a bullet travelling horizontally with a velocity of 750 m/s and weighing 0.25 N. Workout the distance by which the block is displaced from its initial position.

It may be presumed that the bullet after striking the block gets embedded in it.

Solution: Refer Fig. 10.49 which shows the block with various parameters inserted.

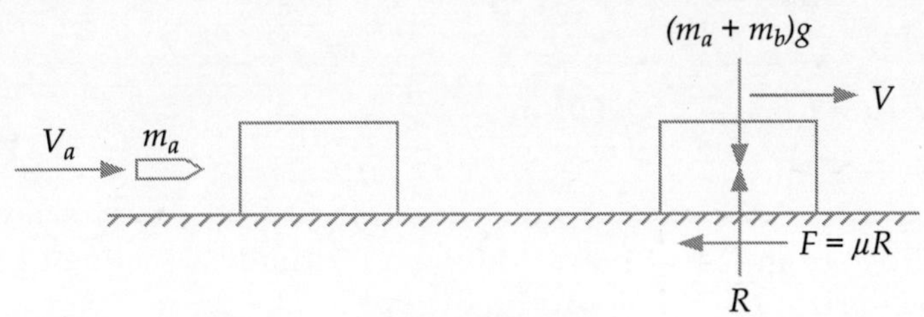

Fig. 10.49

Applying the principle of conservation of momentum,

momentum before impact = momentum after impact

$$m_a V_a + m_b V_b = (m_a + m_b) V$$

or $$\frac{0.25}{9.81} \times 750 + \frac{40}{9.81} \times 0 = \frac{0.25 + 40}{9.81} \times V$$

or $19.11 = 4.103\ V$; $V = \mathbf{4.657\ m/s}$

Normal reaction $R = (m_a + m_b)\, g = \dfrac{0.25 + 40}{g} \times g = 40.25$ N

Force of friction $F = \mu R = 0.35 \times 40.25 = 14.087$ N

Applying the work – energy correlation,

work done to overcome the frictional force = kinetic energy lost by the block with bullet embedded

$$F \times s = \frac{1}{2}(m_a + m_b)\, V^2$$

$$14.087 \times s = \frac{1}{2}\left(\frac{0.25 + 40}{9.81}\right) \times 4.657^2 = 43.632$$

∴ Displacement of block $s = \dfrac{43.632}{14.087} = \mathbf{3.097\ m}$

EXAMPLE 10.58

A box of 60 N weight is pulled up along a 30° plane by applying a 50 N force as shown in Fig. 10.50. The box starts from rest and the surface of the plane is smooth. What speed the box will attain after travelling 3 m up the plane?

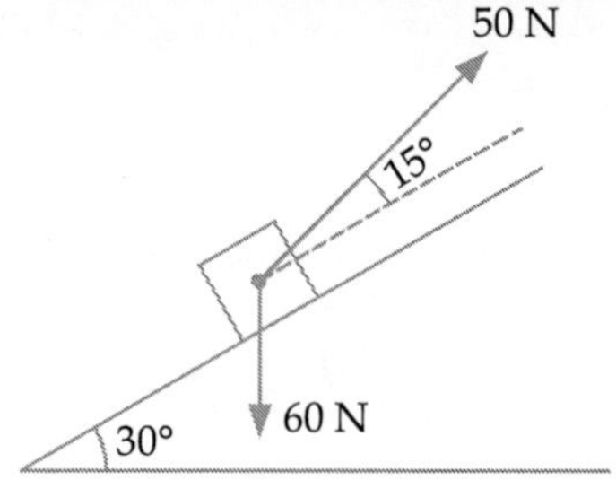

Fig. 10.50

Solution : Using the principle of work and energy

work done = change in kinetic energy

Component of weight parallel to plane

$= 60 \sin 30° = 30 \text{ N } (\leftarrow)$

Component of pull parallel to plane $= 50 \sin 15° = 48.3 \text{ N } (\rightarrow)$

Effective force $= 48.3 - 30 = 18.3$ N

Work done $= 18.3 \times 4 = 732.2$ Nm

Hence $$73.2 = \frac{1}{2} m V_2^2 - \frac{1}{2} m V_1^2 = \frac{1}{2} m V_2^2 \qquad (\because \; V_1 = 0)$$

$$= \frac{1}{2} \times \frac{60}{9.81} V_2^2 = 3.058 \, V_2^2$$

$\therefore$ Speed of box $V_2 = \sqrt{\dfrac{73.2}{3.058}} =$ **4.892 m/s**

EXAMPLE 10.59

A block of 4 kg mass slides from rest at point 1 along a frictionless inclined plane that makes an angle of 30° with the horizontal. What will be the speed of the block at point 2 which lies at a distance of 2.5 m from point?

Obtain your solution by the principle of work and energy or by the principle of conservation of energy.

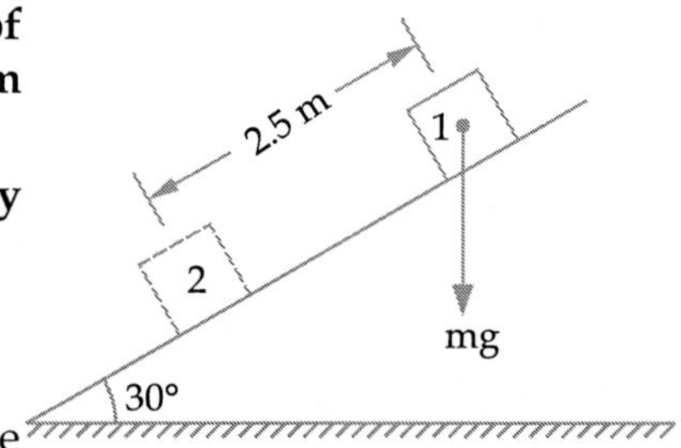

Fig. 10.51

Solution : Using the principle of work and energy

$$W_{12} = (KE)_2 - (KE)_1$$

Component of weight of block acting parallel to the plane

$= mg \sin \theta = (4 \times 9.81) \times \sin 30° = 19.62$ N

Work done during sliding from position 1 to position 2

$$W_{12} = 19.62 \times 2.5 = 49.05 \text{ Nm}$$

Kinetic energy at position 1 $= \frac{1}{2} m V_1^2 = 0 \qquad (\because \; V_1 = 0)$

Kinetic energy at position 2 $= \frac{1}{2} m V_2^2 = \frac{1}{2} \times 4 \times V_2^2 = 2 \, V_2^2$

$\therefore \quad 49.05 = 2 V_2^2 ; \quad V_2 = \sqrt{\dfrac{49.05}{2}} =$ **4.95 m/s**

(*b*) Using the principle of conservation of mechanical energy

$$(KE + PE)_1 = (KE + PE)_2$$

Taking the datum at the level of position 2,

$$0 + mg \times 2.5 \sin 30° = \frac{1}{2} m V_2^2 + 0$$

or $$(4 \times 9.81) \times 2.5 \times 0.5 = \frac{1}{2} \times 4 \times V_2^2$$

$\therefore \quad V_2 = \sqrt{\dfrac{(4 \times 9.81) \times 2.5 \times 0.5}{2}} =$ **4.95 m/s**

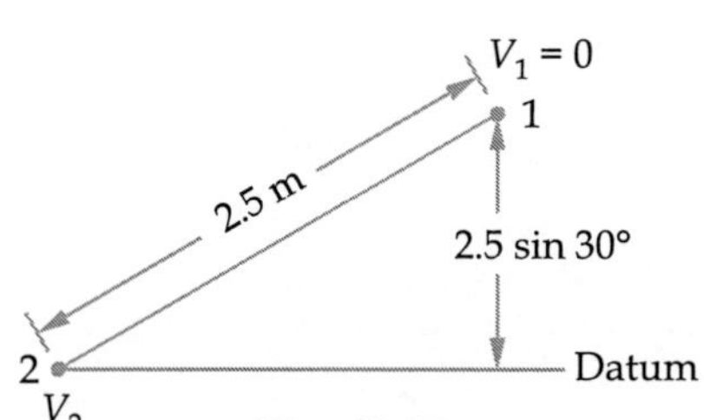

Fig. 10.52

EXAMPLE 10.60

Two bodies weighing 300 N and 450 N are hung to the ends of a rope passing over an ideal pulley. With what acceleration would the heavier body come down? What is the tension in the rope ?

Obtain your solution by using (*i*) the principle of work and energy (*ii*) the principle of impulse and momentum and (*iii*) the D'Alembert's principle.

Solution: Refer Fig. 10.53, for the arrangement of the system and free body diagram for each body.

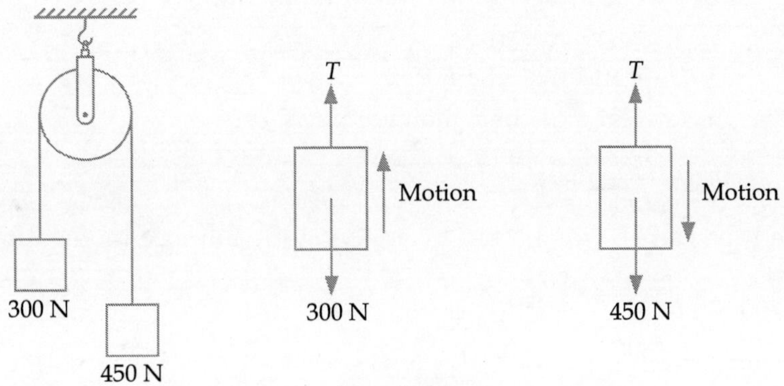

Fig. 10.53

It is to be noted that

(*i*) the acceleration of both the weights is same otherwise the rope would break or become slack.

(*ii*) for a smooth frictionless pulley, the tension induced in the rope will be same throughout.

- **Work-Energy Principle :** Let s be the distance moved by each body. Then

$$\text{work done by 300 N body} = (T - 300)\, s$$
$$\text{work done by 450 N body} = -(T - 450)\, s$$

The – *ve* sign is due to the fact that the direction of motion of 450 N body is opposite to that of 300 N body.

Total work done by the system $= (T - 300)\, s + [-(T - 450)\, s] = 150\, s$

Change in kinetic energy of :

$$\text{300 N body} = \frac{1}{2} \times \frac{300}{9.81}(v^2 - 0^2) = 15.29v^2$$

$$\text{450 N body} = \frac{1}{2} \times \frac{450}{9.81}(v^2 - 0^2) = 22.93v^2$$

Total change in kinetic energy of the system $= 15.29v^2 + 22.93v^2 = 38.22v^2$

From the principle of work and energy,

$$\text{work done} = \text{change in kinetic energy}$$
$$150\, s = 38.22v^2\ ;\quad v^2 = 3.925s$$

Using the kinematic equation, $v^2 - u^2 = 2as$, we have

$$3.925\, s = 2as \qquad (\because u = 0)$$

$\therefore$ Acceleration of the bodies $a = \dfrac{3.925}{2} = \mathbf{1.962\ m/s^2}$

The tension in the rope can be worked out by applying the work-energy principle to any one of the bodies. Considering the 300 N body:

$$(T - 300)\, s = 15.29v^2$$
$$= 15.29 \times 3.925s$$

∴ Tension in the rope $= 300 + (15.29 \times 3.925) = \mathbf{360\ N}$

- **Impulse-momentum principle:** Net force in the direction of motion

$$= (T - 300) - (T - 450) = 150\text{ N}$$

Let this force act for time t seconds. Then from the impulse momentum principle,

impulse = change of momentum of the system

net force × time = final momentum – initial momentum

$$150 \times t = \left(\frac{300}{9.81}v + \frac{450}{9.81}v\right) - 0$$

Initially both bodies are at rest and their momentum is zero

$$150t = \frac{750}{9.81}v\ ; \quad v = 1.962t$$

The tension in the rope can be worked out by applying the impulse-momentum principle to any one of the bodies.

Considering the 300 N body

$$(T - 300)t = \frac{300}{9.81} \times v = \frac{300}{9.81} \times 1.962t$$

∴ Tension in the rope $T = 300 + \dfrac{300}{9.81} \times 1.962 = \mathbf{360\ N}$

Using the kinematic equation $v = u + at$, we have

$$1.962t = 0 + at \qquad (\because u = 0)$$

∴ Acceleration of the bodies $a = \mathbf{1.962\ m/s^2}$

- **D'Alembert's Principle**

Considering motion of 300 N: The weight is brought into dynamic equilibrium by applying inertia force F_i in a direction opposite to that of its motion, which is upwards. Then by D'Alembert's principle

$$T - 300 - F_i = 0 \quad \text{where } F_i = ma$$

$$\therefore \quad T = 300 + F_i = 300 + \frac{300}{9.81}a \qquad \ldots(i)$$

Considering motion of 450 N: The weight moves down wards and accordingly the inertia force will act upwards. Then

$$T - 450 + F_i = 0$$

$$\therefore \quad T = 450 - F_i = 450 - \frac{450}{9.18}a \qquad \ldots(ii)$$

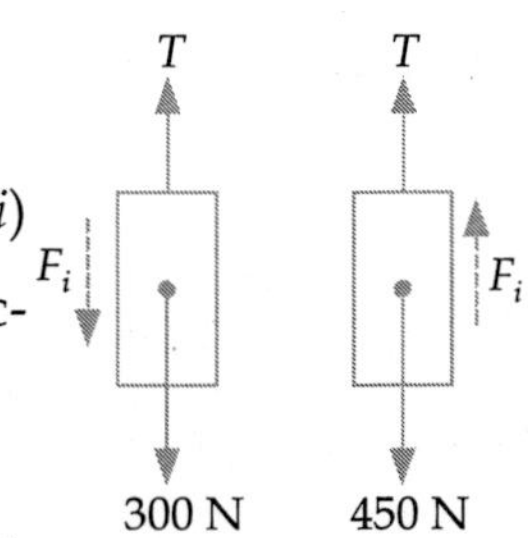

Fig. 10.54

From expressions (i) and (ii),

$$300 + \frac{300}{9.81}a = 450 - \frac{450}{9.81}a$$

or

$$\frac{300 + 450}{9.81}a = 450 - 300 = 150$$

$$\therefore \quad \text{Acceleration of the system } a = \frac{150 \times 9.81}{750} = \mathbf{1.962\ m/s^2}$$

Substituting $a = 1.962\ \text{m/s}^2$ in expression (*i*), we get

$$\text{Tension in the rope } T = 300 + \frac{300}{9.81} \times 1.962 = \mathbf{360\ N}$$

10.8 HAMILTON'S PRINCIPLE

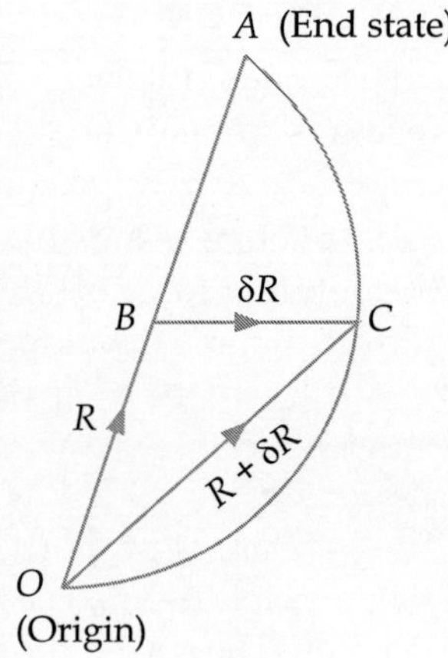

Fig. 10.55

The Hamilton's principle is an important mathematical concept which states that

"the time integral of the difference between the kinetic and potential energies of dynamical system is stationary"

Consider that a force F acts on a particle of mass m and that under the influence of this force, the particle moves from the fixed origin O along the path OBA. If R is the position vector at any time t, then from Newton's second law (force = mass × acceleration)

$$F = m\frac{d^2R}{dt^2}$$

Let the particle now follow path OCA with end position remaining same at A. The variation in path, called virtual displacement, is represented by δR. The work done during virtual displacement δR is

$$\delta W = F \times \delta R = m\frac{d^2R}{dt^2}\,\delta R \qquad ...(i)$$

The kinetic energy of the particle is

$$KE = \frac{1}{2}m\left(\frac{dR}{dt}\right)^2$$

Upon differentiation

$$\delta(KE) = m\left(\frac{dR}{dt}\right)\delta\left(\frac{dR}{dt}\right) = m\frac{dR}{dt}\frac{d}{dt}(\delta R) \qquad ...(ii)$$

Adding expressions (*i*) and (*ii*), we get

$$\delta W + \delta(KE) = m\frac{d^2R}{dt^2} + m\frac{dR}{dt}\frac{d}{dt}(\delta R) = m\frac{d}{dt}\left(\frac{dR}{dt}\delta R\right)$$

Integrating both sides within the time limits from t_0 to t_1

$$\int[\delta W + \delta(KE)]dt = \left| m\frac{dR}{dt}\,\delta R \right|_{t_0}^{t_1}$$

Since $\delta R = 0$ at t_0 and t_1, the above identity may be written as

$$\int[\delta W + \delta(KE)]dt = 0$$

It may recalled that when the force field is conservative, there exists a potential PE (potential energy) such that $W = -(PE)$

The above identity then takes the form

$$\int[\delta(KE) - \delta(PE)]dt = 0 \quad \text{or} \quad \delta\int[(KE - PE)]dt = 0$$

$$\text{or} \qquad \int(KE - PE)dt \text{ is stationary}$$

The Hamilton's principle is valid for a particle. However, its derivation can be extended to a system of particles by summation and to a rigid body by integration. As such the Hamilton's principle is true for any dynamical system.

The integral $\int (KE - PE)dt$ is minimum along the natural path in comparison with that along any other path that joins O to A.

The energy difference (KE – PE) is called the kinetic potential or the Lagrangian function.

10.9 WORK DONE BY A SPRING

Consider a body attached to fixed support by a spring. Initially when the body is at position A_0, the spring is undeformed. When the body is acted upon by a force, the spring extends and the body shifts to a new position which depends upon the magnitude of applied force

Experimental results have shown that when a spring is deformed (compressed or elongated), a corresponding force is developed in the spring. The magnitude of force developed is directly proportional to the displacement of spring from the undeformed position.

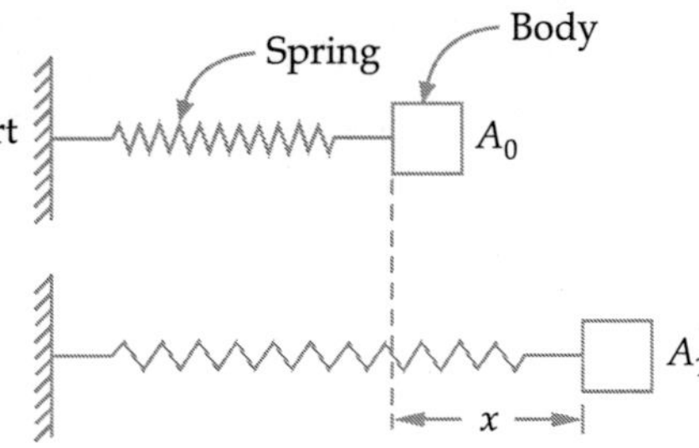

Fig. 10.56

$$F \propto x \text{ or } F = k\,x$$

The constant of proportionality is called the spring constant and it is defined as the force required for unit deformation. The spring constant is expressed in N/m.

Elementary work done by the spring force.

$$= F\,dx = -\,kx\,dx$$

The negative sign stems from the fact that the force induced in the spring acts in a direction opposite to that of displacement.

$$W = -\int_0^x kx\,dx = -\frac{1}{2}kx^2$$

If displacement of body is from x_1 to x_2, then

$$W = -\frac{1}{2}k(x_2^2 - x_1^2) = \frac{1}{2}k\left(x_1^2 - x_2^2\right)$$

The work done by the spring force is represented by the shaded area as shown in Fig. 10.57.

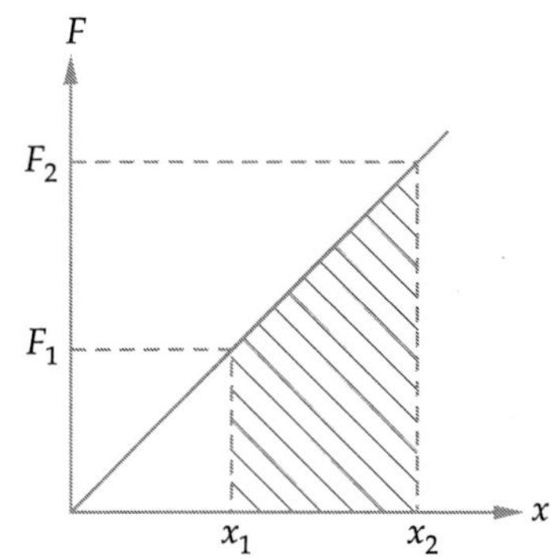

Fig. 10.57

$$W = -\frac{F_2 + F_1}{2} \times (x_2 - x_1)$$

$$W = -\frac{kx_2 + kx_1}{2} \times (x_2 - x_1) = \frac{1}{2}k\left(x_1^2 - x_2^2\right)$$

The work done by spring force is positive when $x_2 < x_1$, *i.e.,* when the spring is returning to its original position.

EXAMPLE 10.61

Determine the length of cord AC in Fig. 10.58 so that 10 kg lamp is suspended in the position shown. The undeformed length of the spring AB is 0.6 m and the spring stiffness is 325 N/m.

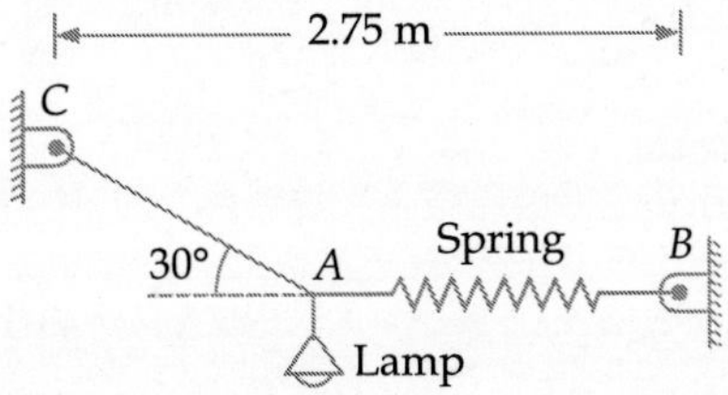

Fig. 10.58

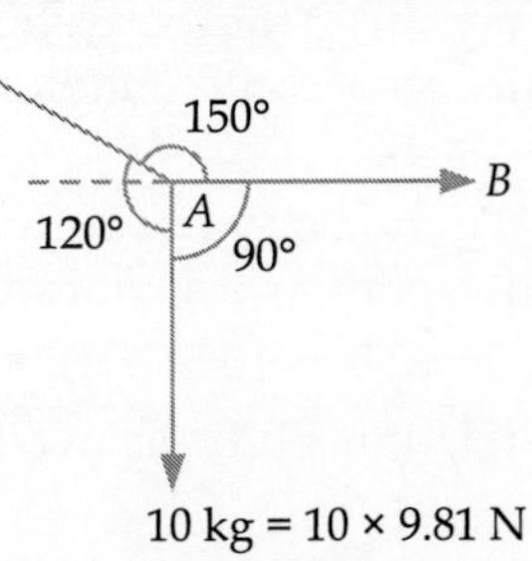

Fig. 10.59

Solution: With reference to free body diagram of junction A, the Lami's theorem gives

$$\frac{F_{ab}}{\sin 120^\circ} = \frac{(10 \times 9.81)}{\sin 150^\circ}$$

$$F_{ab} = (10 \times 9.81)\frac{\sin 120^\circ}{\sin 150^\circ} = (10 \times 9.81) \times \frac{0.866}{0.5} = 169.9 \text{ N}$$

$$\text{Corresponding extension of spring} = \frac{169.9}{325} = 0.523 \text{ m}$$

∴ Deformed length of the spring = undeformed length + extension of spring

$$= 0.6 + 0.523 = 1.123 \text{ m}$$

From the given geometry : $AC \cos 30° = 2.75 - 1.123 = 1.627$

$$\therefore \text{ length of cord } AC = \frac{1.627}{\cos 30^\circ} = \mathbf{1.879 \text{ m}}$$

EXAMPLE 10.62

The spring in a gun is held in vertical position and compressed by 0.25 m. A ball of 40 N weight is placed on the compressed spring which is subsequently released. Calculate the height to which the ball will rise and the velocity it will attain at a height of 2 m. It may be presumed that the spring has a stiffness of 5000 N/mm.

Solution: When the spring is compressed, it has a potential energy stored in it which is given by

$$= \frac{1}{2}kx^2 = \frac{1}{2} \times 5000 \times (0.25)^2 = 200 \text{ Nm}$$

The potential energy of the spring gets transformed into potential energy and kinetic energy of the ball.

(*a*) Let the ball rise to a maximum height of ***h*** m. At maximum height, the velocity of the ball will be zero and so will be its kinetic energy. Therefore

$$200 = 40\,h + 0; \quad h = \mathbf{5 \text{ m}}$$

(*b*) The law of conservation of energy can again be used to find the velocity of ball at a height of 2 m.

$$200 = 40 \times 2 + \frac{1}{2} \times \left(\frac{40}{9.81}\right) V^2 = 80 + 2.039\, V^2; \quad V = \mathbf{7.67 \text{ m/s}}$$

EXAMPLE 10.63

A body of mass 10 kg is made to fall 3 cm height on a spring of stiffness 120 N/cm. Find the displacement of spring. Use the concept that total energy of the mass-spring system remains constant.

Solution : Refer Fig. 10.60 for the system and let the energies be evaluated with respect to undeformed position of the spring as datum.

At **position (1)**, the body has only potential energy and the spring has no energy as it is not deformed. Accordingly

$$E_1 = mgh = 10 \times 9.81 \times 3 = 294.3 \text{ N cm}$$

At position (2), the body as well as the spring have potential energy.

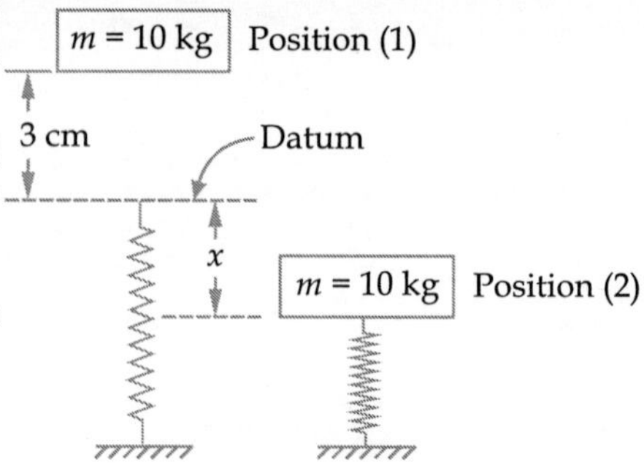

Fig. 10.60

$$E_2 = -mgx + \frac{1}{2}kx^2$$

where x is the compression of the spring in cm

$$E_2 = -10 \times 9.81 \times x + \frac{1}{2} \times 120 \times x^2 = -98.1x + 60x^2$$

Thus using the law of conservation of energy, we can write : $E_1 = E_2$

$$294.3 = -98.1x = 60x^2 \quad \text{or} \quad x^2 - 1.635x - 4.905 = 0$$

$$\therefore \quad x = \frac{-(-1.635) \pm \sqrt{(-1.635)^2 - 4 \times 1 \times}}{2 \times 1} = \frac{1.635 \pm 4.721}{2}$$

Neglecting the negative value which is not possible, we get

$$x = \frac{1.635 + 4.721}{2} = \mathbf{3.178 \text{ cm}}$$

EXAMPLE 10.64

A collar of 5 kg mass slides without friction along a vertical rod as shown in Fig. 10.61. The spring attached to the collar has a spring constant of 250 N/m and its undeformed length is 15 cm. The collar is released from rest at position *A* and slides 20 cm downward to position *B*. What will be the velocity of the collar at position *B*?

Solution : Extension of spring

$$x = BC - AC = \sqrt{(AC)^2 + (AB)^2} - AC$$

$$= \sqrt{15^2 + 20^2} - 15 = 10 \text{ cm}$$

∴ Forced exerted by spring

$$F_s = kx = 250 \times 0.1 = 25 \text{ N}$$

Refer Fig. 10.62 for the free body diagram of the collar when it is at position B

$$\tan\theta = \frac{AC}{AB} = \frac{15}{20} = 0.667\,; \quad \theta = 33.70^\circ$$

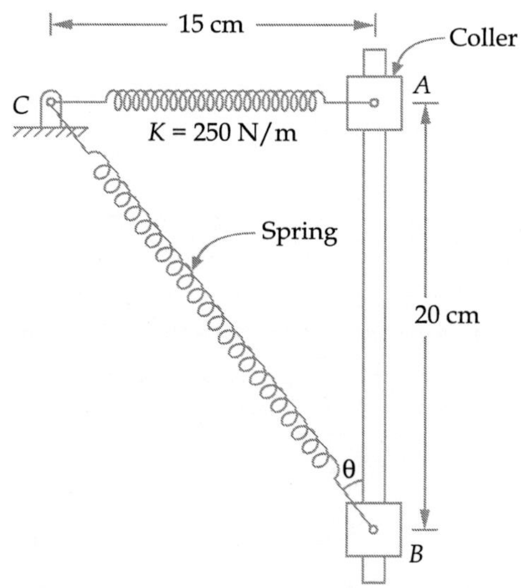

Fig. 10.61

From the equation of equilibrium : $\Sigma F_y = ma$, we have

$$mg - F_s \cos\theta = ma$$

or $5 \times 9.81 - 25 \cos 33.70° = 5 \times a$

$\therefore$ Acceleration of collar motion, $a = \dfrac{5 \times 9.81 - 25 \times 0.832}{5} = 5.65 \text{ m/s}^2$

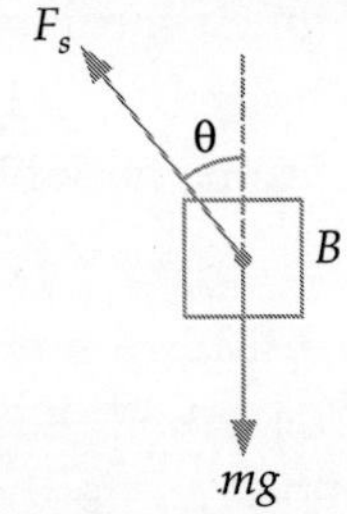

Fig. 10.62

From the kinematic equation of uniformly accelerated motion,

$$v^2 - u^2 = 2as$$

or $v = \sqrt{2as} = \sqrt{2 \times 5.65 \times 0.2}$ $(\because\ u = 0)$

$= \mathbf{1.50\ m/s}$

EXAMPLE 10.65

A block of 50 kg mass is released from rest and slides down a 30° inclined plane as shown in Fig. 10.63. After sliding 1.5 m down the plane, the block hits a spring of spring constant 25 N/mm. If the coefficient of friction between the block and plane is 0.2, make calculations for the maximum deformation of the spring and maximum velocity of the block. Also find the distance the block will move up the plane due to rebound.

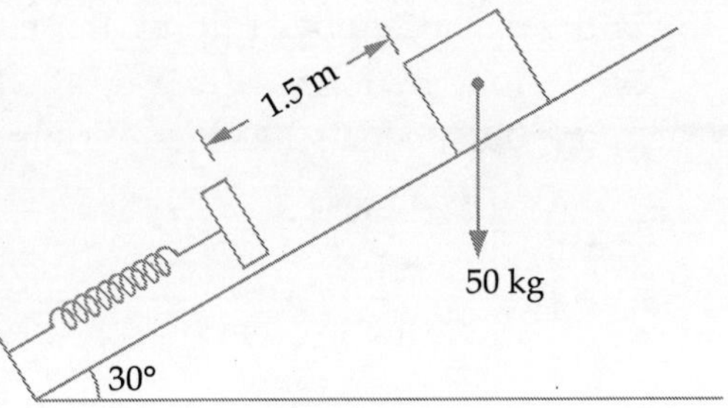

Fig. 10.63

Solution : Refer Fig. 10.64 for the various forces acting on the block.

Normal reaction $R = W \cos\theta = (50 \times 9.81) \cos 30° = 424.78$ N

Force of friction $F = \mu R = 0.2 \times 424.78 = 84.96$ N

This frictional force acts in a direction opposite to that of motion.

Then the net force on the block when it starts moving downwards the plane is

$$= W \sin\theta - F = 50 \times 9.81 \times \sin 30° - 84.96$$
$$= 160.29 \text{ N}$$

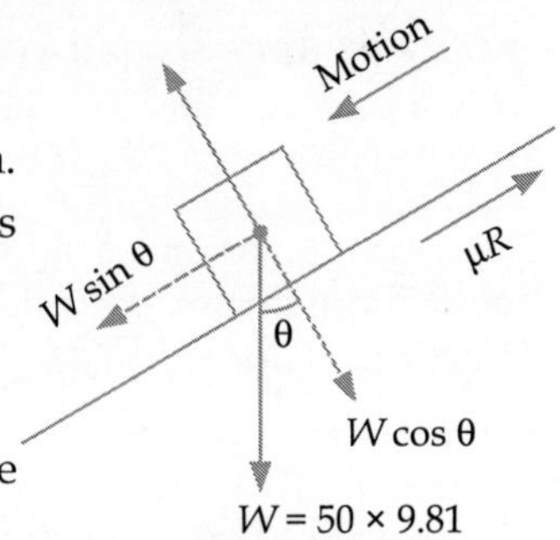

Fig. 10.64

If x is the deformation (compression) of the spring, then distance moved by the block

$$= (1.5 + x)$$

Work done by the block $= 160.29 \times (1.5 + x)$

Work done by the spring $= -\dfrac{1}{2}kx^2 = -\dfrac{1}{2} \times (25 \times 10^3)x^2 = -12.5 \times 10^3 x^2$

$\therefore$ Total work done by the system $= 160.29 \times (1.5 + x) - 12.5 \times 10^3 x^2$

The block starts from rest and finally comes to rest when there is maximum deformation of the spring. Obviously change in kinetic energy of the system is zero. Accordingly from the work-energy principle,

work done = change in kinetic energy

$$160.29 \times (1.5 + x) - 12.5 \times 10^3 x^2 = 0$$

or $12.5x^2 - 0.16x - 0.24 = 0$

$$x = \frac{-(-0.16) \pm \sqrt{(-0.16)^2 - 4(12.5)(}}{2 \times 12.5} = \frac{0.16 \pm 3.47}{25} = 0.145 \text{ m} \quad \text{or} \quad -0.132 \text{ m}$$

The value $x = 0.145$ m is positive and hence acceptable

(*b*) According to Newton's second law $F = ma$

$$160.29 = 50 \times a; \quad a = 3.21 \text{ m/s}^2$$

From the kinematic relation $v^2 - u^2 = 2as$, we have

$$v^2 = 2 \times 3.21 \times 1.5 \qquad (\because u = 0)$$

$\therefore$ Maximum velocity of block $v = \sqrt{2 \times 3.21 \times 1.5} = 3.10$ m/s

(*c*) When the block moves up the plane due to rebound, both the component of weight of block along the plane and the frictional force act in the same direction. Therefore

$$\text{force on the block} = W \sin\theta + \text{frictional force}$$
$$= 50 \times 9.81 \sin 30^\circ + 84.96 = 330.21 \text{ N}$$

If s is the distance moved up by the block, then

work done by forces acting on the block = $330.21 \times s$

The work done equals the energy stored in the spring. Accordingly

$$\frac{1}{2} kx^2 = 330.21 \times s$$

or
$$\frac{1}{2} \times (25 \times 10^3) \times (0.145)^2 = 330.21 \times s; \; s = \mathbf{0.796 \text{ m}}$$

Thus the distance of rebound of block on the plane is 0.796 m.

REVIEW AND SUMMARY

1. From Newton's second law,

$$F = m\frac{dV}{dt} = \frac{d}{dt}(mV) = \frac{mV_2 - mV_1}{t}$$

The product of mass and velociy (mV) is called *momentum*. Apparently the change in linear momentum per unit time is proportional to impressed force and takes place in the direction of force.

2. The product of force and time during which it acts ($F \times t$) is called *impulse*. Impulse equals the change in linear momentum.

3. *Work* is said to be done when the point of application of force moves in the direction of force, and work done equals the product of force and the resulting displacement in the direction of force.

Work done = $F \times x$ when force and displacement are in the same direction
= $F \cos\theta \times x$ when force acts at an angle θ with the direction of displacement

Work done by a force is zero if either the displacement is zero or the force acts normal to displacement.

4. *Power* is the rate of doing work and is obtained by dividing the work by time taken to do that work. *Energy* is defined as the capacity to do work.

5. The energy possessed by a body by virtue of its velocity is known as kinetic energy. Potential energy is the energy by virtue of position of a body with respect to any given datum.

$$KE = \frac{1}{2} mV^2 \quad \text{and} \quad PE = mgh$$

The kinetic energy and potential energy individually change during the motion of body but their sum always remains the same. This is the essence of *law* of *conservation* of *energy*. The sum of *KE* and *PE* of a body is called the *mechanical energy*.

6. From work-energy principle, the work done on a body equals the change in its kinetic energy. Further, if the work of force in moving an object between two positions is independent of tha path followed by the object and can be expressed as change in its potential energy, then such a force is called *conservative force*.
7. Consider a lift (elevator or cage) carrying a load W and moving with uniform acceleration a. There will be reaction of lift which will equal the tension T in the cable supporting the lift. When the lift moves upwards,

$$\text{Net force} = \text{mass} \times \text{acceleration}$$

$$T - W = \frac{W}{g}a$$

$$\text{That gives: } T = W\left(1 + \frac{a}{g}\right)$$

Likewise when the lift moves downwards

$$T = W\left(1 - \frac{a}{g}\right)$$

8. D'Alembert's principle states that "the system of forces acting on a body in motion is in dynamic equilibrium with inertia force of the body".
Let R be the resultant of a system of forces acting on the body which makes the body move along its direction with an acceleration a. Then for dynamic equilibrium of the body.

$$\Sigma F = 0\ ; \quad R - ma = 0 \quad \text{or} \quad R = ma$$

The force ($-\ ma$) is termed as the inertia force or the reverse effective force.
9. The sum of kinetic energy and potential energy of a body remains constant under the action of gravity forces

$$(KE + PE)_1 = (KE + PE)_2$$

10. When a spring is deformed (compressed or elongated), a force is developed the magnitude of which is directly proportional to displacement x of spring from the undeformed position

$$W = -\int_0^x kx\,dx = -\frac{1}{2}kx^2$$

The –ve sign stems from the fact that the force induced in the spring acts in a direction opposite to that of displacement.
If displacement of spring is from x_1 to x_2, then

$$W = -\frac{1}{2}k\left(x_2^2 - x_1^2\right) = \frac{1}{2}\left(x_1^2 - x_2^2\right)$$

REVIEW QUESTIONS

A. Conceptual and conventional questions :

1. State Newton's law that governs the force acting on a body.
2. Define force, momentum and impulse
3. State the principle of conservation of linear momentum
4. Write the impulse-momentum equation and mention its applications
5. What is dynamic equilibrium?

6. State D'Alembert's principle. How it is applied in solving problems related to dynamics?
7. Define energy. What is its relationship with work?
8. Define kinetic energy, potential energy and total mechanical energy.
9. Explain the principle of conservation of mechanical energy.
10. Differentiate between work and power. Mention their units in SI system.
11. Define work. Is it a scalar or vector quantity?
12. State whether any work is done by a man pressing against a vehicle which does not move. Explain.
13. What is the work done by the weight of a body if it is moved horizontally? Explain.
14. State the principle of work and energy.
15. What is a conservative force?
16. Write units for the following: impulse, momentum, work and power
17. A motorised boat of mass 25 tonnes drifts in towards a dock at a speed of 1.8m/s after its engine has been shut off. If the water offers a resistance of 40 N per tonne, what distance the boat will travel before coming to rest? **(4.5 m)**
18. A bullet of mass 10 gm leaves the barrel of a gun with a velocity of 350m/sec. Determine the impulse produced by the discharge of bullet. If the impulse lasts for 0.002 second, what will be average impulsive force? **(14 kgm/s; 7 kN)**
19. A body of mass *M* moves in outer space with velocity *V*. The body breaks into two parts so that mass of one part is $\frac{1}{10}$th of the total mass. After explosion, the heavier part comes to rest while the lighter part continues to move along the original direction of motion. Determine the velocity of smaller part in terms of velocity *V*. **(v = 10V)**
20. A body of 400 kg mass starts from rest and moves along a straight line under the action of a force which varies as the square of time from start. If the magnitude of force equals 400 N at 15 seconds from the start, make calculations for the velocity of body at 15 seconds from the start.

Hint: $F = k\,t^2 \;;\; 400 = k\,(15)^2 \;;\; k = \frac{16}{9}$

$$\therefore \quad 400\,a = \frac{16}{9}t^2 \;;\quad a = \frac{1}{225}t^2$$

$$\int_0^v dv = \frac{1}{225}\int_0^{15} t^2\,dt \qquad \left(\because a = \frac{dv}{dt}\right)$$

$$v = \frac{1}{225}\times\frac{15^3}{3} = 5\text{m/s}$$

21. A running man has half the kinetic energy that a boy of half the mass has. The man has to increase his speed by 1 m/s if he wants to acquire the kinetic energy equal to that of the boy. Make calculations for the original speed of the man and the boy. **(2.4m/s, 4.8m/s)**
22. A man pulls a hand roller on a cricket pitch during 15 minutes recess for tea with an average speed of 0.2 m/s. What work will be done by the man if he applies a pull of 125 N at an upward angle of 30° to the horizontal? Neglect frictional effects, if any. **(19.5 kJ)**
23. A vehicle of 1000 kg mass moving with 48 km/hr collides with the back of another vehicle of 650kg mass moving in the same direction at 24 km/hr. Find (*a*) their common speed after coupling, (*b*) loss in kinetic energy. **(10.71 m/s, 873 J)**

24. A pile driving hammer of mass 250 kg falls 3 metres from rest on a pile of mass 1200 kg. There is no rebound and the pile is driven 20 cm into the ground. Make calculations for the common velocity after impact and the average resistance offered by the ground.

(1.32 m/s, 20.55 kN)

25. A spring of unstretched length 100 mm extends to a length of 125 mm when a load is applied to it. When the load is removed, the spring has to do 2 Joules of work in returning to its original position. Calculate the spring stiffness and the maximum load applied to the spring. **(6.4 N/mm, 160 N)**

B. Complete the following statements with most appropriate word/words :

1. Momentum and impulse are .... quantities with unit ....
2. Work done by a force is zero if the force and displacement are ....
3. Power is the .... of doing work and is obtained by dividing the work done by ....
4. The work done on a body equals the change in .... of the body.
5. The sum of kinetic energy and potential energy of a body is called the ....
6. If the work of force in moving an object between two positions is independent of the path followed by the object, and can be expressed as a change in its potential energy, then such a force is called ....
7. The system of forces acting on a body in motion is in dynamic equilibrium with the inertia force of the body. This is the essence of ....
8. When a bullet is fired from a gun, it recoils in the backward direction. It is due to ....
9. A lift carries a load W and moves up with uniform acceleration a. The tension T in the cables supporting the lift is ....

Answers :

1. vector, kg m/s 2. at right angles 3. rate, time 4. kinetic energy 5. mechanical energy

6. conservative force 7. D'Alembert's law 8. Conservation of momentum 9. $W\left(1+\frac{a}{g}\right)$

C. Multiple choice questions :

1. A man weighing 60 N jumps off a railway train running on horizontal rails at 20 km/hr with a packet weighing 100 N in his hand. The thrust of packet on his hand is
   (*a*) zero (*b*) 100 N (*c*) 500 N (*d*) 700 N
2. A truck of mass 3000 kg is moving at 10 m/s and is acted upon two forces: a forward force of 1000 N due to engine and a retarding force of 400 N due to friction. The truck then gains speed at the rate
   (*a*) 0.2 m/s^2 (*b*) 0.02 m/s^2 (*c*) 0.4 m/s^2 (*d*) 0.33 m/s^2
3. A bullet of mass 0.04 kg moving with a speed of 90 m/s enters a heavy wooden block and is stopped after a distance of 60 cm. The average resistive force exerted by the block on the bullet is
   (*a*) 180 N (*b*) 220 N (*c*) 270 N (*d*) 320 N
4. Two particles with masses in the ratio 1 : 4 are moving with equal kinetic energies. The magnitude of their linear momentums will conform to the ratio
   (*a*) 1 : 8 (*b*) 1 : 2 (*c*) $\sqrt{2}$: 1 (*d*) 4 : 1

5. Which of the following statements is wrong ?
   (*a*) impulse equals the change in momentum
   (*b*) action and reaction are equal and opposite and hence cancel each other
   (*c*) the momentum of a system of two bodies is conserved when there is no external force acting on either body
   (*d*) the work done on a particle must equal change in its kinetic energy.
6. A 150 gm cricket ball moving at 20 m/s is caught by a player and the catching process is completed in 0.1 second. The force of blow exerted by the ball on the hands of player is
   (*a*) 3 N (*b*) 8 N (*c*) 24 N (*d*) 30 N
7. When a bullet is fired from a gun, it is recoiled in the backward direction. It is due to
   (*a*) impulse (*b*) inertia
   (*c*) conservation of momentum (*d*) conservation of energy
8. A force f acts for one second on a body of mass 1 kg moving with an initial velocity u. Then which of the following statements is/are not true ?
   (*a*) body covers a distance $(u + f/2)$
   (*b*) final velocity of the body is $(u + f)$
   (*c*) change in kinetic energy of body is $\frac{1}{2} m f^2$
   (*d*) momentum of the body increases by f
9. A constant force $F = (20\,i + 30\,j + 10\,k)$ kN moves a particle from position $r_1 = (10\,i + 20\,j)$ m to $r_2 = (10\,i + 20\,j + 30\,k)$ m
   The work done by the force in kJ is
   (*a*) zero kJ (*b*) 300 kJ (*c*) 800 kJ (*d*) 1100 kJ
10. A bullet of mass 0.03 kg moving with a speed of 400 m/s penetrates 12 cm into a fixed block of wood. The average force exerted by the wood on the bullet is
   (*a*) 10 kN (*b*) 20 kN (*c*) 25 kN (*d*) 30 kN
11. A man stands on a spring weighing scale in a lift which carries him upwards with an acceleration. The reading on the weighing scale will be
   (*a*) true weight of the man (*b*) lower than the true weight
   (*c*) greater than the true weight (*d*) zero
12. The tension in the cable supporting a lift moving upwards is twice the tension when the lift moves downwards. The acceleration of the lift is equal to
   (*a*) g (*b*) $g/2$ (*c*) $g/3$ (*d*) $g/4$
13. A boy weighing 625 N stands on a spring weighing machine inside a lift. The lift commences its downward journey and the machine reads 500 N. At that instant, the lift is descending with an acceleration of
   (*a*) 1.25 m/s^2 (*b*) 2 m/s^2 (*c*) 8 m/s^2 (*d*) 12 m/s^2
   Take g = 10 m/s^2.
14. Two objects of masses 3 kg and 5 kg respectively are connected by a light cord passing over a smooth weighless pulley
   (*i*) The acceleration with which the masses moves is
      (*a*) $g/4$ (*b*) $g/2$ (*c*) $2g$ (*d*) $4g$

(*ii*) The tension in the string would be

(*a*) 37.5 N (*b*) 80 N (*c*) 150 N (*d*) 0.5 N

Take $g = 10 \text{ m/s}^2$

(*iii*) The pressure on the pulley would be

(*a*) 37.5 N (*b*) 75 N (*c*) 50 N (*d*) 20 N

15. A horizontal force F pulls two masses m_1 and m_2 (lying on a smooth frictionless table) connected by a light inextensible string as shown in Fig. 10.65.

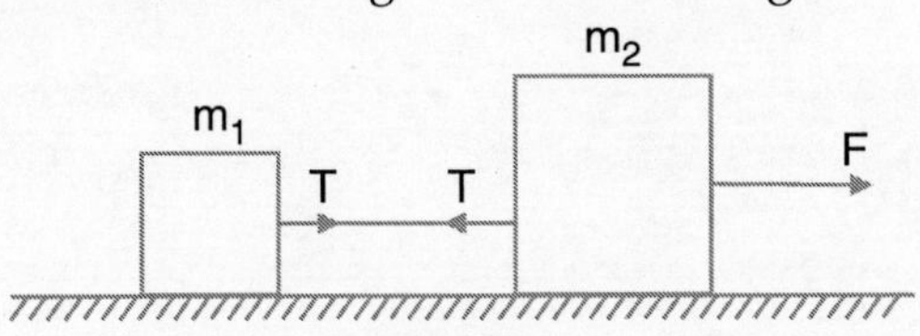

Fig. 10.65.

The tension in the string would be

(*a*) $\dfrac{1}{m_1 + m_2}F$ (*b*) $\dfrac{m_1}{m_1 + m_2}F$ (*c*) $\dfrac{m_2}{m_1 + m_2}F$ (*d*) $\dfrac{m_2}{m_1 + m_2}F$

Answers :

1. (*a*)	2. (*a*)	3. (*c*)	4. (*b*)	5. (*b*)	6. (*d*)
7. (*c*)	8. (*c*)	9. (*b*)	10. (*b*)	11. (*c*)	12. (*c*)
13. (*b*)	14. *i*(*a*), *ii*(*a*), *iii*(*b*)		15. (*b*)		

❑❑❑

CHAPTER

11

KINETICS OF ROTARY MOTION

Learning objectives : The contents of this chapter will make the reader familiar with the following aspects of kinetics of rotary motion of a rigid body

- *rotary motion and laws of rotary motion*
- *moment of momentum, torque and angular momentum*
- *kinetic energy of rolling motion and plane motion*
- *work-energy principle and principle of conservation of energy*
- *analogy between different parameters pertaining to linear and rotary motion*

The subject matter dealt with in this chapter pertains to kinetics of rigid bodies and relates the forces acting on the rigid body, the mass of the body and the motion produced. The analysis, however, has been restricted to plane motion, *i.e*, to a motion in which each particle of the body remains at a constant distance from a fixed reference plane.

11.1 ROTARY MOTION AND TORQUE

- **Rotary motion:** motion of a rigid body in which the particles move in circular paths. The particles continue repeating the same path and have a fixed centre of rotation.
- **Torque:** turning moment of a force on a body on which it acts. Torque equals the product of force and perpendicular distance between the line of action of force and the point of rotation.

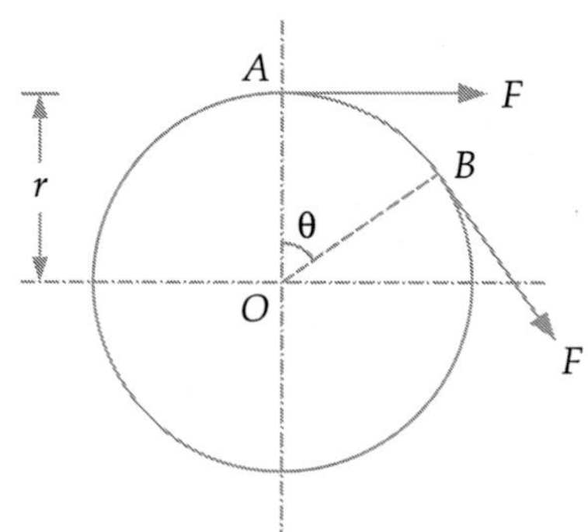

Fig. 11.1

With reference to Fig. 11.1,

$$\text{Torque } T = F \times r$$

If F is in newtons (N) and r is in metres (m), then the unit of torque will be Nm.

When the body pivoted at point O rotates through a small angle θ (measured in radians), the point A shifts to B.

Work done by the tangential force F in rotating the body from A to B is

$$\begin{aligned} &= \text{force} \times \text{distance moved} \\ &= F \times \text{length of arc } AB \\ &= F \times r\theta = (F \times r)\,\theta = T\theta \end{aligned}$$

Thus the work done by torque is given by the product of torque and the angular displacement.

In one rotation $\theta = 2\pi$. If the body turns N revolutions per minute, then the angular displacement per second is $\frac{2\pi N}{60}$

$$\text{Work done} = T \times \frac{2\pi N}{60} \text{ Nm/s}$$

$$\therefore \quad \text{Power of the system} = \frac{2\pi NT}{60} \text{ Watt}$$

11.2 MOMENT OF MOMENTUM

It may be recalled that linear momentum for a body moving along a straight line is the product of the mass of the body and its velocity

With reference of Fig. 11.2, a rigid body of mass m rotates about point O with constant angular velocity.

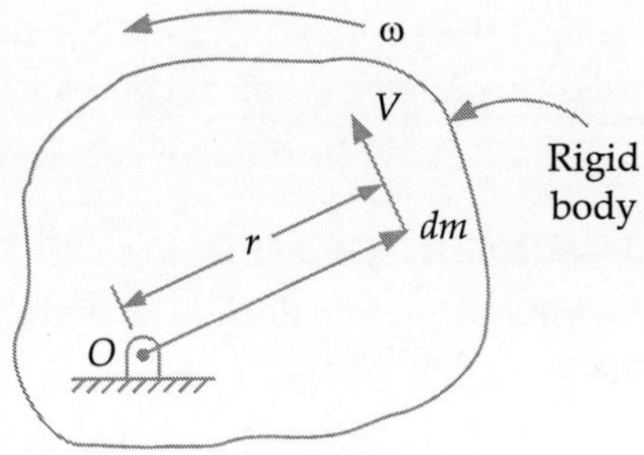

Fig. 11.2

Momentum of elementary mass about O

$$= dm \times V = dm \times \omega r \qquad (\because V = \omega r)$$

Moment of momentum of the elementary mass about O

$$= (dm \times \omega r) \times r = (dm \times r^2) \times \omega$$

For the entire mass, the moment of momentum about O

$$= \int (dm \times r^2) \times \omega = \omega \int (dm \times r^2)$$

The term $\int (dm \times r^2)$ represents the moment of inertia of the whole mass about O. Then

Moment of momentum of the whole body about O

$$= I\omega$$

Thus the moment of momentum may be defined as "*the product of mass moment of inertia and the angular velocity of a rotating body*"

Moment of momentum is also known as *angular momentum*

If the external torque acting on a system is zero, the angular momentum of the system remains constant

11.3 LAWS OF ROTARY MOTION

Similar to laws of linear motion, Newton's laws for rotary motion may be stated as follows:

First law: A body remains at rest or continues to rotate about an axis with constant angular velocity unless it is disturbed by an external torque that changes its state.

Second law: The rate of change of angular momentum of body that rotates about an axis is proportional to the external torque that acts on the body, and the change in angular moment takes place in the direction of the torque.

Third law: To every torque, there is always an equal and opposite torque.

11.4 TORQUE AND ANGULAR MOMENTUM

The product of mass moment of inertia and the angular velocity of a rotating body is known as moment of momentum or angular momentum.

$$\text{Angular momentum} = I\omega$$

Further, the rate of change of angular momentum of a rotating body is proportional to the external torque applied on the body and takes place in the direction of the torque.

$$T \propto \text{rate of change of angular momentum}$$

$$\propto \frac{d}{dt}(I\omega) \propto I\frac{d\omega}{dt} = k\,I\,\alpha$$

where $\alpha = \frac{d\omega}{dt}$ = angular acceleration and k is the constant of proportionality.

Take unit torque ($T = 1$) to produce unit angular acceleration ($\alpha = 1$) in a body of unit mass moment of inertia ($I = 1$), we get $k = 1$ and accordingly,

$$T = I\,\alpha \qquad ...(11.1)$$

The expression $T = I\,\alpha$ may be written as

$$T = I\frac{d\omega}{dt}$$

When the torque T is applied to the body for infinite small interval of time dt, the torque is called an *impulsive* or *angular torque.*

$\therefore$ Impulsive torque $T_i = I\,d\omega$ = change of angular momentum

11.5 MOMENT OF INERTIA AND RADIUS OF GYRATION

The moment of inertia I of a body is the second moment of all elemental areas (or masses) comprising the body.

For a circular lamina of mass m and radius r : $I = \frac{1}{2}mr^2$

For a solid sphere of mass m and radius r : $I = \frac{2}{5}mr^2$

For a uniform bar of mass m and length l : $I = \frac{ml^2}{12}$

The radius of gyration k is defined as the distances of the point, from the axis of reference, where the whole mass of the body is assumed to be concentrated so as the moment of inertia about the given axis does not change.

The moment of inertia of a body in terms of radius of gyration is given by

$$I = mk^2 \quad \text{or} \quad k = \sqrt{\frac{I}{m}}$$

For a circular lamina : $k = \sqrt{\frac{mr^2}{2m}} = \frac{r}{\sqrt{2}} = 0.707\,r$

For a solid sphere : $k = \sqrt{\frac{2}{5}\frac{mr^2}{m}} = \sqrt{\frac{2}{5}}\,r = 0.632\,r$

11.6 KINETIC ENERGY DUE TO ROTATION

With reference to Fig. 11.2, a rigid body of mass m rotates about point O with constant angular velocity ω.

KE of an elementary mass

$$= \frac{1}{2}\,dm\,V^2 = \frac{1}{2}\,dm\,(\omega r)^2 = \frac{1}{2}\omega^2 r^2 dm$$

KE of the whole body,

$$= \int \frac{1}{2}\omega^2 r^2\,dm = \frac{1}{2}\omega^2 \int r^2 dm$$

The quantity $\int r^2\,dm$ represents the moment of inertia of the body about the axis of rotation.

$$\therefore \quad KE \text{ of a rotating body} = \frac{1}{2} I\,\omega^2 \qquad \text{...(11.2)}$$

Work done equals the change in kinetic energy and therefore

$$\text{Work done} = T\,\theta = \frac{1}{2} I(\omega_2^2 - \omega_1^2) \qquad \text{...(11.3)}$$

The plane motion of a rigid body consists of rectilinear motion and a rolling motion without slipping. Then

Total kinetic energy of the body,

= KE in translation + KE in rotation

$$= \frac{1}{2}\,mV^2 + \frac{1}{2}I\omega^2 = \frac{1}{2}\,mV^2 + \frac{1}{2}\,(mk^2) \times \left(\frac{V}{r}\right)^2 = \frac{mV^2}{2}\left[1 + \frac{k^2}{r^2}\right] \qquad \text{...(11.4)}$$

The following analogy exists between different parameters pertaining to linear and rotary motion:

S.No.	Linear motion	Rotary motion
1.	Mass m	Moment of inertia I
2.	Force F	torque T
3.	Linear velocity v	Angular velocity ω
4.	Linear acceleration a	Angular acceleration α
5.	Linear momentum mv	Angular momentum $I\omega$
6.	Force equation $F = ma$	Torque equation $T = I\alpha$
7.	Linear kinetic energy $1/2\,mv^2$	Rotational kinetic energy $1/2\,I\omega^2$
8.	Work done $F\,s$	Work done $T\,\theta$

11.7 WORK-ENERGY PRINCIPLE AND PRINCIPLE OF CONSERVATION OF ENERGY

The work-energy principle connects the work done with change in kinetic energy and be stated as

Work done by a rigid body undergoing general plane motion

= change in KE of the rigid body due to linear motion

+ change in KE of the rigid body due to rotary motion

The work done results both from linear and angular displacement from the initial and final positions.

Then
$$Fs + T\theta = \frac{1}{2}\,m\,(V_2^2 - V_1^2) + \frac{1}{2}\,I\,(\omega_2^2 - \omega_1^2)$$

The principle of conservation of energy states that the sum of potential energy and kinetic energy of a rigid body moving under the influence of conservative forces is always a constant. That is

$$(PE + KE)_1 = (PE + KE)_2$$

11.8 EQUATION OF MOTION

For a rigid body in plane motion (combination of translation and rotation) the equations of motion are.

$$\left.\begin{aligned}\Sigma F_x &= m a_x \\ \Sigma F_y &= m a_y\end{aligned}\right\} \quad \text{Translatory motion}$$

$$\text{and} \quad \Sigma M_G = I_G \alpha \quad \} \quad \text{Rotary motion}$$

where ΣM_G is resultant moment about the mass centre of the rigid body, α is angular acceleration of the rigid body, and I_G is the mass moment of inertia of the rigid body about its mass centre (axis of rotation)

EXAMPLE 11.1

A uniform homogeneous cylinder of 125 mm radius has a mass of 0.15 kg. This cylinder rolls without slipping along a level horizontal surface with a translational velocity of 20 cm/s. Determine its total kinetic energy.

Solution: Kinetic energy due to translation $(KE)_t$

$$= \frac{1}{2} mV^2 = \frac{1}{2} \times 0.15 \times 0.2^2 = 3 \times 10^{-3} \text{ Nm}$$

Kinetic energy due to rotation $(KE)_r = \frac{1}{2} I\omega^2$

$$\omega = \text{angular velocity} = \frac{V}{r} = \frac{0.2}{0.125} = 1.6 \text{ rad/s}$$

$$I = \text{mass moment of inertia } I = \frac{mr^2}{2} = \frac{0.15 \times (0.125)^2}{2} = 1.172 \times 10^{-3} \text{ kgm}^2$$

That gives: $(KE)_r = \frac{1}{2} \times (1.172 \times 10^{-3}) \times 1.6^2 = 1.5 \times 10^{-3}$ Nm

$\therefore$ Total kinetic energy $= (KE)_t + (KE)_r = 3 \times 10^{-3} + 1.5 \times 10^{-3} = \mathbf{4.5 \times 10^{-3}}$ **Nm**

EXAMPLE 11.2

An angular acceleration $\alpha = (0.4\, t^2 + 0.6)$ rad/s² where t is in seconds is imparted to disc A by a motor. If initially the disc has an angular velocity of 5 rad/s, make calculations for the velocity and acceleration of block B at the time instant $t = 1.5$ seconds (Fig. 11.3).

Solution: Angular acceleration $\alpha = 0.4\, t^2 + 0.6$

$$\frac{d\omega}{dt} = 0.4t^2 + 0.6$$

Upon integration within appropriate limits, we get

$$\int_{\omega_0 = 5}^{\omega} d\omega = \int_0^t (0.4t^2 + 0.6)\, dt$$

$$\omega - 5 = 0.4 \times \frac{t^3}{3} + 0.6t$$

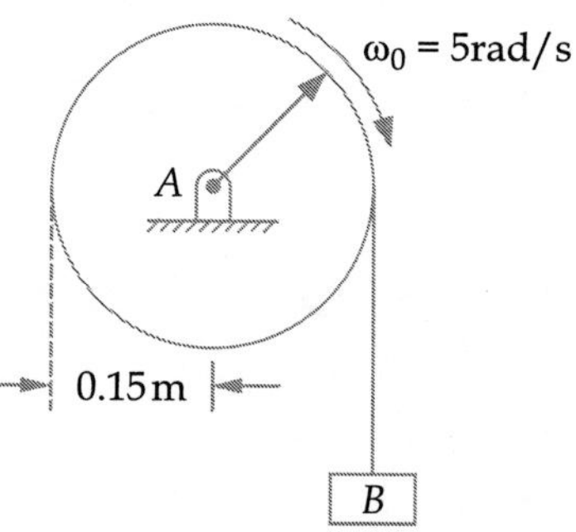

Fig. 11.3

At $t = 1.5$ seconds: $\omega = 0.4 \times \frac{(1.5)^3}{3} + 0.6 \times 1.5 + 5 = 6.35$ rad/s

$$\alpha = 0.4 \times 1.5^2 + 0.6 = 1.5 \text{ rad/s}^2$$

Then velocity V and acceleration α of block B are:

$$V = \omega r = 6.35 \times 0.15 = \mathbf{0.9525\ m/s^2\ (\downarrow)}$$

$$a = \alpha r = 1.5 \times 0.15 = \mathbf{0.225\ m/s^2\ (\downarrow)}$$

EXAMPLE 11.3

A bar pivoted at one end and moving clockwise at 4 rad/s is subjected to a constant angular deceleration. After a certain time, the bar has angular displacement of 6 radians anticlockwise

and has moved through total angle of 16 radians. Determine the angular velocity of the bar at the end of this time interval.

Solution: Refer Fig. 11.4. Considering motion between position OA and OA_1

$$\omega_1^2 - \omega_0^2 = 2\,\alpha\theta_1$$

At position A_1, the bar has travelled through 10 radians and comes to rest.

$$\therefore \quad 0 - (4)^2 = 2\alpha \times 10$$

$$\alpha = -0.8 \text{ rad/s}^2$$

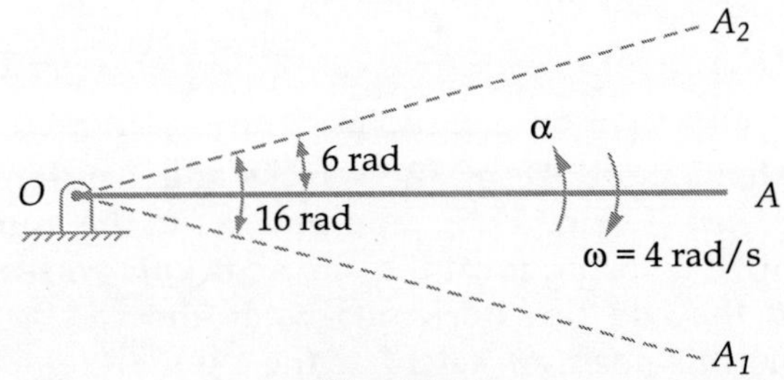

Fig. 11.4

Considering motion between position OA_1 and OA_2

$$\omega_2^2 - \omega_1^2 \times 2\,\alpha\,\theta$$

$$\omega_2^2 - 0^2 = 2 \times 0.8 \times 16$$

$$\therefore \quad \omega_2 = \sqrt{2 \times 0.8 \times 16} = \mathbf{5.06} \text{ rad/s at the end of time interval.}$$

EXAMPLE 11.4

A string of 5 m length has been wrapped round the shaft of a wheel to determine its moment of inertia about its axis. A constant pull of 80 N is applied to the string and observations indicate that the wheel turns 180 rev/min at the instant the string leaves the axle. Calculate the moment of inertia of the wheel.

Solution: Work done in pulling the string

$$= \text{force applied} \times \text{distance moved} = 80 \times 5 = 400 \text{ Nm}$$

Angular velocity of the wheel $\omega = \dfrac{2\pi N}{60} = \dfrac{2\pi \times 180}{60} = 18.84$ rad/s

Then from the work-energy correlation of a rotating system,

work done = kinetic energy of the wheel

$$400 = \frac{1}{2} I\omega^2 = \frac{1}{2} I \times (18.84)^2$$

$\therefore$ Moment of inertia of the wheel $I = \dfrac{400 \times 2}{(18.84)^2} = \mathbf{2.254\ kgm^2}$

EXAMPLE 11.5

A 40 Nm couple moment is applied to a wheel that rests on a plane and has 25 kg mass and 0.4 m radius as shown in Fig. 11.5. If the coefficient of friction between the wheel and plane at A is 0.25, determine:

(*a*) acceleration of the mass centre of wheel, and

(*b*) frictional force.

Presume that radius of gyration for the wheel is 0.20 cm

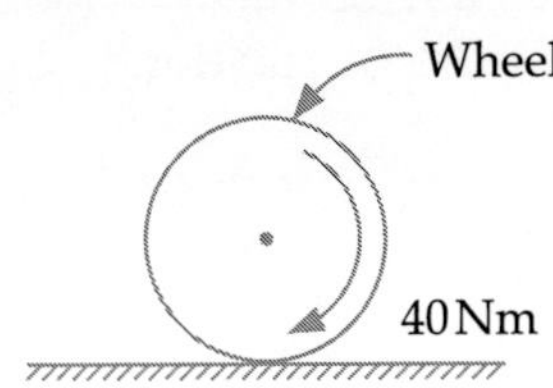

Fig. 11.5

Solution: $I_G = m\,K_G^2 = 25 \times (0.2)^2 = 1 \text{ kg m}^2$

Applying parallel axis theorem,

$$I_A = I_G + m\,h^2 = 1 + 25 \times (0.4)^2 = 5 \text{ kg m}^2$$

Recalling that for a rigid body in plane motion

$$\Sigma M_A = I_A\,\alpha\ ;\ 40 = 5\,\alpha$$

$\therefore$ Angular acceleration of the wheel, $\alpha = \dfrac{40}{5} = 8 \text{ rad/s}^2$

Acceleration of mass centre, $a = \alpha r = 8 \times 0.4 = \mathbf{3.2\ m/s^2}$

(*b*) Frictional force $F = \mu R$

The normal reaction equals the weight of the wheel

$$\therefore \quad F = \mu W = 0.25 \times (25 \times 9.81) = \mathbf{61.31\ N}$$

EXAMPLE 11.6

A slender rod *AB* of 12 kg mass and 1 m length is pivoted at point *O* which is 0.2 m from the end *B* (Fig. 11.6). The end *A* of the rod presses against a spring until the spring is compressed 20 mm and the rod lies horizontally. If the rod is released from this position, calculate the angular velocity when the rod has rotated through 60°.
Take spring constant *K* = 250 kN/m

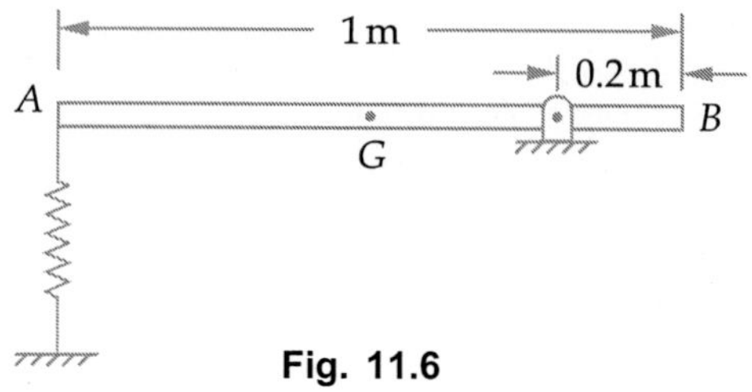

Fig. 11.6

Solution: Magnitude of force in the spring,

$$F = kx = 250 \times (0.02) = 5\text{ kN} = 5000\text{ N}\ (\downarrow)$$

Let I be the moment of inertia of the rod about the pivot point. Then from parallel axis theorem

$$I = I_G + Mh^2 = \frac{ml^2}{12} + mh^2 = \frac{12 \times 1^2}{12} + 12\left(\frac{1}{2} - 0.2\right)^2 = 4.6\text{ kg m}^2$$

For a rigid body in plane motion: $\Sigma M = I\alpha$

$$\therefore \quad 5000 \times 0.8 = 4.6\,\alpha\ ; \quad \alpha = 869.57\text{ rad/s}^2$$

From the kinematic relation $\omega^2 - \omega_0^2 = 2\,\alpha\,\theta$ and noting that $\omega_0 = 0$ and $\theta = 60° = \pi/3$ radian, we have

$$\omega = \sqrt{2 \times 869.57 \times \pi / 3} = \mathbf{42.66\ rad/s}$$

EXAMPLE 11.7

Consider a composite system consisting of a rod *AB* and sphere *BC* fixed rigidly to the rod at *B* as shown in Fig. 11.7. Determine the angular acceleration of the system at the instant when it is released from the horizontal position.

Take mass of rod m_1 = 20 kg,
mass of sphere m_2 = 5 kg

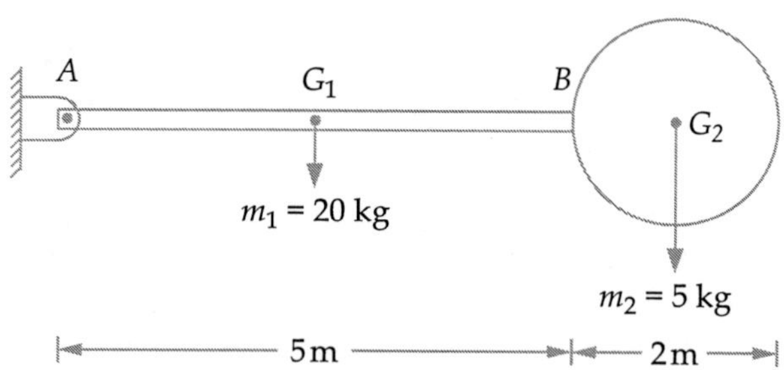

Fig. 11.7

Solution: Moment of inertia of the rod about *A* (centre of rotation),

$$I_1 = \frac{1}{3}ml^2$$

$$= \frac{1}{3} \times 20 \times 5^2 = 166.67\text{ kg m}^2$$

Moment of inertia of sphere about *A*,

$$I_2 = I_{G_2} + mh^2 \quad \text{(parallel axes theorem)}$$

$$= \frac{2}{5}mr^2 + mh^2 = \frac{2}{5} \times 5 \times 1^2 + 5 \times (5+1)^2 = 2 + 180 = 182\text{ kg m}^2$$

$\therefore$ Moment of inertia of the entire system

$$I_a = I_1 + I_2 = 166.67 + 182 = 348.67\text{ kg m}^2$$

For rigid body undergoing rotary motion,

$$\Sigma M = I\alpha \; ; \quad M_a = I_a \times \alpha$$

∴ Angular acceleration of the system,

$$\alpha = \frac{M_a}{I_a} = \frac{(20 \times 9.81 \times 2.5) + (5 \times 9.81 \times 6)}{348.67} = \frac{490.5 + 294.3}{348.67} = \mathbf{2.251 \text{ rad/s}^2}$$

EXAMPLE 11.8

A slender rod *ABCD* of length *l* and mass *m* is supported in a horizontal position as shown in Fig. 11.8 given below:

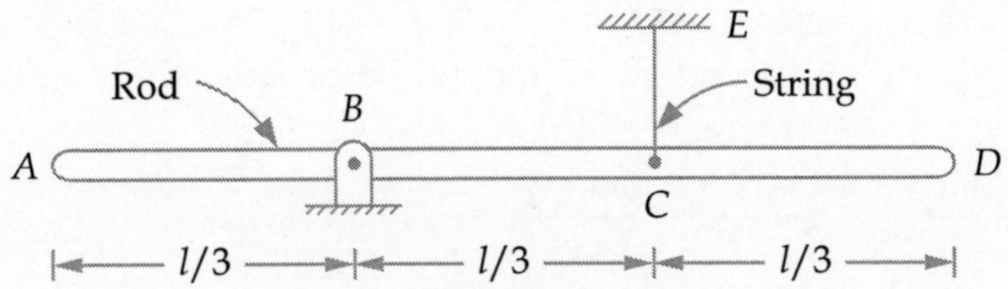

Fig. 11.8

Subsequently the supporting string *CE* is cut. Compute the instantaneous reaction at *B* immediately after string has been cut.

Solution: Refer Fig. 11.9 given below

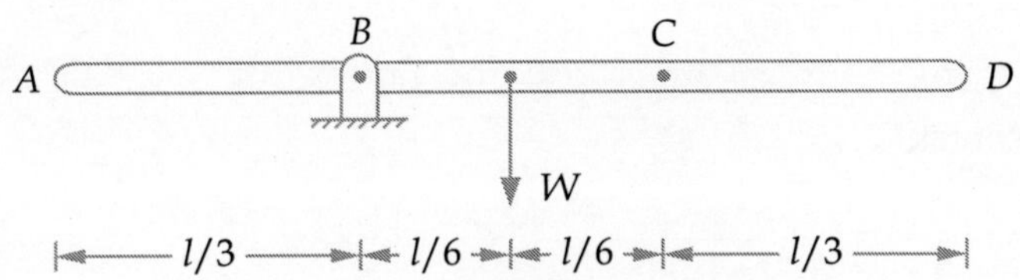

Fig. 11.9

Moment of inertia of the rod about the centre of rotation *B*,

$$I_b = I_G + mh^2 \qquad \text{(parallel axes theorem)}$$

$$I_b = \frac{ml^2}{12} + m\left(\frac{l}{6}\right)^2 = \frac{ml^2}{9}$$

Moment about point $B : M_b = W \times \frac{l}{6} = mg \times \frac{l}{6}$

The from the relation $\Sigma M = I\alpha$ that applies to rigid body rotation, we have

$$mg \times \frac{l}{6} = \frac{ml^2}{9} \times \alpha \; ; \quad \alpha = \frac{3}{2}\frac{g}{l}$$

∴ Linear acceleration at centre of gravity of the rod,

$$a = \alpha \times R = \frac{3}{2}\frac{g}{l} \times \frac{l}{6} = 2.45 \text{ m/s}^2$$

For a rigid body undergoing translatory motion,

net force = mass × acceleration

$$mg - R_b = ma$$

∴ Vertical reaction at suspension point *B*,

$$R_b = m(g - a) = m\,(9.81 - 2.45) = 7.36\,m$$

Since there is no horizontal force acting on the system, the horizontal reaction at the suspension point *B* would be zero.

EXAMPLE 11.9

A bearing carries a shaft that supports a disc of 50 cm radius and 25 kg mass. The shaft is normal to the plane of the disc and is coaxial with the disc as shown in Fig. 11.10. The disc is made to rotate at 1500 rpm and then it decelerates because of bearing friction. If the disc comes to rest in 145 seconds, make calculations for the bearing torque and the number of revolutions the disc turns before it comes to rest. It may be assumed that the bearing friction is independent of the speed of the disc.

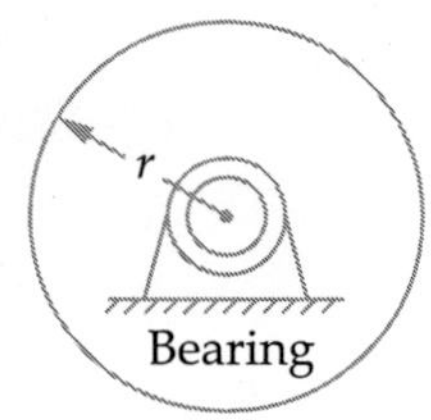

Fig. 11.10

Solution: $\omega_1 = \dfrac{2\pi \times 1500}{60} = 157$ rad/s, and $\omega_2 = 0$

Since the bearing friction is independent of the speed of disc, the angular acceleration is constant and it is given by

$$\alpha = \frac{d\omega}{dt} = \frac{\omega_2 - \omega_1}{t} = \frac{0 - 157}{145} = -1.083 \text{ rad/s}$$

Mass moment of inertia $I = \dfrac{1}{2} mr^2 = \dfrac{1}{2} \times 25 \times 0.5^2 = 3.125 \text{ kg m}^2$

(*a*) Bearing torque $T = I\alpha = 3.125 \times (-1.083) = -3.384$ Nm

The negative sign implies that the frictional torque of the bearing resists the rotation of the disc.

(*b*) From the kinematic relation $\omega_2^2 - \omega_1^2 = 2\alpha\theta$ and noting that $\omega_2 = 0$, we have

Total angular displacement

$$\theta = -\frac{\omega_1^2}{2\alpha} = \frac{-157^2}{2 \times (-1.083)} = 11380 \text{ radians}$$

1 revolution = 2π radians

$\therefore$ Number of revolutions turned by the disc $= \dfrac{11380}{2\pi} = \mathbf{1812}$

EXAMPLE 11.10

A flywheel of mass 4000 kg and radius of gyration 0.8 m loses its speed from 360 rpm to 250 rpm in 2 minutes. Make calculations for (*i*) the torque acting on the wheel, (*ii*) change in the kinetic energy and angular momentum of the wheel during the given period.

Solution: Initial angular velocity $\omega_0 = \dfrac{2\pi N_0}{60} = \dfrac{2\pi \times 360}{60} = 37.68$ rad/s

Final angular velocity $\omega = \dfrac{2\pi N}{60} = \dfrac{2\pi \times 250}{60} = 26.17$ rad/s

Then from the correlation $\omega = \omega_0 + \alpha t$,

angular acceleration $\alpha = \dfrac{\omega - \omega_0}{t} = \dfrac{26.17 - 37.68}{2 \times 60} = -0.0959 \text{ rad/s}^2$

Torque $T = I\alpha = mk^2\alpha = 400 \times (0.8)^2 \times (-0.0959) = \mathbf{-245.5 \text{ Nm}}$

The negative sign means that the torque is retarding torque.

(*ii*) Change in kinetic energy $= \dfrac{1}{2} I\omega^2 - \dfrac{1}{2} I\omega_0^2 = \dfrac{1}{2} I(\omega^2 - \omega_0^2) = \dfrac{1}{2} mk^2 (\omega^2 - \omega_0^2)$

$= \dfrac{1}{2} \times (4000 \times 0.8^2)(26.17^2 - 37.68^2) = \mathbf{-9.407 \times 10^5 \text{ Nm}}$

The negative sign implies decrease in kinetic energy.

(*iii*) Change in angular momentum $= I\omega - I\omega_0 = I(\omega - \omega_0) = mk^2(\omega - \omega_0)$

$= 4000 \times 0.8^2 \times (26.17 - 37.68) = -$ **29466 Nm/s**

The negative sign implies decrease in angular momentum.

EXAMPLE 11.11

A flywheel of 1000 kg mass and 0.8 m radius of gyration is rotating at 180 rev/min. Determine its (*a*) kinetic energy, (*b*) torque and average power required to give the wheel a speed of 180 rev/min in 75 revolutions.

Solution: Angular velocity of wheel $\omega = \frac{2\pi N}{60} = \frac{2\pi \times 80}{60} = 18.84$ rad/s

Moment of inertia of flywheel $I = mk^2 = 1000 \times 0.8^2 = 640$ kgm^2

(*a*) Kinetic energy of the fly wheel $= \frac{1}{2} I\omega^2 = \frac{1}{2} \times 640 \times 18.84^2 = 113582$ Nm

(*b*) Let T be the torque required to give a speed of 180 rev/min in 75 revolutions, *i.e.*, $2\pi \times 75$ radians.

Then from the work-energy correlation of a rotating system,

change in kinetic energy = work done = torque × angular displacement

$$113582 = T \times (2\pi \times 75)\,; \quad T = \mathbf{241.15\ Nm}$$

Average speed of the flywheel $= \frac{0 + 180}{2} = 90$ rev/min

Time taken to make 75 revolutions $= \frac{75}{90} \times 60 = 50$ sec

Work done/sec $= \frac{113582}{50} = 2271.64$ Nm/s

Hence average power = 2271.64 W or **2.272 kW**

EXAMPLE 11.12

A mine cage of 500 kg mass is hung from the drum of hoisting engine which has a mass of 800 kg. The drum is of 150 cm diameter and its radius of gyration is 60 cm. Neglecting frictional effects, determine the velocity the cage will attain when allowed to fall for a period of 4 seconds.

Solution: For drum : Moment of inertia of the drum $I = mk^2 = 800 \times 0.6^2 = 288$ kgm^2

angular impulse = change in angular momentum

$$(T \times 0.75) \times 4 = I\omega_2 - I\omega_1 \quad \text{or} \quad 3T = I\omega_2 \qquad (\because \omega_1 = 0)$$

where T is the tension in the hoisting rope.

Further, the linear velocity and the angular velocity are connected by the relation : $v = \omega r$. That gives

$$\omega_2 = \frac{v}{0.75}$$

$$\therefore \quad T = \frac{I}{3} \times \frac{v}{0.75} = \frac{288}{3} \times \frac{v}{0.75} = 128\,v \qquad ...(i)$$

For 500 kg mass: linear impulse = change in linear momentum

$$(500 \times 9.81 - T) \times 4 = 500\,v \quad \text{or} \quad 4905 - T = 125\,v \qquad ...(ii)$$

From expressions (*i*) ad (*ii*),

$$4905 - 128\,v = 125\,v$$

$\therefore$ Velocity attained by the cage $v = \dfrac{4905}{253} = $ **19.38 m/s**

EXAMPLE 11.13

A cylindrical roller of 400 kg mass and 250 mm radius is at rest on a rough horizontal surface. The roller is pushed by a force of 500 N inclined at 25° to the horizontal and passing through its centre of gravity. The roller then starts rolling without slipping. Calculate the distance moved by the roller by the time it attains a speed of 1.8 m/s.

Solution : Refer Fig. 11.11 for the system of roller and a push being applied to it.

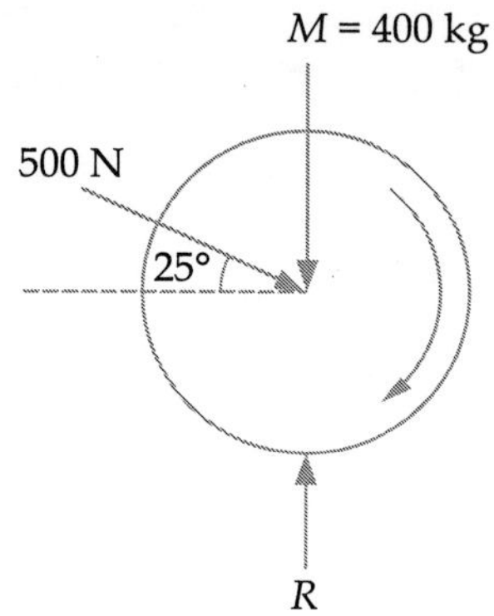

Fig. 11.11

It is to be noted that:

(*i*) the roller moves in the horizontal direction and as such the weight of the roller and the reaction at the rough surface donot contribute towards work.

(*ii*) the velocity at the point of contact is zero and accordingly no work will be done by the frictional force

(*iii*) the work will be done only by the horizontal component of 500 N push.

If S is the distance moved by the roller, then

$$\text{Work done} = (500 \times \cos 25^\circ) \times S = 453.15\, S$$

The roller starts from rest ($u = 0$) and accordingly initial kinetic energy for the roller is zero ($KE_1 = 0$).

Final kinetic energy = kinetic energy due to linear motion + kinetic energy due to rotation

$$KE_2 = \frac{1}{2} mV^2 + \frac{1}{2} I\omega^2 = \frac{1}{2} mV^2 + \frac{1}{2} \times \frac{mr^2}{2} \times \left(\frac{v}{r}\right)^2$$

$$= \frac{1}{2} \times 400 \times 1.8^2 + \frac{1}{2} \times \frac{400 \times 0.25^2}{2} \times \qquad = 648 + 324 = 972 \text{ Nm}$$

From the work-energy principle, the work done equals the change in kinetic energy. Thus

$$453.15\, S = 972 - 0 = 972$$

$\therefore$ Distance moved by the roller $S = \dfrac{972}{453.15} = $ **2.14 m**

EXAMPLE 11.14.

A solid sphere, a cylinder and a hoop are released from rest on an incline at the same elevation. In each case, the mass is *m* and the maximum radius is *R*. Determine the velocity of each body after it has rolled down the incline through a distance *s* as shown in Fig. 11.12.

Solution: It is to be noted that:

(*i*) motion takes place in the gravitational field of the earth

(*ii*) no work is done by the frictional forces is rolling motion

Accordingly the mechanical energy of the system is conserved. That is

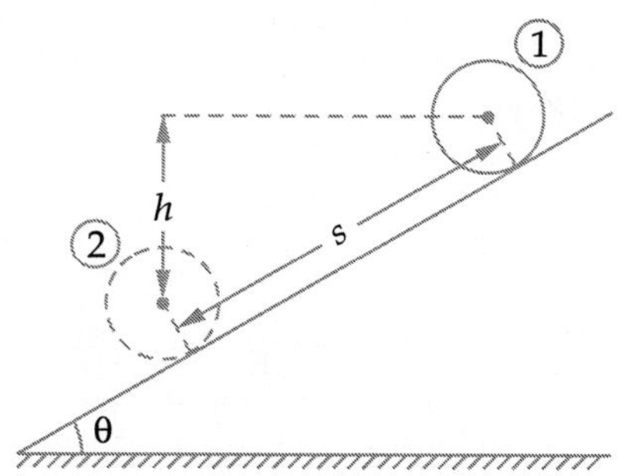

Fig. 11.12

$$(KE + PE)_1 = (KE + PE)_2 \qquad ...(i)$$

Noting that:

$KE_1 = 0$ as the body starts from rest

$PE_1 = 0$ when the initial position is taken as datum

$$KE_2 = \frac{1}{2}mV_2^2 + \frac{1}{2}I\omega^2 = \frac{1}{2}mV_2^2 + \frac{1}{2}I\left(\frac{V_2}{R}\right)^2$$

$$PE_2 = -mgh = -mg\,s\sin\theta$$

Substitution of the above noted values in expression (*i*), we get

$$0 = \frac{1}{2}mV_2^2 + \frac{1}{2}I\left(\frac{V_2}{R}\right)^2 - mg\,s\sin\theta$$

Solution gives: $$V_2 = \frac{\sqrt{2gs\sin\theta}}{\sqrt{1 + I/mR^2}}$$

Substituting the appropriate values of moment of inertia for the given body, the following expressions will appear for the velocity of each body.

- For solid sphere $I = \frac{2}{5}mR^2$ and $V_2 = 0.845\sqrt{2gs\sin\theta}$
- For solid cylinder $I = \frac{1}{2}mR^2$ and $V_2 = 0.816\sqrt{2gs\sin\theta}$
- For hoop $I = mR_2$ and $V_2 = 0.707\sqrt{2gs\sin\theta}$

EXAMPLE 11.15

A homogeneous cylinder of mass 50 kg and 0.5 m in radius is having initial velocity of 6 m/s down a 30° inclined plane. Calculate the velocity of cylinder when it has reached 10 m down the plane from starting point. It may be presumed that the cylinder rolls without sliding.

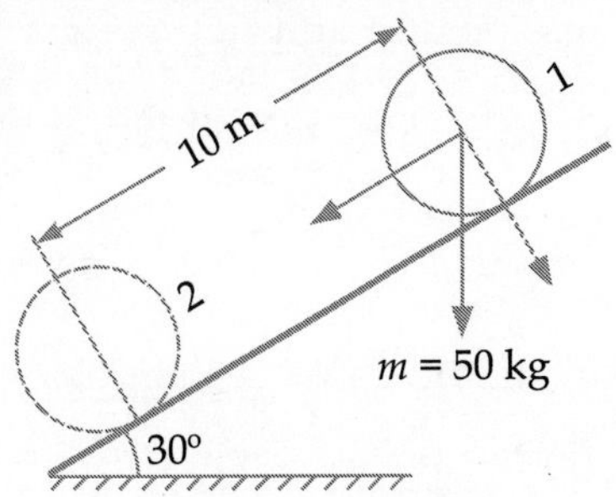

Fig. 11.13

Solution: Component of wight along the plane

$$= mg\sin 30° = 50 \times 9.81 \times 0.5 = 245.25 \text{ N}$$

Work done in moving 10 m down the plane

$$= 245.25 \times 10 = 2452.5 \text{ Nm}$$

Change in kinetic energy of the cylinder

$$\Delta KE = \frac{1}{2}m\left(V_2^2 - V_1^2\right) + \frac{1}{2}I\left(\omega_2^2 - \omega_1^2\right)$$

Noting that $I = \frac{1}{2}mr^2$, $\omega_1 = \frac{V_1}{r}$ and $\omega_2 = \frac{V_2}{r}$, we have

$$\Delta KE = \frac{1}{2} \times 50\left(V_2^2 - 6^2\right) + \frac{1}{2} \times \left(\frac{1}{2} \times 50 \times 0.5^2\right)\left[\left(\frac{V_2}{0.5}\right)^2 - \left(\frac{6}{0.5}\right)^2\right]$$

$$= 25\left(V_2^2 - 36\right) + 3.125\left(4V_2^2 - 144\right)$$

Since work done equals the change in kinetic energy, we have

$$2452.5 = 25\left(V_2^2 - 36\right) + 3.125\left(4V_2^2 - 144\right) = 25V_2^2 - 900 + 12.5V_2^2 - 450$$

$$\therefore \quad V_2 = \sqrt{\frac{2452.5 + 900 + 450}{37.5}} = \mathbf{10.07\ m/s}$$

EXAMPLE 11.16

A solid cylinder and a thin hoop of equal mass m and radius R and connected by a bar G_1G_2 as shown in Fig. 11.14. The system rolls down on the inclined plane without slipping. Determine the acceleration of the system and the tension in the bar.

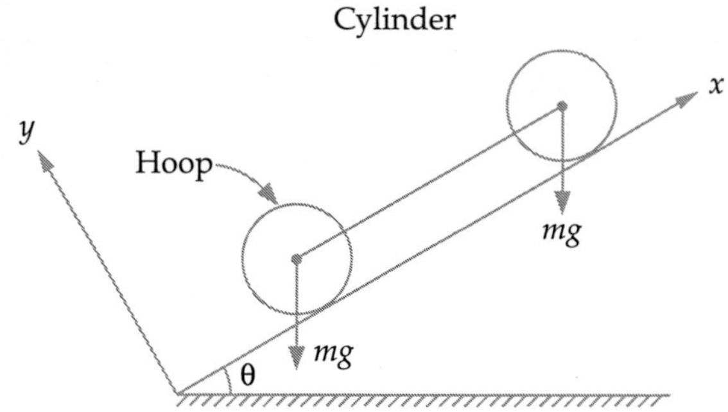

Fig. 11.14

Solution: For the cylinder, the equations of motion are:

$$\Sigma F_x = ma_x: \quad mg \sin\theta - F_1 + T = ma$$

$$F_1 = mg \sin\theta + T - ma \qquad ...(i)$$

where F_1 is the force of Friction on the cylinder and T is the tension in the connecting bar G_1G_2

$$\Sigma M_{G_1} = T_{G_1}\alpha\ ; \quad F_1R = mR^2\alpha\ ; \quad F_1 = mR\,\alpha \qquad ...(ii)$$

The corresponding expressions for hoop would be

$$F_2 = mg \sin\theta - T - ma \qquad ...(iii)$$

$$F_2R = \frac{mR^2}{2}\alpha\ ; \quad F_2 = \frac{mR}{2}\alpha \qquad ...(iv)$$

Adding expression (*i*) and (*iii*) and substituting for F_1 and F_2 as determined above vide expressions (*ii*) and (*iv*), we get

$$mR\,\alpha\left(1 + \frac{1}{2}\right) = 2m\,(g \sin\theta - a) \qquad (\alpha R = a)$$

$$\text{or} \quad \frac{3}{2}a = 2g \sin\theta - 2\alpha\ ; \quad a = \frac{4}{7}g\sin\theta$$

Thus the given system has a linear acceleration equal to $\frac{4}{7}g\sin\theta$

(*b*) From expression (*i*)

$$T = F_1 + ma - mg \sin\theta = mR\,\alpha + ma - mg \sin\theta$$

$$= mR\frac{a}{R} + ma - mg\sin\theta = 2ma - mg \sin\theta$$

$$= 2m \times \frac{4}{9}g \sin\theta - mg \sin\theta = \frac{mg}{7}\sin\theta$$

Thus the connecting bar will be subjected to tension equal to $\frac{mg}{7}\sin\theta$.

EXAMPLE 11.17

A 12 kg plank is supported on two indentical uniform discs; each of mass 5 kg and radius 10 cm. The plank is pulled by 25 N force and its moves 6 m from the rest position. Assuming

that the disc rolls without slipping, make calculations for the velocity with which the plank moves.

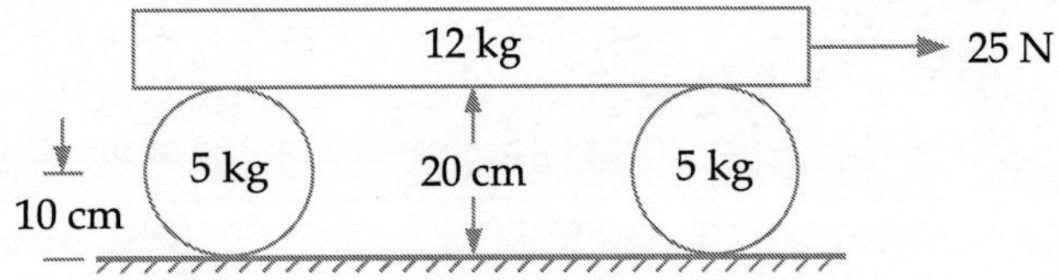

Fig. 11.15

Solution: Work done = force × distance moved = 25 × 6 = 150 Nm

Let the plank move with velocity V m/s towards right and the discs move clockwise with angular ω rad/s.

Initially: The kinetic energy of the plank and discs is zero since initially the system is at rest.

$$\textit{Finally:}\ \text{Kinetic energy of plank} = \frac{1}{2}\,mV^2 = \frac{1}{2} \times 12 \times V^2 = 6\,V^2\,\text{Nm}$$

$$\text{Kinetic energy of two discs} = 2\left[\frac{1}{2}\,m\,V_G^2 + \frac{1}{2}\,I_G\,\omega^2\right]$$

$$\omega = \frac{V}{0.2} = 5\,V;\quad V_G = \omega r = 5\,V \times 0.1 = 0.5\,V$$

$$I = \frac{1}{2}\,mr^2 = \frac{1}{2} \times 5 \times 0.1^2 = 0.025\,\text{kgm}^2$$

$$\therefore\ KE \text{ of two discs} = 2\left[\frac{1}{2} \times 5 \times (0.5V)^2 + \frac{1}{2} \times 0.025 \times (5\right.$$

$$= 2\,[0.625\,V^2 + 0.3125\,V^2] = 1.875\,V^2$$

Total kinetic energy of the system = $6\,V^2 + 1.875\,V^2 = 7.875\,V^2$

This also equals the change in kinetic energy of the system as initial kinetic energy of the system is zero.

Since work done equals the change in kinetic energy, we have

$$150 = 7.875\,V^2;\quad V = \sqrt{19.05} = \mathbf{4.36\ m/s}$$

Thus the plank moves with 4.36 m/s velocity towards right.

EXAMPLE 11.18

A uniform disc of mass 20 kg and diameter 2 m is supported by two pins *A* and *B* as shown in Fig. 11.16. At a particular instant the pin *B* is suddenly removed. Determine the angular acceleration of the disc and the reaction at pin *A* immediately after the removal of pin *B*.

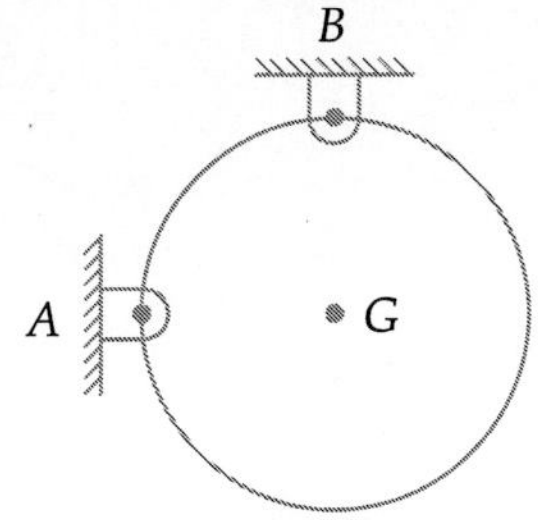

Fig. 11.16

Solution: Refer Fig. 11.17, which shows the various forces acting on the disc when the pin B is removed.

For a rigid body undergoing rotary motion,

$$\Sigma M_a = I_a\alpha$$

or

$$mg \times r = (I_G + mh^2)\,\alpha = \left(\frac{1}{2}mr^2 + mr^2\right)\alpha = \frac{3}{2}mr^2 \times \alpha$$

$$\therefore\ \text{Angular acceleration } \alpha = \frac{2}{3}\,\frac{mg \times r}{mr^2}$$

$$= \frac{2}{3}\frac{g}{r} = \frac{2}{3} \times \frac{9.81}{1} = \mathbf{6.54\ rad/s^2}$$

Further, for translatory motion,

$$\text{net force} = \text{mass} \times \text{acceleration}$$

$$mg - V_a = ma = m\,(r\,\alpha)$$

∴ Vertical reaction at pin A,

$$V_a = mg - m\,r\,\alpha = 20 \times 9.81 - 20 \times 1 \times 6.54 = \mathbf{65.4\ N}$$

Horizontal reaction at pin A = 0

This is because there is no other horizontal component.

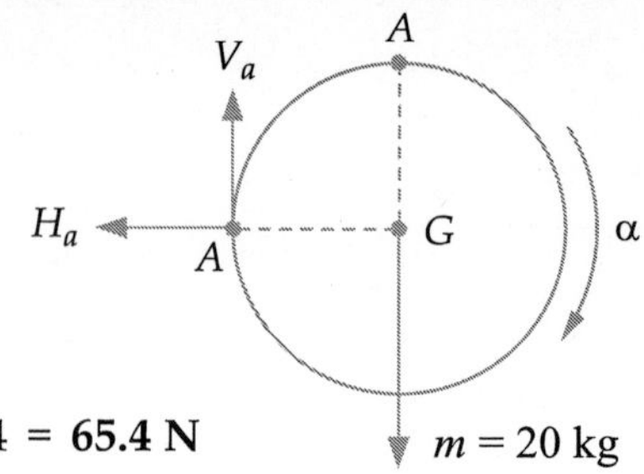

Fig. 11.17

EXAMPLE 11.19

A thin uniform bar of mass *m* and length *l* is suspended by two vertical inextensible strings (Fig. 11.18). If the right hand string *BD* is cut, set up expressions for the angular acceleration of the bar and the tension induced in the left string *AC*.

Solution: When the string BD is cut, the bar AB rotates about its end A.

The bar is acted upon by the following set of forces:

(*i*) weight of the bar $W = mg$ that acts at the point of its centre of gravity G.

(*ii*) tension T induced in the string AC.

For a rigid bar in plane motion (combination of translation and rotation), the equations of motion are

$$\Sigma F_x = ma_x \quad : \text{Here no force acts in the } x\text{-direction}$$

$$\therefore \quad a_x = 0$$

$$\Sigma F_y = ma_y \quad : mg - T = ma \qquad \ldots(i)$$

$$\Sigma M_G = I_G \alpha \quad : T \times \frac{l}{2} = I_G a$$

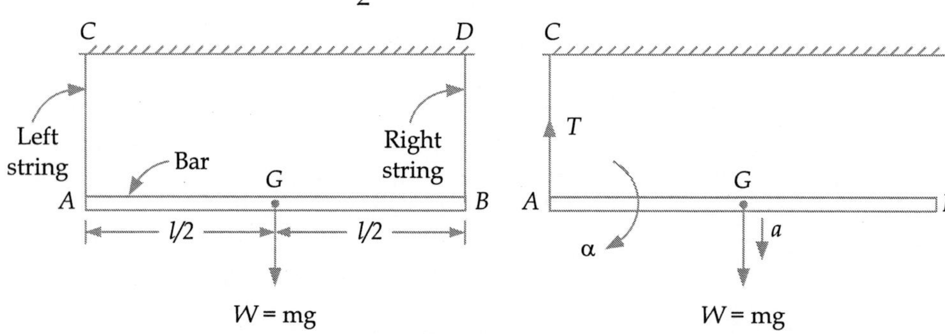

Fig. 11.18

For a thin uniform bar I_G (moment of inertia about the mass centre) $= \dfrac{ml^2}{12}$ and in the absence of slip, the angular acceleration α and linear acceleration a are connected as $a = \alpha r = \alpha l/2$. That gives

$$T \times \frac{l}{2} = \frac{ml^2}{12} \times \frac{2a}{l}; \quad T = \frac{1}{3}\,ma \qquad \ldots(ii)$$

Eliminating T from expressions (*i*) and (*ii*) we have

$$mg - \frac{1}{3}\,ma = ma \quad \text{or} \quad a = \frac{3}{4}\,g \quad \text{and} \quad \alpha = \frac{2a}{l} = \frac{3}{2}\,gl$$

Further : $$T = m\,(g - a) = m \times \left(g - \frac{3}{4}g\right) = \frac{mg}{4}$$

EXAMPLE 11.20

A thin uniform bar of mass m and length l is held at an angle θ with the horizontal by means of two vertical inextensible strings, one at each end as shown in Fig. 11.19. Suddenly the string at the right end breaks, leaving the bar to swing. What will be the angular acceleration of the bar and tension induced in the string at the left end immediately after the string breaks?

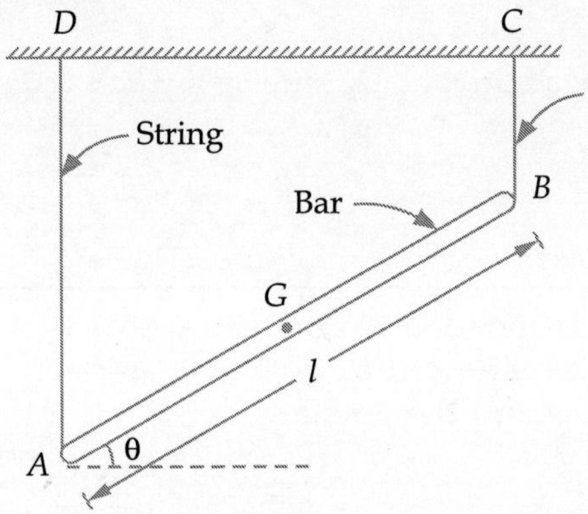

Fig. 11.19

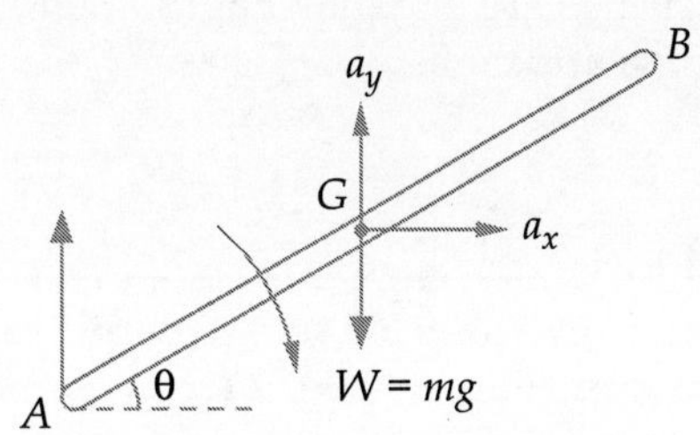

Fig. 11.20

Solution: When the string BC breaks, the bar AB rotates about its end A.

The bar is acted upon by the following set of forces:

(*i*) Weight of the bar $W = mg$ that acts at the point of its centre of gravity G.

(*ii*) Tension T induced in the string AD.

For a rigid bar in plane motion (combination of translation and rotation), the equations of motion are:

$\Sigma F_x = ma_x$; Here no force acts in the x-direction

$\therefore \quad a_x = 0$

$$\Sigma F_y = mg_y \;;\quad mg - T = ma_y \qquad ...(i)$$

$$\Sigma M_G = I_G\,\alpha \;;\; T \times \frac{l}{2}\cos\theta = I_G\,\alpha \qquad ...(ii)$$

For a thin uniform bar I_G (moment of inertia about the mass centre = $ml^2/12$) and in the absence of slip, the angular acceleration α and linear acceleration a are connected as

$$a_y = \alpha r = \alpha\,\frac{l}{2}\cos\theta$$

The from expression (*i*) and (*ii*)

$$mg - T = m\,\alpha\,\frac{l}{2}\cos\theta \qquad ...(iii)$$

and $$T\frac{l}{2}\cos\theta = \frac{ml^2}{12}\alpha \;;\quad T = \frac{ml}{6\cos\theta}\,\alpha \qquad ...(iv)$$

Substituting the value of T as defined by expression (*iv*) into expression (*iii*), we get

$$mg - \frac{ml}{6\cos\theta}\alpha = m\,\alpha\,\frac{l}{2}\cos\theta$$

or
$$mg = \alpha ml\left(\frac{\cos\theta}{2} + \frac{1}{6\cos\theta}\right) = \alpha ml\left(\frac{3\cos^2\theta + 1}{6\cos\theta}\right)$$

The expression for angular velocity α then appears as follows

$$\alpha = \frac{6g\cos\theta}{l(1+3\cos^2\theta)}$$

Then from expression (*iv*)

$$T = \frac{ml}{6\cos\theta} \times \frac{6g\cos\theta}{l(1+3\cos^2\theta)} = \frac{mg}{1+3\cos^2\theta}$$

Note: If the strings are of equal length, then the bar hangs horizontally and $\theta = 0$. The expressions for α and T then appear as

$$\alpha = \frac{3}{2}gl \quad \text{and} \quad T = \frac{mg}{4}$$

EXAMPLE 11.21

A right circular cylinder of mass *m* and radius *r* is suspended from a chord that is wound around its circumference (Fig. 11.21). If the cylinder is allowed to fall freely, determine the acceleration of its mass centre *G* and the tension induced in the chord.

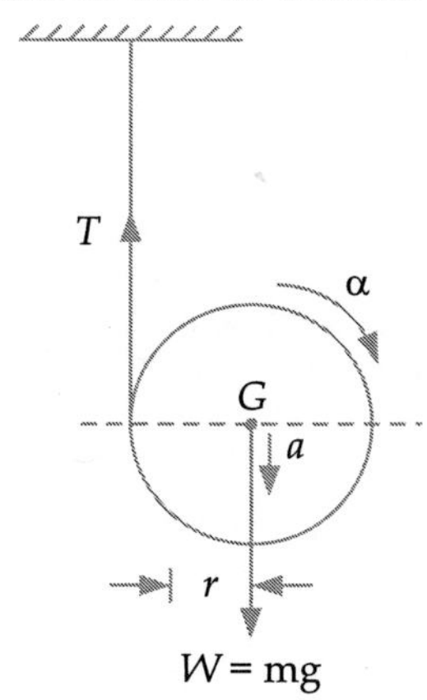

Fig. 11.21

Solution: The cylinder is acted upon by the following set of forces:

(*i*) weight of cylinder $W = mg$ that acts at the point of its centre of gravity G.

(*ii*) tension T induced in the chord.

For a rigid body in plane motion (combination of translation and rotation), the equation of motion are:

$\Sigma F_x = ma_x$: Here no force acts in the x-direction

$\therefore \quad a_x = 0$ (*i*)

$\Sigma F_y = ma_y$: $mg - T = ma$

$\Sigma M_G = I_G\alpha$: $Tr = I_G\alpha$

For a cylinder I_G (moment of inertia about the mass centre) $= \frac{1}{2}mr^2$ and in the absence of slip, the angular acceleration α and linear acceleration a are connected as $a = \alpha r$. That gives:

$$Tr = \frac{1}{2}mr^2 \times \frac{a}{r}; \quad T = \frac{1}{2}ma \qquad \text{...}(ii)$$

Eliminating T from expression (*i*) and (*ii*), we get

$$mg - \frac{1}{2}ma = ma; \quad mg = \frac{3}{2}ma$$

$$\therefore \quad a = \frac{2}{3}g$$

Further : $$T = m(g - a) = m\left(g - \frac{2}{3}g\right) = \frac{mg}{3}$$

EXAMPLE 11.22

A body of weight 8 N is suspended by a light rope wound round a pulley of weight 60 N and radius 30 cm. The other end of the rope is fixed to the periphery of the pulley. If the weight is moving downwards, make calculations for the acceleration of 8 N weight and tension in the string.

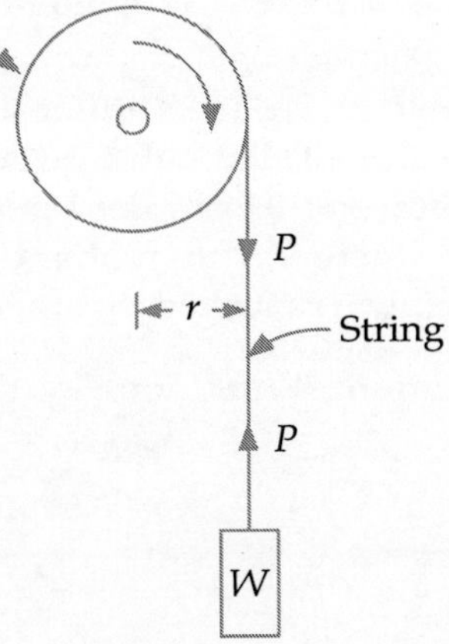

Fig. 11.22

Solution: Refer Fig. 11.22 for the arrangement

Consider the motion of weight (coming downward):

Net downward force on the body

$$= (W - P)$$

If the body moves downward with an acceleration a, then from the equation $F = ma$, we have

$$W - P = \frac{W}{g} a \qquad ...(i)$$

Consider the motion of pulley:

Torque on the pulley $T = I\alpha$

Noting that $T = P \times r$; $\alpha = \frac{a}{r}$, we get

$$P \times r = I \times \frac{a}{r}; \quad P = I \times \frac{a}{r^2}$$

If pulley is considered as a solid disc, then

$$I = \frac{W_0}{g}\frac{r^2}{2} \qquad \text{where } W_0 \text{ is the weight of pulley}$$

$$\therefore \quad P = \frac{W_0}{g}\frac{r^2}{2} \times \frac{a}{r^2} = \frac{W_0}{2g} a \qquad ...(ii)$$

From expressions (*i*) and (*ii*),

$$W - \frac{W_0}{2g} a = \frac{W}{g} a; \quad W = \frac{a}{g}\left(W + \frac{W_0}{2}\right)$$

That gives: $$a = \frac{gW}{W + \frac{W_0}{2}}$$

Then from expression (*i*),

$$W - P = \frac{W}{g} \times \frac{gW}{W + \frac{W_0}{2}} = \frac{2W^2}{2W + W_0}$$

$$\text{or} \quad P = W - \frac{2W}{2W + W_0} = \frac{2W^2 + WW_0 - 2W^2}{2W + W_0} = \frac{WW_0}{2W + W_0}$$

Substituting the given data, we obtain

Acceleration $$a = \frac{9.81 \times 8}{8 + \frac{60}{2}} = \mathbf{2.065\ m/s^2}$$

Tension induced in the string $$P = \frac{8 \times 60}{2 \times 8 + 60} = \mathbf{6.31\ N}$$

EXAMPLE 11.23

An inextensible cord going around a homogeneous cylinder of mass 125 kg and radius 1 m holds a plate *B* of negligible mass.

A collar of 35 kg mass is released from rest in the position shown in Fig 11.23 and it drops on the plate. What will be the velocity of the collar after it has descended an additional distance of 0.5 m after striking the plate. It may be presumed that there is no rebound, *i.e.,* the plate and collar move downwards locked together and the chord remains taut.

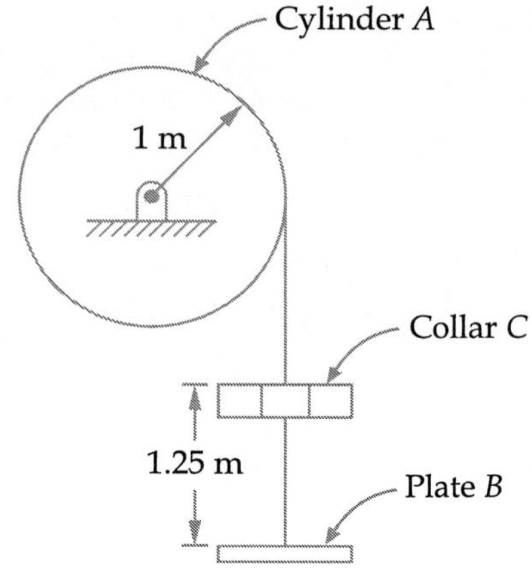

Fig. 11.23

Solution : Work done by the system during fall of collar

$$= mgh = 35 \times 9.81 \times (1.25 + 0.5)$$

$$= 600.86 \text{ Nm}$$

Change in the kinetic energy due to translatory motion of collar

$$= \frac{1}{2} m\left(V_2^2 - V_1^2\right) = \frac{1}{2} mV_2^2 \qquad (\because V_1 = 0)$$

$$= \frac{1}{2} \times 35 \times V_2^2 = 17.5\, V_2^2$$

Change in the kinetic energy due to rotary motion of the cylinder

$$= \frac{1}{2} I(\omega_2^2 - \omega_1^2) = \frac{1}{2} I\omega_2^2 \qquad (\because \omega_1 = 0)$$

$$= \frac{1}{2}\left(\frac{1}{2} mr^2\right) \times \left(\frac{V_2}{r}\right)^2 = \frac{1}{2}\left(\frac{1}{2} \times 125 \times 1^2\right) \times \left(\frac{V_2}{1}\right)^2 = 31.25\, V_2^2$$

According to work – energy principle,

work done by the system = total change in kinetic energy of the system

$$600.86 = 17.5\, V_2^2 + 31.25\, V_2^2 = 48.75\, V_2^2$$

∴ Velocity attained by the collar

$$V_2 = \sqrt{\frac{600.86}{48.75}} = \mathbf{3.51 \text{ m/s}}$$

EXAMPLE 11.24

A homogeneous solid cylinder of weight 100 N rotates about its horizontal axis in frictionless bearings under the action of the weight of 10 N block which is carried by a rope wrapped round the cylinder as shown in Fig. 11.24. What will be the angular velocity of the cylinder two seconds after the motion starts? Assume diameter of cylinder as 100 cm.

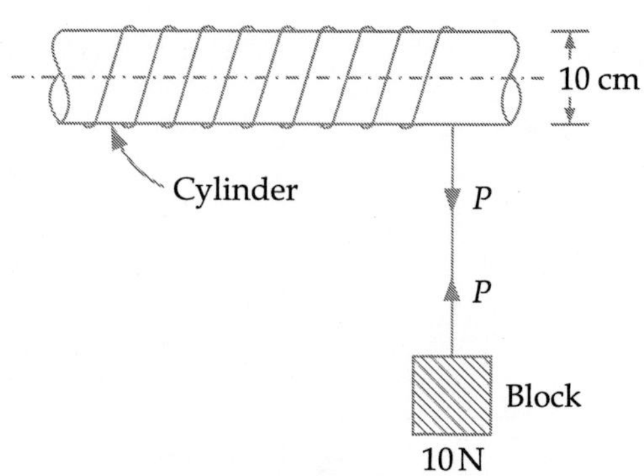

Fig. 11.24

Solution: When the weight moves downward, the cylinder starts rotating. The angular acceleration α of the cylinder and linear acceleration *a* of the system are correlated by the expression.

$a = \alpha R$ where *R* is the radius of the cylinder

∴ $a = 0.5\,\alpha$

Considering downward motion of the block of 10 N weight and applying Newton's second law of motion, we have

$$\text{net force} = \text{mass} \times \text{acceleration}$$

$$(10 - P) = \frac{10}{9.81} \times a = \frac{10}{9.18} \times 0.5\alpha = 0.5097\,\alpha \qquad ...(i)$$

Considering the motion of cylinder,

$$\text{Mass moment of inertia } I = \frac{1}{2} mR^2 = \frac{1}{2} \times \frac{100}{9.81} \times (0.5)^2 = 1.274 \text{ kg m}^2$$

$$\text{Torque on the cylinder } T = P \times R = 0.5\,P$$

From Newton's second law: $T = I\alpha$

$$0.5\,P = 1.274 \times \alpha\,; \quad P = 2.548\,\alpha$$

Substituting this value of P in expression (*i*), we get

$$10 - 2.548\,\alpha = 0.5097\,\alpha$$

∴ Angular acceleration of rotation of cylinder,

$$\alpha = \frac{10}{2.548 + 0.5097} = 3.27 \text{ rad/s}^2$$

From the kinematic relation $\omega = \omega_0 + \alpha t$ and noting that initial angular velocity of cylinder $\omega_0 = 0$, we obtain

$$\omega = 0 + 3.27 \times 2 = \mathbf{6.54 \text{ rad/s}}$$

EXAMPLE 11.25.

A stepped cylinder shown in Fig. 11.25 has the dimensions R_1 = 50 cm and R_2 = 100 cm. The mass of the cylinder is 125 kg with radius of gyration 55 cm. Blocks *A* and *B* of masses 80 kg and 100 kg respectively are connected to the cylinder through the ropes. If the system is released from rest, determine the time required for block *A* to move by 2.5 meters.

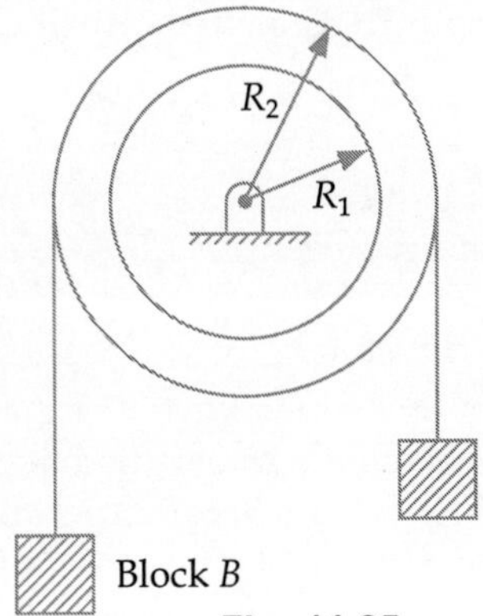

Fig. 11.25

Solution: Since block *B* is heavier than block *A*, the cylinder would rotate in the anti-clockwise direction and the block *A* would get lifted upwards

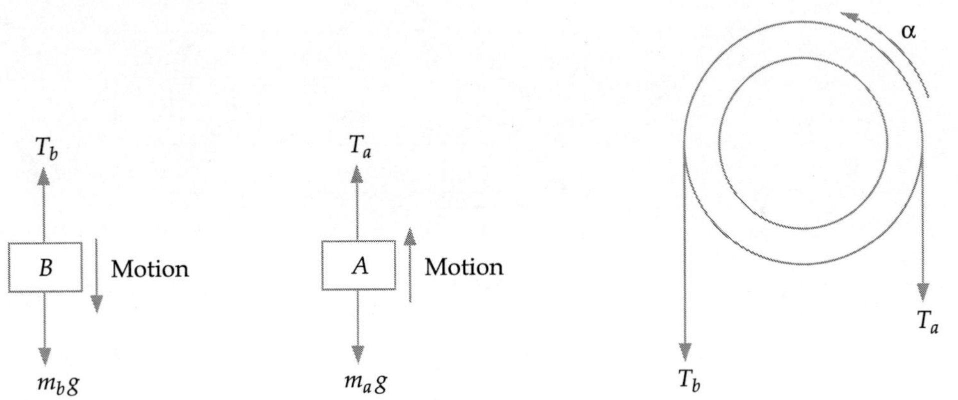

Fig. 11.26

Then with reference to free body diagrams shown in Fig. 11.26

For block *A*:

$$T_a - m_a g = m_a a_a\,; \quad T_a = m_a (a_a + g)$$

$\therefore \quad T_a = 80\,(1 \times \alpha + 9.81) = 80\alpha + 784.8$

For block *B*:

$$T_b - m_b\,g = -\,m_b a_b\,; \quad T_b = m_b\,(g - a_b)$$

$\therefore \quad T_b = 100\,(9.81 - 0.5\alpha) = 981 - 50\,\alpha$

For cylinder:

$$I = mk^2 = 125 \times (0.55)^2 = 37.81 \text{ kg m}^2$$

From Newton's second law; $T = I\alpha$

$$T_b R_1 - T_a R_2 = I\alpha$$

Substituting the appropriate values

$$(981 - 50\alpha) \times 0.5 - (80\alpha + 784.8) \times 1 = 37.81\,\alpha$$

$$(490.5 - 25\,\alpha) - (80\,\alpha + 784.8) = 37.81\,\alpha$$

$\therefore$ Angular acceleration of the cylinder $\alpha = -\,2.06 \text{ rad/s}^2$

Corresponding to 2.5 m upward movement of block *A*, the angular displacement is $2.5/R_2 = 2.5$ radian

Then from the kinematic relation $\theta = \omega t + \dfrac{1}{2}\alpha\, t^2$, we have

$$2.5 = 0 \times t + \frac{1}{2} \times (2.06) \times t^2 \qquad (\omega = 0)$$

$\therefore$ Time required $t = \sqrt{\dfrac{2.5 \times 2}{2.06}} =$ **1.56 sec**

EXAMPLE 11.26

Two blocks of masses $m_1 = 25$ kg and $m_2 = 20$ kg are connected by a light inextensible string which passes over a 25 cm diameter pulley of 2.5 kg mass as shown in Fig. 11.27. Neglecting friction, work out the acceleration of the system and the tension in the string when the masses are released from rest. Assume radius of gyration of the pulley to be equal to its radius.

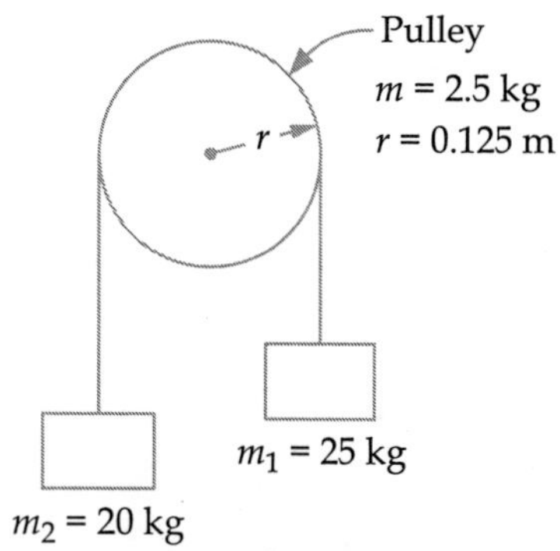

Fig. 11.27

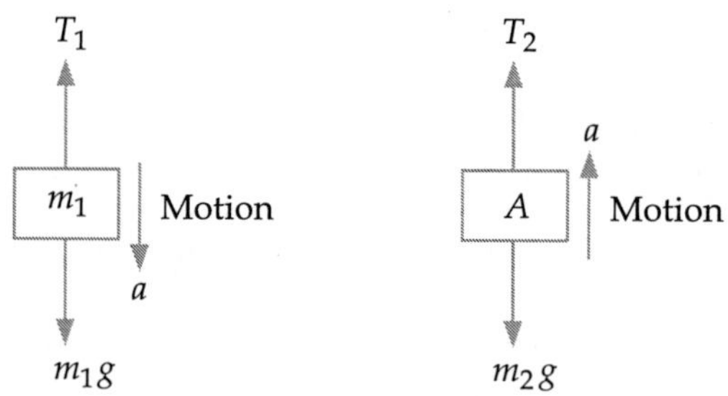

Fig. 11.28

Solution: Considering the motion of masses and applying Newton's second law ($F = ma$), we have

$$m_1 g - T_1 = m_1 a$$

or $\quad 25\,g - T_1 = 25\,a \qquad \ldots(i)$

and

$$T_2 - m_2 g = m_2 a$$

or $\quad T_2 - 20\,g = 20\,a \qquad \ldots(ii)$

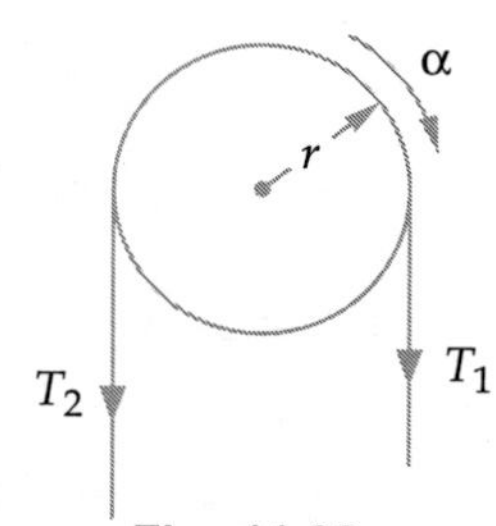

Fig. 11.29

In the context of pulley

Resultant torque $T = (T_1 - T_2)\,r$

$= (T_1 - T_2) \times 0.125$

Mass moment of inertia $I = mk^2 = 2.5 \times (0.125)^2$

$= 0.0391 \text{ kg m}^2$

Angular acceleration $\alpha = \frac{a}{r} = \frac{a}{0.125}$

From Newton's second law: $T = I\alpha$

$$(T_1 - T_2) \times 0.125 = 0.0391 \times \frac{a}{0.125}$$

or $(T_1 - T_2) = 2.5\,a$ ...(*iii*)

Substituting the values of T_1 and T_2 from expression (*i*) and (*ii*) into expression (*iii*),

$$25g - 25a - (20a + 20g) = 2.5a; \quad 47.5\,a = 5\,g$$

$\therefore$ Acceleration of the system $a = \frac{5 \times 9.81}{47.5} = \mathbf{1.033\ m/s^2}$

Then from expression (*i*), $T_1 = 25\,(g - a) = 25\,(9.81 - 1.033) = \mathbf{219.42\ N}$

and from expression (*ii*), $T_2 = 20\,(g + a) = 20\,(9.18 + 1.033) = \mathbf{216.86\ N}$

EXAMPLE 11.27

Two bodies of weight 600 N and 400 N respectively are connected to the two ends of a chord which passes over a pulley of 80 cm diameter. The pulley may be assumed as a solid disc having mass 100 kg and 40 cm radius of gyration about its axis of rotation.

Make calculations for the torque which need to be applied to the pulley to raise the 600 N body with an acceleration of 1.25 m/s².

Neglect friction in the bearings of the pulley.

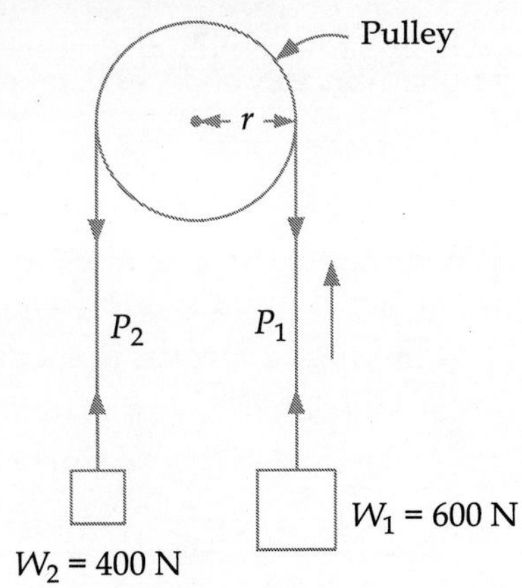

Fig. 11.30

Solution: Refer Fig. 11.30 for the arrangement.

Consider the motion of weight W_1:

The weight is being raised upwards with an acceleration *a*.

Net upward force on weight $= P_1 - W_1$

From the equation $F = ma$,

$$P_1 - W_1 = \frac{W_1}{g} \times a$$

$$\therefore \quad P_1 = \frac{W_1}{g}\,(g + a)$$

$$= \frac{600}{9.81}\,(9.81 + 1.25) = 676.45 \text{ N}$$

Consider the motion of weight W_2:

The weight is being lowered with an acceleration *a*.

Net downward force on weight $= W_2 - P_2$

From the equation $F = ma$

$$W_2 - P_2 = \frac{W_2}{g} \times a$$

$$\therefore \quad P_2 = \frac{W_2}{g}(g - a) = \frac{400}{9.81}(9.81 - 1.25) = 349.03 \text{ N}$$

Consider the motion of pulley:

Moment of inertia of the pulley $I = mk^2 = 100 \times (0.4)^2 = 16 \text{ kgm}^2$

Angular acceleration of pulley $\alpha = \frac{a}{r} = \frac{1.25}{0.4} = 3.125 \text{ rad/s}$

Torque due to rotation of pulley $T_1 = I\alpha = 16 \times 3.125 = 50 \text{ Nm}$

Torque acting on pulley due to difference in pulls in the two ropes,

$$T_2 = (P_1 - P_2)\, r = (6.76.45 - 349.03) \times 0.4 = 130.97 \text{ Nm}$$

∴ Torque which must be applied to the pulley

$$= 50 + 130.97 = \mathbf{180.97\ Nm}$$

REVIEW AND SUMMARY

1. Rotary motion is the motion of a rigid body in which the particles move in a circular path. The particles continue repeating the same path and have a fixed centre of rotation.
2. The product of mass moment of inertia and angular velocity of the body is called the moment of momentum or angular momentum

   Angular momentum $= I\omega$
3. Torque is the turning moment of force on a body on which it acts.

   The rate of change of angular momentum of a rotating body equals the external torque applied on the body and takes place in the direction of the torque

   $$T = I\frac{d\omega}{dt} = I\alpha$$

   When the torque T is applied on the body for infinite small interval of time dt, the torque is called an impulsive or angular torque

   Impulsive torque $T_i = I\, d\omega$ = change of angular momentum
4. Total kinetic energy of a rotating body

   $$= KE \text{ in translation} + KE \text{ in rotation}$$

   $$= \frac{1}{2}mV^2 + \frac{1}{2}I\omega^2$$

   $$= \frac{1}{2}mV^2 + \frac{1}{2}(mk^2)\times\left(\frac{V}{r}\right)^2 = \frac{mV^2}{2}\left[1\right.$$
5. The work done principle connects the work done with change in kinetic energy. Further, work done results both from linear and angular displacement from the initial and final positions. Thus

   $$Fs + T\theta = \frac{1}{2}m\left(V_2^2 - V_1^2\right) + \frac{1}{2}I\left(\omega_2^2 - \omega_1^2\right)$$

6. The principle of conservation of energy states that the sum of potential energy and kinetic energy of a rigid body moving under the influence of conservative forces is always a constant. That is

$$(PE + KE)_1 = (PK + KE)_2$$

REVIEW QUESTIONS

A. Conceptual and conventional questions :

1. Define rotary motion and explain the concept of torque.
2. Define moment of momentum (or angular momentum).
3. Set up an expression for the work done by torque.
4. State the law of conservation of angular momentum.
5. State Newton's laws of motion for angular motion
6. What is the kinetic energy of rotation?
7. Write the equations of motion for a rigid body in plane motion (combination of translation and rotation)
8. State the work-energy principle for a rigid body in plane motion
9. State the principle of conservation of energy for a rigid body.
10. A rotor of mass 20 kg and radius of gyration 1 m is rotating at a speed of 1440 revolutions per minute. If the rotor comes to rest in 2.5 minutes, determine the friction torque at the bearing. **(20.1 Nm)**
11. A uniform homogeneous cylinder of 0.01 kg mass and 8 cm radius rolls without slip along a horizontal level surface. If the translational velocity of the cylinder is 15 cm/s, calculate its total kinetic energy. **(0.000236 Nm)**
12. A flywheel of mass 5000 kg and having a radius of gyration 1 m loses its speed from 400 rpm to 280 rpm in 2 minutes. Calculate (a) the retarding torque acting on it (b) change in the kinetic energy during the above period and (c) change in its angular momentum during the period. **(523.6 Nm, 2236 kNm, 62835 Nm/s)**
13. Two blocks weighing 80 N and 30 N are supported at the ends of rope passing over the rough surface of a pulley mounted on a horizontal axle. The pulley may be assumed to be a solid disc with a weight of 50 N. Determine the tension on a two parts of the rope and the linear acceleration of the blocks. Neglect friction in the bearings of the pulley and the weight of the rope. **(50.37 N, 41.11 N, 3.63 m/s^2)**

B. Complete the following statements with most appropriate word/words :

1. The product of mass moment of inertia and the angular velocity of rolating body is known as ....
2. The .... of a rigid body consists of rectilinear motion and a rolling motion without slipping.
3. The torque applied to a body for infinite small interval of time is called ....
4. If the external torque acting on a system is zero, the .... of the system remains constant.
5. For a solid sphere of mass m and radius r, the radius of gyration is ....

6. The work-energy principle connects the .... with change in ....
7. The work done by torque is given by the product of .... and ....
8. Corresponding to force equation $F = ma$ in linear motion, the torque equation in rotary motion is ....
9. In the expression $1/2\ I\omega^2$ for the kinetic energy of a rigid body in rotation, I represents, the .... moment of inertia of the body about the ....
10. A flywheel has a mass moment of inertia of 10 kgm^2 about the axis of rotation. If it runs with angualr velocity of 100 rad/s, its kinetic energy in kNm would be ....

Answers :

1. angular momentum or moment of momentum 2. plane motions 3. impulsive torque 4. angular momentum 5. 0.632 r 6. Work done and kinetic energy 7. torque, angular displacement 8. $T = I\alpha$ 9. mass, axis of rotation 10. 50

❏❏❏

CHAPTER

12

VIRUTAL WORK

Learning objectives : After the study of this chapter, the reader will able to :

- *define the terms virtual displacement and virtual work*
- *understand the forces that need to be omitted or considered while evaluating the virtual work*
- *express the various virtual displacements in terms of a single virtual displacement*
- *write the virtual-work equation and solve it for unknown force, couple etc.*

When a system or a body is in equilibrium under the action of a set of forces, there is no displacement and accordingly no work in done. However if the system is imagined to undergo any arbitrary but small displacement, some work can be imagined to have been done.

The imaginary small displacement given to the system is called virtual displacement. The product of the net resultant force and the virtual displacement in the direction of force is called virtual work.

The principle of virtual work may be stated as: "*The work done on a rigid body or a system of rigid bodies in equilibrium, when acted upon a set of forces, is zero for any virtual displacement compatible with the constraints on the system.*"

Conversely: "*If the algebraic sum of the virtual work for every virtual displacement is zero, the system is in equilibrium.*"

The concept of virtual work is useful while solving problems related to static equilibrium of beams, framed structures and lifting machines.

Illustration: Consider a rod AB which can rotate about a fulcrum O. A vertical upward force P has been applied at end A and it is required to find the vertical downward force that need to be applied at end B to keep the rod horizontal.

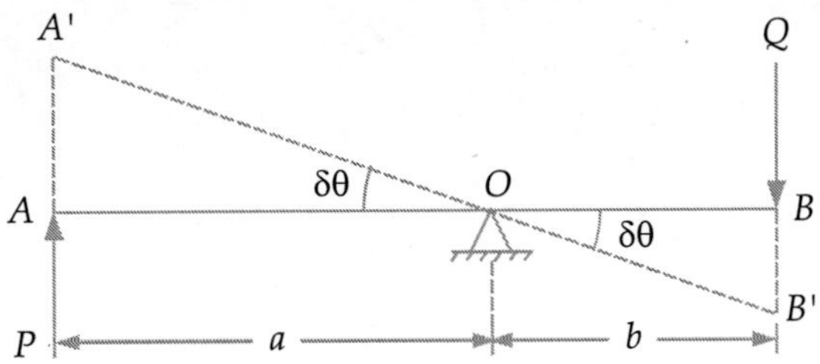

Fig. 12.1

Let it be imagined that the rod rotates through an angle $\delta\theta$ and takes the position $A'B'$. The displacement of A is AA' and that of B is BB'. For small angular rotation of the rod, these displacements can be taken perpendicular to AB and accordingly

$$AA' = a\,\delta\theta \quad \text{and} \quad BB' = b\,\delta\theta$$

$$\text{Work done by force } P = P \times a\,\delta\theta$$

$$\text{Work done by force } Q = -Q \times b\,\delta\theta$$

$$\text{Total work} = Pa\,\delta\theta - Qb\,\delta\theta$$

According to the principle of virtual work, the total work done has to be zero.

$$\therefore \quad Pa\,\delta\theta - Qb\,\delta\theta = 0$$

$$\text{That gives: } Q = P \times \frac{a}{b}$$

The following forces do no work and are omitted while applying the principle of virtual work:

(*i*) tension in a light inextensible string

(*ii*) forces normal to the direction of displacement

(*iii*) reaction between smooth bodies; this is because such a reaction is perpendicular to the direction of displacement.

(*iv*) reaction at smooth pins and hinges which do not move

(*v*) action and reaction between two bodies whose equilibrium is being considered together

(*vi*) reaction at a point or an axis fixed in space around which the body is constrained to turn.

Further, the following sign conventions are adopted while writing the expressions for virtual work:

1. Upward force, forces acting towards right and the forces in the clockwise direction are considered positive.
2. Downward forces, forces acting towards lift and the forces in the anticlockwise direction are taken negative.

The virtual work becomes positive when the displacement is in the same direction as the force and negative when displacement is opposite to the direction of force.

Furthermore, it is generally assumed that there is no friction in hinges, bearings or sliding surfaces.

EXAMPLE 12.1

A beam has been loaded and supported as shown in Fig. 12.2. Use the method of virtual work to determine the reactions at the end supports.

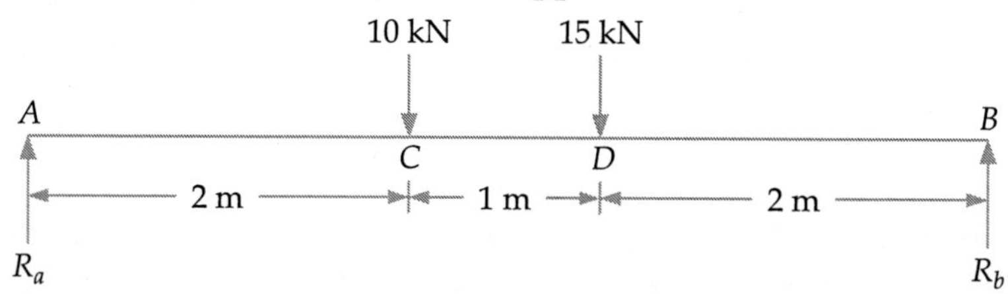

Fig. 12.2

Solution: Let the beam be given a vertical virtual displacement δ_b at end *B* holding the end *A* in position.

From the similarity of triangles

$$\frac{\delta_c}{2} = \frac{\delta_d}{3} = \frac{\delta_b}{5}$$

$$\therefore \quad \delta_c = 0.4\,\delta_b \quad \text{and} \quad \delta_d = 0.6\,\delta_b.$$

$$\text{Total work} = R_b \times \delta_b - 15 \times \delta_d - 10 \times \delta_c + R_a \times 0$$

While writing the above expression, the upward forces have been considered positive, and the down forces as negative.

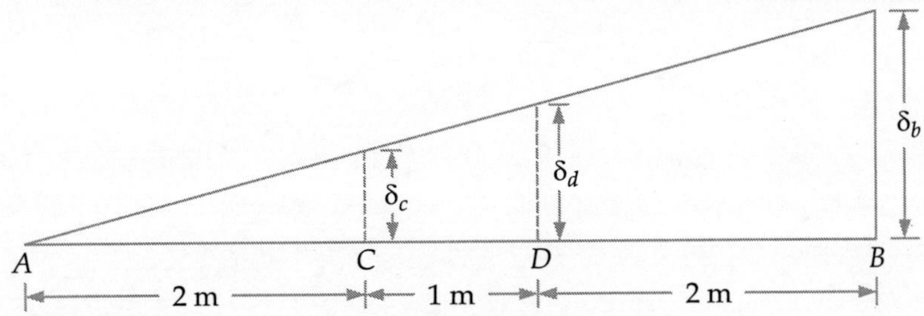

Fig. 12.3

According to the principle of virtual work, the total work done by all the forces must be zero. That gives

$$R_b \times \delta_b - 15 \times \delta_d - 10 \times \delta_c + R_a \times 0 = 0$$

or $$R_b \times \delta_b - 15 \times 0.6\,\delta_b - 10 \times 0.4\,\delta_b = 0$$

or $$R_b - 9 - 4 = 0\ ;\ R_b = \mathbf{13\ kN}$$

$$R_a = (10 + 15) - 13 = \mathbf{12\ kN}$$

EXAMPLE 12.2

Two beams *AC* and *CD* of lengths 9 m and 10 m respectively are hinged at *C*. These are supported on rollers on the ends *A* and *D*, and a hinged support at *B* which lies at distance 7 m from *A*. The beam is acted upon by a concentrated load of 700 N at point *E* located 6 m from end *A*. Determine the reactions at the hinge point and at the support *B*. Use the principle of virtual work.

Solution: Refer Fig. 12.4 for the given system of beams.

Let the system be given a vertical upward virtual displacement y_c at the point *C* holding the ends *A* and *D* in position.

From the geometrical configuration of the system after virtual displacement,

$$y_b = y_c \times \frac{AB}{AC} = \frac{7}{9}\,y_c$$

$$y_e = y_c \times \frac{ED}{CD} = \frac{6}{10}\,y_c$$

Consider beam *CD* alone and assuming R_c acting upward.

Work done by $R_c = R_c \times y_c$

Work done by $R_e = -R_e \times y_e$

Work done by R_a is zero because there is no virtual displacement at *A*

$\therefore$ Total work $= R_c y_c - R_e y_e$

Using the principle of virtual work,

$$R_c y_c - R_e y_e = 0 \quad \text{or} \quad R_c y_c - 700 \times \frac{6}{10}\,y_c = 0; \quad R_c = 420\ \text{kN}$$

The positive sign with R_c suggests that the assumption of R_c acting upward on beam *CD* is correct.

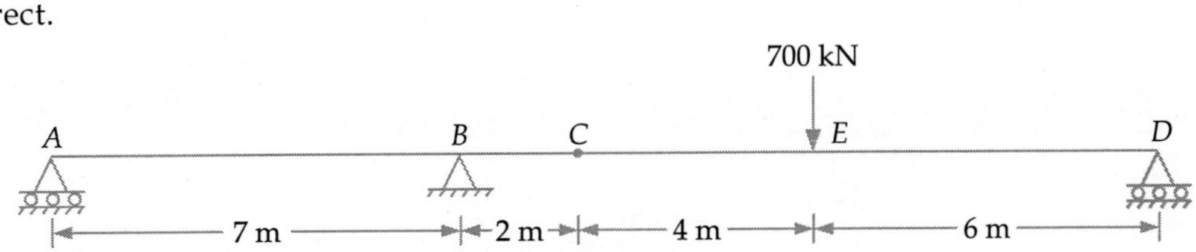

Fig. 12.4

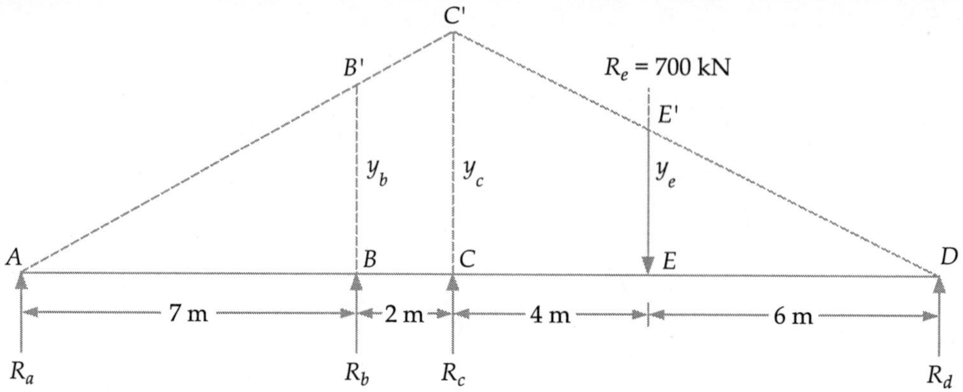

Fig. 12.5

Consider beam *AC* alone:

Since R_c is upward on *CD*, it must be downward on *AC*

Work done by $R_b = R_b \times y_b$

Work done by $R_c = -R_c \times y_c$

Work done by R_d is zero since there is no virtual displacement at *D*

$\therefore$ Total work $= R_b y_b - R_c y_c$

Using the principle of work,

$$R_b y_b - R_c y_c = 0 \quad \text{or} \quad R_b \times \frac{7}{9} y_c - 420 \times y_c = 0; \quad R_b = \mathbf{540\ kN}$$

EXAMPLE 12.3

A beam of span 5 m is supported at *A* and *B* and is subjected to a load system shown in the Fig. 12.6.

Using the principle of virtual work, find the reactions at *A* and *B*.

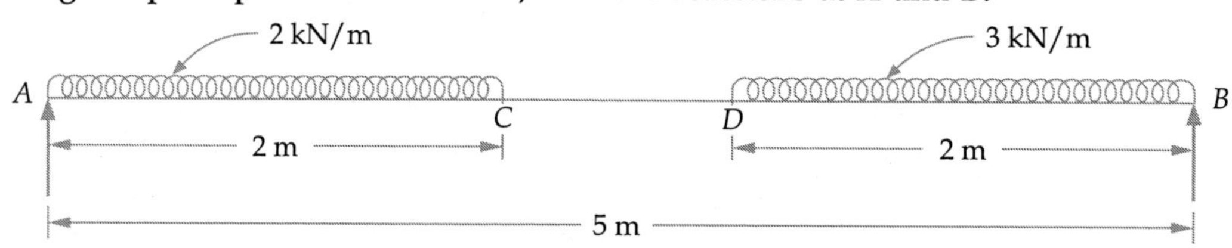

Fig. 12.6

Solution: The given load of 2 kN/m is uniformly distributed for 2 m length between *AC*. Its equivalent load is 2 × 2 = 4 kN and it acts at its centre of gravity at a distance of 1 m form end *A*. Likewise the equivalent load corresponding to *u dl* of 3 kN/m between *DB* is 6 kN and it acts at 1 m distance from end *B*.

The equivalent load diagram for the given beam is as then shown below:

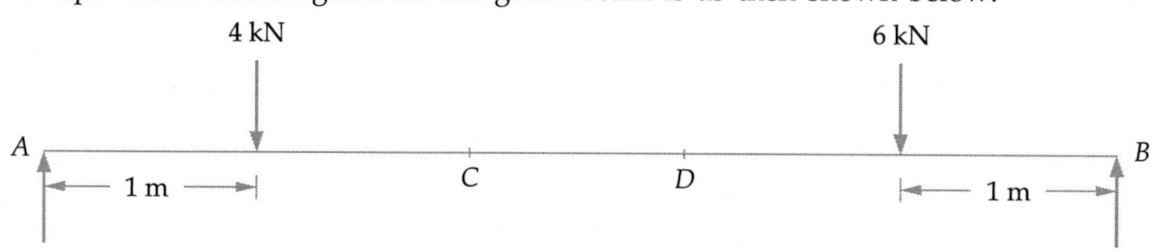

Let the beam be given a vertical virtual displacement δ_b at end B holding the end A in position. The corresponding displacements under the 4 kN and 6 kN load are δ_1 and δ_2 respectively.

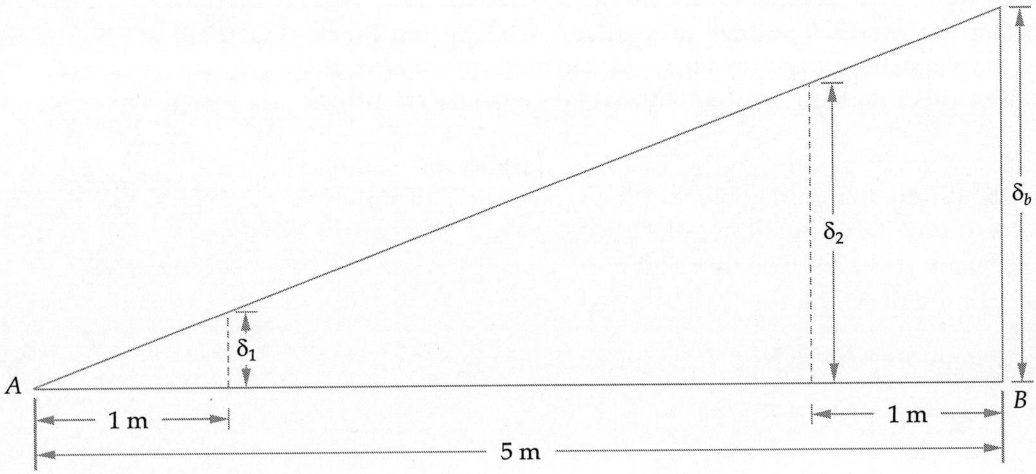

From the similarity of triangles

$$\frac{\delta_1}{1} = \frac{\delta_2}{4} = \frac{\delta_b}{5}$$

$$\therefore \quad \delta_1 = 0.2\,\delta_b \text{ and } \delta_2 = 0.8\,\delta_b$$

$$\text{Total work} = R_b \times \delta_b - 6\delta_2 - 4\delta_1 + R_a \times 0$$

While writing the above expression, the upward forces have been taken positive, and downward forces as negative.

According to the principle of virtual work, the total work done by all the forces is zero. That gives

$$R_b \times \delta_b - 6\delta_2 - 4\delta_1 + R_a \times 0 = 0$$

or $$R_b \times \delta_b - 6 \times 0.8\,\delta_b - 4 \times 0.2\,\delta_b = 0$$

or $$R_b - 4.8 - 0.8 = 0; \quad R_b = \mathbf{5.6\ kN}$$

and $$R_a = (4 + 6) - 5.6 = \mathbf{4.4\ kN}$$

EXAMPLE 12.4

A weight W of 1000 N is to be raised by a system of pulleys as shown in Fig. 12.7. Using the principle of virtual work, find the value of force P which can hold the system in equilibrium.

Solution: Let δy be the virtual vertical downward displacement of force P. Then from the geometry of the system of pulleys, the load will get lifted upwards through distance $\delta y/2$.

$$\text{Work done by force } P = -P \times \delta y$$

$$\text{Work done by load} = W \times \delta y/2$$

While writing the above expressions, the upward displacements have been taken as positive and the down ward displacements as negative.

$$\text{Total work} = -P\,\delta y + W\,\delta y/2$$

According to the principle of virtual work, the total work done by all the forces must be zero. That gives

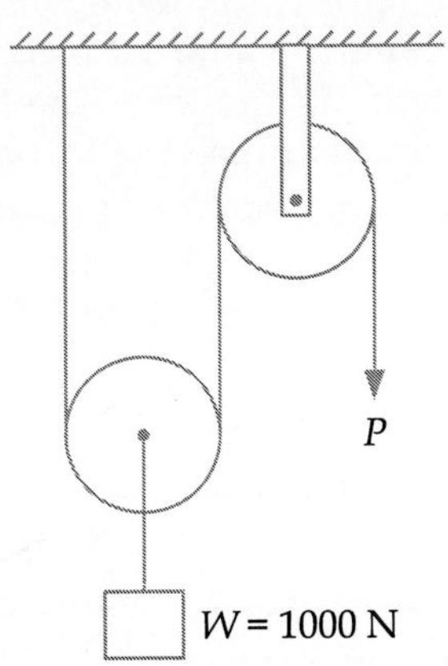

Fig. 12.7

$$-P\,\delta y + W\,\delta y/2 = 0 \quad \text{or} \quad P = \frac{W}{2} = \frac{1000}{2} = \mathbf{500\ N}$$

EXAMPLE 12.5

The diameters of the two pulleys of a differential pulley block (Fig. 12.8) are 30 cm and 25 cm respectively. Using the principle of virtual work, calculate the value of effort *P* to lift a load of 3000 N. Neglect any frictional effects.

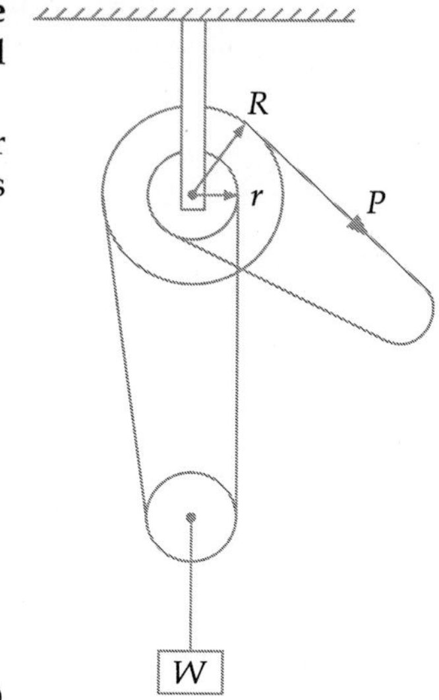

Fig. 12.8

Solution: Let R and r denote the radii of the larger and smaller pulleys of the differential pulley block. When the system of pulleys is imagined to undergo a small angular displacement $\delta\theta$, then

displacement at the effort end = $R\,\delta\theta$

distance through which the load is lifted

$$= \frac{1}{2}\text{(length of chain wound − length of chain unwound)}$$

$$= \frac{1}{2}(R\,\delta\theta - r\,\delta\theta)$$

$$= \frac{1}{2}(R - r)\,\delta\theta$$

That gives:

work done during downward movement of effort = $-P \times R\,\delta\theta$

work done by upward movement of load = $W \times \frac{1}{2}(R - r)\,\delta\theta$

Algebraic sum of the work done = $W \times \frac{1}{2}(R - r)\,\delta\theta - P \times R\,\delta\theta$

According to the principle of virtual work,

$$W \times \frac{1}{2}(R - r)\,\delta\theta - P \times R\,\delta\theta = 0$$

$$\therefore \quad P = \frac{1}{2}W\frac{R-r}{R} = \frac{1}{2} \times 3000 \times \frac{15-12.5}{15} = \mathbf{250\ N}$$

EXAMPLE 12.6

The pulley arrangement shown in Fig. 12.9 is used for hoisting a load *W*. Find the applied force *P* in terms of load *W* for equilibrium of the system. The radii of the two steps of the pulley are r_1 and r_2. Neglect friction and use the virtual work principle.

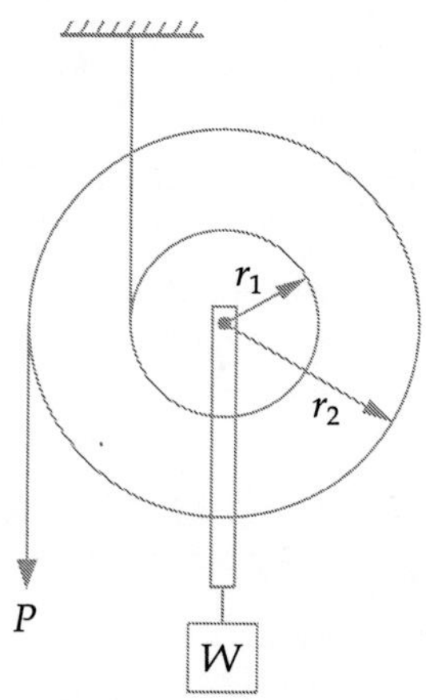

Fig. 12.9

Solution: Let $\delta\theta$ be the small virtual angular displacement of the larger pulley (radius r_2) when pull P is applied. By this action, the smaller pulley (radius r_1) also rotates by angle $\delta\theta$. Then

$$\text{displacement of } W = r_1\,\delta\theta$$

$$\text{displacement of } P = r_2\,\delta\theta - r_1\,\delta\theta = (r_2 - r_1)\,\delta\theta$$

By using the principle of virtual work: $\Sigma W = 0$

$$P(r_2 - r_1)\,\delta\theta - W\,r_1\,\delta\theta = 0$$

The negative sign with the second term is due to the fact that weight acts downward and its lift is upwards (weight force and direction of its movement are in opposite directions)

$$P = \frac{W r_1 \,\delta\theta}{(r_2 - r_1)\delta\theta} = W \frac{r_1}{r_2 - r_1}$$

EXAMPLE 12.7

A uniform rod of length 2 l rests in equilibrium against a smooth vertical wall and upon a smooth peg P at distance a from the wall (Fig. 12.10). Show that, in the position of equilibrium, the inclination of rod with the wall should be $\theta = \sin^{-1}\left(\frac{a}{l}\right)^{1/3}$.

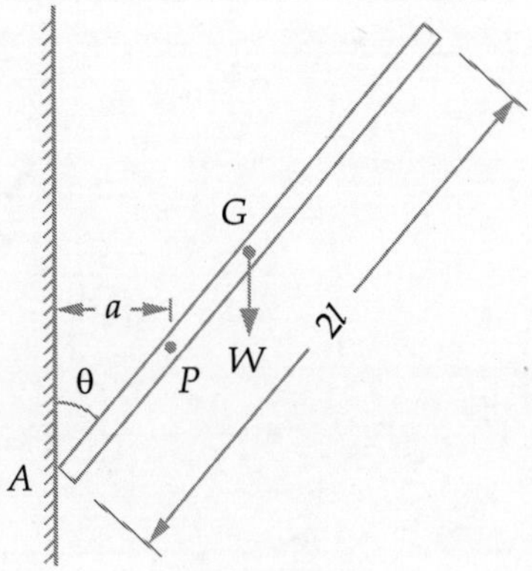

Fig. 12.10

Solution: Let P be the peg and G be the centre of gravity CG of the rod. The weight W of the rod acts vertically downwards at its CG.

The vertical height of G above the peg is given by

$$y = AG \cos\theta - a \cot\theta = l \cos\theta - a \cot\theta$$

Let the system be imagined to rotate through small angle $\delta\theta$ so that θ becomes $(\theta + \delta\theta)$.

$$dy = -l \sin\theta\,\delta\theta + a \operatorname{cosec}^2\theta\,\delta\theta$$

Work done by weight W of the rod,

$$= -W\,dy \qquad \text{(– ve sign due to load acting dowawards)}$$

$$= W(l \sin\theta\,\delta\theta - a \operatorname{cosec}^2\theta\,\delta\theta)$$

The reaction at A and that at P do not contribute towards work. Then from the principle of virtual work

$$W(l \sin\theta\,\delta\theta - a \operatorname{cosec}^2\theta\,\delta\theta) = 0$$

or $\quad l \sin\theta = a \operatorname{cosec}^2\theta$;

or $\quad l \sin\theta = a \times \frac{1}{\sin^2\theta}$; $\quad \sin^3\theta = \frac{a}{l}$

$$\therefore \quad \theta = \sin^{-1}\left(\frac{a}{l}\right)^{1/3}$$

EXAMPLE 12.8

Each of the two uniform hinged bars has a weight W and length l, and is supported and loaded as shown in Fig. 12.11. Using the principle of virtual work, determine the angle θ for a given force P.

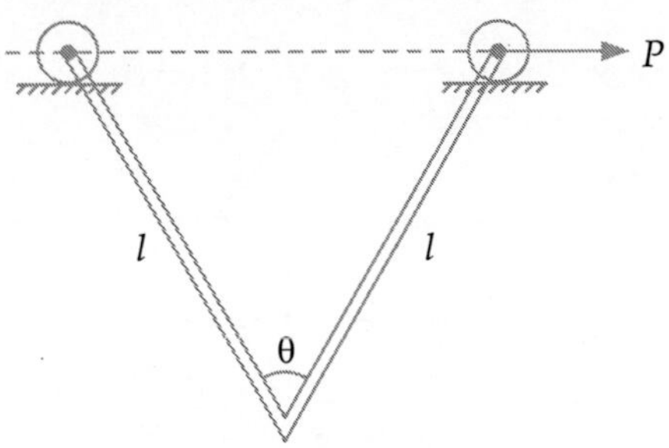

Fig. 12.11

Solution: Taking A as origin (Fig. 12.10), the position and displacement at points D and E are

$$y_d = -\frac{l}{2}\cos\frac{\theta}{2} \;:\; \delta y_d = \frac{l}{4}\sin\frac{\theta}{2}\delta\theta$$

$$y_e = -\frac{l}{2}\cos\frac{\theta}{2} \;:\; \delta y_e = \frac{l}{4}\sin\frac{\theta}{2}\delta\theta$$

D and E are the points of centre of gravity of the bars.

$$y_c = AC = 2l \sin\frac{\theta}{2}$$

$$\delta y_c = l \cos\frac{\theta}{2}\,\delta\theta$$

The virtual displacement δy_c of point C is in the direction of force and accordingly the virtual work done by force P will be positive. Further, when the point C moves towards right, the points D and E move will upwards *i.e.*, opposite to the direction of loads and as such their work will be negative. Then by the principle of virtual work.

$$-W\,\delta y_d - W\,\delta y_e + P\,\delta y_c = 0$$

or
$$-W \times \frac{l}{4}\sin\frac{\theta}{2}\delta\theta$$
$$- W \times \frac{l}{4}\sin\frac{\theta}{2}\delta\theta + p \times l\frac{\cos\theta}{2}\delta\theta$$

or
$$-W\frac{l}{2}\sin\frac{\theta}{2} = Pl\cos\frac{\theta}{2}$$

That gives: $\tan\frac{\theta}{2} = \frac{2l}{W}$

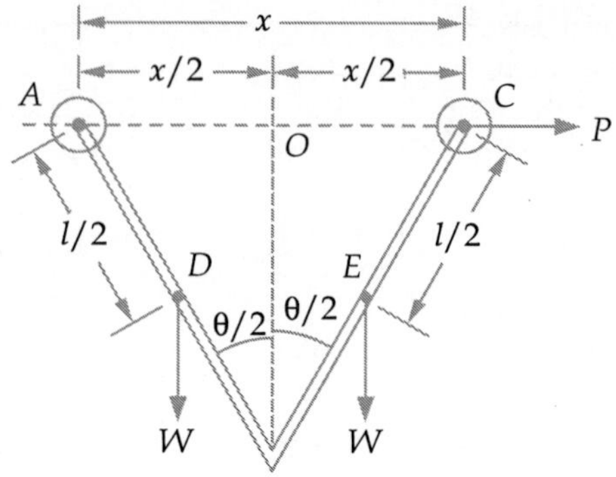

Fig. 12.12

EXAMPLE 12.9

The arrangement shown in Fig. 12.13. is required to remain in state of equilibrium. Set up an expression for tension in the cable in terms of θ and load *W*. Use the method of virtual work.

Solution: It is to be noted that

(*i*) Reaction R_b at the roller support is perpendicular to the plane of support.

(*ii*) Reaction R_a at the hinged support has a component R_{ah} in the horizontal direction and a component R_{av} in the vertical direction.

Since the hinged support has no translatory motion and there is no virtual displacement in the direction of reaction R_b, the reactions at the supports will have no contribution towards work. As such the work will be done only by the forces acting at points C, D and E.

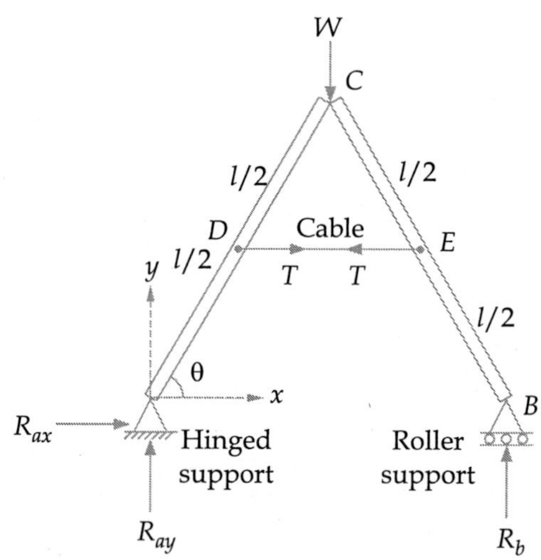

Fig. 12.13

For point C : $y = l\sin\theta$; $\delta y = l\cos\theta\,\delta\theta$

For point D : $x = \frac{l}{2}\cos\theta$; $\delta x = -\frac{l}{2}\sin\theta\,\delta\theta$

For point E : $x = 2l\cos\theta - \frac{l}{2}\cos\theta = \frac{3}{2}l\cos\theta$

$$\delta x = -\frac{3}{2}l\sin\theta\,\delta\theta$$

Total virtual work done

$$= -W\delta y + T \times \delta x \text{ (at } D) - T \times \delta x \text{ (at } E)$$

$$= -W \times l\cos\theta\,\delta\theta + T\left(-\frac{l}{2}\sin\theta\right)\delta\theta - T\left(-\frac{3l}{2}\sin\theta\right)\delta\theta$$

$$= -Wl\cos\theta\,\delta\theta + Tl\sin\theta\,\delta\theta$$

By the principle of virtual work

$$-Wl\cos\theta\,\delta\theta + Tl\sin\theta\,\delta\theta = 0$$

$$W\cos\theta = T\sin\theta\ ;\ T = W\cot\theta$$

EXAMPLE 12.10

A uniform ladder of 300 N weight rests against a smooth vertical wall and a rough horizontal floor making an angle of 60° with the horizontal. Use the method of virtual work to find the frictional force between the foot of ladder and the rough horizontal floor.

Solution: Refer Fig. 12.14 for the ladder AB leaning against a wall and various forces acting on it.

If l is the length of the ladder, then height of G above base line AC,

$$y = AG\sin\theta = \frac{l}{2}\sin\theta$$

Further, let $x = AC = l\cos\theta$

When the end A move leftwards through very small distance δx, the weight of ladder goes downwards through a small distance δy.

$$\delta x = -l\sin\theta\,\delta\theta$$

and

$$\delta y = \frac{l}{2}\cos\theta\,\delta\theta$$

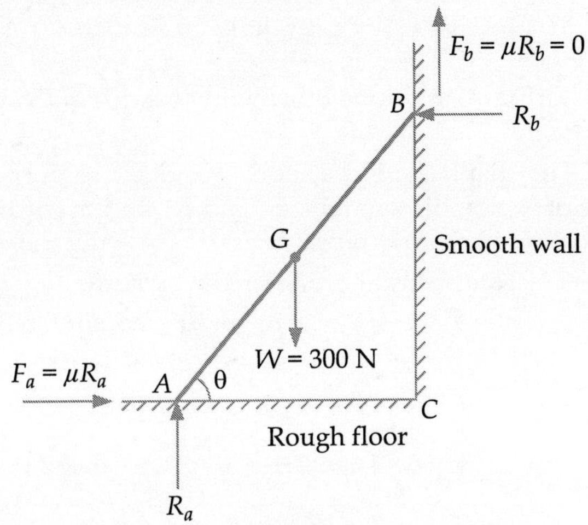

Fig. 12.14

With end A moving left wards, the friction force F_a would act towards right as shown.

Work done by frictional force at $A = F_a \times \delta x$

Work done by weight of ladder $= W \times \delta y$

The displacements at end A and B are perpendicular to the line of action of reactions R_a and R_b respectively and accordingly do not contribute towards work.

$$\text{Total work} = F_a\,\delta x + W\,\delta y = F_a \times (-l\sin\theta\,\delta\theta) + W \times \frac{l}{2}\cos\theta\,\delta\theta$$

Using the principle virtual work,

$$F_a \times (-l\sin\theta\,\delta\theta) + \frac{W}{2}\,l\cos\theta\,d\theta = 0$$

$$\therefore \quad F_a = \frac{W}{2}\cot\theta = \frac{300}{2}\cot 60° = \mathbf{86.6\ N}$$

EXAMPLE 12.11

A uniform ladder of weight 250 N and length 6 m rests on a smooth floor and against a smooth wall as shown in Fig. 12.15. The slipping of ladder is prevented by connecting a horizontal rope from its mid point to the wall. Use the principle of virtual work to calculate the tension induced in the rope.

Solution: Refer Fig. 12.15 for the ladder AB resting on smooth floor and leaning against a smooth vertical wall.

Length of the ladder

$$AB = \sqrt{(AC)^2 + (BC)^2} = \sqrt{3^2 + 4^2} = 5\text{ m}$$

The horizontal rope DE joins the mid points of ladder (D) and wall (E). The weight of the ladder acts at point D which is its centre of gravity.

Let the co-ordinates of point D be (x, y) and the ladder be given a small displacement such that its lower end A moves towards left.

$$x = AD \cos\theta = 2.5 \cos\theta\,; \quad \delta x = -2.5 \sin\theta\,\delta\theta$$
$$y = AD \sin\theta = 2.5 \sin\theta\,; \quad \delta y = 2.5 \cos\theta\,\delta\theta$$

Virtual work done by tension T in the rope $= T(-2.5 \sin\theta\,\delta\theta)$

Virtual work done by weight of ladder (250 N) $= 250\,(2.5 \cos\theta\,\delta\theta)$

The displacement at end A and B are perpendicular to line of action of reactions R_a and R_b and accordingly do not contribute towards work.

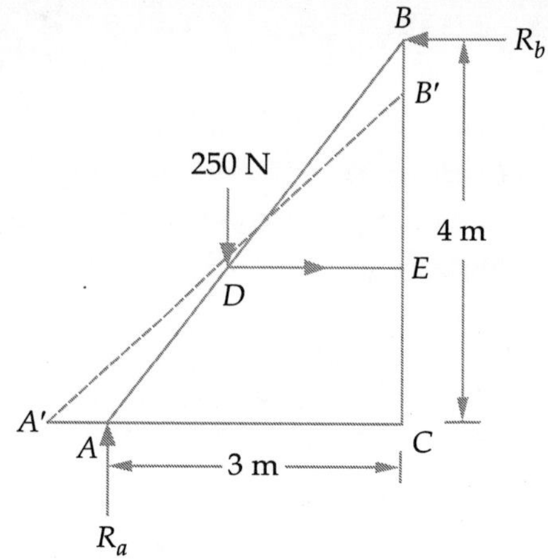

Fig. 12.15

$\therefore$ Total virtual work of the system

$$= T(-2.5 \sin\theta\,\delta\theta) + 250(2.5 \cos\theta\,\delta\theta)$$

Then from the principle of virtual work

$$T(-2.5 \sin\theta\,\delta\theta) + 250\,(2.5 \cos\theta\,\delta\theta) = 0$$

$$T = 250 \cot\theta = 250 \times \frac{3}{4} = \mathbf{187.5\ N}$$

EXAMPLE 12.12

A block of weight 2000 N rests on a smooth inclined plane that makes an angle of 30° with the horizontal. This block is supported by load *P* lying on another smooth plane of inclination 60° as shown in Fig. 12.16. The block and the load have been connected by an inelastic string. Determine the value of load *P* by using the method of virtual work.

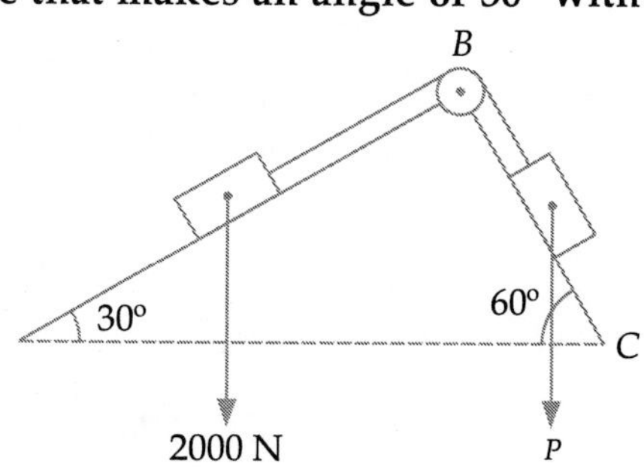

Fig. 12.16

Solution: It is to be noted that:

(*i*) Small displacement δx of 2000 N parallel to plane AB equals the displacement δy of load P along the plane BC.

(*ii*) When the load 2000 N moves down the plane, the load P moves up the plane.

(*iii*) The resolved parts of 2000 N parallel and perpendicular to plane AB are:

$$2000 \sin 30^\circ \quad \text{and} \quad 2000 \cos 30^\circ$$

The resolved parts of P parallel and perpendicular to plane BC are:

$$P \sin 60^\circ \quad \text{and} \quad P \sin 60^\circ$$

Only the resolved parts along the planes will contribute towards work.

Total virtual work $= (2000 \sin 30^\circ) \times \delta x - (P \sin 60^\circ) \times \delta y$

The – *ve* sign in the above identity stems from the fact that movement of the two loads are in opposite directions.

Then from the principle of virtual work,

$$(2000 \sin 30^\circ) \times \delta x - (P \sin 60^\circ) \times \delta y = 0$$

$$\therefore \quad P = 2000 \times \frac{\sin 30^\circ}{\sin 60^\circ} = 2000 \times \frac{0.5}{0.866} = \mathbf{1154.73\ N}$$

EXAMPLE 12.13

Two spheres P and Q of weights 5 N and 10 N respectively can slide freely along the bars AC and BC which are arranged as shown in Fig. 12.17. An extensible string DE connects these spheres. Calculate the value of θ which defines the position of equilibrium. Use the principle of virtual work.

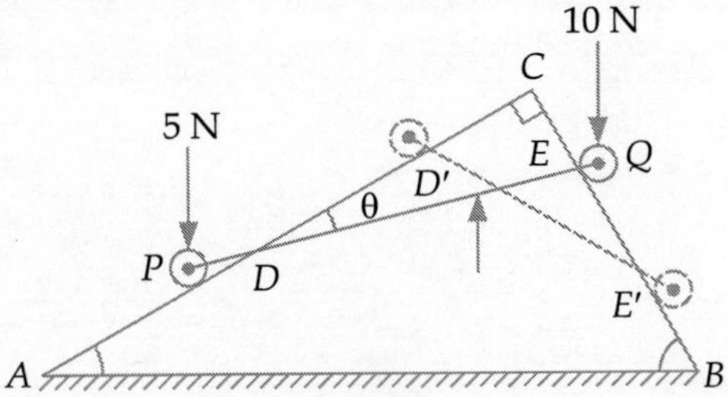

Fig. 12.17

Solution: If l is the length of string ; then

$$DE = l \cos\theta\,; \qquad CE = l \sin\theta$$

Accordingly when a virtual displacement $d\theta$ is given to the string DE, we have

$$\delta_{DE} = -l \sin\theta\,\delta\theta \quad \text{and} \quad \delta_{CE} = l \cos\theta\,\delta\theta$$

The resolved parts of the weights along the planes are

$$5 \sin 30^\circ \quad \text{and} \quad 10 \sin 60^\circ$$

Only these resolved parts contribute towards work.

Work done at point $D = (5 \sin 30^\circ) \times \delta_{DE}$

$= 5 \sin 30^\circ \times (-l \sin\theta\,\delta\theta) = -2.5l \sin\theta\,\delta\theta$

Work done at point $E = 10 \sin 60^\circ \times l \cos\theta\,\delta\theta = 8.66l \cos\theta\,\delta\theta$

Total virtual work $= -2.5l \sin\theta\,\delta\theta + 8.66l \cos\theta\,\delta\theta$

Then from the principle of virtual work,

$$-2.5l \sin\theta\,\delta\theta + 8.66l \cos\theta\,\delta\theta = 0$$

or

$$\tan\theta = \frac{8.66}{2.5} = 3.464; \quad \theta = \mathbf{73.9^\circ}$$

EXAMPLE 12.14

A beam AB of length 1.5 m and negligible weight is hinged at end A and carries a 200 N weight at its free end. The beam is supported at its mid point C by a wire CD as shown in Fig. 12.18. Use the method of virtual work to determine the tension induced in the wire.

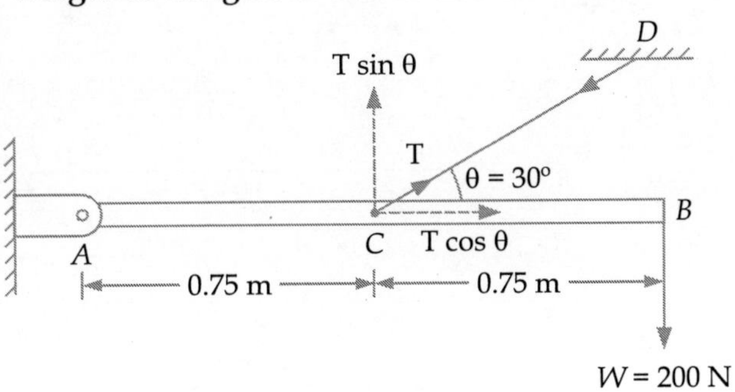

Fig. 12.18

Solution: Let the beam be given a vertical virtual displacement y_b at end B holding the hinged end A in position. The corresponding displacement at C would be

$$y_c = y_b \times \frac{AC}{AB} = y_b \times \frac{0.75}{1.5} = \frac{y_b}{2}$$

If T is the tension in the string, then its vertical and horizontal components are

$$T_v = T \sin\theta \quad \text{and} \quad T_h = T \cos\theta$$

Work done by load W at end $B = -W \times y_b = -200\, y_b$

(– ve sign due to load acting downwards)

Work done by component T_v of tension in the string

$$= T \sin\theta \times y_c = T \sin\theta \times \frac{y_b}{2} = \frac{T y_b}{2} \sin\theta$$

The horizontal component $T \cos\theta$ will have no contribution towards work done because its line of action is normal to the direction of displacement.

Algebraic sum of work done $= \frac{T y_b}{2} \sin\theta - 200\, y_b$

According to the principle of virtual work, the total work done by all the forces must be zero. That gives

$$\frac{T y_b}{2} \sin\theta - 200\, y_b = 0$$

$$T = \frac{200 \times 2}{\sin 30°} = \frac{200 \times 2}{0.5} = \mathbf{800\ N}$$

EXAMPLE 12.15

In a framed structure shown in Fig. 12.19, the length of all the members is equal. Determine the force in the top member *DE* by the method of virtual work.

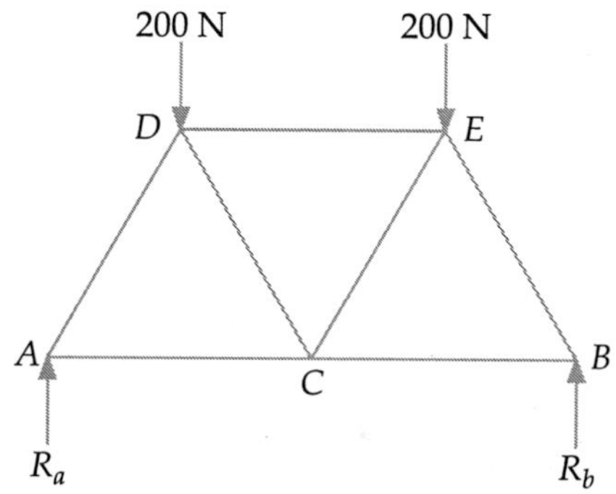

Fig. 12.19

Solution: By symmetry of truss and loading

$R_a = R_b = 200$ N

The member *DE* is cut and replaced by an axial compressive force *P* in it. The system is then made movable by imagining the portion *BCE* to rotate through an angle $\delta\theta$. Then *B* gets displaced to *B*' and *E* to *E*'.

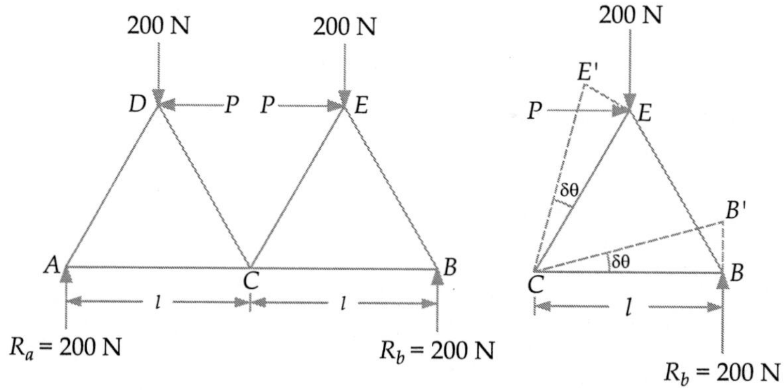

Fig. 12.20

Vertical displacement of $B = BB' = l\,\delta\theta$

Work done by reaction at $B = R_b\, l\,\delta\theta = 200\, l\,\delta\theta$

The horizontal and vertical displacements at *E* are:

$EE' \sin 60° = l\,\delta\theta \times 0.866$ (towards left)

and $EE' \cos 60° = l\,\delta\theta \times 0.5$ (upwards)

Work done by axial force $P = -P\, l\,\delta\theta \times 0.866$

Work done by load of 200 N at $E = -200\, l\,\delta\theta \times 0.5$

Total work done = $200\,l\,\delta\theta - 0.866\,P\,l\,\delta\theta - 100\,l\,\delta\theta$

According to the principle of virtual work,

$$200\,l\,\delta\theta - 0.866\,P\,l\,\delta\theta - 100\,l\,\delta\theta = 0$$

or $0.866\,P = 100$;

$$P = \frac{100}{0.866} = \mathbf{115.47\ N} \quad \text{(Compression)}$$

EXAMPLE 12.16

Using the method of virtual work, determine the force in the top member of the simple truss consisting of equilateral triangles as shown in Fig. 12.21.

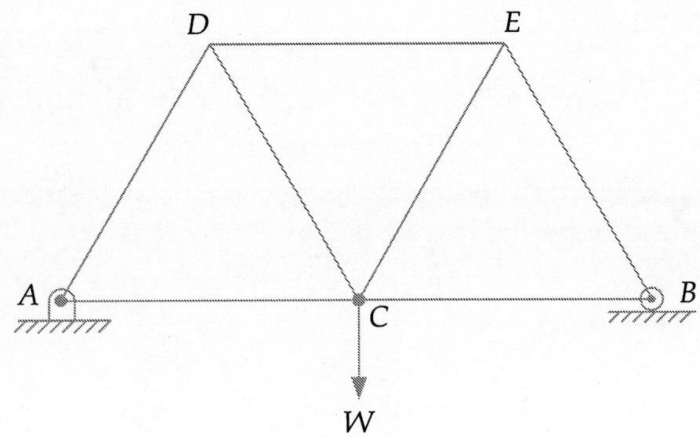

Fig. 12.21

Solution: By symmetry of truss and loading

$$R_a = R_b = \frac{W}{2}$$

The member DE is cut and replaced by an axial compressive force P in it. The system is then made movable by giving a virtual angular displacement of $\delta\theta$ about C. Then B gets displaced to B' and E to E'.

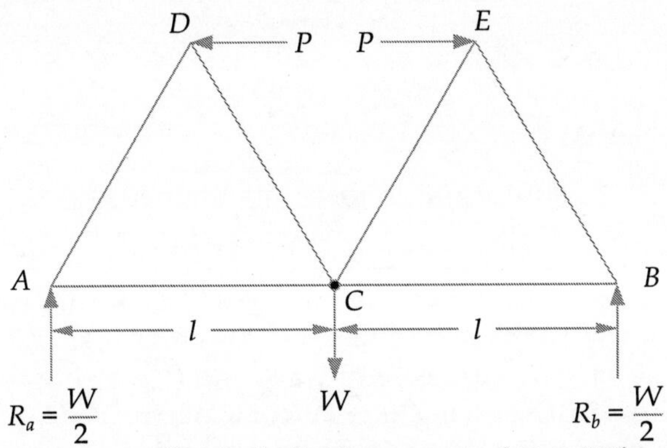

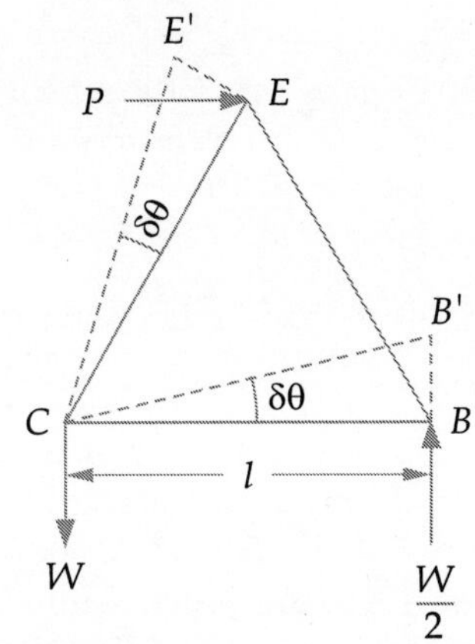

Vertical displacement of $B = BB' = l\,\delta\theta$ (upwards)

Work done by reaction at $B = \dfrac{W}{2} \times l\,\delta\theta$

Horizontal displacement of point $E = EE' \times \sin 60°$

$= l\,\delta\theta \times 0.866$ (towards left)

Work done by axial force $P = -P \times l\,\delta\theta \times 0.866$

The negative sign stems from the fact that force P and displacement of point E are opposite to each other.

According to the principle of virtual work, the work done by all the forces must by zero. That gives

$$\frac{W}{2} \times l\,\delta\theta - P\,l\,\delta\theta\,0.866 = 0$$

or $$P = \frac{W}{2 \times 0.866} = \mathbf{0.577\ W}$$

EXAMPLE 12.17

A number of bars are hinged together to form three identical rhombuses as shown in Fig. 12.22. If a horizontal force of 250 N is applied at *B*, determine the force *P* which must be applied at *A* to preserve equilibrium.

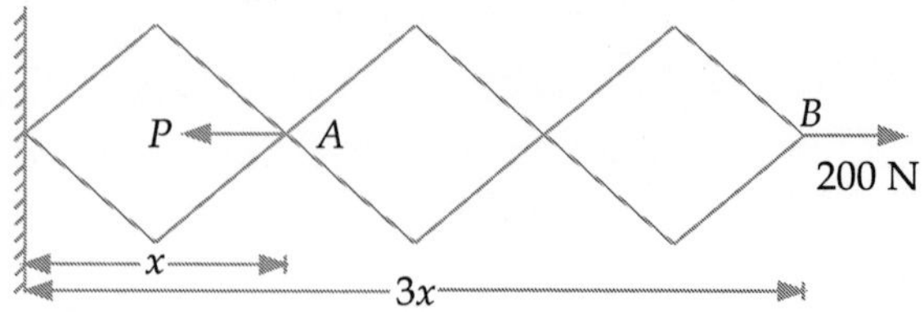

Fig. 12.22

Solution: Let each side of the rhombus be inclined at an angle θ to the horizontal. The horizontal distance of point *A* from the hinged support

$$x = 2\,l\cos\theta$$

Horizontal distance of point *B* from the hinged support

$$3x = 6\,l\cos\theta$$

Accordingly when an imaginary displacement δθ is given to the bars, then corresponding displacements at *A* and *B* are

$$\delta x = -2\,l\sin\theta\,\delta\theta \quad \text{and} \quad 3\,\delta x = -6\,l\sin\theta\,\delta\theta$$

That gives:

$$\text{Work done at } A = -P \times (-2\,l\sin\theta\,\delta\theta)$$

$$\text{Work done at } B = 250 \times (-6\,l\sin\theta\,\delta\theta)$$

Algebraic sum of work done

$$= 2\,Pl\sin\theta\,\delta\theta - 1500\,l\sin\theta\,\delta\theta$$

According to principle of virtual work,

$$2\,Pl\sin\theta\,\delta\theta - 1500\,l\sin\theta\,\delta\theta = 0$$

$$\therefore \quad P = \frac{1500}{2} = \mathbf{750\ N}$$

Thus a force of 750 N must be applied at *A* if equilibrium is to be maintained.

EXAMPLE 12.18

A hexagonal frame *ABCDE* is made up of six bars of equal length and weight. One of the bars is fixed on a horizontal plane and the system lies in the vertical plane as shown in Fig. 12.23. The mid points of the two upper non-horizontal bars are connected by a string *PQ*. Using the principle of virtual work, show that the tension in the string is 6 *W* cot θ where *W* is the weight of each bar and θ is the inclination of the inclined bars with the horizontal.

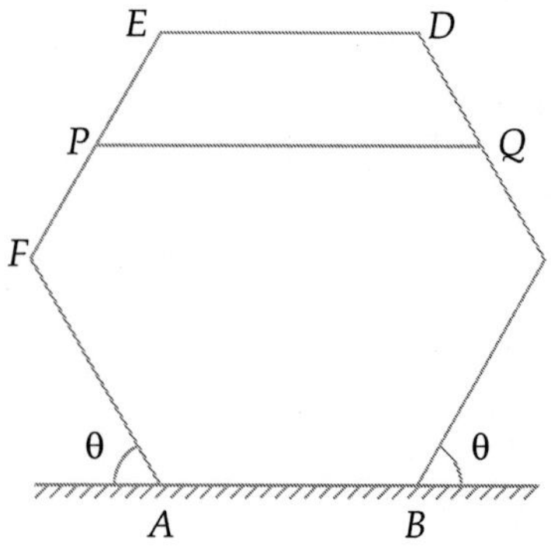

Fig. 12.23

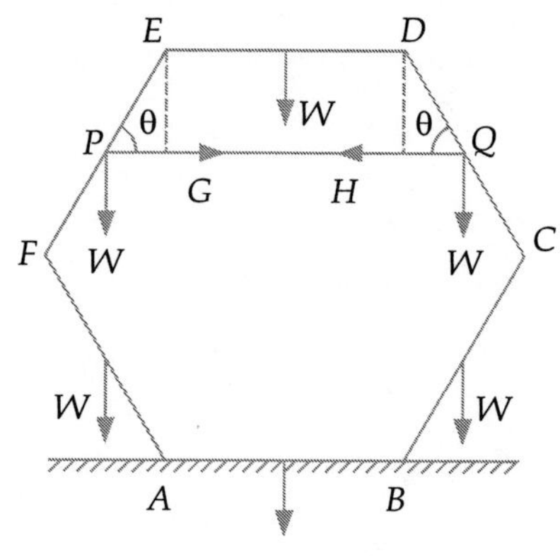

Fig. 12.24

Solution: Let $2a$ be the length of each side of the hexagonal frame.

Length of string $PQ = PG + GH + HQ = a\cos\theta + 2a + a\cos\theta = 2a + 2a\cos\theta$

$\therefore \quad \delta_{PQ} = -2a\sin\theta\,\delta\theta$

The weight of each bar acts vertically downwards at its mid point. Let x, y, z and h be the heights of the mid points of AB, BC and AF, CD and FE, and ED as measured above AB.

Then $x = 0$; $\quad \delta x = 0$

$y = a\sin\theta$; $\quad \delta y = a\cos\theta\,\delta\theta$

$z = 3a\sin\theta$; $\quad \delta z = 3a\cos\theta\,\delta\theta$

$h = 4a\sin\theta$; $\quad \delta h = 4a\cos\theta\,\delta\theta$

By the principle of virtual work,

$$W \times \delta x + 2W \times \delta y + 2W\,\delta z + W\,\delta h + T \times \delta_{PQ} = 0$$

or $\quad W \times 0 + 2W \times a\cos\theta\,\delta\theta + 2W \times 3a\cos\theta\,\delta\theta + W \times 4a\cos\theta\,\delta\theta + T \times (-2a\sin\theta\,\delta\theta) = 0$

or $\quad 12Wa\cos\theta\,\delta\theta = 2Ta\sin\theta\,\delta\theta$

$\therefore$ Tension in the string PQ

$$T = \frac{12Wa\cos\theta\,\delta\theta}{2a\sin\theta\,\delta\theta} = 6W\frac{\cos\theta}{\sin\theta} = 6W\cot\theta$$

EXAMPLE 12.19

Four uniform rods, each of length *a* and weight *W* are hinged together at the extremities and form a rhombus. A string of length *b* joins the shorter diagonal of the rhombus so formed. If one of the rods is supported in the horizontal position, show that the tension *T* of the string is

$$T = \frac{2W(2a^2 - b^2)}{a\sqrt{4a^2 - b^2}}$$

Solution: Refer Fig. 12.25 which shows the rhombus formed by four uniform rods. The side AB is supported in horizontal position and the shorter diagonal BD is joined by a string. The centre of gravity of the four rods will be at point E which is the point of intersection of the diagonals AC and BD of the rhombus.

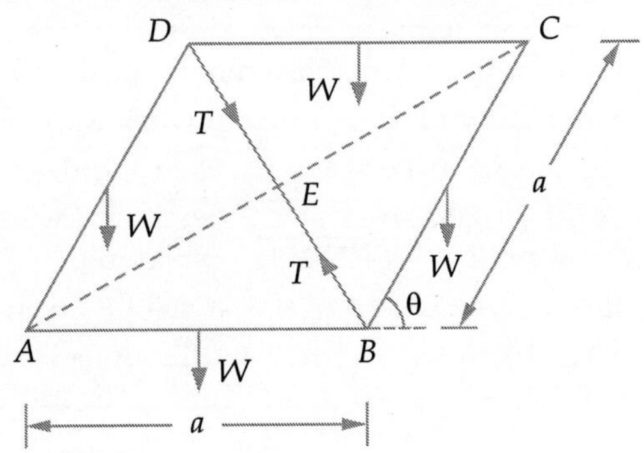

Fig. 12.25

Let angle $BCD = \theta$. Then position of points C and E with respect to rod AB are;

$$y_c = BC\sin\theta = a\sin\theta$$

$$y_e = \frac{1}{2}a\sin\theta$$

$$\delta y_e = \frac{1}{2}\alpha\cos\theta\,\delta\theta$$

Denoting length BD as x, we have

$$x = BD = 2\,BE = 2\,BC\sin\frac{\theta}{2} = 2b\sin\frac{\theta}{2}$$

$$\delta x = b\cos\frac{\theta}{2}\,\delta\theta$$

Algebraic sum of the virtual work done by tension T in the rod BD, and by the weights of the four rods is

$$= -T\,\delta x + 4W\,\delta y_e$$

By the principle of virtual work,

$$-T\,\delta x + 4W\,\delta y_e = 0$$

or

$$T = 4W\frac{\delta y_e}{\delta x} = 4W \times \frac{\frac{1}{2}a\cos\theta\,\delta\theta}{b\cos\frac{\theta}{2}\,\delta\theta} = \frac{2W\cos\theta}{\cos\frac{\theta}{2}}$$

From the geometry of the rhombus,

$$\sin\frac{\theta}{2} = \frac{BE}{BC} = \frac{b/2}{a} = \frac{b}{2a}$$

From trigonometric relations,

$$\cos\frac{\theta}{2} = \sqrt{1-\sin^2\frac{\theta}{2}} = \sqrt{1-\frac{b^2}{4a^2}} = \sqrt{\frac{4a^2-b^2}{2a}}$$

$$\cos\theta = 1-2\sin^2\frac{\theta}{2} = 1-2\frac{b^2}{4a^2} = \frac{2a^2-b^2}{2a^2}$$

$$\therefore \quad \text{Tension } T \text{ in the string} = 2W \times \frac{(2a^2-b^2)/2a^2}{\sqrt{4a^2-b^2}/2a} = \frac{2W(2a^2-b^2)}{a\sqrt{4a^2-b^2}}$$

which is the required expression.

REVIEW AND SUMMARY

1. A machine or mechanism has no tendency to move when it is in a state of equilibrium. However, a small displacement may be just imagined to have been caused. Since this displacement does not actually occur, it is referred to as virtual displacement.
2. Virtual work is the work done by a force or couple during a virtual displacement which is consistent with the constraints imposed on the system. While computing the virtual work, the various virtual displacements are expressed in terms of a single virtual displacement.
3. The following forces do no work and are omitted in forming the equation for virtual work :
   (*i*) reaction between the body and any smooth surface with which it is in contact
   (*ii*) internal forces at connections (mutual action and reaction between two bodies whose equilibrium is being considered)
   (*iii*) forces exerted by inextensible cords and cables.
4. The principle of virtual work states that : The total virtual work done by all the forces and couples acting on a rigid body for any virtual displacement is zero.

REVIEW QUESTIONS

A. Conceptual and conventional questions :

1. State the principle of virtual work. Explain it giving a suitable example.
2. Mention the forces which are generally omitted while applying the principle of virtual work.
3. A load *W* is being lifted by a pulley system as shown in Fig. 12.26, by applying a pull at the end of a cord supporting the pulleys having the same diameter. Using the principle of virtual, set up the relationship between *P* and *W*.

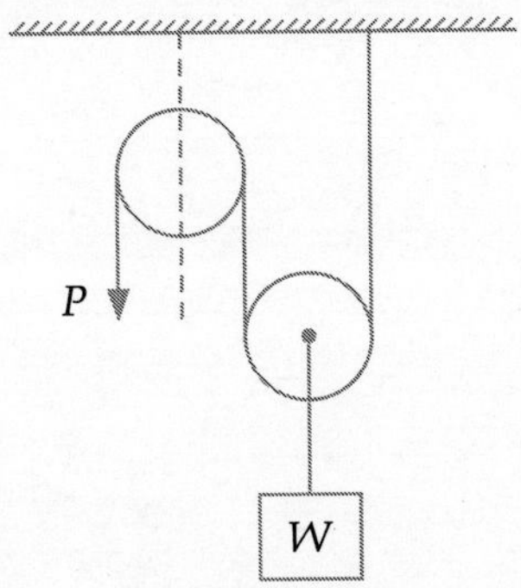

Fig. 12.26

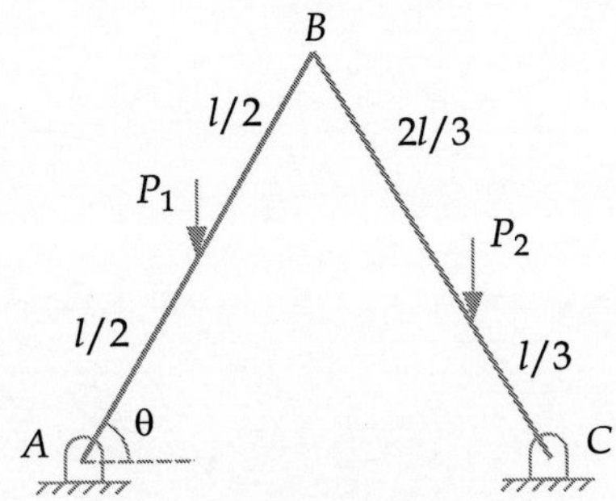

Fig. 12.27

4. Using the method of virtual work, setup the following expression for horizontal component of reaction at C in respect of the frame supported and loaded as shown in Fig. 12.27.

$$R_{ch} = \frac{P_1 P_2}{24}\cot\theta$$

How this expression would be affected if force P_2 acts horizontally to the right instead of vertically?

$$\left(\frac{P_1}{4}\cot\theta + \frac{5}{6}P_2\right)$$

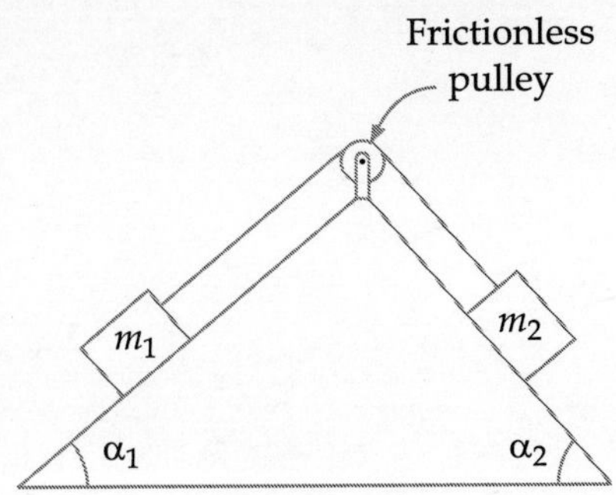

Fig. 12.28

5. A friction less double-incline with angle α_1 and α_2 as shown in Fig. 12.28, carries a set of sliding masses m_1 and m_2 connected with an inextensible string and passing over a frictionless pulley. By using the principle of virtual work, obtain the relationship between α_1 and α_2 in terms of m_1 and m_2 and hence determine α_2 if $\alpha_1 = 30°$ and $m_1 = 2m_2$

6. A regular hexagonal frame *ABCDEF* is made up of six uniform identical bars each weighing *W*, joined together (Fig. 12.29). The frame rests in a vertical plane and the bar *AB* is fixed in horizontal position. The mid points of *AB* and *DE* are connected by a light string. Using the method of virtual work, find tension in the string.

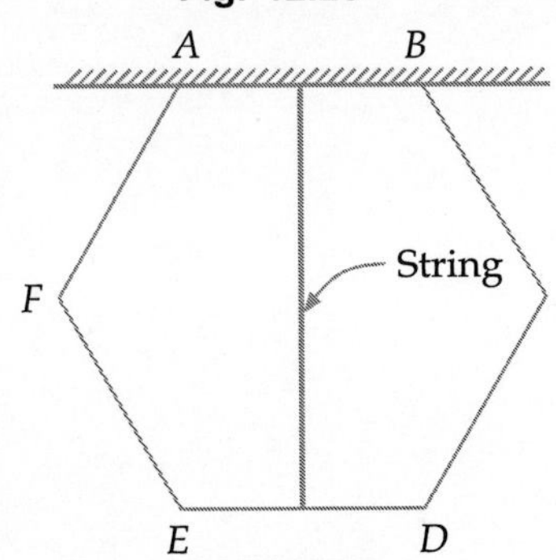

Fig. 12.29

Hint: Let $2a$ be the length of each side of the hexagonal frame. The weight of each bar acts vertically downwards at its mid point. Let x, y, z and h be the heights of the mid points of *AB*, *BC* and *AF*, *CD* and *FE*, and *ED* as measured below *AB*. Then

$$x = 0; \quad \delta x = 0$$
$$y = a\sin\theta; \quad \delta y = a\cos\theta\,\delta\theta$$
$$z = 3a\sin\theta; \quad \delta z = 3a\cos\theta\,\delta\theta$$
$$h = 4a\sin\theta; \quad \delta h = 4a\cos\theta\,\delta\theta$$

Length of the string $l = 4a\sin\theta$; $\delta l = 4a\cos\theta\,\delta\theta$

By the principle of virtual work,

$$W \times 0 + 2W \times a\cos\theta\,\delta\theta + 2W \times 3a\cos\theta\,d\theta$$
$$+ W \times 4a\cos\theta\,\delta\theta - T \times 4a\cos\theta\,\delta\theta = 0$$

$\therefore$ Tension in the string, $T = \dfrac{12Wa\cos\theta\,\delta\theta}{4a\cos\theta\,\delta\theta} = \mathbf{3\,W}$

❏❏❏

EXPERIMENTS

LIST OF PRACTICALS

1. To determine the reaction at supports in case of a simply supported beam.
2. To verify the polygon law of co-planar (con-current) force system.
3. To verify the forces in the different members of a jib crane.
4. To determine the moment of inertia of a flywheel.
5. To determine the coefficient of friction between different surfaces on an inclined plane.
6. To determine the coefficient of friction between the pulley and a belt (flat belt, V-belt).
7. To determine the velocity ratio, mechanical advantage and efficiency of a worm and worm wheel for lifting loads.
8. To determine the velocity ratio, mechanical advantage and efficiency of a simple screw jack.
9. To compare the experimentally determined values of forces in the struts and tie of a basic roof truss with the analytically determined values.
10. Study of gear trains.

EXPERIMENT-1

OBJECT

To determine the reactions at the supports of a simply supported beam.

APPARATUS

Simply supported beam, weights, metre scale

THEORY

A beam is a structural member, whose longitudinal dimension is large compared to the transverse dimension. The beam is supported along its length and is acted upon by a system of loads at right angles to its axis. A simply supported beam freely rests on its end supports which may be knife edges or rollers. The end supports exert only forces but no moments on the beams. The horizontal distance between the supports is called span.

Reactions in a simply supported beam are determined by using the principle of moments which states that :

A body acted upon by a number of coplanar forces will be in equilibrium, if the algebraic sum of moments of all the forces about a point lying in the same plane is zero. Mathematically

$$\Sigma M = 0$$

i.e., clockwise moments = anticlockwise moments

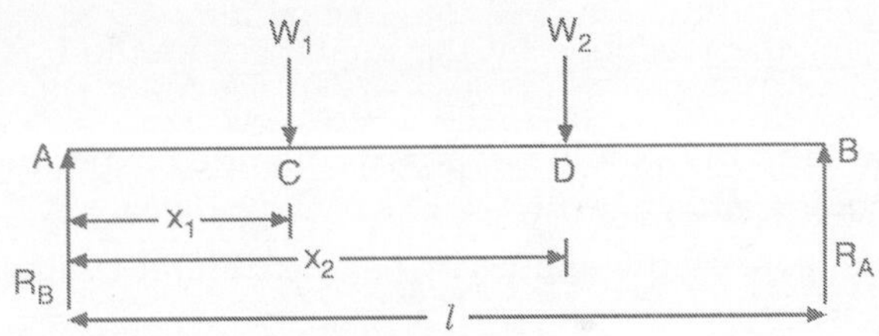

Fig. 1

With reference to the arrangement shown above, *AB* is a simply supported beam supported at *A* and *B*. The loads W_1 and W_2 are applied at points *C* and *D* which are located at distances x_1 and x_2 respectively from end *A*.

Applying conditions of equilibrium,

(*a*) $\Sigma F_y = 0$; The algebraic sum of the vertical components of different forces is zero. That is

$$R_A + R_B - W_1 - W_2 = 0 \ ; \qquad R_A + R_B = W_1 + W_2 \qquad ...(i)$$

(*b*) $\Sigma M = 0$; The algebraic sum of the moments of all forces about a point lying in the same plane is zero. Then taking moments about the support end *A* and equating it to zero, one gets

$$W_1 x_1 + W_2 x_2 - R_B l = 0 \qquad ...(ii)$$

$$\therefore \qquad R_B = \frac{W_1 x_2 - W_2 x_2}{l}$$

and $$R_A = W_1 + W_2 - R_B$$

PROCEDURE

Refer Fig. 2 given below which shows the apparatus pertaining to a simply supported beam.

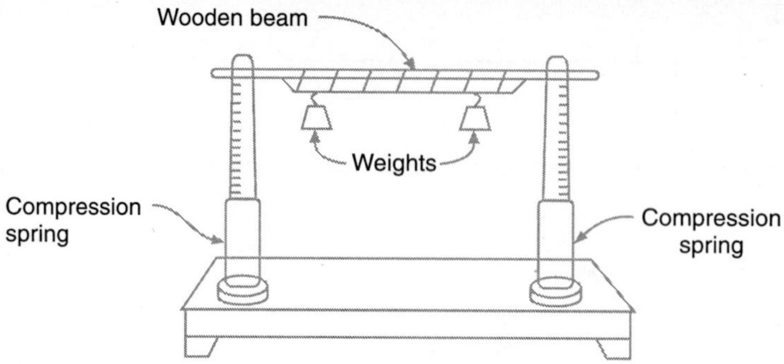

Fig. 2

(i) Measure distance l which denotes the span of the given beam.

(ii) Note the initial readings S_A and S_B of the spring balances provided at the end supports A and B.

(iii) Place the loads W_1 and W_2 on the slider hook and measure the distances of their locations x_1 and x_2 from the end support A.

(iv) Note the final readings S_A' and S_B' of the spring balances.

Then the reactions at the end supports are then given by the difference in the final and initial readings of the spring balances. That is

$$R_A = S_A' - S_A \quad \text{and} \quad R_B = S_B' - S_B$$

(v) Make calculations for the reactions by the analytical method as outlined in the theory stated above.

(vi) The reactions at the supports should be same as obtained experimentally and analytically. If not, calculate the percentage error.

(vii) Repeat the experiment atleast three times by having different values of W_1 and W_2 installed at different locations.

OBSERVATIONS

$l =$; $S_A =$; $S_B =$

S.No.	*Weights on beams and their distances*				*Final spring balance readings*		*Reactions observed*		*Reactions calculated*		*%age error*
	W_1	x_1	W_2	x_2	S_A'	S_B'	$R_A = S_A' - S_A$	$R_B = S_B' - S_B$	R_A	R_B	
1.											
2.											
3.											

PRECAUTIONS

(i) Make accurate measurements of the distances l, x_2 and x_2

(ii) Note accurately the readings of the spring balances.

EXPERIMENT-2

OBJECT

To state and prove the polygon law of forces.

APPARATUS

Wooden board, white sheat of paper and drawing pins, pulleys with clamps, threads/strings, pans with weights and mirror strips

THEORY

The polygon law of forces states that :

"*If a number of concurrent forces acting simultaneously on a body are represented in magnitude and direction by the sides of a polygon, taken in order, then the resultant is represented in magnitude and direction by the closing side of the polygon, taken in opposite order.*"

Consider forces P_1, P_2, P_3 and P_4 acting on the body at point O. Line AB is drawn to represent force P_1, line BC to represent P_2, line CD to represent P_3 and line DE to represent P_4. The polygon is completed by drawing the closing line AE. This closing line AE represents the resultant of the given system in magnitude, line of action and direction.

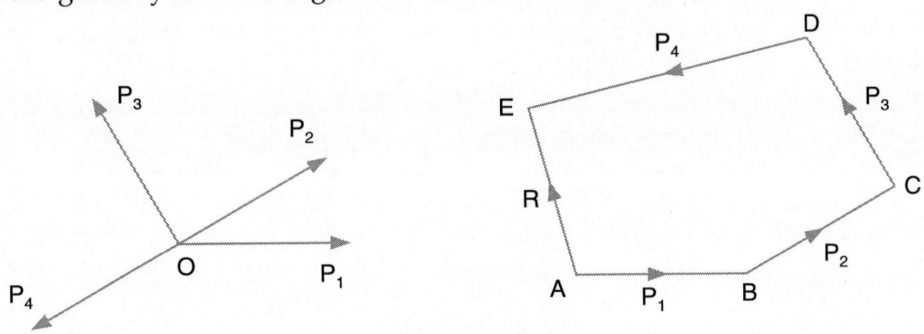

Fig. 1

Quite often the polygon law of forces is stated as follows :

"If any number of forces acting at a point can be represented in magnitude and direction by the sides of a polygon taken in order, then the system of forces is in equilibrium.

PROCEDURE

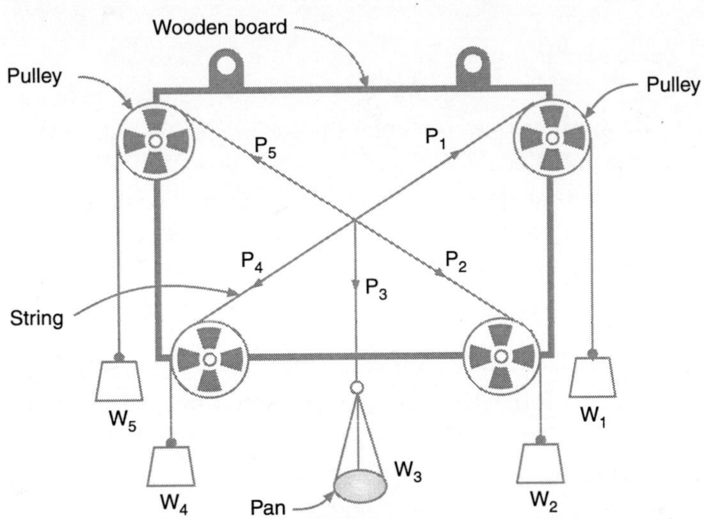

Fig. 2

With reference to the force polygon apparatus shown above

(*i*) Attach the white sheet of paper on the wooden force table with the help of drawing pins.

(*ii*) Fix the pulleys to the board using clamps.

(*iii*) Take a string and pass over two pulleys.

(*iv*) Take another string and tie the middle of this string to the middle of the first string.

(*v*) Pass the second string over the other two pulleys.

(*vi*) Attach the known weights to the free ends of the strings.

(*vii*) Take a third string and fasten it to the knot made between the first and second strings. Attach a weight pan to the free end of this string.

(*viii*) Put some known weights in the pan in such a way that the knot comes almost at the centre of paper.

(*ix*) Allow the system to come to rest.

(*x*) Mark the direction of strings by drawing straight lines on the paper by placing the mirror strips under the strings.

(*xi*) Note and write down the magnitude of weights near the corresponding string line on the paper.

(*xii*) Remove the paper from the board and produce the lines of force at point *O*.

This gives the position (space) diagram showing the various forces (or loads) along with their magnitude and line of action.

(*xiii*) Name the forces in the space diagram by using Bow's notation by which each force is named by two capital letters placed on its either side.

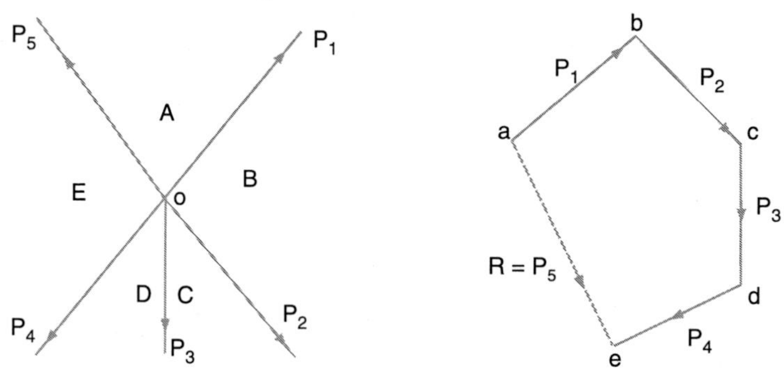

Space diagram

Force polygon

Fig. 3

(*xiv*) Construct the force diagram from a convenient point. Go on adding all the forces vectorially one by one to some suitable scale.

In the context of the force system indicated, draw vector *ab* equal to force P_1 to some suitable scale and parallel to the line of action of P_1. From point *b*, draw vector *bc* to represent P_2 in magnitude and direction. From point *c*, draw vector *cd* to represent force P_3 in magnitude and direction. Then from point draw vector *de* equal and parallal to force P_4.

(*xv*) Join *ae* and measure its length. Check whether it equals P_5 or not. As per polygon law of forces, *ae* (converted into force *R*) should represent force P_5 both in magnitude and direction.

If *R* is not equivalent to P_5, then calculate the error.

(*xvi*) Repeat the steps atleast three times by changing the loads.

OBSERVATIONS

S.No.	*Weight suspended or forces*					*Calculated resultant* (R)	% Error in 'R' $\frac{R-P_5}{R} \times 100$
	$P_1 = W_1$	$P_2 = W_2$	$P_3 = W_3$	$P_4 = W_4$	$P_5 = W_5$		
1							
2							
3							

PRECAUTIONS

(*i*) Make the pulleys frictionless by proper lubrication.
(*ii*) Ensure that the weights and the pan donot touch any surface.
(*iii*) The strings should not overlap the pulleys.
(*iv*) Direction of strings should be marked carefully with the help of mirror strips.
(*v*) The addition of weights in the pan should be gradual and in gentle way.

EXPERIMENT-3

OBJECT

To verify the forces in the different members of a jib crane.

APPARATUS

Jib crane apparatus, spring balance, weights, metre scale, etc.

THEORY

The jib crane is a hoisting mechanism used for lifting heavy loads. The unit essentially consists a tie member and a jib which are attached to a vertical post. The string carrying the load to be lifted is fastened to the point where are connected the tie and jib members. During the act of lifting load, the tie member is under tension and the jib member is under compression.

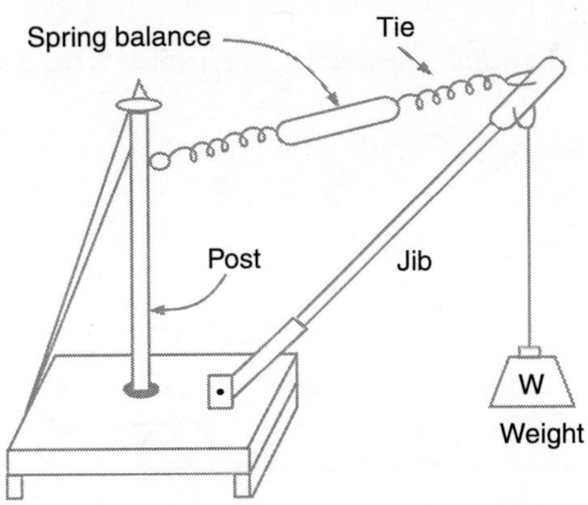

Fig. 1

The experiment is based on the triangle law of forces which may be stated as:

"If two forces acting simultaneously on a body are represented by the sides of a triangle taken in order, their resultant is represented by the closing side of the triangle taken in the opposite order".

Consider two forces P and Q acting on the body as shown :

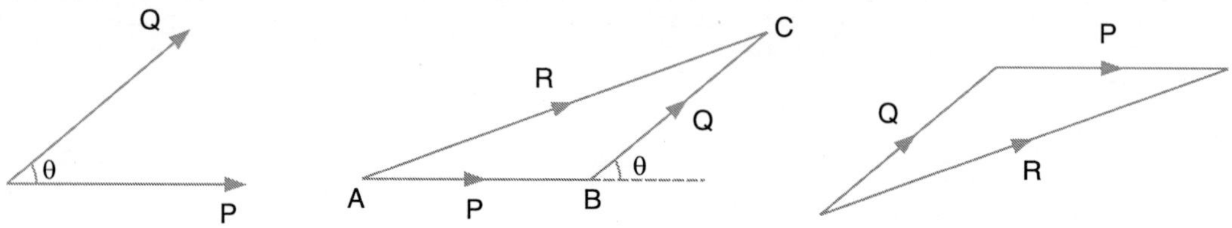

Fig. 2

Line AB be drawn to represent force P and BC to represent Q. The triangle ABC is completed by drawing the closing line AC. Line AC represents the resultant in magnitude, line of action and direction.

It is to be noted that addition of forces P and Q in any order gives the same resultant R.

The triangle law of forces may also be stated as :

"If a system of three forces acting upon a body can be represented in magnitude and direction by the sides of a triangle, taken in order, then the system will be in equilibrium."

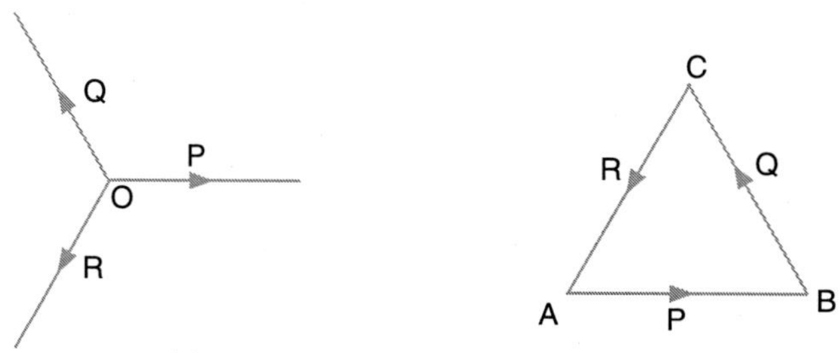

(a) (b)

Fig. 3

With reference to the figure shown above, the three forces P, Q and R acting at point O have been represented by the sides AB, BC and CA of the triangle ABC.

PROCEDURE

(*i*) Note the initial readings S_t and S_j of the spring balances provided in the tie member and jib member.

(*ii*) Attach a load W to the string, fastened to the point where are joined the tie and jib members.

(*iii*) Note the final readings S_t' and S_j' of the spring balances.

The forces F_t and F_j are then given by the difference in the final and initial readings of the spring balances. That is

$$F_t = S_t' - S_t \quad \text{and} \quad F_j = S_j' - S_j$$

(*iv*) Measure the lengths of jib, tie and post after the weight W has been attached.

(*v*) Draw the space diagram to some convenient scale by using the measured lengths of jib, tie and post. The space diagram represents the forces in jib, tie, post and the weight W.

(*vi*) Name the forces in the space diagram by using Bow's notation by which each force is named by two capital letters placed on its either side.

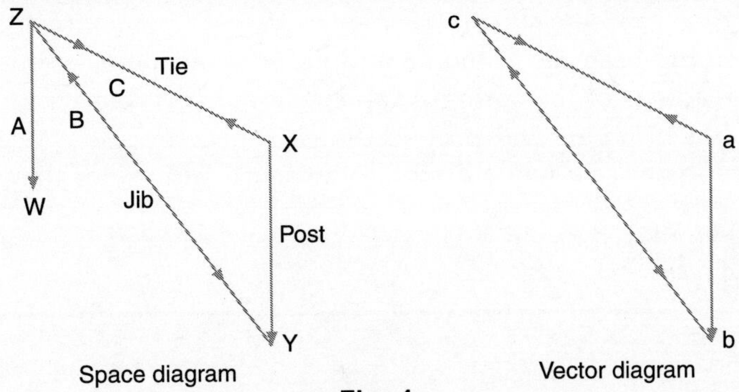

Fig. 4

(*vii*) Construct the force diagram from a convenient point. Go on adding all the forces vectorically one by one to some suitable scale.

In the context of force diagram indicated, draw vector *ab* vertically equal to weight *W*. Line *ac* is drawn parallel to *XZ* and line *bc* is drawn parallel to *YZ* of the space diagram.

(*viii*) Measure the lengths *ac* and *bc* and convert these lengths to weight (force) as per the chosen scale.

This gives the calculated values for the forces in the tie and jib members.

(*ix*) Repeat steps atleast three times by changing the weight.

OBSERVATIONS

S.No.	Weight W	Length			Initial readings		Final readings		Observed force		Graphically calculated force		% error	
		Tie	Jib	Post	Tie	Jib	Tie	Jib	Tie F_t	Jib F_j	Tie F_{tc}	Jib F_{jc}	Tie	Jib
1														
2														
3														

CALCULATIONS AND RESULT

$$\text{Percentage error in tie member } e_t = \frac{F_t - F_{tc}}{F_t} \times 100$$

$$\text{Percentage error in jib } e_j = \frac{F_j - F_{jc}}{F_j} \times 100$$

$$\text{Mean percentage error in tie member} = \frac{e_{t1} + e_{t2} + e_{t3}}{3}$$

$$\text{Mean percentage error in jib} = \frac{e_{j1} + e_{j2} + e_{j3}}{3}$$

PRECAUTIONS

(*i*) Ensure that the spring balances in the tie and jib are functioning properly.
(*ii*) Initial and final readings of the spring balances should be taken with great care.
(*iii*) Measure the lengths of jib, tie and post accurately.
(*iv*) Choose the appropriate scale while drawing the space and vector diagrams.

EXPERIMENT-4

OBJECT

To determine the moment of inertia of flywheel.

APPARATUS

Flywheel, load hanger, cord, stop watch, tape, weights, string etc.

THEORY

A flywheel is a heavy wheel which when used in a machine acts as an energy accumulator. It stores energy during the period when the supply of energy is more than demand, and releases it during the period when the requirement becomes more than the supply. In an internal combustion engine, the energy is developed only during the expansion (power) stroke and this is much more than the engine load. The excess energy is stored in the flywheel, and is released to the crankshaft during other strokes (suction, compression and exhaust) where no energy is developed. These actions make the engine crankshaft to rotate at constant speed. The flywheels are also used with machines where the operation is intermittent such as punching, shearing and rivetting machines. In these machines, the flywheel stores energy from the power source during the greater portion of the operating cycle and releases it during a small period of the cycle.

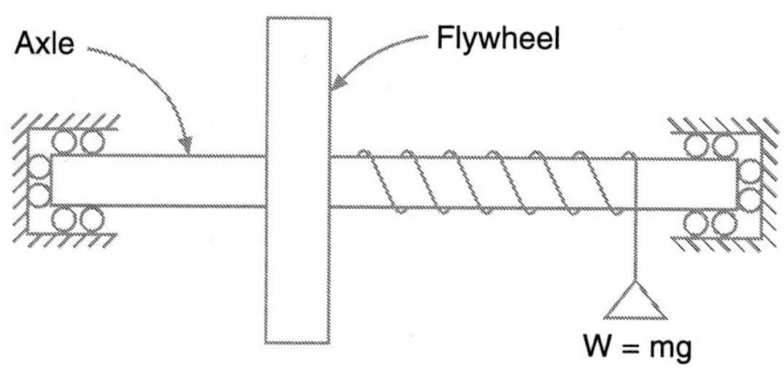

Fig. 1

The experimental setup has the flywheel mounted on the axle which is supported on small bearings. Let

N_1 = number of revolutions made by the flywheel when mass m falls through height h
N_2 = number of revolutions made by the flywheel before coming to rest on detachment of mass
t = time taken to complete revolutions N_2

Then moment of inertia I of the flywheel is worked out by using the correlation

$$I = m\frac{N_1}{N_1 + N_2}\left[\frac{ght^2 - r^2}{2\pi^2 N_2^2}\right]$$

where r is radius of flywheel.

PROCEDURE

1. Attach the string having its length equal to distance between the horizontal peg and floor, and wind it.
2. Attach a known mass to the free end of string.
3. Measure distance between floor and mass body, and release the wheel.
4. Note the revolutions made by the wheel during the period the mass is falling.
5. Start stop watch when the mass gets detatched and note the time taken for the wheel to come to rest.
6. Note the revolutions made by wheel after mass is detached and wheel comest to rest.
7. Repeat the procedure for different masses.
8. Measure the radius of flywheel.

OBSERVATIONS

Thickness of string t_1 = ...

Effective radius of axle $r = \frac{d}{2} + \frac{t_1}{2}$ = ...

S.No.	*Mass* m	*Distance h through which mass falls*	*No. of revolutions* N_1 *made by wheel when mass falls*	*No. of revolutions* N_2 *made by wheel after mass is detached*	*Time* t
1					
2					
3					

CALCULATIONS

Moment of inertia of flywheel, $I = m\frac{N_1}{N_1 + N_2}\left[\frac{ght^2 - r^2}{2\pi^2 N_2^2}\right]$

RESULT

The average moment of inertia of flywheel is $I = \frac{I_1 + I_2 + I_3}{3}$

PRECAUTIONS

(*i*) Ensure that there is no overlapping of the turns of string on the axle.
(*ii*) Start and stop the watch at correct time.
(*iii*) Adopt proper units while making calculations for the moment of inertia.

EXPERIMENT-5

OBJECT

To determine the coefficient of friction between different surfaces on an inclined plane.

APPARATUS

An inclined plane, block, scale pan and weights, pulley and string

The long steel plane has a ground surface and can be locked in any angular position which is indicated on a semi-circular protractor scale.

THEORY

Consider an inclined plane with angle of inclination α. When the body (trolley) just begins to move up, then the force of friction on the body acts down the plane. The various forces which keep the body in equilibrium are :

(*i*) Weight W of the block

(*ii*) Reaction R of the plane

(*iii*) Tractive force P, and

(*iv*) Friction force $F = \mu R$

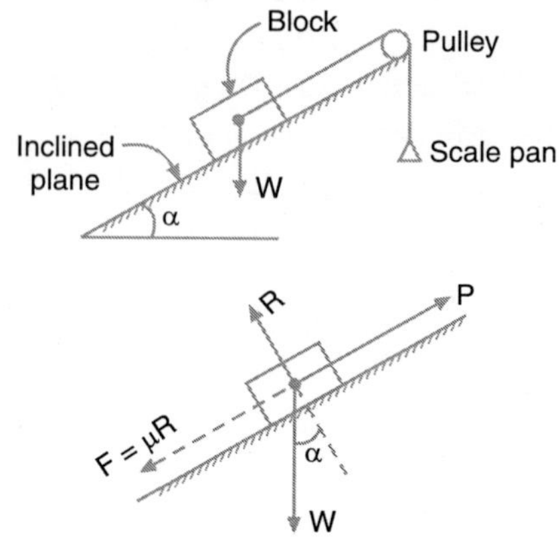

Fig. 1

Resolving these forces along and perpendicular to the plane, we get

$$P = \mu R + W \sin \theta$$

and $$R = W \cos \alpha$$

From these two identities, the coefficient of friction μ works out as

$$P = \mu W \cos \alpha + W \sin \alpha$$

or $$\mu = \frac{P - W \sin \alpha}{W \cos \alpha}$$

This expression enables to determine the coefficient of friction from different values of P, W and α.

PROCEDURE

1. Place the trolley on the inclined plane and connect it to one end of the string
2. Pass the string over the pulley and attach scale pan at the free end of the string
3. Fix the angle of inclination α of the inclined plane
4. Add some weights in the trolley and make a record of the weights added
5. Put weights slowly in the scale pan till it just begins to move up the plane. The magnitude of these weights represents the effort P.
6. Repeat the steps three/four times by changing the load and effort
7. Calculate the value of coefficient of friction by using the relation

$$\mu = \frac{P - W \sin \alpha}{W \cos \alpha}$$

OBSERVATIONS

S.No.	Angle α	*Load W*			*Effort P*			μ
		Wt. of trolley W_1	*Wt. put in the trolley* W_2	*Total weight* $W = W_1 + W_2$	*Wt. of scale pan* P_1	*Wt. put in the scale pan* P_2	*Total weight* $P = P_1 + P_2$	
1								
2								
3								

RESULTS

The average coefficient of friction μ between the two surrace is

$$\mu = \frac{\mu_1 + \mu_2 + \mu_3}{3}$$

PRECAUTIONS

(*i*) The string should be parallel to the plane
(*ii*) There should not be any knot in the string
(*iii*) The pulley should be frictionless
(*iv*) The weights should be put gently and slowly in the scale pan.

EXPERIMENT-6

OBJECT

To determine the coefficient of friction between the pulley and a belt (flat velt, V-bolt)

APPARATUS

The experimental setup consists of a wall mounted pulley with a loaded belt. The pulley is made of cast iron and has three machined grooves to suit a flat belt, V-belt and a rope. The apparatus has the facility to change the position of peg in different holes and this allows to get different positions of belt on pulley, *i.e.*, different lap angles.

THEORY

The transmission of power in factories from one rotating shaft to another that lies at a considerable distance is achieved through belts and ropes. The shafts are fitted with pulleys, the belt is wrapped round the pulleys and its ends are connected to form and endless connector. The belt and the pulley remain in contact by frictional grip.

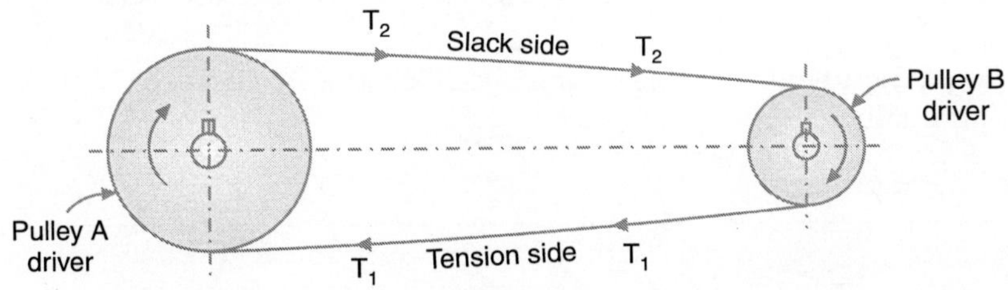

Fig. 1

The pulley A which is connected to rotating shaft is called the driver, and the pulley B that needs to be driven is termed as follower. When the driver rotates, it carries the belt because of friction that exists between the pulley and the belt. The frictional resistance develops all along the contact surfaces. That makes the belt carry the follower which too starts rotating. The driving pulley pulls the belt from one side (called tension side) and delivers it to the other side (called slack side). The tension T_1 in the belt on the tension side is more than tension T_2 on the slack side. Further, only a part of the belt is in contact with the pulley. The angle subtended at the centre of the pulley by the position of belt in contact with it is called the angle of contact or the angle of lap.

For a flat belt :

$$\frac{T_1}{T_2} = e^{\mu\theta} \qquad ...(i)$$

For a V-bolt or a rope running into a grooved pulley

$$\frac{T_1}{T_2} = e^{(\mu \operatorname{cosec} \alpha)\theta} \qquad ...(ii)$$

where θ is the angle of contact, μ is the coefficient of friction between belt and pulley/groove and α is half the angle of groove.

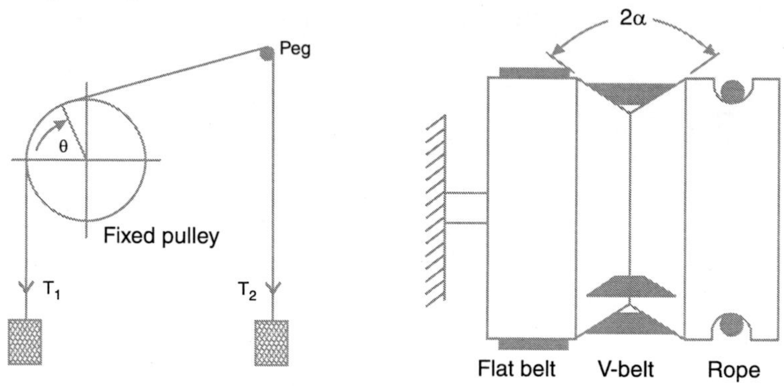

Fig. 2

PROCEDURE

1. Select the type of belt and pulley
2. Add weights to the two hangers suspended from the belt
3. Fix the peg in the board to obtain the desired angle of lap θ
4. Note the tension (weights) T_1 and T_2 when the motion is impending (just to take place) between the belt and pulley.
5. Use the relation (*i*) or (*ii*) (as the case may be) to determine the value of coefficient of friction μ

While conducting experiment, the peg should be fixed tightly to avoid slip, and the loads should be applied gradually.

OBSERVATIONS

Pulley	T_1	T_2	μ
Flat			
120°–V			
90°–V			
60°–V			

RESULTS

The calculated values of coefficient for different belts are :

Flat belt =

120° V belt =

90° V belt =

60° V belt =

EXPERIMENT-7

OBJECT

To determine the velocity ratio, mechanical advantage and efficiency of a worm and worm-wheel.

APPARATUS

Worm and worm-wheel set up, strings, a set of weights, scale pans and metre scale

THEORY

The worm and worm-wheel unit essentially consists of :

(*i*) a square threaded screw called worm, and

(*ii*) a toothed wheel called the worm wheel.

Both the worm and wheel are geared together and their axes are at right angles to each other. A wheel is attached to the worm and the effort *P* is applied to one end of the rope which passes over this wheel. A load durm is securely mounted on the worm wheel and the load *W* is connected to the string wound around the load drum.

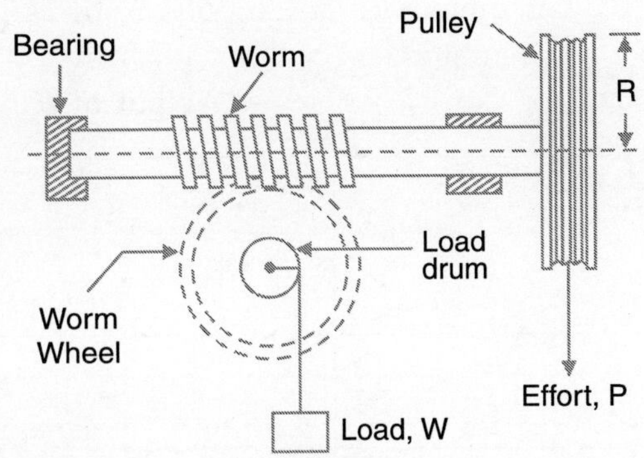

Fig. 1

Let one complete revolution be given to the effort wheel.

Then the effort will move through distance $2\pi R$ where *R* is radius of the wheel.

If the worm is *single threaded,* then for one revolution of the effort wheel, the worm would push the worm wheel through one teeth. If the number of teeth on the worm wheel is *T*, then a push of one teeth implies that the load drum will move through $\frac{1}{T}$ revolution. Therefore,

$$\text{Distance moved by load} = 2\pi r \times \frac{1}{T} = \frac{2\pi r}{T} \quad \text{where } r \text{ is the radius of load drum.}$$

$$\text{Velocity ratio } (VR) = \frac{\text{distance moved by effort}}{\text{distance moved by load}} = \frac{2\pi R}{2\pi r / T} = \frac{RT}{r}$$

$$\text{Mechanical advantage (MA)} = \frac{\text{weight lifted}}{\text{effort applied}} = \frac{W}{P}$$

$$\text{Efficiency } \eta = \frac{\text{mechanical advantage}}{\text{velocity ratio}} = \frac{MA}{VR}$$

$$= \frac{W/P}{RT/r} \times 100 = \frac{Wr}{PRT} \times 100$$

PROCEDURE

(*i*) Measure the circumference of the load drum and the pulley rim, and there from calculate the radius of the pulley rim *R* and the load drum *r*.

(*ii*) Count the number of teeth *T* corresponding to the given apparatus of worm and worm-wheel unit, and calculate the velocity ratio of the set up $\left(VR = \frac{RT}{r}\right)$

(*iii*) Wrap a string round the effort pulley and attach a pan to its free end.

(*iv*) Wrap another string round the load drum and suspend a load W to the free end of this string.

(*v*) Place some weights in the pan attached to the string passing around the effort pulley. Go on increasing these weights till the load W just starts moving (impending motion).

(*vi*) Note down the weights in the pan. We get the value of effort P when the weight of pan is added to weights in the pan.

(*vii*) Repeat the steps three/four times by changing the load and effort.

OBSERVATIONS

Number of teeth in the worm-wheel T =

Circumference of the pulley rim $2\pi R$ =

Circumference of the load drum $2\pi r$ =

Weight of pan =

S.No.	Load lifted W	Effort P		$MA = \frac{W}{P}$	$\eta = \frac{MA}{VR} \times 100$
		Weight in the pan	Weight of pan + weight in the pan		
1					
2					
3					

RESULTS

Velocity ratio VR =

$$\text{Mechanical advantage MA} = \frac{(MA)_1 + (MA)_2 + (MA)_3}{3}$$

$$\text{Efficiency } \eta = \frac{\eta_1 + \eta_2 + \eta_3}{3}$$

PRECAUTIONS

(*i*) Lubricate the pulleys to make them frictionless.

(*ii*) The string should not overlap on the pulley or on the drum.

(*iii*) The load to be lifted, and the pan should not touch any surface.

(*iv*) Circumference of the pulley and load drum should be measured with great care.

(*v*) The weights for effort should be placed gradually and gently in the pan.

EXPERIMENT-8

OBJECT

To determine the velocity ratio, mechanical advantage and efficiency of a simple screw jack.

APPARATUS

Simple screw jack, strings, a set of weights, scale pans, threads and metre scale

THEORY

Screw jack is a simple machine used for lifting heavy loads, through short distances, with the help of small effort applied at its handle. The most common application is the raising of the front or rear portion of a vehicle for the purpose of changing the wheel or tyre.

The unit consists of a base which is in the shape of the frustrum of a hollow cone. The frustrum has a hole which is threaded to receive a strong square threaded screw. A collar is fitted onto the top of the screw in such a manner that the screw can rotate freely in it. The load to be raised or lowered is placed on the top of the collar. A lever or handle, often called tommy bar, can be inserted into a hole provided in the collar. When an effort is applied at the end of the lever, the top of the screw goes either up or down depending on the direction in which the lever is turned.

When one rotation is given to the handle,

$$\text{distance moved by the effort} = 2\pi l$$

where l is the length of lever arm from the centre of screw.

$$\text{distance through which load is lifted} = \text{pitch of the thread} = p$$

$$\therefore \text{ Velocity ratio } VR = \frac{\text{distance moved by effort}}{\text{distance moved by the load}} = \frac{2\pi l}{p}$$

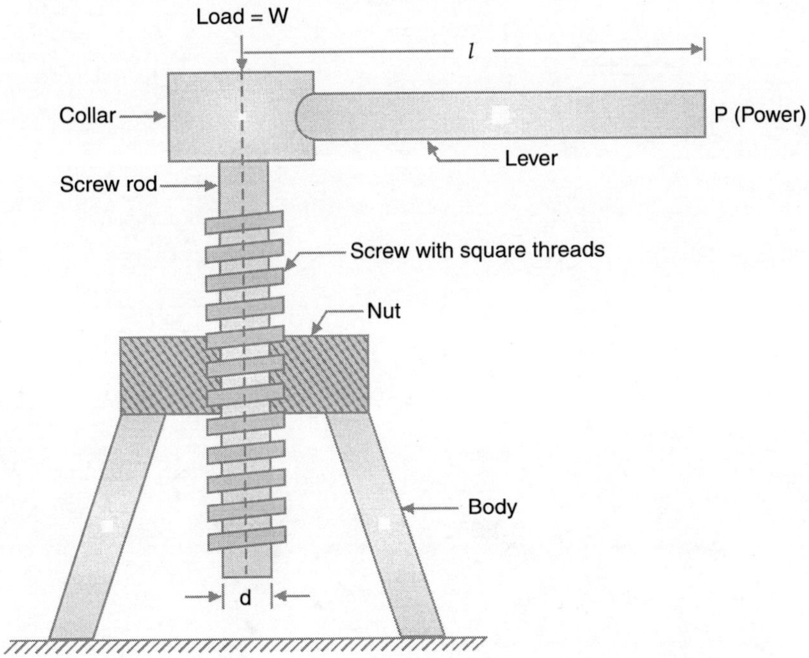

Fig. 1

If P is the effort applied at the end of the lever of the screw jack and W is the weight lifted or lowered, then

$$\text{Mechanical advantage } MA = \frac{W}{P}$$

$$\text{Efficiency } \eta = \frac{\text{mechanical advantage}}{\text{velocity ratio}} = \frac{MA}{VR}$$

In the experimental setup, there is a double pulley platform system that enables to apply the effort for lifting and lowering the load.

PROCEDURE

(*i*) Note the pitch p of the screw from the manual corresponding to the given screw jack.

(*ii*) Measure the circumference of the double pulley platform with the help of a thread (circumference = $2\pi r$).

(*iii*) Calculate the velocity ratio of the setup $\left(VR = \dfrac{2\pi r}{p}\right)$.

(*iv*) Place some known weight W on the screw jack.

(*v*) Put some weights in the scale pan and increase this weight till the load starts to move up (impending motion).

(*vi*) Note the weight of effort pan w and the weights added W_1 for the impending motion.

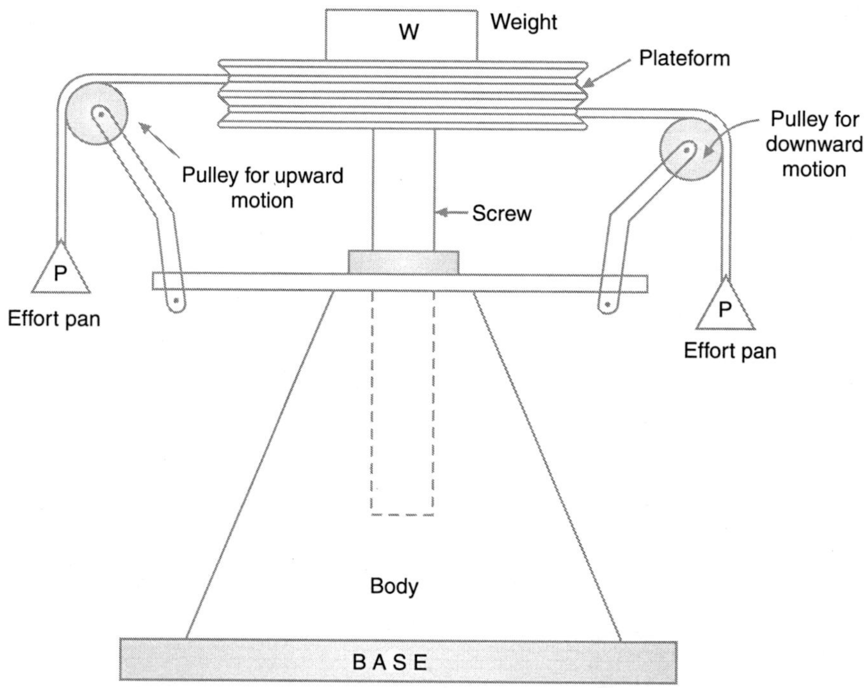

Fig. 2

OBSERVATIONS

S.No.	*Weight on the screw jack W*	*Effort P*		$MA = \dfrac{W}{P}$	$\eta = \dfrac{MA}{VR} \times 100$
		Weight in the pan W_1	*Total weight* $P = W_1 + W$		
1					
2					
3					

RESULTS

$$\text{Velocity ratio } VR =$$

$$\text{Mechanical advantage } MA = \frac{(MA)_1 + (MA)_2 + (MA)_3}{3}$$

$$\text{Efficiency } \eta = \frac{\eta_1 + \eta_2 + \eta_3}{3}$$

PRECAUTIONS

(*i*) There should not be any overlap of strings

(*ii*) Circumference of double pulley platform should be measured carefully

(*iii*) The weight should be placed gradually and gently in the pan.

EXPERIMENT-9

OBJECT

To compare the experimentally determined values of forces in the struts and tie of a basic roof truss with the analytically determined values.

APPARATUS

A segment of plane truss for experimental purpose

THEORY

A framed structure is an assemblage of a number of bars or rods joined together in such a way to form a rigid framework; this structure is designed to resist geometrical distortion under any applied system of loading. The ends are joined by riveting, welding or by nuts and bolts. The bars used in the framed structure are slender (cross-sectional dimensions quite small compared to length) and are usually of angle section, *T*-section, *I*-section etc. The bars are called two-force members because they are either in tension or in compression. The framed structure finds application in the roofs of sheds at railway platforms, workshops and industrial buildings, and bridge trusses, etc.

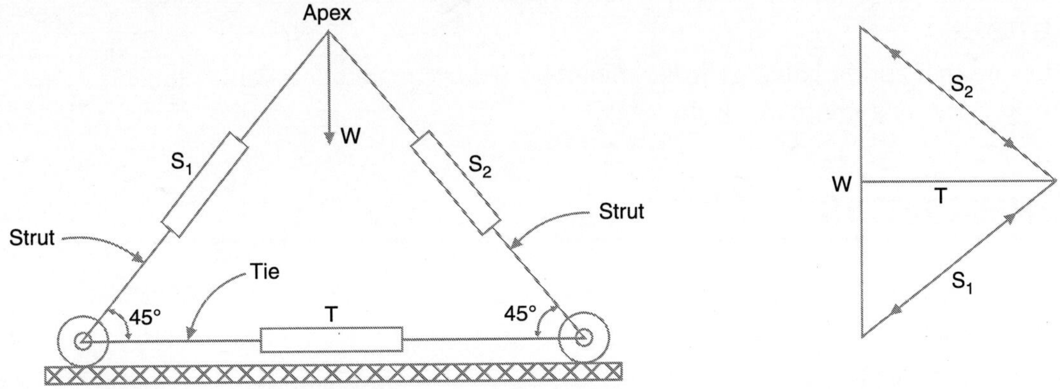

(a) Plane truss

(b) Force polygon

Fig. 1

The experimental roof truss consists of two rafters and struts and a restraining tie. Both struts are pivoted at their apex. The other end of one of the rafters is pivoted to a free standing base, while the remaining strut end runs on ball bearings along a track. When a load is hung from the apex, the free end of the strut moves side ways and restrained by a spring balance.

Both the struts include spring balances and as such all internal loads can be directly measured.

PROCEDURE

(*i*) Set the apparatus and then apply the weight at weight hanger positioned at the apex.

(*ii*) Note the spring balance readings which give the values of internal forces in the struts and tie.

(*iii*) Draw the force diagram and determine therefrom the values of forces setup in the struts and the tie.

(*iv*) Repeat the above steps for different loads.

OBSERVATIONS

S.No.	*Applied weight*	*Spring balance readings*		
		Strut-1 S_1	*Strut-2* S_2	*Tie* (*T*)
1				
2				
3				

Comparison of Values Obtained from Observation and Force Diagram

	Spring balance readings	*Values from force diagram*
S_1 S_2 T		

PRECAUTIONS

(*i*) Ensure that spring balances work smoothly, *i.e.*, frictional effects are negligible

(*ii*) Apply the load gradually at the apex.

EXPERIMENT-10

OBJECT

Study of gear trains

THEORY

Gear train is any combination of gear wheels by means of which power and motion is transmitted from one shaft to another.

Various types of gear trains are :

1. Simple gear train
2. Compound gear train
3. Reverted gear train
4. Epicyclic gear train

The nature of the train used depends upon the velocity ratio required and the relative position of the axes of the shafts.

• **Simple gear train :** Refer Fig. 1 for a simple gear train in which each shaft carries one wheel only. Between the driving gear 1 and the driven or follower gear 4, two intermediate gears 2 and 3 have been provided.

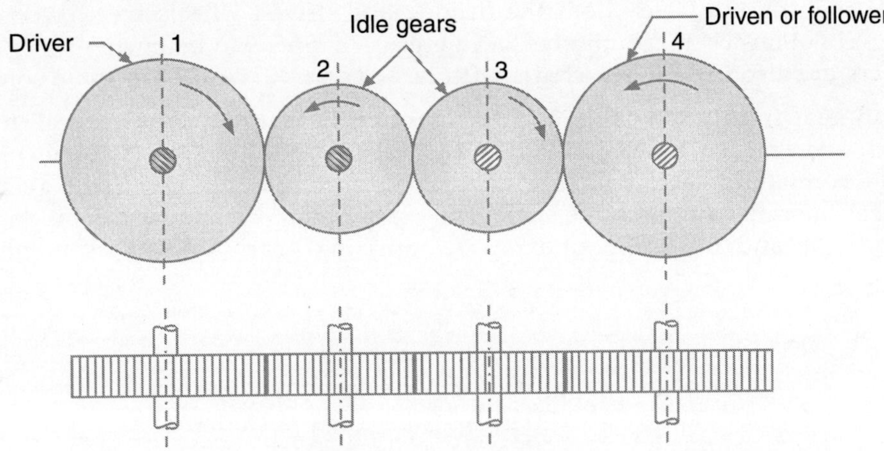

Fig. 1. Simple gear train

Let,

N_1, N_2, N_3 and N_4 be the rotational speeds of gears 1, 2, 3 and 4

T_1, T_2, T_3 and T_4 be the number of teeth on gears 1, 2, 3 and 4

The driving gear 1 is in mesh with the intermediate gear 2, and the speed ratio for these gears is

$$\frac{N_1}{N_2} = \frac{T_2}{T_1} \quad \text{...}(i)$$

Likewise the intermediate gears 2 and 3 are in mesh and accordingly

$$\frac{N_2}{N_3} = \frac{T_3}{T_2} \quad \text{...}(ii)$$

Further, the intermediate gear 3 is in engagement with the driven (follower) gear, and therefore

$$\frac{N_3}{N_4} = \frac{T_4}{T_3} \quad \text{...}(iii)$$

The speed ratio N_1/N_4 for the gear assembly can be obtained by multiplying identities (*i*), (*ii*) and (*iii*).

$$\frac{N_1}{N_2} \times \frac{N_2}{N_3} \times \frac{N_3}{N_4} = \frac{T_2}{T_1} \times \frac{T_3}{T_2} \times \frac{T_4}{T_3} \quad \text{or} \quad \frac{N_1}{N_4} = \frac{T_4}{T_1}$$

$$\therefore \text{ Speed or velocity ratio} = \frac{\text{speed of the driving wheel}}{\text{speed of the driven wheel}}$$

$$= \frac{\text{number of teeth on the driven w}}{\text{number of teeth on the driving v}}$$

Train value of a gear train is defined as the reciprocal of the speed ratio of gear train.

$$\text{Train value} = \frac{\text{speed of the driven wheel}}{\text{speed of the driving wheel}} = \frac{\text{number of teeth on the driving v}}{\text{number of teeth on the driven w}}$$

Comments :

(*i*) The speed ratio and the train value are independent of the size and number of intermediate gears.

(*ii*) The intermediate gears help to obtain the desired direction of motion of the driven gear. With odd number of idle gears, the direction of rotation of the driven wheel will be same as that of the driving wheel. Both the driving and driven wheels will have opposite direction of motion when the number of idle gears happens to be even.

The idle gears are also used when shafts with large centre distance are to be connected.

• **Compound gear train :** Refer Fig. 2 for a compound gear train in which all the gears donot mesh with one another. Between the driving gear 1 and the driven (follower) gears 6, two intermediate compound gears have been provided. A compound gear is the gear which carries two wheels mounted on the same shaft. The gears 2 and 3 constitute a compound gear mounted on shaft *M*, and the compound gear comprising gears 4 and 5 is mounted on shaft *P*.

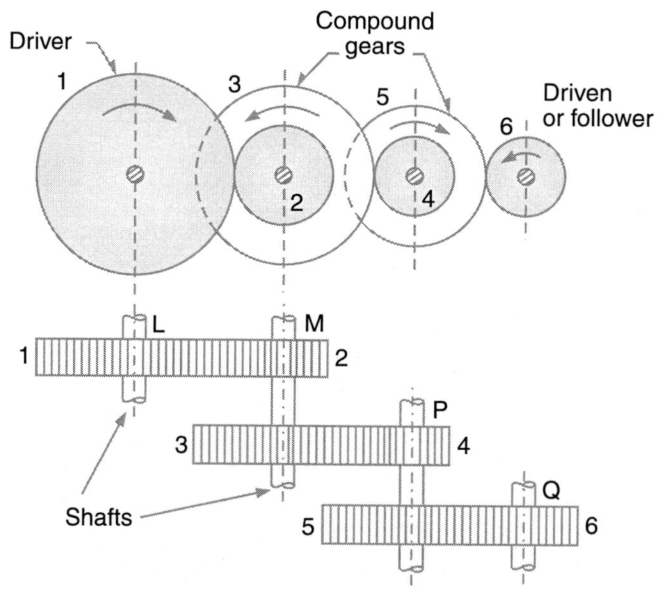

Fig. 2. Compound gear train

In the compound gear train, gear 2 is follower to the driving gear 1, and gear 3 is the driving wheel for gear 4. Obviously gears 1, 3 and 5 are the drivers, and gears 2, 4 and 6 are the drivens.

Let, N_1, N_2, N_3, N_4, N_5 and N_6 be the speeds of the respective gears

T_1, T_2, T_3, T_4, T_5 and T_6 are the number of teeth on respective gears.

Gears 1 is in mesh with gear 2 and therefore its speed ratio is

$$\frac{N_1}{N_2} = \frac{T_2}{T_1} \qquad ...(a)$$

Gear 3 is in mesh with gear 4 and therefore the speed ratio is

$$\frac{N_3}{N_4} = \frac{T_4}{T_3} \qquad ...(b)$$

Similarly for the gears 5 and 6 which engage each other, speed ratio is

$$\frac{N_5}{N_6} = \frac{T_6}{T_5} \qquad ...(c)$$

The speed ratio N_1/N_6 for the gear assembly can be worked out by multiplying identities (*a*), (*b*) and (*c*)

$$\frac{N_1}{N_2} \times \frac{N_3}{N_4} \times \frac{N_5}{N_6} = \frac{T_2 \times T_4 \times T_6}{T_1 \times T_3 \times T_5} \qquad ....(d)$$

Gears 2 and 3 are mounted on the same shaft *M* and therefore $N_2 = N_3$. Again gears 4 and 5 are mounted on the same shaft *P* and therefore $N_4 = N_5$. With these substitutions, the expression (*d*) transforms to

$$\text{Speed or velocity ratio} = \frac{N_1}{N_6} = \frac{T_2 \times T_4 \times T_6}{T_1 \times T_3 \times T_5}$$

$$\text{That is} \quad \frac{N_1}{N_6} = \frac{\text{speed of the first driver}}{\text{speed of the last driven}}$$

$$= \frac{\text{product of the number of teeth o}}{\text{product of the number of teeth c}} \qquad ....(e)$$

A compound gear train helps to obtain a large speed reduction from the first to the last shaft with an assembly of small gears. Usually for a speed reduction of the order of 8 : 1, compound gear train or worm gearing is employed.

- **Reverted gear train :** A reverted gear train manifests when the first driving gear and the last driven gear are on the same axis; axes are coincident and coaxial.

Refer Fig. 3 for the arrangement of a reverted gear train.

The gear 1 is the first driving gear (clockwise motion); it meshes with gear 2 and makes it turn in the opposite direction (anitclockwise motion). Gears 2 and 3 are mounted on the same shaft and constitute a compound gear. The motion of gear 3 is like to that of gear 2 (anticlockwise). The gear 3 acts as driver for the gear 4. The direction of rotation of gear 4 will be clockwise, *i.e.*, opposite to that of gear 3. Apparently in the reverted gear train, the motion of the first gear and the last gear is like.

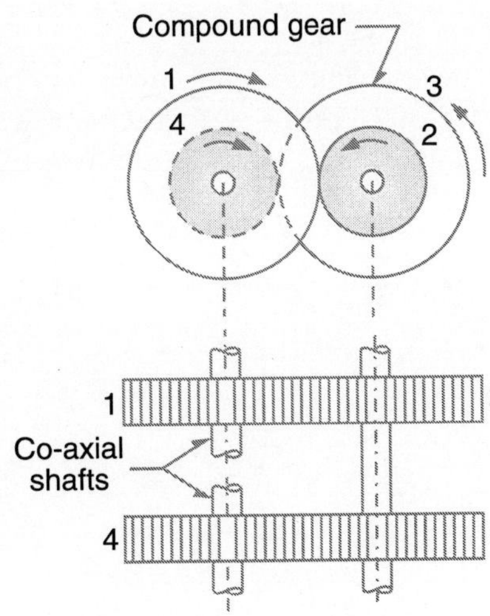

Fig. 3. Reverted gear train

Let D_1, D_2, D_3 and D_4 be the pitch circle diameters of the respective wheels, and their corresponding speeds be N_1, N_2, N_3 and N_4.

The distance between shaft centres of gears 1 and 2, and that between gears 3 and 4 is same. That gives

$$\frac{D_1 + D_2}{2} = \frac{D_3 + D_4}{2} \quad \text{or} \quad D_1 + D_2 = D_3 + D_4$$

Presuming that the circular pitch or module of all the gear wheels is same, the number of teeth on each gear is proportional to its circumference or diameter.

That gives

$$T_1 + T_2 = T_3 + T_4$$

Now, Speed ratio $$\frac{N_1}{N_4} = \frac{\text{number of teeth on drivens}}{\text{number of teeth on drivers}} = \frac{T_2 \times T_4}{T_1 \times T_3} \qquad ...(f)$$

The reverted gear gear train is used to connect hour hand to minute hand in a clock mechanism. Other applications lie in automotive transmissions, lathe back gears and industrial speed reducers.

• **Epicyclic gear train :** In the simple and compound gear trains, the axis of the wheels remain fixed relative to one another. The epicyclic gear train is a special type of gear train in which axis of rotation of one or more of the wheels is carried on an arm and this arm is free to rotate about the axis of rotation of one or the other wheels in the train.

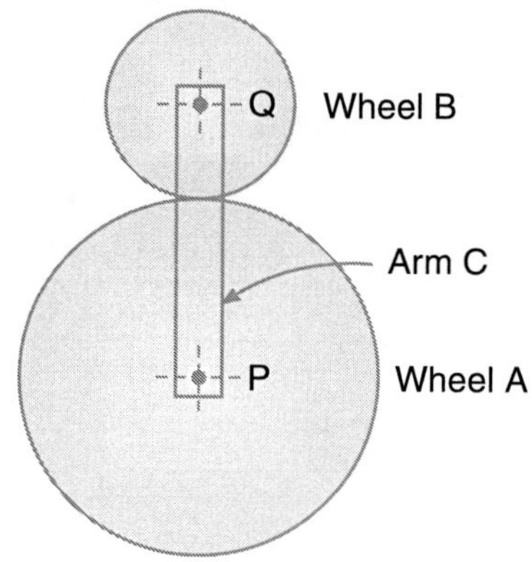

Fig. 4. Epicyclic gear train

Refer Fig. 4 for the arrangement of epicyclic gear train. The gear *A* and the arm *C* can rotate about the axis at *P*, the gear *B* meshes with gear *A* and has its axis of rotation on the arm at *Q*. When the gear *A* is fixed and the arm *C* is made to rotate about *P*, the gear *B* would be forced to roll around the outside of the gear *A*.

With N_C revolutions of the arm, the gear wheel *B* rotates about its axis by speed N_B given by

$$\frac{N_B}{N_C} = 1 + \frac{T_A}{T_B} \qquad ...(g)$$

The epicyclic gearing can bring about extensive speed reduction, and is capable of transmitting high velocity ratios with gears of moderate size in a comparatively short space. These aspects render it ideally suitable for use in differential gears of automobiles, wrist watches, hoists and pulley blocks etc.

❏❏❏

INDEX

INDEX

K

L

M

N

P

R

S

T

❑❑❑

Win Prizes

ATTENTION READERS

Win Prizes

We request for your frank assessment regarding various aspects of the book as given below :

A TEXTBOOK OF ENGINEERING MECHANICS

By: Dr. D.S. Kumar

Please Fill up the Sheet in Neat and Clean Writing

(i) What topics of your syllabus, important for your examination point of view, are not covered in this book ?

....

....

....

....

(ii) What are the chapters wherein the treatment of the subject matter is not systematic or organised or updated ?

....

....

....

....

(iii) Have you come across misprints/mistakes/factual inaccuracies in the book? Please specify the chapters and the page numbers.

....

....

....

....

(iv) What other books on the same subject have you found/heard better than the present book? Please specify in terms of price and quality

....

....

....

....

(v) Further suggestions and comments for the improvement of this book.

....

....

....

....

(PS: If need be please attach a separate sheet to write your views)

OTHER DETAILS

(i) Who recommended you this book ? (Please tick (✓) in the box near the option relevant to you.

☐ Teacher ☐ Friends ☐ Bookseller

....

....

....

....

(ii) Name of the recommending teacher, his designation and address :

....

....

....

....

(iii) Name and address of the bookseller you purchased the book from :

....

....

....

....

(iv) Name and address of your institution (Please mention the University or Board as the case may be)

....

....

....

....

(v) Your name and complete postal address :

....

....

....

....

(vi) Course you are enrolled in :

....

....

....

....

The best assessment will be awarded each month: The award will be in the form of our publications as decided by the Editorial Board.

Please mail the filled up coupon at your earliest to :

S.K. KATARIA & SONS®
4885/109, Prakash Mahal, Dr. Subhash Bhargav Lane,
Opposite Delhi Medical Association, Daryaganj, New Delhi-110002 (INDIA)
Phone: +91-11-43551243; Mobile: +91-9871775858, 7678425209
e-mail: katariabook@gmail.com, katariabooks@yahoo.com
Website: www.skkatariaandsons.com